pH AND BRAIN FUNCTION

pH AND BRAIN FUNCTION

Edited by

KAI KAILA
and
BRUCE R. RANSOM

WILEY-LISS

A JOHN WILEY & SONS, INC., PUBLICATION

New York • Chichester • Weinheim • Brisbane • Singapore • Toronto

This book is printed on acid-free paper. ♾

Library of Congress Cataloging-in-Publication Data:

pH and brain function / edited by Kai Kaila and Bruce R. Ransom.
p. cm.
"A Wiley Liss publication."
Includes bibliographical references and index.
ISBN 0-471-11838-9 (cloth : alk. paper)
1. Neurochemistry. 2. Hydrogen-ion concentration. 3. Nervous system—Pathophysiology. 4. Neurons. 5. Homeostasis. I. Kaila, K. II. Ransom, Bruce R.
[DNLM: 1. Brain—physiology. 2. Hydrogen-Ion Concentration. WL 300 P532 1998]
QP356.3.P497 1998
573.8′4—dc21
DNLM/DLC
for Library of Congress 97-40568

Printed in the United States of America.

10 9 8 7 6 5 4 3 2 1

CONTENTS

PREFACE

The activity patterns of brain cells are highly sensitive to the influence of a vast number of chemical factors, both exogenous (e.g. drugs and toxins) and intrinsic ones, such as neurotransmitters and hormones. In terms of concentration, one of the most potent intrinsic neuromodulators within the brain is the H^+ ion, which can exert powerful actions on neuronal function with fluctuations at nanomolar levels. The central role of pH in the control of brain function is based on the high H^+ sensitivity of a large number of diverse protein molecules, which include various types of voltage- and ligand-gated channels as well as ion pumps and carriers, constituents of second-messenger pathways, and enzymes that control cellular energy metabolism.

It has also become evident that pH in the brain varies with neural activity. Indeed, H^+ can function as a signal or messenger in the brain, either within cells or between them. The obvious relevance of these insights to an understanding of how the brain works under both physiological and pathophysiological conditions is clear.

Over the last 20 years or so, an enormous amount of exciting new information has become available on this topic. The story about how brain cells regulate pH and are affected by pH has now become clear enough, and is important enough, to justify a book dedicated to this subject. Most of the primary research information about the physiology of pH in the nervous system is widely scattered in the scientific literature and there has been no book or monograph that effectively integrates these emerging concepts in a way which would be useful to the general neuroscience community. pH and Brain Function was assembled to address this shortcoming. Our goal was to create a book that would make detailed information, as well as general concepts, accessible to an audience that has no special training in this field. With this in mind we incorporated an extensive introductory section that covers the fundamental concepts of acid-base biology. This section also describes the multitude of current techniques used for direct monitoring of pH and the physicochemical and physiological processes that affect it. At the same time, we wanted the book to be authoritative and to contain lucid summaries of pertinent experimental work. To achieve this, we persuaded many of the most influential and respected scientists in this field to participate as authors. We were happy not only about their enthusiasm to join us in this endeavor, but also about the fact that these individuals felt as we did, that a comprehensive summary of this topic was long overdue.

Another goal we had in mind was set in the spirit of integrative biology (which, parenthetically, can be taken as a modern expression for what was formerly known simply as physiology): the exponentially growing output of current research at the molecular and cellular levels has to be put into a functional context or it has no real meaning. We have, therefore, written a brief chapter (Chapter 1) which addresses some of the relevant teleological, evolutionary and whole-organism aspects of pH physiology in an attempt to provide a general framework for some of the more specific data presented elsewhere in this volume (*). In addition, and more importantly, the general structure of the book is based on chapters that address the topic of pH and brain function at various levels of biological organization, ranging from cells, organelles, and isolated tissue preparations to the whole brain *in vivo*.

Multi-author books inevitably reflect the differing writing styles and perspectives of the contributors. Realizing that this can lead not only to repetition and redundancy but also to serious omissions, we exercised considerable editorial prerogative on first drafts of the chapters. In addition, we requested and received comments about each chapter from other contributors. This extra effort provided by our authors was beyond the call of duty but was rendered graciously, nevertheless. We thank them for their patience and support, their good advice, and their fine contributions.

The importance of pH as a crucial variable in the complex functioning of the brain was the real driving force for the authors and editors to create this book. As research expands our knowledge about this topic, we predict it will spawn interest in a progressively larger audience of scientists and clinicians. Our hope is that *pH and Brain Function* by providing a comprehensive view of what we currently know, facilitates the development of new ideas to inspire further experimental work in the laboratory.

Kai Kaila
Bruce R. Ransom

(*)We would like to express our sincere thanks to Dr. George Somero (Stanford University) for helpful discussions and for comments on an early version of Chapter 1.

CONTRIBUTORS

ALESSANDRA AMATO, Department of Physiology, University College London, London, United Kingdom

DAVID ATTWELL, Department of Physiology, University College London, London, United Kingdom

KLAUS BALLANYI, Physiologisches Institut, Universitat Gottingen, Humboldtallee 23, Gottingen, D-37073, Germany

STEVEN BARNES, Neuroscience Research Group, University of Calgary, Faculty of Medicine, Calgary, Alberta, Canada

STUART BEVEN, Novartis Institute for Medical Sciences, London, WC1E 6BN, United Kingdom

MARK O. BEVENSEE, Department of Cellular and Molecular Physiology, Yale University School of Medicine, New Haven, CT 06520

BRIAN BILLUPS, Department of Physiology, University College London, London, United Kingdom

WALTER F. BORON, Department of Cellular and Molecular Physiology, Yale University School of Medicine, New Haven, CT 06520

MURIEL BOUVIER, Department of Physiology, University College London, London, United Kingdom

MITCHELL CHESLER, Department of Physiology and Neuroscience and Department of Neurosurgery, New York University Medical Center, New York, NY 10016

JOACHIM W. DEITMER, Abteilung fur Allgemeine Zoologie, FB Biologie, Universität Kaiserslautern, D-67653 Kaiserslautern, Germany

LIZ HOPKINSON, Novartis Institute for Medical Sciences, London, WC1E 6BN, United Kingdom

KAI KAILA, Department of Biosciences, Division of Animal Physiology, University of Helsinki, FIN-00014 Helsinki, Finland

KEN-ICHIRO KATSURA, The Department of Internal Medicine, Nippon Medical School, 1-1-5 Sendagi, Bunkyo-ku, Tokyo 113, Japan

RISTO A. KAUPPINEN, A. I. Virtanen Institute, University of Kuopio, Neulaniementie 2, FIN-7021, Kuopio, Finland

HAROLD KIMELBERG, Departments of Surgery, Pharmacology and Neuroscience, Albany Medical College, Albany, NY 12208

RICHARD P. KRAIG, Departments of Neurology and Pharmocological and Physiological Sciences, Committee on Neurobiology, The University of Chicago, Chicago, IL 60637

JOSEPH C. LAMANNA, Department of Neurology, Case Western Reserve University, School of Medicine, Cleveland, OH 44106

KAREN L. LAURO, Department of Neurology, Case Western University, School of Medicine, Clevend, OH 44106

PEKKA MELLERGÅRD, Department of Neurosurgery, University Hospital, S-221 85 Lund, Sweden

OGNEN A. C. PETROFF, Departments of Neurology and Medicine, Yale University, New Haven, CT 06520

JAMES W. PRICHARD, Departments of Neurology and Medicine, Yale University, New Haven, CT 06520

BRUCE R. RANSOM, Department of Neurology, University of Washington, Seattle, WA 98195-6465

YVONNE RIDDERSTRÅLE, Department of Animal Physiology, Box 7045 5-750 07, Swedish University of Agricultural Sciences, Uppsala, Sweden

CHRISTINE R. ROSE, Institute of Physiology, University of Saar Landes, 66421 Hamburg, Germany

DOUGLAS L. ROTHMAN, Departments of Neurology and Medicine, Yale University, New Haven, CT 06520

ELIANA SCEMES, Department of Physiology, University of Sao Paulo, Sao Paulo, Brazil

CHRISTOF J. SCHWIENING, Department of Physiology, University of Cambridge, Downing Site, Cambridge CB2 3EQ, United Kingdom

BO K. SIESJÖ, Laboratory for Experimental Brain Research, Wallenberg Neuroscience Center, University Hospital, S-221 85 Lund, Sweden

GEORGE G. SOMJEN, Department of Cell Biology, Duke University Medical Center, Durham, NC 27710

DAVID C. SPRAY, Department of Neuroscience, Albert Einstein College of Medicine, Bronx, New York 10461

EVA SYKOVÁ, Department of Neuroscience, Charles University and Institute of Experimental Medicine, Academy of Science of the Czech Republic, UEM AVCR, 142 20 Prague 4, Czech Republic

MAREK SZATKOWSKI, Department of Physiology, University College London, London, United Kingdom

ROGER C. THOMAS, Department of Physiology, University of Cambridge, Downing Site Cambridge, CB2 3EQ, United Kingdom

GEOFFREY C. TOMBAUGH, Department of Neuroscience and Cell Biology, University of Medicine and Dentistry of New Jersey, Piscataway, NJ 08854

STEVEN TRAYNELIS, Department of Pharmacology, Emory University, Atlanta, GA

JUHA VOIPIO, Department of Biosciences, Division of Animal Physiology, University of Helsinki FIN-00014, Helsinki, Finland

PER J. WISTRAND, Department of Ophthalmology, Uppsala University Hospital, Uppsala, Sweden

SECTION I

Introduction to pH

CHAPTER 1

Concept of pH and its Importance in Neurobiology

KAI KAILA AND BRUCE R. RANSOM
Department of Biosciences, Division of Animal Physiology, University of Helsinki, FIN-00014 Helsinki, Finland (K.K.); Department of Neurology, University of Washington, Seattle, Washington 98195-6465 (B.R.)

HYDROGEN IONS AND pH

What is an H^+ Ion

The hydrogen ion (H^+) is a single proton, and hence, it is the smallest independent chemically reactive particle in the universe. H^+ ions are a product of the ionization of water; but water is a very weak acid, so that, within an organism, only one of about 100–1000 million water molecules is, in fact, ionized. However, despite their small size and sparse occurrence, H^+ ions are of enormous biological importance. This is largely due to the presence in protein molecules of charged groups capable of binding and releasing H^+ ions. Very often, these interactions give rise to alterations in protein conformation, which manifest themselves as functional changes at various levels of biological organization. Within neurobiology, the mechanisms and consequences of H^+-mediated effects are currently being studied by experimental approaches that range from exploiting the techniques of molecular biology, such as the construction of point-mutated or chimeric proteins, to imaging of the human brain *in vivo* (see Chapters 6–9, 22, and 30).

Quantifying H^+: the Various Definitions of pH

The effective concentration of H^+ is usually quantified in terms of pH, which is defined by the relationship $pH = -\log_{10}a_H$, where a_H is the activity (or thermodynamically effective concentration) of the hydrogen ions (Chapters 2, 6, and 10). However, the concept of pH is perhaps not as straightforward as one might hope and, in fact, there are two main kinds of definitions of pH, theoretical and practical (Chapter 6). The one given above is based on the single-ion thermodynamical activity of the H^+ ion and falls into the theoretical category. Practical definitions rely on the use of a few standard pH solutions based on the operation of the hydrogen electrode. Such standard solutions (or derivatives thereof)

pH and Brain Function, Edited by Kai Kaila and Bruce R. Ransom
ISBN 0-471-11838-9

are routinely used in the calibration of laboratory pH meters. Virtually all biological experimentation uses the practical definition of pH.

It is sometimes argued that the logarithmic nature of the pH scale can obscure the fact that the physiologically relevant pH range (from, say 6.5 to 8.0) corresponds to a wide H^+ activity range at very low absolute values (from 10 to 300 nM). While this may be so, it should be noted that the H^+ sensitivities of various proteins are usually expressed in terms of the apparent pK values of their functionally critical ionizable groups. Therefore, a conversion of physiological pH shifts to the molar scale obscures how they might affect the function of putative H^+-sensitive proteins. Nevertheless, however one quantitatively expresses the effective H^+ concentration, the H^+ ion is an extremely potent modulator of diverse protein molecules (Chapters 10, 17, and 22–27). In addition, the properties of membrane lipids and are sensitive to small changes in pH (e.g., Hazel et al., 1992), and some pH effects on membrane proteins are indirect in that they are mediated by pH-induced changes in the fluidity of the lipid matrix.

HYDROGEN IONS IN AN ORGANISM

Why Are pH Values What They Are?

The above subheading has been borrowed from the thought-provoking monograph by Hochachka and Somero (1984), who raise the interesting evolutionary point that most metabolites of intracellular biochemical pathways have to be ionized in order to be effectively "trapped" within the cell. This implies maintaining an intracellular pH (pH_i) level at which there will be a net charge on metabolic intermediates which, in their neutral form, would tend to escape across the cell membrane. Another fundamental aspect of pH regulation is that physiological pH values are close to the p*K* values of histidine imidazole groups. Consequently, imidazole groups are uniquely suited for reversible protonation-deprotonation steps that can influence protein functions. These include structural transitions of proteins, protein-ligand interactions, and pH-buffering reactions.

Ultimately, one must answer the question, what is, in fact, being regulated when pH is kept under homeostatic control? Important insights can be gained from comparative studies on nonmammalian species. It has been known for a long time that "cold-blooded" animals such as amphibians and fish have higher blood pH values than humans and other mammalians and, as demonstrated by Reeves (1972; see also Cameron, 1989; Hochachka and Somero, 1984), pH regulation in animal tissues at varying temperatures is aimed at maintaining a constant fractional dissociation of histidine-imidazole side chains, termed "alpha-imidazole". Hence, this kind of pH homeostasis was called "alphastat regulation," in contrast to the alternative possibility, "pHstat regulation," the maintenance of a fixed pH (which would, of course, result in varying levels of imidazole dissociation with varying temperatures).

A central aspect of alphastat regulation is that this strategy facilitates reversibility of enzymatic reactions. A good example of this is provided by lactate dehydrogenase (LDH), which has a substrate-binding histidine at its active site. Lactate can bind to the critical imidazole side chain only if the latter is deprotonated, whereas pyruvate can bind only to the protonated site. With the active site histidine in a state where either increases or decreases in pH_i can affect the binding of either substrate, the enzyme is able to alter its direction of net function (lactate dehydrogenase to produce pyruvate; or pyruvate reductase to produce lactate) in response to changes in pH_i. It should be emphasized that this type of reversibility pertains for any histidine-based process.

Beyond the conceptual importance of distinguishing between alphastat and pHstat regulation, there are also, for the experimentalist, practical implications in this distinction. It is not an infrequent practice to conduct *in vitro* experiments on isolated mammalian cells and tissues at temperatures much lower than those prevailing *in vivo*. Implications of alphastat regulation should also be taken into account in clinical procedures that have an effect on the patient's temperature, as with hypothermia for certain types of surgery. At temperatures lower than normal in mammals, keeping extracellular pH at its standard physiological level (e.g., at 7.4) is, in functional terms, equivalent to imposing a significant acidosis.

pH Regulation Is Synonymous With Acid-Base Regulation

Taken literally, pH regulation might seem to imply the regulation of H^+ ions only. However, from a physiological point of view, this type of definition would make no sense, as it is clear that at all levels of biological organization (organism, tissue, cell, organelle) homeostatic mechanisms involving H^+ ions are inseparable from those controlling the production, consumption, and clearance of CO_2 and HCO_3^- as well as weak-acid end products of energy metabolism (e.g., lactate). Hence, in biological systems, pH regulation is synonymous with acid-base regulation.

At the whole-organism level, the key elements in pH regulation are the lungs or, depending on the animal species, other structures with respiratory epithelia, where the regulation of ventilation controls the partial pressure of CO_2 (P_{CO_2}) in the blood. The kidneys, on the other hand, are responsible for the net trafficking of other important acid-base species, notably HCO_3^- and NH_4^+. From a neurophysiological point of view, it is interesting that the central role of the brain in the regulation of breathing (Chapters 20, 26, 27) leads to a unique control loop: as aptly put by Balestrino and Somjen (1988), "since the brain regulates respiration and, through respiration, the pH and CO_2 content in its own fluids, the brain also exercises control over its own excitability."

Two unique mechanisms provide the brain with additional capacity to manage its pH affairs. The blood-brain barrier (Chapters 17, 32) prevents charged acid-base species in blood from having free access to brain extracellular (interstitial) fluid. The second mechanism is the choroid plexus, sometimes referred to as the brain's kidney, which produces cerebrospinal fluid (CSF) by an active secretory process. CSF is in slow equilibrium with interstitial fluid. The enhanced pH-regulating ability provided by the blood-brain barrier and the choroid plexus underscores the fact that pH is a crucial factor vis a vis brain function.

Intracellular and Extracellular pH Regulation

All (nucleated) animal cells, including neurons and glial cells, maintain their pH_i at a resting level that is more alkaline than what would be predicted on the basis of a thermodynamically passive plasmalemmal distribution of H^+ ions. This, together with extensive studies on cellular pH regulation, indicates the presence of plasmalemmal mechanisms that extrude acid (Chapters 10–15). In addition to the cellular plasma membrane, subcellular membranes (e.g., those of mitochondria, lysosomes, and transmitter vesicles) are also able to generate and maintain an H^+ gradient, which implies compartmentalization of the cell interior with respect to pH. Therefore, it should be noted that the term intracellular pH (pH_i) refers to the pH of the nonorganellar cytoplasm (the cytosol).

Homeostasis of pH within brain cells is based on the operation of a number of membrane transport mechanisms (Chapters 10–15). It is not surprising, therefore, that many acid-transporting ion-exchange mechanisms are controlled by H^+-induced allosteric effects, a feedback design that is teleologically easy to understand. A classical example of this kind of a feedback mechanism is acid extrusion by the Na^+/H^+ exchanger, which is activated by low pH_i. The Na^+/H^+ exchanger is a secondary active transport mechanism, where the extrusion of protons is powered by the steep inward Na^+ gradient built up by Na^+/K^+ ATPase. However, brain cells have a number of different acid-extruding ion-transport mechanisms, as will be discussed in various chapters in this book.

On the other hand, there are also ionic mechanisms that serve to take up acid in response to intracellular alkaline loads (Chapters 10 and 13–14). The simultaneous presence in a cell of acid extruders and uptake mechanisms makes it possible to have a much more tightly controlled pH_i than would be the case with a system where pH_i would be set by extrusion mechanisms working against variable loads.

Intracellular acid loads can be generated by a large number of biophysical and biochemical processes, including thermodynamically downhill-oriented transmembrane movements of acid-base species, as well as the acid loads associated with both aerobic and anaerobic energy metabolism (Chapters 5, 10, 28, and 30). However, energy metabolism does not, necessarily and always, simply represent a load for the pH regulatory systems of the brain. There is much evidence that adjustments of energy metabolism may serve a homeostatic role in pH dynamics (Chapter 5). For instance, a tissue alkalosis caused by hyperventilation (a net loss of CO_2) might well be counteracted by an increased glycolytic production of lactate that is triggered by a rise in pH_i. This kind of coupling is probably mediated by a well-known example of a pH-sensitive protein, phosphofructokinase (PFK), one of the principal flux-governing enzymes of glycolysis. The pH sensitivity of PFK in glial cells may, in fact, become the focus of much interest because, in at least some kinds of neurons, oxidative energy metabolism is largely fuelled by lactate that is produced by glycolysis within glial cells.

How to Obtain Simultaneous Homeostasis of Intracellular and Extracellular pH?

Both neurons and glial cells contain a variety of pH-sensitive receptors, channels, carriers, and enzymes. Since the pH-sensitive sites of these proteins are found in both the extracellular and intracellular compartments, the physiologically intriguing question arises as to which compartment is the primary one to be regulated. Indeed, all the acid extruded from a cell will end up in the interstitial space, and during major deviations from a steady-state situation, homeostasis of pH_i must to some extent be constrained by what happens to extracellular pH (pH_o), and vice versa. This situation is perhaps even more problematic when considering that CO_2 is the only acid-base species that can readily cross the blood-brain barrier.

In the light of the present information, it seems that the heterogeneity of pH in various compartments (both intracellular and extracellular) provides a partial solution to the above problem. For instance, there is evidence that, depending on their molecular mechanisms of generation (Chapters 3 and 16–19), transient changes in pH_o may be confined to distinct microdomains within the interstitial space. There are also mechanisms of pH_o "spatial buffering" provided by some types of glial cells (Newman, 1996). Such cells are capable of rapidly transporting acid-base equivalents away from their site of origin within the brain parenchyma, thus alleviating any instantaneous acid load generated by local cellular activity.

SIGNALLING IN THE BRAIN: ROLE OF H^+ IONS

Slight changes in intracellular or extracellular pH can produce conspicuous effects on the biochemical, ion-regulatory, or electrical machinery of nerve and glial cells. An interesting aspect of the very low physiological concentration range of protons is that it applies to both the intracellular and extracellular compartments. In this respect, the situation differs from that of the Ca^{2+} ion, a well-established signalling factor, which is (in agreement with its second-messenger function) kept at a very low level only *within* cells. By contrast, H^+ ions may have a signalling role both within and between brain cells, as discussed in various chapters of this volume (Chapters 4, 16, 17, and 20–27).

Generation of H^+ Transients

The mechanisms that generate pH transients within brain tissue belong to two basic categories: those involving *de novo* production of acid-base species and those caused by net transmembrane fluxes. All of the former kind are of metabolic origin, in most cases resulting from changes related to energy metabolism (Chapters 5, 28, and 32). The latter are caused by the operation of plasmalemmal channels and carriers that transport acid-base species (Chapters 10–19). The most thoroughly studied examples of such mechanisms are ion-transport mechanisms involved in cellular pH homeostasis, as well as ion channels permeable to HCO_3^- and H^+.

Targets of H^+ Modulation

All those familiar with electrophoresis techniques are aware that the ionization of protein side groups (and hence the net charge of the molecule) can be dramatically influenced by ambient pH. This does not mean, however, that all proteins will show a structural/functional sensitivity to changes in pH in the physiological (or even pathophysiological) range. In fact, a look at various closely related proteins, such as, distinct members of ionotropic receptor families gated by glutamate and γ-aminobutyric acid (GABA), reveals a high (both qualitative and quantitative) variation in pH sensitivity (Chapter 22). This diversity is hardly likely to reflect random selection during the evolution of the receptor families.

A pH-sensitive site without a biological function would only confer a possible source of functional instability on a protein molecule. In fact, it is not unreasonable to assume that the presence in any protein of allosterically important amino-acid side chains prone to pH modulation in the physiological range is likely to point to functional H^+ regulation of that protein. In other words, some proteins (or sometimes only certain isoforms thereof, see Chapter 22), have been "tuned" in the course of evolution to be extremely sensitive to the slightest changes in ambient pH.

It is worth noting that the pK of an amino-acid residue in a protein is strongly affected by the microenvironment offered by adjacent residues, and that conformational changes at one site may influence the pK of some distant site (e.g., Antosiewicz et al., 1994). These factors provide a high number of degrees of freedom in the evolutionary design and, perhaps, in the developmental expression patterns of protein isoforms with distinct sensitivities to pH.

A further factor that adds to the spectrum of H^+-mediated actions at the cellular level is the profound effect that modest changes in pH have on the fluidity of lipid membranes. While it has been known for a long time that protein function in biological membranes is sensitive to changes in lipid fluidity, there is not much information available on fluidity effects mediated by pH in nervous tissue.

pH Modulation of Brain Function

The peripheral nervous system, as well as the CNS, contain neuronal elements involved in the regulation of breathing, whose excitability is altered by changes in pH_i (Chapters 26–27). This enables them to function as CO_2 sensors, and thus to monitor respiratory efficacy. Neural excitability in most regions of the brain is suppressed by acidosis and enhanced by alkalosis (Chapters 16–17, and 20–25). This dependency of excitability on pH appears to be due to a large number of factors that act on both intrinsic neuronal excitability, as well as on both excitatory and inhibitory synaptic transmission. Moreover, the sites of H^+ action are both intracellular and extracellular.

A large number of proteins immediately involved in the control of neuronal excitability appear to be prone to pH modulation. These include not only various kinds of ligand-gated channels as already mentioned, but also voltage-sensitive Ca^{2+} and K^+ channels (Chapter 21). The literature on H^+ effects on channels span an extremely wide pH range, but one must note that the purpose of much of this work has been to probe ion-channel structure and function, using pH variations far outside the physiological range as an experimental tool (Hille, 1992), rather than to look for a possible modulatory role of physiological or pathophysiological pH shifts.

An intriguing idea that has emerged from studies of pH-dependent alterations in excitability is that highly localized pH transients might play a signalling role in neuronal communication. For instance, the fast extracellular alkaline shifts that are associated with excitatory transmission may act on postsynaptic N-methyl-D-aspartate (NMDA) receptors (and perhaps on voltage-gated Ca^{2+} channels) to enhance excitability and Ca^{2+} uptake in the postsynaptic neuron. Evidence of this kind raises the possibility that such "intrinsic proton modulation" of neuronal function could play a central role in synaptic plasticity (Chapter 17). Recent evidence for a tight coupling between intraneuronal regulation of Ca^{2+} and H^+ through a plasmalemmal Ca^{2+}/H^+ ATPase (Chapters 15 and 16) further underscores this idea. In fact, it is quite possible that changes in neuronal excitability and synaptic plasticity, hitherto solely attributed to intracellular Ca^{2+} transients, may include a significant component mediated by pH shifts.

When considering the actions of pH on brain function, a picture emerges where pH has effects at progressively more complex levels of organization. The effects of activity-induced pH shifts on receptors, channels, and carriers at the molecular level give rise to changes in neuronal excitability and communication that may even persist on longer time scales. This will, of course, have an influence on neuronal network function and, in the living organism, on behavioral and other observable expressions of the workings of the brain.

Even though its molecular basis is not completely understood, breathing is a nice example of "higher level" phenomena involving modulation of nervous function by pH and

CO_2. The coupling of brain excitability to breathing is a very tight one, and it is common knowledge that hyperventilation leads to hyperexcitability, while hypoventilation has an opposite effect (Chapter 20). In fact, it appears that controlling one's state of mind by voluntary control of breathing has been used for millennia in diverse meditation techniques developed in oriental cultures.*

pH AND BRAIN PATHOPHYSIOLOGY

It is a truism that studies unraveling novel physiologically important phenomena are bound to uncover novel factors with a pathophysiological role of their own. This, of course, applies also to the regulation and modulation of pH in the brain (Chapters 28–33). Although the topic of pH and brain pathophysiology is not well understood, it is of great potential importance in fathoming how diseases affect nervous system function. Moreover, such knowledge may have relevance for the treatment of these diseases.

A major area of interest is the role of pH-mediated effects in the molecular and cellular mechanisms controlling neuronal apoptosis and necrosis. Early work in this field suggested that brain acidosis augments cell death under various pathophysiological conditions, especially stroke (Chapters 28–31). However, more recently it has become clear that, at least under certain experimental conditions, a slight acidosis may, in fact, be neuroprotective against, for instance, excitotoxic insults associated with an excessive activity of glutamate-gated ion channels. The details of this H^+-mediated protective action may be complex and probably involve more than just an inhibition of glutamate-gated receptors of the NMDA type. A clinically relevant question that emerges here is, whether it would be possible to design therapeutic interventions to reduce the degree of brain cell injury. An intriguing recent finding is that some endonucleases are strongly activated by a fall in intracellular pH (Barry and Eastman, 1993), indicating a possible direct involvement of pH in the triggering of programmed cell death. This takes on added significance in the light of the fact that apoptosis is now realized to play a role in stroke-related cell loss.

Epilepsy is a very common form of abnormal brain function that appears to be influenced by pH. The factors that provoke and halt seizure discharge are still not well understood, but, as previously mentioned, changes in CO_2 partial pressure produced by voluntary changes in breathing pattern can affect seizure likelihood. Specifically, hyperventilation, which removes CO_2, can precipitate a seizure in a susceptible person. More work on the role of pH in epilepsy is needed, and it may turn out to have great practical importance in developing treatment strategies. For example, ketogenic diets can be beneficial in certain forms of childhood epilepsy and appear to help by inducing a systemic acidosis. One wonders whether drugs capable of inducing a mild brain acidosis would also have an antiepileptic action.

Finally, there are general questions about how brains afflicted by chronic degenerative diseases (e.g., Parkinson's disease, amyotrophic lateral sclerosis, Alzheimer's disease) manage their pH regulation. Glial cells may be important in this issue because they are

*A provocative recent study reports an apparent link between brain cell pH and IQ (Rae et al., 1996). Using 31P magnetic resonance spectroscopy to measure intracellular pH in 42 boys, a significant positive correlation was found between brain pH and performance in various tests of intelligent behavior, such as those used to determine the WISC-III Full Scale IQ.

often present in higher numbers in diseased brains. Microglial cells tend to proliferate at sites of injury or disease, and astrocytes become altered to form what is known as astrogliosis. Do these cellular changes alter regional pH behavior in the brain, and does this affect the disease process? Another interesting point worth noting is that the vast majority of primary neoplastic diseases of the CNS involve glial cells. Since there is suggestive work to indicate that changes in pH_i can strongly influence cellular proliferation, the possibility emerges that therapeutic maneuvers to control glial pH_i might be useful in the treatment of brain tumors.

REFERENCES

Antosiewicz J, McCammon JA, Gilson MK (1994): Prediction of pH-dependent properties of proteins. J Mol Biol 238: 415–436.

Balestrino M, Somjen GG (1988): Concentration of carbon dioxide, interstitial pH and synaptic transmission in hippocampal formation of the rat. J Physiol 396: 247–266.

Barry MA, Eastman A (1993): Identification of deoxyribonuclease II as an endonuclease involved in apoptosis. Arch Biochem Biophys 300: 44–450.

Cameron JN (1989): Acid-base homeostasis: Past and present perspectives. Physiol Zool 62: 845–865.

Hazel JR, McKinley SJ, Williams EE (1992): Thermal adaptation in biological membranes: Interacting effects of temperature and pH. J Comp Physiol B 162: 593–601.

Hille B (1992): Ionic Channels of Excitable Membranes, 2nd ed. Sunderland, MA: Sinauer.

Hochachka PW, Somero GN (1984): Biochemical Adaptation. Princeton, NJ: Princeton University Press.

Newman EA (1996): Acid efflux from retinal glial cells generated by sodium bicarbonate cotransport. J Neurosci 16: 159–168.

Rae C, Scott RB, Thompson CH, Kemp GJ, Dumughn I, Styles P, Tracey I, Radda GK (1996): Is pH a biochemical marker of IQ? Proc R Soc Lond B 263: 1061–1064.

Reeves RB (1972): An imidazole alphastat hypothesis for vertebrate acid-base regulation: Tissue carbon dioxide and body temperature in bullfrogs. Respir Physiol 14: 219–236.

CHAPTER 2

Principles and Practical Aspects of pH Buffering

MITCHELL CHESLER
Department of Physiology and Neuroscience and Department of Neurosurgery N.Y.U. Medical Center New York, NY 10016

RELEVANCE OF pH BUFFERING

The principles of pH buffering are important to neurobiologists for both practical and theoretical reasons. In the experimental setting, a given stable pH is often expected. However, the pH of extracellular or intracellular fluid may be far different from the value assumed. This is often the result of endogenous acid production, the permeability of membranes to neutral form of acids and bases, and the generation of pH shifts by a variety of membrane-based transport systems. To understand how these factors contribute to the regulation and modulation of pH, a quantitative appreciation of buffering is required, as this determines the size and speed of a pH shift. Moreover, for a given pH change, the magnitude of the underlying physiological processes can be interpreted only if the buffering capacity is known. The purpose of this chapter is to outline the principles of pH buffering and raise issues of particular concern in the context of experimental neurobiology. This discussion begins with a brief outline of classical buffering theory, then ventures into the concerns and pitfalls in the experimental setting.

Classical Buffering Concepts

In principle, the change in pH resulting from a perturbation in acid-base status can be expressed in a variety of ways. Thus, "buffering capacity" can be formalized as the change in pH resulting from either a change in CO_2 tension, a shift in nonvolatile weak acid or base content, or a change in strong acid or base concentration (Roos and Boron, 1981; Chesler, 1990). The latter conceptualization is the most general, and has found greatest use in the biological setting. The buffering power (β) of an aqueous solution is therefore defined as the change in pH arising per incremental concentration change of strong base or acid. Thus,

$$\beta = \frac{db}{d\text{pH}} \tag{1}$$

pH and Brain Function, Edited by Kai Kaila and Bruce R. Ransom
ISBN 0-471-11838-9

or,

$$\beta = -\frac{da}{dpH} \tag{2}$$

where *db* and *da* represent incremental changes in the concentration of strong base or strong acid, respectively. Note, that because pH is dimensionless, the buffering capacity is expressed in terms of concentration. The total concentration of "strong" acid or base in solution is sometimes referred to as the "strong ion difference" or "SID."

Koppel and Spiro (1914) first quantified the buffering capacity of weak acid and weak base solutions, using the above definition (an English translation of their classic work is provided by Roos and Boron (1980)). For a simple weak acid HA, which dissociates into A^- and H^+ with a dissociation constant K_a,

$$\beta = \frac{db}{dpH} = \frac{2.3[H^+][A_t][K_a]}{([H^+] + K_a)^2} \tag{3}$$

where $[A_t] = [HA] + [A^-]$. For such a system, the maximum buffering occurs when $pH = pK_a$. Equation 3 then reduces to

$$\beta = 0.575\ [A_t] \tag{4}$$

An equivalent pair of equations can be derived for a simple weak base. For a 10 mM HEPES-buffered solution where the pH is equal to the pK_a of 7.55, Eq. 4 predicts a buffering power of 5.75 mM. At a pH of 7.0, the buffering power calculated from Eq. 3 would be 3.95 mM. For a more complex solution containing a number of weak acids and bases, the total buffering capacity of the solution (β_T) is simply the algebraic sum of the buffering capacities of the individual components. Thus, for a solution containing weak acids a–j, the total, buffering capacity (β_T) is given by

$$\beta_T = \beta_a + \beta_b + \ .\ .\ . + \beta_j \tag{5}$$

where β_a–β_j are the respective buffering powers contributed by each weak acid–conjugate base pair.

In the intracellular compartment, a significant fraction of the total buffering power is provided by imidazole groups on the histidine residues of proteins and by the free and bound forms of phosphate (Burton, 1978). A large component of biological buffering is also provided by carbon dioxide and bicarbonate. Buffering occurs through the equilibria with carbonic acid (H_2CO_3)

$$CO_2 + H_2O \overset{K_1}{\rightleftharpoons} H_2CO_3 \overset{K_2}{\rightleftharpoons} \underset{\displaystyle H^+}{\underset{+}{HCO_3^-}} \overset{K_3}{\rightleftharpoons} CO_3^{2-} + H^+ \tag{6}$$

where K_1, K_2, and K_3 are the equilibrium constants for the respective hydration and dissociation reactions. Van Slyke (1922) noted that in many biological settings, where local

CO_2 is in rapid equilibrium with a large fixed source, the $[CO_2]$ can be treated as a constant (i.e., the system is "open" with respect to CO_2). Under these conditions the buffering power provided by CO_2 and HCO_3^- (βb) is much greater than that of an equivalent concentration of weak acid or base in a closed system, and

$$\beta b = 2.3\,[HCO_3^-] + 4.6[CO_3^{2-}] \tag{7}$$

(Roos and Boron, 1981). At physiological pH, the contribution of CO_3^{2-} is negligible and βb can be approximated by

$$\beta b = 2.3\,[HCO_3^-] \tag{8}$$

Comparing Eqs. 4 and 8 indicates that for equivalent concentrations of buffer, a CO_2–HCO_3^- system open to CO_2 will provide roughly four times the buffering capacity of a closed-system weak-acid buffer at its optimum pH. It is also important to note that at a fixed P_{CO_2}, βb will increase markedly as fluid becomes more alkaline. This is apparent upon rearrangement of the Henderson-Hasselbalch equation:

$$[HCO_3^-] = \frac{s \times P_{CO_2}}{10^{pK_a}} \times 10^{pH} \tag{9}$$

where s is the CO_2 solubility constant. It is evident that the $[HCO_3^-]$ will increase exponentially as a function of pH if the CO_2 tension is held constant.

Although the buffering capacity of a CO_2–HCO_3^- system can be considerable, the assumptions inherent in the above formulation are not always valid. Under conditions where the $[CO_2]$ is not fixed, the buffering capacity provided by the CO_2–HCO_3^- equilibria will be considerably less than that predicted by Eq. 8. A marked decrease in the effectiveness of CO_2–HCO_3^- buffering may also be evident if the acid-base challenge is extremely rapid. This can occur because the hydration of CO_2 is a slow reaction, with a time constant on the order of 20–30 seconds (Maren, 1967). In many biological settings, the rate of this reaction is speeded by the enzyme carbonic anhydrase (Chapter 3). However, where this enzyme is absent, insufficient, or inhibited, the ability to buffer rapid acid-base challenges becomes compromised.

INTRACELLULAR BUFFERING POWER

Definitions

The notion that a cell type has a single intracellular buffering capacity is an oversimplification. In an idealized sense, the intracellular buffering power of a cell is usually taken to mean the physicochemical buffering capacity of the cytosol. However, determination of the isolated buffering capacity of this fluid is experimentally elusive. When analyzing the intracellular pH shifts arising from imposed acid-base challenges, it is often difficult to discern the contribution of metabolism, or the effects of acid transport across the plasmalemma or the membranes of organelles.

Thus, intracellular buffering capacity would ideally refer only to the buffering capacity of the cytoplasm (Katsura et al., 1993). However, the additional intracellular processes that might influence cytoplasmic pH remain ill-defined, and are experimentally difficult to exclude. Operational definitions of intracellular buffering capacity, as emphasized by Roos and Boron (1981), are therefore unavoidable. These authors suggested that for practical purposes, the "intracellular buffering" would refer to all of the processes that mitigate a cytoplasmic acid-base challenge, *exclusive of plasma membrane transport.* This definition allows for experimental approaches to intracellular buffering capacity, although the exclusion of plasmalemmal transporters may be difficult to achieve in practice.

Since the buffering capacity of a system is given by the summed contribution of the constituent buffers, cytoplasmic buffering is often conceptually divided into its CO_2/HCO_3^--dependent component, and the remaining contributors. These second factors constitute the "intrinsic buffering capacity" of the cell. Thus, for cytoplasm, the total buffering capacity (β_T) is given by

$$\beta_T = \beta_b + \beta_i \tag{10}$$

where β_b and β_i refer to the bicarbonate-dependent and the intrinsic components. Ideally, these components would refer to the pure physicochemical buffering power of the cytoplasmic water. Thus, β_i would derive largely from the titratable amino acids and phosphates, whereas β_b would refer to the buffering provided by CO_2 and HCO_3^-. In neurons and glial cells, β_i typically ranges between 10 and 20 mM, whereas β_b, at a pH_i of 7.0 and Pco_2 of 40 mmHg, is approximately 23 mM. Thus the total intracellular buffering capacity (β_T) of neurons and glial cells is on the order of 30–50 mM.

In practice, the determination of these buffering capacities involves the imposition of quantified acid or base loads while measuring the change in intracellular pH. Most methods involve exposure of a cell to a weak acid (e.g., CO_2 or propionic acid) or a weak base (e.g., NH_3 or trimethylamine). The concepts and pitfalls associated with these techniques are discussed in a number of sources (Roos and Boron, 1981; Thomas, 1984; Chesler, 1990). Alternative methods involve the direct titration of tissue or cell homogenates using NaOH or HCl. A cogent discussion of this approach and its drawbacks is provided in a recent approach contribution by Siesjo and colleagues (Katsura et al., 1993).

Why Measure Intracellular Buffering Capacity?

The modulation of intracellular pH by acid transporters involves a small net change in H^+ activity (several nanomolar range). In the face of considerable intracellular buffering capacity, however, the necessary fluxes of H^+ equivalents amount to millimoles per liter. To account for this acid transport, therefore, requires measurement of the intracellular buffering capacity. Typically, the molar flux of acid equivalents (J) across a cell is determined from the rate of a pH change and the intracellular buffering power. Thus, in general

$$J = \frac{dpH}{dt} \times \beta_T \tag{11}$$

where J is expressed in millimoles liter^{-1} seconds^{-1}, and β_T is the total intracellular buffering capacity. It is evident that if the buffering capacities under two conditions are not the same, then identical rates of pH change will not represent equivalent rates of transport. In practice, this consideration arises when comparing intracellular pH shifts in

HCO_3^- versus HCO_3^--free media. For example, if the rate of pH recovery from an acid load is identical in HCO_3^- versus HEPES-buffered media, one must conclude that the molar flux was greater in the former condition, since β_T would be larger in the presence of intracellular bicarbonate.

EXTRACELLULAR BUFFERING POWER

The extracellular space of excitable tissue is subject to large, rapid changes in pH. These shifts result from normal electrical activity, and become exaggerated under pathological conditions, such as spreading depression or ischemia (Chesler, 1990; Chesler and Kaila, 1992; reviewed in Chapter 5 and 28). The speed and size of these pH changes will determine their impact on pH-sensitive processes, such as N-methyl-D-aspartate (NMDA) receptor gating (see Chapter 22). However, the magnitude of these pH shifts, and therefore their functional impact, ultimately depend on the buffering properties of the extracellular compartment.

Since interstitial buffering in the nervous system is provided almost exclusively by CO_2 and HCO_3^-, the buffering capacity at equilibrium would be given by Eq. 8, provided the P_{CO_2} was constant. However, microelectrode studies of CO_2 have demonstrated that the extracellular P_{CO_2} changes during repetitive neural activity. In hippocampal slices, a rapid fall of a few millimeters of mercury was followed by a prolonged elevation of the P_{CO_2}, accounting for the late fall in extracellular pH (Voipio and Kaila, 1993). Because the P_{CO_2} does not remain clamped, the effective buffering power must be less than that predicted by Eq. 8 (see Chapter 4).

Whether CO_2 and HCO_3^- remain in equilibrium may also be questioned. While there is evidence for extracellular carbonic anhydrase in brain slices (Chen and Chesler, 1992a, b; Kaila et al., 1992), it is unclear whether the enzyme is present in sufficient concentration to insure equilibration. Studies with CO_3^{2-}- and pH-sensitive microelectrodes have indicated that the rise in HCO_3^- during stimulus-evoked alkaline shifts was far less than anticipated if the buffering system were to remain in equilibrium (Chesler and Chen, 1993). Moreover addition of as little as 10^{-7} M carbonic anhydrase to the superfusing Ringer's solution approximately halved extracellular alkaline transients in brain slices (Huang et al., 1995).

EXPERIMENTAL CONSIDERATIONS

Extracellular pH in Brain Slice Preparations

Measurements with pH microelectrodes have established that the extracellular pH of brain slices is lower than that of the superfusate by as much as several tenths of a pH unit (Balestrino and Somjen, 1988; Walz, 1989; Chen and Chesler, 1992a). To normalize extracellular pH, one can modestly raise the bicarbonate concentration of the Ringer's solution (Chen and Chesler, 1992a). However, the interstitial acidosis in the hippocampal slice is almost entirely due to the endogenous generation of CO_2 by the tissue (Voipio and Kaila, 1993; Chesler et al., 1994). Therefore, the more appropriate Ringer's solution may be one equilibrated with less than the typical 5% CO_2. For example, using a superfusate with a P_{CO_2} of 40 mmHg ($\sim$ 5%), the P_{CO_2} of the slice may be approximately 60 mmHg. The superfusate P_{CO_2}

might then be reasonably lowered to 20 mmHg (2.5%) to achieve a P_{CO_2} of approximately 40 mmHg within the slice. While the pH of the Ringer's solution would then rise to roughly 7.7, the pH in the slice would be far closer to the *in vivo* range of 7.25–7.35 (Cragg et al., 1977; Javaheri et al., 1983; Kraig et al., 1983). Slices in submersion chambers may have a greater accumulation of tissue CO_2 owing to more limited exchange with the surround. Consequently, compensation for endogenous acidosis might require lowering the P_{CO_2} of the Ringer's solution to a value below that required for an interface chamber. In practice, however, it is probably necessary to first measure the extracellular pH in one's own preparation, since the generation and dispersal of CO_2 will depend upon the condition and thickness of the tissue, as well as the exchange rate of the superfusate.

Consideration of Temperature

Temperature is an additional factor when considering the pH of an *in vitro* preparation. In many brain slice preparations the temperature is maintained considerably below 37°C. Acutely dissociated cells are often studied at room temperature. An obvious effect of cooling is to slow the metabolic production of CO_2, which, in the brain slice, would be expected to considerably lessen the endogenous acidosis. However, this effect may be somewhat countered by the temperature-dependence of CO_2 solubility; as the temperature falls, more CO_2 enters solution, causing the pH to fall. A decrease in temperature can also be expected to slow the rate of acid transporters, particularly the ATP-coupled mechanisms. The net effect of temperature on acid transport may be complex, however, due to the assortment of acid loading and acid extruding mechanisms among cells (see Chapter 10).

Effects of Endogenous CO_2 on Intracellular Buffering Capacity

The accumulation of CO_2, in addition to causing an endogenous acidosis, can dramatically increase the intracellular buffering power. For instance, if an intracellular pH of 7.0 were maintained in the face of twice the normal (5%) P_{CO_2}, then the intracellular bicarbonate concentration would have doubled from roughly 10 to 20 mM. Accordingly, the bicarbonate contribution to buffering would have increased from 23 to 46 mM. In view of the high P_{CO_2} in slices, changes in intracellular pH arising from neural activity may therefore be blunted in these preparations, compared with the *in vivo* condition.

Use of HEPES Buffer in Brain Slice Preparations

Although uncommon, *in vitro* CNS preparations have occasionally been used with non-bicarbonate buffers such as N-2-hydroxyethylpiperazine-N′-2-ethanesulfonic acid HEPES (Pittman et al., 1981, Chesler and Chan, 1988; Chesler and Rice, 1991; Church, 1992; Gottfried and Chesler, 1994). Use of HEPES has sometimes been associated with a greater extracellular acidosis (Chesler and Rice, 1991), most likely due to insufficient buffer concentration. Using 26 mM HEPES in 100% O_2, Gottfried and Chesler (1994) noted that the extracellular pH of hippocampal slices was comparable to that of slices in standard 26 mM bicarbonate Ringer's solution (5% CO_2, 95% O_2). In this context, it is important to point out that the comparison of pH buffering in the slice interstitium is not analogous to the quantitative descriptions of buffering in a beaker, since all buffer species are free to shuttle between tissue and the superfusate (see Chapter 4).

Ringer's solution buffered with HEPES is often used to study a preparation under nominally bicarbonate-free conditions. However, in a brain slice, the endogenous production of CO_2 will insure that the tissue is far from bicarbonate-free, despite the lack of bicarbonate in the Ringer's solution. Assuming the interstitial acidosis is due mainly to CO_2 production, it is likely that a bicarbonate concentration of a few millimolar is generated in the extracellular and intracellular compartments. Since intracellular pH regulation is heavily reliant on bicarbonate transport (see Chapters 10–14), the persistence of bicarbonate may promote slice viability. The generation of bicarbonate will also augment both the extracellular and intracellular buffering capacity.

While organic buffers such as HEPES are usually assumed to be inert, associated changes in membrane properties have been occasionally reported. Such effects are discussed in a useful contribution by Deitmer and colleagues (Schmidt et al., 1996), who noted the induction of a Cl^- current when leech neurons were exposed to some of these agents. As these authors point out, caution is warranted when using non-bicarbonate buffers, particularly since the side effects of these compounds are unlikely to be uniform across different preparations.

Microelectrode Filling Solutions—Fine Micropipettes

High-resistance micropipettes used for conventional intracellular recording are typically filled with potassium salts. To avoid leakage of Cl^- into cells, and to provide favorable current passing characteristics, a popular choice has been potassium acetate, typically used in concentrations of 3–4 M. Although the pK_a of acetic acid (~4.7) is far from physiological pH, molar concentrations have a considerable buffering capacity. However, the intracellular acetate concentration is likely to only rise by 10–100 mM at most, making the buffering capacity and the pH of the filling solution a small concern.

Of greater importance, may be the ability of the neutral acid to shuttle hydrogen ions across the plasma membrane. Inclusion of millimolar acetate in Ringer's solution will cause an intracellular acidification, due to the permeation of neutral acetic acid (Roos and Boron, 1981; Thomas, 1984). By the same mechanism, a leakage of acetate into cytoplasm would result in the efflux of acetic acid from the cell, with a resultant increase in intracellular pH (Fig. 1A).

An intracellular pH record using a reference electrode containing 3 M potassium acetate is shown in Fig. 1B. After penetration of a giant lamprey neuron by the electrode the intracellular pH was about 7.5, near the average value for these cells (Chesler, 1986). Over the next 5 minutes, the pH gradually alkalinized towards 7.9. A similar alkaline drift was noted using 3 M lithium acetate, but was never seen when using either 3 M KCl, or low concentrations of acetate as the reference filling solution (Chesler, 1986).

Although extremely fine pipettes are used in typical intracellular recording situations, leakage of electrolyte can be significant, particularly when recording from small cells. In these instances, use of 4 M potassium acetate may place a considerable base load on the cell, leading to an unphysiological rise in intracellular pH. By utilizing a salt of a strong acid and strong base, intracellular alkalinization can be avoided. For example, solutions of potassium sulfate or potassium methanesulfonate will have negligible concentrations of neutral acid, due to their low pK_a values. While the former solution saturates at concentrations of only 0.6–0.7 M, molar solutions of potassium methanesulfonate are attainable.

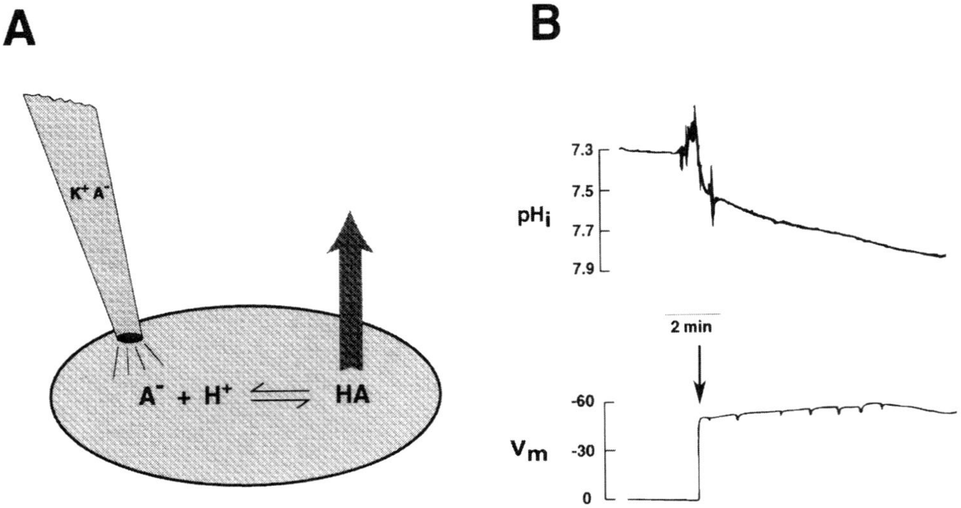

Figure 1. Effect of acetate filling solution on intracellular pH. **A.** A mechanism of intracellular alkalinization. Efflux and disappearance of neutral acetic acid (HA) leads to loss of intracellular H^+. **B.** Use of 3 M potassium acetate in the reference side of a pH microelectrode was associated with a progressive intracellular alkalinization in lamprey reticulospinal neurons (M. Chesler, unpublished observation).

Microelectrode Filling Solutions—Whole Cell Pipettes

In the whole-cell recording configuration, pipettes are typically filled with a buffered electrolyte solution. Commonly, the pH of the filling solution is set in the range of 7.0–7.4, using 10 mM HEPES titrated with KOH or CsOH. Given a pK_a of 7.5, the buffering capacity of such a solution is roughly 5 mM. As this may be one fourth to one half the intrinsic buffering capacity of the cell, which is provided by largely nonmobile (and nondialyzable) amino acid residues, the cell interior is unlikely to be clamped to the pH of the filling solution. Higher concentrations of HEPES have found recent use, as these solutions seemed to diminish the run down of high-threshold calcium currents (Hockberger and Nam, 1994). While the basis of this effect is unclear, these observations suggest that the pH stability is less than optimal in most whole-cell experiments.

When whole-cell recording is performed in CO_2-HCO_3^--buffered media, control over internal pH is more unlikely. In brain slice recordings, for example, the ambient P_{CO_2} is roughly 60–80 mmHg, representing a virtually infinite CO_2 source from the standpoint of the cell. In the face of this CO_2 load, the internal pH is likely to deviate considerably from the pH of the filling solution. In addition, the diffusion of CO_2 into the cell interior will cause the generation of bicarbonate, due to the HEPES and the intrinsic buffers, altering the ionic composition and the assumed buffering properties of the cell interior.

Although dialyzed with filling solution, the cell interior may also be subject to changes in pH due to membrane transport. The effectiveness of plasma transport mechanisms in the whole-cell configuration is not clear. However, any alteration of the intracellular pH from its normal value can be expected to be countered by the variety of acid transporters discussed in Chapters 10–14.

CONCLUDING REMARKS

In the classic study of cellular pH regulation, experimental design depended on the understanding of acid-base chemistry, including a quantitative appreciation of buffering power, and an awareness of buffering kinetics. These topics remain critical in the merging of acid-base physiology and neurobiology brought about in this text. Studies described in subsequent chapters demonstrate that modest changes in H^+ activity can affect diverse cellular processes, and that local pH can change rapidly in neural tissue. It is therefore evident that the dynamics of pH and pH buffering should be important considerations in the design and interpretation of many neurobiological experiments.

REFERENCES

Balestrino M, Somjen GG (1988): Concentration of carbon dioxide, interstitial pH and synaptic transmission in hippocampal formation of the rat. J Physiol 396: 247–266.

Burton RF (1978): Intracellular buffering. Resp Physiol 33: 51–58.

Chen JCT, Chesler M (1992a): Modulation of extracellular pH by glutamate and GABA in rat hippocampal slices. J Neurophysiol 67: 29–36.

Chen JCT, Chesler M (1992b): pH transients evoked by excitatory synaptic transmission are increased by inhibition of extracellular carbonic anhydrase. Proc Natl Acad Sci USA 89: 7786–7790.

Chesler M (1986): The regulation of intracellular pH in reticulospinal neurones of the lamprey, *Petromyzon marinus*. J Physiol 381: 241–261.

Chesler M (1990): The regulation and modulation of pH in the nervous system. Prog Neurobiol 34: 401–427.

Chesler M, Kaila K (1992): Modulation of pH by neuronal activity. Trends in Neurosciences 15: 396–402.

Chesler M, Rice ME (1991): Extracellular alkaline-acid pH shifts evoked by iontophoresis of glutamate and aspartate in turtle cerebellum. Neuroscience 41: 257–267.

Chesler M, Chan CY (1988): Stimulus-induced extracellular pH transients in the in vitro turtle cerebellum. Neuroscience 27: 941–948.

Chesler M, Chen JCT (1993): Evoked shifts in extracellular pH, bicarbonate and CO_2 in rat hippocampal slices. Soc Neurosci Abstr 19: 1522.

Chesler M, Chen JCT, Kraig RP (1994): Determination of extracellular bicarbonate and carbon dioxide concentrations in brain slices using carbonate and pH-selective microelectrodes. J Neurosci Methods 53: 129–136.

Church J (1992): A change from HCO_3^--CO_2-to hepes-buffered medium modifies membrane properties of rat CA1 pyramidal neurones in vitro. J Physiol 455: 51–71.

Cragg P, Patterson L, Purves MJ (1977): The pH of brain extracellular fluid in the cat. J Physiol 272: 137–166.

Gottfried JA, Chesler M (1994): Endogenous H^+ modulation of NMDA receptor-mediated EPSCs revealed by carbonic anhydrase inhibition in rat hippocampus. J Physiol 478: 373–378.

Hockberger PE, Nam SC (1994): High-voltage-activated calcium current in developing neurons is insensitive to nifedipine. Eur J Physiol 426: 402–411.

Huang, W, Smith, SE, Chesler, M (1995): Addition of carbonic anhydrase augments extracellular pH buffering in rat cerebral cortex. J Neurophysiol 74: 1806–1809.

Javaheri S, De Hemptinne A, Vanheel B, Leusen I (1983): Changes in brain ECF pH during metabolic acidosis and alkalosis: A microelectrode study. J Appl Physiol 55: 1849–1853.

Kaila K, Paalasmaa P, Taira T, Voipio J (1992): pH transients due to monosynaptic activation of GABA A receptors in rat hippocampal slices. Neuro report 3: 105–108.

Katsura K, Mellergard P, Theander S, Ouyang Y, Siesjo BK (1993): Buffer capacity of rat cortical tissue as well as of cultured neurons and astrocytes. Brain Res 618: 283–294.

Koppel M, Spiro K (1914): Uber die Wirkung von Moderatoren (Puffern) bei der Werschiebung des Saurebasengleichge wichtes in biologischen Flussigkeiten. Biochem Z 65: 409–439.

Kraig RP, Ferreira-Filho CR, Nicholson C (1983): Alkaline and acid transients in cerebellar microenvironment. J Neurophysiol 49: 831–850.

Maren TH (1967): Carbonic anhydrase: Chemistry, physiology, and inhibition. Physiol Rev 47: 595–781.

Pittman QJ, Hatton JD, Bloom FE (1981): Spontaneous activity in perfused hypothalamic slices: Dependence on calcium content of the perfusate. Exp Brain Res 42: 49–52.

Roos A, Boron WF (1980): The buffer value of weak acids and bases: Origin of the concept, and first mathematical derivation and application to physico-chemical systems—The work of M. Koppel and K. Spiro (1914). Resp Physiol 40: 1–32.

Roos A, Boron WF (1981): Intracellular pH. Physiol Rev 61: 296–433.

Schmidt J, Mangold C, Deitmer JW (1996): Membrane responses evoked by organic buffers in identified leech neurones. J Exp Biol 199: 327–335.

Stewart PA (1978): Independent and dependent variables of acid-base control. Resp Physiol 33: 9–26.

Thomas RC (1984): Experimental displacement of intracellular pH and the mechanism of its subsequent recovery. J Physiol 354: 3P–22P.

Van Slyke DD (1922): On the measurement of buffer values and on the relationship of buffer value to the dissociation constant of the buffer and the concentration and the reaction of the buffer solution. J Biol Chem 52: 525–570.

Voipio J, Kaila K (1993): Interstitial P_{CO_2} and pH in rat hippocampal slices measured by means of a novel fast CO_2/H^+-sensitive microelectrode based on a pVC-gelled membrane. Pflugers Arch 423: 193–201.

Walz W (1989): pH shifts evoked by neuronal stimulation in slices of rat hippocampus. Can J Physiol Pharmacol 67: 577–581.

CHAPTER 3

Carbonic Anhydrase Isoforms in the Mammalian Nervous System

YVONNE RIDDERSTRÅLE AND PER J WISTRAND
Department of Animal Physiology, Swedish University of Agricultural Sciences, Uppsala, Sweden (Y.R.); and Department of Ophthalmology, Uppsala University Hospital, Uppsala, Sweden (P.J.W.)

INTRODUCTION

Carbonic anhydrases (CA; E.C.4.2.1.1) are zinc metalloenzymes that occur in mammalian tissues as seven (CA I–VII) genetically distinct isozymes, all of which catalyze the reversible hydration of carbon dioxide:

$$CO_2 + H_2O \rightleftharpoons HCO_3^- + H^+$$

All known physiological functions of CA are believed to result from this CO_2/HCO_3^- interconversion (Maren, 1977).

Some isozymes also have esterase activities, but there is as yet no known endogenous substrate (Chapman and Maren, 1978). The isozymes differ kinetically with respect to catalytic activities and sensitivity to inhibition by sulfonamides and anions. Their genes are differently expressed in most tissues, either as soluble (CA I, II, III, V), membrane-associated (CA IV) or secreted (VI, VII) forms. Their amino acid and DNA sequences have been determined and many have been purified to homogeneity. The genes for CA I–VII have been located to chromosomes 1, 8, 16, and 17 (cf. Tashian, 1989; Sly and Hu, 1995).

Carbonic anhydrase activity in the nervous system was first discovered by van Goor in 1940, and then studied by Ashby and her colleagues, who analyzed the distribution of CA activity in normal and pathological human brains, and in brains of several animal species. Ashby and Cobbs (Ashby, 1944) were the first to suggest that the activity originates from oligodendroglia. This was confirmed by Giacobini (1961), who found high CA catalytic activity in isolated single glial cells of the rat brain. Somewhat later, CA activity was also detected in the peripheral nervous system (cf. Riley et al., 1984).

Since then, at least three different isozymes of CA have been discovered in the nervous system. The functional implication of these isozymes with regard to the various postulated

pH and Brain Function, Edited by Kai Kaila and Bruce R. Ransom
ISBN 0-471-11838-9

roles for CA in the nervous system is not known (see also chapter 17). The CA activity in the brain is said to facilitate transport of fluid and ions out of myelin sheaths (Bourke and Kimelberg, 1975), and to deliver bicarbonate ions for biosynthetic synthesis (Cammer, 1991a). It modulates pH in the extracellular fluid in brain parenchyma (Chesler and Kaila, 1992) and has roles in cerebrospinal fluid formation (Vogh, 1980), in the respiratory response to CO_2 (Hanson et al., 1981), and in seizure activity (Anderson et al., 1984).

Carbonic anhydrase inhibitors reduce glial swelling, cerebrospinal fluid pressure, and seizure activity (Millichap, 1958), and have favorable therapeutic effects in hyperventilation and mountain sickness. Congenital CA II deficiency results in mental retardation in certain patients (cf. Sly and Hu, 1995).

In view of the many diverse roles of CA and the effects of inhibitors on brain function, it is appropriate to review the occurrence and properties of the various isoforms of CA, recently found in the central (CNS) and peripheral (PNS) nervous system.

TISSUE DISTRIBUTION AND PROPERTIES OF CA ISOFORMS

Properties of the isoforms are given in Table 1.

The function of CA I, next to hemoglobin the most abundant protein (150 μM) in human erythrocytes, is not known. Genetic lack of CA I in the human erythrocytes does not impair CO_2 transport or other functions (Kendall and Tashian, 1977).

Carbonic anhydrase II has a maximal CO_2 hydration turnover of 2×10^6, under physiological conditions (Wistrand, 1981). Its K_m values for CO_2 and HCO_3^- are well above the tissue concentrations of these substrates. The cytosolic form CA II and the membrane-associated form CA IV appear often to coexist in cells that secrete H^+ or HCO_3^- ions or transport CO_2. CA IV is located either at the basolateral or luminal plasma membranes. The catalytic activities, the hydration of CO_2 in the cytosol by CA II, and the dehydration of H_2CO_3 at the membrane exterior by CA IV, are thought to provide an adequate supply of H^+ and HCO_3^- ions for exchange with other ions for membrane transport or for rapid response to perturbations of the CO_2/HCO_3^- equilibrium in the cytosol. A CA II deficiency syndrome has been discovered by Sly et al. (cf. Sly and Hu, 1995), with clinical manifestations such as mental retardation, cerebral calcification, osteopetrosis, renal tubular acidosis, anemia, and hepatosplenomegaly. The reason for these various symptoms in the patients is not known. Mice deficient in CA II in all tissues show growth retardation, renal tubular acidosis, and vascular calcifications, but not osteopetrosis or cerebral calcifications (Brechue et al., 1991; Spicer et al., 1989).

The function of CA III, the most abundant soluble protein (240 μM) in the red, "slow" muscle fibers, is not known, although it has been suggested that CA III facilitates the diffusion of CO_2 from muscle to blood (Gros, 1991).

Carbonic anhydrase is found in some mitochondria (Lynch et al., 1995; Dodgson, 1991) but not in brain. Its precise location is not known, but it appears to be a soluble protein localized between the mitochondrial membranes. It seems to have a function in the intermediary metabolism by supplying HCO_3^- ions for the mitochondrial carboxylases in lipid, urea, and glucose synthesis.

Carbonic anhydrase VI and CA VII are probably confined to the salivary glands. CA VI is secreted into the saliva, where its presumed function is to buffer salivary pH (Aldred et al., 1991).

TABLE 1 Some Properties of Mammalian CA Isozymes

Isozyme	k_{cat} (sec^{-1})	K_m (mM)	Tissue Distribution	Function
CA I	29,000	4.4	Cytoplasmic in erythrocytes, certain epithelia	Unknown
CA II	236,000 5.5U/μg	10	Cytoplasmic in erythrocytes, secretory epithelia, oligodendroglia, certain neurons, astrocytes, microglia, choroid plexus epithelium	CO_2 and ion transfer
Membrane-associated CA II	5.5U/μg		Associated with membranes of oligodendroglia? Not fully characterized	Compaction of myelin?
Myelin CA II	Lower than CA II		Associated with myelin in CNS; less so in PNS; not fully characterized	Myelin compaction and synthesis?
CA III	1,000	40	Cytoplasmic in red muscle fibres, adipocytes, erythrocytes, choroid plexus epithelium? Microglia	Unknown CO_2 transfer?
CA IV	170,000	10	Bound to plasma membranes of secretory epithelia and vascular endothelium. Choroid plexus epithelium, cerebral capillaries, axons.	CO_2 and ion transfer
CA V	2,000 at pH 7.2 30,000 at pH 9.2	10	Soluble in certain mitochondria, not in brain	Unknown Gluconeogenesis? Ureagenesis
CA VI	6,000		Found only in salivary glands; secreted into saliva.	pH regulation of saliva?
CA VII			Cytosolic mRNA in salivary glands; has not yet been isolated and characterized.	

Isozymes in bold are found in the nervous system. References are found in the text or below the table.
Data were obtained at 0–1°C at pH 7.4. Data on human CA, I, II, and IV, and on cat CA III, are from Maren et al. (1993); those on murine CA V from Heck et al. (1994), and those on human VI from Murakami and Sly (1987). Data on membrane- and myelin-associated CA in brain are from Sapirstein et al. (1984), Sapirstein et al. (1983), and Sapirstein and Lees (1978).

Many cells, like renal tubular cells (Wistrand, 1980) and erythrocytes (Wistrand, 1981), appear to have a large, 20- to 50-fold excess of CA activity in relation to physiological needs, and it would therefore seem unnecessary for the cell to regulate its CA activity. However, there is some evidence that the activity of CA I, II, III, and IV may be regulated either rapidly, involving the cAMP system, or slowly, involving protein synthesis. Such up- or downregulation has been observed in the kidney and brain after changes in hormone levels or acid base balance (Brion et al., 1994a,b; Schwartz et al., 1994).

CHARACTERIZATION AND LOCALIZATION TECHNIQUES

Biochemical Methods

Biochemical methods have been reviewed in detail by Chegwidden (1991).

Affinity chromatography is the best method for purifying the isozymes, since they all bind to sulfonamide inhibitors. However, it is rare that this yields more than 95% purity. The purified enzyme should be assessed for homogeneity by methods suitable for the native, non-SDS-denatured enzyme (SDS-sodium dodecyl sulfate). Substrate (CO_2 and esters) and inhibitor (sulfonamides, anions and SDS) kinetic activities should be measured, because such data can be used to differentiate between various isozymes. The possibility of contamination by other isozymes originating from erythrocytes (CA I, II and III) or capillary endothelium (CA IV and perhaps CA I) should be ruled out.

CA II and CA III have been produced in *Escherichia coli* from cDNA clones (Forsman et al., 1988).

Histochemical Methods

These are all based on the formation of a complex cobalt-phosphate-carbonate precipitate at the enzyme site (Ridderstråle, 1991). They localize the activities of all isozymes, provided they exhibit catalytic activities and are present in high enough tissue concentrations. Model experiments have indicated that tissue concentrations of 0.1 μM or more are needed to stain the high-activity CA II; the corresponding value for the low-activity form CA III is 0.1 mM. This would explain why CA III has been detected by the histochemical methods only in red fibers of striated muscles, where the concentration of CA III is high, about 200 μM. The high-activity forms CA II and CA IV have a concentration of 1–10 μM in the cytoplasm and plasma membranes (Wistrand, 1980) of most cells, and are therefore easily visualized histochemically.

Fixatives inactivate the isozymes to different degrees, which could be crucial if the isozymes exist in low concentrations. Fixation with glutaraldehyde and embedding in water-soluble resin is preferred, due to excellent morphology and high resolution of stain localization (Ridderstråle, 1991). Freeze sectioning preserves more of the enzyme activity but does not regularly allow visualization of the staining of cell membranes or capillary endothelium.

The histochemical precipitate can be visualized by light or electron microscopy. It should be emphasized, however, that even the latter technique does not allow differentiation between a precipitate formed through the activity of a cytoplasmic enzyme, close to or adhering to the inside of a cell membrane, and a precipitate formed from the activity of a membrane-bound enzyme. Neither can the location of the enzymes' active site in the membrane be ascertained.

Carbonic anhydrase III is relatively insensitive to inhibition by sulfonamides. Incubation of tissue slices with 10 μM acetazolamide will inhibit all isozymes but CA III, which therefore can be localized only if present in high enough tissue concentrations, like in certain striated muscles. Sodium dodecyl sulfate inhibits CA I, II, and III, but not CA IV. Incubation of tissue slices with SDS could therefore be tried to preferentially stain CA IV.

Immunocytochemical Methods

Immunocytochemical methods have been described by Gay (1991). All the isozymes are antigenically different and allow the production of antibodies sufficiently selective for use in various immunoassays. However, cross-reactions may occur if another isozyme is simultaneously present in the same cell in higher concentrations than the isozyme to be determined. Monoclonal antibodies are now available. Use of paraformaldehyde, Bouin's fluid, and Carnoy's fixative does not remove antigenicity, and are useful, particularly if perfusion fixation is used for delicate tissues like brain and spinal cord. Lipids are removed from brain tissue prior to the tissue being embedded in paraffin to enable the reagents to permeate the myelinated fibers. However, Carnoy's fixative has been shown to remove myelin proteins (Cammer et al., 1985).

Cryostat sections (Matsumoto and Schwartz, 1992) avoid fixation, dehydration, and embedding solutions. However, as previously noted for histochemistry, this gives a less well preserved morphology.

The peroxidase-antiperoxidase (PAP) complex methods and their newer variants are at least 100 times more sensitive than the catalytic methods for detection of the high-activity CA II and IV in a given tissue. For example, biotin-amplified immunostaining techniques have a sensitivity of 0.2 ng/ml for CA I and II. These immunohistochemical methods can be adapted to the ultrastructural level using electron microscopy (EM) (Ogawa et al., 1993), but it has not yet been possible to localize the membrane-associated isozymes to the interior or exterior part of the plasma membrane.

There is a need to validate these techniques when they are used to localize membrane-associated isozymes, since many results regarding the presence of CA IV in a given tissue have been at variance with those obtained by physiological, histochemical, and biochemical techniques. It is also advantageous to use the histochemical and immunocytochemical methods simultaneously in order to prove the presence of a given isozyme.

Other Methods to Localize CA Isozymes

Other techniques include autoradiography with isotopically labelled inhibitor (Gay et al., 1981) and the use of fluorochrome sulfonamide histochemical probes (Dermietzel et al., 1985). However, these methods are less sensitive than those mentioned above.

Isotopically-labelled acetazolamide could be used in scanning experiments to localize CA isozymes in brain glia and gliomas, after intravenous administration of the inhibitor to the patient (Le Bars et al., 1988).

Inhibitors Used in the Study of Physiological Functions

Some inhibitors that are useful for studies of brain CA are listed in Table 2.

Maren (1977), has outlined the concepts of how to use inhibitors for studying the function of carbonic anhydrase. Isozyme selective inhibitors have been discussed by Lindskog and Wistrand (1989).

There are many strong inhibitors available but they are not discriminatory enough to enable the selective inhibition of CA I, II, V, and CA VI. However, the membrane-associated isozymes, with their active sites facing the outside of the plasma membrane, can be selectively inhibited by membrane-impermeant inhibitors applied to the outside of the

TABLE 2 Inhibitors Useful for Studying Physiological Functions of CA Isozymes in the Nervous System

Inhibitor	Isozyme	Ki (nM)	Comment
Methazolamide	CA II	8	Membrane permeant
	CA III	10,000	
	CA IV	56	
Dorzolamide	CA II	2	Membrane permeant. Inactive against CA I.
	CA III	80,000	
	CA IV	33	
Chlorzolamide	CA II	1	Membrane permeant. Relatively active against CA III.
	CA III	300	
	CA IV	19	
Benzolamide	CA II	1	Relatively membrane impermeant anion, but may be transported into the cell by organic anion transport system
	CA III	3,000	
	CA IV	20	
Quartenized benzolamide	CA II	2.5	Membrane impermeant small cation, but may be transported into the cell by organic cation transport system
	CA IV	60	
Quartenized ammonium sulfanilamide	CA II	200	Membrane impermeant small cation, but may be transported into the cell by organic cation transport system
	CA IV	1,800	
Dextran-coupled prontosil	CA II	200	Mol wt 5kDa for interstitial and 100 kDa for intravascular use. Membrane impermeant. Anaphylaxis?
Dextran-coupled amino-benzolamide	CA II	0.47μg/ml	Mol wt 5–30 kDa. Membrane impermeant. May cause anaphylactic reactions.
	CA IV	0.5μg/ml	
Polyethylene-coupled aminobenzolamide	CA II	29	Mol wt 3.35 kDa. Membrane-impermeant. Can be given intravenously.
	CA IV	690	

References for the impermeant inhibitors are given in the text. K_i:s for the permeant inhibitors are from Maren et al. (1993); they were measured at 0–1°C.

membranes. They can be constructed by linking strong sulfonamide inhibitors of an appropriate length to macromolecules of different sizes, such as dextrans (Karlmark et al., 1979; Tinker et al., 1981; Geers et al., 1985) or polyethylene (Conroy et al., 1996) via nonhydrolyzable spacer arms. Rigorous controls must be used to ascertain that the linkage does not break to release free inhibitor. Ionized small inhibitors in the form of anions or cations at physiological pH can also be used, if it can be established that these inhibitors are not taken up by the relevant cells through organic ion transport systems. Experiments have been conducted with benzolamide and quarternized sulfanilamide (Karlmark et al., 1979; Henry, 1987) and benzolamide (Supuran and Clare, 1995).

Selective inhibition of CA IV gives a larger physiological response than inhibition of CA II in the renal proximal tubular cell (Lucci et al., 1983). In general, more than 95% inhibition of the CA II activity is needed to evoke physiological effects on CO_2 and ion transport in various organs (Maren, 1977; Wistrand, 1980, 1981), including brain (Millichap, 1958). Whether this is the case also for the other isozymes is not known.

Recently, a transferrin-like protein found in the plasma of certain mammals, has been shown to strongly inhibit CA II (K_i = nM) and CA IV (K_i = μM), and has been used to selectively inhibit membrane-bound pulmonary vascular CA activity (Heming et al., 1993).

Antisense RNA and DNA molecules targeted against CA II are now available to block the expression of this protein in vitro (Laitala and Väänänen, 1994).

OCCURRENCE AND FUNCTION OF CARBONIC ANHYDRASE ISOFORMS IN THE NERVOUS SYSTEM

Comprehensive reviews on CA in CNS and PNS have been written by Cammer (1991b; functions of CA in glial cells and myelin); Neubauer (1991; location and function of neuronal CA), Sapirstein (1984; biosynthesis and function of glial CA), Lees and Sapirstein (1984; myelin-associated CA), Torrance (1993; CA in chemoreceptors). Some properties of brain CAs are given in Table 2.

Ashby's (1944) suggestion and Giacobini's (1961) original finding of the localization of most brain CA activity to glial cells has been confirmed by biochemical analysis of enriched oligodendrocyte fractions (cf. Ghandour et al., 1980). The activity reflects the density of oligodendrocytes in the particular region.

The soluble enzyme in the homogenates of rat brain was shown to be similar to erythrocyte CA II (Filippi et al., 1978; Sapirstein and Lees, 1978). However, Karler and Woodbury (1960) also found membrane-associated CA activity in the particulate fractions of rat brain homogenates. Actually, in the adult human and rodent brain, more than 50–60% of the CA activity is membrane-associated (cf. Sapirstein, 1984). This high percentage of membrane-associated enzyme is unique to the brain and seems to occur in all areas of the CNS. The membrane-associated form had the same molecular weight (30 kDa) and isoelectric point (7.2), as the soluble brain CA II, but was different antigenically. Antibodies directed against soluble CA II might therefore not be suited for immunohistochemical localization of the membrane-associated form. The cellular origin of the membrane-associated CA II has not yet been clarified.

The soluble and membrane-associated form of CA II are synthesized by two independent mechanisms on free and membrane-bound polysomes, perhaps using two different mRNAs (Sapirstein et al., 1983). The synthesis of membrane-associated CA II on rough endoplasmic reticulum would direct the enzyme to the plasma membranes of distant cellular processes. A membrane-associated form of CA II has also been reported to be present in pancreas (Mahieu et al., 1994).

In addition to this membrane-associated CA II, Brion et al. (1994b) found CA activity in the supernatants of mice brain homogenates; this activity was resistant to inhibition by SDS, indicating the presence of CA IV in these brains. The cellular origin of this CA IV-type of activity will be discussed below in the section neurons (p. 32) and cerebral vessels (p. 33).

It is intriguing that Kimura and MacLeod (1975) found that CA I exists in the particulate fraction of the pituitary gland, because results from immunoassay and inhibitor kinetics on brain homogenates of mouse (Ghandour et al., 1989), rat (Nogradi, 1993), and man (Funakoshi and Deutsch, 1971) indicate that the brain does not contain CA I.

Recently, Krueger and Saito (1992) reported on a tyrosine phosphatase in the brain having an amino acid sequence highly homologous to CA, but without CO_2 catalytic activity. Kato (1990) has found a CA-related polypeptide in Purkinje cells. Whether these forms are antigenically similar to known CA isozymes in not known. Such similarities could obscure immunohistochemical findings.

Various techniques can localize the type of CA isozyme to various glial cells, microglia, and to several different populations of neurons. However, the cells are sometimes difficult to identify in histological preparations, a situation that could explain some of the contradictory results in the literature. In addition, species differences exist with regard to the presence of CA in CNS and PNS (cf., the discussion of the difference between sheep and man with respect to CA activity in astrocytes (p. 31) and oligodendrocytes (see below). Problems with the use of immunocytochemistry for glial cell identification have been described by Korr et al. (1994). Immunohistochemistry for the localization of CA is the most sensitive method, but is apt to produce false results, as described above. With these precautions in mind, we shall now discuss CA activity in different cells in the CNS and PNS, as found by different methods.

Oligodendrocytes

Occurrence of CA isozymes in cytoplasm and membranes. Using the Cartesian diver technique on isolated oligodendroglia, Giacobini (1961) found that these cells contained one fifth of the catalytic activity of erythrocytes, corresponding to a concentration of 4 μM of CA II in the oligodendrocyte cytoplasm, a concentration similar to that in renal tubular cells. CA activity was later found histochemically in oligodendrocytes in the brain of cat (Ridderstråle and Hanson, 1985), rat (Nogradi and Mihaly, 1990), and mouse (Ridderstråle and Wistrand, 1998). The cell membranes and the myelin sheaths were not stained in these studies, but the mouse oligodendrocytes were found to be surrounded by a filamentous staining, in the areas of white matter, probably localized to membranes of the axons (see p. 32).

The CA activity originates from CA II, found immunohistochemically inside the cell body and its cytoplasmic projections of oligodendrocytes of several species (Delaunoy et al., 1977; Agnati et al., 1995), including man (Kumpulainen et al., 1983). Also, hybridization (cDNA probe) studies have demonstrated that CA II is exclusively restricted to oligodendroglia (Ghandour and Skoff, 1991). However, sheep (Jeffrey et al., 1991) and goat (Ridderstråle, personal communication) oligodendrocytes lack CA II, illustrating that species differences do occur.

Occurrence of CA isozymes in myelin. Myelin consists of several concentric layers of the glial plasma membrane. During myelin formation, the oligodendrocyte wraps itself turn by turn around the axon, pushing the cytoplasm and nucleus towards the periphery (Fig. 1). The space between two glial cells, where the axon is left uncovered, constitutes Ranvier's node. Unmyelinated nerve fibers are covered by a single layer of a glial cell process. The proximity of these structures accounts for the difficulties in resolving the staining of morphological techniques, and perhaps also the difficulties in separating them biochemically.

A considerable part (7–20%) of the membrane-associated enzyme is associated with myelin in the CNS and even more so in the PNS (cf., Lees and Sapirstein, 1984; Cammer and Tansey, 1987). The white matter, with more myelin, thus has a higher proportion of myelin-associated CA activity than the grey matter.

The myelin-type of enzyme was the first membrane-bound CA isozyme to be purified (Sapirstein and Lees, 1978). It was shown to be similar to the erythrocyte CA II, with respect to molecular weight (30 kDa), isoelectric point, sensitivity to inhibition by acetazolamide (Cammer et al., 1976), and lack of sensitivity to inhibition by chloride (Sapirstein

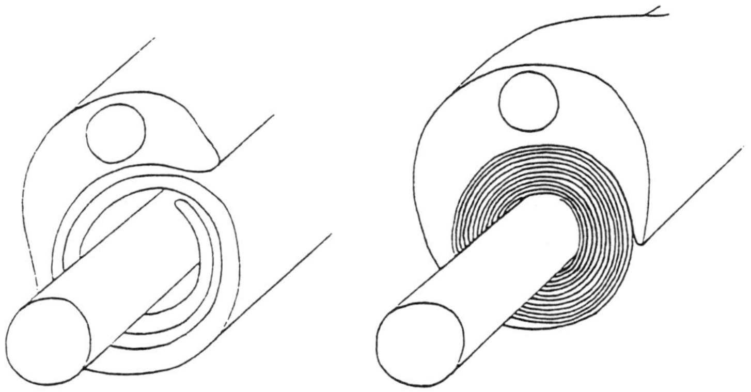

FIGURE 1. Schematic presentation of myelin formation by the glial cell. The myelin sheath is composed of concentric layers of cell membranes, formed by the glial cell. The glial cell wraps itself around the axon, while its cytoplasm is pressed towards the periphery.

and Lees, 1978). However, its amino acid composition and relatively low specific activity were different from those of CA II, presumably due to the association to myelin. How this association affects its immunolocalization should be further evaluated. Also, the question how this myelin-associated CA II is related to the membrane-associated CA II should be clarified.

Biochemical analysis has shown that different myelin fractions, have different specific CA activities (Cammer et al., 1976). The highest activity was found in the myelin-like and heavy myelin fractions, and the lowest in light myelin, indicating that the enzyme is concentrated towards the outside of the myelin sheath. Recent studies by Sapirstein et al. (1993) have shown that, within myelin, CA is localized to the periaxolemmal myelin and there, perhaps, to the paranodal domain, known to be engaged in regulation of ion and fluid transport.

Several immunohistochemical studies have shown CA II to be present in myelin. On the other hand, histochemical studies indicate that CA activity is absent in the myelin sheath. The different results can perhaps be explained by difficulties in discriminating the axonal cytoplasm and membrane from the myelin sheath and the surrounding thin rim of oligodendrocyte cytoplasm (see Figs. 1,2)

Recently a mutation in mice has been described (Lewis et al., 1988) in which CA II is lacking biochemically (Brechue et al., 1991) and immunologically (Spicer et al., 1989; Ghandour et al., 1989) in all tissues tested, including the brain and its myelin fractions.

Brion et al. (1994a) found that brain myelin extracts, as well as spinal cord homogenates and pellets from brain of such CA II-deficient mice, contained small amounts of SDS-resistant CA activities, indicating the presence of CA IV in these fractions. This was confirmed by Western blot analysis. Immunocytochemical evidence for the presence of membrane-bound CA IV in oligodendrocytes and myelinated tracts of mouse brains was also reported by Cammer et al. (1995). However, Ghandour et al. (1992), using an immunohistochemical technique, did not find CA IV in glia, myelin, or neurons. These authors instead found CA IV on the luminal side of the endothelium of the rat cerebral vessels. Thus, there is controversy about the presence of CA IV in glia. Contamination of glial cell fractions with CA IV from capillary endothelium and the use of different antibodies against CA IV, could perhaps explain the varying results.

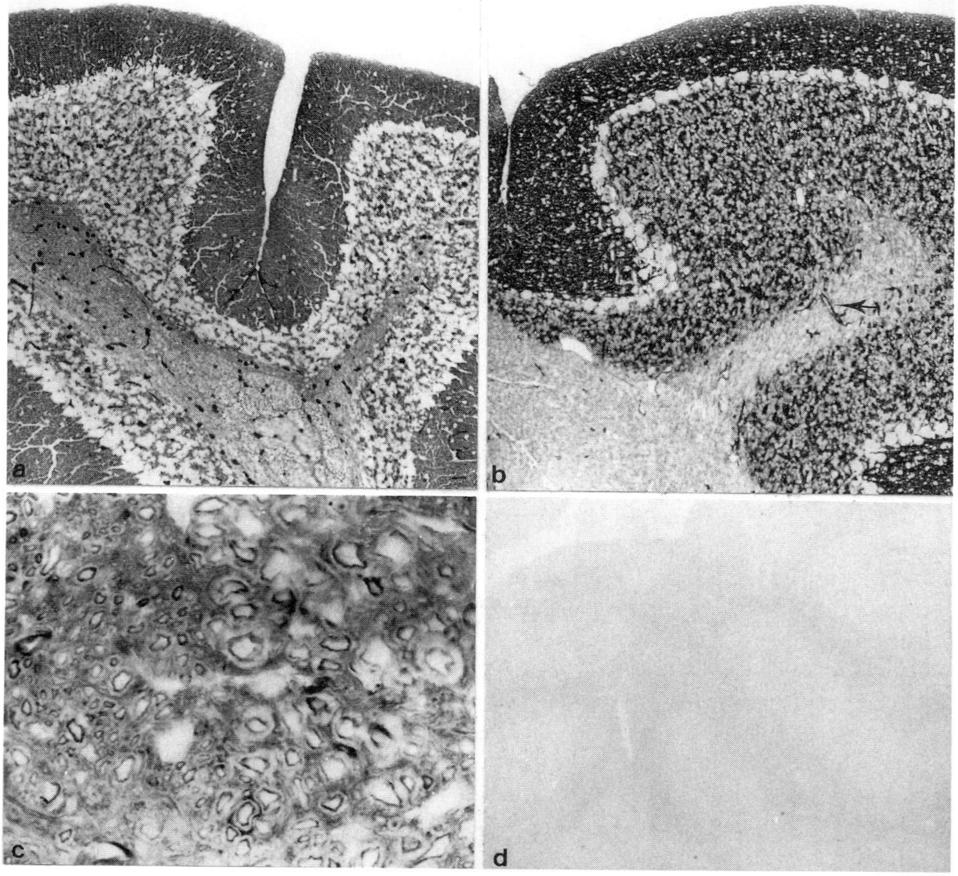

FIGURE 2. Histochemical localization of carbonic anhydrase in cerebellum of normal and CA II-deficient mouse.

a. Normal mouse. Incubation time 6 min. Magnification 150 ×.
b. CA II-deficient mouse. Incubation time 6 min. Magnification 150 ×.
c. CA II-deficient mouse, white matter. Incubation time 6 min. Magnification 1200 ×.
d. CA II-deficient mouse. Incubated 6 min with 10^{-5} M acetazolamide included in the medium. Magnification 1200 ×.

In both animals membrane-bound CA is indicated by intense staining in the neuropil of the molecular layer, in a fine network surrounding the granular cells, and in the capillaries (arrow). Also the axonal membrane in myelinated fibers is stained, while the myelin is unstained (c). Only in the normal animal (a) are stained oligodendrocytes seen especially in the white matter.

In the CA II-deficient mice, histochemistry shows that axons and probably also dendrites belonging to cerebellar neurons have membrane-associated CA activity, perhaps originating from CA IV (Fig. 2b,c). Therefore the CA IV found by Brion et al. (1994a) in the myelin fractions could perhaps originate from contaminating axons as well as capillary endothelium.

Function of CA in oligodendrocytes. CA may participate in the transfer of fluid and ions across the myelin sheaths to the extracellular fluid and thus function in the compaction of myelin. Lack of CA could therefore result in myelin swelling (cf. Cammer et al., 1995). CA in myelin can also provide fast delivery of bicarbonate ions for biochemical reactions. In the oligodendrocytes, acetyl CoA carboxylase and fatty acid synthase were found to be colocalized with CA II, and this isozyme therefore has been thought to provide bicarbonate for the synthesis of fatty acids in these cells. However, in spite of the postulated dependence on CA II for the synthesis of myelin lipids, chronic inhibition studies (Maren, 1985) and genetic deficiency of CA II in mice have not revealed any large structural abnormality of myelin in these animals. These mutants have a normal morphology of the brain (Ghandour et al., 1989) at both light and electron microscopic level, although Cammer et al. (1993) reported that the oligodendrocytes in these mice appeared shrinking in both gray and white matter, even though their myelin and myelin protein was normal, with compact myelin. The different results could perhaps be due to the use of different strains of CA II-deficient mice.

The pathology of the mental retardation in patients with congenital CA II deficiency has not yet been described.

It has been suggested that CA II is involved in the fixation of iron to transferrin via its production of bicarbonate (Griot and Vandevelde, 1988). It is therefore interesting that transferrin immunostaining has been found in the oligodendrocytes of the gerbil brain (Hafidi et al., 1994).

Astrocytes

Presence of CA isozymes in cytoplasm and membranes. In addition to immature astrocytes, two forms of mature astrocytes have been described, the fibrous and the protoplasmic (Cammer and Zhang, 1992).

Histochemical studies have shown either lack of, or very weak, CA activity in the astrocytes of rat (Riley et al., 1984; Nogradi and Mihaly, 1990) and cat (Ridderstråle and Hanson, 1985) but the cells were not clearly identified. Faintly stained astrocytes were found in the developing rat brain (Nogradi and Mihaly, 1991). This activity probably originates from CA II, since immunohistochemistry has located this isozyme to astrocytes and their precursors in neonatal and young rats (Cammer and Zhang, 1992). Also primary astroglial cultures of neonatal rodent brain have CA II (Chow et al., 1991). Some astrocytes in the adult brain also contain CA II, as shown for rats (Roussel et al., 1979; Agnati et al., 1995), sheep (Jeffrey et al., 1991) and goat (Ridderstråle; personal communication). Several authors have denied that astrocytes contain CA II and have suggested that erroneous identification of cell type, antisera contamination, pinocytosis of CA II, or the developmental stage of the astrocytes could explain the positive staining (Ghandour et al., 1989; Anderson et al., 1984; Snyder et al., 1983; Kumpulainen et al., 1983; Ghandour et al., 1980). Also the astrocytes in the adult human brain seem to lack CA II (Kumpulainen et al., 1983).

Function of CA activity in astrocytes. When astrocytes in rat cortex and hippocampus (Pappas and Ransom, 1994) are depolarized by extracellular potassium ions in HCO_3^--buffered solutions, intracellular pH is increased (Chapter 14). Astrocyte swelling induced by K^+, and dependent on HCO_3^-, is abolished by acetazolamide, indicating that swelling is controlled by CA (Bourke et al., 1975). Astrocytes in primary cultures exhibit three different mechanisms to regulate intracellular pH; NA^+/H^+ exchange, Na^+-independent $C1^-/HCO_3^-$-exchange, and Na^+/HCO_3^- cotransport (Chow et al., 1991, 1992). Incubation of the astrocytes with the membrane-permeant inhibitor acetazolamide decreased the intracellular pH, indicating that CA activity in these cells is involved in the pH regulation. However, Kaila (personal communication) saw no effect of acetazolamide on the CO_2-induced decrease of intracellular pH of rat astrocytes in primary culture.

Recently, Cammer (1991a) reported the colocalization of CA, glutamine synthetase, and carbamoylphosphate synthase II in astrocytes, which would suggest that these enzymes are involved in the synthesis of pyrimidine by furnishing bicarbonate and glutamine. Pyrimidine could then, in turn, be used in the DNA synthesis.

In view of the need for CA activity to carry out ion transport and pyrimidine synthesis it is bewildering that there is such controversy about its presence in astrocytes, as found in morphological and functional studies. Species differences or type of astrocyte studied, may explain some, but not all, of the different results.

Microglia

Presence of CA Isozymes in Cytoplasm and Membranes. The ameboid and reactive microglial cells belong to the brain macrophages and probably originate from blood. They can be distinguished by various cell markers detected immunocytochemically.

Histochemical CA activity was found in the ameboid reactive microglia, distributed throughout the white matter of the cerebrum and cerebellum and the gray matter of the brain stem (Nogradi, 1993). Visualization by EM of the histochemical precipitate showed it to be present in the cytoplasm.

Nogradi (1993) first reported the presence of CA II and CA III in the reactive microglia, and faint immunoreactivity for CA III in the resting microglia.

Function of CA Activity in Microglia. Macrophages of different tissues have been shown (Spicer and Schulte, 1991) to have varying contents of CA I, II, and III. The function of CA II in brain microglia is probably similar to that in other macrophages, that is, to buffer the cytoplasm during the acidosis brought about by the catabolic activities of these cells. However, this remains to be studied, as does the function of CA III in microglia.

Neurons

Presence of CA Isozymes in Cytoplasm and Membranes. Korhonen et al. (1964) found vigorous histochemical staining in the stratum glomerulosum and gelatinosum of the mouse olfactory bulb, and recently Nogradi and Mihaly (1991) reported on histochemical CA activity in rat granule cells of the olfactory tubercle, cells thought to have a neuroendocrine function. The histochemical staining was found in the cytoplasm and in the nuclei of these neurons. In another study, Nogradi and Mihaly (1990) found axonal staining in rat medulla oblongata and spinal cord. In the cat, Ridderstråle and Hanson (1985) saw histochemical evidence for the presence of CA in neurons, predominantly in the cytoplasm, but sometimes also associated with the cell membranes of some large neurons in the medulla oblongata, where the chemosensitive area is located. The Purkinje cells of cerebellum showed cytoplasmic and nuclear staining. Intensely stained axons were frequently present in the cerebellar white matter, in medulla oblongata, and spinal cord. Also in the CA II-deficient mouse, histochemistry showed intense staining in the axons of the white matter, in a fine network surrounding the granular cells and in the neuropil of the cerebellar molecular layer (Fig. 2). Part of the filamentous staining found in the cortical layers may emanate from axons and dendrites. Immunohistochemistry showed that this activity was not due to the presence of CA I, CA II, or CA III. Therefore the membrane-associated CA activity at these locations should originate from a membrane-associated CA, perhaps CA IV (Ridderstråle and Wistrand, 1998). The results of these studies indicate that most of the neuronal cell bodies in the CNS do not have CA activity and moreover lack CA immunoreactivity (Kumpulainen et al., 1983; Cammer and Zhang, 1992; Ghandour et al., 1992). However, functional evidence for the presence of intracellular CA activity in isolated rat hippocampal neurons has been obtained by Pasternack et al. (1993).

In the PNS, Cammer and Tansey (1987) found CA II in the axoplasm of subpopulations of myelinated fibers in rat and mouse sciatic nerve, spinal root, and cranial nerve, and in the perikarya of the ganglion cells and satellite cells. Other investigations show the presence of both CA II and histochemical activity in subpopulations of ganglion cells, satellite cells and axons, and in peripheral nerves (Riley et al., 1984; Wong et al., 1983; Kazimierczak et al., 1986; Aldskogius et al., 1988). Sommer et al. (1985) have related the CA staining to several morphological and cytochemical features. Interestingly, dorsal root ganglion cells with histochemical CA activity have electrophysiological properties different from those lacking such activity (Prabhakar and Lawson, 1995).

In conclusion, morphological and functional studies indicate that some CNS neurons, particularly those with sensory functions, and certain sensory ganglion cells, contain CA. In the PNS, CA appears to be located primarily in medium and large sensory neurons but not in the motoneurons (cf. Cammer and Tansey, 1987).

Function of CA in neurons. In medullary chemoreceptors and in receptors of the carotid body, the cytoplasmic CA activity is needed for the receptors to respond rapidly to a change in P_{CO_2} (Hanson et al., 1981; Torrance, 1993). However, it could not be ruled out that also a membrane-associated CA has a role in this response.

During excitation, the intracellular pH of neurons decreases, primarily due to production of CO_2 and lactate, and entry of protons from the extracellular fluid through various channels. Thus, there would seem to be a need for an intracellular CA in these neurons to catalyze the hydration of CO_2 in the cytoplasm, delivering H^+ and HCO_3^- ions for buffering and exchange with other ions during membrane transport.

Cerebral Blood Vessels

Presence of CA isozymes. In all parts of the brain, except for the meninges, histochemical investigations (Nogradi et al., 1993; Ridderstråle and Hanson, 1985) have localized CA activity to the endothelium of the capillaries, but not of the large blood vessels. Also CA II-deficient mice have histochemical staining in small cerebral vessels (Fig. 2b), which are sensitive to acetazolamide inhibition, indicating that the CA activity in the brain capillaries originates from a high-activity type of CA isozyme, shown to be CA IV (Ghandour et al., 1992).

Functional studies using membrane-impermeant inhibitors in the blood vessels of ventral medulla (Hanson et al., 1981) have indicated that CA activity is indeed located on the luminal surface of the endothelial cell membrane and thus must be membrane-associated.

Nogradi and Mihaly (1990) found that the walls of rat cerebral blood vessels have pericyte-like cells, with CA staining.

Function of CA activity in blood vessels. Hanson et al., (1981) suggested that the CA activity lining the endothelium of the capillaries passing through the central chemoreceptors brings the plasma and red cells of the blood to CO_2 equilibrium, much like the role of CA IV in the lung capillaries (Bidani et al., 1978; Heming et al., 1994).

Carbonic anhydrase activity in small vessels (located at cell borders of the capillary fenestrated endothelium) is found in many organs but not in all. It has been suggested (Lütjen-Drecoll et al., 1985) that here the function of CA is to regulate the size of the fenestrate. However, this would not agree with its location in tight cerebral capillaries lacking fenestrae.

Systemic administration of CA inhibitors causes cerebral vessels to dilate and increases cerebral blood flow, as does administration of CO_2 (Mithoefer et al., 1957). The effect could be due to brain acidification by the inhibition of CA in erythrocytes, giving a respiratory acidosis, and/or by inhibition of CA in cerebral capillaries and brain. The anticonvulsant effect and the favorable effect of CA inhibitors on the syndrome of high-altitude sickness have also been attributed to brain acidification. The relative importance of CA inhibition in brain and erythrocytes for these effects should be analyzed.

Choroid Plexus

Presence of CA isozymes in cytoplasm and plasma membranes. Fisher and Copenhaver (1959) were the first to report on biochemical CA activity in choroid epithelial cells, and Giacobini (1961) found that the cells had one-third of the activity of erythrocytes. Histochemistry in rodents by Korhonen et al. (1964), and Masuzawa et al. (1981), and in dog by Nakamura et al. (1991), using EM visualization, showed the staining to be located in the cytoplasm and along the cell borders, on apical and basolateral membranes. CA II-deficient mice show strong histochemical CA staining of the plasma membranes of these epithelial cells, indicating that the activity originates from a membrane-bound enzyme (Ridderstråle and Wistrand, 1998).

Roussel et al. (1979), Ghandour et al. (1992), and Nogradi (1993) reported that the choroid plexus cells in the rat had CA II immunoreactivity. However, there was no immunochemical evidence for CA IV in the choroid plexus, which is surprising in view of the positive findings in the histochemical studies. This discrepancy needs to be clarified.

Nogradi et al. (1993) found CA III, in addition to CA II, in the cytoplasm and microvilli of the choroid epithelial cells from rabbits, rats, and humans. However, in homogenates of mouse brain there was no acetazolamide-resistant CA activity and therefore no indication for the presence of CA III (Brion et al., 1994a). These discrepancies as to the presence of CA III in brain need further clarification.

Korhonen et al. (1964) and Roussell et al. (1979) found no histochemical activity or immunoreactive CA II in the ependymal cells, also thought to be involved in secretion of cerebrospinal fluid.

Monamine activity from sympathetic nerve stimulation (Lindvall et al., 1978) or from local application of noradrenaline or serotonin (Nakamura et al., 1991) to the choroid plexus has been found to reduce CA activity of the choroid plexus, as measured morphometrically or biochemically. Since the membrane-associated activity is relatively small compared to that of CA II, this effect on CA activity probably reflects a reduction of the CA II activity.

Function of CA activity in choroid plexus. The colocalization of CA II in the cytoplasm, and of histochemical CA activity on the plasma membranes of the choroid epithelial cells, is similar to that in other secretory epithelia. By the hydration of CO_2 in the cytoplasm and dehydration of H_2CO_3 at the cells' outside, intracellular buffering and membrane transport of H^+ and HCO_3^- ions is facilitated. Inhibition results in a reduced secretion of cerebrospinal fluid (CSF) (Maren and Broder, 1970), and inhibition of the membrane-associated CA activity is probably responsible for most of the effect, as was shown to be the case for the renal proximal tubular cells (Lucci et al., 1983). This hypothesis should be tested.

Interstitial Space in Brain Parenchyma

Presence of CA activity. The best evidence for CA activity in the interstitial space comes from functional studies. By applying membrane-impermeate dextran-coupled inhibitors (Kaila et al., 1992; Chen and Chesler, 1992) to brain slices from the hippocampal pyramidal region, and observing the effects of these inhibitors on activity-induced pH shifts in the extracellular space, it could be shown that CA activity modulates pH in this space. The boundary of the space is formed by neurons, glial cells, microglia, and cerebral vessels. So far, localization studies, referred to above, indicate that the function CA activity could be located on certain neuronal and axonal membranes. Glial cells and microglia seem to lack membrane-associated CA activity. The cerebral capillaries have membrane-bound enzyme, CA IV, on the luminal side, but probably also on the basal cell membrane. Whether this enzyme is facing the extracellular space is not known.

Function of the membrane-associated CA isozyme in the interstitial space. Neuronal and glial cell depolarization acidify the extracellular space, either by metabolic activity, by producing CO_2 or lactic acid, or by active acid secretion. The space may also be alkalinized during neuronal activity, due to efflux of base or entry of protons, through nonspecific or ligand-gated channels. At the same time, acidification and alkalinization of the space, i.e., shifts in proton concentrations, have been shown to affect synaptic currents (Taira et al., 1993; Gottfried and Chesler, 1994). Since CA activity has been shown to modulate extracellular pH transients, CA can be said to be involved in neuronal activity (Chapter 17).

Velisek et al. (1995) have found that CA II-deficient mice are resistant to experimental seizures, which could be due to interstitial acidosis; however, whether this is due to the systemic acidosis, or lack of CA II in the brain of these mice, is not known.

DEVELOPMENTAL AND REGULATORY ASPECTS OF CA SYNTHESIS

Carbonic anhydrase in brain is a marker of postnatal glial differentiation. Biochemical studies (Sapirstein et al., 1978b; Snyder et al., 1983) on isolated rat oligodendroglia, brain homogenates, and optic nerve (Davis et al., 1987) have demonstrated an increase of CA activity during postnatal development. On the first postnatal day, CA II was already found in precursors of the oligodendroglia (Cammer and Zhang, 1992). The number of astrocytes and oligodendrocytes also increased during this period. The membrane-associated CA of rat brain was reported to be fully expressed by the 16th postnatal day (Sapirstein et al., 1978a). In humans, adult values of CA activity in brain are reached by the age of $3^1/_2$

years, with the membrane-bound activity amounting to 50–75% of the total activity (Lees et al., 1980). Histochemical studies have confirmed the biochemical findings (Nogradi and Mihaly, 1990). Thus CA staining of the rat oligodendroglia and pericytes was seen on the first day after birth and had reached adult levels after 2 weeks.

Rat cerebellar Purkinje neurons were intensely CA active a few days after birth, but interestingly enough, had lost their activity at 1 month after birth. In the mouse, neuronal CA activity is first detected during the perinatal stage in the root ganglia. It could function as a marker for functional activity of the proprioceptive neurons (Mayeux et al., 1993).

The reactive ameboid microglia of the rat brain, found before birth in the cingulum, periaqueductal region, the roof of the lateral ventricles, and in the white matter of the cerebellum, had disappeared 2 weeks after birth. The epithelial cells of the rat choroid plexus were already stained at 4 days after birth (Nogradi and Mihaly, 1990).

In brains with myelin deficiency, as found in the optic nerve of newborn rats (Davis et al., 1987) or in the adult "Jimpy mutant mouse" and Quaking mouse, CA activity and CA II are low (Delaunoy et al., 1978; Sapirstein et al., 1980). A loss of CA in patients with leukodystrophies like Krabbe's and adrenoleukodystrophy was found and expected, since oligodendroglia containing CA II are fewer in the brains of these patients (Lees et al., 1980). Thyroid hormones can stimulate myelinogenesis in the CNS. In hypothyroid mice, the CA activity of subcellular fractions, including myelin, was relatively low (Li and Chow, 1994). On the other hand, in hyperthyroid DBA mice, the overproduction of myelin is associated with an increased CA activity (Engström et al., 1984).

Human cerebrospinal fluid contains low concentrations of CA I and II (Kumpulainen et al., 1985; Parkkila et al., 1994), but CA II was found in higher concentrations in patients with multiple sclerosis, perhaps indicating a leakage from degenerating oligodendroglia.

In CA II-deficient mice, Brion et al. (1994a,b) and Cammer et al. (1995) found higher CA IV activities in soluble and membrane-associated fractions of brain homogenates and in pure myelin. They suggested that this was a compensatory measure—an upregulation of CA IV—in response to the lack of CA II in these animals. However, the increase was small, and the data need to be confirmed.

SUMMARY AND CONCLUSIONS

The nervous system appears to have five forms of carbonic anhydrase, the cytoplasmic isozymes CA II and CA III, the myelin- and membrane-associated isoforms of CA II, and the membrane-associated isozyme CA IV. Each form is probably expressed by its own gene, except for the myelin-associated and membrane-associated forms of CA II, which could be expressed by the same gene as CA II, but be other transcripts (Sapirstein et al., 1983). There is an urgent need to characterize biochemically the myelin-associated and membrane-associated forms of CA II, and the CA IV isozyme, and to identify more precisely their localization in myelin and various cell membranes. The myelin-associated CA II seems to be situated in the high-molecular-weight outer lamellar part of the myelin sheaths and the region near the node, but it has been hard to localize with available morphological techniques. One likely function of the enzyme at the node, be it a myelin-type of enzyme or an axonal enzyme, would be to facilitate ion transport by furnishing H^+ and HCO_3^- ions for fluid and ion transport. Another suggested role for the myelin-associated

CA II is to deliver HCO_3^- ions as building blocks for the production of myelin. However, this theory has not been validated by inhibition studies, since animals on chronic large doses of strong, lipid-soluble inhibitors show no sign of structurally abnormal myelin. Also, CA II-deficient mice seem to have normal or near-normal myelin in their brains in spite of their lack of CA II in the myelin. The role of myelin-associated CA II therefore needs further clarification. If, as stated, CA IV is associated to myelin-like CA II, one wonders why myelin should contain two membrane-associated isozymes, with very similar CO_2 kinetics.

By analogy to other ion-transporting epithelia, one would expect to find membrane-bound CA IV in the cell membranes of ion-transporting glia and of epithelial cells of the choroid plexus. Available techniques have not been able to localize CA activity to astrocyte or oligodendrocyte cell membranes. However, like the axons and dendrites of certain neurons and like cerebral capillary endothelium; choroid plexus cells show strong membrane-associated histochemical staining. It is, therefore, surprising that there is immunohistochemical evidence for CA IV only in capillary endothelium. Discrepancies between histochemical and immunohistochemical membrane staining need to be resolved. However, both techniques agree as to the presence of CA IV in the capillary endothelium. Moreover, functional studies have confirmed the presence of a capillary endothelial CA IV enzyme, with its active site facing the blood lumen.

Functional studies, applying membrane-impermeant inhibitors in the brain extracellular fluid, have shown that a membrane-associated CA activity is necessary for the modulation of pH gradients in the extracellular space, brought about by neuronal and metabolic activities. This CA activity is probably associated with the membranes of certain neurons and their processes, since astrocytes, oligodendroglia, and microglia appear to have no such CA activity.

The CA II isozyme is ubiquitous and found in the cytoplasm of choroid plexus epithelial cells, oligodendrocytes, certain neurons and ganglion cells, and occasionally in microglia and astrocytes. Its function there is to facilitate the buffering of the intracellular fluid and deliver H^+ and HCO_3^- ions for the ion exchange systems. This has been verified in studies on many cell types. However, other roles ascribed to this isozyme, such as its involvement in pyrimidine synthesis in the astrocytes need to be validated.

The recent reports on the presence of CA III in the cytoplasm of reactive microglia is interesting. Why macrophages express CA III only when they become stimulated and reactive should be studied further. The function would not appear to be the catalysis of CO_2 hydration, since this function is carried out more effectively by CA II, also present in the cytoplasm of these cells, be they active or nonactive. Unfortunately there is, as yet, no CA III-selective inhibitor available to study its physiological function.

REFERENCES

Agnati LF, Tinner B, Staines WA, Väänänen K, Fuxe K (1995): On the cellular localization and distribution of carbonic anhydrase II immunoreactivity in the rat brain. Brain Res 676: 10–24.

Aldred P, Fu P, Barrett G, Penschow JD, Wright RRD, Coghlan JP, Fernley RT (1991): Human secreted carbonic anhydrase: cDNA cloning, nucleotide sequence, and hybridization histochemistry. Biochemistry 30: 569–575.

Aldskogius H, Arvidsson J, Hansson P (1988): Carbonic anhydrase enzyme histochemistry of cranial nerve primary sensory afferent neurons in the rat. Histochemistry 88: 151–154.

Anderson RE, Engstrom FL, Woodbury DM (1984): Localization of carbonic anhydrase in the cerebrum and cerebellum of normal and audiogenic seizure mice. Ann NY Acad Sci 429: 502–504.

Ashby W, Cobb S (1944): A parallelism between the quantitative incidence of carbonic anhydrase and functional levels of the central nervous system. J Biol Chem 152: 235–240.

Bidani A, Crandall ED, Forster RE (1978): Analysis of postcapillary pH changes in blood in vivo after gas exchange. J Appl Physiol 44: 770–781.

Bourke RS, Kimelberg HG (1975): The effect of HCO_3^- on the swelling and ion uptake of monkey cerebral cortex under conditions of raised extracellular potassium. J Neurochem 25: 323–328.

Brechue WF, Kinne Saffran E, Kinne RK, Maren TH (1991): Localization and activity of renal carbonic anhydrase (CA) in CA-II deficient mice. Biochim Biophys Acta 1066: 201–207.

Brion LP, Suarez C, Zhang H, Cammer W (1994a): Up-regulation of carbonic anhydrase isozyme IV in CNS myelin of mice genetically deficient in carbonic anhydrase II. J Neurochem 63: 360–366.

Brion LP, Zavilowitz BJ, Suarez C, Schwartz GJ (1994b): Metabolic acidosis stimulates carbonic anhydrase activity in rabbit proximal tubule and medullary collecting duct. Am J Physiol 266: F185–F195.

Cammer W (1991a): Carbonic anhydrases in myelin and glial cells in the mammalian central nervous system. In Dodgson SJ, Tashian RE, Gros G, Carter ND (eds): The Carbonic Anhydrases. Cellular Physiology and Molecular Genetics. New York: Plenum, pp 325–332.

Cammer W (1991b): Immunostaining of carbamoylphosphate synthase II and fatty acid synthase in glial cells in rat, mouse, and hamster brains suggests roles for carbonic anhydrase in biosynthetic processes. Neurosci Lett 129: 247–250.

Cammer W, Tansey FA (1987): Immunocytochemical localization of carbonic anhydrase in myelinated fibers in peripheral nerves of rat and mouse. J Histochem Cytochem 35: 865–870.

Cammer W, Zhang H (1992): Carbonic anhydrase in distinct precursors of astrocytes and oligodendrocytes in the forebrains of neonatal and young rats. Brain Res Dev Brain Res 67: 257–263.

Cammer W, Fredman T, Rose AL, Norton WT (1976): Brain carbonic anhydrase: activity in isolated myelin and the effect of hexachlorophene. Neurochem 27:165–171.

Cammer W, Sacchi R, Sapirstein V (1985): Immunocytochemical localization of carbonic anhydrase in the spinal cords of normal and mutant (shiverer) adult mice with comparisons among fixation methods. J Histochem Cytochem 33: 45–54.

Cammer W, Zhang H, Cammer M (1993): Glial cell abnormalities in the CNS of the carbonic anhydrase II deficient mutant mouse. J Neurol Sci 118: 1–9.

Cammer W, Zhang H, Tansey FA (1995): Effects of carbonic anhydrase II (CAII) deficiency on CNS structure and function in the myelin-deficient CAII-deficient double mutant mouse. J Neurosci Res 40: 451–457.

Chapman SK, Maren TH (1978): A search for the function of human carbonic anhydrase B. Biochim Biophys Acta 527: 272–276.

Chegwidden WR (1991): Purification of the carbonic anhydrases. In Dodgson SJ, Tashian RE, Gros G, Carter ND (eds): The Carbonic Anhydrases. Cellular Physiology and Molecular Genetics. New York: Plenum Press, pp 101–118.

Chen JC, Chesler M (1992): pH transients evoked by excitatory synaptic transmission are increased by inhibition of extracellular carbonic anhydrase. Proc Natl Acad Sci USA 89: 7786–7790.

Chesler M, Kaila K (1992): Modulation of pH by neuronal activity. Trends Neurosci 15: 396–402.

Chow SY, Yen Chow YC, White HS, Woodbury DM (1991): pH regulation after acid load in primary cultures of mouse astrocytes. Brain Res Dev Brain Res 60: 69–78.

Chow SY, Yen Chow YC, Woodbury DM (1992): Studies on pH regulatory mechanisms in cultured astrocytes of DBA and C57 mice. Epilepsia 33:775–784.

Conroy CW, Wynns GW, Maren TM (1996): Synthesis and properties of two new membrane impermeant high molecular weight carbonic anhydrase inhibitors. Bioorganic Chemistry 24: 262–272.

Davis PK, Carlini WG, Ransom BR, Black JA, Waxman SG (1987): Carbonic anhydrase activity develops postnatally in the rat optic nerve. Brain Res 428: 291–298.

Delaunoy JP, Roussel G, Mandel P (1977): Localisation immunohistochimique de l'anhydrase carbonique forme C dans le systéme nerveux central du rat. (Immunohistochemical localization of carbonic anhydrase C in the rat central nervous system). CR Acad Sci 285: 801–804.

Delaunoy JP, Filippi D, Laurent G, Mandel P (1978): Quantitative measurements of carbonic anhydrase (CAII) in the central nervous system of neurological mutant mice "Jimpy" by radioimmunoassay. Brain Res 155: 201–204.

Dermietzel R, Leibstein A, Siffert W, Zamboglou N, Gros G (1985): A fast screening method for histochemical localization of carbonic anhydrase. Application to kidney, skeletal muscle, and thrombocytes. J Histochem Cytochem 33: 93–98.

Dodgson SJ (1991): Liver mitochondrial carbonic anhydrase (CA V), gluconeogenesis, and ureagenesis in the hepatocyte. In Dodgson SJ, Tashian RE, Gros G, Carter ND (eds): The Carbonic Anhydrases. Cellular Physiology and Molecular Genetics. New York: Plenum Press, pp 297–306.

Engström FL, Kemp JW, Woodbury DM (1984): Subcellular distribution of carbonic anhydrase and Na^+, K^+ and HCO_3^- ATPases in brains of DBA and C57 mice. Epilepsia 25: 759–764.

Filippi D, Sciaky M, Limozin N, Laurent G (1978): Anhydrase carbonique du systéme nerveux central dur rat. Isolement et propriétés. (Carbonic anhydrase in the central nervous system of the rat. Isolation and properties). Biochimie 60: 99–102.

Fisher RG, Copenhaver JH (1959): The metabolic activity of the choroid plexus. J Neurosurg 16: 167–176.

Forsman C, Behravan G, Osterman A, Jonsson BH (1988): Production of active human carbonic anhydrase II in E. coli. Acta Chem Scand 42: 314–318.

Funakoshi S, Deutsch HF (1971): Human carbonic anhydrases, VI. Levels of isozymes in old and young erythrocytes and in various tissues. Biol Chem 246: 1088–1092.

Gay CV, Shanabrook VM, Sharkey DJ (1981): Development of a procedure for autoradiographic localization of carbonic anhydrase at the electron microscope level. J Histochem Cytochem 29: 415–419.

Gay VC (1991): Localization of the carbonic anhydrases by immunocytochemistry. In Dodgson SJ, Tashian RE, Gros G, Carter ND (eds): The Carbonic Anhydrases. Cellular Physiology and Molecular Genetics. New York: Plenum, pp 145–151.

Geers C, Gros G, Gartner A (1985): Extracellular carbonic anhydrase of skeletal muscle associated with the sarcolemma. J Appl Physiol 59: 548–558.

Ghandour MS, Langley OK, Zhu XL, Waheed A, Sly WS (1992): Carbonic anhydrase IV on brain capillary endothelial cells: A marker associated with the blood-brain barrier. Proc Natl Acad Sci USA 89: 6823–6827.

Ghandour MS, Skoff RP (1991): Double-labeling in situ hybridization analysis of mRNAs for carbonic anhydrase II and myelin basic protein: expression in developing cultured glial cells. GLIA 4: 1–10.

Ghandour MS, Skoff RP, Venta PJ, Tashian RE (1989): Oligodendrocytes express a normal phenotype in carbonic anhydrase II-deficient mice. J Neurosci Res 23: 180–190.

Ghandour MS, Langley OK, Vincendon G, Gombos G, Filippi D, Limozin N, Dalmasso D, Laurent G (1980): Immunochemical and immunohistochemical study of carbonic anhydrase II in adult rat cerebellum: A marker for oligodendrocytes. Neuroscience 5: 559–571.

Giacobini E (1961): Localization of carbonic anhydrase in the nervous system. Science 134: 1524–1525.

Goor van H (1940): Die Verbreitung und Bedeutung der Carboanhydrase. Enzymologia 8: 113–128.

Gottfried JA, Chesler M (1994): Endogenous H^+ modulation of NMDA receptor-mediated EPSCs revealed by carbonic anhydrase inhibition in rat hippocampus. J Physiol (Lond) 478: 373–378.

Griot C, Vandevelde M (1988): Transferrin, carbonic anhydrase C and ferritin in dissociated murine brain cell cultures. J Neuroimmunol 18: 333–340.

Gros G (1991): The role of carbonic anhydrase within the tissues, with special reference to mammalian striated muscle. Soc Exp Biol Semin Ser 41: 35–54.

Hafidi A, Sanes DH, Hillman DE, Kedeshian P (1994): Structural and molecular heterogeneity of astrocytes and oligodendrocytes in the gerbil lateral superior olive. Neuroscience 60: 503–519.

Hanson MA, Nye PC, Torrance RW (1981): The location of carbonic anhydrase in relation to the blood-brain barrier at the medullary chemoreceptors of the cat. J Physiol (Lond) 320: 113–125.

Heck RW, Tanhauser SM, Manda R, Tu C, Laipis PJ, Silverman DN (1994): Catalytic properties of mouse carbonic anhydrase V. J Biol Chem 269: 24742–24746.

Heming TA, Vanoye CG, Stabenau EK, Roush ED, Fierke CA, Bidani A (1993): Inhibitor sensitivity of pulmonary vascular carbonic anhydrase. J Appl Physiol 75: 1642–1649.

Heming TA, Stabenau EK, Vanoye CG, Moghadasi H, Bidani A (1994): Roles of intra- and extracellular carbonic anhydrase in alveolar-capillary CO_2 equilibration. J Appl Physiol 77: 697–705.

Henry RP (1987): Quaternary ammonium sulfanilamide: A membrane-impermeant carbonic anhydrase inhibitor. Am J Physiol 252: R959–R965.

Jeffrey M, Wells GA, Bridges AW (1991): Carbonic anhydrase II expression in fibrous astrocytes of the sheep. J Comp Pathol 104: 337–343.

Kaila K, Paalasmaa P, Taira T, Voipio J (1992): pH transients due to monosynaptic activation of $GABA_A$ receptors in rat hippocampal slices. Neuroreport 3: 105–108.

Karler R, Woodbury DM (1960): Intracellular distribution of carbonic anhydrase. Biochem J 75: 538–543.

Karlmark B, Agerup B, Wistrand PJ (1979): Renal proximal tubular acidification. Role of brush-border and cytoplasmic carbonic anhydrase. Acta Physiol Scand 106: 145–150.

Kato K (1990): Sequence of a novel carbonic anhydrase-related polypeptide and its exclusive presence in Purkinje cells. FEBS Lett 271: 137–140.

Kazimierczak J, Sommer EW, Philippe E, Droz B (1986): Carbonic anhydrase activity in primary sensory neurons, I. Requirements for the cytochemical localization in the dorsal root ganglion of chicken and mouse by light and electron microscopy. Cell Tissue Res 245: 487–495.

Kendall AG, Tashian RE (1977): Erythrocyte carbonic anhydrase, I. Inherited deficiency in humans. Science 197: 471–472.

Kimura H, MacLeod RM (1975): Evidence of existence of two isozymes of carbonic anhydrase in the anterior pituitary gland of female rats. J Biol Chem 250: 1933–1938.

Korhonen LK, Näätänen E, Hyyppä M (1964): A histochemical study of carbonic anhydrase in some parts of the mouse brain. Acta Histochem 18: 336–347.

Korr H, Horsmann C, Schurmann M, Delaunoy JP, Labourdette G (1994): Problems encountered when immunocytochemistry is used for quantitative glial cell identification in autoradiographic studies of cell proliferation in the brain of the unlesioned adult mouse. Cell Tissue Res 278: 85–95.

Krueger NX, Saito H (1992): A human transmembrane protein-tyrosine-phosphatase, PTP zeta, is expressed in brain and has an N-terminal receptor domain homologous to carbonic anhydrases. Proc Natl Acad Sci USA 89: 7417–7421.

Kumpulainen T, Dahl D, Korhonen LK, Nyström SH (1983): Immunolabeling of carbonic anhydrase isoenzyme C and glial fibrillary acidic protein in paraffin-embedded tissue sections of human brain and retina. J Histochem Cytochem 31: 879–886.

Kumpulainen T, Rajaniemi H, Myllylä VV, Korhonen LK (1985): A single-step solid phase radioimmunoassay for quantifying human carbonic anhydrase I and II in cerebrospinal fluid. Clin Chim Acta 150: 205–212.

Laitala T, Väänänen HK (1994): Inhibition of bone resorption in vitro by antisense RNA and DNA molecules targeted against carbonic anhydrase II or two subunits of vacuolar H(+)-ATPase. J Clin Invest 93: 2311–2318.

LeBars D, Luthra SK, Pike VW, Swenson ER (1988): The radiosynthesis of a "C-labelled carbonic anhydrase inhibitor-[*carbonyl*-"C] acetazolamide. Appl Radiat JSOT 39: 371–675.

Lees MB, Sapirstein VS (1984): Myelin-associated enzymes. In Lajtha A (ed): Handbook of Neurochemistry, 2nd ed. New York: Plenum Press, pp 435–460.

Lees MB, Sapirstein VS, Reiss DS, Kolodny EH (1980): Carbonic anhydrase and 2′,3′-cyclic nucleotide-3′-phosphohydrolase activity in normal human brain and in demyelinating diseases. Neurology 30: 719–725.

Lewis SE, Erickson RP, Barnett LB, Venta PJ, Tashian RE (1988): N-ethyl-N-nitrosourea-induced null mutation at the mouse Car-2 locus: An animal model for human carbonic anhydrase II deficiency syndrome. Proc Natl Acad Sci USA 85: 1962–1966.

Li J, Chow SY (1994): Subcellular distribution of carbonic anhydrase and Na^+, $K(^+)$-ATPase in the brain of the hyt/hyt hypothyroid mice. Neurochem Res 19: 83–88.

Lindskog S, Wistrand PJ (1989): Carbonic anhydrase inhibitors. In Sandler M, Smith J (eds): Design of Enzyme Inhibitors as Drugs. Oxford: Oxford University Press, pp 698–723.

Lindvall M, Edvinsson L, Owman C (1978): Sympathetic nervous control of cerebrospinal fluid production from the choroid plexus. Science 201: 176–178.

Lucci MS, Tinker JP, Weiner IM, DuBose Jr TD (1983): Function of proximal tubule carbonic anhydrase defined by selective inhibition. Am J Physiol 245: F443–F449.

Lüjten-Drecoll E, Eichhorn M, Bárány EH (1985): Carbonic anhydrase in epithelia and fenestrated juxtaepithelial capillaries of Macaca fascicularis. Acta Physiol Scan 124: 295–307.

Lynch CJ, Fox H, Hazen SA, Stanley BA, Dodgson S, Lanoue KF (1995): Role of hepatic carbonic anhydrase in de novo lipogenesis. Biochem J 310: 197–202.

Mahieu I, Hollande E, Carter N (1994): Membrane targeting of carbonic anhydrase II (CAII) in human pancreatic ductal Capan I cells in culture. Biochem Soc Trans 22: 4385.

Maren TH (1977): Use of inhibitors in physiological studies of carbonic anhydrase. Am J Physiol 232: F291–F297.

Maren TH (1985): Carbonic anhydrase [editorial]. N Engl J Med 313: 179–181.

Maren TH, Broder LE (1970): The role of carbonic anhydrase in anion secretion into cerebrospinal fluid. J Pharmacol Exp Ther 172: 197–202.

Maren TH, Wynns GC, Wistrand PJ (1993): Chemical properties of carbonic anhydrase IV, the membrane-bound enzyme. Mol Pharmacol 44: 901–905.

Masuzawa T, Shimabukuro H, Sato F, Saito T (1981): Ultrastructural localization of carbonic anhydrase activity in the rat choroid plexus epithelial cell. Histochemistry 73: 201–209.

Matsumoto T, Schwartz GJ (1992): Novel method for performing carbonic anhydrase histochemistry and immunocytochemistry on cryosections. J Histochem Cytochem 40: 1223–1227.

Mayeux V, Tafti M, Baldy Moulinier M, Valmier J (1993): Developmental regulation of carbonic anhydrase expression in mouse dorsal root ganglia. Brain Res Dev Brain Res 71: 201–208.

Millichap GJ (1958): Seizure patterns in young animals. Significance of brain carbonic anhydrase. Proc Soc Exp Biol Med 97: 606–611.

Mithoefer JC, Mayer PW, Stocks JF (1957): Effect of carbonic anhydrase inhibition on the cerebral circulation of the anesthetized dog. Fed Proc 16: 88–89.

Murakami H, Sly WS (1987): Purification and characterization of human salivary carbonic anhydrase. J Biol Chem 262: 1382–1388.

Nakamura S, Sasaki J, Tsubokawa T (1991): Serotonergic effects on carbonic anhydrase activity in the choroid plexus. Childs Nerv Syst 7: 442–447.

Neubauer J (1991): Carbonic anhydrase and sensory function in the central nervous system. In Dodgson SJ, Tashian RE, Gros G, Carter N (eds): The Carbonic Anhydrases. Cellular Physiology and Molecular Genetics. New York: Plenum, pp 319–329.

Nógrádi A (1993): Differential expression of carbonic anhydrase isozymes in microglial cell types. Glia 8: 133–142.

Nógrádi A, Mihály A (1990): Light microscopic histochemistry of the postnatal development and localization of carbonic anhydrase activity in glial and neuronal cell types of the rat central nervous system. Histochemistry 94: 441–447.

Nógrádi A, Mihály A (1991): Expression and quantitative changes of carbonic anhydrase in developing neurones of rat central nervous system. Int J Devl Neurosci 9: 555–561.

Ogawa Y, Hong SS, Toyosawa S, Kuwahara H, Shimazaki M, Yagi T (1993): Immunoelectron microscopy of carbonic anhydrase isozyme VI in human submandibular gland: Comparison with isozymes I and II. J Histochem Cytochem 41: 343–351.

Pappas CA, Ransom BR (1994): Depolarization-induced alkalinization (DIA) in rat hippocampal astrocytes. J Neurophysiol 72: 2816–2826.

Parkkila AK, Parkkila S, Serlo W, Reunanen M, Vierjoki T, Rajaniemi H (1994): A competitive dual-label time-resolved immunofluorometric assay for simultaneous detection of carbonic anhydrase I and II in cerebrospinal fluid. Clin Chim Acta 230: 81–89.

Pasternack M, Voipio J, Kaila K (1993): Intracellular carbonic anhydrase activity and its role in GABA-induced acidosis in isolated rat hippocampal pyramidal neurones. Acta Physiol Scand 148: 229–231.

Prabhakar E, Lawson SN (1995): The electrophysiological properties of rat primary afferent neurones with carbonic anhydrase activity. J Physiol (Lond) 482: 609–622.

Ridderstråle Y (1991): Localization of the carbonic anhydrases by chemical reactions. In Dodgson SJ, Tashian RE, Gros G, Carter N (eds): The Carbonic Anhydrases. Cellular Physiology and Molecular Genetics. New York: Plenum Press, pp 133–144.

Ridderstråle Y, Hanson MA (1985): Histochemical study of the distribution of carbonic anhydrase in the cat brain. Acta Physiol Scand 124: 557–564.

Ridderstråle Y, Wistrand PJ (1998): Membrane-associated carbonic anhydrase activity in the brain of the CA II-deficient mouse (submitted).

Riley DA, Ellis S, Lang DH, Bain JL (1984): Examination of carbonic anhydrase activities in the peripheral nervous systems of humans, rhesus monkeys, cats, and rats. Ann NY Acad Sci 429: 408–411.

Roussel G, Delaunoy JP, Nussbaum JL, Mandel P (1979): Demonstration of a specific localization of carbonic anhydrase C in the glial cells of rat CNS by an immunohistochemical method. Brain Res 160: 47–55.

Sapirstein V, Trachtenberg M, Lees MB, Koul O (1978a): Regional developmental and fractional studies on myelin and other carbonic anhydrases in rat CNS. Adv Exp Med Biol 100: 55–69.

Sapirstein VS (1984): Carbonic anhydrase. In Lajtha A (ed): Handbook of Neurochemistry, 2nd ed. Enzymes in the Nervous System. New York: Plenum Press, vol 4, pp 385–402.

Sapirstein VS, Lees MB (1978): Purification of myelin carbonic anhydrase. J Neurochem 31: 505–511.

Sapirstein VS, Lees MB, Trachtenberg MC (1978): Soluble and membrane bound carbonic anhydrases from rat CNS: Regional development. J Neurochem 31: 283–287.

Sapirstein VS, Flynn C, Lees MB (1980): Developmental changes in carbonic anhydrase and adenylate cyclase in quaking mice. Brain Res 185:373–383.

Sapirstein VS, Strocchi P, Wesolowski M, Gilbert JM (1983): Characterization and biosynthesis of soluble and membrane-bound carbonic anhydrase in brain. J Neurochem 40: 1251–1261.

Sapirstein VS, Strocchi P, Gilbert JM (1984): Properties and function of brain carbonic anhydrase. Ann NY Acad Sci 429: 481–493.

Sapirstein VS, Durrie R, Nolan CE, Marks N (1993): Identification of membrane-bound carbonic anhydrase in white matter coated vesicles: The fate of carbonic anhydrase and other white matter coated vesicle proteins in triethyl tin-induced leukoencephalopathy. J Neurosci Res 35: 83–91.

Schwartz GJ, Brown D, Mankus R, Alexander EA, Schwartz JH (1994): Low pH enhances expression of carbonic anhydrase II by cultured rat inner medullary collecting duct cells. Am J Physiol 266: C508–C514.

Sly WS, Hu PY (1995): Human carbonic anhydrases and carbonic anhydrase deficiencies. Annu Rev Biochem 64: 375–401.

Snyder DS, Zimmerman Jr TR, Farooq M, Norton WT, Cammer W (1983): Carbonic anhydrase, 5′-nucleotidase, and 2′,3′-cyclic nucleotide-3′-phosphodiesterase activities in oligodendrocytes, astrocytes, and neurons isolated from the brains of developing rats. J Neurochem 40: 120–127.

Sommer EW, Kazimierczak J, Droz B (1985): Neuronal subpopulations in the dorsal root ganglion of the mouse as characterized by combination of ultrastructural and cytochemical features. Brain Res 346: 310–326.

Spicer SS, Schulte BA (1991): Biologic role of carbonic anhydrase assessed immunocytochemically. In Botrà F, Gros G, Storey BT (eds): Carbonic Anhydrase. From Biochemistry and Genetics to Physiology and Clinical Medicine. Weinheim: VCH, pp 208–225.

Spicer SS, Lewis SE, Tashian RE, Schulte BA (1989): Mice carrying a CAR-2 null allele lack carbonic anhydrase II immunohistochemically and show vascular calcification. Am J Pathol 134: 947–954.

Supuran CT, Clare BW (1995): Carbonic anhydrase inhibitors, 24. A quantitative structure-activity relationship study of positively charged sulfonamide inhibitors. Eur J Med Chem 30: 687–696.

Taira T, Smirnov S, Voipio J, Kaila K (1993): Intrinsic proton modulation of excitatory transmission in rat hippocampal slices. NeuroReport 4 :93–96.

Tashian RE (1989): The carbonic anhydrases: Widening perspectives on their evolution, expression and function. Bioessays 10: 186–192.

Tinker JP, Coulson R, Weiner IM (1981): Dextran-bound inhibitors of carbonic anhydrase. J Pharmacol Exp Ther 218: 600–607.

Torrance RW (1993): Carbonic anhydrase near central chemoreceptors. In Data PG: Neurobiology and Cell Physiology of Chemoreception. Plenum Press, New York, pp 235–239.

Velisek L, Moshe SL, Stanton PK (1995): Resistance of hippocampal synaptic transmission to hypoxia in carbonic anhydrase II-deficient mice. Brain Res 671: 245–253.

Vogh BP (1980): The relation of choroid plexus carbonic anhydrase activity to cerebrospinal fluid formation: study of three inhibitors in cat with extrapolation to man. J Pharmacol Exp Ther 213: 321–331.

Wistrand PJ (1980): Human renal cytoplasmic carbonic anhydrase. Tissue levels and kinetic properties under near physiological conditions. Acta Physiol Scand 109: 239–248.

Wistrand PJ (1981): The importance of carbonic anhydrase B and C for the unloading of CO_2 by the human erythrocyte. Acta Physiol Scand 113: 417–426.

Wong V, Barrett CP, Donati EJ, Eng LF, Guth L (1983): Carbonic anhydrase activity in first-order sensory neurons of the rat. J Histochem Cytochem 31: 293–300.

CHAPTER 4

Diffusion and Buffering Aspects of H^+, HCO_3^-, and CO_2 Movements in Brain Tissue

JUHA VOIPIO
Department of Biosciences, Division of Animal Physiology, University of Helsinki, FIN-00014 Helsinki, Finland

INTRODUCTION

Brain tissue is effectively perfused through its dense network of capillaries. This circulation provides, among its other functions, a distributed mechanism for gas exchange, bringing oxygen in and carbon dioxide out and thereby removing metabolically produced acid from the brain. With respect to processes involving acid-base, it is essential to understand how different buffer systems operate in concert under varying load conditions at the cellular level, and how diffusion of acid equivalents takes place in the neuronal microenvironment. This is important, not only in understanding the brain *in vivo,* but also in the design and interpretation of *in vitro* experiments.

Buffers do not simply buffer (i.e., attenuate) pH changes. Depending on the types of buffers present, a large variation is found in the time course and shape of pH responses to different kinds of acid-base loads. In certain cases, buffers are needed in the generation of pH transients.

Concentration gradients give rise to diffusion. Although a pH gradient is associated with a net flux of acid, the movement of free protons plays but little role under most physiological conditions. Instead, protonated and nonprotonated forms of mobile buffers diffuse down their concentration gradients, thereby facilitating proton diffusion. Therefore, all mobile buffers present in a system, and their duffusion characteristics, are involved in the net flux of acid associated with a pH gradient.

BUFFERING

The main buffer in the brain is the CO_2/HCO_3^- buffer system. It is usually considered to be an open buffer, assuming maintenance of a constant partial pressure of CO_2 ($P\text{CO}_2$). However, this assumption is not necessarily valid, especially when considering local conditions

pH and Brain Function, Edited by Kai Kaila and Bruce R. Ransom
ISBN 0-471-11838-9

within a short timescale. For instance, direct measurements of $P\text{CO}_2$ in rat hippocampal slices have revealed both a steady-state gradient and activity-induced transient changes in $P\text{CO}_2$ (Voipio and Kaila, 1993). Unless catalyzed by the enzyme carbonic anhydrase (CA), the buffering action of the CO_2/HCO_3^- system is slow, the half-time of equilibration of the CO_2 hydration-dehydration reaction being roughly 15–30 seconds depending on temperature and pH. There is increasing evidence for the presence of CA activity in the main compartments of the brain parenchyma (see Chapter 3): in various neurons, in glial cells, the interstitial space, and also in the cerebrospinal fluid (CSF) and in endothelial cells of the blood-brain barrier. Therefore, rapid transmembrane movements of acid equivalents or changes in metabolic activity may give rise to significant transient changes in the concentrations of the CO_2/HCO_3^- buffer species.

Although the basic theory of pH buffering is well established (see Chapter 2; Roos and Boron, 1981), there are certain cases where the operation of the CO_2/HCO_3^- buffer may still appear somewhat confusing. To understand such phenomena, one has to consider different types of acid-base loads in the presence of a second, independent buffer system. Thereby it becomes evident that even a fairly small non-CO_2/HCO_3^- buffer capacity does, indeed, play an important role in shaping pH responses. The buffer species carbonic acid and carbonate cannot be ignored, as is often implicitly assumed, especially if carbonic anhydrase is not present, or its activity has been blocked.

Factors Shaping pH Responses

The striking variety of different possible shapes of pH responses, which can be generated in a CO_2/HCO_3^- buffer system, can most clearly be demonstrated by considering a simple one-compartment system. The results can then be extended to more complex systems like those encountered in nervous tissue, where transmembrane fluxes of acid equivalents and changes in metabolic activity cause pH shifts in compartments bordered by cellular membranes. In fact, manipulation of buffer systems can be used, and has been used, as a tool in resolving mechanisms underlying pH signals in the nervous tissue (see Chapter 17; Taira et al., 1995; Voipio et al., 1995).

The system to be considered is illustrated in Fig. 1A. This system may be assumed to be either closed or open with respect to CO_2; in the latter case CO_2 tends to equilibrate between the system and its surroundings. With respect to all other buffer species, the system is closed. It may contain CA or a non-CO_2/HCO_3^- buffer, or both.

In what follows, only responses to instantaneous and uniform addition of H^+ (or OH^-), HCO_3^-, CO_3^{2-}, or CO_2 will be considered. The immediate pH response is obtained by assuming immediate equilibration of reactions between different buffer species within the system, including carbonate and carbonic acid (see Fig. 1A). However, the hydration-dehydration step between CO_2 and H_2CO_3 is included in these reactions only if it is catalyzed by CA. Equilibration of CO_2 between the system and its surroundings involves a net movement of CO_2, and it takes anywhere from milliseconds to minutes in real preparations, depending on the dimensions involved. Therefore, over a very short timescale, even an open system behaves as a closed system. The final pH level is obtained by considering equilibration of all those reactions shown in Fig. 1A that are relevant to the case under study. In all cases, the immediate and final shifts in pH may be calculated from groups of equations that are given in Appendix C. The equations include not only those defining equilibrium or dissociation constants, since their number is smaller than that of

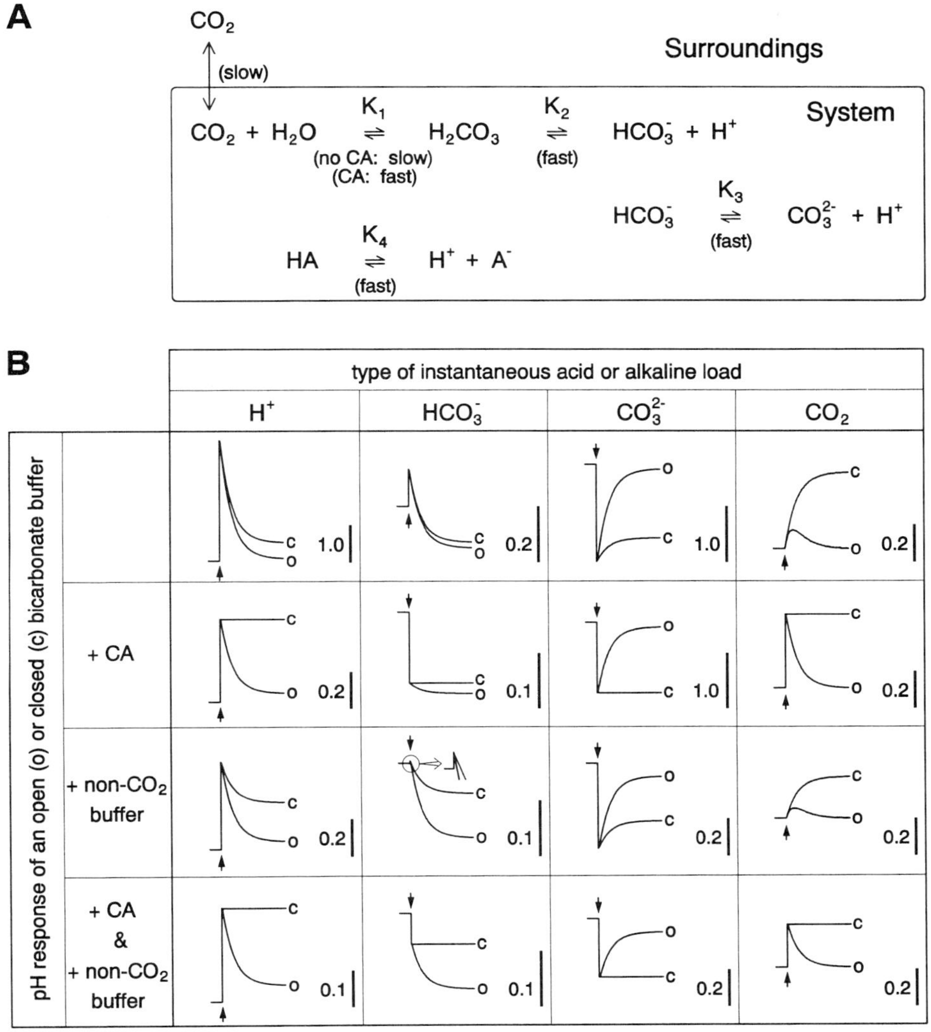

FIGURE 1. A schematic representation of pH responses in a single-compartment model system upon sudden addition of an acid or alkaline load. **A.** The model system consists of a single compartment with a CO_2/HCO_3^- buffer. The system may be either open or closed with respect to CO_2, and it may also have a non-CO_2/HCO_3^- buffer, which is represented by a weak acid, HA. Note that there are four reactions within the system which need to be considered. **B.** The amplitude and qualitative time course of pH responses that are generated in the model system shown in (A) upon uniform addition (indicated by arrow) of H^+ (2.5 mM), HCO_3^- (10 mM), CO_3^{2-} (2.5 mM) or CO_2 (2.5 mM). Traces are sketched for cases with and without CA in open (o) and closed (c) systems. The non-CO_2/HCO_3^- buffer, if present, corresponds to 10 mM HEPES buffer. The initial HCO_3^- concentration in the system is 25 mM and pH is 7.4. Further details are given in Appendix C. Upward deflections indicate a fall in pH. The time course of responses is shown on a relative timescale and the figure is not intended for any quantitative temporal comparisons between traces.

unknown variables, but also equations that are obtained by writing conditions for conservation of matter (total carbon, total H^+). The calculations are easy and straightforward, although somewhat laborious.

Attaining the final pH level is instantaneous only if the system is closed and contains CA. Otherwise time is required for equilibration of either the hydration-dehydration step or for the flux of CO_2 across the system's boundaries, or both.

Figure 1B illustrates pH responses calculated for a conventional CO_2/HCO_3^- buffer system (25 mM HCO_3^-, pH 7.4; see Appendix C for details). The pH responses upon uniform addition of H^+ (2.5 mM), HCO_3^- (10 mM), CO_3^{2-} (2.5 mM), or CO_2 (2.5 mM) are sketched for cases with and without CA and/or a non-CO_2/HCO_3^- buffer (10 mM weak acid with a pK_a of 7.4). The time courses of pH changes are shown in Fig. 1B only qualitatively, on a relative timescale. The mechanisms involved in the shaping of individual responses are discussed in more detail on p 48–50.

H^+-induced changes in pH. In the absence of CA activity, instantaneous addition of H^+ (or OH^-) gives rise to a rapid pH transient, the amplitude of which depends mainly on the non-CO_2/HCO_3^- buffer capacity of the solution. As illustrated in Fig. 1B, if no non-CO_2/HCO_3^- buffer exists in the system, the amplitude of the immediate pH peak is very high, because only small amounts of protons are involved in the large relative changes of the low concentrations of carbonic acid and carbonate, or, in other words, because pH is much higher than the logarithmic dissociation constant of carbonic acid ($-\log_{10} K_2$, cf. Fig. 1A) and much lower than that of bicarbonate ($-\log_{10} K_3$). The time course of the subsequent pH recovery reflects the slow net dehydration reaction and, in an open system, also CO_2 transfer across the system's boundaries.

In the presence of CA activity, all reactions within the system equilibrate immediately. Accordingly, the change in pH is stepwise if the system is closed. In an open system, an immediate shift of identical amplitude is observed, because the time-consuming O_2 exchange between the system and its surroundings has no effect on the instantaneous pH transient.

Another possibility would be addition of OH^-, which is equivalent to removing H^+ ions. In this case, the amount of free CO_2 initially present in the system may be insufficient for rapid buffering of a large OH^- load.

CO_2-induced changes in pH. After suddenly adding a given amount of CO_2 to an open system, the concentration of free CO_2 ($[CO_2]$) returns to its original level and there will be no steady-state change in pH. However, a pH transient will be generated depending on the kinetics of CO_2 reequilibration and of hydration-dehydration. In the presence of CA activity, the pH transient peaks rapidly (i.e., CA is needed in the generation of the rapid pH shift), whereas in the absence of CA the maximum change in pH is smaller and is attained with a delay. It should also be noted that a maintained change in $[CO_2]$ causes a steady-state shift in pH, which develops either rapidly or slowly depending on the presence or absence of CA activity (not illustrated). In each case, a non-CO_2/HCO_3^- buffer has an attenuating effect on the pH changes.

In a closed system, addition of CO_2 results in a maintained fall in pH, as all reactions within the system equilibrate. The response is slow in the absence of CA.

HCO_3^--induced changes in pH. If the acid-base load is due to bicarbonate ions, interesting qualitative differences are found in the pH responses compared with those described above.

In the absence of CA activity, bicarbonate will take part in two immediate reactions: a fraction of HCO_3^- ions will dissociate to CO_3^{2-} ions by releasing a proton, while formation of carbonic acid by proton binding will occur as well (cf. Fig. 1A). In the cases depicted in Fig. 1B, the logarithmic acid dissociation constant of H_2CO_3 ($-\log_{10} K_2$) is roughly four units lower than pH, which is, on the other hand, less than three units lower than the logarithmic acid dissociation constant of HCO_3^- ($-\log_{10} K_3$). Therefore, the immediate release of protons by dissociation of HCO_3^- masks the less pronounced binding of protons in H_2CO_3 formation, and a transient acidification results. However, since the initial concentrations of H_2CO_3 and CO_3^{2-} are very small (1.6 μM and 50 μM, respectively), the net amount of protons released is small and, consequently, even a small amount of a non-CO_2/HCO_3^- buffer can attenuate the immediate pH transient effectively.

If the hydration-dehydration reaction of CO_2 is catalyzed by CA, addition of HCO_3^- causes an immediate alkaline shift in pH under all physiologically relevant conditions. This is because the net dehydration of HCO_3^- to CO_2 dominates over the effects related to H_2CO_3 and CO_3^{2-} (note that the moderate relative changes in the millimolar concentrations of CO_2 and HCO_3^-, which are required for attaining a new equilibrium, are associated with a much larger number of protons than corresponding relative changes in the micromolar concentrations of H_2CO_3 and CO_3^{2-}).

For the same reason, the final response is always an alkalosis, regardless of the kinetics of the process. These kinetics depend again on CA activity and on CO_2 equilibration, as illustrated in Fig. 1B. Note also that the final pH shift has a larger amplitude in an open system than in a closed system, which is contrary to what is seen upon addition of H^+, CO_2, or CO_3^{2-}. This is because the H^+-consuming net dehydration, caused by HCO_3^- addition, is enhanced if CO_2 can leave the system.

The total amount of CO_2 that is produced deserves a few words. In the new equilibrium state, in a system devoid of a non-CO_2/HCO_3^- buffer, the only marked absolute concentration change is that in HCO_3^-, and this is almost equivalent to the change that is brought about directly by the amount of HCO_3^- added. Therefore, only a very small amount of CO_2 is produced by net dehydration, and the open and closed system responses are nearly similar (see Fig. 1B). With regard to making physiological solutions, the HCO_3^- concentration that prevails in a CO_2/HCO_3^--buffered solution following equilibration with a CO_2-containing gas mixture, is almost equivalent to the concentration of HCO_3^- that was added when making the solution from salts and water. However, if the solution contains a non-CO_2/HCO_3^- buffer, this buffer should be titrated separately, because otherwise a large net hydration-dehydration takes place, giving rise to a significant change in the HCO_3^- concentration. In other words, the final pH would not correspond to that predicted by the CO_2 partial pressure and the initial HCO_3^- concentration. (In this respect, the effect of 1 mM $H_2PO_4^-$, that is common in salines, is quite small and can usually be ignored.)

With respect to the transient acidosis seen in the absence of CA, it may be worth pointing out that adding $NaHCO_3$ from a stock solution (e.g., when titrating solutions) is not equivalent to adding pure (dry) $NaHCO_3$, since the stock solution contains also small (CO_2-dependent) amounts of carbonate and carbonic acid.

CO_3^{2-}-induced changes in pH. The kinetics of pH changes induced by sudden addition of CO_3^{2-} may appear surprising when compared to those induced by HCO_3^-. The immediate response is, of course, a transient increase in pH brought about by the H^+-

consuming formation of HCO_3^- from the added CO_3^{2-}. The resulting rise in the concentration of HCO_3^- does not, however, cause a net reaction towards H_2CO_3 and CO_2, (cf. Fig. 1A) as one might (erroneously) expect on the basis of what has been described in the case of HCO_3^- addition above. Instead, although the (initially 25 mM) HCO_3^- concentration increases, its production from the added (2.5 mM) CO_3^{2-} is coupled to an one-to-one consumption of the free (initially 0.04 μM !) H^+, and the resulting large *relative* fall in H^+ fully masks the effect of the small relative increase in HCO_3^- on the reaction between HCO_3^- and H_2CO_3. Therefore, net dissociation of H_2CO_3 occurs, which gives rise to net hydration of CO_2, and a recovery from the peak alkalosis is observed. If the system is open, more CO_2 is available and the recovery plateaus at a level that is less alkaline than in a closed system. If a non-CO_2/HCO_3^- buffer is present, the pH shifts are qualitatively similar but have smaller amplitudes. Only if the system is closed and equipped with CA, do all reactions equilibrate immediately and there is a step alkalosis without further changes in pH. The large difference in the final open and closed system responses is attributable to depletion of CO_2 in the closed system.

Carbonate ion can bind two protons, whereas HCO_3^- can bind only one. This 2 : 1 ratio is seen in open systems, where similar steady-state alkaline shifts are caused by a given amount of CO_3^{2-} and twice as much HCO_3^-.

Differences in Non-CO_2/HCO_3^- Buffer Capacity: Their Role in a Multicompartment System

Let us apply the above principles in a more complex system. For example, consider a simple two-compartment system. Both compartments contain carbonic anhydrase, but only one is equipped with a non-CO_2/HCO_3^- buffer, and the compartments are separated with a gas-permeable membrane. As mentioned above, addition of bicarbonate to a CO_2/HCO_3^--buffered solution devoid of a non-CO_2/HCO_3^- buffer gives rise to an alkaline shift with little CO_2 production. If a given amount of HCO_3^- is transferred from the compartment with the non-CO_2/HCO_3^- buffer to the one without, the alkalosis in the latter one will not be associated with significant CO_2 generation. However, provided that the non-CO_2/HCO_3^- buffer capacity is very high, there is hardly any pH change in the first compartment, and the loss of HCO_3^- has to be covered by net hydration of CO_2 in order to attain a new equilibrium of the hydration-dehydration reaction. What results is consumption of CO_2 and a fall of the amount of free CO_2 (or P_{CO_2}) in the whole system, whereas the total HCO_3^- content increases.

It is evident that this result can be generalized: If bicarbonate is translocated in a multicompartment system between regions of different non-CO_2/HCO_3^- buffer capacity, significant changes in P_{CO_2} may arise. If we consider bicarbonate movements between intracellular and extracellular compartments in the brain, it is worth noting that the small amount of free CO_2 makes it prone to significant relative changes. For instance, the possible physiologically relevant consequences may be related to mechanisms of spatial pH buffering and nonsynaptic communication. Interestingly, an activity-induced transient fall in P_{CO_2} that was enhanced upon inhibition of extracellular carbonic anhydrase activity (which is indicative of an intracellular sink of CO_2) has been recorded in rat hippocampal slices (Voipio and Kaila, 1993). The possible role of this type of differences in intrinsic buffering power with respect to H^+ signalling within the nervous tissue have not yet been studied.

DIFFUSION

Buffering is not the only physicochemical factor that shapes pH responses in living tissue. In order to understand the dynamics of acid-base shifts in the brain, the diffusion of various buffer species must be considered.

Fick's Laws of Diffusion

Diffusion results from random movement of mobile particles. If a dissolved substance is not uniformly distributed, a concentration gradient exists and the nonoriented stochastic movement of individual particles results in a net transfer of the substance in the direction of decreasing concentration. In an isotropic medium, the flux is proportional to the concentration gradient of the diffusing substance and to a substance-specific factor of proportionality, which is called the diffusion coefficient. This dependence is called Fick's first law of diffusion, and it is often used to calculate the net flux of a substance in a system at steady state.

Diffusion is driven by a virtual force. On the other hand, an electric field exerts a "true" force on individual ions, and thereby gives rise to another flux component. The simultaneous effects of a concentration gradient and an electric field on charged particles are expressed in the so-called Nernst-Planck electrodiffusion equation (see, e.g., Hille, 1992), which electrophysiologists get acquainted with when deriving the well known Goldman-Hodgkin-Katz equations. In this chapter, however, the discussion will be limited to cases where electric fields do not play a role or may be ignored.

The fundamental equation of diffusion is called Fick's second law, and it can be easily derived on the basis of Fick's first law (see Crank, 1956). This equation is useful when the system is not at steady state, since it relates local accumulation of a diffusing substance to the shape of its concentration gradient and to its production and consumption in chemical reactions.

The mathematical formulation of Fick's laws is given in Appendix A.

Diffusion in the Brain Extracellular Space

If we consider ionic diffusion through brain extracellular space, Fick's laws are valid only on a microscale, where the moving particles do not meet any diffusion barriers such as cell membranes. On a larger scale, a straight line between two points repeatedly crosses cell membranes, and the flux of impermeant ions takes place along tortuous extracellular routes. However, diffusion can still be treated as if it were occurring along a straight line, if we consider, in addition to each point on the line, also the space in its vicinity. This is accomplished mathematically by averaging variables within a volume that is large with respect to cell size, but small with respect to macroscopic dimensions of the system (the approach follows that for diffusion in porous media and is discussed in detail by Nicholson and Phillips, 1981; see also Whitaker, 1969).

In addition to volume averaging, two new variables are needed. The reduced free space is described by the ratio of extracellular volume to total volume, called volume fraction (α), and the increase in the path length is represented by tortuosity (λ), which manifests itself as a decrease in the diffusion coefficient (see Nicholson, 1988–1989). Fick's first and second laws can then be modified (see Appendix A) to describe extracellular diffusion as if it occurred in a homogeneous and isotropic continuum (Nicholson and Phillips, 1981; see also Nicholson and Rice, 1986).

The validity of these new equations has been examined on a theoretical basis (Nicholson and Phillips, 1981), but the most convincing evidence is the excellent fit of experimental data to theoretical predictions. This has been a consistent finding in numerous studies (for references, see below), where the most common probe ion has been tetramethylammonium (TMA^+). In general, the results are similar with cations and anions, and also with the divalent cation Ca^{2+}. However, data on K^+ movements do not fit the modified equation, most likely because of spatial buffering by the glial syncytium (cf. Gardner-Medwin, 1983).

Numerical values of α and λ can be obtained by injecting an impermeable probe ion into brain interstitial space and following the time course of its concentration changes at a known distance from the point source (for the method, see Nicholson, 1985, 1992, 1993). *In vivo* and *in vitro* studies have been performed on several regions of the CNS using mainly rat and turtle. The data are reviewed in Nicholson and Rice (1986), and in Nicholson (1992, 1993); some recent studies are cited below.

In general, these studies have revealed typical α and λ values of 0.2 and 1.6, respectively. Extreme values of α have been observed in the molecular layer of isolated turtle cerebellum (0.31; Rice et al., 1993) and in the CA1 region of rat hippocampus (0.12; McBain et al., 1990). The common assumption that extracellular diffusion is isotropic may be true in many brain regions, but anisotropic diffusion (i.e., orientation dependence of tortuosity) has been reported in the molecular layer of turtle isolated cerebellum (Rice et al., 1993) as well as in rat corpus callosum after development of myelination (Vorisek and Syková, 1997a). During early postnatal development, extracellular volume fraction is relatively large, but maturation is not reflected in the values of tortuosity (Lehmenkühler et al., 1993; see also Voipio and Ballanyi, 1997; but see Vorisek and Syková, 1997a). Neuronal activity can be coupled to a significant reduction of extracellular space (Dietzel et al., 1980; Ransom et al., 1985; Svoboda and Syková, 1991), and the effect is even larger during pathophysiological conditions like anoxia, ischemia, spreading depression, or epileptic activity (Lux et al., 1986; Syková et al., 1994; Perez-Pinzon et al., 1995; Vorisek and Syková, 1997b; earlier work reviewed by Hansen, 1985).

Taken together, although extracellular diffusion is restricted by the complex geometry and limited volume of the available space, a concentration gradient of a substance (e.g., a mobile impermeant buffer) will give rise to a net flux and the flux can have a substantial value. Diffusion is less hindered during early stages of maturation than in the adult and, in a resting preparation, diffusion takes place more freely than under states of intense or pathophysiological activity.

Local boundary conditions are different if the diffusing substance is membrane-permeant. This is particularly true for CO_2, which easily permeates through plasma membranes (see Gutknecht et al., 1977). However, the path of CO_2 is not completely independent of limiting geometrical factors, since the intracellular density of macromolecular structures is not negligible (cf. Gros and Moll, 1971). On the other hand, within individual compartments, facilitated diffusion of CO_2 (Longmuir et al., 1966; Gros et al., 1976) may occur. In a recent study (Voipio and Ballanyi, 1997), the diffusion of free CO_2 was found to be similar in saline (cf. Ratcliff and Holdcroft, 1963) and in neonatal rat brainstem tissue, suggesting that the hindering and facilitating factors within the tissue play either minor or mutually cancelling roles. However, it remains for future work to determine the relative contributions of the two effects in different brain areas and at different stages of postnatal development.

Proton Diffusion

Single-ion diffusion coefficients can be calculated from experimentally determined equivalent conductivities (Robinson and Stokes, 1959; see also Hille, 1992). The diffusion coefficient of protons in water ($D_H = 9.3 \times 10^{-5}$ cm^2/s, $T = 25°C$) is roughly an order of magnitude higher than that of inorganic cations and anions. However, diffusion of free protons results in negligible flux rates at (patho)physiologically relevant pH levels, because the flux across a distance is proportional to D_H and the corresponding concentration difference (not ratio, see Eq. A1), which, being in the nanomolar range, is close to zero.

Proton fluxes may be affected by electrostatic factors in the vicinity of cell membranes. Assuming no transmembrane fluxes and considering a pH gradient in bulk water phase, a negative surface potential might cause an increase in the surface concentration of free H^+ sufficient to induce a noticeable enhancement in the flux of free H^+ along the membrane surface. While we can only speculate about the condition where there is no transmembrane flux, more is known about proton migration along membrane surfaces when protons are released at the surface. In this case, protons can diffuse laterally over relatively long distances, not because of a high diffusion coefficient, but rather because of a slow surface-to-bulk transfer rate (Heberle et al., 1994; see also Gabriel and Teissié, 1996), which can be increased by mobile buffers (cf. Gutknecht and Tosteson, 1973; Engasser and Horvath, 1974; Gutknecht et al., 1977). These mechanisms may play an important role in the coupling of processes generating pH shifts to H^+-sensitive ion channels, receptors, and transporters (cf. Taira et al., 1993 and Chapters 17, 21, and 22).

Reversible binding to buffer molecules can dramatically change the effects mediated by H^+ diffusion. For instance, especially if the buffer molecules are fixed, relaxation of a suddenly induced pH gradient proceeds very slowly, because displacement of a large amount of acid has to be mediated by the flux of free protons. A simple general modification of Fick's second law (Crank, 1956; see Appendix B) describes such an effect in terms of free H^+ concentration and an effective diffusion coefficient ($D_{H_{eff}}$), which is always smaller than D_H. This approach has been extended to cases where both mobile and fixed buffers contribute (Junge and McLaughlin, 1987; Irving et al., 1990; see also Al-Baldawi and Abercrombie, 1992). The finding that $D_{H_{eff}}$ is always smaller than D_H, even when the buffers are mobile, may appear to be at variance with the well-known facilitation of H^+ diffusion by mobile buffers. The explanation is, that $D_{H_{eff}}$ must be used in Fick's first law together with the *total* concentration of H^+, although free concentration is used in Fick's second law (see Appendix B). The use of total H^+ concentration is not convenient if the aim is to estimate fluxes related to a certain pH gradient.

The diffusion of protonated and nonprotonated mobile buffer molecules down their concentration gradients, as a result of pH gradient, can facilitate the net flux of protons by orders of magnitude at physiologically relevant pH levels. For example, facilitated proton transfer by phosphate diffusion can give rise to proton fluxes that are 10,000 times larger than rates of free diffusion (Gros et al., 1976). This mechanism is essential for the net flux of acid down a pH gradient within brain tissue. The focus of this discussion must therefore be shifted to the diffusion of acid equivalents such as CO_2 and HCO_3^-.

Acid-Base Movement by Diffusion in the Brain Parenchyma

Measurements carried out using ion-sensitive microelectrodes in brain slice preparations have consistently revealed an acidic extracellular pH (pH_o) with respect to the pH of superfusing solution. A comparable pH difference also prevails between arterial blood and brain tissue *in vivo*. In addition to this steady-state pH_o gradient, activity-related acid transients are a common observation in various preparations. These changes in pH_o have been attributed to cellular metabolism, especially to the production of lactic acid (for references, see Chapter 17), but direct measurements with CO_2/H^+-sensitive microelectrodes (see Chapter 6) indicate that the major cause, at least in rat hippocampal slices maintained in an interface chamber, is accumulation of CO_2 (Voipio and Kaila, 1993). This result indicates that substantial $P\text{CO}_2$ gradients can exist within brain tissue, despite the high permeability of cell membranes to CO_2.

It seems likely, however, that during neuronal activity, as well as in preparations where the O_2 supply is less than sufficient, anaerobic energy metabolism will indirectly give rise to a significant CO_2 production when the H^+ load is buffered by bicarbonate. With regard to the kinetics of buffering, it is important that functionally active carbonic anhydrase is present in many compartments of the brain parenchyma (see Chapter 3). Whether the rapid buffering reaction takes place in the interstitial space or in intracellular compartments (utilizing active bicarbonate uptake mechanisms), it is coupled to a net consumption of HCO_3^-. Indeed, in the submerged neonatal rat brainstem preparation, which has large dimensions compared to conventional brain slices, high $P\text{CO}_2$ levels were coupled to a low concentration of extracellular HCO_3^- (Voipio and Ballanyi, 1997). This suggests that anaerobic metabolism is involved in the generation of the pH_o gradient in this preparation.

It should be emphasized that the facilitated diffusion of CO_2 is based on parallel diffusion of CO_2 and HCO_3^- (see Kreuzer and Hoofd, 1987), which requires that the concentration gradients of CO_2 and HCO_3^- are oriented in the same direction. This mechanism is not effective in translocating acid unless mobile buffers facilitate a parallel flux of protons (Gros et al., 1976). In contrast, when HCO_3^- is consumed in buffering an H^+ load, supplementary HCO_3^- diffuses into this region of tissue, while CO_2 diffuses out. The resulting shuttle provides a mechanism of facilitated H^+ diffusion.

Such a CO_2/HCO_3^- shuttle mechanism has gained convincing support from direct measurements of pH_o and $P\text{CO}_2$ gradients, which were performed on the neonatal rat brainstem preparation and which also allowed indirect monitoring of the extracellular HCO_3^- concentration (Voipio and Ballanyi, 1997). The results of this and a previous study (Voipio and Kaila, 1993) are consistent with the following three conclusions, which concern passive movement of acid from a tissue area of low pH_o to its surroundings (Fig. 2):

1. CO_2 produced by oxidative metabolism causes an increase in tissue $P\text{CO}_2$ and in the CO_2 flux down its concentration gradient. Depending on the non-CO_2/HCO_3^- buffer capacity, a certain amount of this CO_2 is (reversibly) converted to HCO_3^- plus H^+.
2. The CO_2/HCO_3^- shuttle illustrated schematically in Fig. 2 is the main mechanism for removing acid that is produced by anaerobic energy metabolism. It is reasonable to assume that electroneutrality is maintained, at least partly, by the flux of lactate anions.
3. Proton transport by phosphate diffusion may contribute to some extent, but it should be noted that the phosphate concentrations used in the demonstration of facilitated proton transfer by Gros et al. (1976) were orders of magnitude higher than those present in typical extracellular solutions. Other mobile buffers may act as proton carriers in a similar diffusion loop, if present in sufficient concentration.

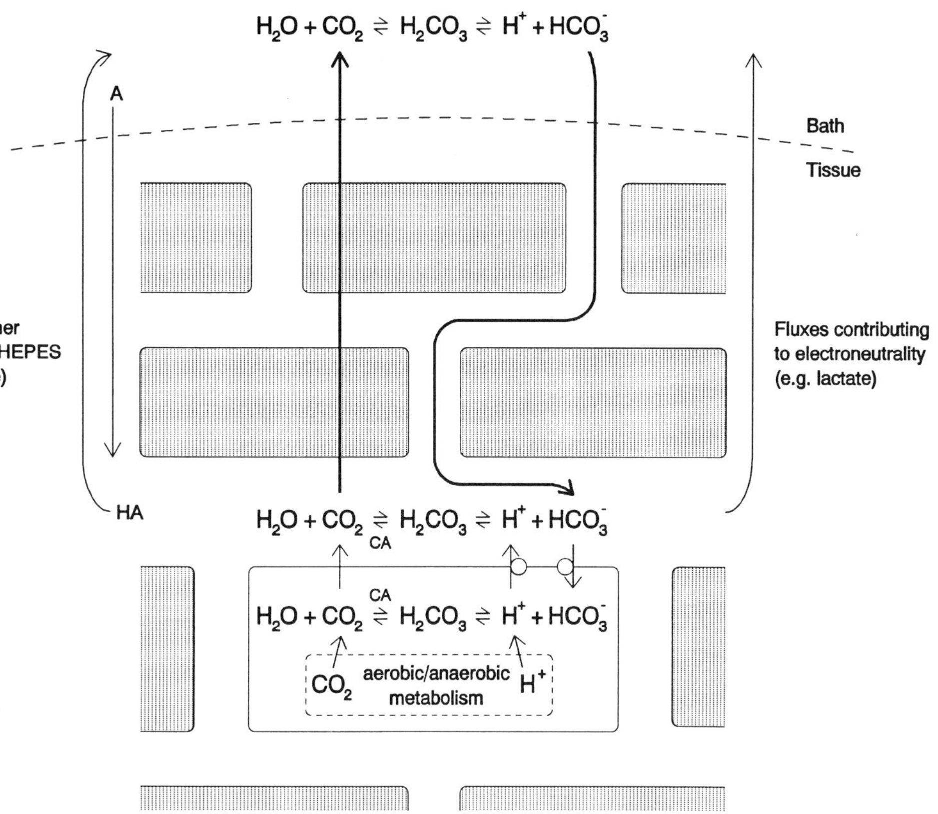

FIGURE 2. Diffusion of acid equivalents in an *in vitro* brain tissue preparation. CO_2 may be generated within a preparation directly by aerobic metabolism or indirectly when an H^+ load of anaerobic metabolism titrates HCO_3^-. This gives rise to a flux of HCO_3^- from the bath along the extracellular space as indicated by the long arrow around individual cells (rectangles). Since CO_2 molecules readily cross cell membranes, the path of their efflux is drawn as a straight line. A shuttle of protonated (HA) and nonprotonated (A) species of other buffers may contribute as indicated. (From Voipio and Ballanyi, 1997, with permission.)

The above view is based on data obtained using *in vitro* preparations, and it assumes that diffusion over long distances of charged buffer species takes place along the extracellular space, whereas diffusion of CO_2 is not significantly hindered by cell membranes. The same mechanisms are likely to be similarly involved in the brain *in vivo*. However, intracellular diffusion of charged buffer species (cf. Al-Baldawi and Abercrombie, 1992) coupled to active plasmalemmal transporters may be important when diffusion distances are short. The shape of astrocytes is particularly striking in this respect and, recently, glial cells have been shown to be capable of spatial buffering of pH_o (Newman, 1996).

3.5 Free Diffusion Coefficients of HCO_3^- and CO_2

Bicarbonate is a fairly small polyatomic anion. Its limiting equivalent conductivity (Robinson and Stokes, 1959) gives a diffusion coefficient of 1.18×10^{-5} cm²/s in water (25°C). Taking into account the effect of ionic strength (see Appendix in Gros and Moll, 1974), and assuming a temperature dependence similar to that reported for NO_3^- (Robinson and Stokes, 1959) yields a $D_{HCO_3^-}$ of 1.10×10^{-5} cm²/s in a typical mammalian saline at 35°C. Making corresponding corrections (cf. Ratcliff and Holdcroft, 1963; Gros and Moll, 1971), the diffusion coefficient of CO_2 is about 2.3×10^{-5} cm²/s at 35°C.

The diffusion coefficient of larger particles can be estimated by using the Stokes-Einstein relation (Einstein-Stokes formula; see Robinson and Stokes, 1959; Hille, 1992), which states that D is inversely proportional to the particle radius. The dependence of D on $(mw)^{-1/3}$ (where mw is molecular weight) follows directly, if constant density and spherical particle shape are assumed. This result can be used as a simple rule of thumb when comparing the flux of large buffer molecules (or macromolecules in general) down a given concentration gradient.

Brain Tissue *In Vitro:* Maintaining a Physiological Acid-Base Balance

In vitro brain tissue preparations are commonly maintained in a superfusion chamber where the oxygen and glucose supply is provided by diffusion into the tissue. Tissue is usually sliced to a thickness of 300–500 μm, temperature is often reduced by a few degrees from 37°C, and the superfusing solution is gassed with 95% O_2. This practice helps to avoid hypoxia within the preparation, as does the use of interface-type chambers.

Systematic observations of acidic extracellular pH gradients in brain slices have led to the suggestion that bicarbonate concentrations in superfusing solutions should be increased. However, in the light of the above discussion, conditions within an *in vitro* slice preparation, which would correspond to 5% CO_2 and pH 7.4, could be better effected by simply reducing the CO_2 content of the gas mixture. It is hard to give any general recommendations, but using 4.0–4.5% CO_2 and 25 mM HCO_3^- might be adequate for 400 μm slices maintained in an interface chamber between 32°C and 34°C. If the preparation has larger dimensions, a more pronounced reduction in CO_2 and an increase in $[HCO_3^-]$ are required (cf. Voipio and Ballanyi, 1997). It is also interesting to note that, in interface chambers, the gas phase dominates in setting P_{CO_2} within slices (K. Kaila, P. Paalasmaa, and J. Voipio, unpublished observations), which suggests that this chamber type is less prone to cause hypercapnic conditions than submerged chambers.

Finally, it must be emphasized that, when changing buffers in superfusing solutions, neither a constant buffer capacity nor a constant buffer concentration guarantees the maintenance of a constant extracellular pH within a preparation. This is, of course, due to the dependence of the steady-state pH_o on the diffusion characteristics of each buffer species available.

CONCLUSIONS

The main focus in current research on acid-base dynamics in the brain is largely on transmembrane fluxes and cellular generation and/or consumption of acid equivalents. As is evident from the considerations presented in this chapter, understanding pH dynamics requires quantitative knowledge about two kinds of additional factors—those that govern the net diffusion of acid-base species and those that govern their buffering.

While buffering has received a fair amount of attention in some of the published work, the fact that the presence or absence of various "intrinsic" and exogenous buffers can affect pH transients not only in quantitative, but also in qualitative ways, has largely been ignored. As illustrated in Fig. 1, pH shifts with a wide spectrum in their amplitudes and time courses can be evoked in response to a given kind of acid-base load by variations in the buffer systems. Clearly, a division of these systems simply into "open" and "closed" ones is a rather crude approximation that fails to provide a sound basis for an analysis of pH shifts.

In addition, most experimental studies on pH shifts in brain tissue have not explicitly recognized that buffering in all living tissues is strongly affected by the diffusional properties of acid-base species. The factors to be considered here fall into two categories, the physicochemical ones and those related to the geometrical properties of the interstitial and intracellular compartments.

In view of the complex and heterogeneous structure of nervous tissue, it is obvious that present measurement techniques may impose a serious limitation to carrying out truly quantitative work on the mechanisms governing pH dynamics at the level of H^+-sensitive targets. Perhaps even more important, there are numerous conceptual limitations in current experimental approaches that easily lead to ill-defined experimental strategies. Nevertheless, whatever the methodological and conceptual approaches may be, understanding the mechanisms of acid-base buffering and diffusion forms a cornerstone for experimental studies within this field.

ACKNOWLEDGMENTS

The original research work of the author has been funded by the Academy of Finland.

APPENDIX A. DIFFUSION EQUATIONS

Fick's first law of diffusion states that, in an isotropic medium, the rate of transfer per unit area, or flux J_s, of substance S is proportional to the gradient of its concentration C_s (e.g., Crank, 1956):

$$J_s = -D_s \, dC_s/dx \tag{A.1a}$$

for diffusion in the direction of x, or in a more general form

$$\bar{J}_s = -D_s \nabla C_s \quad \text{(A.1b)}$$

where D_s is the (free) diffusion coefficient of S, ∇ is the gradient operator, and $\bar{J}_s$ is the vector form of J_s.

Local accumulation of S results if the concentration gradient of S is not constant and/or if there are sources and sinks of S. If q denotes the difference of source and sink densities of S, and t denotes time, Fick's second law is written as (e.g., Crank, 1956):

$$\partial C_s/\partial t = D_s\, \partial^2 C_s/\partial x^2 + q \quad \text{(A.2a)}$$

or

$$\partial C_s/\partial t = D_s\, \nabla^2 C_s + q \quad \text{(A.2b)}$$

Fick's second law must be modified before it can be applied to diffusion along the geometrically complex brain extracellular space. Using volume averaging, and introducing two new variables (volume fraction α, and tortuosity λ), Fick's first and second laws can be restated as (Nicholson and Phillips, 1981; see also Nicholson and Rice, 1986)

$$J_s = -\alpha(D_s/\lambda^2)\, dC_s/dx \quad \text{(A.3a)}$$

or

$$\bar{J}_s = -\alpha(D_s/\lambda^2)\, \nabla C_s \quad \text{(A.3b)}$$

and

$$\partial C_s/\partial t = (D_s/\lambda^2)\, \partial^2 C_s/\partial x^2 + q/\alpha \quad \text{(A.4a)}$$

or

$$\partial C_s/\partial t = (D_s/\lambda^2)\, \nabla^2 C_s + q/\alpha \quad \text{(A.4b)}$$

where D_s is the free, and D_s/λ^2 the apparent, diffusion coefficient of S, C_s is the free ion concentration within the extracellular space, and q is the volume average of the net source density of S.

APPENDIX B. EFFECTIVE DIFFUSION COEFFICIENT(S) OF H^+

A fixed buffer F (weak acid HF) releases and binds H^+ in a reversible and immediate reaction

$$HF \rightleftharpoons H^+ + F^-$$

If this reaction has a $K_F \gg [H^+]$, the concentration of HF ([HF]) is directly proportional to the free H^+ concentration, $[H^+]$ (i.e., $[HF] = R[H^+]$) where R is a constant. This is evident from

$$K_F = \frac{[H^+][F^-]}{[HF]} = [H^+]\frac{[F]_{tot} - [HF]}{[HF]} \tag{B.1}$$

from which

$$[HF] = \frac{[F]_{tot}}{K_F + [H^+]}[H^+] \cong \frac{[F]_{tot}}{K_F}[H^+] = R[H^+] \tag{B.2}$$

where $[F]_{tot} = [F^-] + [HF]$. The source term in Fick's second law (q in Eq. A.2b) is now $-\,\partial[HF]/\partial t$, and a substitution for [HF] from Eq. B.2 gives (cf. Crank, 1956; Junge and McLaughlin, 1987; see also Irving et al., 1990)

$$\begin{aligned}\partial[H^+]/\partial t &= \frac{D_H}{R+1}\nabla^2[H^+] \\ &= \frac{D_H}{1 + [F]_{tot}/K_F}\nabla^2[H^+] \\ &= D_{H_{eff}}\nabla^2[H^+]\end{aligned} \tag{B.3}$$

In this equation, the consequences of reversible binding to an immobile buffer are described as a reduction of D. The new diffusion coefficient $D_{H_{eff}}$ has been named as the modified (Crank, 1956), effective (Junge and McLaughlin, 1987), or apparent (Irving et al., 1990) diffusion coefficient of H^+; "effective" is used here to avoid confusion with the apparent diffusion coefficient D/λ^2 of Eqs. A.3 and A.4. Note that the D_H of free protons does not change.

Making assumptions similar to those above for both a fixed and a mobile buffer (F and M, respectively), Junge and McLaughlin (1987) have derived an effective diffusion coefficient $D_{H_{eff}}$ for Fick's second law

$$\partial[H^+]/\partial t = D_{H_{eff}}\nabla^2[H^+]$$

where

$$D_{H_{eff}} = \frac{D_H + D_{HM}[M]_{tot}/K_M}{1 + [M]_{tot}/K_M + [F]_{tot}/K_F} \tag{B.4}$$

and Irving et al. (1990) have extended the validity of this equation beyond the assumptions made above. Again, the effects of binding to both buffers and the effects of H^+ movements in the mobile form HM are all included in $D_{H_{eff}}$ and no source-sink terms appear.

Since it is very evident that $D_{HM} < D_H$, $D_{H_{eff}}$ is always smaller than D_H. When compared to a fixed buffer, facilitation of H^+ diffusion by a mobile buffer is seen as a smaller decrease in $D_{\mathrm{H_{eff}}}$. Therefore even mobile buffers, despite facilitating net proton fluxes, slow down relaxation of evoked H^+ gradients.

May the above forms of $D_{\mathrm{H_{eff}}}$ be used in Fick's first law? Recalling the assumption $K_M \gg [H^+]$, Eq. B.2 can be written for [HM] as well, and the total flux of protons ($J_{\mathrm{H_{tot}}}$) is

$$\begin{aligned} J_{\mathrm{H,tot}} &= J_{\mathrm{H}} + J_{\mathrm{HM}} = D_{\mathrm{H}}\nabla[\mathrm{H}^+] + D_{\mathrm{HM}}\nabla[\mathrm{HM}] \\ &\cong D_{\mathrm{H}}\nabla[\mathrm{H}^+] + D_{\mathrm{HM}}\frac{\nabla[\mathrm{M}]_{\mathrm{tot}}[\mathrm{H}^+]}{K_{\mathrm{m}}} \\ &= (D_{\mathrm{H}} + D_{\mathrm{HM}}[\mathrm{M}]_{\mathrm{tot}}/K_{\mathrm{M}})\,\nabla[\mathrm{H}^+] \\ &= D^*_{\mathrm{H_{eff}}}\,\nabla[\mathrm{H}^+] \end{aligned} \tag{B.5}$$

$D^*_{\mathrm{H_{eff}}}$ of this equation is the numerator in the formula of $D_{\mathrm{H_{eff}}}$. Keeping the assumption $K_M \gg [H^+]$, the denominator in Eq. B.4 is the ratio of total to free protons, and therefore Fick's first law may also be written

$$J_{\mathrm{H_{tot}}} = D_{\mathrm{H_{eff}}}\,\nabla[\mathrm{H}^+]_{\mathrm{tot}} \tag{B.6}$$

where $[H^+]_{tot} = [H^+] + [HM] + [HF]$.

APPENDIX C

When acid or base is added to a system such as that shown in Fig. 1A, the immediate pH responses can be calculated by considering only the rapidly equilibrating reactions; similarly, final pH responses can be calculated by considering both rapidly and slowly equilibrating reactions. There are, however, more variables to be solved for than there are linearly independent equations defining equilibrium or dissociation constants. The required additional equations are obtained by writing conditions for conservation of matter (total carbon, total H^+).

In the equations below, H^+ concentration ($[H^+]$) is used instead of H^+ activity (a_H). This is a bold oversimplification, and it should be kept in mind that there are two definitions of pH (see Chapter 6); the theoretical one defines pH as $-\log_{10} a_H$. Accordingly, a_H should be used in equations defining dissociation constants (cf. Eqs. C.2–C4 below), since otherwise the pK_a of a weak acid is not the pH of maximum buffer capacity.

Using the notation of Fig. 1A, the equilibrium constant K_1 and the three dissociation constants K_2, K_3, and K_4 are expressed as

$$K_1 = \frac{[\mathrm{H_2CO_3}]}{[\mathrm{CO_2}]} \tag{C.1}$$

$$K_2 = \frac{[\mathrm{HCO_3^-}][\mathrm{H^+}]}{[\mathrm{H_2CO_3}]} \tag{C.2}$$

$$K_3 = \frac{[\mathrm{CO_3^{2-}}][\mathrm{H^+}]}{[\mathrm{HCO_3^-}]} \tag{C.3}$$

$$K_4 = \frac{[\mathrm{H^+}][\mathrm{A^-}]}{[\mathrm{HA}]} \tag{C.4}$$

These conditions must always be fulfilled, with the exception that Eq. C.1 is ignored when calculating immediate pH responses of systems devoid of carbonic anhydrase activity (cf. Fig. 1A).

Since it was assumed that only CO_2 may leave or enter the system shown in Fig. 1A, no volume changes occur, and the total concentration of the non-CO_2/HCO_3^- buffer remains constant within the system. Thus, at any time

$$[HA] + [A^-] = \text{constant} \tag{C.5a}$$

and, in the absence of a non-CO_2/HCO_3^- buffer

$$[HA] = [A^-] = 0 \tag{C.5b}$$

In a closed system, the total concentration of carbon can change only if the acid-base load is due to addition (or removal) of a carbon-containing buffer species. If $\Delta[C]$ denotes the concentration of carbon added to the system, the total concentration of carbon $[C]_{tot}$ within the system is

$$\begin{aligned}[C]_{tot} &= [CO_2]^* + [H_2CO_3]^* + [HCO_3^-]^* + [CO_3^{2-}]^* + \Delta[C] \\ &= [CO_2] + [H_2CO_3] + [HCO_3^-] + [CO_3^{2-}]\end{aligned} \tag{C.6}$$

where concentrations before and after addition of the acid-base load are indicated by [ ]* and [ ], respectively. Equation C.6 must be fulfilled at any time in a closed system, as well as in an open system immediately after addition of the load. $[CO_2]^*$ and $[CO_2]$ are equal and cancel out in calculations concerning immediate pH responses of systems devoid of carbonic anhydrase.

Similarly, the total concentration of protons $[H]_{tot}$ is constant in a closed system. In addition to free protons and those bound to the non-CO_2/HCO_3^- buffer (HA), each HCO_3^- ion is associated with one proton, while both H_2CO_3 and CO_2 (or "CO_2/H_2O pair") are associated with two protons. Therefore

$$\begin{aligned}[H]_{tot} &= 2[CO_2]^* + 2[H_2CO_3]^* + [HCO_3^-]^* + [HA]^* + [H^+]^* + \Delta[H] \\ &= 2[CO_2] + 2[H_2CO_3] + [HCO_3^-] + [HA] + [H^+]\end{aligned} \tag{C.7}$$

where $\Delta[H]$ is the increase in the total proton concentration brought to the system by the acid-base load. Note that $\Delta[H]$ is zero, or one or two times the load concentration when the load consists of CO_3^{2-}, HCO_3^-, or CO_2, respectively.

Because equilibration of CO_2 between the system and its surroundings was assumed not to be immediate, the final concentration of free CO_2 in an open system attains its original value:

$$[CO_2] = [CO_2]^* \tag{C.8}$$

In an open system, net efflux of CO_2 from the system causes a decrease in the total carbon concentration. On the other hand, removing one CO_2 molecule corresponds to removing two protons. Therefore, the changes caused by the flux of CO_2, which brought the system back to open state equilibrium, can be written as a decrease in the total carbon concentration, or as a twofold larger decrease in the total proton concentration, namely,

$$\begin{aligned} & [HCO_3^-] + [CO_3^{2-}] - ([HCO_3^-]^* + [CO_3^{2-}]^* + \Delta[C]) \\ &= \tfrac{1}{2}\{[HCO_3^-] + [HA] + [H^+] - ([HCO_3^-]^* + [HA]^* + [H^+]^* + \Delta[H])\} \end{aligned} \quad \text{(C.9)}$$

where the terms $\Delta[C]$ and $\Delta[H]$ have the same meaning as in Eqs. C.6 and C.7, and the differences $[CO_2] - [CO_2]^*$ as well as $[H_2CO_3] - [H_2CO_3]^*$ are not shown, since they equal zero.

Equations C.1–C.9 provide in each case a number of linearly independent equations sufficient for calculating the immediate and final pH responses. In most cases, the set of equations can be solved most easily by first eliminating variables and then using numerical methods to find $[H^+]$.

The responses shown in Fig. 1B were calculated with the following assumptions: Equilibrium and dissociation constants

$$\begin{aligned} K_1 &= 1.0 \times 10^{-3}\ (pK_1 = -\log_{10} K_1 = 3.0) \\ K_2 &= 0.63 \times 10^{-3}\ (pK_2 = 3.2) \\ K_3 &= 79 \times 10^{-12}\ (pK_3 = 10.1) \\ K_4 &= 40 \times 10^{-9}\ (pK_4 = 7.4) \end{aligned}$$

Note that the commonly used first dissociation constant of CO2

$$pK_{1(CO_2)} = -\log_{10} \frac{[H^+][HCO_3^-]}{[CO_2]}$$

would now be

$$pK_{1(CO_2)} = pK_1 + pK_2 = 6.2.$$

Initial concentrations

$$\begin{aligned} [H^+]^* &= 40\ \text{nM}\ (pH^* = 7.4) \\ [CO_2]^* &= 1.5\ \text{mM} \\ [HCO_3^-]^* &= 25\ \text{mM} \\ [H_2CO_3]^* &= 1.5\ \mu\text{M} \\ [CO_3^{2-}]^* &= 50\ \mu\text{M} \end{aligned}$$

Finally, it may be worth a reminder that the concentration of free CO_2 in a solution equilibrated with humidified 5% CO_2 gas is calculated as follows:

$$\begin{aligned} [CO_2] &= S_{CO_2} P_{CO_2} \\ &= S_{CO_2} \frac{5\%}{100\%} (760\ \text{mmHg} - P_{H_2O,sat}) \end{aligned} \quad \text{(C.10)}$$

where S_{CO_2} is the solubility of CO_2, $P_{H_2O,sat}$ is the partial pressure of saturated water vapor, and the ambient atmospheric pressure is assumed to be 760 mmHg. Numerical values of S_{CO_2} for cerebrospinal fluid and plasma at temperatures from 10°C to 40°C are given in Severinghaus (1965).

REFERENCES

Al-Baldawi NF and Abercrombie RF (1992): Cytoplasmic hydrogen ion diffusion coefficient. Biophys J 61: 1470–1479.

Crank J (1956): The Mathematics of Diffusion. London: Oxford University Press.

Dietzel I, Heinemann U, Hofmeier G, Lux HD (1980): Transient changes in the size of the extracellular space in the sensorimotor cortex of cats in relation to stimulus-induced changes in potassium concentration. Exp Brain Res 40: 432–439.

Engasser J-M, Horvath C (1974): Buffer-facilitated proton transport pH profile of bound enzymes. Biochim Biophys Acta 358: 178–192.

Gabriel B, Teissié J (1996): Proton long-range migration along protein monolayers and its consequences on membrane coupling. Proc Natl Acad Sci USA 93: 14521–14525.

Gardner-Medwin AR (1983): A study of the mechanisms by which potassium moves through brain tissue in the rat. J Physiol 335: 353–374.

Gros G, Moll W (1971): The diffusion of carbon dioxide in erythrocytes and hemoglobin solutions. Pflügers Arch 324: 249–266.

Gros G, Moll W (1974): Facilitated diffusion of CO_2 across albumin solutions. J Gen Physiol 64: 356–371.

Gros G, Moll W, Hoppe H, Gros H (1976): Proton transport by phosphate diffusion—a mechanism of facilitated CO_2 transfer. J Gen Physiol 67: 773–790.

Gutknecht J, Tosteson DC (1973): Diffusion of weak acids across lipid bilayer membranes: Effects of chemical reactions in the unstirred layers. Science 182: 1258–1261.

Gutknecht J, Bisson MA, Tosteson FC (1977): Diffusion of carbon dioxide through lipid bilayer membranes. Effects of carbonic anhydrase, bicarbonate and unstirred layers. J Gen Physiol 69: 779–794.

Hansen AJ (1985): Effect of anoxia on ion distribution in the brain. Physiol Rev 65: 101–148.

Heberle J, Riesle J, Thiedemann G, Oesterhelt D, Dencher NA (1994): Proton migration along the membrane surface and retarded surface to bulk transfer. Nature 370: 379–382.

Hille B (1992): Ionic Channels of Excitable Membranes, 2nd ed. Sunderland, MA: Sinauer.

Irving M, Maylie J, Sizto NL, Chandler WK (1990): Intracellular diffusion in the presence of mobile buffers. Application to proton movement in muscle. Biophys J 57: 717–721.

Junge W, McLaughlin S (1987): The role of fixed and mobile buffers in the kinetics of proton movement. Biochim Biophys Acta 890: 1–5.

Kreuzer F, Hoofd L (1987): Facilitated diffusion of oxygen and carbon dioxide. In AP Fishman (ed): Handbook of Physiology, Section 3: The Respiratory System, vol. IV, Gas Exchange. Bethesda, American Physiological Society, pp. 89–111.

Lehmenkühler A, Syková E, Svoboda J, Zilles K, Nicholson C (1993): Extracellular space parameters in the rat neocortex and subcortical white matter during postnatal development determined by diffusion analysis. Neuroscience 55: 339–351.

Longmuir IS, Forster RE, Woo CY (1966): Diffusion of carbon dioxide through thin layers of solution. Nature 209: 393–394.

Lux HD, Heinemann U, Dietzel I (1986): Ionic changes and alterations in the size of the extracellular space during epileptic activity. Adv Neurol 44: 619–639.

McBain CJ, Traynelis SF, Dingledine R (1990): Regional variation of extracellular space in the hippocampus. Science 249: 674–677.

Newman EA (1996): Acid efflux from retinal glial cells generated by sodium bicarbonate cotransport. J Neurosci 16: 159–168.

Nicholson C (1985): Diffusion from an injected volume of a substance in brain tissue with arbitrary volume fraction and tortuosity. Brain Res 333: 325–329.

Nicholson C (1988–1989): Issues involved in the transmission of chemical signals through the brain extracellular space. Acta Morphologica Neerlando-Scandinavica 26: 69–80.

Nicholson C (1992): Quantitative analysis of extracellular space using the method of TMA^+ iontophoresis and the issue of TMA^+ uptake. Can J Physiol Pharmacol 70: S314–S322.

Nicholson C (1993): Ion-selective microelectrodes and diffusion measurements as tools to explore the brain cell microenvironment. J Neurosci Meth 48: 199–213.

Nicholson C, Phillips JM (1981): Ion diffusion modified by tortuosity and volume fraction in the extracellular microenvironment of the rat cerebellum. J Physiol 321: 225–257.

Nicholson C, Rice ME (1986): The migration of substances in the neuronal microenvironment. Ann NY Acad Sci 481: 55–71.

Perez-Pinzon MA, Tao L, Nicholson C (1995): Extracellular potassium, volume fraction, and tortuosity in rat hippocampal CA1, CA3, and cortical slices during ischemia. J Neurophysiol 74: 565–573.

Ransom BR, Yamate CL, Connors BW (1985): Activity-dependent shrinkage of extracellular space in rat optic nerve: A development study. J Neurosci 5: 532–535.

Ratcliff GA, Holdcroft JG (1963): Diffusivities of gases in aqueous electrolyte solutions. Transactions of the Institution of Chemical Engineers 41: 315–319.

Rice ME, Okada YC, Nicholson C (1993): Anisotropic and heterogeneous diffusion in the turtle cerebellum: Implications for volume transmission. J Neurophysiol 70: 2035–2044.

Robinson RA, Stokes RH (1959): Electrolyte Solutions. London: Butterworths.

Roos A, Boron WF (1981): Intracellular pH. Physiol Rev 61: 296–433.

Severinghaus JW (1965): Blood gas concentrations. In WO Fenn and H Rahn (eds): Handbook of Physiology, Section 3: Respiration, vol. II. Washington, D.C., American Physiological Society pp. 1475–1487.

Svoboda J, Syková E (1991): Extracellular space volume changes in the rat spinal cord produced by nerve stimulation and peripheral injury. Brain Res 560: 216–224.

Syková E, Svoboda J, Polak J, Chvatal A (1994): Extracellular volume fraction and diffusion characteristics during progressive ischemia and terminal anoxia in the spinal cord of the rat. J Cereb Blood Flow Metab 14: 301–311.

Taira T, Smirnov S, Voipio J, Kaila K (1993): Intrinsic proton modulation of excitatory transmission in rat hippocampal slices. Neuroreport 4: 93–96.

Taira T, Paalasmaa P, Voipio J, Kaila K (1995): Relative contributions of excitatory and inhibitory neuronal activity to alkaline transients evoked by stimulation of Schaffer collaterals in the rat hippocampal slice. J Neurophysiol 74: 643–649.

Voipio J, Kaila K (1993): Interstitial $P\text{co}_2$ and pH in rat hippocampal slices measured by means of a novel fast CO_2/H^+-sensitive microelectrode based on a PVC-gelled membrane. Pflügers Arch 423: 193–201.

Voipio J, Paalasmaa P, Taira T, Kaila K (1995): Pharmacological characterization of interstitial pH transients evoked by selective synaptic and exogenous activation of AMPA, NMDA and $GABA_A$ receptors in the rat hippocampal slice. J Neurophysiol 74: 633–641.

Voipio J, Ballanyi K (1997): Interstitial $P\text{co}_2$ and pH and their role as chemostimulants in the isolated respiratory network of neonatal rats. J Physiol 499: 527–542.

Vorisek I and Syková E (1997a): Evolution of anisotropic diffusion in the developing rat corpus callosum. J Neurophysiol 78: 912–919.

Vorisek I and Syková E (1997b): Ischemia-induced changes in the extracellular space diffusion parameters, K^+, and pH in the developing rat cortex and corpus callosum. J Cereb Blood Flow Metab 17: 191–203.

Whitaker S (1969): Fluid motion in porous media. Industrial and Engineering Chemistry 61: 14–28.

CHAPTER 5

Cerebral Energy Metabolism and pH

PEKKA MELLERGÅRD AND BO K. SIESJÖ
Department of Neurosurgery, University Hospital, S-221 85 Lund, Sweden (P.M.);
Laboratory for Experimental Brain Research, Wallenberg Neuroscience Center, University Hospital, S-221 85 Lund, Sweden (B.K.S.)

ENERGY METABOLISM OF THE BRAIN; COUPLING TO PRODUCTION AND CONSUMPTION OF H^+

As reviewed by Krebs et al. (1975) and by Alberti and Cuthbert (1982), mammalian cellular metabolism is a proton-producing process, with H^+ being generated in a multitude of metabolic reactions (see also Hochachka and Mommsen, 1983). Indeed, the turnover of H^+ ions is greater than that of any other metabolite. There is, though, a balance between production and elimination of H^+. Thus, the major metabolic end product is CO_2, which is eliminated through the lungs, preventing accumulation of H_2CO_3 and release of H^+. Furthermore, small amounts of nonvolatile acids are excreted by the kidneys. However, the major part of the H^+ produced during metabolism is intrinsically consumed in reactions related to energy metabolism. For example, H^+ is generated when ATP is hydrolyzed and consumed when it is re-formed (Hochachka and Mommsen, 1983). Before discussing these reactions in more detail, we will discuss some general concepts of cerebral energy metabolism (for reviews, see Clarke and Sokoloff, 1996; Erecinska and Silver, 1989; Siesjö 1978, 1985b; Siesjö et al., 1993).

The basic chemical currency of energy in all living cells, including neurons and glia, are the two high-energy phosphate bonds contained in adenosine triphosphate (ATP). To a lesser extent other purine and pyrimidine nucleotides, notably guanosine, cytidine, uridine, and inosine triphosphates (GTP, CTP, UTP, and ITP, respectively) can serve as energy sources. Since these systems are recharged by ATP, such reactions are indirectly driven by ATP energy.

When work is done (i.e., when the energy contained in ATP is utilized) the molecule is degraded to adenosine diphosphate (ADP) and orthophosphate (Eq. 1). Further hydrolysis of ADP to adenosine monophosphate (AMP) and inorganic phosphate (Pi) releases addi-

pH and Brain Function, Edited by Kai Kaila and Bruce R. Ransom
ISBN 0-471-11838-9

tional energy (Eq. 2). It should be noted that both of these reactions are associated with release of protons. The equations are usually written as

$$[ATP^{4-}] + [HOH] \rightarrow [ADP^{3-}] + [HPO_4^{2-}] + [H^+] \quad (1)$$

$$[ADP^{3-}] + [HOH] \rightarrow [AMP^{2-}] + [HPO_4^{2-}] + [H^+] \quad (2)$$

However, as discussed below the number of H^+ released depends on Mg^{2+} concentration and pH (Hochachka and Mommsen, 1983; Krebs et al., 1975).

Cytosolic ATP, ADP, and AMP constitute a pool of adenine nucleotides, the members of which are interrelated by a reversible reaction catalyzed by adenylate kinase.

$$[ATP^{4-}] + [AMP^{2-}] \rightleftharpoons [ADP^{3-}] + [ADP^{3-}] \quad (3)$$

The total amounts of ATP, ADP, and AMP in the mammalian brain are under physiological conditions, approximately 3.0, 0.3, and 0.03 μmol/g (wet weight), respectively, so that the ratios of ATP/ADP and ATP/AMP are about 10 and 100. However, most of the ADP and AMP appear to be compartmentalized—with a major fraction being localized to the mitochondria, and another one bound to various cellular constituents. Therefore, the free cytosolic concentration of ADP is only 10–20 μM, and that of AMP less than 1 nM (Erecinska and Silver, 1989; Veech et al., 1979). The concentration of inorganic phosphate, which can be measured by ^{31}P NMR, is 1.5–2.0 mM. There appears to be no major differences in the concentration of ATP or related metabolites, among various mammalian species (Erecinska and Silver, 1989).

The adenine nucleotide pool is related to the phosphocreatine-creatine (PCr-Cr) pool in which PCr serves as a readily available source of ATP according to the reaction

$$PCr + ADP + H^+ \rightarrow Cr + ATP \quad (4)$$

This reaction suggests that whenever the ADP concentration rises, or pH falls, ATP will be formed at the expense of the PCr stores. We will return to these relationships later in our discussion.

The brain differs from other organs in having unusually high energy requirements. Although constituting less than 2% of the body weight, the human brain receives approximately 25% of the cardiac output at rest and utilizes around 20% of the body's total energy production (Siesjö, 1978; Clarke and Sokoloff, 1996). Under normal conditions there is only one substrate, namely glucose, that can support cerebral ATP production. The only important exception to this occurs in conditions like diabetic ketoacidosis or starvation, during which the blood concentrations of ketone bodies are increased. In these conditions, acetoacetate and β-hydroxybutyrate can supply up to 50% of the substrate for cerebral oxidative metabolism. The brain contains no triglycerides, and fatty acids are not utilized as energy sources in the mature brain. However, the brain may, under certain circumstances, utilize endogenous carbohydrate metabolites and amino acids to supply some energy. This, and the potentially important role of lactate as an energy substrate will be discussed subsequently.

Under physiological conditions the oxidation of glucose is tightly coupled to ATP production according to the overall equation.

$$\text{Glucose} + 6\,O_2 + 36\,ADP + 36\,Pi \rightarrow 6\,CO_2 + 44\,H_2O + 36\,ATP \quad (5)$$

The heat loss in this reaction is around 30%, which seems to be a price nature is ready to pay to ensure that the reaction is always spontaneous, and to secure ATP formation under most conditions encountered (for further discussion, see Siesjö, 1978).

As shown by Eq. 6, CO_2 is a potential source of protons

$$CO_2 + H_2O \rightleftharpoons H_2CO_3 \leftrightarrow HCO_3^- + H^+ \tag{6}$$

Being rapidly diffusible gas, CO_2 normally equilibrates with blood and is transported away for elimination via the lungs. However, accumulation of CO_2 may contribute to stimulus-related acidification (Voipio and Kaila, 1993; and in Chapters 16–17). In ischemia, when H^+ is produced by other reactions, the reaction in Eq. 6 is driven to the left (as written), causing P_{CO_2} to rise to values of 200–300 mmHg (Ekholm et al., 1991; Katsura et al., 1992; Kraig et al., 1985, 1986).

Transport and Use of Glucose

Glucose transport over the blood-brain barrier. The transport of glucose over the blood-brain barrier (BBB) occurs by facilitated diffusion down a normally large concentration gradient, the brain intracellular glucose concentration generally being about 30% of the plasma concentration. This transport is mediated by the carrier GLUT1, a member of a supergene family of Na^+-independent glucose transporters, which may also transport other hexoses, like galactose and fructose (Pardridge and Boaddo, 1993). Under normal conditions brain glucose utilization is limited by glucose phosphorylation, not glucose transport (Lund-Andersen, 1979). However, under pathological conditions, notably hypoglycemia, the intracellular glucose concentration may be so low that glucose transport over the BBB becomes rate-limiting. It is, therefore, of interest that GLUT1 is asymmetrically distributed, with 40% in the cytoplasmatic compartment, and the remaining 60% being distributed so that the number of immunoreactive GLUT1 glucose transporters on the abluminal membrane is three to four times that on the luminal membrane (Farrell and Pardridge, 1991). Such an asymmetric distribution of GLUT1 may provide a mechanism whereby glucose transport through the BBB can be immediately upregulated by selective translocation of glucose transporter from the cytosol to the luminal membrane.

GLUT1 has also been detected in neuronal and glial preparations, and may mediate glucose transport into these cells. However, the major transporter of glucose into neurons is probably another isomer, the so called GLUT3 carrier (Nagamatsu et al., 1990; Pardridge and Boaddo, 1993).

Utilization of glucose. Once in the brain intracellular space, the glucose molecules enter the glycolytic pathway by being phosphorylated to glucose-6-phosphate (G-6-P), a reaction which is catalyzed by hexokinase and occurs at the expense of ATP energy (see Fig. 1). From here the molecule may be metabolized along three different paths: to CO_2 via the hexose monophosphate shunt (HMP), to glycogen, or to pyruvate, by a series of steps constituting the glycolytic pathway.

The hexose monophosphate shunt (also called the pentose phosphate or the phosphogluconate pathway) constitutes a complex series of reactions. In the present context it is sufficient to mention that HMP contributes to approximately 1–3% of the glucose flux, the important net result being the formation of pentose sugar for nucleotide synthesis, and NADPH for lipid synthesis and/or glutathione reduction (Clarke and Sokoloff, 1996; Siesjö, 1978).

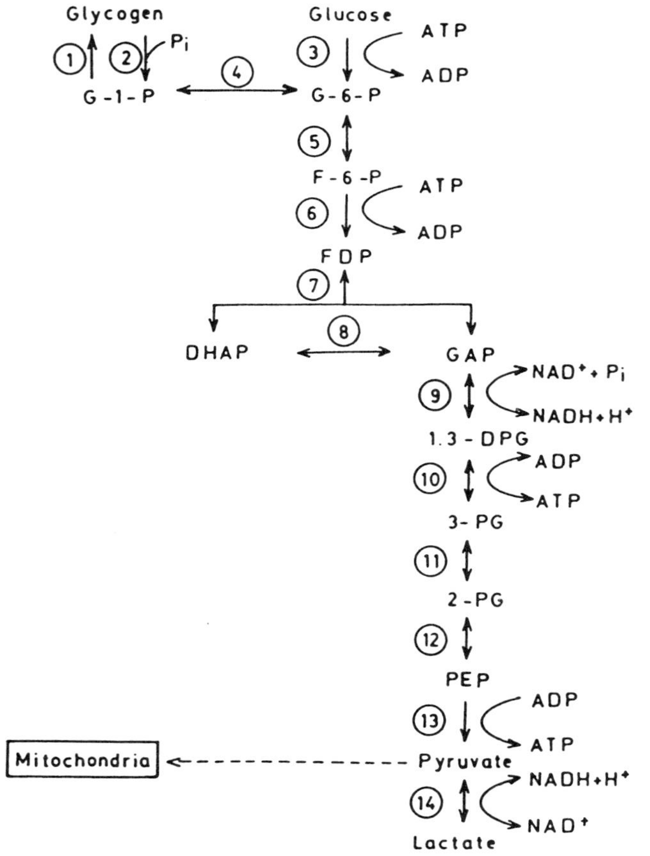

FIGURE 1. Individual reactions of the glycolytic pathway. G-6-P = glucose-6-phosphate, F-6-P = fructose-6-phosphate, Fructose-1,6-diphosphate, DHAP = dihydroxyacetone phosphate, GAP = glyceraldehyde-3-phosphate, DPG = diphosphoglycerate, PG = phosphoglycerate, PEP = phosphoenol pyruvate. The trivial names of the catalyzing enzymes are as follows. 1 = glycogen synthetase, 2 = phosphorylase, 3 = hexokinase, 4 = phosphoglucomutase, 5 = glucosephosphate isomerase, 6 = phosphofructokinase, 7 = aldolase, 8 = triosephosphate isomerase, 9 = glyceraldehyde-3-phosphate dehydrogenase, 10 = phosphoglycerate kinase, 11 = phosphoglyceromutase, 12 = enolase, 13 = pyruvate kinase, 14 = lactate dehydrogenase. Modified from Siesjö (1978).

Glycogen is a glucose polymer, which serves as a readily available storage form of glucose and forms approximately 1% of the total energy stores in the body. Although present in the brain, its concentration is only 2–4 μmol/g, which is one and two orders of magnitude less than the concentration in muscle and liver, respectively. Thus, if the glucose supply to the brain is interrupted, the glycogen stores of cortical tissue can only support normal brain activity for a few minutes. Even so, glycogen is the largest energy reserve available in the CNS. The glycogen of the brain is primarily found in astrocytes, where glycogen phosphorylase (the main glycogenolytic enzyme) is also predominantly localized (Cataldo and Broadwell, 1986; Reinhart et al., 1990). This may not be surprising, considering that the astrocytic endfeet surround the capillaries that are the source of glucose. The glycogen of the brain has a relatively fast turnover, and is closely regulated by different neurohormones. As we will discuss shortly, recent findings indicate that astrocytic glycogen might be more important as a source of lactate than of glucose.

Production of Energy

Anaerobic metabolism of glucose. The glucose destined for ATP production enters the anaerobic sequence of glycolysis reactions or the Embden-Meyerhof pathway (Fig. 1). Thus, after the conversion of G-6-P to fructose-6-phosphate (F-6-P) there follows a second phosphorylation to fructose-1,6-diphosphate. This reaction is catalyzed by phosphofructokinase (PFK), is irreversible, and is sensitive to pH changes (see below). These reactions can be summed up in Eq. 6.

$$\text{Glucose} + 2\text{Pi} + 2\text{ ADP} + 2\text{ NAD} \rightarrow 2\text{ Pyruvate}^- + 2\text{ ATP} + 2\text{ NADH} \tag{7}$$

As observed, glycolysis thus produces 2 moles of ATP per mole of glucose. When glycogen is the precursor, 3 moles of ATP are formed, since the energy-requiring hexokinase step is bypassed.

Under anaerobic conditions, pyruvate is reduced to lactate in a reversible reaction catalyzed by lactate dehydrogenase (LDH). As shown in Eq. 8, this reaction does not yield additional ATP, but permits glycolysis to proceed, since it regenerates the oxidized form of the coenzyme nicotine adenine nucleotide (NAD^+).

$$\text{Lactate}^- + \text{NAD}^+ \rightleftharpoons \text{Pyruvate}^- + \text{NADH} + \text{H}^+ \tag{8}$$

However, under normal conditions (i.e., with an unlimited oxygen supply), lactate production accounts for less than 5% of the glucose metabolized by the brain. Instead, pyruvate is further oxidized to CO_2 and water by mitochondrial enzymes.

It is still commonly believed that glycolysis is a major source of hydrogen ions, and that these reflect the formation of lactic acid. However, it has been realized that release of H^+ mainly occurs when ATP is hydrolyzed, and that the formation of lactate in the LDH reaction is not necessarily accompanied by a corresponding release of H^+ (Alberti and Cuthbert, 1982; Hochachka and Mommsen, 1983; Krebs et al., 1975).

ATP and the other adenine nucleotides have phosphate groups that are only partly ionized at physiological pH values. The adenine nucleotides are also to an appreciable extent complexed with Mg^{2+}. The phosphates differ in their tendency to form complexes, ATP having a high, ADP an intermediate, and AMP a low affinity for Mg^{2+}. This means that there are different forms of these phosphates present in the cytosol. If ATP is taken as an example, one can write the following equations

$$[\text{Mg}^{2+}] + [\text{HATP}^{3-}] \rightleftharpoons [\text{MgHATP}^-] \tag{9}$$

$$[\text{Mg}^{2+}] + [\text{HATP}^{4-}] \rightleftharpoons [\text{MgHATP}^{2-}] \tag{10}$$

Thus, there will be at least four different forms of ATP present, and it is clear that both the concentration of free Mg^{2+} and H^+ enter as important variables, determining the relative proportions of these different forms. Since the pK values of the different forms of adenine nucleotides are different, the amount of H^+ formed during glycolysis depends on pH (Krebs et al., 1975; Hochachka and Mommsen, 1983). To illustrate this, the following equations can be written, covering two extreme cases (Hochachka and Mommsen, 1983):

$$\text{Glucose} + 2\ \text{HADP}^{2-} + 2\ \text{H}_2\text{PO}_{4-} \rightarrow 2\ \text{Lactate}^- + 2\ \text{H}_20 + 2\ \text{HATP}^{3-} + 2\text{H}^+ \quad (11)$$

$$\text{Glucose} + 2\ \text{ADP}^{3-} + 2\ \text{HPO}_4^{2-} \rightarrow 2\ \text{Lactate}^- + 2\ \text{H}_2\text{O} + 2\ \text{ATP}^{4-} \quad (12)$$

For a given [Mg^{2+}], the lower the pH, the greater the amount of H^+ formed. Thus, Eq. 11 is only valid at pH values below 6.0, and if pH is high enough (>8.0), no H^+ is released (Eq. 12). At pH 7.0 (i.e., close to the intracellular pH of most brain cells), approximately 1 mole of H^+ is formed for every 2 moles of lactate (Hochachka and Mommsen, 1983). Nevertheless, anaerobic glycolysis is still associated with the production of equal amounts of lactate and H^+. This is because H^+ is released during hydrolysis of ATP. As pointed out by Krebs et al. (1975), this is true for most, but not for all, of the ATP hydrolyzed. The main intracellular factors determining the release of H^+ during ATP utilization are again Mg^{2+} and pH. Interestingly, here the dependence on pH is reciprocal, compared to the H^+ production from glycolysis. Thus, the lower the pH, the fewer moles of H^+ are formed per mole of ATP hydrolyzed (Hochachka and Mommsen, 1983; Krebs et al., 1975). Because of these opposite pH dependencies of H^+ production by glycolysis and the associated ATP hydrolysis, the total amount of H^+ generated is independent of pH, and is always the same. Thus, the amount of H^+ formed per mole of glucose metabolized to lactate, plus the moles of H^+ released during hydrolysis of the ATP formed, always adds up to 2. Interestingly, the same stochiometry of net H^+ production prevails in species where glucose is fermented to other end products, like succinate (Hochachka and Mommsen, 1983).

In essence, this means that at constant [ATP], anaerobic glycolysis is always associated with the production of equal amounts of lactate and H^+. If there is hydrolysis of preexisting ATP stores (i.e., a lowering of the total concentration of ATP), additional H^+ will be released. This occurs in pathological situations, like ischemia. To our knowledge an increase in the total ATP store has never been documented (see also pp. 75). Thus, the fall in pH during ischemia should be at least as large as that corresponding to the amount of lactate formed.

Aerobic metabolism of glucose. In most animal cells, glycolysis serves to generate pyruvate, which is then oxidized by the mitochondria, where the final aerobic metabolism of glucose takes place. This oxidative phosphorylation or "respiration" starts with the formation of acetyl coenzyme A (acetyl CoA) through decarboxylation of pyruvate. The first step in the citric acid cycle involves condensation of acetyl CoA with oxaloacetate to form citrate. Through a cyclic series of reactions, involving different acidic intermediates such as α-ketoglutarate and malate, the carbons of acetyl CoA appear as CO_2, and oxaloacetate is regenerated (see Fig. 2).

The citric acid cycle provides reducing equivalents in the form of NADH and $FADH_2$ (the reduced form of flavoprotein FAD), which constitute the "fuel" for oxidative phosphorylation. Their electrons are delivered to the respiratory chain (also called the electron transport chain). The free energy released during electron transfer is used to pump protons into the intermembrane space of the mitochondrion, thereby creating an electrochemical proton gradient across the inner mitochondrial membrane (Mitchell, 1966). The backflow of protons down this gradient is in turn used to drive the membrane-bound ATP-synthetase, which catalyzes the conversion of ADP and Pi to ATP in a reaction that is es-

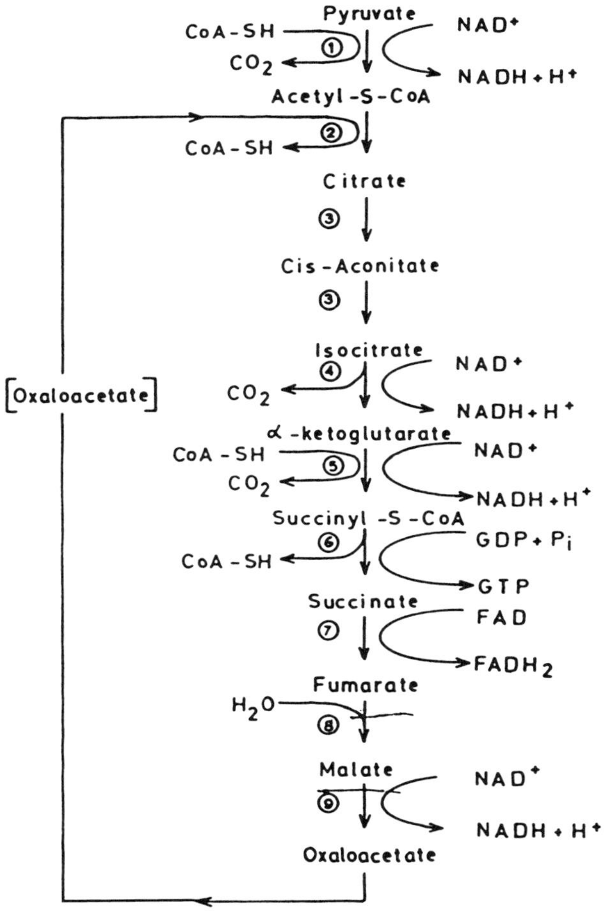

FIGURE 2. Individual reactions of the citric acid cycle. Reactions leading to generation of NADH, $FADH_2$, or GTP have been displayed to the right of the pathway arrows. The trivial names of the catalyzing enzymes are as follows. 1 = pyruvate dehydrogenase, 2 = citrate synthetase (condensing enzyme), 3 = aconitase, 4 = isocitrate dehydrogenase, 5 = α-ketoglutarate dehydrogenase, 6 = succinate thiokinase, 7 = succinate dehydrogenase, 8 = fumarase, 9 = malate dehydrogenase. Modified from Siesjö (1978).

sentially a reversal of Eq. 1. The process of oxidative phosphorylation is thus completed, each mole of glucose having yielded 36 moles of ATP. For the present discussion, it is important to observe that for each mole of ATP formed 1 mole of H^+ is consumed (Mitchell, 1966).

Lactate utilization. Lactate is transported over the BBB by the same monocarboxylic acid carrier that is responsible for the transport of ketone bodies (Pardridge, 1993; Wozniak et al., 1993). However, the transporter has a limited capacity; hence plasma lactate cannot substitute for glucose as a metabolic substrate (Pardridge, 1993). However, a specific lactate transporter has been described in neurons, suggesting that lactate may still play a role as a metabolic substrate in the CNS (Dringen et al., 1993b). Additional evidence supports such a hypothesis. Lactate may serve as a substrate for cells like oligodendrocytes and cerebellar Purkinje cells, which contain low activities of hexokinase, and is therefore likely to have a low capacity to convert glucose to glucose-6-phosphate (Kao-Jen and Wilson, 1980; Snyder and Wilson, 1983). It has also been shown that synaptic activity can be maintained in rat hippocampal slices in the absence of glucose, with lactate and pyruvate as the sole energy substrates (Schurr et al., 1988). Moreover, neurodegeneration induced by glucose deprivation is prevented in the presence of lactate (Izumi et al., 1993). *In vitro* utilization of lactate in nervous tissue has been demonstrated also in the presence of glucose (Larrabee, 1983).

As previously mentioned, astrocytes contain almost all the glycogen of the brain. Astrocytes may in fact constitute a source of lactate that can be utilized by neurons as an energy substrate. It has been shown that rat astrocytes in primary culture lack fructokinase activity and do not release glucose, but only lactate, following degradation of their glycogen store (Dringen et al., 1993a; Bergbauer, 1996). Furthermore, in metabolically compartmentalized nervous systems such as the honeybee retina, glycolysis occurs in glial cells, while oxidative metabolism takes place in the neurons, the necessary carbohydrate substrates being supplied from the glia (Tsacopoulos et al., 1988). Similar findings were reported in guinea pig retina (Poitry-Yamate and Tsacopoulos, 1991). Recently Pellerin and Magistretti (1994) reported that glutamate stimulates lactate production in mouse astrocytes. A relationship between glutamate uptake and glycolysis would provide a simple and direct mechanism to tightly couple neuronal activity to glucose utilization.

Redox state of the cell. In the mitochondrion, electrons from the substrates used are translocated along a series of redox systems to be finally accepted by oxygen. The oxidation-reduction status (the so called "redox state") of the tissue, which can be defined as the balance between production of reducing equivalents and the acceptance of electrons by oxygen, is an important variable in discussions of cell energy metabolism. The redox state is normally relatively constant, but may change during pathological conditions. The ratio of reduced to oxidized form of the proximal electron carrier nicotinamide adenine dinucleotide ($NADH/NAD^+$) is often regarded as a convenient measure of the redox state. One method for estimating this $NADH/NAH^+$ ratio *in vivo* is to measure the tissue concentrations of substrate couples that are in equilibrium with the mitochondrial or cytosolic NAD pool. In many types of tissue the hydroxybutyrate/acetoacetate ratio can be used to indicate the mitochondrial redox status, but not so in the brain. This is because the substrates are normally not present in detectable concentrations. Instead the cytosolic redox state can be derived from the LDH reaction (see Eq. 8).

This reaction shows that the $NADH/NAD^+$ ratio is a function of both the lactate/pyruvate ratio and pH. The reaction also serves to illustrate the fact that reduction of NAD^+ (or $NADP^+$) leads to release of H^+. However, these H^+ ions will be consumed when the reduced forms of the redox couples are re-oxidized.

Clearly the lactate/pyruvate ratio can only serve as a true indicator of the cytosolic $NADH/NAD^+$ ratio if the cytosolic pH remains constant. If pH is reduced, the lactate/pyruvate ratio will rise at a constant $NADH/NAD^+$ ratio (Siesjö et al., 1972). For example, a fall in cytosolic pH of 0.3 units (i.e., doubling of the H^+ concentration) would result in a doubling of the lactate/pyruvate ratio without any change in the redox state (as determined by the $NADH/NAD^+$ ratio).

Energy Consumption and Its Regulation

Energy requirements. Under normal conditions brain tissues consume 22–32 μmol · g^{-1} · min^{-1} of ATP. This ATP utilization corresponds to a rate of cerebral glucose consumption (CMR_{glu}) between 0.25 and 0.8 μmol g^{-1} min^{-1}, and a rate of cerebral oxygen metabolism (CMR_{O2}) that varies from 1.5 to 5 μmol g^{-1} min^{-1}. A fair estimation, based mainly on measurements of CMR_{O2} and CMR_{glu} under conditions where ion movements were blocked, is that the maintenance of ionic gradients normally requires 12–16 μmol ATP/g (i.e., 50–60% of ATP utilization). The major fraction of this ATP is used by Na^+/K^+-ATPase (Astrup, 1982; Erecinska and Silver, 1989). One may expect that, in the brain, a sizable fraction of the remaining energy use can be attributed to neurotransmitter metabolism and to phosphorylation reactions, apart from more general cellular work like protein and lipid synthesis, and cell maintenance and repair.

With a rate of ATP utilization of around 30 μmol g^{-1} min^{-1}, the available store of ATP is only sufficient for 10–15 seconds. There is, though, a storage form of ATP in the form of phosphocreatine (PCr). Thus, the fall in ATP can be retarded by means of phosphorylation of ADP via the reversible creatine kinase (CK) reaction, as described in Eq. 13. As seen, this reaction involves uptake or release of protons.

$$[PCr^{2-}] + [ADP^{3-}] + [H^+] \rightleftharpoons [Cr] + [ATP^{4-}] \qquad (13)$$

This equation differs from Eq. 4 in only one respect; it shows that the reaction is reversible. Since the tissue concentration of PCr is about 5 μmol/g it can only provide ATP for an additional 10 seconds of normal work, thus emphasizing the need for an uninterrupted oxidative metabolism. The creatine present adds approximately 5 μmol/g to the total creatine pool. Although widely distributed in both neurons and glia, the CK activity may be increased in regions where energy expenditure is particularly large (Erecinska and Silver, 1989). Due to the high equilibrium constant, CK will favor formation of ATP, so that any ADP formed will immediately be rephosphorylated to ATP (Veech et al., 1979). A consequence of this is that under conditions of increased ATP utilization (or decreased ATP synthesis) a fall in PCr preceeds that in ATP, and only after PCr has decreased to a low level will ADP and AMP rise appreciably (Lowry et al., 1964; Ekholm et al., 1993).

Higher rates of CMR_{glu} and CMR_{O2} are observed in smaller animals like rats, where the neuron packing density is relatively higher; the lower value applies to man. By analogy CMR_{glu} and CMR_{O2} are also higher in gray matter compared to white matter (Sokoloff, 1977; Clarke and Sokoloff, 1996). However, there is support for the assumption that energy metabolism also varies between different regions of the brain, between different cell types, and even among different parts of the same cell (Sokoloff, 1977; Clarke and Sokoloff, 1996). Indirect evidence for this comes from the observed difference in sensitivity to hypoxia between different cell types (Siesjö, 1985b). Enzymes of the glycolytic pathway and citric acid cycle are also unevenly distributed in brain (Lai and Cooper, 1986). More direct evidence for variation in energy metabolism is provided by studies of 2-deoxyglucose uptake and determination of enzyme activity. To summarize such investigations, it can be said that neurons in general have greater oxidative metabolic activity than glia (however, see Hertz et al., 1992), that the activity seems to be higher in neurons receiving a strong excitatory input, and that higher rates of oxidative metabolism are seen in dendrites and synaptic terminals (Erecinska and Silver, 1989; Siesjö, 1978; Sokoloff, 1977).

Regulation of energy production: relationship to acid-base changes. The observations discussed above raise questions as to how energy production is regulated, not only in different regions of the brain, but also at the cellular level. At the cellular level, the rate of energy conversion (i.e., ATP production) and work (i.e., ATP hydrolysis) must be balanced to achieve steady state. Such a balance is also observed *in vivo*. The classical model to explain this interaction predicts that the concentrations of ADP and Pi (i.e., the products of ATP hydrolysis) in themselves constitute the cytosolic feedback mechanisms. Since the concentration of these metabolites can be assumed to be proportional to the rate of ATP hydrolysis, the cytosolic concentrations of ADP and Pi would provide an excellent feedback signal to regulate oxidative phosphorylation during changes in energy utilization. It has, however, been difficult to show any significant changes in the concentrations of ADP and Pi associated with large increases in oxidative phosphorylation (Balaban, 1990; Erecinska and Silver, 1989). The small concentration changes of these metabolites suggest a system that relies on other mechanisms for metabolic control. The delivery of reducing equivalents into the respiratory chain could represent such a regulatory step: this delivery is in turn being influenced by, for example, increased activity of pyruvate dehydrogenase, secondary to an increase in intracellular Ca^{2+}. However, although this and other possibilities exist, the primary controlling factor regulating oxidative phosphorylation is still unknown (Balaban, 1990). In fact, considering the coupling of ATP production to the proton electrochemical gradient across the mitochondrial membrane, a modification of this gradient (e.g., by a changed proton permeability) could be an effective modulator of ATP production (Gunter et al., 1994).

For a long time it has been accepted that in the brain there is a close regional coupling between metabolic rate and blood flow, with local modulations in neuronal activity inducing similar modulations in local metabolism that, in turn, will alter local blood flow to supply the necessary oxygen and energy substrates. Clearly, regional CMR_{glu} and cerebral blood flow (CBF) are increased during focal increases in neuronal activity (Sokoloff, 1977; Clarke and Sokoloff, 1996). However, this does not necessarily imply an increase in glucose oxidation, or a response to oxygen depletion. In fact, recent measurements using positron emission tomography have shown that increases in neuronal activity cause, at lest transiently, a tissue uptake of glucose in excess of that consumed by oxidative metabolism (Fox et al., 1998; Frostig et al., 1990). It has been suggested that focal physiological cortical activation results in a metabolic uncoupling, meaning that the increase in blood flow and glucose utilization are not matched by a commensurate increase in oxygen consumption. Furthermore, both ^1H NMR spectroscopy and *in vivo* microdialysis have revealed focally increased lactate levels after physiological stimulation (Fellow et al., 1993; Prichard et al., 1991). Thus, transient increases in neuronal activity may preferentially induce glycolysis, at least initially.

Since increased neuronal activity is usually associated with glutamate release, it is of considerable interest that glutamate uptake in astrocytes is associated with a stimulation of glycolysis (Pellerin and Magistretti, 1994), thus providing a direct mechanism for coupling neuronal activity to glucose utilization in the brain (see also Bittar et al., 1996). We note that the glutamate carrier in astrocytes also seems to generate an alkalinization outside, and an acidification inside glial cells, perhaps providing a mechanism for neuron–glia interaction via H^+ exchange (Bouvier et al., 1992).

Estimation of energy balance. Since the energy requirements cannot be measured, measurements of oxygen or glucose consumption (i.e., CMR_{O_2} and CMR_{glu}) can never give accurate information on whether or not energy production is sufficient to meet the requirements of the tissue. This is obvious if one compares energy needs of the brain under conditions of hypothermia and seizures. Thus, at 22°C, an energy production of 25% of the normal will cover the needs, while during status epilepticus a two- to threefold increase in normal energy production is just barely sufficient. Obviously, energy balance can only be assessed from measurements of tissue metabolites. Since the size of the adenine nucleotide pool may vary appreciably, measurements of ATP only are not sufficient. A convenient measure of the balance between production and utilization of energy is the adenylate energy charge (E.C.), defined by

$$\text{E.C.} = \frac{[\text{ATP}] + 0.5[\text{ADP}]}{[\text{ATP}] + [\text{ADP}] + [\text{AMP}]} \tag{14}$$

This expression was coined by Atkinson (1968). It is still a valid measure of the charge of the "energy battery" of cells.

Energy Metabolism and pH

Production and consumption of protons during normal energy metabolism. The most frequent single reaction in aerobic cells is the conversion of ATP into ADP and Pi. The preceeding discussion has hopefully made it clear that utilization of energy (i.e., the hydrolysis of ATP and other nucleoside triphosphates) is accompanied by production of H^+. Although up to 30% of the ATP hydrolyzed is not accompanied by production of H^+ ions (Krebs et al., 1975) ATP hydrolysis is still the most important contributor to the vast amounts of H^+ ions that are constantly being formed in the brain.

Additional protons are formed in the mitochondrial respiratory chain when hydrogen is donated to Fe^{3+} (e.g., from a flavoprotein) according to the equation

$$2\,Fe^{3+} + 2\,H^+ \rightarrow 2\,Fe^{2+} + 2\,H^+ \tag{15}$$

As remarked above, another important source of H^+ are reactions in which NAD^+ or $NADP^+$ are reduced. One example of this is the step in glycolysis where glyceraldehyde-3-phosphate (GAP) is oxidized to 1,3-diphosphoglycerate (DPG):

$$\text{GAP} + NAD^+ + HPO_4^{2-} \rightarrow \text{DPG} + \text{NADH} + H^+ \tag{16}$$

However, since the consumption and production of energy are normally closely balanced, the net acid production from these reactions is usually negligible. We have already pointed out that when ATP is resynthesized via the phosphorylation of ADP, there is a consumption of H^+ (see Eq. 1). Similarly, the replenishment of PCr^{2-} by reversal of the creatine kinase reaction (see Eq. 13) consumes the H^+ formed when PCr^{2-} is hydrolyzed. The reoxidation of NADH or NADPH consumes protons on a mole for mole basis. Since the hydrogen concentration is a stoichiometric factor in the reaction, changes in pH would be expected to influence the creatine kinase equilibrium (Lawson and Veech, 1979). In fact, attempts have been made to utilize this for the determination of changes in cytosolic pH (Siesjö et al., 1972).

The question is then: When or under what conditions, is acid actually formed? Before addressing this question in some more detail, we need to make a few comments on how changes in pH may influence processes involved in energy metabolism.

Effect of pH changes on energy metabolism. A disturbance of the pH homeostasis may affect both glycolysis and oxidative metabolism, and may also have an indirect effect on ATP production by influencing energy-requiring cellular processes. In the following we will point out some examples of this.

Oxidative phosphorylation is influenced by pH. For example, in isolated mitochondria and cell suspensions, acidosis reduces respiration (Hillered et al., 1985; Nioka et al., 1987; Tobin et al., 1972). Possible mediators of such H^+-dependent effects are the NADH-ubiquinone oxidoreductase (Rouslin, 1991), or the 2-oxo-glutarate dehydrogenase complex, which has a remarkably narrow optimum pH range of 7.2–7.4 (Lai and Cooper, 1986).

A number of different types of enzymes exhibit a pH optimum for their activity, which means that many cellular reactions are influenced by changes in pH. To take only one example, it has been shown that the activity of Na^+/K^+-ATPase exhibits a pH optimum that is dependent on the concentration of ATP. The enzyme activity shifts to a higher value when the ATP levels fall, from an optimum pH of 7.1–7.4 at normal ATP concentrations to 7.6–7.8 at micromolar concentrations of ATP (Skou, 1979). Acidosis could thus be expected to inhibit Na^+/K^+-ATPase activity, particularly if accompanied by a lowering of [ATP].

A classical example of a pH-dependent enzyme is phosphofructokinase (PFK), the enzyme responsible for the third (and rate limiting) step in glycolysis (see Fig. 4). Thus, PFK is fully active at pH values just 0.1 units higher than the pH at which it is completely inactive (Trivedi and Danforth, 1966). It has also been shown that glycolysis is markedly pH-dependent, with acidosis having an inhibitory and alkalosis a stimulatory effect (Ueda, 1979). However, at the same time there are many reports of large increases in lactate under anaerobic conditions in spite of acidosis, showing that glycolysis may well continue despite a lowering of pH. This nicely illustrates the complexity of the problems with which we are dealing. Thus, PFK is not only very sensitive to variations in pH, it is also influenced (and activated) by ADP, AMP, and Pi as well as by cAMP and NH_4^+ (Lowry and Passonneau, 1966). During conditions when such stimulators abound, as in cerebral ischemia, their effect may overcome the inhibitory effect of a decreased pH.

INFLUENCE OF CHANGES IN pH ON CELLULAR ENERGY METABOLISM

Changes in pH trigger metabolic responses whose purpose seems to be to compensate for the alterations in intracellular pH (i.e., a metabolic buffering). The primary observations supporting this concept were obtained in experiments on induced hypocapnia and hypercapnia. Additional information was obtained in experiments on ammonia intoxication.

Hypocapnia

In a buffered fluid maintained at a certain P_{CO_2}, a decrease in CO_2 tension is accompanied by a rise in pH (i.e., by alkalosis) whose magnitude depends on two variables: the decrease in P_{CO_2}, and the buffer capacity of the system. *In vivo,* a lowering of P_{CO_2} (hypocapnia) does not lead to the expected rise in pH (Kjällquist et al., 1969; MacMillan and Siesjö, 1972), probably because the rise in pH triggers a compensatory production of metabolic acids (Siesjö, 1973). Very likely, this production is largely due to stimulation of PFK, with an ensuing rise in glycolytic rate and, thereby, in production of lactate plus H^+.

In the brain, hypocapnia is not only accompanied by a rise in the concentrations of lactate and pyruvate but also of many other anions of metabolic acid, including those formed in the citric acid cycle (MacMillan and Siesjö, 1973). Figure 3 shows the relationship between P_{CO_2} and the tissue concentrations of lactate and pyruvate, as well as of changes in pH_i. In the figure, the changes in pH_i derived have been compared to those expected from the change in P_{CO_2}, and the buffer capacity of the system. The latter was assumed to be 25 mM l^{-1} pH^{-1} (Siesjö, 1973). This value is close to that subsequently measured in brain homogenates, neurons, and glial cells (Katsura et al., 1996).

Figure 3 shows that, at all reduced P_{CO_2} values, the change in pH_i was smaller than that expected from the physiochemical buffer capacity of the system. This suggests that

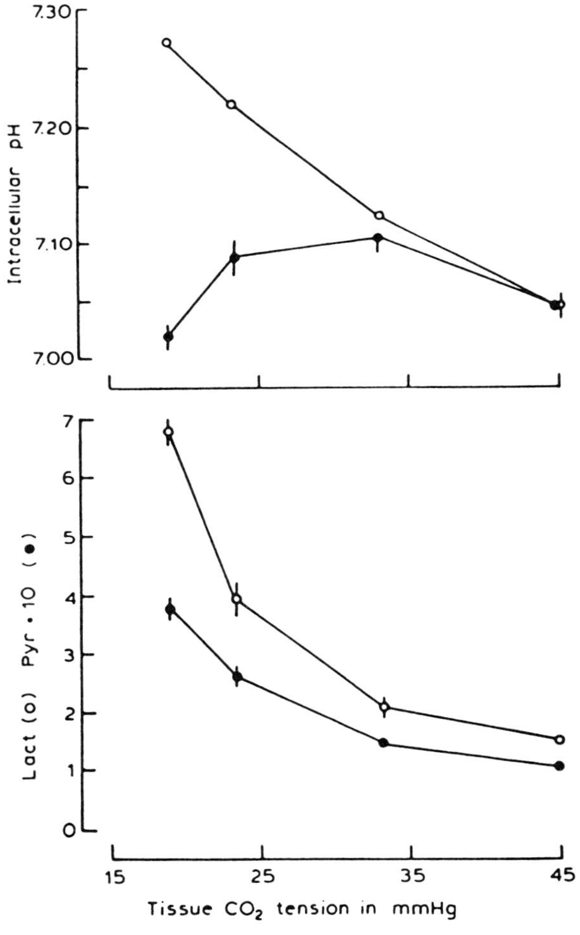

FIGURE 3. Regulation of pH_i during hypocapnia by acid production. The upper panel compares actual pH_i values, derived for an extra cellular fluid volume of 15%, and those calculated on the assumption that physiochemical buffering alone would regulate pH_i. The difference between the two curves reflects a decrease in buffer base (BB) concentration. The lower panel demonstrates that one likely mechanism is production of metabolic acids. The intracellular lactate and pyruvate concentrations were calculated from the tissue, cerebrospinal fluid, and blood contents. Similar decreases were observed in the tissue concentrations of citrate, α-ketoglutamate, and malate. Data from (MacMillan and Siesjö, 1973) reproduced with permission from Siesjö (1985a).

metabolic acids were formed when pH was raised in response to the lowering of P_{CO_2}. Actually, the "regulation" of pH can be quantitatively explained by the production of metabolic acids (see below).

One remarkable feature of the data in Fig. 3 is that, at very low P_{CO_2} values, the pH_i derived returns to normal and even subnormal values. In essence, this means that, at these very low P_{CO_2} values, pH is "overcompensated." Probably this is because, in extreme hypocapnia, CBF becomes low enough to trigger anaerobic production of lactate plus H^+.

Hypercapnia

Induction of hypercapnia in the absence of hypoxia is probably the best model available to study the effect of acidosis on energy metabolism. Since CO_2 readily passes brain cellular membranes, an increase in CO_2 tension gives rise to a rapid fall in intracellular pH—the magnitude being 0.3–0.4 units when P_{CO_2} is increased from 40 to 140 mmHg (Siesjö, 1978). However, the pH_i response to hypercapnia *in vivo* is biphasic, that is, the initial pH_i shift is transient and is followed by a slow shift towards normal, indicating that pH_i is regulated (Messetter and Siesjö, 1971; Nioka et al., 1987).

Acidosis due to hypercapnia is accompanied by distinct changes in brain energy metabolism. These encompass a reduction in CMR_{glu} in the absence of a corresponding decrease in CMR_{O_2} (Folbergrová et al., 1975; Miller and Veech, 1975), a transient increase in G-6-P and F-6-P, a reduction in glycolytic metabolites downstream from F-6-P, a reduction in levels of citric acid cycle metabolites (except succinate), and a decrease in the amino acid pool, with a fall in glutamate and a (transient) increase in aspartate concentration (Folbergrová et al., 1972a, 1974, 1975; Miller and Veech, 1975). The principal changes are illustrated in Fig. 4.

Even if hypercapnia is extreme, it does not reduce the ATP concentration, or the adenylate energy charge (Folbergrová et al., 1992, 1993, 1995); nor does it alter the $NADH/NAD^+$ ratio (Siesjö et al., 1972). Thus, although stepwise increases in P_{CO_2} (and decreases in pH_i) increase the lactate/pyruvate ratio, this seems to reflect a pH-dependent rise in the ratio, at a constant $NADH/NAD^+$ ratio. All these results can be interpreted to show that, in acidosis, endogenous glycolytic substrates are being oxidized, partly replacing exogenous glucose as substrates. This is not because glucose supply is limited but because its metabolism is inhibited, particularly at the PFK step. Since the metabolites normally exist as anions of moderately strong acids, their oxidation should lead to removal of a stoichiometric amount of H^+, explaining the regulation of pH_i (Siesjö, 1973). Very probably, transamination with conversion of glutamate to aspartate, and a decrease in the pool of dicarboxylic acids, serve the same purpose, that is, removing H^+ from the intracellular fluids.

The concept of regulation of pH_i by production (hypocapnia) and consumption (hypercapnia) of metabolic acid provides a plausible explanation of metabolic pH_i control (Siesjö, 1973). In fact, this may be the dominant mechanism in the control of pH during the first 1 to 3 hours of hypo- and hypercapnia. Figure 5 illustrates this fact by comparing changes in pH_i observed after 45 minutes of hypo- and hypercapnia to those expected from the changes in buffer base caused by production and consumption of metabolic acids, respectively. Figure 5 also illustrates how the actual pH_i differs from that expected, should pH_i be regulated by physiochemical buffering only. Clearly, the initial regulation of pH_i in hypo- and hypercapnia can be quantitatively explained by the production and consumption of metabolic acids. This does not, of course, exclude a role for active

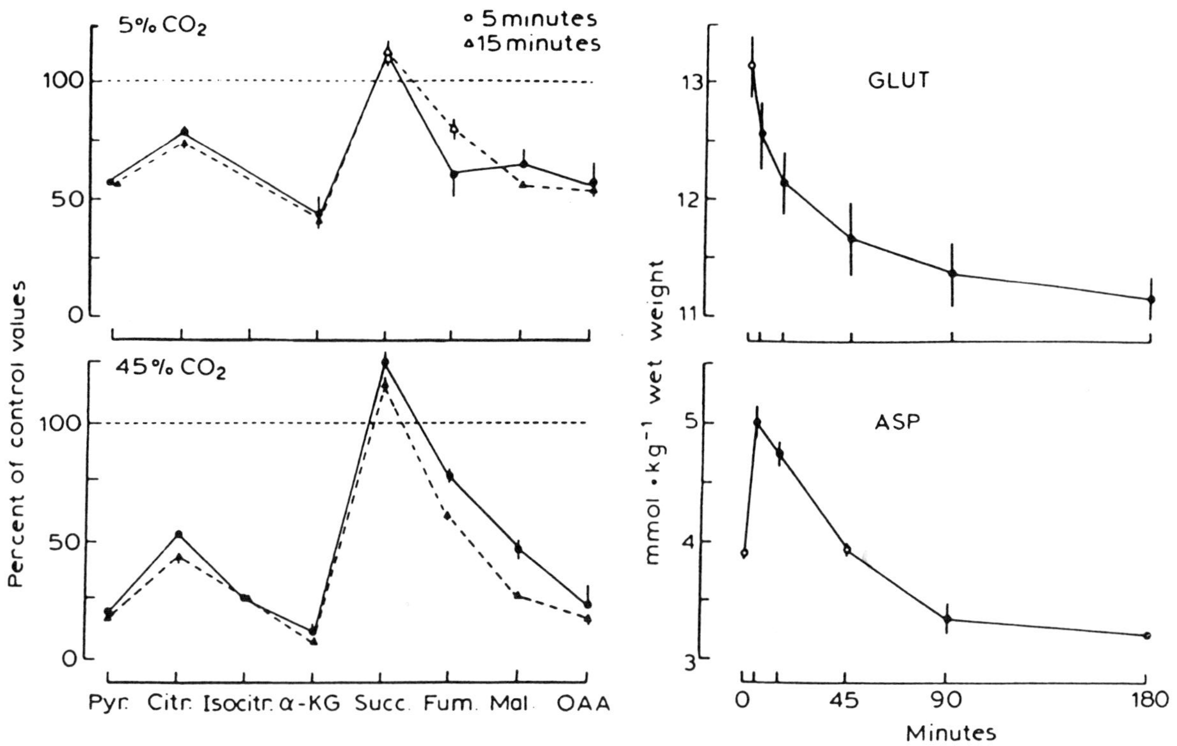

FIGURE 4. Reduction of concentrations of metabolic acids during hypercapnia. The panels to the left demonstrate reductions in tissue concentrations of pyruvate and of all citric acid cycle intermediates except succinate. The "acid consumption," which is proportional to the degree of hypercapnia, is maximal within the first 5–15 minutes. The panels to the right demonstrate a slower reduction in the tissue concentrations of diprotic amino acids (see page 142). Data from Folbergrová et al., (1972a, 1975).

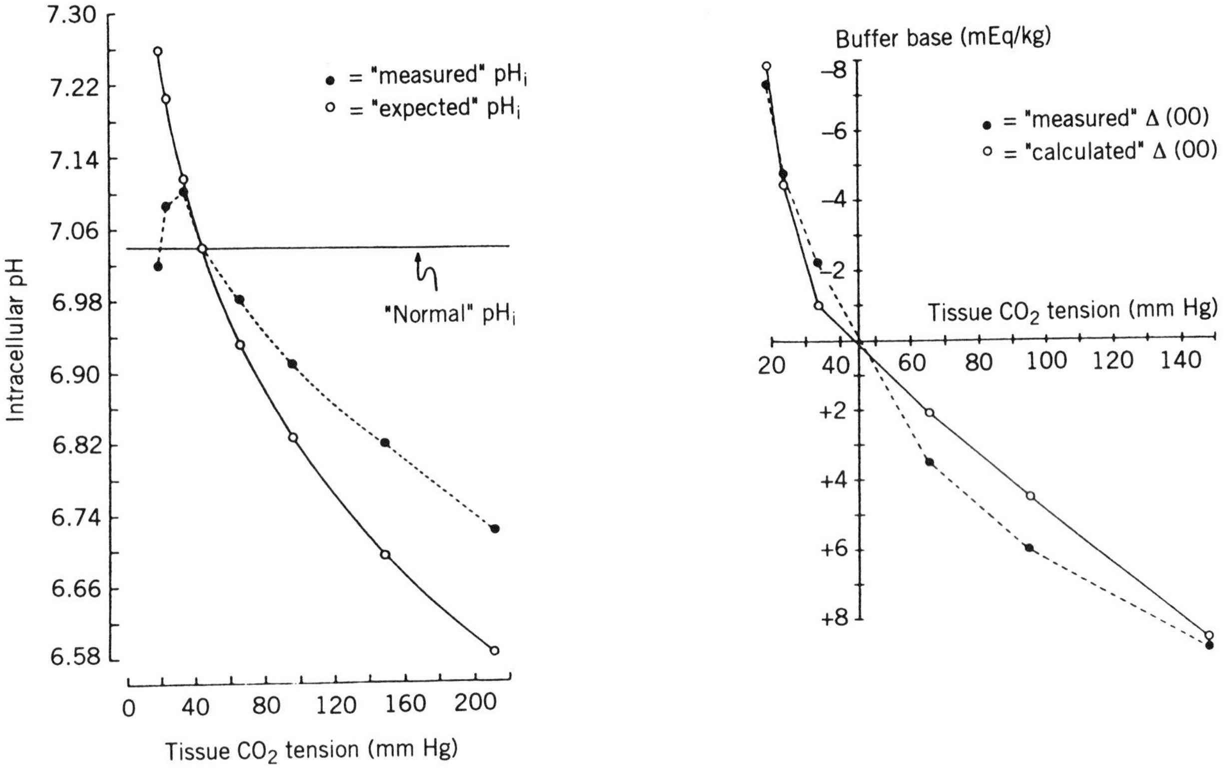

FIGURE 5. The left panel of the figure illustrates changes in pH with Pco_2 in a solution that simulates the intracellular buffer capacity (open circles), as well as the pH_i changes obtained *in vivo* when the Pco_2 was either lowered for 45 minutes or increased for the same period (filled circles). The pH_i changes were less for any given change in Pco_2 than what corresponds to the physicochemical buffer capacity. The difference in pH at any given Pco_2 represents a change in buffer base concentrations (BB) (see Siesjö, 1973). The right part of the figure compares changes in buffer base (ΔBB) calculated from the measured changes in tissue metabolites to those derived from a theoretical equation that mimicks the buffer capacity of the system, and into which measured Pco_2 and pH_i values are inserted. The correspondence between the measured and calculated ΔBB suggests that acute pH_i regulation in hypo- and hypercapnia is due to production and consumption of metabolic acids, respectively. Reproduced from Siesjö, (1973).

transport of H^+ over cell membranes. However, regulation of pH_i is initially slow, occurring at the expense of the regulation of pH_e (Katsura et al., 1994). This indicates that regulation of pH_i (and pH_e) by H^+/HCO_3^- transport across the BBB is slow (see also Chapter 32).

Ammonia Intoxication

Ammonia intoxication accompanies liver disease. However, in this and related conditions, a rise in blood ammonia concentration is only part of a complex metabolic cascade of events. From an acid-base point of view, special attention should be given to experimentally induced ammonia intoxication. This is because diffusion of NH_3 from blood to brain is one way of carrying basic equivalents into the tissue, with a resulting increase in pH_i and pH_e (Roos and Boron, 1981).

Ammonia intoxication, thus induced, is accompanied by an array of symptoms, ranging from subtle neuropsychiatric symptoms to coma. These changes probably reflect alterations in synaptic transmission (Raabe, 1993) and, at least in part, alterations in the bioenergetic state of the cells. The change in energy metabolism could be related to a protein kinase C-mediated phosphorylation of Na^+/K^+-ATPase that is secondary to activation of N-methyl-D-aspartate (NMDA) receptors; this would lead to an increased activity of the ATPase and, therefore, to an increased consumption of ATP (Kosenko et al., 1994).

From the present perspective, though, we are concerned with the effects of hyperammonemia on pH_e and pH_i. Injection or infusion of ammonia salts are found to cause a rise in tissue pH. However, the rise in pH is of very moderate degree, probably because the pH change triggers a compensatory increase in glycolytic rate and lactic acid production; in fact, at very high blood (and tissue) ammonia concentrations (injected dose 0.78 mM/100 mg of body weight) the tissue lactate concentrations increase from about 1–2 mM/kg to values of 10–12 mM kg (Hindfelt and Siesjö, 1971). Ammonia intoxication is thus another example of metabolic pH regulation in conditions when pH_i increases above normal. In fact, since cells react so promptly to alkalosis by production of acids a sizable increase in pH_i (alkalosis) is never observed.

In summary, when cells are challenged with externally imposed changes that threaten to upset their pH homeostasis, they respond by increasing the production of metabolic acids (alkalosis) or by oxidizing anions of such acids (acidosis). The result is a finely tuned regulation of pH_i and, thereby, of pH_e as well.

INFLUENCE OF CHANGES IN ENERGY METABOLISM ON INTRA- AND EXTRACELLULAR pH

Conditions involving a primary increase in acid production or consumption encompass ischemia/hypoxia, epileptic seizures, and hypoglycemia. The most dramatic example of a relationship between a perturbed energy metabolism and deranged acid-base homeostasis is ischemia/hypoxia, that is, conditions in which a shift from aerobic to anaerobic metabolism is accompanied by production of metabolic acids, notably lactate plus H^+. However, since acid-base changes in ischemia are the subject of Chapter 28 in this volume, we will concentrate on changes in epileptic seizures and hypoglycemia.

Epileptic Seizures

It has been known for decades that epileptic seizures are accompanied by acidification of extra- and intracellular fluids. The documentation was based on recordings of pH_e (Caspers and Speckmann, 1972; Wang and Sonnenschein, 1955), and on derivation of pH_i with indirect techniques (Duffy et al., 1975). The latter findings were subsequently confirmed and extended by research workers using ^{31}P NMR techniques (Petroff et al., 1986).

Results obtained during the last 10 years have clarified the relationship between production of acid and decrease in pH_i/pH_e, the "regulation" of pH_i and the relationship between pH_i and pH_e in such regulation.

The induction of status epilepticus by bicuculline (or flurothyl) leads to a dramatic increase in metabolic rate to 200–300% of control (Borgström et al., 1976; Meldrum and Nilsson, 1976). Initially, PCr is reduced by 30–40% and ATP falls moderately, with corresponding increases in ADP and AMP; however, ATP concentration is subsequently increased to about 95% of control and is maintained at these levels even if the seizures are sustained for 2 hours (Chapman et al., 1977). During the first few minutes of seizure activity, the lactate content is increased from about 1.5 mmol/kg to values of 7–10 mmol/kg, and the values are subsequently maintained at these levels until the seizures are terminated (Chapman et al., 1977).

Figure 6 shows changes in intracellular pH during and following a 60-minute period of sustained seizures (Siesjö et al., 1985). Soon after seizure induction, pH_i decreases by 0.2–0.3 units, but it then slowly returns to normal values. The results suggest that the initial production of lactate plus H^+ reduces pH_i by about 0.25 units, and pH_e by 0.3–0.4 units.

The results obtained during seizures thus demonstrate an initial acidification, with subsequent regulation of pH_i, and secondary intracellular alkalinization upon termination of the seizures. In all probability, the initial acidification is related to the increase in lactate con-

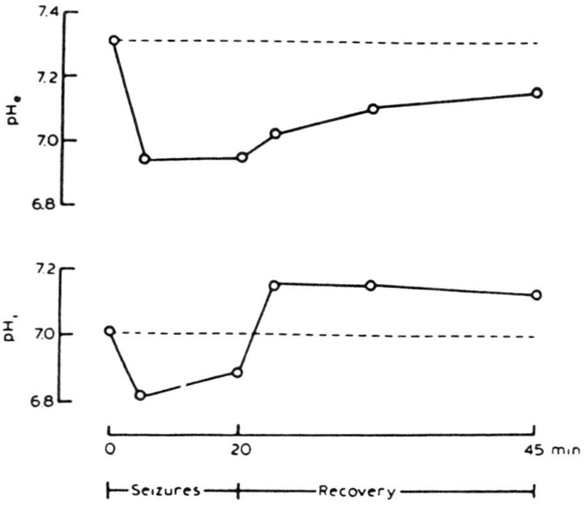

FIGURE 6. Regulation of pH_i and pH_e during and following a period of status epilepticus lasting 20 minutes and induced by flurothyl. Note regulation of pH_i during status and overshoot of pH_i during recovery, at constantly low pH_e values. Reproduced from Siesjö (1985a).

centration, with a corresponding release of H^+. Thus, the changes occur in spite of a constant P_{CO_2}, and a near-normal ATP concentration. During ongoing seizures, pH_i seems to be regulated back towards normal values and, when the seizures are terminated, it increases to supranormal values, suggesting active transport regulation that occurs at the expense of pH_e regulation. We tentatively conclude that seizures lead to an initial acidosis, mediated by lactic acid, and that pH_i is regulated by mechanisms that depend on the continuous generation of ATP and that involve translocation of H^+ from intra- to extracellular fluids.

Hypoglycemia

Since the brain is so highly dependent on glucose, it is not surprising that pronounced hypoglycemia is accompanied by cerebral symptoms ranging from drowsiness to convulsions and coma (Auer and Siesjö, 1988; Siesjö, 1988). When sufficiently prolonged, hypoglycemic coma leads to irreversible damage to brain cells (Auer et al., 1984a,b). In this respect, hypoglycemia resembles ischemia and/or hypoxia. Common to the two conditions is cerebral bioenergetic failure (Chapman et al., 1981; Feise et al., 1976; Lewis et al., 1974b; Norberg and Siesjö, 1976) and loss of ion homeostasis, the latter involving release of K^+ from cells, and uptake of Ca^{2+} (Harris et al., 1984; Wieloch et al., 1984). Hypoglycemia is also accompanied by extensive lipolysis, with release of free fatty acids (Agardh et al., 1991); the onset of coma leads to release of excitatory amino acids (Sandberg et al., 1986; Wieloch et al., 1986). However, hypoglycemia differs from ischemia and/or hypoxia in several important respects (Auer and Siesjö, 1988): First, the blood flow and oxygen delivery rates are normal or increased (Abdul-Rahman et al., 1979), redox couples are oxidized rather than reduced (Lewis et al., 1974a; Norberg and Siesjö, 1976), and pH_i is either unchanged or increased above normal (Behar et al., 1985; Feise et al., 1976; Lewis et al., 1974a; Pelligrino and Siesjö, 1981).

The difference in the acid-base response between hypoglycemia and ischemia and/or hypoxia is a fundamental one, and may explain why hypoglycemia, when severe and long-lasting, leads to selective neuronal necrosis rather than to pan-necrosis (i.e., infarction; Auer and Siesjö, 1988). The reasons why acidosis does not develop in hypoglycemia are rather straightforward: Since the substrate of glycolysis is short in supply, metabolic acids are not produced in the energy-depleted tissue; rather, they are used as metabolic fuels (Siesjö, 1988). This is reflected in the mismatch between CMR_{glu} (which is markedly reduced) and CMR_{O_2} (which, at least initially, is upheld), suggesting consumption of endogenous substrates (for literature, see Siesjö, 1988). These substrates proved to encompass glycolytic metabolites, citric acid cycle intermediates, and associated amino acids (Agardh et al., 1978; Feise et al., 1976; Lewis et al., 1974a; Norberg and Siesjö, 1976), with hydrolysis of lipids playing an unknown role (Agardh et al., 1981).

When hypoglycemia is sufficiently severe to induce "coma," with cessation of EEG activity, ATP and other nucleoside triphosphates are extensively hydrolyzed, and ion homeostasis is lost. In spite of this, lactate is not produced but rather consumed, together with anions of many other metabolic acids. This probably explains why pH_i is unchanged or increased: Any H^+ formed by hydrolysis of nucleoside triphosphates, or by hydrolysis of lipids, is balanced by consumption of H^+ during oxidation of metabolic acids.

In summary, the combination of bioenergetic failure, loss of cellular ion homeostasis, and lack of lactic acidosis makes hypoglycemic coma a unique insult, which, when compared with conditions of ischemia and/or hypoxia allows us to analyze the role played by acidosis in the cellular damage caused by bioenergetic failure and loss of calcium homeostasis.

REFERENCES

Abdul-Rahman A, Dahlgren N, Johansson BB, Siesjö BK (1979): Increase in local cerebral blood flow induced by circulating adrenaline: Involvement of blood-brain barrier dysfunction. Acta Physiol Scand 107: 227–232.

Agardh C-D, Folbergrova J, Siesjö BK (1978): Cerebral metabolic changes in profound insulin-induced hypoglycemia and in the recovery period following glucose administration. J Neurochem 31: 1135–1142.

Agardh C-D, Kalimo H, Olsson Y, Siesjö BK (1981): Hypoglycemic brain injury: Metabolic and structural findings in rat cerebellar cortex during profound insulin-induced hypoglycemia and in the recovery period following glucose administration. J Cereb Blood Flow Metab 1: 71–84.

Agardh C-D, Zhang H, Smith M-L, Siesjö BK (1991): Free radical production and ischemic brain damage: Influence of postischemic oxygen tension. Int J Devl Neuroscience 9: 127–138.

Alberti K, Cuthbert C (1982): The hydrogen ion in normal metabolism: A review. In Porter R, Lawrenson G (eds): Metabolic Acidosis. Bath: The Pitman Press, pp 1–19.

Astrup J (1982): Energy-requiring cell functions in the ischemic brain. J Neurosurg 56: 482–497.

Atkinson D (1968): The energy charge of the adenylate pool as a regulatory parameter. Interaction with feedback modifiers. Biochemistry 7: 4030–4034.

Auer RN, Siesjö BK (1988): Biological differences between ischemia, hypoglycemia and epilepsy. Ann Neurol 24: 699–707.

Auer RN, Wieloch T, Olsson Y, Siesjö BK (1984a): Distribution of hypoglycemic brain damage: Relationship to white matter and cerebrospinal fluid pathways. Acta Neuropathol (Berl) 64: 177–191.

Auer RN, Wieloch T, Olsson Y, Siesjö BK (1984b):Hypoglycemic brain injury in the rat: Correlation of density of brain damage with the EEG isoelectric time. A quantitative study. Diabetes 33: 1090–1098.

Balaban R (1990): Regulation of oxidative phosphorylation in the mammalian cell. Am J Physiol 258: C377–C389.

Behar K, den Hollander J, Petroff O, Hetherington H, Prichard J, Shulman R (1985): Effect of hypoglycemic encephalopathy upon amino acids, high-energy phosphates, and pHi in the rat brain in vivo: Detection by sequential ^{1}H and ^{31}P NMR spectroscopy. J Neurochem 44: 1045–1055.

Bergbauer K, Dringen R, Verleysdonk S, Gebhardt R, Hamprecht B, Wiesinger H (1996): Studies on fructose metabolism in cultured astroglial cells and control hepatocytes: lack of fructokinase activity and immunoreactivity in astrocytes. Dev Neurosci 18(5-6): 371–9.

Bittar PG, Charnay Y, Pellerin L, Bouras C, Magistretti PJ (1996): Selective distribution of lactate dehydrogenase isoenzymes in neurons and astrocytes of human brain. J Cereb Blood Flow Metab 16(6): 1079–89.

Borgström L, Norberg K, Siesjö BK (1976): Glucose consumption in rat cerebral cortex in normoxia, hypoxia and hypercapnia. Acta Physiol Scand 96: 569–574.

Bouvier M, Szatkowski M, Amato A, Attwell D (1992): The glial cell glutamate uptake carrier countertransport pH-changing anions. Nature 360: 471–474.

Caspers J, Speckmann E (1972): Cerebral P_{O_2}, P_{CO}, and pH: Changes during convulsive activity and their significance for spontaneous arrest of seizures. Epilepsia 13: 699–725.

Cataldo A, Broadwell R (1986): Cytochemical identification of cerebral glycogen and glucose-6-phosphatase activity under normal and experimental conditions: Neurons and glia. J Electron Microsc Tech 3: 413–437.

Chapman A, Meldrum B, Siesjö B (1977): Cerebral metabolic changes during prolonged epileptic seizures in rats. J Neurochem 28: 1025–1035.

Chapman A, Westerberg E, Siesjö B (1981): The metabolism of purine and pyrimidine nucleotides in rat cortex during insulin-induced hypoglycemia and recovery. J Neurochem 36: 179–189.

Clarke D, Sokoloff L (1996): Circulation and energy metabolism of the brain. In: Siegel G, Agranoff B, Albers R, Molinoff P (eds): Basic Neurochemistry. New York: Raven Press, pp 645–680.

Dringen R, Gebhardt R, Hamprecht B (1993a): Glycogen in astrocytes: Possible function as lactate supply for neighboring cells. Brain Res 623: 208–214.

Dringen R, Wiesinger H, Hamprecht B (1993b): Uptake of L-lactate by cultured rat brain neurons. Neurosci Lett 163: 5–7.

Duffy T, Howse D, Plum F (1975): Cerebral energy metabolism during experimental status epilepticus. J Neurochem 24: 925–934.

Ekholm A, Katsura K, Kristián T, Liu M, Folbergrová J, Siesjö BK. Coupling of cellular energy state and ion homeostasis during recovery following brain ischemia. Brain Res 604(1–2): 185–91, 1993 Feb 26.

Ekholm A, Kenichiro K, Siesjö BK (1991): Tissue lactate content and tissue P_{CO_2}: Implications for compartmentation of H^+. Neurol Res 13: 74–76.

Erecinska M, Silver I (1989): ATP and brain function. J Cereb Blood Flow Metab 9: 2–19.

Farrell C, Pardridge W (1991): Blood brain glucose transporter is assymetrically distributed on brain endothelial luminal and abluminal membranes: An electron microscopic immunogold study. Proc Natl Acad Sci USA 88: 5779–5783.

Feise G, Kogure K, Busto R, Scheinberg P, Renmuth OM (1976): Effect of insulin hypoglycemia upon cerebral energy metabolism and EEG activity in the rat. Brain Res 126: 263–280.

Fellows L, Boutelle M, Fillenz M (1993): ATP-sensitive potassium channels and local energy demands in the rat hippocampus: an in vivo study. J Neurochem 60: 1258–1263.

Folbergrová J, MacMillan V, Siesjö BK (1972a): The effect of hypercapnic acidosis upon some glycolytic and Krebs cycle-associated intermediates in the rat brain. J Neurochem 19: 2507–2517.

Folbergrová J, MacMillan V, Siesjö BK (1972b): The effect of moderate and marked hypercapnia upon the energy state and upon the cytoplasmic $NADH/NAD^+$ ratio of the rat brain. J Neurochem 19: 2497–2505.

Folbergrová J, Pontén U, Siesjö BK (1974): Patterns of changes in brain carbohydrate metabolites, amino acids and organic phosphates at increased carbon dioxide tensions. J Neurochem 22: 1115–1125.

Folbergrová J, Norberg K, Quistorff B, Siesjö BK (1975): Carbohydrate and amino acid metabolism in rat cerebral cortex in moderate and extreme hypercapnia. J Neurochem 25: 457–462.

Folbergrová J, Memezawa H, Smith M-L, Siesjö BK (1992): Focal and perifocal changes in tissue energy state during middle cerebral artery occlusion in normo- and hyperglycemic rats. J Cereb Blood Flow Metab 12: 25–33.

Folbergrová J, Kiyota Y, Pahlmark K, Memezawa H, Smith M-L, Siesjö BK (1993): Does ischemia with reperfusion lead to oxidative damage to proteins in the brain? J Cereb Blood Flow Metab 13: 145–152.

Folbergrová J, Zhao Q, Katsura K, Siesjö B (1995): *N-tert*-butyl-α-phenylnitrone improves recovery of brain energy state in the rats following transient focal ischemia. Proc Natl Acad Sci USA 92: 5057–5061.

Fox P, Raichle M, Mintun M, Dence C (1988): Nonoxidative glucose consumption during focal physiological neural activity. Science 241: 462–464.

Frostig R, Lieke E, Tsó D, Grinvald A (1990): Cortical functional architecture and local coupling between neuronal activity and the microcirculation revealed by in vivo high-resolution optical imaging of intrinsic signals. Proc Natl Acad Sci USA 87: 6082–6086.

Gunter T, Gunter K, Sheu S, Gavin C (1994): Mitochondrial calcium transport: Physiological and pathological relevance. Am J Physiol Soc 267: C313–C339.

Harris R, Wieloch T, Symon L, Siesjö BK (1984): Cerebral extracellular calcium activity in severe hypoglycemia: Relation to extracellular potassium activity and energy state. J Cereb Blood Flow Metab 4: 187–193.

Hertz L, Code W, Sykova E (1992): Ions, water, and energy in brain cells: A synopsis of interrelations. Can J Physiol Pharmacol 70: S100–S106.

Hillered L, Smith M-L, Siesjö BK (1985): Lactic acidosis and recovery of mitochondrial function following forebrain ischemia in the rat. J Cereb Blood Flow Metab 5: 259–266.

Hindfelt B, Siesjö BK (1971): Cerebral effects of acute ammonia intoxication, I. The influence on intra- and extracellular acid-base parameters. Scand J Clin Lab Invest 28: 353–364.

Hochachka P, Mommsen T (1983): Protons and Anaerobiosis. Science 219: 1391–1397.

Izumi Y, Benz A, Zorumski C, Olney J (1993): Effects of lactate and pyruvate on glucose deprivation in rat hippocampal slices. Neuroreport 5: 617–620.

Kao-Jen J, Wilson J (1980): Localization of hexokinase in neuronal tissue: Electron microscopic studies of rat cerebellar cortex. J Neurochem 35: 667–678.

Katsura K, Ekholm A, Siesjö BK (1992): Tissue P_{CO_2} in brain ischemia related to lactate content in normo- and hypercapnic rats. J Cereb Blood Flow Metab 12: 270–280.

Katsura K, Kristián T, Nair R, Siesjö BK (1994): Regulation of intra- and extracellular pH in the brain in acute hypercapnia: A reappraisal. Brain Res 651: 47–56.

Katsura K, Folbergrová J, Siesjö B (1996): Changes in labile energy metabolites, redox state and intracellular pH in postischemic brain of normo- and hyperglycemic rats. Brain Res 726: 57–63.

Kjällquist Å, Nardini M, Siesjö BK (1969): The regulation of extra- and intracellular acid-base parameters in the rat brain during hyper- and hypocapnia. Acta Physiol Scand 76: 485–494.

Kosenko E, Kaminsky Y, Grau E, Minana M-D, Marcaida G, Grisolia S, Felipo V (1994): Brain ATP depletion induced by acute ammonia intoxication in rats is mediated by activation of the NMDA receptor and NA^+, K^+-ATPase. J Neurochem 63: 2172–2178.

Kraig RP, Pulsinelli WA, Plum F (1985): Hydrogen ion buffering during complete brain ischemia. Brain Res 342: 281–290.

Kraig RP, Pulsinelli WA, Plum F (1986): Carbonic acid buffer changes during complete brain ischemia. Am J Physiol 250: R348–R357.

Krebs H, Woods H, Alberti K (1975): Hyperlactataemia and lactic acidosis. Essays Med Biochem 1: 81–103.

Lai J, Cooper A (1986): Brain α-ketoglutarate dehydrogenase complex: Kinetic properties, regional distribution, and effects of inhibitors. J Neurochem 47: 1376–1386.

Larrabee M (1983): Lactate uptake and release in the presence of glucose by sympathetic ganglia of chicken embryos and by neuronal and nonneuronal cultures prepared from these ganglia. J Neurochem 40: 1237–1250.

Lawson JW, Veech RL (1979): Effects of pH and free Mg^{2+} on the K_{eq} of the creatine kinase reaction and other phosphate hydrolyses and phosphate transfer reactions. J Biol Chem 254: 6528–6537.

Lewis LD, Ljunggren B, Norberg K, Siesjö BK (1974a): Changes in carbohydrate substrates, amino acids and ammonia in the brain during insulin-induced hypoglycemia. J Neurochem 23: 659–671.

Lewis LD, Ljunggren B, Ratcheson RA, Siesjö BK (1974b): Cerebral energy state in insulin-induced hypoglycemia, related to blood glucose and to EEG. J Neurochem 23: 673–679.

Lowry O, Passonneau J (1966): Kinetic evidence for multiple binding sites on phosphofructokinase. J Biol Chem 241: 2268–2279.

Lowry OH, Passoneau JV, Hasselberger FK, Schulz DW (1964): Effect of ischemia on known substrates and cofactors of the glycolytic pathway in the brain. J Biol Chem 239: 18–30.

Lund-Andersen H (1979): Transport of glucose from blood to brain. Physiol Rev 59: 305–352.

MacMillan V, Siesjö B (1972): Intracellular pH of the brain in arterial hypoxemia, evaluated with the CO_2 method and from the creatine phosphokinase equilibrium. Scand J Clin Lab Invest 30: 117–125.

MacMillan V, Siesjö BK (1973): The influence of hypocapnia upon intracellular pH and upon some carbohydrate substrates, amino acids and organic phosphates in the brain. J Neurochem 21: 1283–1299.

Meldrum B, Nilsson B (1976): Cerebral blood flow and metabolic rate early and late in prolonged seizures induced in rats by bicuculline. Brain 99: 523–544.

Messeter K, Siesjö B (1971): The intracellular pH in the brain in acute and sustained hypercapnia. Acta Physiol Scand 83: 210–219.

Miller A, Veech R (1975): Decreased rate of glucose utilization by rat brain in vivo after exposure to atmospheres containing high concentrations of CO^2. J Neurochem 25: 553–558.

Mitchell P (1966): Chemiosmotic coupling in oxidative and photosynthetic phosphorylation. Biol Rev 41: 445–502.

Nagamatsu S, Kornhauser J, Burant C, Seino S, Mayo KE, Bell GI (1990): Glucose transporter expression in brain cDNA sequence of mouse GLUT3, the brain facilitative glucose transporter isoform, and identification of sites of expression by in situ hybridization. J Biol Chem 267: 467–472.

Nioka S, Chance B, Hilberman M, Subramanian H, Leigh JJ, Veech R, Forster R (1987): Relationship between intracellular pH and energy metabolism in dog brain as measured by ^{31}P-NMR. J Appl Physiol 62:5: 2094–2101.

Norberg K, Siesjö BK (1976): Oxidative metabolism of the cerebral cortex of the rat in severe insulin-induced hypoglycemia. J Neurochem 26: 345–352.

Pardridge W (1993): Transport of insulin-related peptides and glucose across the blood-brain barrier. Ann NY Acad Sci 692: 126–137.

Pardridge W, Boaddo R (1993): Molecular cloning and regulation of gene expression of blood-brain glucose transporter. The blood-brain barrier: Cellular and molecular biology. In Pardridge W, (ed): Cellular and Molecular Biology of the Blood-Brain Barrier. New York: Raven Press, pp 395–340.

Pellerin L, Magistretti P (1994): Glutamate uptake into astrocytes stimulates aerobic glycolysis: A mechanism coupling neuronal activity to glucose utilization. Proc Natl Acad Sci USA 91(22) 10625–10629.

Pelligrino D, Siesjö BK (1981): Regulation of extra- and intracellular pH in the brain in severe hypoglycemia. J Cereb Blood Flow Metab 1: 85–96.

Petroff O, Prichard J, Ogino T, Avison M, Alger J, Shulman R (1986): Combined ^{1}H and ^{31}P nuclear magnetic resonance spectroscopic studies of bicuculline-induced seizures in vivo. Ann Neurol 20: 185–193.

Poitry-Yamate C, Tsacopoulos M (1991): Glial (Muller) cells take up and phosphorylate [3H]-2-deoxy-D-glucose in mammalian retina. Neurosci Lett 122: 241–244.

Prichard J, Rothman D, Novotny E, Petroff O, Kuwabara T, Avison M, Howseman A, Hanstock C, Schulman R (1991): Lactate rise detected by ^{1}H NMR in human visual cortex during physiologic stimulation. Proc Natl Acad Sci USA 88: 5829–5831.

Raabe W (1993): Effects of hyperammonemia on neuronal function: NH4+, IPSP and Cl(-)-extrusion. Adv Exp Med Biol 341: 71-82.

Reinhart P, Pfeiffer B, Spengler S, Hamprecht B (1990): Purification of glycogen phosphorylase from bovine brain and immunocytochemical examination of rat primary cultures using monoclonal antibodies raised against this enzyme. J Neurochem 54: 1471–1483.

Roos A, Boron W (1981): Intracellular pH. Physiol Rev 61: 296–434.

Rouslin W (1991): Effects of acidosis and ATP depletion on cardiac muscle electron transfer complex. J Mol Cell Cardiol 23: 1127–1135.

Sandberg M, Butcher SP, Hagberg H (1986): Extracellular overflow of neuroactive amino acids during severe insulin-induced hypoglycemia: In vivo dialysis of the rat hippocampus. J Neurochem 47: 178–184.

Schurr A, West C, Rigor B (1988): Lactate-supported synaptic function in the rat hippocampal slice preparation. Science 240: 1326–1328.

Siesjö BK (1973): Metabolic control of intracellular pH. (Editorial). Scand J Clin Lab Invest 32: 97–104.

Siesjö B (1978): Brain Energy Metabolism. John Wiley & Sons. New York.

Siesjö BK (1985a): Acid-base homeostasis in the brain: Physiology, chemistry, and neurochemical pathology. In: Kogura K, Hossman KA, Siesjö BK, Welsh FA (eds): Progress in Brain Research, Elsevier, pp 121–154, New York.

Siesjö BK (1985b): The biology of cerebral ischemia. Part 1: Pathology and pathophysiology. Part 2: Neurochemical Pathology. Neurology and Neurosurg. 6(15 and 16): 1–8 in both issues.

Siesjö BK (1988): Hypoglycemia, brain metabolism, and brain damage. Diabetes Metab Rev 4: 113–144.

Siesjö BK, Folbergrová J, MacMillan V (1972): The effect of hypercapnia upon intracellular pH in the brain, evaluated by the bicarbonate-carbonic acid method and from the creatine phosphokinase equilibrium. J Neurochem 19: 2483–2495.

Siesjö BK, von Hanwehr R, Nergelius G, Nevander G, Ingvar M (1985): Extra- and intracellular pH in the brain during seizures and in the recovery period following the arrest of seizure activity. J Cereb Blood Flow Metab 5: 47–57.

Siesjö BK, Katsura K, Mellergård P, Ekholm A, Lundgren J, Smith M-L (1993): Acidosis-related brain damage. In Kogure K, Hossmann K-A, Siesjö BK (eds): Neurobiology of Ischemic Brain Damage. Amsterdam: Elsevier Science, pp 23–48.

Skou J (1979): Effects of ATP on the intermediary steps of the reaction of the (Na^+-K^+)-ATPase. Effects of ATP on K0.5 for Na^+ and hydrolysis at different pH and temperature. Biochim Biophys Acta 567: 421–435.

Snyder C, Wilson J (1983): Reactive levels of hexokinase in isolated neuronal, astrocytic and oligodendroglial fractions from rat brain. J Neurochem 40: 1178–1181.

Sokoloff L (1977): Relation between physiological functions and energy metabolism in the central nervous system. J Neurochem 29: 13–26.

Tobin R, Mackerer C, Mehlman M (1972): pH effects on oxidative phosphorylation of rat liver mitochondria. Am J Physiol 223: 83–88.

Trivedi B, Danforth W (1966): Effect of pH on the kinetics of frog muscle phosphofructokinase. J Biol Chem 241: 4110–4114.

Tsacopoulos M, Evequoz-Mercier V, Perrottet P, Buchner E (1988): Honeybee retinal glial cells transform glucose and supply the neurons with metabolic substrate. Proc Natl Acad Sci USA 85: 8727–8731.

Ueda K (1979): Phosphofructokinase. Adv Enzymol 48: 194–244.

Veech R, Lawson J, Cornell N, Krebs H (1979): Cytosolic phosphorylation potential. J Biol Chem 254: 6538–6547.

Voipio J, Kaila K (1993): Interstitial PCO_2 and pH in rat hippocampal slices measured by means of a novel fast CO2/H(+)-sensitive microelectrode based on a PVC-gelled membrane. Pflugers Arch, 423(3-4): 193–201.

Wang R, Sonnenschein R (1955): pH of cerebral cortex during induced convulsions. J Neurophysiol 18: 130–137.

Wieloch T, Harris RJ, Symon L, Siesjö BK (1984): Influence of severe hypoglycemia on brain extracellular calcium and potassium activities, energy charge and phospholipid metabolism. J Neurochem 43: 160–168.

Wieloch T, Koide T, Westerberg E (1986): Inhibitory neurotransmitters and neuromodulators as protective agents against ischemic brain damage. In Krieglstein K (ed): The Pharmacology of Ischemic Brain Damage. Amsterdam: Elsevier, pp 191–197.

Wozniak M, Rydzewski B, Baker S, Raizada M (1993): The cellular and physiological actions of insulin in the central nervous system. Neurochem Int 22: 1–10.

SECTION II

Methods of Studying pH

CHAPTER 6

Ion-Sensitive Microelectrodes

JUHA VOIPIO
Department of Biosciences, Division of Animal Physiology, University of Helsinki, FIN-00014 Helsinki, Finland

INTRODUCTION

Ion-sensitive microelectrodes (ISMEs) were first used by physiologists in the 1950s (Caldwell, 1954; Hinke, 1959; Lev and Buzhinsky, 1961). Since then ISMEs have provided a powerful means for direct monitoring of ion activities in live cells and tissue both *in vitro* and *in vivo*. Much of the information related to processes such as cellular pH regulation and ion shifts during neuronal activity or pathophysiological phenomena has been collected using ISMEs.

Other methods for ion measurements have also evolved. Nowadays, fluorescent indicators (see Chapter 8) are often the method of choice when recording intracellular pH, especially when cells are small. However, the two techniques are complementary rather than competing, since their fields of application differ. For instance, ISMEs can be used to measure not only H^+ but also some other physiologically important acid-base species. The present chapter is not intended to give an introduction to the art of constructing ISMEs. Instead, it provides a brief account of the scope and limitations of the ISME techniques. The emphasis is on liquid-membrane microelectrodes, since glass-membrane ISMEs are rarely used today.

ION ACTIVITY, FREE ION CONCENTRATION, AND pH

Strong Electrolytes

In an infinitely dilute electrolyte solution, free ion concentration and ion activity are equal. At higher concentrations, interionic distances are small, and individual ions are not independent of each other due to their mutual interactions. This is manifested as a lower "effective concentration" (i.e., ion activity) compared to free ion concentration. The factor relating these two is called the activity coefficient, which approaches unity in dilute solutions. Mean activity coefficients are unambiguously defined and measurable, whereas only quasi-empirical estimates of single-ion activity coefficients can be obtained (see

pH and Brain Function, Edited by Kai Kaila and Bruce R. Ransom
ISBN 0-471-11838-9

Ammann, 1986). Typical values of single-ion activity coefficients for monovalent and divalent ions in physiological electrolyte solutions are around 0.7 and 0.3, respectively.

Although electrodes sense activity, ion measurements are sometimes presented in terms of free ion concentration. Which of these two terms is used depends on electrode calibration and uncertainties about activity coefficients in biological fluids. Because calibration solutions are made on the basis of fixed concentrations of inorganic salts, the concept of free concentration is perhaps intuitively more clear. However, quantifying signals measured from a preparation in terms of free ion concentration rests on the assumption that activity coefficients are identical in the fluids of the preparation and in calibration solutions. (This is the reason why ISMEs are calibrated at a physiological ionic strength.) Uncertainty about this equality makes the use of activities preferable to using free ion concentrations, especially in the case of intracellular measurements. This topic has been discussed in more detail by Voipio et al., (1994).

H^+ and pH

If we then consider the so called "free protons" and pH measurements in aqueous solutions, we must first ask: What does "pH" actually mean? Too many textbooks, though correctly recognizing that H^+ ions in aqueous solutions appear as H_3O^+, define pH as the negative logarithm of H^+ concentration. A thermodynamically much more sensible *theoretical definition* equates pH to the negative logarithm of H^+ activity (a_H), that is, $\mathrm{pH} \equiv -\log_{10} a_H$ (see e.g. Atkins, 1978). From this we must distinguish the *operational definition* of pH, which connects pH to the electromotive force of the hydrogen electrode. This definition can be found in slightly differing forms (Robinson and Stokes, 1959; Bates, 1981; Weast, 1989) and frequently includes the statement that a pH glass electrode makes a good approximation of the cumbersome hydrogen electrode. The operational definition is completed by designating certain buffer solutions as pH standards (Bates, 1981; Weast, 1989; see also Robinson and Stokes, 1959). It should be noted that the definition does not ignore but includes the liquid junction potential across a saturated KCl bridge.

A profound difference is buried in the above two definitions of pH. The theoretical definition is an exact statement, whereas the operational one defines pH as a relative concept. It is this relative pH that people work with in the laboratory. In theoretical considerations, like those on the thermodynamics of membrane transport, one deals with proton activity.

If we have two definitions, do we have also two pHs? Over the pH range 4–9, carefully measured pH may be equated to $-\log_{10} a_H$ (Robinson and Stokes, 1959; see also Bates, 1981). At higher and lower pH levels the correspondence fails and the difference (which appears in the second decimal place) can be approximated by simple equations (Robinson and Stokes, 1959; Weast, 1989). As pointed out in the same references, these corrections have been incorporated in certain tables and should be subtracted for (operational) pH. When relating pH to $-\log_{10} a_H$ at any pH level, accuracy is limited also by the fact that somewhat differing pH values are measured, depending on which of the accepted pH standards are chosen for electrode calibration (see Bates, 1981).

The fairly high ionic strength of physiological electrolyte solutions may cause an additional error amounting up to 0.05 pH units, which can, however, be partially eliminated by modification of the pH standards (Bates et al., 1978). For all these reasons, although small pH changes in a sample can be measured with high accuracy, absolute pH values should in practice be given with no more than two decimal places.

Constant errors and uncertainties in measured pH values do not always cause a corresponding error when interpreting pH as $-\log_{10}a_H$. A typical example is given by the equilibrium potential of H^+, which is calculated from the difference of measured intracellular and extracellular pH values (see Chapter 10). In this case subtraction eliminates any constant offsets of the pH scale. In contrast, experimental determination of H^+ dissociation constants relies on measured absolute pH values (for the concepts of thermodynamic, apparent, and classical dissociation constants, see Bates, 1981).

Taken together, confusion could be avoided by devoting the term pH to its operational meaning. Accordingly, when dealing with the theoretical (thermodynamic) pH concept, using $-\log_{10}a_H$ does not leave a chance for misinterpretation. Other symbols and definitions exist (see Bates, 1981), but they are most likely of little importance from a physiologist's point of view.

An alternative approach to pH and acid-base analysis has been developed by Stewart (1981), based on the so-called strong-ion difference (SID). This approach does not, however, provide any advantages and is therefore not considered here. I agree with the explicit criticism on SID given by Cameron (1989).

Laboratory Practice. The routine of calibrating pH microelectrodes in buffer solutions that are titrated using a standard laboratory pH meter follows the operational definition of pH. Therefore, it is preferable to present (at least) primary data in terms of pH, not as a_H. It is hard to find any need or basis for converting physiological pH signals to free proton concentration.

If such a conversion were done for some reason, an estimate of the H^+ activity coefficient (γ_{H^+}) would be needed. Figure 5.4 in Ammann (1986) plots predicted single-ion activity coefficients as a function of ionic strength. Such a plot shows that γ_{H^+} is about 4% higher than γ_{Na^+} and equals 0.77 at an ionic strength of 0.16–0.17 (see also Bates et al., 1978).

A significant source of error can arise from fixing a more or less precise relative pH scale to an incorrect absolute pH level. It appears that physiologists have adopted salines containing 25 mM HCO_3^- and equilibrated with 5% CO_2 as some kind of a secondary standard of pH 7.4. This convention is very practical, but one should be aware of its weaknesses with respect to both reproducibility and accuracy of absolute pH values. The fact that the pH of this "standard" is not exactly 7.4 complicates the interpretation of pH values. However, this is not the main problem, as is evident from the following: First, salines frequently contain other buffers, such as HEPES (N-[2-Hydroxyethyl]piperazine-N′-[2-ethanesulfonic acid]), as well. If such a non-CO_2/HCO_3^- buffer is not adjusted by acid or base addition to the pH that would result from the concentrations of CO_2 and HCO_3^-, the two buffers will titrate each other, which results in changes in the HCO_3^- concentration and in pH. Second, the solubility of CO_2 is strongly dependent on temperature (Severinghaus, 1965). Therefore, in order to maintain a constant pH in solution, the HCO_3^- concentration should be increased from 25 mM to 30 mM when reducing the temperature from near 37°C to room temperature (22°C). Third, it is not uncommon that the CO_2 content of a 5% CO_2 gas deviates significantly from its nominal value. One should therefore prefer qualified gas suppliers who can provide an analysis for each gas bottle they deliver. And fourth, changes in the ambient atmospheric pressure give rise to corresponding shifts in the CO_2 partial pressure and may cause pH changes exceeding $\pm$ 0.02 pH units.

In addition, it is perhaps worth reminding the reader, in this context, that superfusion solutions should be equilibrated at the experimental temperature (or only slightly above it, which is sometimes necessary in order to avoid bubble formation in tubing and bath) using humidified gas. (Note that the partial pressure of saturated water vapor decreases that of CO_2; see Appendix C in Chapter 4). In addition, the superfusion system should be made of materials having a low gas permeability (e.g., Tygon is acceptable but silicone is not.)

BASICS OF ISMES

ISME Structure and Operation

Ion-sensitive microelectrodes are made using glass capillary micropipettes filled with a stable reference solution. Ion sensitivity is achieved by sealing the pipette tip with a membrane (often termed a "sensor") that is selectively permeable (or at least behaves in that manner) to the ion species of interest. When the tip is immersed into a sample solution of unknown ion content, a potential difference develops rapidly across the membrane. This potential is measured against a reference electrode, which is usually a conventional microelectrode, and the two can be constructed to make one single electrode by using double-barrelled glass tubing.

The electrical response of ion-sensitive electrodes is described by two equations (see Fig. 1A) (Ammann, 1986; Voipio et al., 1994). The Nicolsky-Eisenman equation takes into account not only the primary ion but also interfering ions, which enter with corresponding selectivity constants. Interference is often ignored, and the electrode response is assumed to obey the Nernst equation. It follows that the electrode response is logarithmically related to changes in ion activity, and therefore, the response of H^+ electrodes is directly related to pH.

Electrical Properties of ISMEs

The small dimensions of microelectrode tips and the low electrical conductivity of ion-specific membranes result in an extremely high electrode resistance (up to a few hundred thousand $M\Omega$). This requires specifically designed electrometer amplifiers and causes sensitivity to capacitive noise coupling. However, these are not serious limitations of the technique but rather are method-specific features that can be easily dealt with (see Voipio et al., 1994).

Practical ISME Construction

ISMEs offer a cheap and straightforward method for ion measurements. Readers interested in the construction and use of ISMEs in practice are referred to other sources: Liquid-membrane microelectrodes (Voipio et al., 1994) are easy to make, and ready-to-use membrane solutions are available for most relevant ions. ISMEs based on the use of H^+ or Na^+-sensitive glass (Thomas, 1978) are more tricky to make but offer superior selectivity against lipophilic substances. The book by D. Ammann (1986) is recommended as a comprehensive presentation of the theory of (liquid-membrane) ISMEs.

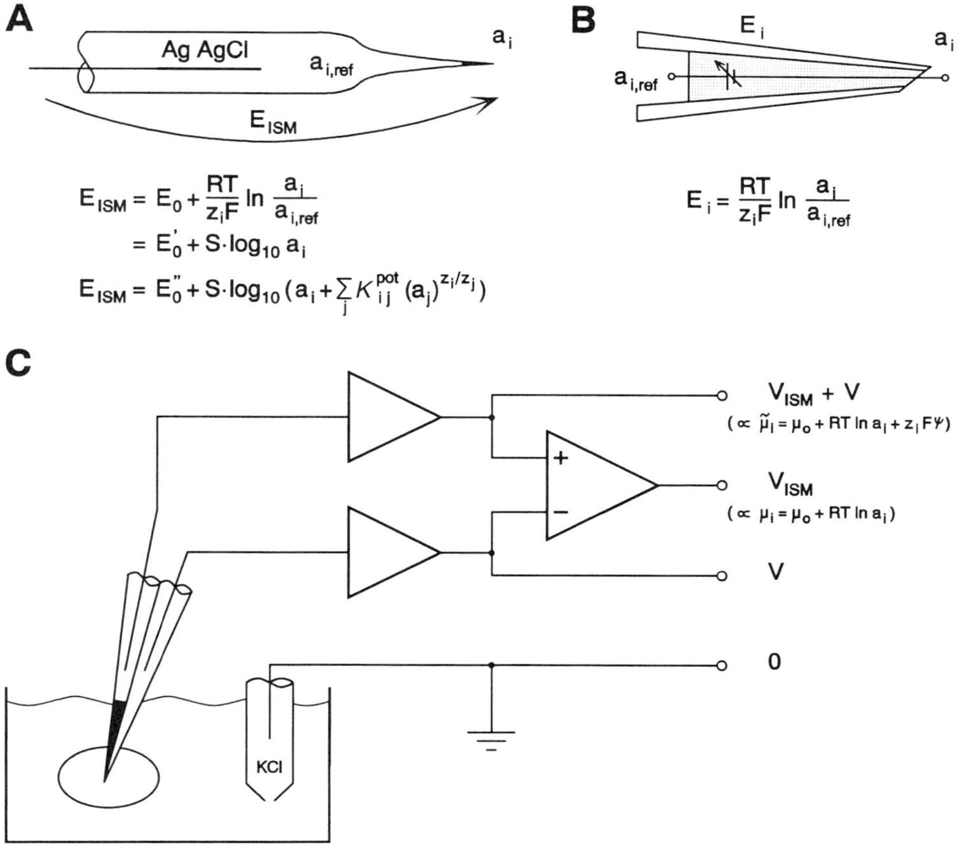

FIGURE 1. **A.** An ion-sensitive electrode generates a signal (E_{ISM}), which is recorded from an Ag|AgCl wire with respect to the potential of the sample solution. For an ideally selective electrode, E_{ISM} is given by the Nernst equation where R, T, F, and z have their usual meaning. Since the activity in the electrode filling solution of the ion to be sensed ($a_{i,\text{ref}}$) is constant, its effect is usually included, together with the electrode potential of the Ag|AgCl wire, in the constant E_0'. In practice, the observed "slope" (S) of the electrode response to the primary ion's activity in the sample (a_i) is often a few percent lower than that predicted by the Nernst equation. The effect of interfering ion species is described by the Nicolsky-Eisenman equation (see Ammann, 1986; or Voipio et al., 1994), where K_{ij}^{pot} is the potentiometric selectivity coefficient of the interfering ion(s) (j) with respect to the primary ion (i).

B. A column of liquid-membrane solution, taken into the micropipette tip by suction, behaves, despite its length, as an ion-sensitive membrane. Therefore, in the ideal case, the potential difference across the column is given by the Nernst equation.

C. When measuring from a preparation, an ion-sensitive electrode records not only ion activity but also the extracellular or intracellular potential (V). This is recorded separately by a conventional microelectrode, which may be constructed as the reference barrel in a double-barrelled micropipette configuration. Subtracting the two signals gives a signal (V_{ISM}), which depends on ion activity only. Note that V_{ISM} is proportional to the chemical potential of the ion i (μ_i), whereas the ion-barrel signal $V_{ISM} + V$ is proportional to the ion's electrochemical potential ($\tilde{\mu}_i$); μ_0 is the standard state (electro)chemical potential and Ψ denotes electrical potential.

ADVANTAGES AND LIMITATIONS OF ISMEs

From a technical point of view, ISMEs are fancy microscopic devices that enable reliable monitoring of ion activities in extremely small samples. Fluorescent probes offer another advanced technique for the same purpose. These two means cannot be compared without considering their use in live preparations. It appears that the choice between the two methods is often governed by conditions set by the preparation and by the physiological phenomenon of interest.

ISMEs are easy to calibrate and they are very stable, even in experiments lasting for many hours. Multi-ion measurements are simple, provided the preparation tolerates several ISMEs. In these respects ISMEs score well when compared to optical indicators. In addition, fluorescent probes suffer from photo-bleaching and may alter buffering. Fluorescent dyes may also be phototoxic, whereas ISMEs are invasive, a feature that totally prohibits their use in smallest cells (see p. 101). On the other hand, measurements from deep inside the tissue are possible with ISMEs because they are invasive, and ISMEs can be easily used in *in vivo* work. Furthermore, since only the very tip of ISMEs is sensitive, they record local signals and can be readily used to measure surface activities, ion gradients, etc.

The transmembrane driving force of an ion species is a function of both membrane potential and ion distribution across the membrane. For example, in order to study the thermodynamic basis of cellular pH regulation, knowledge of intracellular pH is not sufficient, and one needs to know the membrane potential as well. It is therefore a clear advantage of the intracellular ISME technique that a reference microelectrode is used to measure the membrane potential. However, even without a reference electrode, an ISME would faithfully record the thermodynamic driving force of the ions that it senses.

Response Time of ISMEs

ISMEs respond much more slowly than optical indicators. Typical estimates of liquid-membrane ISME response times are on the order of a few seconds.

In theory, a distorted signal can be used to reconstruct the primary signal if the type of distortion is known. This method has been used to reduce the apparent response time of liquid-membrane pH microelectrodes (Mückenhoff et al., 1994). However, I am doubtful about the applicability of this approach when recording from a live preparation, since there the recorded signal consists not only of the ion signal, but also of changes in electrical potential and of interference coupled from surrounding noise sources. As the various electrode input signals are distorted in a different way and the relative contribution of each signal cannot be known precisely, a genuine reconstruction is not possible.

In practice, there is little one can do about the properties of liquid-membrane solutions in order to speed up the electrode response (see Ammann, 1986). However, it is possible to reduce electric RC factors, formed by electrode resistance (R) and stray capacitance (C), which are often the main cause of slow responses. Bevelling of micropipette tips reduces ISME resistance (Kaila and Voipio, 1985), and using short membrane columns (see Vaughan-Jones and Kaila, 1986; Voipio et al., 1994) allows the capacitance compensation circuit to function optimally. When properly adjusted, this circuit provides most of the charge needed to charge the stray capacitance across the micropipette wall during changes in electrode potential. This can be rapidly accomplished only if the access resistance to the stray capacitance is low. Since membrane solutions have high specific resistances, a residual capacitance remains. This cannot be rapidly compensated for but can be reduced by making short-column electrodes. In addition, if one uses only freshly made electrodes, it is easy to obtain a significant improvement in ISME response times.

USING ISMEs AT THE CELLULAR LEVEL

Recording of Intracellular Ion Activities

Ion-sensitive microelectrodes are suitable for recording intracellular ion activities in a large number of invertebrate preparations including classical research models such as squid axons, crustacean neurons and muscle fibers, snail neurons, leech neurons and glial cells, etc. It should be born in mind that many of the mechanisms involved in synaptic transmission and active regulation of intracellular pH, now known to be functionally important in the mammalian brain, were originally characterized in invertebrate preparations (see Chapter 11; Roos and Boron, 1981; Thomas, 1984; Chesler, 1990; Kaila, 1994). Many of these preparations are outstanding in that they allow simultaneous intracellular recording of one or more ions and the use of voltage clamp in experiments lasting for hours (e.g., Kaila et al., 1989).

Reliable ISME recording of intracellular ion activities from mammalian neurons and glial cells is much more difficult and in many cases impossible. Although these cells seem in general to be too vulnerable to survive impalements even by ISMEs with submicron tips, some reports of such experiments have appeared. For example, intraneuronal calcium has been recorded *in vivo* in different regions of rat brain (Silver and Erecińska, 1990), respiration-related intracellular pH changes have been monitored *in vivo* in cat medullary neurons and glial cells (Ballanyi et al., 1994), and intracellular pH has been measured *in vivo* from rat cortical astrocytes (Chesler and Kraig, 1987, 1989). Although *in vivo* conditions seem to be more favorable for intracellular ISME experiments, stable recording periods are fairly short, lasting typically for only about 10 minutes (Chesler and Kraig, 1989) or less (Silver and Erecińska, 1990).

Recording of Interstitial Ion Activities

The interstitial space amounts to about 20% of the total volume in mammalian brain parenchyma (for references, see Nicholson, 1979; Hansen, 1985; Nicholson and Rice, 1986). Confined by neurons and glial cells, this space consists of clefts that are so narrow (typically 20-nm wide) that even the smallest ISMEs have tip diameters one or two orders of magnitude larger. It is therefore impossible to make direct ISME recordings from the intact interstitial space. However, the space around a microelectrode tip inserted in brain tissue is in direct contact with the surrounding interstitial space. It is generally believed that a diffusion equilibrium prevails in this system and that, therefore, the extracellular ion activities recorded by ISMEs are equal to those in the nearby intact interstitium (Chapter 17).

The lines of evidence supporting this view are as follows: Changes in membrane potential of glial cells upon experimentally induced changes in extracellular K^+ concentration parallel the changes recorded by extracellular K^+ microelectrodes (Lothman and Somjen, 1975; Futamachi and Pedley, 1976). Brain extracellular volume fraction values, measured using ISMEs and local microinjections, are independent of electrode distance and the results are in agreement with those obtained with noninvasive methods (Nicholson and Phillips, 1981). According to a recent report (Gottfried and Chesler, 1996), the fluorescent probe BCECF records activity-induced extracellular pH transients in a slice preparation that are faster than those recorded using pH microelectrodes. However, a detailed comparison of the extracellular use of ISMEs and optical indicators in brain slices has not been done.

The diffusion delay from the intercellular clefts to the dead space created by an ISME tip may attenuate the recorded peak amplitude of rapid interstitial ion transients. Furthermore, local signals are averaged during this diffusion process. The dead space volume, and thereupon the attenuation, is likely to depend on the ISME tip diameter and on the mechanical stability of the ISME with respect to the preparation. Ransom, Carlini, and Yamate (1987) recorded stimulation-induced extracellular increases in $[K^+]$ in rat optic nerve and ended up with an upper limit of 2 μm for the ISME tip diameter. We have observed a similar distortion of rapid K^+ transients recorded in rat hippocampal slices when the ISME tip diameter was over 2 μm (K. Lamsa, K. Kaila, and J. Voipio, unpublished observations).

Tissue damage induced by the ISME might also cause contamination of the interstitial space by intracellular substances. For instance, more than one group of investigators has considered the possibility that extracellular carbonic anhydrase activity may sometimes be artificially created (Huang et al., 1995; Voipio et al., 1995) and might affect recorded extracellular pH signals.

MICROELECTRODES FOR VARIOUS ACID-BASE SPECIES

H^+ Electrodes

Most pH-sensitive microelectrodes fall into two main categories, depending on whether the H^+ sensitive membrane used is glass or liquid. Glass-membrane pH microelectrodes are, in principle, miniaturized versions of standard laboratory pH electrodes. Constructing such electrodes requires training and is described in detail elsewhere (Thomas, 1978). Although liquid-membrane pH microelectrodes have largely superseded their glass-membrane counterparts, the latter are still outstanding in their insensitivity to lipophilic compounds.

In liquid-membrane microelectrodes, the tip of the micropipette is filled with an organic solution that behaves as a selectively permeable membrane (see Fig. 2). Some membrane solutions can be termed ion-exchangers but usually these "membrane cocktails," as they are often called, contain a ligand molecule that is capable of specifically binding one ion species and is dissolved with additives in an organic solvent (Ammann, 1986). Liquid-membrane pH microelectrodes based on the neutral carrier tri-*n*-dodecylamine (Ammann et al., 1981) have been widely used, and two slightly differing ready-to-use membrane cocktails are commercially available spanning a pH range of 5.5–12. Using another hydrogen ionophore, 4-nonadecylpyridine, pH values down to 2.0 can be measured (Chao et al., 1988). This ionophone is also available as a membrane cocktail.

Liquid-membrane pH microelectrodes are often made in a double-barrelled configuration, where one barrel is used as a conventional microelectrode. This is not necessary when working with large invertebrate cells, which can be easily impaled with two separate microelectrodes. If visual positioning of two microelectrode tips is not possible, double-barrelled pH microelectrodes are preferable, even in extracellular use.

Carbonate Electrodes

A group of liquid-membrane solutions, based on the use of quaternary ammonium salts dissolved in an organic solvent system, were originally patented as organic ion exchanger sensors for bicarbonate-sensitive (macro)electrodes (Wise, 1973). However, these membrane solutions were soon found to be sensitive to carbonate instead of bicarbonate (Herman and Rechnitz, 1974, 1975; Greenberg and Meyerhoff, 1982). Later, they were used as carbonate-sensitive microelectrodes (Wietasch and Kraig, 1991; Chesler et al., 1994). If the aim is to calculate the concentration of bicarbonate and CO_2 from measured values of pH and carbonate (cf. Kraig and Cooper, 1986; Wietasch and Kraig, 1991), the nonideal selectivity against other anions (Chesler et al., 1994) and the low concentration of carbonate ions at physiologically relevant pH levels may limit the applicability of these microelectrodes.

Carbon Dioxide Electrodes

Carbon dioxide electrodes make use of H^+-sensitive membranes. The well-known Stow-Severinghause type of CO_2-sensitive electrodes (Stow et al., 1957; Severinghause and Bradley, 1958) is based on a thin gas-permeable membrane (e.g., Teflon), which seals a thin layer of unbuffered electrolyte solution on the active surface of a glass pH electrode. Changes in the partial pressure of CO_2 ($P\text{CO}_2$) in the sample solution give rise to pH changes below the Teflon membrane. These changes are then detected by the pH electrode.

Early measurements of brain $P\text{CO}_2$ were made using a macroelectrode of the Stow-Severinghaus type (Hertz and Siesjö, 1959) gently pressed against tissue surface (Ingvar et al., 1959; Siesjö, 1961). The Stow-Severinghaus electrode type has also been miniaturized to a micro $P\text{CO}_2$ electrode. In these designs, a thin membrane (Caflisch and Carter, 1974; Pucacco and Carter, 1978), or an air gap (Ma, 1990), seals a micropipette tip and allows CO_2 diffusion into a very small volume of bicarbonate solution. The resulting CO_2-dependent pH changes are recorded with a pH microelectrode that is inserted within the pipette.

If the gas-permeable Teflon or rubber membrane is replaced by an H^+-sensitive liquid membrane that is also permeable to CO_2, the signal generated across the membrane depends on two factors (Niedrach, 1975; see also Coon et al., 1976). First, the membrane acts as in an ordinary pH electrode and, second, the CO_2-dependent pH changes of the filling solution are sensed by the inner membrane–solution interface (see Fig. 2). The transmembrane signal depends, therefore, on both pH and $P\text{CO}_2$ of the sample solution. If the pH of the sample is measured separately, subtraction of the two signals gives the desired $P\text{CO}_2$ signal. Bomsztyk and Calalb (1986) combined such two electrodes in a double-barrelled microelectrode configuration, which was therefore able to measure both pH and $P\text{CO}_2$ simultaneously.

A severe drawback of all the CO_2 electrodes described above is their slow response to CO_2. The 90% response time of these electrodes is often a few minutes. This is due to the time required for CO_2 diffusion through the membrane and the subsequent slow hydration reaction in the filling solution.

A CO_2 response time of only a few seconds was more recently obtained by using a polyvinyl chloride-gelled membrane solution and carbonic anhydrase in a double-barrelled microelectrode (Voipio and Kaila, 1993; see also Voipio et al., 1994 see Fig. 2).

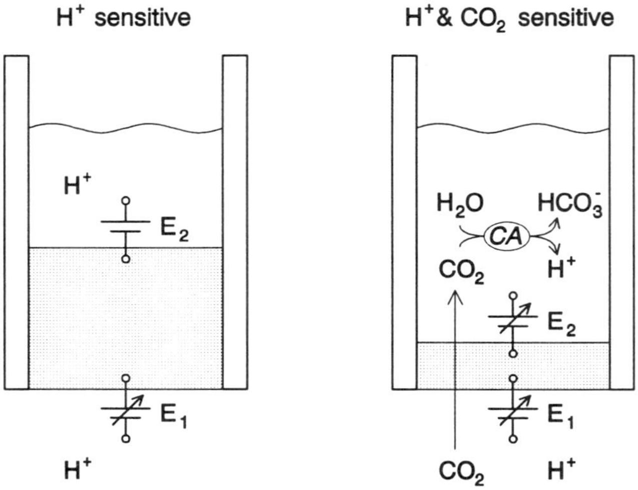

FIGURE 2. An ion-sensitive membrane has two active interfaces. In conventional ion-sensitive electrodes, as in the pH electrode shown on the left, the potential difference E_2 across the inner membrane-solution interface remains constant, since the filling electrolyte solution is well buffered. The outer interface at the tip senses ion activity in the sample and generates the ion-sensitive signal component E_1.

As schematically illustrated on the right, if the filling solution of a liquid-membrane pH electrode is not buffered, CO_2 diffusing through the membrane can alter the pH at the inner membrane solution interface, thereby making E_2 sensitive to CO_2. Since E_1 behaves as in a pH electrode, the electrode output is the sum of pH and CO_2 sensitive signals. If the former is recorded with a pH-sensitive barrel of a double-barrelled microelectrode, subtracting the two signals results in a solely CO_2-sensitive output. The CO_2 response is made fast by using very short membrane columns (i.e., a short diffusion path) and by adding carbonic anhydrase (CA) in the filling solution to speed up the otherwise slow hydration-dehydration reaction of CO_2. For further discussion, see text.

These microelectrodes measure pH and $P\text{CO}_2$ simultaneously, and they have proved to be reliable in experiments on rat hippocampal slices (Voipio and Kaila, 1993), optic nerve (Ransom et al., 1994), and brainstem (Voipio and Ballanyi, 1997). Preliminary experiments on leech neurons have indicated that intracellular recordings are also possible (C. Rose and J. Voipio, unpublished data).

Bicarbonate Electrodes

As already mentioned, it has been shown that a group of anion exchanger electrodes that were initially claimed to be sensitive to bicarbonate are in fact carbonate electrodes. This group also includes the bicarbonate microelectrode of Khuri et al. (1974). Since the rapidly equilibrating dissociation reaction of bicarbonate produces a proton and a carbonate ion, simultaneously measured pH and carbonate signals can be converted to a bicarbonate signal. Accordingly, a triple-barrelled structure, combining carbonate, pH, and voltage-sensitive barrels in a single microelectrode, has been used for indirect recording of bicarbonate (Wietasch and Kraig, 1991; see also Chesler et al., 1994).

A minielectrode based on a CO_2-permeable H^+-sensitive membrane of the type already discussed (p 103) has been reported as a bicarbonate electrode (Funck et al., 1982; Mostert et al., 1984). As the type of membrane suggests, the electrode is not directly bicarbonate-sensitive but its operation is based on sensing pH and $P\text{CO}_2$, hence an equilibrium of the CO_2 hydration reaction is a prerequisite for this electrode to function properly. This rules out bicarbonate measurements in those live preparations where carbonic anhydrase activity does not exist, or is pharmacologically blocked. In addition, at least in my hands, this type of a membrane can be reliably used in microelectrodes only in the CO_2/H^+ microelectrode configuration, which allows off-line calculation of bicarbonate from the recorded pH and $P\text{CO}_2$ (see Voipio and Ballanyi, 1997).

Taken together, all electrode measurements of bicarbonate activity have so far been indirect and based on direct sensing of H^+ and CO_3^{2-}, or of CO_2 and H^+. A breakthrough in developing a directly bicarbonate-sensitive liquid-membrane solution suitable for use in microelectrodes is therefore keenly anticipated.

Electrodes for Other Acid-Base Species

A liquid-membrane solution is commercially available for making NH_4^+ microelectrodes, but the reported logarithmic selectivity coefficient with respect to potassium is only −0.6 (Bührer et al., 1988). However, NH_4^+ microelectrodes have also been applied intracellularly (Fresser et al., 1991).

No lactate-sensitive liquid-membrane solutions exist. Lactate oxidase could be used as the enzyme in amperometric microdialysis biosensors, but even the smallest sensors available are far too big to be called microelectrodes. Nevertheless, such sensors have been used to measure transmitter release in rat brain *in vivo* (Walker et al., 1995). If sensors based on oxidase enzymes could be reduced to microelectrode size, they would evidently be useful in direct monitoring metabolic products such as pyruvate and lactate.

CONCLUSIONS

Ion-sensitive microelectrodes provide a quantitatively superior means for direct ion measurements. Their stability is excellent and precise calibrations are easily made. Depending on the experimental design and the aim of the work, the point-measurement nature can be considered a disadvantage, but in most cases it has proved to be very useful. This is especially obvious in a structurally heterogeneous environment, such as that encountered in measurements of interstitial pH in hippocampal slices (e.g., Taira et al., 1995; Voipio et al., 1995).

Knowing the thermodynamical forces involved in transmembrane as well as interstitial ion movements is essential for an understanding of the mechanisms underlying various neurobiological processes. ISMEs are sensitive to the electrochemical potential of an ion species. One therefore obtains both the chemical and electrical components of the forces driving ionic net movements directly from the data, which is a great advantage compared with other methods.

Ion-sensitive microelectrodes have been, and are currently being, used in a wide range of preparations both *in vivo* and *in vitro*. The potential applications of the ISME techniques are far from being exhausted.

ACKNOWLEDGMENTS

The original research work of the author has been funded by the Academy of Finland.

REFERENCES

Ammann D, Lanter F, Steiner RA, Schulthess P, Shijo Y, Simon W (1981): Neutral carrier based hydrogen ion selective microelectrode for extra- and intracellular studies. Anal Chem 53: 2267–2269.

Ammann D (1986): Ion-selective Microelectrodes. Springer-Verlag: Berlin.

Atkins PW (1978): Physical Chemistry. Oxford: Oxford University Press.

Ballanyi K, Mückenhoff K, Bellingham MC, Okada Y, Scheid P, Richter DW (1994): Activity-related pH changes in respiratory neurones and glial cells of cats. Neuroreport 6: 33–36.

Bates RG, Vega CA, White DR Jr (1978): Standards for pH measurements in isotonic saline media of ionic strength I = 0.16. Anal Chem 50: 1295–1300.

Bates RG (1981): The modern meaning of pH. CRC Crit Rev Anal Chem 10: 247–278.

Bomsztyk K, Calalb MB (1986): A new microelectrode method for simultaneous measurement of pH and P_{CO_2}. Am J Physiol 251: F933–F937.

Bührer T, Peter H, Simon W (1988): NH_4^+ ion selective microelectrode based on the antibiotics nonactin/monactin. Pflügers Arch 412: 359–362.

Caflisch CR, Carter NW (1974): A micro P_{CO_2} electrode. Anal Biochem 60: 252–257.

Caldwell PC (1954): An investigation of the intracellular pH of crab muscle fibres by means of micro-glass and micro-tungsten electrodes. J Physiol 126: 169–180.

Cameron JN (1989): Acid-base homeostasis: past and present perspectives. Physiol Zoology 62: 845–865.

Chao P, Ammann D, Oesch U, Simon W, Lang F (1988): Extra- and intracellular hydrogen ion-selective microelectrode based on neutral carriers with extended pH response range in acid media. Pflügers Arch 411: 216–219.

Chesler M (1990): The regulation and modulation of pH in the nervous system. Prog Neurobiol 34: 401–427.

Chesler M, Chen JCT, Kraig RP (1994): Determination of extracellular bicarbonate and carbon dioxide concentrations in brain slices using carbonate and pH-selective microelectrodes. J Neurosci Methods 53: 129–136.

Chesler M, Kraig RP (1987): Intracellular pH of astrocytes increases rapidly with cortical stimulation. Am J Physiol 253: R666–R670.

Chesler M, Kraig RP (1989): Intracellular pH transients of mammalian astrocytes. J Neurosci 9: 2011–2019.

Coon RL, Lai NCJ, Kampine JP (1976): Evaluation of a dual-function pH and P_{CO_2} in vivo sensor. J Appl Physiol 40: 625–629.

Fresser F, Moser H, Mair N (1991): Intra- and extracellular use and evaluation of ammonium-selective microelectrodes. J Exp Biol 157: 227–241.

Funck RJJ, Morf WE, Schulthess P, Ammann D, Simon W (1982): Bicarbonate-sensitive liquid membrane electrodes based on neutral carriers for hydrogen ions. Anal Chem 54: 423–429.

Futamachi KJ, Pedley TA (1976): Glial cells and extracellular potassium: Their relationship in mammalian cortex. Brain Res 109: 311–322.

Gottfried JA, Chesler M (1996): Temporal resolution of activity-dependent pH shifts in rat hippocampal slices. J Neurophysiol 76: 2804–2807.

Greenberg JA, Meyerhoff ME (1982): Response properties, applications and limitations of carbonate-selective polymer membrane electrodes. Anal Chim Acta 141: 57–64.

Hansen AJ (1985): Effect of anoxia on ion distribution in the brain. Physiol Rev 65: 101–148.

Herman HB, Rechnitz GA (1974): Carbonate ion-selective membrane electrode. Science 184: 1074–1075.

Herman HB, Rechnitz GA (1975): Preparation and properties of a carbonate ion-selective membrane electrode. Anal Chim Acta 76: 155–164.

Hertz CH, Siesjö B (1959): A rapid and sensitive electrode for continuous measurement of pCO_2 in liquids and tissue. Acta Physiol Scand 47: 115–123.

Hinke JAM (1959): Glass microelectrodes for measuring intracellular activities of sodium and potassium. Nature 184: 1257–1258.

Huang W, Smith SE, Chesler M (1995): Addition of carbonic anhydrase augments extracellular pH buffering in rat cerebral cortex. J Neurophysiol 74: 1806–1809.

Ingvar DH, Siesjö B, Hertz CH (1959): Measurement of tissue pCO_2 in the brain. Experientia 15: 306–308.

Kaila K (1994): Ionic basis of $GABA_A$ receptor channel function in the nervous system. Prog Neurobiol 42: 489–537.

Kaila K, Pasternack M, Saarikoski J, Voipio J (1989): Influence of GABA-gated bicarbonate conductance on membrane potential, current and intracellular chloride in crayfish muscle fibres. J Physiol 416: 161–181.

Kaila K, Voipio J (1985): A simple method for dry bevelling of micropipettes used in the construction of ion-selective micro-electrodes. J Physiol. 369: 8P.

Khuri RN, Bogharian KK, Agulian SK (1974): Intracellular bicarbonate in single skeletal muscle fibres. Pflügers Arch 349: 285–294.

Kraig RP, Cooper AJL (1986): Bicarbonate and ammonia changes in brain during spreading depression. Can J Physiol Pharmacol 65: 1099–1104.

Lev AA, Buzhinsky EP (1961): On the method for determining the activity of K^+-sensitive glass microelectrodes. Tsitologiia 3: 614–621.

Lothman EW, Somjen GG (1975): Extracellular potassium activity, intracellular and extracellular potential responses in the spinal cord. J Physiol 252: 115–136.

Ma Y (1990): A carbon dioxide air-gap microelectrode based on a neutral hydrogen ion exchanger pH microelectrode. Anal Biochem 186: 74–77.

Mostert IA, Morf WE, Simon W (1984): Optimization of a bicarbonate-selective liquid membrane electrode. Mikrochim Acta 3: 425–432.

Mückenhoff K, Schreiber S, De Santis A, Okada Y, Scheid P (1994): Ion-sensitive microelectrode system with short response time. J Neurosci Methods 51: 147–153.

Nicholson C (1979): Brain-cell microenvironment as a communication channel. In Schmitt FO, Warden FG (eds) The Neurosciences: Fourth Study Program. Cambridge, MA: MIT Press, pp 457–473.

Nicholson C, Phillips JM (1981): Ion diffusion modified by tortuosity and volume fraction in the extracellular microenvironment of the rat cerebellum. J Physiol 321: 225–257.

Nicholson C, Rice ME (1986): The migration of substances in the neuronal microenvironment. Ann N Y Acad Sci 481: 55–71.

Niedrach LW (1975): Bicarbonate Ion Electrode and Sensor. U.S. Patent 3,898,147.

Pucacco LR, Carter NW (1978): An improved pCO_2 microelectrode. Anal Biochem 90: 427–434.

Ransom BR, Carlini WG, Yamate CL (1987): Tip size of ion-exchanger based K^+-selective microelectrodes. Can J Physiol Pharmacol 65: 894–897.

Ransom BR, Voipio J, Kaila K, Paalasmaa P (1994): Mechanisms of activity-dependent extracellular acid shifts in white matter of the mammalian brain. Soc Neurosci Abstr 20: 242.11.

Robinson RA, Stokes RH (1959): Electrolyte Solutions. London: Butterworths.

Roos A, Boron WF (1981). Intracellular pH. Physiol Rev 61: 296–433.

Severinghaus JW, Bradley AF (1958): Electrodes for blood pO_2 and pCO_2 determination. J Appl Physiol 13: 515–520.

Severinghaus JW (1965): Blood gas concentrations. In Fenn WO, Rahn H (eds) Handbook of Physiology, Section 3: Respiration, Vol. II, pp 1475–1487.

Siesjö BK (1961): A method for continuous measurement of the carbon dioxide tension on the cerebral cortex. Acta Physiol Scand 51: 297–313.

Silver IA, Erecińska M (1990): Intracellular and extracellular changes of $[CA^{2+}]$ in hypoxia and ischemia in rat brain in vivo. J Gen Physiol 95: 837–866.

Stewart PA (1981): How to Understand Acid-Base. New York: Elsevier.

Stow RW, Baer RF, Randall BF (1957): Rapid measurement of the tension of carbon dioxide in blood. Arch Phys Med Rehabil 38: 646–650.

Taira T, Paalasmaa P, Voipio J, Kaila K (1995): Relative contributions of excitatory and inhibitory neuronal activity to alkaline transients evoked by stimulation of Schaffer collaterals in the rat hippocampal slice. J Neurophysiol 74: 643–649.

Thomas RC (1978): Ion-sensitive Intracellular Microelectrodes. London: Academic Press.

Thomas RC (1984): Experimental displacement of intracellular pH and the mechanism of its subsequent recovery. J Physiol 354: 3P–22P.

Vaughan-Jones RD, Kaila K (1986): The sensitivity of liquid sensor, ion-selective microelectrodes to changes in temperature and solution level. Pflügers Arch 406: 641–644.

Voipio J, Kaila K (1993): Interstitial Pco_2 and pH in rat hippocampal slices measured by means of a novel fast CO_2/H^+-sensitive microelectrode based on a PVC-gelled membrane. Pflügers Arch 423: 193–201.

Voipio J, Pasternack M, MacLeod K (1994): Ion-sensitive microelectrodes. In Ogden D (ed): Microelectrode Techniques—The Plymouth Workshop Handbook, 2nd ed. The Company of Biologists, Cambridge: pp 275–316.

Voipio J, Paalasmaa P, Taira T, Kaila K (1995): Pharmacological characterization of interstitial pH transients evoked by selective synaptic and exogenous activation of AMPA, NMDA and $GABA_A$ receptors in the rat hippocampal slice. J Neurophysiol 74: 633–641.

Voipio J, Ballanyi K (1997): Interstitial Pco_2 and pH and their role as chemostimulants in the isolated respiratory network of neonatal rats. J Physiol 499: 527–542.

Walker MC, Galley PT, Errington ML, Shorvon SD, Jeffreys JGR (1995): Ascorbate and glutamate release in the rat hippocampus after perforant path stimulation: A "dialysis electrode" study. J Neurochem 65: 725–731.

Weast RC (ed) (1989): CRC Handbook of Chemistry and Physics, 69th ed. CRC Boca Raton, FL: CRC Press.

Wietasch K, Kraig RP (1991): Carbonic acid buffer species measured in real time with an intracellular microelectrode array. Am J Physiol 261: R760–R765.

Wise WM (1973): Bicarbonate Ion Sensitive Electrode. U.S. Patent 3,723,281.

CHAPTER 7

Absorbance pH Indicators

JOSEPH C. LAMANNA AND KAREN L. LAURO
Department of Neurology, Case Western Reserve University, School of Medicine, Cleveland, OH

INTRODUCTION

The use of colorimetric indicators is one of the oldest methods available for pH estimation. The essential characteristics of the colorimetric technique are familiar to anyone who has taken any elementary science course in chemistry or biology. Qualitatively, litmus paper can be used to discriminate solutions as above or below neutrality. Somewhat more quantitative pH analyses can be done using dye-impregnated papers that can be quickly used to estimate solution pH to within 0.1 pH unit by direct visual comparison to a color chart. These dye molecules have chromophores, that is, groups of constituent atoms containing π (bonding) electrons. Absorption of photons that have wavelengths in the visible spectrum involve electronic transitions within the chromophores.

Absorption dyes have a reasonable potential for use as colorimetric pH indicators in living cells and tissues, although to be useful any given dye must be able to gain access to the relevant cellular compartments of interest and, once there, have some stability. The indicator should have minimal toxicity and have a large enough molar absorption coefficient so that color changes can be detected using dye concentrations that will not affect the buffer capacity of the tissue. The dye should have a p*K* in the physiologically relevant range (i.e., near neutrality). Optical absorption dyes that meet the above criteria, by their inherent properties, can provide quantitative estimates of pH with high sensitivity and excellent temporal and spatial resolution.

The development of colorimetric methods for quantitative applications to physiological questions has followed the refinement of optical techniques and improved computer analyses. Photodetectors, diffraction gratings, lenses, filters, and other optical and mechanical components now provide commercial and custom instrumentation systems with high sensitivity and with broad spectral response. The cost of computers that have the speed and performance to analyze the data produced by such instruments has fallen dramatically. There has been a large growth in the consumer and scientific demand for video (CCD) cameras, frame-grabber and video capture boards, and both commercial and

pH and Brain Function, Edited by Kai Kaila and Bruce R. Ransom
ISBN 0-471-11838-9

public domain image-processing software. These advances have allowed the translation and application of quantitative spectrophotometric assay principles to intact cells and tissues.

PRINCIPLES OF pH MEASUREMENT BY ABSORPTION SPECTROPHOTOMETRY

Beer-Lambert-Bouguer Law

The quantitation of pH colorimetric indicators is based fundamentally on the principles of Beer's law (based also on the work of Lambert and Bourguer, who were eighteenth century astronomers). The law relates the attenuation of light as it passes through a system, to the distance travelled through the medium, and the concentration of absorber and, thus, essentially defines the equivalence of mass and pathlength with respect to absorption. The familiar equation is $A = \epsilon CL$; where ϵ is the molar absorption coefficient (also called molar absorptivity or molar extinction coefficient), L is the pathlength in centimeters, and A is absorbance in optical density units. In the usual applications of this law, L and ϵ are constants that depend on the optical instrument and absorbing molecule, respectively; A is the measured quantity; and C is the unknown absorber concentration. Absorption can be determined from the relation $A = -\log T$, where T is the transmission, and $T = I/I_0$. I is the intensity of the light that passes through the absorber; I_0, the intensity of the light passing through a reference devoid of the absorber (a "blank"). Absorbance (A) is the official SI term for what had been previously called optical density (OD).

The law is an ideal law and is derived theoretically for perfectly transparent solutions at infinite dilution. In practice, the linearity expressed in the equation is valid over a restricted range of absorption representing fairly dilute systems, where the measurement error due to nonideal conditions is small up to 1 or 2 A units. For most practical considerations the maximum reliability will be in the range of 0.2 to 0.8 A units (15–65%T). The method is used to determine the unknown concentration, C, of an absorbing molecule for which the extinction coefficient is known or can be determined. To determine the extinction coefficient, it is only necessary to measure the absorbance (A) of a sample of known concentration (C). Thus, $\epsilon = A/CL$. Yet, it is better to estimate the mean ϵ using a series of samples with a range of known concentrations that span the expected range of the unknown concentrations. This assumes that the relationship of A to C is linear over the range of C. The graphical plot of A as a function of C can be used to assess linearity, or it can be used to define the specific relationship of A to C and serve as a calibration curve. The outcome of the process is that an optical signal is used to measure the unknown concentration of a light-absorbing molecule.

Henderson-Hasselbalch Equation

Colorimetric pH indicators are absorption dyes that are weak acids or weak bases and change color with dissociation. Thus, these dyes are amphiprotic and can exist in a protonated (acid) form or an uprotonated (base) form. The dissociation is described by the Henderson-Hasselbach equation:

$$\mathrm{pH} = \mathrm{p}K - \log \frac{[acid]}{[base]} \tag{1}$$

where [acid] and [base] are the concentrations of the protonated and unprotonated forms of the dye molecules, and the constant, p*K*, is the pH at which half the molecules are protonated and half are unprotonated. The pH of a solution containing a colorimetric pH indicator can be determined from this relationship, if the optical pathlength is kept constant and if the p*K* and total dye concentration ([acid] + [base]) are known, by spectrophotometrically measuring the concentration of the chromophore of one of the two amphiprotic forms.

This means that only one of the two forms needs to have a chromophore, the other form can be colorless. Phenol red is an example of such a dye. It is red in its base form and colorless in its acid form. A simple, single-wavelength spectrophotometric instrument can be used to measure the concentration of the colored form of the dye. The pH versus dye concentration curve then resembles a standard titration curve. The absorption midpoint of the curve is the p*K*. Because absorption scales linearly with concentration, the change in absorption signal for a given change in pH gets smaller the further the pH is from the p*K*. The greatest sensitivity is achieved ± 1 pH unit from the p*K*. When the pH is greater than ± 2 pH units from the p*K* the method is essentially useless, another dye must be found, with a more appropriate p*K*, to cover the new pH range of interest. Thus, colorimetric indicators are only useful over a narrow range of pH, and the p*K* of an indicator is an important characteristic to consider in dye selection. All pH_i measurement methods that depend on the Henderson-Hasselbalch relationship will suffer from this limitations.

Single-wavelength, single-chromophore systems have been used to estimate pH_i in biological preparations such as skeletal muscle (Sun et al., 1993; Hollingworth and Baylor, 1990; Baylor and Hollingworth, 1990) and invertebrate neurons (Ahmed and Connor, 1980). Advantages of this method are its simplicity, low cost, and speed. This method is useful where measurement response time is an important consideration. Unfortunately, the p*K* of a dye *in vivo* is not always known, or easy to determine, because the "apparent" p*K* may be different from the p*K* of the dye as measured in a test solution. Furthermore, it is not always possible to know or control the total dye content.

Macdonald Equation

The conditions that make single-wavelength spectrophotometry useful are rarely applicable to any experimental protocols used for brain tissue *in vivo* or *in vitro*. In these preparations, however, absorption dyes with dual chromophores, (i.e., dyes where both the protonated and unprotonated forms are colorimetric indicators) can be used. When dual-chromophore dyes are used, then neither the total (protonated and unprotonated) dye concentration, nor the pathlength have to be known or remain constant. In this case, the pH signal is proportional to the ratio of the absorbance measured at two wavelengths, one at the wavelength of maximum absorption of the acid form of the dye and the other at the characteristic wavelength of the base form of the dye. The quantitative relationship was expressed by Macdonald, et al. (1977).

$$\frac{A_{a\lambda}}{A_{b\lambda}} = \frac{\epsilon_{a\lambda}{}^{a} \cdot [\mathrm{H}^{+}] + K\epsilon_{a\lambda}{}^{b}}{\epsilon_{b\lambda}{}^{a} \cdot [\mathrm{H}^{+}] + K\epsilon_{b\lambda}{}^{b}} \tag{2}$$

The equation above was derived from combination of Beer's law of absorption (A) at the acid wavelength (a_λ) and base wavelength (b_λ) with the Henderson-Hasselbalch equation.

The equation shows that the ratio of absorption at the two wavelengths is proportional to the hydrogen ion concentration and five constants. Four of the five constants are the extinction coefficients of the two forms at each of the two wavelengths, and the fifth is the dissociation equilibrium constant. The ratio method is inherently more sensitive than single absorption measurements because of the reciprocal relationship of the concentrations of the two dye forms. Thus, the numerator and denominator of the ratio are changing simultaneously in opposite directions.

These constants are "apparent" constants, which are affected by variables that affect the electronic environment of the chromophore, such as polarization of the solvent and chromophore, and the dielectric constant of the medium. Deviations from Beer's law due to the effects of light scattering mean that optical geometry and physical instrumentation characteristics will also affect the values of these constants. For these reasons, a calibration curve is required for each *in vivo* and *in vitro* preparation, and for each optical instrumentation layout.

Instrumentation

We have previously addressed many of the pertinent aspects of optical instrumentation as they relate to *in vitro* and *in vivo* absorption measurements (Sick and LaManna, 1995; LaManna, 1992). Nevertheless, there are a few important points that should be emphasized to facilitate the evaluation and comparison of the different instrumentation and analytical approaches that have been used for the measurement of intracellular pH (pH_i) by absorption methods.

Transmission versus reflection spectrophotometry. The optical properties of both *in vitro* and *in vivo* preparations are dominated by light scattering. The two types of spectrophotometry differ in that the light source and detector may be on the same side of the tissue (reflection mode) or on opposite sides of the tissue (transmission mode). In the reflection mode, most of the light detected will be back-scattered, diffusely reflected light. In the transmission mode, most of the detected light will be forward scattered. Because of this difference, the coefficients of the calibration curves and the wavelengths of maximum absorption will differ between the two modes.

Scanning versus multiwavelength spectrophotometry. In scanning spectrophotometry, broad spectral data are collected and recorded, usually in the form of a graph that plots light intensity as a function of wavelength. Preservation of full spectral data allows corrections for baseline and absorption peak shifts induced by light scattering. Multiwavelength (usually dual) spectrophotometry depends upon either theoretical or experimental determination of the appropriate wavelengths of absorption peaks and isosbestic or reference wavelengths. Dual-wavelength spectrophotometry is used where temporal response is an issue, although methods that make use of neural nets may allow faster response from instruments scanning the full spectrum (Lin et al., 1992).

Imaging versus non-imaging. The detector used may be imaging or nonimaging. The imaging detector, video or CCD array, preserves spatial resolution, but usually at the cost of poor temporal resolution. Nonimaging detectors can be used where homogeneous responses are expected or where rapid response is required.

Data analysis. The most important aspect of data analysis, often overlooked, is numeric resolution. In optical absorption studies, absorbance measurements can routinely be in the range of 0.01 to 1 A units. This minimally requires 10-bit, and preferably 12-bit analog to digital (A/D) converter resolution. The impact of this requirement is even more evident in imaging instrumentation. The usual video capture boards produce 8-bit image arrays, but this limitation can be overcome through various tricks during data acquisition. Ten- and 12-bit CCD proprietary imaging systems, which avoid using the low-resolution capture boards, are becoming available at reasonable costs.

A more important limitation stems from analyses that require ratio calculations of absorption images, that are then used to calculate pH from linear or nonlinear calibration curves. If this is done with 8-bit resolution there will be significant round-off errors introduced. Desktop computer programs, such as Alice™ (formerly DIPstation; Hayden Image Processing Group, Boulder, CO) and NIH Image (version 1.56 onward) are available with the required resolution.

PROPERTIES OF ABSORPTION DYES SUITABLE AS COLORIMETRIC INDICATORS OF pH

Historical Perspective

There are hundreds of thousands of absorption dyes, many of which have pH-sensitive chromophores. The number of dyes that have been used for pH measurement in living tissue is a tiny fraction of this number. An extensive body of literature covers the use of voltage-sensitive absorption dyes for membrane potential measurement (Grinvald et al., 1988). A few pH indicator dyes have been used in biological cell and tissue preparations (for review see Roos and Boron, 1981; Caldwell, 1956; LaManna, 1987), but the use of absorption dyes for intracellular pH measurements in brain and neural tissue has been limited. Ahmed and Connor (1980) tested a number of pH-sensitive absorption dyes in their invertebrate nerve preparation. The dyes tested included bromthymol blue, bromcresol purple, brilliant yellow, neutral red, cresol red, and phenol red. They concluded that phenol red and bromcresol purple were the best indicators in this application, on the basis of absorbance signal amplitude and uniformity of intracellular distribution. They pressure-microinjected phenol red into molluscan neurons, and observed cytoplasmic acidification with calcium influx, or in response to trains of action potentials.

Applications of absorption dyes to pH_i measurements in brain slices and in intact brain have been limited solely to one dye—neutral red. The historical background for neutral red has been given previously (LaManna, 1987, 1992). Neutral red is a vital dye, that is, it can be used in living cells and tissues. Many observations concerning the interaction of the dye with biological tissue were made by the early histologists. The overall conclusions from an analysis of this literature are that neutral red was considered the least toxic of the vital stains and suitable for many physiological studies. The dye was observed to stain intracellular granules, such as mitochondria and Golgi bodies, and is excluded from metaphase nuclei and the nuclear membrane (Schmidtmann, 1925). Much of this early literature was reviewed by Koehring (1930).

Over the last 20 years, based primarily on the work of Macdonald (Macdonald et al., 1977; Macdonald and Jöbsis, 1976), neutral red has been used as a quantitative pH_i indicator in toad heart ventricular strips (Snow, 1986; Lutaya et al., 1981; Snow et al., 1982; LaManna et al., 1978, 1980), and hamster diaphragm muscle (Zhang et al., 1990).

Optical Properties of Neutral Red

Neutral red [color index (C.I.) 50040; 3-amino-7-dimethylamino-2-methylphenazine hydrochloride; MW = 289] (see Fig. 1) is available from Sigma Chemical Company in a purified form (N 7005, > 90% dye content). The molecule is a cationic weak base. The base form of the molecule is uncharged and the protonated, acid form is positively charged. Neutral red is a relatively nontoxic vital dye and a sensitive pH indicator in the range of 6–8. In water, the absorption maximum of the acid form is at a wavelength of 530 nm and the molar extinction coefficient is $\epsilon = 2.75 \times 10^4$. The absorption maximum of the base form is at 452 nm, and $\epsilon = 1.39 \times 10^4$. The p*K* is 6.81 (Walz et al., 1975). When the dye is in tissue, the effects of the chemical microenvironment, and the effects of light scattering result in a red shift of the absorption peaks and an alkaline shift of the p*K*. The wavelength of the acid form is 550–560 nm and of the base form is 450–480 nm. The p*K* shifts to about 7 (LaManna and McCracken, 1984).

Biological Properties of Neutral Red

Tissue uptake. Neutral red has a lipid:water partition coefficient of about 30 and, primarily as the uncharged free base, is rapidly taken up by cells. Metabolically competent brain slices concentrate neutral red to 100 times the solution concentration (McIlwain and Grinyer, 1950). In addition, neutral red becomes bound to the intracellular organelle membranes and remains trapped. *In vivo,* the dye can be administered intraperitoneally or intravenously. Because the dye is accumulated by the cells and bound, an effective intracellular dye content can be amassed with relatively low exposure concentrations. The time course of the brain uptake of neutral red after intraperitoneal administration of 1 ml of a 2% dye solution occurs over 2 hours, with a peak at about 30 minutes (LaManna and McCracken, 1984). Almost no neutral red can be found in the cerebrospinal fluid (Kogure et al., 1980).

Intracellular location of the dye. Differential centrifugation (LaManna and McCracken, 1984) shows almost all of the recovered dye to be in the particulate fractions (88%) with the highest specific activity in the P3 fraction. *In vitro* studies have shown that the dye is in the organelles such as mitochondria (Wikström, 1984) and Golgi bodies (Barr et al., 1984). There is little association of the dye with the nucleus or nuclear membranes, and there is no interaction of the dye with myelin fractions.

In its protonated form, this dye is associated, on both sides of the membrane, with the negative charges of the membrane surface. Other cations such as Mg^{2+} displace the molecule away from the surface, such that the displacement distance is proportional to the

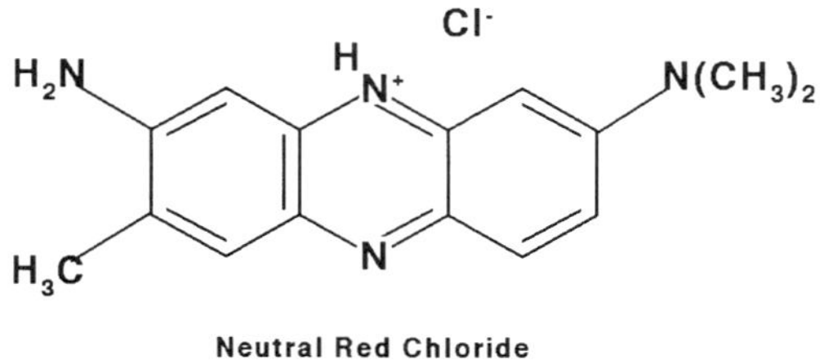

FIGURE 1. Neutral red chloride.

cation concentration. This environment may not be in rapid and complete communication with the bulk phase. This is significant because charged buffer molecules such as HEPES N-[2-Hydroxyethyl]piperazine-N′-[2-ethane sulphonic acid] may not have access to this surface space (Van Walraven et al., 1984).

Disappearance of the dye. *In vivo,* the dye disappears from the tissue after about 2 hours, but lasts much longer in the brain slice. The loss of staining color represents a metabolic change rather than a leak, because in the slice preparation, once the slices are stained, they are put into fresh buffer and that buffer does not get red. The disappearance is most likely due to metabolism of the molecule to a leuco form with consequent loss of the chromophore (Prince et al., 1981).

Physiological Properties of Neutral Red

Neutral red is a vital dye, considered to have low toxicity. Nevertheless, biological effects from both acute and chronic exposure have been reported.

Cytotoxicity. Neutral red will bind to DNA and is mutagenic in high concentrations. In addition, mainly through its cationic form, neutral red can influence the activity of some enzymes and has a tendency to self-aggregate and bind ATP.

In vivo toxicity. The main effect that has been systematically observed after neutral red administration is mild systemic hypotension (LaManna and McCracken, 1984). Millimolar concentrations of neutral red have been reported to stimulate glycolysis (McIlwain and Grinyer, 1950). No effects of micromolar concentrations of neutral red on cerebral energy metabolism have been observed.

Long-term treatment with neutral red has been reported to injure pancreatic glucagon-secreting cells in rats. Chronic exposure to neutral red has been reported to affect carbohydrate metabolism in an invertebrate preparation. This literature has been reviewed (LaManna, 1987, 1992).

Pitfalls and Limitations

Neutral red can be a useful pH_i indicator *in vivo* and *in vitro*. Nevertheless, there are certain pitfalls that should be considered and which could limit its use in some experimental preparations.

1. The range of maximum sensitivity is limited to pH 6–8. Although this range can be extended somewhat, reliability begins to fall off rapidly outside this range.
2. Cellular identification cannot be made on the basis of dye uptake, but if the cells have different pH_i they may be distinguishable up to the limits of light microscopy optical resolution.
3. The intracellular compartment may not be completely identifiable. Actively dividing cells may have nuclear stains, and so may dead cells, in which the nuclear membrane may be compromised.
4. The sensitivity of absorption dyes is relatively low compared to fluorescence dyes. This means that a larger volume of stained tissue is required to provide a measurable signal. Thus, neutral red is not a good indicator for white matter studies

because myelin, which doesn't stain, and the axoplasm, which does stain, comprise a small fraction of the total volume. Fluorescence dyes (Chapter 8) usually provide a greater signal, which can be especially advantageous in smaller systems such as single cells, where absorption dyes would be of limited value. For example, for an absorption dye with a molar absorption coefficient of 50,000, a concentration of 100 μM, would provide a maximum signal of 0.001 A units in a cell 50 μ thick.

MEASUREMENTS OF INTRACELLULAR pH USING ABSORPTION DYES

Advantages of Absorbance Measurements

The use of neutral red as an *in vitro* or *in vivo* pH_i indicator is especially warranted where spatial resolution is an important consideration. Methods based on nuclear magnetic resonance are limited to large tissue volumes and cannot reach the potential resolution of the dye methods. Microelectrode methods can sample small volumes, but cannot provide comprehensive coverage of heterogeneous tissue regions. Fluorescence methods can be useful in some of these circumstances, but cell loading with fluorescent dyes may not be convenient, especially where larger quantities of expensive dyes would be required for sufficient loading. The cost of optical instrumentation for absorption dyes is also much less than for fluorescent dyes. An additional advantage of the absorption dye method is that a reasonable estimate of the intracellular pH can be made by eye by the investigator, even before any signal processing.

Some specific examples of applications where absorption dyes have proved to be useful are given in the following sections.

In Vitro Studies: Brain Slices. Transmission Spectrophotometry in Diffusely Scattering Media

The brain slice preparation provides a useful system to study pH regulation and functional activity. Functionally, the brain slice, especially the hippocampal brain slice, retains synaptic competency such that evoked responses can be elicited. The blood can be perfused out of the brain before making the slice, thereby eliminating hemoglobin, which absorbs heavily in the visible spectrum. The few intrinsic chromophores (e.g., the cytochromes) are present in low concentration. This fact can be verified simply by noting that the brain slice is white, indicating that the optical properties of the brain slice preparation are dominated by light scattering without specific wavelength-dependent absorption. The scattering properties of the slice change with physiological state. Investigators who use brain slice preparations can usually distinguish metabolically competent slices by visual inspection. The light-scattering properties appear to be influenced by changes in intracellular and extracellular volume (Andrew and MacVicar, 1994; Turner et al., 1995; Kreisman et al., 1995). The physical and optical properties of the brain slice preparation are the subject of a recent review (Sick and LaManna, 1995).

The intracellular pH of the hippocampal brain slice measured by scanning transmission spectrophotometry of neutral red, was 7.71 $\pm$ 0.07 ($n = 8$) (Sick et al., 1989), with the slice in a bicarbonate buffer (26 mM, pHb 7.4, 37°C). This was surprisingly more alkaline than the pH of the bath buffer, and much more alkaline than pH_i measured *in vivo*. This puzzling finding led to a comprehensive examination of the dye calibration procedures in an attempt to determine if a calibration error might explain the alkaline measurement.

Calibration. Two primary methods of calibration were used to quantitate pH_i measurements from neutral red. The first method made use of brain pastes. Brain pastes were made from the discarded cerebral hemispheres of the brains used for hippocampal slice preparations. These brains were homogenized with little or no additional fluid in the presence of collagenase and deoxycholate. Neutral red was added to the paste, reserving a small sample of unstained paste to act as spectrophotometric blank. The pH of the paste was altered over a range by addition of concentrated HCl or NaOH and measured with an electrode in a standard laboratory pH meter, or by ^{31}P NMR. A few drops of paste were placed on a glass slide and mounted in the place of the hippocampal brain slice in the chamber. The ratio of absorption at the two characteristic wavelengths was plotted as a function of pH to generate a calibration curve. Over the range of pH 6–8, a linear fit to the calibration curve was considered to provide sufficient accuracy. The resultant linear calibration equation was reported as:

$$pH = \frac{(R - 6.08)}{-0.7} \tag{3}$$

Where $R = A_{550}/A_{450}$ (Sick et al., 1989).

Estimates of statistical reliability showed that the standard error of the estimate of absolute pH for a single measurement was $\pm$ 0.25 which could be improved by averaging multiple measurements. The 95% confidence interval for the relative pH sensitivity (the slope of the calibration curve) was 0.01 pH units.

A common technique for calibrating pH_i is the nigericin method (Thomas et al., 1979). The major advantage of this method is that it can be applied to the intact tissue slice. The technique uses the ionophore nigericin, which serves to equilibrate the combined hydrogen and potassium ion gradients. When external and internal potassium ion concentrations are made equal, then the intracellular pH will be equal to the external buffer pH. Thus, it is necessary to add the nigericin in a high-K^+ buffer. With this method, the known buffer pH can be used to calibrate the dye absorption ratio. In the brain slice preparation, however, the equilibration time for nigericin was found to be on the order of 30–60 minutes, most likely due to the restricted extracellular space. Nevertheless, a one or two point calibration could be made at the end of the experiment, and it was found to agree very closely to the pH calculated from the neutral red brain paste equation (Lin et al., 1996). Significantly, when the nigericin/high-K^+ buffer (pH_o 7.4) was added to the brain slice preparation, the neutral red signal indicated an acid shift of the intracellular compartment, demonstrating that the pH_i had been more alkaline than the extracellular pH (pH_o).

Hippocampal brain slice. With the corroboration of the calibration procedure, it was concluded that the intracellular compartment that contains neutral red in the brain slice preparation has a pH between 7.4 and 7.8 (LaManna et al., 1992; Whittingham et al., 1989; Sick et al., 1989, 1987). Current estimates of pH_i in the brain slice preparation are highly method-specific, for as yet unexplained reasons. Estimates range from 6.9 by the DMO (5,5-dimethyl-2,4-oxazolidinedione method; Hertz et al., 1970), 7.03 by microelectrode (Grichtchenko and Chesler, 1994), 7.08 and 6.95 using BCECF [2′,7′-bis(carboxyethyl)carboxyfluorescein] fluorescence (Amato et al., 1994; Fujiwara et al., 1992), 7.09 by SNARF fluorescence (Lin et al., 1993b), and 7.39 by ^{31}P NMR (Brooks et al., 1989). The explanation for the discrepancies in absolute pH_i estimation must be related to variations in the incubation conditions, buffers, compartmentation, calibration procedures, and intrinsic peculiarities of each method (Lin et al., 1996).

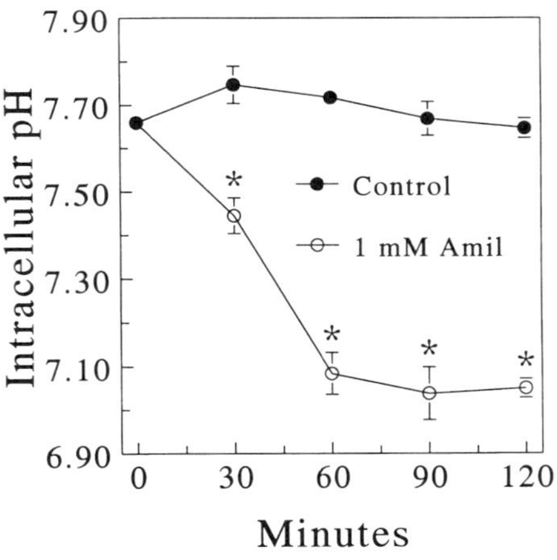

FIGURE 2. The effect of 1 mM amiloride on pH_i in the hippocampal slice. Under control conditions (filled circles) hippocampal brain slices maintain a pH_i more alkaline than the external bicarbonate buffer (pH_o=7.4, 37°C). The pH_i decreases in the presence of the Na^+/H^+ exchange transporter inhibitor amiloride (open circles).

When the hippocampal slice was treated with 1 mM amiloride, a Na^+/H^+ exchange transporter inhibitor, the pH_i measured by neutral red became more acidic (Fig. 2), suggesting that the alkalinization in the compartment associated with neutral red was due to activity of this exchanger. This finding was confirmed and extended using a series of specific amiloride analogues (Lin et al., 1996, 1993a).

Comparison with other brain regions. In a separate series, we compared the hippocampal brain slice pH_i with the pH_i of cortical and cerebellar slices (LaManna and Farrell, 1989). The cerebral cortical brain slice pH_i in bicarbonate buffer at pH 7.4 was 7.57 ± 0.02 (SEM, n = 40). The cerebellar brain slice pH_i was 7.62 ± 0.02 (n = 8), and hippocampal brain slice pH_i was 7.66 ± 0.02 (n = 13). No significant differences between any of the groups were found.

In Vivo Studies

In situ *rat brain: diffuse reflection spectrophotometry.* Optical studies of the intact brain *in situ* are complicated by the presence of hemoglobin. Not only does hemoglobin absorb strongly in the visible portion of the spectrum, but its absorption changes with oxygen saturation. Nevertheless, it was possible to obtain an estimate of cerebral cortical pH_i *in situ* by absorption spectrophotometry of neutral red, despite the inability to control for changes in blood volume and hemoglobin saturation (LaManna and McCracken, 1984).

A calibration curve for diffuse reflection spectrophotometry was constructed using the brain paste method. The calibration curve exhibited a greater influence of diffusely backscattered light than the calibration curve generated for transmission mode spectrophotometry used for brain slices. This is reflected in the coefficients of the linear equation:

$$pH = \frac{(R - 10.5)}{-1.3} \tag{4}$$

Although it was not possible to determine *in vivo* pH_i from this equation, it was possible to estimate *in vivo* pH_i from the shape of the differential reflectance spectrum from the brain surface. The method involved detecting the shift in the overall peak of the specific neutral red absorption spectrum as a function of pH. In rat anesthetized with chloral hydrates the estimated pH_i of the cerebral cortex ($n = 11$) was 7.04 ± 0.01 (LaManna and McCracken, 1984), which is well within the expected range (Chesler, 1990).

Color film histophotometry of frozen rat brain. To avoid the problems caused by hemoglobin, to take advantage of correlative biochemical assays, and to achieve some degree of spatial resolution, we developed the method of color film histophotometry of frozen rat brain (Griffith et al., 1992; LaManna et al., 1986). For this method, neutral red was administered intravenously 45 minutes before the anesthetized rat brain was frozen *in situ* with liquid nitrogen (Pontén et al, 1973). The frozen brain was removed in a glove box, and sectioned with a microtome. Periodically, 35-mm photographic color slides were made of the block face as the brain was sectioned. The sections were lyophilized and used for quantitative histochemistry. The color slides were analyzed by making monochromatic images of the slide at 450 nm and 550 nm, the absorption peaks of the base and acid forms of neutral red. (These wavelengths were also near the absorption maxima of two of the three dyes in the Fujichrome Velvia ASA 50 film that was used.) An image analysis system was used to create an image that was the ratio of these two monochromatic images. The ratio image was then scaled by a calibration curve to produce a grayscale or pseudocolor image where intensity was proportional to pH_i (LaManna et al., 1992). The calibration curve was constructed from photographic slide images of brain paste homogenates stained with neutral red in which pH was varied over a range. A nonlinear least squares fitting routine was used to estimate the constants of the following equation:

$$\mathrm{pH} = \mathrm{p}K + \log\left[\frac{\epsilon^{a}_{550\lambda} - \epsilon^{a}_{450\lambda} \cdot R}{\epsilon^{b}_{450\lambda} \cdot R + \epsilon^{b}_{550\lambda}}\right] \tag{5}$$

The constants were found to be $\epsilon^{a}_{550\lambda} = 46094$, $\epsilon^{a}_{450\lambda} = 15203$, $\epsilon^{b}_{450\lambda} = 18532$, $\epsilon^{b}_{550\lambda} = 536$; the p$K$ was 7.15. Very similar pH values, within the range of 6 to 8, can be calculated from the linear approximation (LaManna and McCracken, 1984), which we have used in the past (LaManna et al., 1992; Griffith et al., 1992) for reflection mode measurements.

In Fig. 3, the top image is a monochromatic (450 nm) image of the block face of a sagittally sectioned frozen rat brain. The image was collected through a black and white CCD camera (Javelin Ultrachip) and capture board (Data Translation) using NIH Image on a Macintosh IIci. Extraneous parts of the image were edited out. Various shades of neutral red staining can be seen in the image, with little or no staining of white matter. Anatomical structures are readily apparent.

The lower image in Fig. 3 is the processed pseudocolor (grayscale) pH_i image. The figure demonstrates a degree of pH_i heterogeneity that is probably physiologically significant. The image has been clipped to show only the locations with valid signal (i.e., optical densities above 0.05 A units in the monochrome images) and where the pH is between 6.5 and 7.5. All other image pixels are set to black. Within the analysis program, regions of interest can be identified and the frequency histogram of the pH_i of each pixel in the region of interest recorded (LaManna et al., 1992). The pH_i observations reported below were derived from such analyses and are presented as the mean for the region of interest, plus or minus the

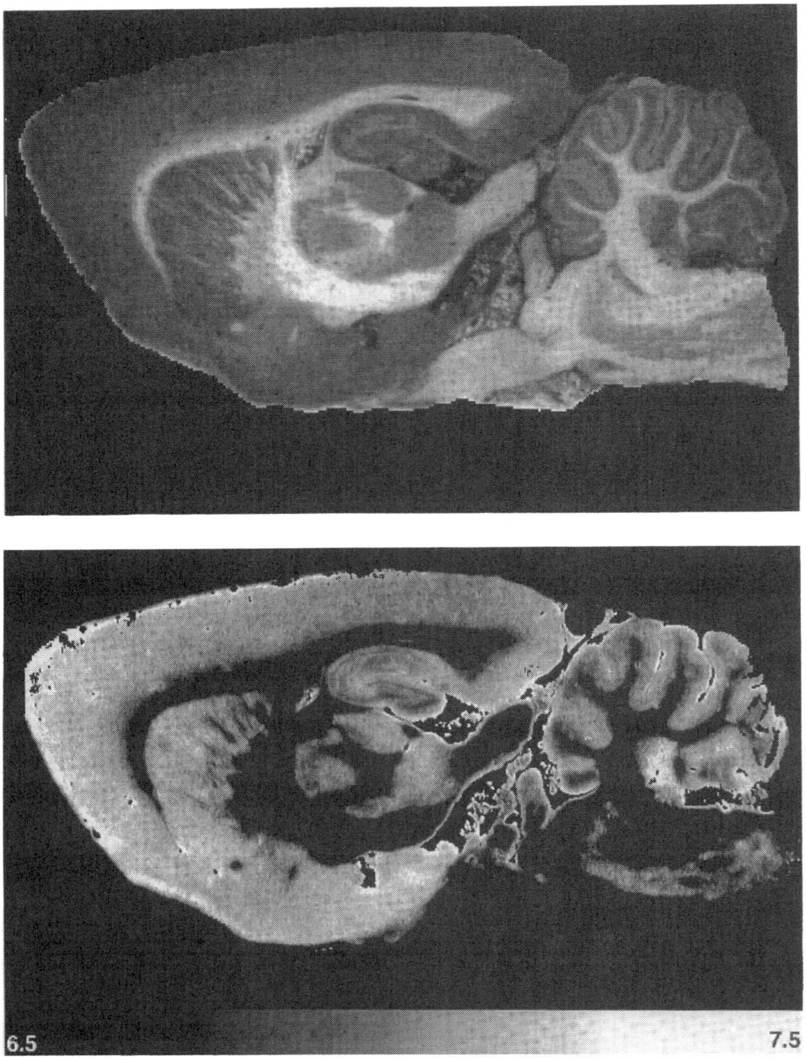

FIGURE 3. *Top:* Digitized monochromatic (450 nm) image of frozen rat brain treated with neutral red. The block face was sectioned sagittally. *Bottom:* Grayscale, pH_i image corresponding to top image. Lighter grays indicate alkaline pH_i, darker grays indicate acid. Calibration is shown on the continuous tone color bar at the bottom of the figure.

standard deviation of the distribution. The mean pH_i of the frontal cortex was 7.16 ± 0.09. The corresponding value for the parietal cortex was 7.01 ± 0.03, and for the occipital cortex was 7.08 ± 0.19. The mean pH_i in the striatum was 7.11 ± 0.04. The mean hippocampal pH_i was 7.01 ± 0.15. There appeared to be fairly uniform distribution of pH_i within the cerebral cortex, with no obvious microregional structure at this level of spatial resolution. This was not true in other, subcortical brain regions such as the hippocampus, cerebellum, and brainstem, where there were clear indications of spatial heterogeneity.

A series of sagittal sections was used to determine the pH_i of various specific nuclei in the brainstem of control rats. Representative values are shown in the table below. Regional changes in brainstem pH_i and energy metabolites with moderate hypoxia have been reported (LaManna et al., 1996).

Brainstem Region	pHi (mean ± SEM)
Medial vestibular nucleus	7.23 ± 0.08
Nucleus tractus solitarius	7.17 ± 0.08
Parapyramidal nucleus	7.12 ± 0.07
Nucleus ambiguus	7.03 ± 0.08
Spinal trigeminal nucleus	6.99 ± 0.10

A higher magnification image of the hippocampus is shown in Fig. 4. The top image is the thionin-stained section of the hippocampus shown in the grayscale pH_i image at the bottom. The stratum oriens (pH_i=7.01 ± 0.11) and stratum moleculare (pH_i=7.05 ± 0.07) are usually slightly more alkaline than the stratum pyramidale (pH_i=6.88 ± 0.13). The stratum radiatum is usually more heterogeneous (pH_i=6.92 ± 0.15). It appears that neuronal cell body layers, in general, are slightly more acidic than the rest of the parenchyma.

Most methods of pH_i determination report average brain tissue pH_i to be between 7.0 and 7.2 (Chesler, 1990; and Chapters 12, 14, 16). The evidence for pH_i compartmentation is more ambiguous and may depend on many uncontrolled physiological variables. Evidence from high-resolution ^{31}P NMR (Gilboe et al., 1993) suggests the possibility of multiple pH_i compartments in brain that could correspond to values we observed in the neuronal cell body regions of the hippocampus and cerebellum.

Pathophysiological applications. Where spatially heterogeneous pH_i responses to pathophysiological challenges can be expected, the neutral red methods are especially useful. Three examples of such applications are given below:

1. During complete brain ischemia, such as occurs with cardiac arrest, pH_i becomes more acidotic, and the degree of acidification is correlated with glycolytic metabolism (Chapters 5 and 28). The neutral red method demonstrated heterogeneous acidosis in the cerebral cortex during total cerebral ischemia (Crumrine et al., 1991). This heterogeneity increased with increasing acidosis (Griffith et al., 1992). Furthermore, a rapid recovery of pH_i was shown in the rat after cardiac arrest and resuscitation (LaManna et al., 1992), and after transient bilateral carotid occlusion in the gerbil (Hoffmanet al., 1994).
2. The pattern of pH_i in focal stroke is quite distinctive (Fig. 5). The central infarct becomes acidotic. The penumbra can maintain pH_i homeostasis for as much as an hour, despite elevated tissue lactate content. However, after 1.5 hours the penumbra may be as acidotic as the core (Yao et al., 1995). The contralateral and medial cerebral cortex tend to be slightly more alkaline than controls.
3. Spreading depression (SD) is another interesting brain metabolic state that can be studied with the neutral red method (Gault et al., 1994). For these studies, the

spreading depression wave was initiated in the frontal cortex, and then after some time interval the brain was frozen. Thus, the SD wave was "trapped" in the parietal cortex, such that the frontal cortex was recovering from the passage of the wave and the occipital cortex acted as control tissue where the SD wave had not yet reached. The rate of travel of the electrical SD wavefront was about 3.0 mm/min.

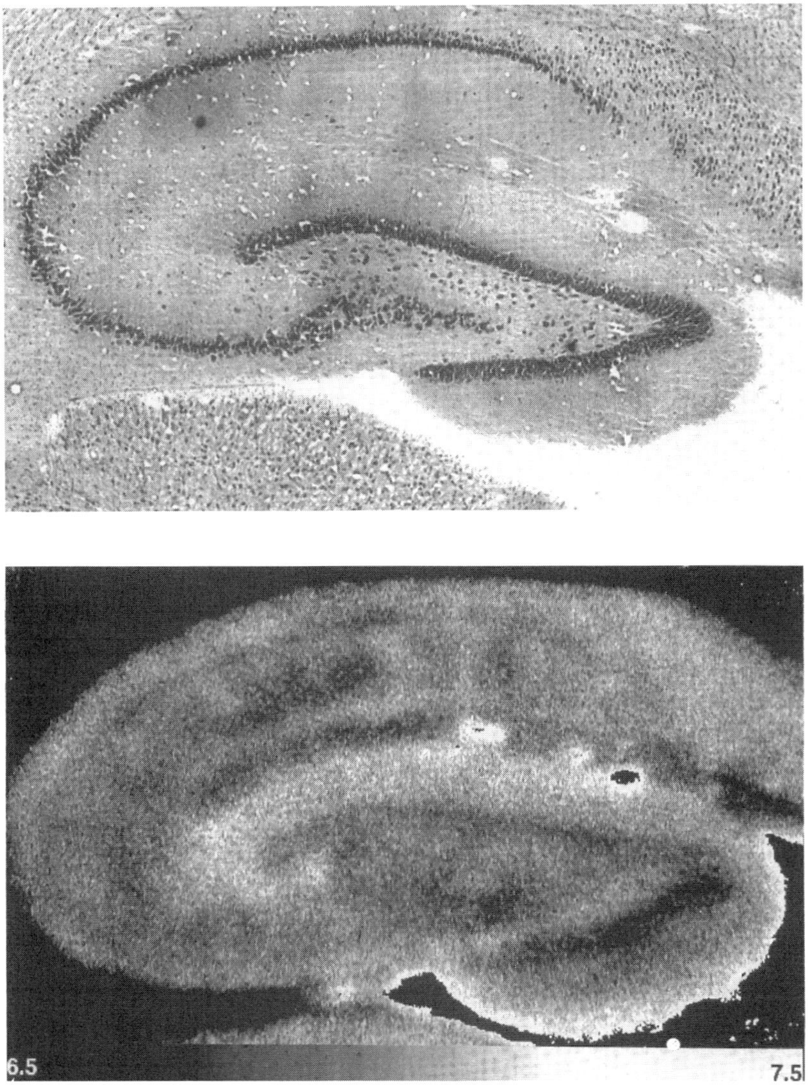

FIGURE 4. *Top:* Thionin-stained 20-μm section of the hippocampus from a frozen rat brain stained with neutral red. *Bottom:* Grayscale pH_i image of the blockface from which the top section was cut. Lighter grays are alkaline and darker grays acidic. Calibration is shown on the continuous tone color bar at the bottom of the figure. Neuronal cell body layers appear more acidic than the other layers.

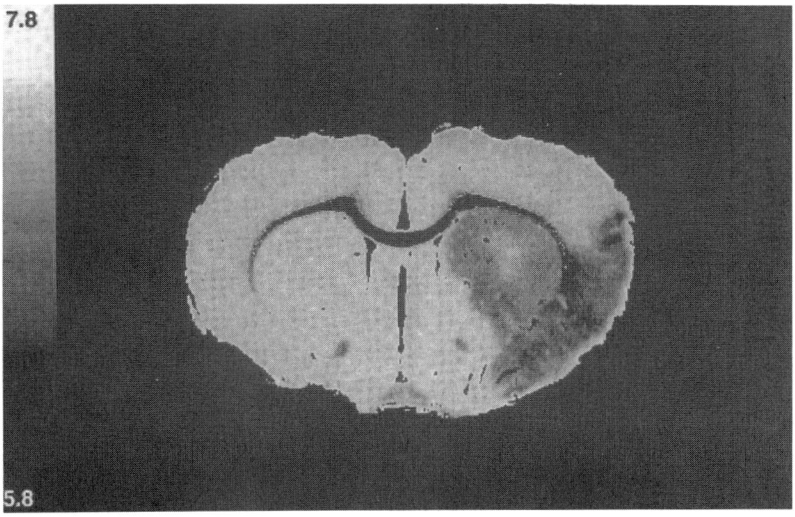

FIGURE 5. Grayscale pH_i image of a frozen rat brain stained with neutral red. The brain is from a hyperglycemic rat 1 hour after permanent middle cerebral artery occlusion. Darker grays in the core of the ischemic zone indicate regions of significant acidosis. Calibration is shown on the continuous tone color bar at the bottom of the figure.

Figure 6 demonstrates the results obtained from a rat brain frozen during the passage of a spreading depression wave. The pH_i grayscale image is shown at the bottom. The track for the needle used to microinject concentrated KCl to initiate spreading depression is clearly visible in the frontal cortex. The top trace is a plot of the electrical signal recorded from a parietal cortical electrode that has been scaled and temporally inverted so that the horizontal dimension is linear distance in millimeters. The plot was placed over the cortex where it was estimated that the SD electrical wave was trapped at the time of freezing. The middle tracing is a column average of pH_i pixel values through the cerebral cortex and shows an initial alkalinization at the wavefront, followed by a transient acidification. The three images are aligned vertically. The alkalinization at the wavefront may reflect the response of astrocytes, which are known to undergo transient alkalinization during activation such as that occurring in spreading depression (Chesler and Kraig, 1987, 1989).

SUMMARY AND CONCLUSIONS

Neutral red can be used as a quantitative pH indicator with potential for excellent temporal and spatial resolution. Application of the method is still rudimentary; there is much basic information still to be obtained and much development necessary. Nevertheless, methods that use absorption dyes, and specifically neutral red, to measure brain pH are clearly capable of providing insight into physiological and pathophysiological mechanisms.

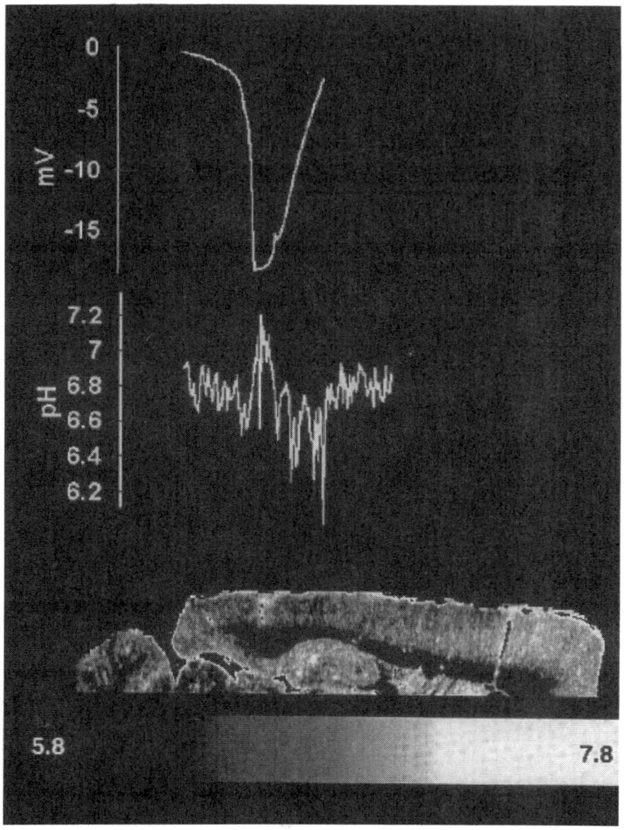

FIGURE 6. Spreading depression. At the bottom of the figure, just above the continuous tone calibration bar, is a grayscale pH_i image of a sagittal section through the cortex and part of the cerebellum (on the left) of a frozen rat brain. The rat brain was frozen during propagation of a spreading depression wave that was initiated in the frontal cortex (observe the needle track). The middle trace represents the column (vertical) average of the pH_i of each pixel in the cortex directly below. A zone of alkalinization is apparent at the point in the cortex where it is presumed that the peak of the electrical depolarization occurs. (The upper trace is the time-reversed, distance-scaled electrical response from a microelectrode in the parietal cortex). The alkaline transient was followed by an transitory acidification

ACKNOWLEDGMENTS

The studies reviewed in this chapter were supported by NIH grants NS22077, NS22571, HL25830, and HL42215. Figure 2 was based on the data originally presented as an abstract (LaManna et al., 1987). Figure 5 is from a study originally reported as an abstract (Griffith et al., 1991). The brain section used for Fig. 6 comes from the study by Gault et al., (1994). S. Foss was instrumental in manuscript preparation.

REFERENCES

Ahmed Z, Connor JA (1980): Intracellular pH changes induced by calcium influx during electrical activity in molluscan neurons. J Gen Physiol 75: 403–426.

Amato A, Ballerini L, Attwell D (1994): Intracellular pH changes produced by glutamate uptake in rat hippocampal slices. J Neurophysiol 72: 1686–1696.

Andrew RD, MacVicar BA (1994): Imaging cell volume changes and neuronal excitation in the hippocampal slice. Neuroscience 62: 371–383.

Barr R, Safranski K, Sun IL, Crane FL, Morre DJ (1984): An electrogenic proton pump associated with the golgi apparatus of mouse liver driven by NADH and ATP. J Biol Chem 259: 14064–14067.

Baylor SM, Hollingworth S (1990): Absorbance signals from resting frog skeletal muscle fibers injected with the pH indicator dye, phenol red. J Gen Physiol 96: 449–471.

Brooks KJ, Kauppinen RA, Williams SR, Bachelard HS, Bates TE, Gadian DG (1989): Ammonia causes a drop in intracellular pH in metabolizing cortical brain slices. A [^{31}P]-and [^{1}H]nuclear magnetic resonance study. Neuroscience 33: 185–192.

Caldwell PC (1956): Intracellular pH. Int Rev Cytol 5: 229–277.

Chesler M (1990): The regulation and modulation of pH in the nervous system. Prog Neurobiol 34: 401–427.

Chesler M, Kraig RP (1987): Intracellular pH of astrocytes increases rapidly with cortical stimulation. Am J Physiol 253: R666–R670.

Chesler M, Kraig RP (1989): Intracellular pH transients of mammalian astrocytes. J Neurosci 9: 2011–2019.

Crumrine RC, LaManna JC, Lust WD (1991): Regional changes in intracellular pH determined by neutral red histophotometry and high energy metabolites during cardiac arrest and following resuscitation in the rat. Metab Brain Dis 6: 145–155.

Fujiwara N, Abe T, Endoh H, Warashina A, Shimoji K (1992): Changes in intracellular pH of mouse hippocampal slices responding to hypoxia and/or glucose depletion. Brain Res 572: 335–339.

Gault LM, Lin C-W, LaManna JC, Lust WD (1994): Changes in energy metabolites, cGMP, and intracellular pH during cortical spreading depression. Brain Res 641: 176–180.

Gilboe DD, Kintner D, Anderson ME, Fitzpatrick JH, Emoto SE, Markley JL (1993): Inorganic phosphate compartmentation in the normal isolated canine brain. J Neurochem 60: 2192–2203.

Grichtchenko II, Chesler M (1994): Depolarization-induced alkalinization of astrocytes in gliotic hippocampal slices. Neuroscience 62: 1071–1078.

Griffith JK, LaManna JC, Crumrine RC, Lust WD, Selman WR (1991): Rapid loss of metabolic viability in the penumbra following MCA occlusion in spontaneously hypertensive rats (SHR). J Cereb Blood Flow Metab 11 (Suppl): S549 (Abstract).

Griffith JK, Cordisco BR, Lin C-W, LaManna JC (1992): Distribution of intracellular pH in the rat brain cortex after global ischemia as measured by color film histophotometry of neutral red. Brain Res 573: 1–7.

Grinvald A, Frostig RD, Lieke E, Hildesheim R (1988): Optical imaging of neuronal activity. Physiol Rev 68: 1285–1366.

Hertz L, Schousboe A, Weiss GB (1970): Estimation of ionic concentrations and intracellular pH in slices from different areas of rat brain. Acta Physiol Scand 79: 506–515.

Hoffman TL, LaManna JC, Pundik S, Selman WR, Whittingham TS, Ratcheson RA, Lust WD (1994): Early reversal of acidosis is a first step to metabolic recovery following ischemia. J Neurosurg 81: 567–573.

Hollingworth S, Baylor SM (1990): Changes in phenol red absorbance in response to electrical stimulation of frog skeletal muscle fibers. J Gen Physiol 96: 473–491.

Koehring V (1930): The neutral red reaction. J Morphol Physiol 49: 45–137.

Kogure K, Alonso OF, Martinez E (1980): A topographic measurement of brain pH. Brain Res 195: 95–109.

Kreisman NR, LaManna JC, Liao S-C, Yeh ER, Alcala JR (1995): Light transmittance as an index of cell volume in hippocampal slices: Optical differences of interfaced and submerged positions. Brain Res 693: 179–186.

LaManna JC (1987): Intracellular pH determination by absorption spectrophotometry of neutral red. Metab Brain Dis 2: 167–182.

LaManna JC (1992): Determination of intracellular pH in tissues by quantitative spectrophotometry of neutral red. In Frank K, Kessler M (eds): Quantitative Spectroscopy in Tissue. Frankfurt: pmi Verlagsgruppe GmbH, pp 84–97.

LaManna JC, Farrell ER (1989): Intracellular pH in brain slices from rat cerebral cortex, cerebellar cortex and hippocampus. Proceedings of the International Congress on Physiological Science 17: 370 (Abstract).

LaManna JC, McCracken KA (1984): The use of neutral red as an intracellular pH indicator in rat brain cortex *in vivo*. Anal Biochem 142: 117–125.

LaManna JC, Saive J-J, Macdonald VW, Jöbsis FF (1978): Simultaneous monitoring by optical techniques of respiratory chain and intracellular pH in toad ventricle strips. Experientia 34: 203–204.

LaManna JC, Saive J-J, Snow TR (1980): The relative time course of early changes in mitochondrial function and intracellular pH during hypoxia in the isolated toad ventricle strip. Circ Res 46: 755–763.

LaManna JC, McCracken KA, Whittingham TS, Lust WD (1986): Determination of intracellular pH by color film histophotometry of frozen *in situ* rat brain. In Longmuir IS (ed): Oxygen Transport to Tissue, VIII; Advances in Experimental Medicine and Biology, vol. 200. New York: Plenum, pp 253–259.

LaManna JC, Assaf H, Sick TJ, Whittingham TS (1987): Amiloride reversal of alkaline intracellular pH in hippocampal slices. Soc Neurosci Abstr 13: 126 (Abstract).

LaManna JC, Griffith JK, Cordisco BR, Lin C-W, Lust WD (1992): Intracellular pH in rat brain *in vivo* and in brain slices. Can J Physiol Pharmacol 70: S269–S277.

LaManna JC, Haxhiu MA, Kutina-Nelson KL, Pundik S, Erokwu B, Yeh ER, Lust WD, Cherniack NS (1996): Decreased energy metabolism in brainstem during central respiratory depression in response to hypoxia. J Appl Physiol: 81: 1772–1777, 1996.

Lin C-W, LaManna JC, Takefuji Y (1992): Quantitative measurement of two-component pH-sensitive colorimetric spectra using multilayer neural networks. Biol Cybern 67: 303–308.

Lin C-W, Whittingham TS, LaManna JC (1993a): Intracellular pH regulation mechanism of hippocampal brain slices-optical measurement by neutral red. Biomedical Engineering, Application, Basis, Communication 5: 759–772.

Lin C-W, Whittingham TS, LaManna JC (1993b): The design and development of a ratio spectral and ratio imaging system for measuring intracellular pH from brain tissues. Biomedical Engineering, Application, Basis, Communication 5: 613–621.

Lin C-W, Kalaria RN, Kroon SN, Bae J-Y, Sayre LM, LaManna JC (1996): The amiloride-sensitive Na^+/H^+ exchange antiporter and control of intracellular pH in hippocampal brain slices. Brain Res: 731: 108–113, 1996.

Lutaya G, Rahmin ZHA, Shuttlewood RJ, Bashford CL, Griffiths JR (1981): Alkalinization of phosphorylase kinase-deficient muscle during tetanic contraction. Biosci Rep 1: 177–182.

Macdonald VW, Jöbsis FF (1976): Spectrophotometric studies on the pH of frog skeletal muscle. J Gen Physiol 68: 179–195.

Macdonald VW, Keizer JH, Jöbsis FF (1977): Spectrophotometric measurements of metabolically induced pH changes in frog skeletal muscle. Arch Biochem Biophys 184: 423–430.

McIlwain H, Grinyer I (1950): Concentration or exclusion of some heterocyclic compounds by brain cortex in relation to their effects on carbohydrate metabolism. Biochem J 46: 620–628.

Pontén U, Ratcheson RA, Salford LG, Siesjö BK (1973): Optimal freezing conditions for cerebral metabolites in rats. J Neurochem 21: 1127–1138.

Prince RC, Linkletter SJG, Dutton PL (1981): The thermodynamic properties of some commonly used oxidation-reduction mediators, inhibitors and dyes, as determined by polarography. Biochim Biophys Acta 635: 132–148.

Roos A, Boron WF (1981): Intracellular pH. Physiol Rev 61: 296–434.

Schmidtmann M (1925): Über die intracellulaire Wasserstoffionenkonzentration unter physiologischen und einigen pathologischen Bedingungen. Zeitschrift für die Gesamte Experimentelle Medizin 45: 714–742.

Sick TJ, LaManna JC (1995): Intrinsic optical properties of brain slices: Useful indices of electrophysiology and metabolism. In Schurr A, Rigor BM (eds): Brain Slices in Basic and Clinical Research. Boca Raton, FL: CRC Press. pp 47–63.

Sick TJ, LaManna JC, Hollinden G, Rosenthal M (1987): Optical studies of metabolism and intracellular ion homeostasis in the hippocampal slice. In Schurr A, Teyler TJ, Tseng MT (eds): Brain Slices: Fundamentals, Applications and Implications. Basel: Karger, pp 70–87.

Sick TJ, Whittingham TS, LaManna JC (1989): Determination of intracellular pH in the in vitro hippocampal slice preparation by transillumination spectrophotometry of neutral red. J Neurosci Methods 27: 25–34.

Snow TR (1986): A study of the effects of substrates on intracellular pH in toad ventricular strips. J Mol Cell Cardiol 18: 723–732.

Snow TR, Saive J-J, Hemstreet TM (1982): A study of the temporal relation between intracellular pH and contractile performance in toad ventricle strips during hypercapnic acidosis. J Mol Cell Cardiol 14: 1–12.

Sun H, Jacquey F, Bernengo JC (1993): A micro-spectrophotometric study on the spectral properties of phenol red injected into cytoplasm for pH_i determination. Biophys Chem 48: 91–100.

Thomas JA, Buchsbaum RN, Zimmiak A, Racker E (1979): Intracellular pH measurements in Ehrlich ascites tumor cells utilizing spectrophotometric probes generated in situ. Biochemistry 18: 2210–2218.

Turner DA, Aitken PG, Somjen GG (1995): Optical mapping of translucence changes in rat hippocampal slices during hypoxia. Neurosci Lett 195: 209–213.

Van Walraven HS, Marvin HJ, Koppenaal E, Kraayenhof R (1984): Proton movements and electric potential generation in reconstituted ATPase proteoliposomes from the thermophilic cyanobacterium Synechococcus 6716. Eur J Biochem 144: 555–561.

Walz FG, Terrena B, Rokince D (1975): Equilibrium studies on neutral red-DNA binding. Biopolymers 14: 825–837.

Whittingham TS, Warman E, Assaf H, Sick TJ, LaManna JC (1989): Manipulating the intracellular environment of hippocampal slices: pH and high energy phosphates. J Neurosci Methods 28: 83–91.

Wikström M (1984): Two protons are pumped from the mitochondrial matrix per electron transferred between NADH and ubiquinone. FEBS Lett 169: 300–304.

Yao H, Ginsberg MD, Eveleth DD, LaManna JC, Watson BD, Alonso OF, Loor JY, Busto R (1995): Local cerebral glucose utilization and cytoskeletal proteolysis as indices of evolving focal ischemic injury in core and penumbra. J Cereb Blood Flow Metab 15: 398–408.

Zhang R, Kelsen SG, LaManna JC (1990): Measurement of intracellular pH in hamster diaphragm by absorption spectrophotometry. J Appl Physiol 68: 1101–1106.

CHAPTER 8

Fluorescence Indicators

MARK O. BEVENSEE AND WALTER F. BORON
Department of Cellular and Molecular Physiology, Yale University School of Medicine, New Haven, CT 06520

HISTORICAL INTRODUCTION

Of the many methods available for measuring intracellular pH (pH_i), the use of fluorescent dyes has become the most popular. Only a quarter of a century ago, those interested in measuring pH_i were limited to either (i) computing pH_i from the transmembrane distribution of weak acids or bases, or (ii) measuring pH_i directly with pH-sensitive microelectrodes. The principle underlying the *weak-acid/base* approach is that the electrically neutral form of a monoprotic weak acid or base readily equilibrates across the cell membrane, so that the intracellular concentration of the charged form depends solely on pH_i. For example, with neutral weak acids such as CO_2 and 5,5-dimethyl-2,4-oxazolidinedione (DMO), the more alkaline the cell interior, the higher the intracellular concentration of HCO_3^- or the anionic form of DMO. HCO_3^- can be assayed chemically, and DMO is available as a radioisotope. Although the weak-acid/base technique has the advantage that pH_i can be measured even in very small cells, it is generally impossible to measure pH_i in a single cell. Furthermore, because the cells are destroyed in the assay, continuous pH_i measurements are impossible.

The *microelectrode technique* for measuring pH_i exploits the pH sensitivity of specialized glasses or of "liquid membranes" containing organic molecules, as discussed in Chapter 6 of this volume. The underlying principle is that the glass or liquid membrane behaves as an H^+-selective barrier, so that the voltage difference across the barrier changes linearly with the pH difference across the barrier. In contrast to the weak-acid/base method, the use of microelectrodes allows one to monitor pH_i in single cells, and to do so continuously. The chief disadvantage of the microelectrode technique is the necessity of using cells large enough to tolerate the impalement, such as squid axons, *Xenopus* oocytes, invertebrate neurons and glial cells, and amphibian epithelial cells.

Since the advent of modern dyes, beginning with 2′,7′-biscarboxyethyl-5,6-carboxyfluorescein (BCECF) in the early 1980s (Rink et al., 1982), pH-sensitive fluorescent indicators have become the method of choice for continuously measuring pH_i in small cells. As discussed in Chapter 7 of this book, pH-sensitive absorbance indicators have also been

pH and Brain Function, Edited by Kai Kaila and Bruce R. Ransom
ISBN 0-471-11838-9

employed to measure pH_i. As described below, absorbance measurements require one to discriminate between light transmitted through a cell in the absence of intracellular dye (a large number) and in the presence of intracellular dye (a smaller number). Because the absorbance of light by the dye is proportional to the length of the light path through the cell, the absorbance signal for intracellular dye tends to be small and noisy in single cells, which are typically thin. In contrast, fluorescence signals tend to be rather easy to measure, even in single cells. Even with low intracellular dye concentrations, it is usually possible to use a photomultiplier tube (or some other device) to amplify the fluorescence signal. Moreover, the background fluorescence (i.e., autofluorescence) at the relevant wavelengths is usually rather low, so that the discrimination between dye fluorescence and autofluorescence is very good.

The first pH-sensitive dyes to be used by biologists were naturally occurring pH-sensitive dyes whose color (i.e., absorbance) changes with pH. For example, De Vries (1871) observed that intracellular alkalinization causes the pigment betacyanin in tissue slices of beets to change in color from red to yellowish brown. Furthermore, Willstätter and Mallison (1915) demonstrated that the pigments of several flowers are anthocyanin dyes that appear red in acid solutions, and blue in alkaline solutions. Nowadays, many synthetic pH-sensitive fluorescence indicators are commercially available. Some of the more commonly used indicators are listed in Table 1 and displayed in Fig. 1. Some indicators are designed to measure pH in the cytosol (pH range: 6.8–7.4), whereas others are designed to measure pH in more acid compartments or organelles (pH range: 4.5–6). The use of fluorescence indicators has become increasingly popular among physiologists and cell biologists. For example, probes are also available for measuring other ions (e.g., Na^+, K^+, Cl^-, Ca^{2+}, Mg^{2+}, Zn^{2+}) and cell membrane potential, and for probing specific cellular components (e.g., receptors, ion channels, signal transduction intermediates).

In this chapter, we will first discuss the photochemistry of fluorescence, and the factors that contribute to the pH sensitivity of a dye molecule. Then, we will consider some of the qualities that make an intracellular dye useful in physiological experiments, such as ease of loading the dye into cells. Next, we will discuss the calibration of intracellular indicators. Finally, we will describe the various techniques used for measuring fluorescence in cells.

TABLE 1 Commonly Used pH-sensitive Fluorescence Indicators

	Excitation Wavelengths (nm)		Emission Wavelengths (nm)		
Indicator	Ex_1	Ex_2	Em_1	Em_2	p*K*
BCECF[1]	439	505	535		~7.0
Carboxy-SNAFL-1[2,3]	515	550	610		~7.8
Carboxy-SNARF-1[4]	488–534		580	640	~7.5
HPTS[5]	405	450	511		~7.3

[1]2′,7′-bis-(2-carboxyethyl)-5,6-carboxyfluorescein.

[2]5′(and 6′)-carboxyseminaphthofluorescein or 5′(and 6′)-carboxy-3,10-dihydroxy-spiro[7*H*-benzo[*c*]xanthene-7,1′(3′*H*)-isobenzofuran]-3′-one.

[3]For both carboxy-SNAFL-1 and carboxy-SNARF-1, the excitation and emission wavelengths given in the table are some of the more commonly used ones. However, many different combinations of excitation and emission wavelengths can be employed when using SNAFL and SNARF indicators to measure pH (Whitaker et al., 1991). Also, these indicators can be used as both dual excitation and dual emission pH sensors.

[4]5′(and 6′)-carboxy-10-dimethylamino-3-hydroxy-spiro[7*H*-benzo[*c*]xanthene-7,1′(3′*H*)-isobenzofuran]-3′-one.

[5]8-Hydroxypyrene-1,3,6-trisulfonic acid or pyranine.

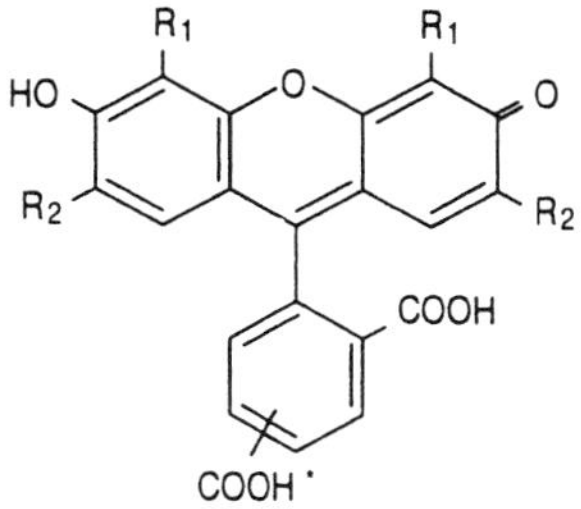

Fluorescein derivative

Indicator	R_1	R_2	pK
fluorescein (COOH* absent)	H	H	6.7
5,6-carboxyfluorescein	H	H	6.4
4',5'-dimethyl-5,6-carboxyfluorescein (DMCF)	CH_3	H	7.0
2',7'-biscarboxyethyl-5,6-carboxyfluorescein (BCECF)	H	$(CH_2)_2COOH$	7.0

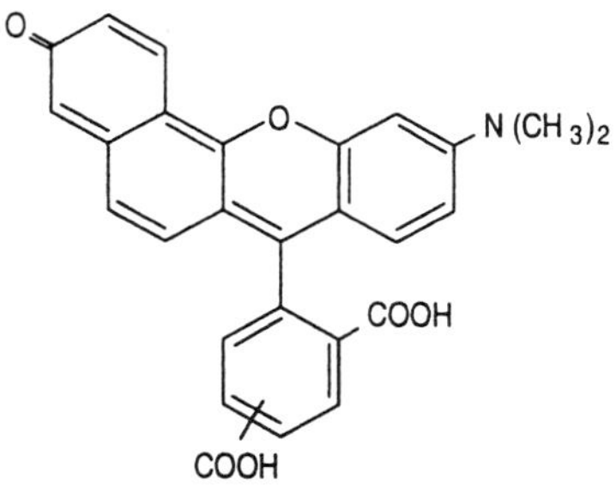

carboxy-SNARF-1
(pK=7.5)

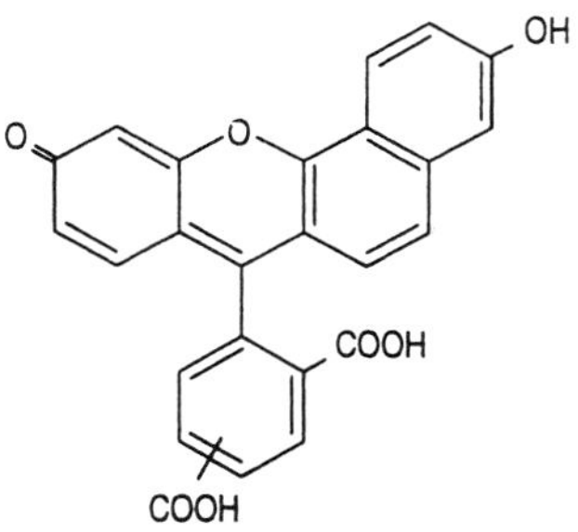

carboxy-SNAFL-1
(pK=7.8)

FIGURE 1. Some commonly used pH-sensitive fluorescent indicators. There are actually six different forms of fluorescein that exist in aqueous solution. Four of these (ionic charge: 0, − 1, − 2, − 3) represent successive protonations, with p*K* values of approximately 2.2, 4.4, and 6.7. The neutral form has two additional isomers. In the physiological pH range, the forms with ionic charges of − 2 (both carboxyls deprotonated) and − 3 (phenolic H dissociated) are most prevalent, and it is the relative fluorescence levels of these (p*K* = 6.7) that are exploited in the ratiometric determination of pH_i.

PRINCIPLES OF FLUORESCENCE

Energy Levels of Molecules

Because it is important to understand the chemistry underlying some of the experimental artifacts (e.g., bleaching and photodynamic damage) that arise with the use of pH-sensitive dyes, we will provide a brief overview of the chemistry of absorbance and fluorescence.

Photochemistry, broadly defined, includes any chemical effect associated with the absorption or emission of electromagnetic radiation. According to the Stark-Einstein law of photochemical equivalence, the key event in a photochemical reaction is the absorption of one photon, or quantum of energy, by each participating molecule. The energy of a quantum of light is

$$E = \mathrm{h}\nu$$

where h is Planck's constant and ν is the frequency of the electromagnetic radiation. The energy of the molecule absorbing the quantum of light is the sum of the electronic energy (i.e., energy of the molecule's electrons), as well as the energies of vibration and rotation:

$$E = E_{\mathrm{e}} + E_{\mathrm{vib}} + E_{\mathrm{rot}}$$

The vibrational component (E_{vib}) includes the stretching of bonds between atoms, as well as the bending of bond angles among three sequential atoms arranged in the shape of a "V". The rotational component (E_{rot}) includes tumbling and spinning of the molecule. Quantum theory permits changes in the electronic, vibrational, and rotational energies by only discrete amounts. In general, the differences in electronic energy levels (ΔE_{e}) are much greater than those of vibrational energy levels (ΔE_{vib}), which are in turn much greater than those of rotational energy levels (ΔE_{rot}):

$$\Delta E_{\mathrm{e}} >> \Delta E_{\mathrm{vib}} >> \Delta E_{\mathrm{rot}}$$

Therefore, when a quantum of light is absorbed by a molecule, the energy is partially consumed in an electronic transition, partly in a vibrational one, and partly in a rotational one. For example, if a molecule is irradiated with light in the far infrared/microwave region of the electromagnetic spectrum (low frequency/long wavelength) in which the energy input is low, only rotational energy transitions would be expected. However, if a molecule is irradiated with light in the near infrared region in which the energy input is higher, vibrational energy transitions would also be expected. Finally, if a molecule is irradiated with light in the visible and ultraviolet region, in which the energy input is the highest, electronic transitions would be expected.

The Fate of an Excited Molecule

As discussed above, if a molecule (M) in its ground state is irradiated with light of the appropriate energy (hν), the molecule will absorb a quantum of light and be raised to an excited state, M*

$$\mathrm{M} + \mathrm{h}\nu \rightarrow \mathrm{M}^*$$

This transition is depicted by path *a* in Fig. 2. In this figure, the two widely separated clusters of closely spaced horizontal lines represent different E_e values (large E transition). Within one of these clusters, the closely spaced horizontal lines represent the different E_{vib} values possible for a given electronic state (intermediate E transitions). Not shown are the different E_{rot} values, which could be represented as even more closely spaced lines (smallest E transitions). In the excited M* state, the molecule may have any of several fates, as outlined below.

Resonance radiation. M* may simply emit a single photon of the same wavelength as the one absorbed, and thereby return to its ground state (path *b*, Fig. 2):

$$M + h\nu \rightarrow M^* \rightarrow M + h\nu$$

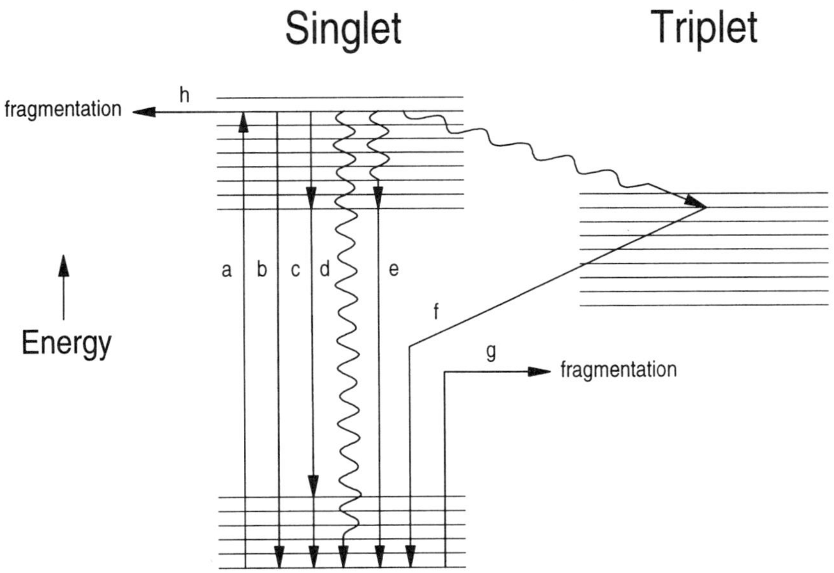

FIGURE 2. Energy transitions for a molecule absorbing a quantum of light. The two widely separated clusters of closely spaced horizontal lines represent different electronic energy levels (E_e), whereas the closely spaced horizontal lines represent the different vibrational energy (E_{vib}) for a given electronic state. The absorption of a photon elevates the electron from the ground state to an excited state (path *a*). Solid lines with downward-pointing arrowheads represent photon emissions, and wavy lines represent energy transitions resulting in heat dissipation. Once in an excited state, a molecule may return to its ground energy state by a number of routes: Emitting a photon of the same energy as the one absorbed (resonance radiation; path *b*); emitting two or more photons of lesser energy (fluorescence, path *c*); colliding with and transferring energy to other molecules in a series of small energy transitions (path *d*); emitting a photon after dissipating some energy in collisions with other molecules (path *e*); or emitting a photon after making a transition to a metastable triplet state (phosphorescence path *f*). Upon excitation, a molecule may fragment into other products, either before being elevated to the excited state (dissociation, path *g*), or after (predissociation, path *h*).

Fluorescence. In the following example, the excitation of M to M* is followed by three transitions to lower energy levels, with a photon emitted in each case:

$$M + h\nu \rightarrow M^*$$
$$M^* \rightarrow M^{*\prime} + h\nu_1$$
$$M^{*\prime} \rightarrow M^{*\prime\prime} + h\nu_2$$
$$M^{*\prime\prime} \rightarrow M + h\nu_3$$

For each transition, the emitted photon will have a different energy: A three-photon transition is shown as path *c* in Fig. 2. The emitted photon always has a greater wavelength (lower energy) than the photon that previously excited the molecule.

Collision. M* may collide with a like molecule, M, or a different molecule or atom, X. Whereas the absorption of a photon requires only $\sim 10^{-15}$ seconds, the M* has a lifetime on the order of 10^{-8} seconds. During this long lifetime of M*, random collisions cause the molecule to fall to lower vibrational states (lower energy states). These random collisions involve several scenarios. In the first scenario, M* may gradually lose all of its excess energy by a series of collisions. In each case, M* falls to a slightly lower energy level (M*') as it transmits a small amount of energy to X, which gains (↑) vibrational, rotational and/or translational energy. Ultimately, the transferred energy is manifested as heat (path *d*, Fig. 2):

$$M^* + X \rightarrow M^{*\prime} + X(\uparrow E_{vib}, E_{rot}, E_{trans}) \rightarrow\rightarrow\rightarrow M + X(\uparrow E_{vib}, E_{rot}, E_{trans})$$

This type of collision event is called *quenching* because it decreases the number of molecules in the excited M* state, and thus decreases the likelihood that M* can fluoresce. Halide ions, such as I^- and Br^- are extremely effective quenchers, and as substituent groups on a molecule, they can substantially decrease the probability of fluorescence. Quenching is to be distinguished from the *inner-filter effect,* in which molecule M, or other molecules in the solution, absorb the exciting and/or emitted light, and thus result in a decrease in measured fluorescence.

A second possibility is that collision reduces the energy of an excited molecule from M* to M*' as a prelude to fluorescence (path *e*, Fig. 2):

$$M^* + X \rightarrow M^{*\prime} + X(\uparrow E_{vib}, E_{rot}, E_{trans})$$
$$M^{*\prime} \rightarrow M + h\nu$$

The probability of emitting a photon of energy $h\nu$ is greatest at the point of minimal vibrational energy in a particular electronic state.

A third possibility is that after collision M* transfers enough energy to raise X to an excited state, from which X* can fluoresce:

$$M^* + X \rightarrow M + X^*$$
$$X^* \rightarrow X + h\nu$$

Phosphorescence. An excited molecule can have one of two states. In the excited "singlet" state of the molecule, which is far more common, the spin of the excited electron is opposite that of the remaining electron in the original orbital. However, in rare cases, an excited electron in the singlet state can make a transition to a metastable "triplet" state (M**) of lower energy. Here, the spin of the excited electron is identical to that of the remaining electron in the original orbital. The lifetime of M** is orders of magnitude longer than that of M*, often on the order of seconds. The emission of the photon from this metastable state is termed *phosphorescence* (path *f* in Fig. 2):

$$M^* \rightarrow M^{**} \rightarrow M + h\nu$$

Dissociation. Upon absorbing a photon, but before reaching an excited state, M may dissociate into two or more products, termed fragments (path *g*, Fig. 2):

$$M + h\nu \rightarrow \text{fragments}$$

This process of fragmentation, which occurs predominantly at lower energy levels, is the reason why not all molecules are capable of fluorescing.

Predissociation. M may dissociate only after being excited to M* (path *h*, Fig. 2):

$$M + h\nu \rightarrow M^* \rightarrow \text{fragments}$$

Enhanced chemical reactivity. A molecule in the M* state may often become more reactive. For example, bleaching of pH-sensitive dyes is a common problem in which the lifetime of a dye molecule becomes greatly reduced with excessive light excitation. Furthermore, bleaching of pH-sensitive dye (e.g., fluorescein) can be exacerbated at higher oxygen levels. Presumably, the interaction of O_2 with the dye is greatly accelerated by exciting the dye to the M* state.

Fluorescence Excitation and Emission Spectra

To define a fluorescence measurement fully, it is necessary to excite the molecule at a specific wavelength, and to monitor the fluorescence emission at a specific wavelength. To determine the best excitation wavelength experimentally, one can generate a *fluorescence excitation spectrum* by exciting the molecule over a range of wavelengths of light, while monitoring the fluorescence emission at a single wavelength. As an example, the excitation spectrum of BCECF free acid is shown on the left in Fig. 3. The dye was excited with light from 400 nm to 550 nm, and the emitted fluorescence was collected at 530 nm. The excitation spectrum thus measures the amount of emitted light as a function of the excitation wavelength. Likewise, to determine the best emission wavelength, one can generate a *fluorescence emission spectrum* by exciting the molecule at a single wavelength of light, and monitoring the fluorescence emission over a range of wavelengths. The emission spectrum of BCECF free acid is shown on the right in Fig. 3. The dye was excited with light at 490 nm, and the emitted fluorescence was collected from 450 nm to 650 nm. The emission spectrum thus measures the amount of emitted light as a function of the emitted wavelength. As discussed above, because emission wavelengths are greater

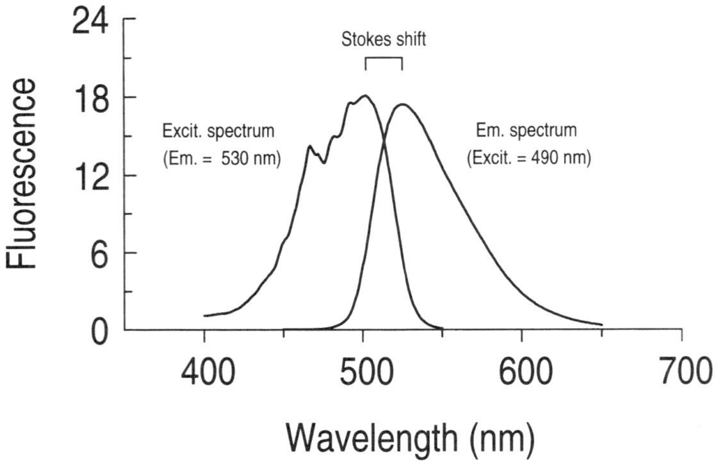

FIGURE 3. Excitation and emission spectra of 200 nM BCECF (free acid) in a HEPES-buffered solution (pH = 7.5). The spectra were obtained at room temperature in a commercial spectrofluorometer (SPEX Fluorolog-2, Model CM1T10E, SPEX Industries, Edison, NJ). The excitation spectrum on the left was obtained by exciting the dye with light from 400 nm to 550 nm (increments of 1 nm) and recording the emitted fluorescence at 530 nm. The emission spectrum on the right was obtained by exciting the dye with light at 490 nm and recording the emitted fluorescence from 450 nm to 650 nm (increments of 1 nm). The Stokes shift, the difference between the peak emission (525 nm) and the peak excitation (502 nm), is 23 nm. Fluorescence on the *y*-axis is plotted in arbitrary units.

than excitation wavelengths, an emission spectrum is always shifted towards higher wavelengths (i.e., "red-shifted") in relation to the excitation spectrum. For such spectra, the difference between the peak emission wavelength and the peak excitation wavelength is often referred to as the Stokes shift. For the BCECF spectra shown in Fig. 3, the Stokes shift is 23 nm (525 nm, peak emission − 502 nm, peak excitation).

In generating an excitation or an emission spectrum, one should be aware of two potential sources of error. First, the intensity of the excitation light from a given light source is somewhat dependent on wavelength. Second, the sensitivity of the emitted-light detector, such as a photomultiplier tube or CCD camera, is also dependent on wavelength. For precise measurements, physicists and chemists use a wavelength-independent thermopile to monitor the heat generated by incident light. For other types of light detectors, one can determine the wavelength dependence empirically using a calibrated light source.

The Chemistry of Fluorescence

Other things being equal, molecules with a greater absorbance obviously tend to have a greater fluorescence. For molecules with low absorbance, fluorescence is more dependent on the concentration of the molecule. For a beam of light shining through a cuvette containing a solution of dye, the Beer-Lambert law (discussed in greater detail in Chapter 7) describes the relationship between absorbance (A), the intensity of the incident light (I_o) and the intensity of the transmitted light (I):

$$A = \epsilon lc = -\log (I/I_o)$$

where ϵ is the molar absorptivity, l is the optical path length through the cuvette, and c is the concentration of the dye molecule. The amount of light absorbed is the difference between I_o and I:

$$\text{absorbed light} = I_o - I = I_o (1 - 10^{-\epsilon lc})$$

Because only a fraction (φ_M) of this absorbed light can be emitted as photons, fluorescence (F) can be described by the following equation:

$$F = \varphi_M I_o (1 - 10^{-\epsilon lc})$$

where φ_M is also known as the *quantum yield,* or fraction of excited dye molecules that emits photons. If ϵlc is very large, then $10^{-\epsilon lc}$ becomes negligible compared to 1, and

$$F = \varphi_M I_o$$

That is, the fluorescence at any particular wavelength is proportional to the excitation light intensity, but invariant of dye concentration. On the other hand, if ϵlc is extremely small, an expansion of the first of the above two equations yields:

$$F = \ln (10) \varphi_M I_o \epsilon lc$$

In other words, F is proportional to both the concentration of the fluorophore and the intensity of the exciting light. In typical biological experiments, in which ϵlc is neither extremely small nor extremely large, F is also proportional to both the concentration of the fluorophore and the intensity of the exciting light.

Molecules with multiple conjugated double bonds tend to absorb light at higher wavelengths (i.e., lower energies), and also to fluoresce. As noted above, quenching substituent groups such as $-$I and $-$Br, as well as $-NO_2$, $-$COOH, and azo groups reduce fluorescence. On the other hand, electron-donating groups such as $-$OH, $-NH_2$ and $-OCH_3$ tend to enhance fluorescence. For pH-sensitive indicators, such as BCECF and SNAFL (see Table 1), fluorescence depends on the level of deprotonation of the molecules. When excited at the appropriate wavelength, these dyes display greater fluorescence under more alkaline conditions (i.e., when the molecules are more deprotonated).

DESIRABLE QUALITIES IN DYES

The following characteristics of a fluorescent dye are necessary if one is to measure pH_i effectively.

1. The dye should be easy to load into the cell.
2. Once inside the cell, the dye molecules should remain in the cell.
3. Inside the cell, the molecules should be uniformly distributed, predominantly in the cytoplasm.
4. The dye molecules should absorb light and fluoresce at convenient wavelengths.
5. The dye should be relatively pH-sensitive at one wavelength, and relatively pH-insensitive (or have the opposite pH sensitivity) at another. By dividing the signal

obtained at one wavelength by that obtained at the other, one obtains a "ratiometric" measurement that is, in principle, insensitive to changes in such parameters as dye concentration and focus.

6. The pK of the dye should be reasonable.
7. Neither the dye, nor any of its degradation products, should affect the physiology of the cell.
8. The dye molecules should have a high quantum yield.
9. The dye should be as resistant as possible to photobleaching.

We shall now discuss each of these characteristics in turn.

Convenient Dye Loading

In general, a dye that readily leaks into a cell can also readily leak out. Therefore, experimenters have devised techniques to maximize the unidirectional transport of dye into cells. For example, one can microinject dextran-linked dye molecules into a cell, thus ensuring that the dye will be confined to the cytoplasm and not leak out of the cell. Several of these dextran-linked dye preparations are commercially available. However, microinjection is technically difficult, and not practical with a population of cells. Electrophysiologists performing whole-cell patch-clamp experiments have often loaded dyes useful for assessing cell viability into cells by diffusion from the patch-pipette solution. Using a similar method, Trapp et al. (1996) and Lückermann et al. (1997) have loaded the free-acid form of BCECF into rat medullary neurons to measure simultaneous changes in pH_i and membrane currents or potentials. Because this form of the dye is charged, it does not readily cross the plasma and organellar membranes, and therefore remains trapped in the cell cytosol.

The most common technique used to load cells with a pH-sensitive dye is to incubate cells in a solution containing an ester derivative of the dye. Ester derivatives of various fluorescein compounds have a low net charge, and usually move readily across cell membranes into cells. Inside the cells, native esterases cleave the ester groups from the precursor, thus releasing the dye. Having a greater net negative charge compared to its precursor, the dye remains trapped inside the cell. For many years, biochemists have used ester derivatives of dyes to stain cells supravitally, and assess viability. The charged form of the dye remains inside cells with healthy cell membranes (positive stain), whereas the dye leaks out of cells with damaged cell membranes (negative stain). Later, when Roger Tsien and his colleagues developed BCECF, probably the most widely used of the fluorescein-based pH-sensitive dyes, they used the esterification technique to load BCECF into cells (Rink et al., 1982). The acetoxymethyl esterification of BCECF is similar to that of another popular dye, Ca^{2+}-sensitive Fura-2, also developed by Roger Tsien. Many ester derivatives of indicators, particularly the high-molecular-weight ones such as Fura-2-AM, are poorly soluble in aqueous solutions. However, dispersing agents such as fetal calf serum, albumin, or mild detergents can be added to the dye-loading medium to minimize precipitation. One commonly used detergent is Pluronic F-127, a mild nonionic surfactant that sequesters dye into micelles (Kao, 1994).

Low Dye-Leakage Rate

Dyes invariably leak, to a greater or lesser extent, out of virtually any cell. Some cells transport dye out of the cell by means of specific transporters. For example, cultured thyroid cells (Gérard et al., 1990) and epithelial kidney and intestinal cells (Allen et al., 1990; Collington et al., 1992) extrude fluorescent dyes by means of a probenecid-sensitive organic anion transporter. In epithelial cells, dye transport generally occurs predominantly across the apical membrane. One can determine the extent of dye loss by monitoring the amount of intracellular dye over time, as can be done with dyes having an isosbestic point (see p. 140 under Ratiometric Quantitation). It is easy to confuse dye leakage with bleaching, inasmuch as both processes result in a gradual decrease in dye concentration. However, because the degree of bleaching depends on the aggregate number of photons absorbed by the dye, leakage and bleaching can be distinguished by comparing the time course of the dye signal with continuous versus discontinuous illumination. In other words, a dye may leak from a cell in the presence or absence of illumination. In contrast, dye bleaching in a cell will occur only in the presence of illumination.

Uniform Cytoplasmic Distribution of Dye

Thomas et al. (1979) noted that fluorescein tends to accumulate in mitochondria, as well as the cytoplasm, and to leak out of the cell rather quickly. Dimethylcarboxyfluorescein, on the other hand, is better confined to and retained in the cytoplasm. A legitimate concern is that the acetoxymethyl-ester derivatives of BCECF and other dyes might accumulate and become de-esterified in a number of cell compartments. In using the ester form of BCECF, our group has observed that BCECF seems to be concentrated in the nucleus of some cells, such as osteoclasts. However, in osteoclasts, nuclear pH is virtually identical to cytoplasmic pH (Ravesloot et al., 1995), suggesting that nuclear dye accumulation is not problematic. In what we would regard as healthy cells, we have observed a *uniform* distribution of BCECF throughout the cytoplasm (and sometimes the nucleus). However, we have observed that excessive illumination of cells, which leads to dye bleaching and photodynamic damage (see below), is often associated with punctuate accumulations of BCECF, blebbing of the cell membrane, a grainy appearance to the cell, and an accelerated rate of dye loss (Bevensee et al., 1995). Slayman et al. (1994) have observed an accumulation of BCECF in vacuoles of *Neurospora,* and have suggested that, in animal cells, BCECF may accumulate inside acidic organelles that are too small to be visualized by light microscopy. However, if a substantial fraction of intracellular BCECF ($pK \cong 7.2$) were confined to organelles with a pH below 6.0, then changes in cytoplasmic pH, even if they were reflected by equal changes in organellar pH, would have very little effect on the dye signal. Experimental dye calibrations are not consistent with this prediction. Thus, we can conclude that BCECF predominantly reports the pH of the cytoplasm. Indeed, as discussed below for leech glial cells, the pH transients measured with BCECF are similar to those measured with pH-sensitive microelectrodes, which presumably measure cytosolic pH (Nett and Deitmer, 1996).

Convenient Wavelengths

The most convenient pH-sensitive fluorescence emission wavelengths are those in the visible range, because special optics are required to allow the transmission of UV light. The pH-sensitive dyes BCECF, SNARF, and SNAFL (see Table 1), can be excited with visible light, and the fluorescence emission is also in the visible range. In principle, one should be able to measure pH and Ca^{2+} with near simultaneity by choosing the correct combination of dyes, so that the excitation or emission spectra of the dyes do not overlap substantially. For example, because the excitation spectrum for carboxy-SNAFL-1 (see Table 1) is substantially more red-shifted (peak excitations at 515 and 550 nm) than the excitation spectrum for Fura-2 (peak excitations at 340 and 380 nm), these two dyes would be ideal candidates for measuring pH and Ca^{2+} simultaneously.

Ratiometric Quantitation

For dyes such as BCECF and SNAFL, both the absorbance and fluorescence excitation spectra are pH sensitive. The pH sensitivity of a spectrum can be reflected by a change in either the magnitude or the shape of the spectrum. With magnitude changes, the complete spectrum could be scaled up or down as a function of pH. However, such a scaling is not optimal, because all wavelengths of the spectrum are scaled by the same factor. Spectral-shape changes are far more useful because a pH increase might cause the spectrum to scale up substantially at some wavelengths, to scale down at others, and to exhibit no pH sensitivity whatsoever at the isosbestic wavelength. This sort of pH sensitivity allows one to determine pH_i ratiometrically and to eliminate pH-independent factors that may contribute to spectral changes (e.g., dye loss and unexpected changes in illumination). In the case of BCECF, a dual-excitation dye, fluorescence emission rises steeply with pH when excited at ~ 500 nm, but is nearly pH insensitive when excited at 440 nm. The ratio of the two intensities (I_{500}/I_{440}) is sensitive to pH, but not to changes in other parameters, such as dye concentration. Because I_{440} is relatively insensitive to pH_i, yet sensitive to dye concentration, the rate of decrease of I_{440} ($-k_{440}$) can represent the degree of leakiness of the cell membrane to the dye. Increased leakiness during an experiment is indicative of a compromised cell membrane (Bevensee et al., 1995). For carboxy-SNARF-1, which can be used as a dual-emission dye, the shape of the emission spectrum is pH sensitive. When excited at a wavelength of 488 nm, the acid form of the dye has an emission peak at 580 nm, the basic form has an emission peak at 640 nm, and the isosbestic wavelength for the emission spectrum is 604 nm. For dual-excitation dyes such as BCECF, one must periodically capture the emitted fluorescence of the dye first with one excitation wavelength, and then with the other, before generating the ratio. However, in using dual-emission dyes such as carboxy-SNARF-1, in which there is only one excitation wavelength, one can capture the two emission wavelengths simultaneously (e.g., by using a beam splitter and two photomultiplier tubes), and thereby obtain a ratio that is less susceptible to time-dependent optical artifacts (e.g., fluctuations in light-source intensity, or movements of the cell).

Convenient p*K*

The sensitivity of the dye to pH changes (i.e., changes in fluorescence per change in pH) is greatest when pH equals p*K*. Therefore, for sensitive pH_i measurements, one should use a dye with a p*K* similar to the pH of the cytoplasm of most cells (e.g., between 7.0 and 7.3). Intracellular BCECF typically has a p*K* of 7.2–7.3, and is not very useful for distinguishing pH_i values much below 6.0 or above 8.5.

Low Toxicity

In general, pH-sensitive dyes do not appear to be toxic to cells. However, dimethylcarboxyfluorescein inhibits excitation-contraction coupling in frog skeletal muscle (Irving et al., 1989). Also, fluorescein analogues and their isothiocyanate derivatives inhibit the calcium pump in the red blood cell (Gatto and Milanick, 1993 and also Chapter 15).

High Quantum Yield

As discussed above, a dye molecule that absorbs a photon can have a number of fates. As noted above, the fraction of excited dye molecules that emits photons is termed the quantum yield (φ_M). To maximize the fluorescence signal during measurements, one should use a dye with as high a quantum yield as possible. Fluorescein dyes generally have a quantum yield that exceeds 90%. Fura-2, on the other hand, has a quantum yield of only 23% when free and 49% when bound to Ca^{2+}.

Low Rate of Bleaching

After a dye molecule has been excited by absorbing a photon, the molecule is in a high-energy, highly reactive state. As noted above, the excited molecule can simply disintegrate into any number of fragments. Such fragmentation results in *bleaching*. The damage done to the cell by the byproducts of bleaching, which include highly reactive free radicals, is termed *photodynamic damage*. In physiological conditions (i.e., in the presence of O_2), BCECF molecules bleach at the rate of one every ~ 36,000 excitations. Not surprisingly, the rate of bleaching is proportional to the illumination intensity and illumination duration (i.e., the cumulative dose of exciting photons). Thus, intense illumination for brief periods and dim illumination for prolonged periods produce similar amounts of bleaching. As discussed above, the rate of bleaching also increases at higher oxygen concentrations. The lifetime of a fluorophore is an order of magnitude greater in the absence of O_2 than in its presence. Rhodamine-based dyes appear to have a lower bleaching rate, albeit a lower quantum yield, than do fluorescein-based dyes.

INTRACELLULAR CALIBRATION

There are four major approaches available for calibrating intracellular pH sensitive dyes: *in vitro* cuvette calibration, independent pH_i measurement, the high-$[K^+]$/nigericin approach, and the null-point weak-acid/base technique.

In Vitro Cuvette Calibration

The easiest technique for calibrating an intracellular dye is to perform the calibration in K^+-based saline that mimics the composition of intracellular fluid. However, the spectral properties of a dye in a cell are probably different from those outside a cell in a cuvette. For example, the *absorbance* properties of dimethylcarboxyfluorescein (DMCF) have been studied in some detail, both *in vitro* and *in vivo* (Chaillet et al., 1985). Compared to the *in vitro* absorbance spectrum obtained at comparable pH and dye concentrations, the *in vivo* absorbance spectrum of DMCF is both broader and red-shifted. Also, the apparent p*K* of the intracellular dye is several tenths of a pH unit more alkaline than the apparent p*K* of the dye in the cuvette. The discrepancies between the dye in different environments probably reflects binding of the dye to one or more intracellular components. Indeed, the spectral properties of the *in vitro* dye in the presence of a protein such as albumin are similar to those of the *in vivo* dye. The discrepancies might also be due to differences in viscosity between the extracellular and intracellular environments. Unfortunately, for dye calibration, it is difficult to predict the best *in vitro* environment that mimics the *in vivo* environment.

Independent pH_i Measurement

A more satisfying approach for calibrating intracellular dye is to use a second method to obtain an independent estimate of pH_i in the same cell, or at least in a comparable cell. pH_i can be estimated by nonfluorometric techniques, such as microelectrodes, weak acid/base distributions, and NMR. However, the technique provides only a rough calibration, and the experiments need to be performed under comparable conditions of temperature and buffer. Ideally, the optical and independent method could be performed on the same cell, or on a randomly selected cell from the same population. For example, microelectrodes have been used to verify the absorbance calibration of DMCF in salamander proximal tubule cells (Chaillet and Boron, 1985), and the fluorescence calibration of BCECF in leech glial cells (Nett and Deitmer, 1996), both in the steady state and during various pH_i transients.

The High-[K^+]/Nigericin Approach

By far the most common method for calibrating intracellular dyes is to expose cells to a solution containing nigericin and K^+ as the predominant cation (Thomas et al., 1979). Nigericin is an ionophore that exchanges H^+ on one side of the membrane for K^+, and Na^+ to a lesser extent, on the other side. In the absence of Na^+ from the calibration cocktail, nigericin becomes a *de facto* K^+/H^+ exchanger and the transmembrane chemical K^+ gradient ($\Delta\mu_K$) approaches the transmembrane chemical H^+ gradient ($\Delta\mu_H$):

$$\Delta\mu_K = RT \ln\left(\frac{[K^+]_i}{[K^+]_0}\right)$$

$$\Delta\mu_H = RT \ln\left(\frac{[H^+]_i}{[H^+]_0}\right)$$

Where R is the universal gas constant, and T is temperature. Thus, if $\Delta\mu_K = \Delta\mu_H$, at equilibrium

$$\frac{[H^+]_i}{[H^+]_0} = \frac{[K^+]_i}{[K^+]_0}$$

In other words, if $[K^+]_0 = [K^+]_i$, then it follows that pH_i is the same as pH_0. In order for the approach to be successful, the effect of nigericin on pH_i must dominate the ability of the cell to regulate pH_i. There are at least three potential problems in using the nigericin approach to calibrate dyes accurately.

First, the accuracy of the high-K^+/nigericin technique requires that the extracellular solution contains a K^+ concentration similar to that inside the cell. The greater the difference between $[K^+]_0$ and $[K^+]_i$, the greater the error in the calibration. Second, even if $[K^+]_0 = [K^+]_i$, the nigericin approach could yield a faulty calibration depending on the pH_i of the cell when exposed to the nigericin solution. For example, if a cell with a pH_i of 6.0 and a $[K^+]_i$ of 105 mM is exposed to a nigericin solution of pH 7.0 and $[K^+]_0$ of 105 mM, then $[K^+]_i = [K^+]_0$, but $[H^+]_0 \ll [H^+]_i$. In this case, nigericin will catalyze the exchange of intracellular H^+ for extracellular K^+, causing $[K^+]_i$, to rise, such that $[K^+]_i$ will never equal $[K^+]_0$. How much $[K^+]_i$ rises depends on (i) other pathways for K^+ transport and (ii) the intracellular buffering power of the cell (β). Cellular mechanisms such as K^+ channels and transporters (e.g., $Na^+/K^+/Cl^-$ cotransporter) may help establish a constant $[K^+]_i$, even with nigericin transporting K^+ into the cell. The higher the β, the more K^+ and H^+ need to exchange before pH_i rises to 7.0, that is, before the nigericin system reaches equilibrium. In theory, if β were high enough, $[K^+]_i$ would continually increase, and the cell would swell and eventually burst. A third potential problem with using the nigericin approach for calibrating intracellular dyes pertains to experiments on cells deep within a tissue (e.g., in a brain slice). Because of unstirred layers, the extracellular pH and $[K^+]$ near the cells deep within the tissue (pH_{deep} and $[K^+]_{deep}$, respectively) may not equal the pH and $[K^+]$ of the bulk solution. In such cases, unless one measures pH_{deep} and $[K^+]_{deep}$, it will not be possible to calibrate an intracellular pH-sensitive dye reliably using the nigericin technique.

In Fig. 4A, we show the results of an experiment in which BCECF in a single neuron acutely dissociated (i.e., isolated on the same day of the experiment) from the CA1 region of the rat hippocampus is calibrated using the high-$[K^+]$/nigericin technique. I_{490}/I_{440} decreased (path *abc*) when the neuron was switched from a HEPES-buffered solution ($pH = 7.4$) to one containing 105 mM K^+ and 10 μM nigericin ($pH = 7.0$). I_{490}/I_{440} changed in a predictable manner when the neuron was then exposed to high-$[K^+]$/nigericin solutions of different pH. The results from 14 other experiments are shown in Fig. 4B, in which the I_{490}/I_{440} values from each experiment are normalized to unity at pH 7.0. The best-fit curve to the data can be described by a modified form of the Michaelis-Menten equation.

The Null-Point Weak-Acid/Base Technique

When a cell is exposed to a weak base such as NH_3, the influx of NH_3 leads to a pH_i increase, the magnitude of which increases in a predictable way at progressively lower initial pH_i values. Conversely, when a cell is exposed to a weak acid such as CO_2, the influx of CO_2 elicits a pH_i decrease, the magnitude of which increases in a predictable way at progressively higher initial pH_i values. Thus, it is possible to choose a combination of weak acid and weak base concentrations, such that the simultaneous application of the acid and base causes no net ΔpH_i. Under these conditions, one can ascertain the initial pH_i (Szatkowski and Thomas, 1986; Eisner et al., 1989). This null-point approach may prove valuable for validating the high-$[K^+]$/nigericin technique. For example, in rabbit vascular smooth muscle cells, the steady-state pH_i calculated by the high-$[K^+]$/nigericin technique is as much as ~ 0.30 higher than that given by the null-point technique (Boyarsky et al., 1996, 1997a,b). The null-point technique is cumbersome because one needs a number of solutions containing different combinations of the weak acid/weak base to acquire more than one calibration point per experiment. Potential weaknesses of

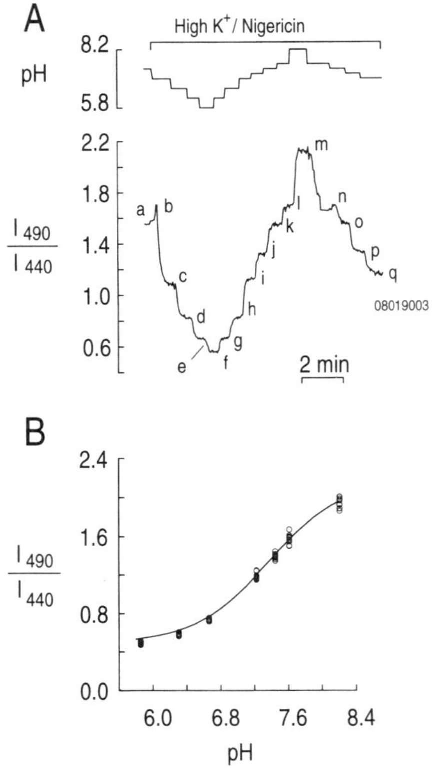

FIGURE 4. The high-[K^+]/nigericin technique for calibrating the pH-sensitive dye BCECF in single neurons acutely isolated from the CA1 region of the rat hippocampus. **A.** When a neuron was switched from a HEPES-buffered solution (pH = 7.4), to a Na^+-free solution containing 105 mM K^+ and 10 μM nigericin (pH = 7.0), I_{490}/I_{440} decreased (*abc*). I_{490}/I_{440} changed in a predictable fashion when the neuron was subsequently exposed to nigericin-containing solutions of the following pHs: 6.6 (*cd*), 6.2 (*de*), 5.8 (*ef*), 6.2 (*fg*), 6.6 (*gh*), 7.0 (*hi*), 7.2 (*ij*), 7.4 (*jk*), 7.6 (*kl*), 8.2 (*lm*), 7.6 (*mn*), 7.4 (*no*), 7.2 (*op*), 7.0 (*pq*). **B.** The pH-dependence of I_{490}/I_{440} from 14 other experiments similar to that in Fig. 4A. The data from each experiment were normalized such that I_{490}/I_{440} = 1.0 at pH = 7.0. The best-fit curve describing the data was obtained using a modified form of the Michaelis-Menten equation (G. Boyarsky, M. O. Bevensee, and W. F. Boron, unpublished).

the null-point technique are that one must know the p*K* values for the weak acid and base, and must assume that neither the weak acid (or its conjugate weak base) nor the weak base (or its conjugate weak acid) are transported or metabolized by the cell.

TECHNIQUES FOR MEASURING FLUORESCENCE IN CELLS

Introduction

The signal-to-noise ratio (*S*/*N*) of a fluorescence signal can be reduced by any of a number of interfering factors, such as mechanical vibrations of the cell due to solution flow and fluctuations in the intensity of the light source. However, if such factors are minimized, the S/N becomes limited by the number of photons that can be detected. Under these conditions,

$$S/N \propto \sqrt{F}$$

where F is the intensity of the emitted fluorescent light (photons/s). For example, if the number of photons per second increases by a factor of 100, the S/N will improve by a factor of 10. Obviously, the S/N of a group of fluorescing cells will be higher than the S/N of any single cell in the group. Moreover, for either a single cell, or a population of cells, the S/N will be greater the longer photons are collected. Thus, there is a square-root trade-off between S/N (i.e., pH resolution), spatial resolution, and temporal resolution: For every four-fold degradation in spatial or temporal resolution, there is a two-fold improvement in S/N.

Fluorescence Measurements in a Population of Cells

The simplest, and most efficient approach for monitoring pH_i in a population of cells attached to a coverslip is to use a commercial, macroscopic spectrofluorometer. In such a device, the excitation light beam typically has an area of several square millimeters and thus illuminates a large number of cells. If one wishes to execute a dual-excitation protocol using a machine with only a single excitation monochromator, then this monochromator must be periodically shifted from the pH-sensitive wavelength to another wavelength (e.g., the isobestic wavelength) to confirm the stability of the dye concentration (Ganz et al., 1988, 1989). This shifting of wavelengths can take several seconds, greatly compromising the time resolution. With a dual-monochromator spectrofluorometer, on the other hand, one can rapidly alternate between two excitation wavelengths (Gupta et al., 1994; Kaplan and Boron, 1994; Martínez-Zaguilán et al., 1994; Sánchez-Armass et al., 1994; Shrode and Putnam, 1994; Nett and Deitmer, 1996). This macroscopic approach, using a dual-beam machine, is technically easy, and is probably the method of choice for monitoring pH_i in a large number of functionally identical and readily available cells. However, it should be noted that, even with a supposedly uniform cell population, the cells are unlikely to be truly identical. For example, they are unlikely to be in the same position in the cell cycle. Moreover, if one were interested in monitoring something analogous to $[Ca^{2+}]_i$ oscillations, this would only be possible using a single-cell approach.

Fluorescence Measurements in Single Cells

For a particular cellular preparation, it may become necessary to monitor pH_i in single cells, especially if the number of available cells is limited, or if the cells are functionally heterogeneous. Such pH_i measurements can be achieved with the use of either an upright or inverted microscope, with the excitation light passing through the objective (epi-illumination). In the case of an upright microscope, one would wish to use a water-immersion objective to avoid the optical noise created by an air-water interface. For a dual-excitation dye such as BCECF, the two excitation wavelengths can be generated by first splitting the light into short and long wavelengths using a dichroic mirror, and then passing the two paths of light through band-pass filters of the appropriate wavelengths. Alternatively, one could position a filter wheel, which houses band-pass filters, in front of the light source, and periodically "chop" between the appropriate wavelengths. Figure 5 is a schematic of

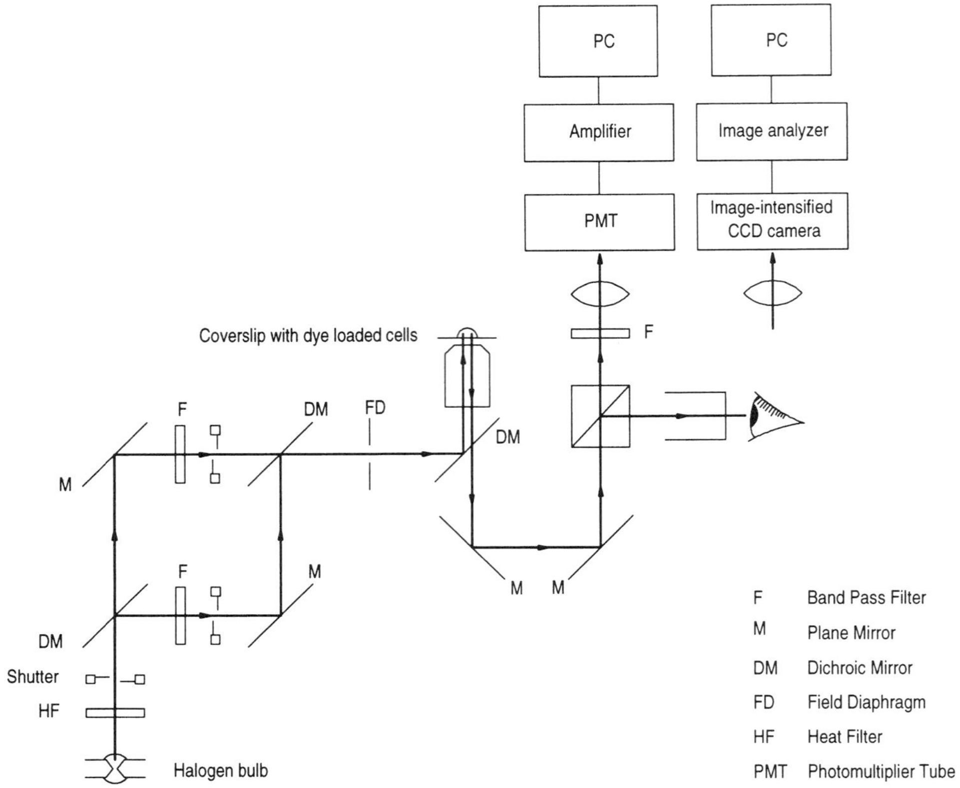

FIGURE 5. A schematic of a single-cell fluorescence apparatus. In two-dimensional fluorescence imaging, the photomultiplier tube (PMT) and amplifier are replaced by an image-intensified television (CCD) camera and image analyzer.

a single-cell fluorescence apparatus used in our laboratory. A small region of a cell (e.g., 10 μm diameter) is excited with light of the appropriate wavelength(s), and the emitted fluorescent light is captured by a photomultiplier tube (PMT). Two major problems, stray light and bleaching and/or photodynamic damage, can be minimized by enclosing the apparatus, and minimizing the period of illumination, respectively. The fluorescence signal may be optimized by programming the amplifier to adjust its gain according to the magnitude of the expected signal at each excitation wavelength. This fluorescence technique has been used with freshly dissociated hippocampal neurons (Schwiening and Boron, 1994; Bevensee et al., 1995, 1996), cultured neurons (Gaillard and Dupont, 1990; Raley-Susman et al., 1991; Nedergaard and Goldman, 1993; Wang et al., 1994; Ou Yang et al., 1995; Amos and Richards, 1996; Amos et al., 1996), cultured astrocytes (Bouvier et al., 1992; Boyarsky et al., 1993; Bevensee et al., 1997a,b), neurons in slices of the brainstem (Trapp et al., 1996; Lückermann et al., 1997) and of the hippocampus (Trapp et al., 1997), and macrophages (Kapus et al., 1993), as well as with isolated rabbit proximal tubules (Geibel et al., 1989).

For monitoring pH_i in multiple cells in a population, or for monitoring apparent pH_i values in different regions of a single cell, two-dimensional fluorescence imaging is the method of choice (Spring, 1985; Paradiso and Machen, 1987). The approach is similar to that used in single-cell fluorescence experiments, except that the photodetector, rather than being a PMT, is some sort of highly sensitive imaging device, such as an image-intensified television camera (Fig. 5). The fluorescent images can be digitally analyzed using a digital image processor. Conventional digital imaging-processing equipment is used to determine light intensities at a video frame rate (~30 frames/s), at typical spatial resolutions of 512×512 pixels (i.e., picture elements) or 1024×1024 pixels. More rapidly responding or more sensitive systems, not based on commercial video systems, are also available, although these may be extremely expensive. Two-dimensional fluorescence imaging of pH_i has been used with cultured hippocampal neurons (Hartley and Dubinsky, 1993; Baxter and Church, 1996), cultured hippocampal astrocytes (Pappas and Ransom, 1993, 1994; Rose and Ransom, 1996), acutely isolated hippocampal neurons (Pasternack et al., 1993), neurons in brainstem slices (Ritucci et al., 1996, 1997), single osteoclasts (Nordström et al., 1995; Ravesloot et al., 1995), the parietal and chief cells in a single isolated perfused gastric gland (Waisbren et al., 1994a,b), and the cells of isolated perfused colonic crypts (Singh et al., 1995).

Even though it is now relatively straightforward to use digital imaging systems to monitor pH_i in single cells, there are serious and insurmountable technical limitations to the useful spatial resolution that one can obtain. Because the limiting *S/N* is proportional to the square root of the number of photons (see above), the *S/N* for a single pixel is extremely low. For example, if one collects 10^6 photons from an entire cell in a sample period, the *S/N* can be as high as 1000:1. If, on the other hand, one images this cell on a 512×512 array (i.e., ~262,000 pixels), the number of photons counted per pixel during the same collection period will average 1,000,000/262,000 = ~4, and the limiting *S/N* would be 2. Thus, the *S/N* for each pixel will be only 1/500 as great as for the entire cell! There are only two ways to avoid such a reduction in *S/N*. First, pixels can be grouped with a corresponding decrease in two-dimensional spatial resolution. Each four-fold grouping of pixels would improve the *S/N* by a factor of 2. Second, the sampling period can be increased with a corresponding decrease in temporal resolution. Each four-fold increase in sampling period also would improve the *S/N* by a factor of 2. It should be clear that there is an inescapable trade-off between pH_i resolution (i.e., *S/N*), spatial resolution, and temporal resolution. Another problem with two-dimensional fluorescence imaging is inhomogeneity of illumination of either the target or the detector. If the inhomogeneities are not wavelength dependent, this problem is minimized with the ratioing technique.

Fluorescence Measurements Using Confocal Microscopy

With laser-scanning confocal microscopy, one achieves extremely good spatial resolution in three dimensions. However, as might be anticipated from the discussion in the previous section, the problem of photon availability can become extremely serious. In two-dimensional imaging, a significant fraction (depending on the objective used) of the photons typically come from above or below the plane of focus. In confocal microscopy, the plane of focus is narrower and the out-of-focus photons are largely eliminated, so that the number of photons may decrease by a factor of 10 or more. As a result, the number of photons available for detection in a confocal system are greatly reduced, with an unavoidable deterioration in signal-to-noise ratio. As discussed above in the previous section, there will be an even larger trade-off between pH_i resolution, spatial resolution (three dimensions), and temporal resolution. If one attempts to increase the power of the laser (thereby increase the rate of photon emission from each dye molecule), the lifetime of the dye molecules will be substantially reduced. The measurement of pH_i with confocal fluorescence microscopy has been used in Madin-Darby canine kidney (MDCK) cells (Zhou et al., 1995), cultured rabbit cardiac myocytes (Chacon et al., 1994), and *Xenopus* yolk platelets (Fagotto and Maxfield, 1994), as well as cells of the cortical collecting tubule (Wang et al., 1990), rat proximal convoluted tubule (Weinlich et al., 1993) and mouse distal colon (Chu and Montrose, 1995).

CONCLUDING REMARKS

The use of fluorescent indicators to measure pH_i has grown markedly over the past 15 years, owing to the development of relatively well-behaved intracellular dyes, as well as techniques for calibrating them. With the optical techniques discussed above, pH-sensitive fluorescent indicators can be used in a wide range of applications. For example, dyes have been used to measure pH_i in neurons and glial cells grown in culture, acutely isolated, or nestled in a slice of brain tissue. Such experiments will make it possible to study the mechanisms that alter or regulate pH_i in brain cells under well-defined conditions (in the case of cultured or acutely isolated cells), and under conditions that closely mimic *in vivo* conditions (in the case of cells in tissue slices). If it should become possible genetically to engineer a pH-sensitive dye that could be expressed in specific brain cells, one could in principle study pH_i *in situ* under conditions that are even closer to physiological.

ACKNOWLEDGMENTS

We thank Dr. Gregory Boyarsky for the data presented in Fig. 4.

REFERENCES

Allen CN, Harpur ES, Gray TJB, Simmons NL, Hirst BH (1990): Efflux of bis-carboxyethyl-carboxyfluorescein (BCECF) by a novel ATP-dependent transport mechanism in epithelial cells. Biochem Biophys Res Commun 172: 262–267.

Amos BJ, Richards CD (1996): Intrinsic hydrogen ion buffering in rat CNS neurones maintained in culture. Exp Physiol 81: 261–271.

Amos BJ, Pocock G, Richards CD (1996): On the role of bicarbonate as a hydrogen ion buffer in rat CNS neurones. Exp Physiol 81: 623–632.

Baxter KA, Church J (1996): Characterization of acid extrusion mechanisms in cultured fetal rat hippocampal neurones. J Physiol (Lond) 493: 457–470.

Bevensee MO, Schwiening CJ, Boron WF (1995): Use of BCECF and propidium iodide to assess membrane integrity of acutely isolated CA1 neurons from rat hippocampus. J Neurosci Methods 58: 61–75.

Bevensee MO, Cummins TR, Haddad GG, Boron WF, Boyarsky G (1996): pH regulation in single CA1 neurons acutely isolated from the hippocampi of immature and mature rats. J Physiol (Lond) 494: 315–328.

Bevensee MO, Weed RA, Boron WF (1997a): Intracellular pH regulation in cultured astrocytes from rat hippocampus. I. Role of HCO_3^-. J Gen Physiol 110: 453–465.

Bevensee MO, Apkon M, Boron WF (1997b): Intracellular pH regulation in cultured astrocytes from rat hippocampus. II. Electrogenic Na/HCO_3 cotransport. J Gen Physiol 110: 467–483.

Bouvier M, Szatkowski M, Amato A, Attwell D (1992): The glial cell glutamate uptake carrier countertransports pH-changing anions. Nature 360: 471–474.

Boyarsky G, Ransom B, Schlue W-R, Davis MBE, Boron WF (1993): Intracellular pH regulation in single cultured astrocytes from rat forebrain. Glia 8: 241–248.

Boyarsky G, Hanssen C, Clyne LA (1996): Superiority of in vitro over in vivo calibrations of BCECF in vascular smooth muscle cells. FASEB J 10: 1205–1212.

Boyarsky G, Hanssen C, Clyne LA (1996): Inadequacy of high K^+/nigericin for calibrating BCECF. I. Estimating steady-state intracellular pH. Am J Physiol 271: C1131–C1145.

Boyarsky G, Hanssen C, Clyne LA (1996): Inadequacy of high K^+/nigericin for calibrating BCECF. II. Intracellular pH dependence of the correction. Am J Physiol 271: C1146–C1156.

Chacon E, Reece JM, Nieminen A-L, Zahrebelski G, Herman B, Lemasters JJ (1994): Distribution of electrical potential, pH, free Ca^{2+}, and volume inside cultured adult rabbit cardiac myocytes during chemical hypoxia: A multiparameter digitized confocal microscopic study. Biophys J 66: 942–952.

Chaillet JR, Boron WF (1985): Intracellular calibration of a pH-sensitive dye in isolated, perfused salamander proximal tubules. J Gen Physiol 86: 765–794.

Chaillet JR, Lopes AG, Boron WF (1985): Basolateral Na-H exchange in the rabbit cortical collecting tubule. J Gen Physiol 86: 795–812.

Chu S, Montrose MH (1995): Extracellular pH regulation in microdomains of colonic crypts: Effects of short-chain fatty acids. Proc Natl Acad Sci USA 92: 3303–3307.

Collington GK, Hunter J, Allen CN, Simmons NL, Hirst BH (1992): Polarized efflux of 2′,7′-bis(2-carboxyethyl)-5(6)-carboxyfluorescein from cultured epithelial cell monolayers. Biochem Pharmacol 44: 417–424.

De Vries H (1871): Sur la permeabilité due protoplasma des Betteraves rouges. Arch Neerl Sci Exactes Nat 6: 118–126.

Eisner DA, Kenning NA, O'Neill SC, Pocock G, Richards CD, Valdeolmillos M (1989): A novel method for absolute calibration of intracellular pH indicators. Pflügers Arch 413: 553–558.

Fagotto F, Maxfield FR (1994): Changes in yolk platelet pH during *Xenopus laevis* development correlate with yolk utilization. A quantitative confocal microscopy study. J Cell Sci 107: 3325–3337.

Gaillard S, Dupont J-L (1990): Ionic control of intracellular pH in rat cerebellar Purkinje cells maintained in culture. J Physiol 425: 71–83.

Ganz MB, Boyarsky G, Boron WF, Sterzel RB (1988): Effects of angiotensin II and vasopressin on intracellular pH of glomerular mesangial cells. Am J Physiol 254: F787–F794.

Ganz MB, Boyarsky G, Sterzel RB, Boron WF (1989): Arginine vasopressin enhances pH_i regulation in the presence of HCO_3^- by stimulating three acid-base transport systems. Nature 337: 648–651.

Gatto C, Milanick MA (1993): Inhibition of the red blood cell calcium pump by eosin and other fluorescein analogues. Am J Physiol 264: C1577–C1586.

Geibel J, Giebisch G, Boron WF (1989): Basolateral sodium-coupled acid-base transport mechanisms of the rabbit proximal tubule. Am J Physiol 257: F790–F797.

Gérard C, Boudier J-A, Mauchamp J, Verrier B (1990): Evidence for probenecid-sensitive organic anion transporters on polarized thyroid cells in culture. J Cell Physiol 144: 354–364.

Gupta A, Schwiening CJ, Boron WF (1994): Effects of CGRP, forskolin, PMA, and ionomycin on pH_i dependence of Na-H exchange in UMR-106 cells. Am J Physiol 266: C1083–C1092.

Hartley Z, Dubinsky JM (1993): Changes in intracellular pH associated with glutamate excitotoxicity. J Neurosci 13: 4690–4699.

Irving M, Maylie J, Sizto NL, Chandler WK (1989): Simultaneous monitoring of changes in magnesium and calcium concentrations in frog cut twitch fibers containing antipyrylazo III. J Gen Physiol 93: 585–608.

Kao JPY (1994): Practical aspects of measuring [Ca^{2+}] with fluorescent indicators. Methods Cell Biol 40: 155–181.

Kaplan DL, Boron WF (1994): Long-term expression of c-H-*ras* stimulates Na-H and Na^+-dependent Cl-HCO_3 exchange in NIH-3T3 fibroblasts. J Biol Chem 269: 4116–4124.

Kapus A, Romanek R, Yi Qu A, Rotstein OD, Grinstein S (1993): A pH-sensitive and voltage-dependent proton conductance in the plasma membrane of macrophages. J Gen Physiol 102: 729–760.

Lückermann M, Trapp S, Ballanyi K (1997): GABA- and glycine-mediated fall of intracellular pH in rat medullary neurons in situ. J Neurophysiol 77: 1844–1852.

Martínez-Zaguilán R, Gillies RJ, Sánchez-Armass S (1994): Regulation of pH in rat brain synaptosomes. II. Role of Cl^-. J Neurophysiol 71: 2249–2257.

Nedergaard M, Goldman SA (1993): Carrier-mediated transport of lactic acid in cultured neurons and astrocytes. Am J Physiol 265: R282–R289.

Nett W, Deitmer JW (1996): Simultaneous measurements of intracellular pH in the leech giant glial cell using 2′,7′-bis-(2-carboxyethyl)-5,6-carboxyfluorescein and ion-sensitive microelectrodes. Biophys J 71: 394–402.

Nordström T, Rotstein OD, Romanek R, Asotra S, Heersche JNM, Manolson MF, Brisseau GF, Grinstein S (1995): Regulation of cytosolic pH in osteoclasts. Contribution of proton pumps and a proton-selective conductance. J Biol Chem 270: 2203–2212.

Ou Yang Y-B, Kristián T, Kristiánová V, Mellergård P, Siesjö BK (1995): The influence of calcium transients on intracellular pH in cortical neurons in primary culture. Brain Res 676: 307–313.

Pappas CA, Ransom BR (1993): A depolarization-stimulated, bafilomycin-inhibitable H^+ pump in hippocampal astrocytes. Glia 9: 280–291.

Pappas CA, Ransom BR (1994): Depolarization-induced alkalinization (DIA) in rat hippocampal astrocytes. J Neurophysiol 72: 2816–2826.

Paradiso AM, Machen TE, Tsien RY (1987): Digital image processing of intracellular pH in gastric oxyntic and chief cells. Nature 325: 447–450.

Pasternack M, Voipio J, Kaila K (1993): Intracellular carbonic anhydrase activity and its role in GABA-induced acidosis in isolated rat hippocampal pyramidal neurones. Acta Physiol Scand 148: 229–231.

Raley-Susman KM, Cragoe EJ Jr, Sapolsky RM, Kopito RR (1991): Regulation of intracellular pH in cultured hippocampal neurons by an amiloride-insensitive Na^+/H^+ exchanger. J Biol Chem 266: 2739–2745.

Ravesloot JH, Eisen T, Baron R, Boron WF (1995): Role of Na-H exchangers and vacuolar H^+ pumps in intracellular pH regulation in neonatal rat osteoclasts. J Gen Physiol 105: 177–208.

Rink TJ, Tsien RY, Pozzan T (1982): Cytoplasmic pH and free Mg^{2+} in lymphocytes. J Cell Biol 95: 189–196.

Ritucci NA, Erlichman JS, Dean JB, Putnam RW (1996): A fluorescence technique to measure intracellular pH of single neurons in brainstem slices. J Neurosci Methods 68: 149–163.

Ritucci NA, Dean JB, Putnam RW (1997): Intracellular pH response to hypercapnia in neurons from chemosensitive areas of the medulla. Am J Physiol 273: R433–R441.

Rose CR, Ransom BR (1996): Mechanisms of H^+ and Na^+ changes induced by glutamate, kainate, and D-aspartate in rat hippocampal astrocytes. J Neurosci 16: 5393–5404.

Sánchez-Armass S, Martínez-Zaguilán R, Martínez GM, Gillies RJ (1994): Regulation of pH in rat brain synaptosomes. I. Role of sodium, bicarbonate, and potassium. J Neurophysiol 71: 2236–2248.

Schwiening CJ, Boron WF (1994): Regulation of intracellular pH in pyramidal neurones from the rat hippocampus by Na^+-dependent Cl^--HCO_3^- exchange. J Physiol (Lond) 475: 59–67.

Shrode LD, Putnam RW (1994): Intracellular pH regulation in primary rat astrocytes and C6 glioma cells. Glia 12: 196–210.

Singh SK, Binder HJ, Geibel JP, Boron WF (1995): An apical permeability barrier to NH_3/NH_4^+ in isolated, perfused colonic crypts. Proc Natl Acad Sci USA 92: 11573–11577.

Slayman CL, Moussatos VV, Webb WW (1994): Endosomal accumulation of pH indicator dyes delivered as acetoxymethyl esters. J Exp Biol 196: 419–438.

Spring KR (1985): The study of epithelial function by quantitative light microscopy. Pflügers Arch 405: S23–S27.

Szatkowski MS, Thomas RC (1986): New method for calculating pH_i from accurately measured changes in pH_i induced by a weak acid and base. Pflügers Arch 407: 59–63.

Thomas JA, Buchsbaum RN, Zimniak A, Racker E (1979): Intracellular pH measurements in Ehrlich ascites tumor cells utilizing spectroscopic probes generated in situ. Biochemistry 18: 2210–2218.

Trapp S, Lückermann M, Brooks PA, Ballanyi K (1996): Acidosis of rat dorsal vagal neurons *in situ* during spontaneous and evoked activity. J Physiol (Lond) 496: 695–710.

Trapp S, Lückermann M, Kaila K, Ballanyi K (1997): Acidosis of hippocampal neurones mediated by a plasmalemmal Ca^{2+}/H^+ pump. Neuroreport 7: 2000–20004.

Waisbren SJ, Geibel J, Boron WF, Modlin IM (1994a): Luminal perfusion of isolated gastric glands. Am J Physiol 266: C1013–1027.

Waisbren SJ, Geibel JP, Modlin IM, Boron WF (1994b): Unusual permeability properties of gastric gland cells. Nature 368: 332–335.

Wang X, Kurtz I (1990): H^+/base transport in principal cells characterized by confocal fluorescence imaging. Am J Physiol 259: C365–C373.

Wang GJ, Randall RD, Thayer SA (1994): Glutamate-induced intracellular acidification of cultured hippocampal neurons demonstrates altered energy metabolism resulting from Ca^{2+} loads. J Neurophysiol 72: 2563–2569.

Weinlich M, Capasso G, Kinne RKH (1993): Intracellular pH in renal tubules in situ: single-cell measurements by confocal laserscan microscopy. Pflügers Arch 422: 523–529.

Whitaker JE, Haugland RP, Prendergast FG (1991): Spectral and photophysical studies of benzo[*c*]xanthene dyes: Dual emission pH sensors. Anal Biochem 194: 330–344.

Willstätter R, Mallison H (1915): Über Variationen der Blutenfarben. Ann Chem 408: 147–162.

Zhou Y, Marcus EM, Haugland RP, Opas M (1995): Use of a new fluorescent probe, seminaphthofluorescein-calcein, for determination of intracellular pH by simultaneous dual-emission imaging laser scanning confocal microscopy. J Cell Physiol 164: 9–16.

CHAPTER 9

Brain pH Measurement by Nuclear Magnetic Resonance Spectroscopy

JAMES W. PRICHARD, DOUGLAS L. ROTHMAN, AND
OGNEN A.C. PETROFF
Departments of Neurology and Medicine, Yale University, New Haven, CT 06520

INTRODUCTION

Hydrogen ion concentration is basic to brain function, as this volume demonstrates. However, it is hard to measure accurately, because any kind of tissue-disruptive sampling procedure changes it, and exogenous indicators may be unevenly distributed, toxic, or both.

Since its demonstration in 1946 (Bloch et al., 1946; Purcell et al., 1946), the nuclear magnetic resonance (NMR) phenomenon has been productive in a wide range of scientific applications. As NMR instruments large enough for measurements on humans became available in the 1980s, magnetic resonance imaging (MRI) based on detection of the strong NMR signal from hydrogen nuclei (protons) of tissue water began to replace x-ray techniques for many purposes and soon emerged as the premier imaging technique for medical diagnosis (Stark and Bradley, 1992).

In addition to the uniquely large water proton signal, many other NMR signals are available from living tissue. Because they are smaller, the technical problems of detecting them are somewhat different, and for that reason study of them is called magnetic resonance spectroscopy (MRS), to distinguish it from MRI. The distinction is an informal one made as a matter of convenience by workers in the field; MRI and MRS use the same NMR physics in different ways. In recent years, a trend to use machines capable of both has started, as the extensive advantages of making many NMR measurements on the same sample in a single machine have become both evident and practical. Tissue pH is an excellent example of a measurement that is valuable in itself, yet gains in value when combined with other measurements, as of lactate and inorganic phosphate, for example (see Chapters 4, 5, and 30). The capability of modern noninvasive NMR methods to measure these and many other variables repeatedly in the same subject is unprecedented in biomedical science. Among the numerous attractive prospects it offers neurobiology is the opportunity to achieve new understanding of how pH is related to function and disease in the nervous system.

pH and Brain Function, Edited by Kai Kaila and Bruce R. Ransom
ISBN 0-471-11838-9

In 1973, nondestructive measurement of erythrocyte intracellular pH (pH_i) by ^{31}P nuclear magnetic resonance spectroscopy (MRS) was reported (Moon and Richards, 1973). The measurement is possible because the ^{31}P signal from inorganic phosphate (Pi) is sensitive to pH. Use of the technique to study pH_i of ischemic muscle followed shortly (Hoult et al., 1974). Over the next decade, measurements of pH_i in the brains of living animals (Ackerman et al., 1980; Chance et al., 1980) and humans (Cady et al., 1983; Younkin et al., 1984) appeared. Numerous ^{31}P spectroscopic studies on pH phenomena in a variety of organisms, tissues, and organelles were done in the same period. This early literature has been well reviewed (Gadian, 1982; Nucitelli and Deamer, 1982; Petroff and Prichard, 1995; Petroff, 1996).

Subsequent work has shown that NMR methods are well suited for pH_i measurements in living tissue. They are noninvasive, and therefore can be used repeatedly on the same sample without disturbing it. The "sample" can be an inorganic solution, a suspension of cells, or a functioning organ of a living human being, including the brain. Many endogenous indicators of pH_i detectable by NMR exist in biological tissues; exogenous ones are available for special tasks. The growing availability of NMR machines has, or soon will have, placed NMR measurements within reach of most neurobiologists whose work can benefit from them. A paramount advantage of brain pH measurement by NMR is the potential for a wide variety of other NMR-based measurements on the same tissue at the same time (Stark and Bradley, 1992; Prichard, 1995; Prichard, 1996a,b).

This chapter describes the basis of NMR pH measurements in brain. Since most published work has been done by ^{31}P MRS, the specific examples given are mainly from that area. Examples from other tissues and other nuclei are mentioned to emphasize the versatility of NMR methods and to indicate existence of an extensive literature that cannot be reviewed here.

THE PHYSICAL BASIS OF BRAIN pH_i MEASUREMENTS BY NMR

Biological NMR measurements are based on the fact that resonant frequencies of nuclear precession can be detected from ^{1}H, ^{31}P, ^{13}C, ^{15}N, ^{23}Na, ^{17}O, and several other magnetic nuclei that are natural constituents of molecules found in biological tissue. All NMR measurements depend on the same few principles of nuclear physics; these are illustrated in the accompanying box, for the benefit of readers who are unfamiliar with them.

Exogenous nuclei with magnetic properties can be used as metabolic tracers by enrichment above their natural abundance, as has been done with compounds containing, ^{13}C, ^{15}N, and ^{17}O. Nuclei not naturally present in biological tissue may also serve as tracers or pH indicators under some conditions (^{19}F is an example, see pps 162-163), or, through their influence on endogenous signals, (e.g., as MRI contrast agents, the most widely used of which contain ^{157}Gd).

Slight differences in resonant frequency from the same nuclear species incorporated into different chemical structures often allow selective measurement of the specific compounds containing the nucleus. Measurements are made by placing the subject in a strong, static magnetic field which aligns magnetic nuclei in an orderly way. This ordering of the nuclei allows magnetic signals from large populations of them to be detected after controlled perturbation by small injections of magnetic energy near the precession frequencies of the nuclei to be observed.

elementary NMR physics

1 Magnetic nuclei are somewhat like small compass needles which spin around their own long axis:

2 In the strong magnetic field of an NMR spectrometer, they line up with or against the flux lines in a Boltzmann distribution, with a few more pointing up than down, and their spins cause them to precess around the flux lines:

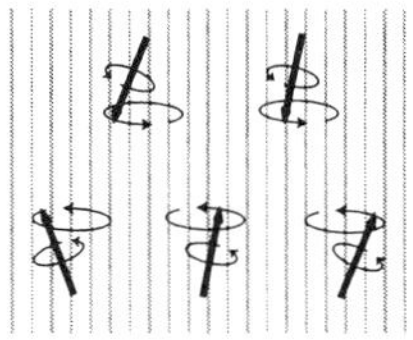

3 The net magnetization of such a spin population is a vector parallel to the flux lines, which run along the magnet's Z axis:

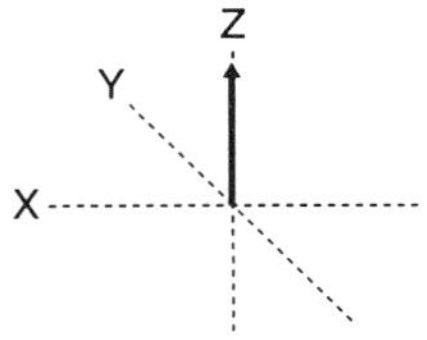

4 For measurements, a small additional magnetic field rotating near the spins' precession frequency is applied to tip the vector into the X-Y plane:

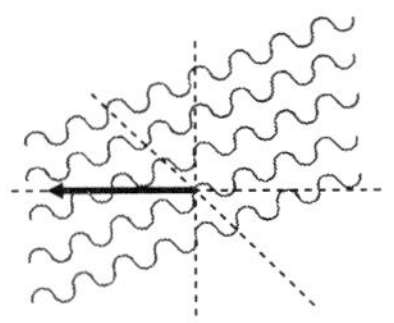

5 Slight differences in precession frequency due to magnetic field inhomogeneities cause spins to dephase, up spins going one way, down spins the other (T2 relaxation), while some return to the Z axis (T1 relaxation):

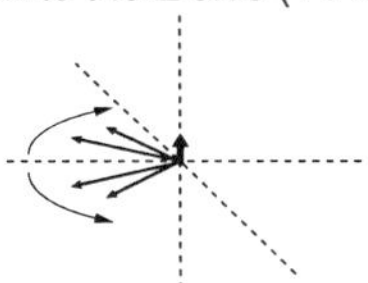

6 Another kind of brief magnetic pulse flips spins still in the X-Y plane by 180°:

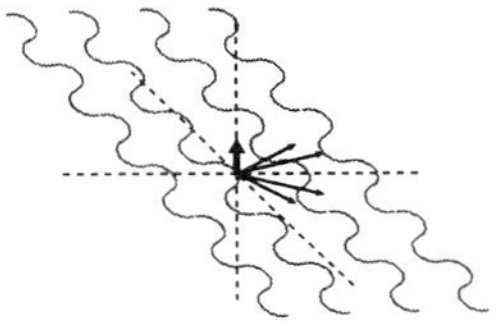

7 Because the spins continue to precess in the same direction, their net magnetization refocuses as a vector along -X:

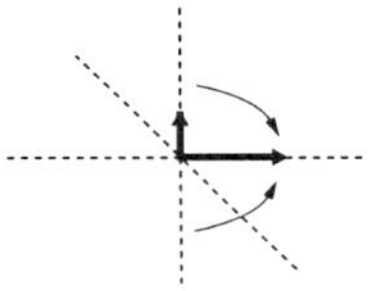

8 A coil in the spectrometer detects the vector in -X as an induced voltage, like this one showing 2 resonant frequencies:

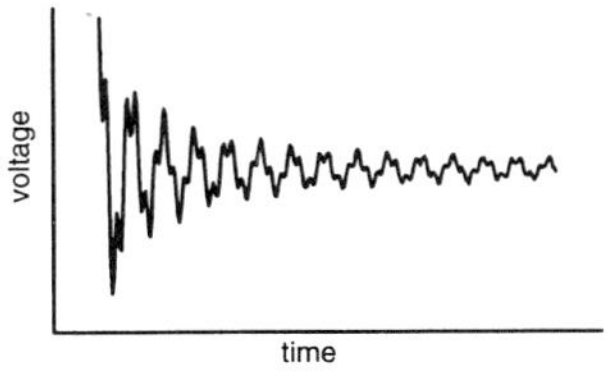

9 Fourier transformation of such a signal into the frequency domain produces a spectrum with two frequency peaks:

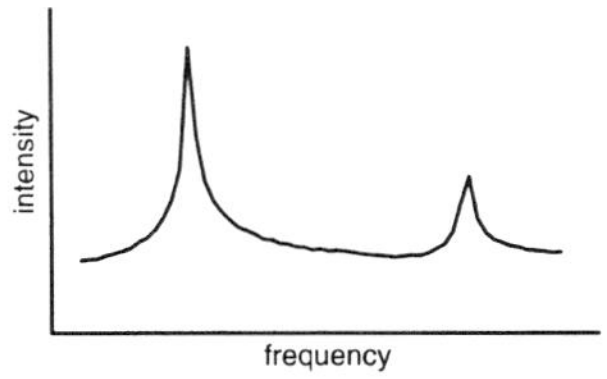

Many different variables can be measured by NMR methods, depending on which nuclei and which compounds containing them are chosen for study. As noted above, the ^1H signal from tissue water is the strongest signal from biological tissues and is the basis of current medical MRI. Weaker signals from non-water ^1H and other nuclei allow measurement of specific compounds such as adenosine triphosphate (ATP), phosphocreatine

(PCr), lactate, some amino acids, and a variety of other small molecules. Some MRS signals are strong enough to be made into maps of the anatomical distribution of the chemicals that produce them. Research is active in this area, and chemically specific maps of dozens of compounds including the ones mentioned above will soon be routinely obtainable from normal humans and patients with brain diseases, an availability that will open a new era in systematic investigation of the brain's chemical development and pathology.

Measurement of tissue pH by NMR is possible because some MRS signals titrate. That is, the resonant frequencies of particular nuclei in some compounds are changed by changes in the pH of the tissue compartment containing them. When the pK of the change is in a region of biological interest and the titrating signal can be resolved well enough from other signals, noninvasive pH measurements can be made at whatever time and space resolutions the strength of the particular signal allows, usually a few minutes and a few cubic centimeters. The considerable potential of pH measurement by NMR indicated by the first decade's work led investigators interested in the brain to evaluate various NMR methods for use in that organ. The Pi signal in ^{31}P spectra emerged clearly as the best indicator under most conditions, although others are available, as described later in this chapter (pps 162-163).

Differences along the frequency axes of NMR spectra are called "chemical shifts" because they are caused mainly by differences among the magnetic microenvironments created by the chemical structure of molecules. Magnetic nuclei bound into a molecule influence the electron clouds around nearby nuclei, and thereby raise or lower the degree of shielding of those nuclei from the main magnetic field of the NMR spectrometer. The resonant frequency of a magnetic nucleus is determined by the net magnetic field that the nucleus "sees"—the main field plus shielding effects. Though slight, the effects of shielding due to molecular structure are detectable in living tissue; they are the basis of chemically specific NMR measurements. Nuclei that are part of the same chemical structure anywhere in the sample will have similar resonant frequencies, which are different from the frequencies of the same nuclear species in other chemical structures. For example, the terminal phosphorus nuclei of all ATP molecules in the same tissue compartment resonate at nearly the same frequency, because they all experience similar shielding from the main magnetic field by intramolecular distortions of electron clouds due to the molecule's other two phosphorus nuclei and its protons; the α and β phosphorus nuclei have different resonant frequencies from the terminal phosphorus and from each other because of their different magnetic microenvironments. The capability of an NMR instrument to detect these differences in frequency is termed its resolution, which is improved by higher and more uniform (homogenous) magnetic fields.

Measurements of pH by NMR depend on detection of the chemical shift of the conjugate acid and base forms of some compound used as a pH indicator. If the two forms exchange protons at a rate that is slow relative to the duration of the NMR measurement, separate signals from each are detected. The ratio of their intensities—corrected if necessary for differences in relaxation properties—can be substituted for the base/acid concentration ratio in the Henderson-Hasselbalch equation

$$\mathrm{pH} = \mathrm{p}K + \log\left(\frac{[\mathrm{base}]}{[\mathrm{acid}]}\right)$$

The precision of the pH measurement depends on how well the intensities of the two conjugates can be measured and determination of a pK value appropriate for the conditions.

If protons exchange between the base and acid forms rapidly relative to the duration of the NMR measurement, a single signal is observed, with a chemical shift that is the concentration-weighted average of the chemical shifts of the acid and base forms of the indicator. In this case, precision of pH estimation depends on the precision with which the chemical shift of the exchange-averaged signal can be measured, and on knowledge of pK and the chemical shifts of both the acid and base forms of the indicator. Like pK, the latter must have been determined separately under conditions considered similar to those of the observed system. For an observed chemical shift (δ) comprised of a weighted average between the shift of a pure base (δ base) and that of its conjugated acid (δ acid), the Henderson-Hasselbalch equation is rewritten as

$$\text{pH} = \text{p}K + \log\left(\frac{(\delta - \delta\ \text{acid})}{(\delta\ \text{base} - \delta)}\right)$$

In many biological tissues, Pi is the best NMR indicator of pH_i, for two reasons: (i) The importance of phosphorus chemistry in energy metabolism ensures that NMR-detectable concentrations of Pi are usually present. Its component conjugate acid ($H_2PO_4^-$) and base (HPO_4^{2-}) are sufficiently concentrated for both to influence the weighted average of the Pi signal. (ii) The pK of this dissociation is 6.77, which places the most sensitive part of its titration curve in a range of principal biological interest. The pKs of the first and third dissociations of phosphoric acid are so far from the pH of most biological tissues that H_3PO_4 and PO_4^{3-} concentrations are too low to affect the chemical shift of the Pi signal.

Thus, for NMR measurements of pH_i based on dissociation of phosphoric acid, the Henderson-Hasselbalch equation is written as

$$\text{pH} = \text{p}K + \log\left(\frac{(\delta\text{Pi} - \delta H_2PO_4^{1-})}{(\delta HPO_4^{2-} - \delta\text{Pi})}\right)$$

where δPi is the observed chemical shift of the concentration-weighted average signal from $H_2PO_4^{1-}$ and HPO_4^{2-}, and $\delta H_2PO_4^{1-}$ and δHPO_4^{2-} are the chemical shifts of the pure compounds known from separate measurements. For measurement of pH_i from Pi in mammalian brain, an appropriate numerical form of the equation is

$$\text{pH} = 6.77 + \log\left(\frac{(\delta\ \text{Pi} - 3.29)}{(5.68 - \delta\ \text{Pi})}\right)$$

the constants of which were determined by *in vitro* titration under conditions similar to the environment of Pi in the brain (Petroff et al., 1985).

In order to avoid misinterpretation of NMR data intended to measure pH, one must remain constantly aware that whatever affects the pK of Pi titration, or the ratio $[H_2PO_4^-]/[HPO_4^{2-}]$, changes the apparent pH. Factors known to affect Pi titration include temperature, ionic strength, and the concentrations of Mg^{2+}, Ca^{2+}, protein, Na^+, K^+, and the inorganic phosphates themselves. Choice of a signal from which to measure the chemical shift of Pi is important, because no signal from the sample itself is frequency invariant over all conditions, and external signals can never reflect exactly the same conditions that affect the Pi signal. These considerations are general ones that apply to any NMR indicator of pH in any system. Fuller treatment of them is readily accessible

(Roberts et al., 1981; Gard and Ackerman, 1983; Seo et al., 1983; Petroff et al., 1985; Kushmerick et al., 1986; Madden et al., 1991; Petroff and Prichard, 1995; Petroff, 1996).

Figure 1 shows simulated ^{31}P spectra as they might appear under conditions associated with widely different pH_i. Such spectra are obtained from volumes of tissue containing millions of cells. Since most NMR-observable Pi is in cytosol, the signal principally reflects the pH of this intracellular compartment.

COMPARTMENTATION OF pH

Living tissues are known to maintain different values of pH in cytosol, organelles, and extracellular spaces. In the brain, naturally occurring differences in pH_i across different kinds of cells and even from one region to another of the same cell are known or suspected; separate variations of pH_i in neurons and glia during brain activation have been documented (Chesler, 1990; Chesler and Kaila, 1992; Chapters 15–18). All of these compartments are well below the spatial resolution of NMR methods that can be used with living brain. Interpretation of NMR measurements of pH must therefore always include consideration of what is known about the compartmental distribution of the nuclei that generate the pH-sensitive signal, and such knowledge is always incomplete. Consequently, tissue compartmentation is responsible for a special kind of uncertainty affecting interpretation of pH_i measurements in the brain, and this kind of uncertainty is different from the measurement errors discussed below. Even if all measurement errors could be eliminated, heterogenous distribution of the indicator across tissue compartments would still mandate a conditional element in biological conclusions based on NMR data from intact tissue. The case of Pi provides an instructive example.

If Pi exchange between two chemical environments having different pH is slow, two Pi signals may be observed (Gadian 1982). The faster the rate of exchange, the closer the resonant frequencies become. This idea is expressed formally as

$$\Delta f = \left[(f_A - f_B)^2 - \frac{2k^2}{\pi^2} \right]^{1/2}$$

where Δf is the observed difference between two Pi signals, f_A and f_B are the different resonant frequencies of each population of Pi nuclei in their separate environments, and k is the rate of exchange between them. As k increases, Δf becomes smaller, so that at any given magnetic field strength (which determines $f_A - f_B$) and level of measurement noise, some value of k exists above which two signals can no longer be observed. Exchange between cytosol and extracellular fluid is usually not fast enough to approach this condition.

If exchange of Pi between two compartments with different pH is negligible, the intensity of the Pi signal from each compartment is proportional to the Pi concentrations and the fraction that each compartment is of the whole volume. In principle, ^{31}P spectroscopy ought to distinguish extra- from intracellular pH, because the difference between them is typically 0.3–0.4 pH units in animal tissues. However, clear Pi signals from both compartments must be detectable. In these early days of pH measurement in living tissue by MRS, that requirement is met only in a few favorable cases. A signal from Pi in the extracellular space can be seen easily in cell suspensions (Navon et al., 1977; Slonczewski et al., 1981; Gillies et al., 1982; Lien et al., 1992; Merle et al., 1992), but clear instances have been reported only infrequently from perfused organs (Garlick et al., 1979; Salhany et al., 1979; Gilboe et al., 1993). An extracellular Pi signal from organs of intact vertebrates was not reported in

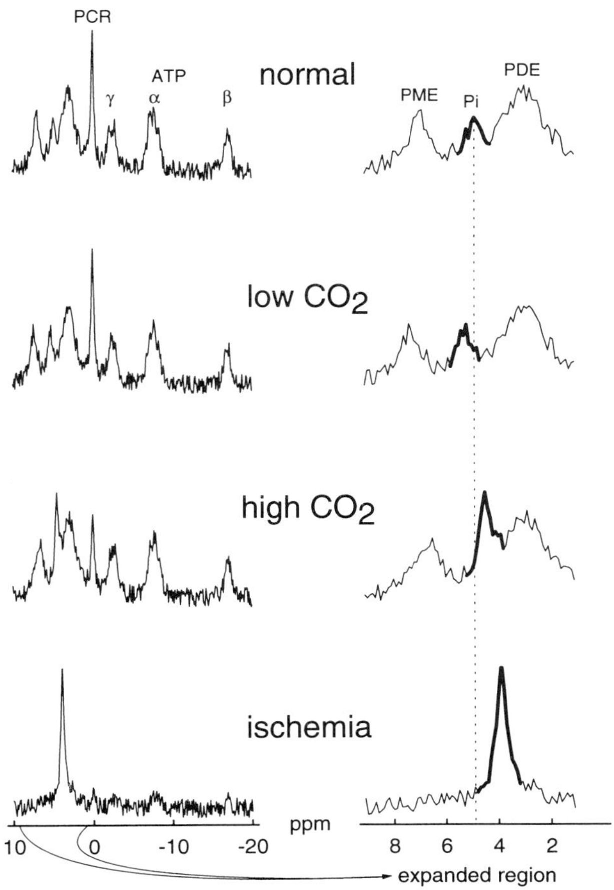

FIGURE 1. Simulated normal ^{31}P spectrum of brain and changes that occur in hypocarbic alkalosis, hypercarbic acidosis, and acidosis with energy failure in ischemia. The spectra were simulated mathematically to allow presentation of important biological features without distraction by unrelated variations in acquisition and processing; the difference from actual data is not detectable by inspection. The peaks in the spectra represent resonant frequencies of ^{31}P nuclei in molecules of living tissue. They are obtained from the brain in an NMR spectrometer by the measurement process outlined in the box "elementary NMR physics".

The normal ^{31}P spectrum is represented at top left, with the region containing the (boldface) signal from inorganic phosphate (Pi) expanded on the right. The other signals are from phosphocreatine (PCr), the α, β, and γ phosphates of adenosine triphosphate (ATP), and compounds with phosphomonoester (PME) and phosphodiester (PDE) groups. The abscissa is a normalized resonant frequency axis with units of parts per million (ppm). Signals obtained at any magnetic field strength can be compared on the ppm scale, which increases to the left by conventions rooted in NMR history. In studies of living tissue, some prominent and narrow internal signal such as PCR is commonly used as the reference frequency.

In *hypocarbia,* the Pi signal shifts to the left, reflecting the higher ratio of dibasic to monobasic phosphate in the alkalotic state. In *hypercarbic acidosis,* the Pi signal shifts in the opposite direction. In addition, Pi signal intensity rises and PCR intensity falls, due to concentration changes caused by the effect of acidosis on the creatine kinase equilibrium. Note that the PME signal also titrates with the changes in P_{CO_2}. In *ischemia* severe enough to cause loss of tissue energy stores and disruption of cell membranes, Pi may be the only signal remaining, shifted far in the acid direction, as shown.

$$\text{ppm} = 10^6 \times \frac{\text{resonant frequency in Hertz} - \text{reference frequency in Hertz}}{\text{reference frequency in Hertz}}$$

several studies that might have detected it (Gadian and Radda, 1981; Taylor et al., 1983; Luyten et al., 1989; Gallis and Canioni, 1992; Gadian et al., 1993), but one group reported observing extra- and intracellular Pi signals from living sheep brain well enough to measure the pH of both compartments, and thereby demonstrate the superior buffering capacity of the intracellular compartment during manipulation of Pco_2 (Portman et al., 1991). Neoplastic tissues often have increased extra- to intracellular volume ratios; nevertheless, pH measured by ^{31}P spectroscopy in tumors is probably a measure principally of pHi (Stubbs et al., 1992).

Despite the difficulties of detecting a Pi signal from any compartment except cytosol, pH heterogeneity measured from multiple Pi signals has been demonstrated in a number of biological systems. These include perfused adrenal gland, platelets, plants, and unicellular organisms (Roberts et al., 1980; Gadian and Radda, 1981; Brenot et al., 1992; Walter et al., 1992). Signals attributed to intramitochondrial pH have been measured in isolated mitochondria (Ogawa et al., 1981), isolated liver cells (Cohen et al., 1978), and perfused heart (Garlick et al., 1983). The mitochondrial matrix space of the brain is only about 5% of the organ's intracellular volume. Pi signals plausibly attributable to mitochondria have not been detected in brain slices or cultured brain cells (Brooks et al., 1989; Lien et al., 1992; Merle et al., 1992), or in intact brain (Luyten et al., 1989; Gadian et al., 1993).

In normal intact brain, arithmetic is against detection of compartmental pH differences. Since many cells have total intracellular Pi concentrations of 2–3.5 mM (Veech et al., 1979; McIlwain and Bachelard, 1985; Connett, 1988; Luyten et al., 1989), one must assume that this range is the average one for brain cells. Cerebrospinal fluid Pi concentration is 0.4–0.6 mM and its pH is approximately 7.3 (Fishman 1992). The pH of extracellular fluid is also about 7.3 (Chesler, 1990), and it is thought also to have a Pi concentration in the range of 0.5 mM. Therefore, if the extracellular fluid space is 20% of an observed volume of brain and intracellular fluid is 80%, the intra- to extracellular ratio of Pi is on the order of 20. This value is too high to allow measurement of both intra- and extracellular pH in the brain by ^{31}P spectroscopy, given the noise level of current instruments. Nor have intracellular compartments with different pH been reliably identified.

In ischemic brain tissue, Pi concentration rises several-fold (Chapters 5, 28, 30), and pH_i moves in the acidic direction, along the most sensitive portion of the Pi titration curve. One might therefore expect mapping of pH_i by ^{31}P spectroscopy to reveal selectively regions of brain distressed by focal ischemia, since in such regions, both the increased Pi and its journey down the steep part of the curve favor detection of acidic pH over the contribution of less affected regions (Gadian et al., 1993). Such a measurement is not possible yet, but it will be available routinely in the future, if the best efforts of the 1990s at mapping ^{31}P metabolite signals onto brain anatomy (Brown, 1992) continue to advance.

NMR pH INDICATORS OTHER THAN Pi

Other ^{31}P Resonances

In addition to Pi, several other metabolites detectable by ^{31}P MRS in biological tissues have resonant frequencies sensitive to pH. Among these are γ-adenosine triphosphate (γ-ATP) (Gadian et al., 1979), phosphorylethanolamine (Corbett et al., 1987, 1988), 2,3-diphosphoglycerate (Moon and Richards, 1973; Stewart et al., 1986), fructose phosphates (Navon et al., 1977; Gonzalez-Mendez et al., 1988), and glucose phosphate (Roberts et al., 1980). The last three of these are too dilute in brain to serve as practical indicators of pH_i. Figure 2 shows the NMR titration curves of the other two compounds on the same scale with the curve of Pi.

Phosphorylethanolamine has a large signal in the ^{31}P spectrum of neonatal brain (Gyulai et al., 1984; Corbett et al., 1987; Burri et al., 1988; Corbett et al., 1988), and its p*K* of 5.63 (Brooks et al., 1989) is favorable for measurement of pH_i in the acidic part of the physiological range. In the adult brain, however, phosphorylcholine is present in similar amounts. Since its resonant frequency is close to that of phosphorylethanolamine, signals from the two compounds cannot be resolved *in vivo;* they are the principal components of a composite phosphomonoester (PME) signal. The titration behavior of this signal is a complex function of magnetic field, as well as of the relative concentrations and the similar, but not identical, titration properties of its two major components. Consequently, accuracy of pH_i estimations using the PME signal is limited at present.

The γ-ATP resonance has a wide titration range but, in the brain, most γ-ATP is complexed with magnesium (Gadian et al., 1979; Ackerman et al., 1980; Madden et al., 1991;

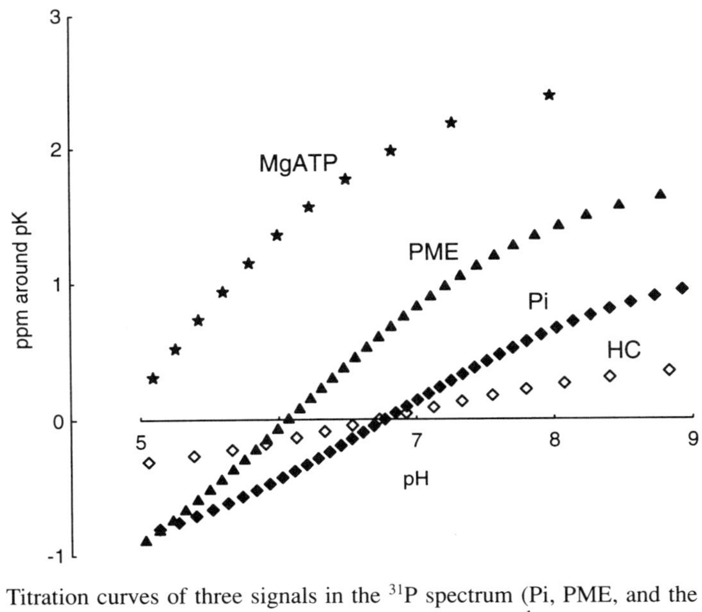

FIGURE 2. Titration curves of three signals in the ^{31}P spectrum (Pi, PME, and the γ phosphate of Mg^{2+}-complexed ATP) and homocarnosine (HC) in the ^{1}H spectrum, normalized to the resonant frequencies of their p*K*s on the ordinate. On each curve, the spacing of points indicates very roughly the separate pH values that might be discriminated in a few cubic centimeters of human brain by a few minutes of continuous data collection under realistic conditions. Actual precision depends not only on volume of tissue and recording time, but also on numerous other variables ranging from magnetic field strength to the distribution across tissue compartments of compounds contributing to the titrating signal. Pi is the best of the ^{31}P indicators because its signal is large and narrow, and its p*K* is near the intracellular pH of brain. Despite the wider chemical shift excursion of its titration, PME is less precise because its p*K* is farther from the physiological pH range, and because it is a broader, composite resonance from compounds that have slightly different titration behavior (see text). MgATP precision is coarse because its p*K* is far from the physiological range. The narrow chemical shift range of HC predicts that its precision may be low, although the greater sensitivity of ^{1}H compared to ^{31}P may result in better relative precision than estimated in the figure.

Halvorson et al., 1992; Gadian et al., 1993). The pK of the complex is 4.85 (Gadian et al., 1979), which is too acidic to allow accurate measurements of pH_i in the range of greatest interest. These features are illustrated in Fig. 2.

Exogenous indicators of pH detectable by NMR would find a place in brain pH measurement if nontoxic compounds with suitable p*K*s and tissue distribution properties were available. Various metabolites with titratable ^{31}P signals have been used to measure pH_i and extracellular pH in other tissues (Stewart et al., 1986; Brenot et al., 1992; Clarke et al., 1993). However, exogenous compounds generally useful for measurement of brain pH have not been described. The principal reason is the basic one of sensitivity. NMR methods can detect few signals from compounds less concentrated than 1 mM *in vivo*. Hence, exogenous indicators must be used in quantities likely to alter brain function. An example is 2-deoxyglucose-6-phosphate, a compound well known to neurobiologists for its utility in estimation of brain glycolytic rate when labeled with ^{14}C for autoradiography or with ^{18}F for positron emission tomography. The phosphate resonance titrates around a p*K* of 6.22 under physiological conditions, and when the compound is sufficiently concentrated, it is detectable by ^{31}P MRS (Navon et al., 1977; Bailey et al., 1981; Brooks et al., 1989; Petroff, 1996). Unfortunately, the concentrations necessary are so high that they cause electroencephalographic slowing and ATP depletion, presumably due to interference with glucose phosphorylation in the brain (unpublished observations in rabbits).

Signals From ^{1}H, ^{13}C, and ^{19}F

Magnetic nuclei other than ^{31}P can yield pH measurements from biological tissues. The ^{1}H spectrum includes signals from several compounds with strong signals sensitive to pH_i. The signal from the histidine moiety of carnosine and anserine has been used for this purpose in muscle (Yoshizaki et al., 1981; Williams et al., 1985; Arus et al., 1986; Lundberg and Vogel, 1987; Pan et al., 1988). Citrate is concentrated in prostate and has been suggested as a pH_i indicator for that tissue (Moore and Sillerud, 1993).

An indicator of brain pH_i in the ^{1}H spectrum would be highly useful. A titrating signal from elevated brain histidine in histidinemic mice has been reported (Gadian et al., 1986), but measurements at normal concentrations of histidine are yet to be made. The ^{1}H signal from carnosine has been used to measure pH_i in human muscle and, during recovery from exercise, it gave values quite similar to Pi in the ^{31}P spectrum (Pan et al., 1988). Carnosine is present in brain, but only at concentrations too low for useful pH_i measurements. Homocarnosine is more concentrated, (0.2–0.8 μmols/g wet weight; Perry, 1982). It is therefore a good candidate as an indicator of pH_i in normal brain, provided that interference from coresonant broad signals can be overcome, as preliminary observations indicate is possible (Rothman et al., 1997). The homocarnosine titration curve is shown in Fig. 2 for comparison with the ^{31}P indicators discussed above.

In the ^{13}C spectrum, signals from carnosine, anserine, citrate, and malate have been used to measure pH_i in muscle, prostate, and plants (Lundberg and Vogel, 1987; Walter et al., 1992). No similar efforts have been reported in brain. Although the resolution of the ^{13}C spectrum is excellent, the relatively low sensitivity of the nucleus and its low natural abundance (1.1%) require acquisition times that are impractically long for brain work.

Compounds of fluorine do not occur naturally in biological tissues, and use of radioactive ^{18}F in positron emission tomography has shown that they are not toxic in trace amounts. In principle, they might be used as indicators of metabolic activity and pH, but the low inherent sensitivity of NMR methods dictates that concentrations far above tracer level would be necessary for practical applications. This exigent requirement has so far limited ^{19}F-based biological pH measurements to cultured tissue (Taylor and Deutsch, 1983; Deutsch and Taylor, 1987), isolated frog skin (Civan et al., 1984), and preliminary efforts in rat brain (Chiles et al., 1990).

VALIDATION AND ERROR RANGE OF BRAIN pH_i MEASUREMENTS

In similar biological systems, pH_i values determined *in vivo* from the Pi signal of ^{31}P spectra show good general agreement with values obtained by other methods (Gadian et al., 1979; Petroff et al., 1985). Direct comparisons made in several laboratories have produced the same result (Gadian and Radda, 1981; Adler et al., 1988; Gonzalez-Mendez et al., 1988; Hoehn-Berlage et al., 1989; Jayasundar et al., 1992).

Experimental manipulations of pH_i are reliably detected by NMR methods *in vivo*. Studies of brain buffering capacity by measurement of the pH_i response to administration of carbon dioxide and acetazolamide are in broad agreement with estimates made by other methods in both animal models (Litt et al., 1985; Petroff et al., 1985; Behar et al., 1987; Adler et al., 1988; Bickler et al., 1988; Corbett et al., 1988; Portman et al., 1991) and in humans (van Rijen et al., 1989; Vorstrup et al., 1989; Bain et al., 1992). The response of brain pH_i to seizures, ischemia, hypoglycemia, and ammonia challenge is as expected from earlier work by other techniques (Prichard et al., 1983; Sessler et al., 1987; Andersen et al., 1988; Rosenberg et al., 1989; Behar, 1993; Gadian et al., 1993).

In other organs, similarly good agreement between pH_i determined from Pi has been the rule, as in studies of acid challenge to heart (Bernard and Cozzone, 1992) and liver (Gallis and Canioni, 1992) and of the relationship of pH to contractile force and fatigue in striated muscle (Weiner et al., 1990; Argov and Bank, 1991; Gallis and Canioni, 1992). When two different NMR indicators were used to measure pH_i in erythrocytes (Stewart et al., 1986) and exercising muscle (Pan et al., 1988), the numbers agreed well.

The precision of pH_i measurements from Pi has been reported as about 0.1 pH unit under practical conditions in studies of liver mitochondria (Ogawa et al., 1981), muscle (Gadian et al., 1979; Taylor et al., 1983), and brain (Alger et al., 1989; Behar et al., 1989). Many factors that influence the limit of precision are common to all NMR methods for measurement of pH_i. The case of Pi illustrates them.

As with pH measurement by any indicator method, the p*K* of the indicator determines the pH range that can be effectively observed (Chapters 7 and 8). Under conditions prevailing in the brain, the Pi titration curve is well situated for monitoring of pH_i changes in the range from 5.5 to 7.5. Distribution of the indicator across tissue compartments is obviously critical for interpretation of pH values inferred from a single signal averaged over volumes of complex tissue that are large compared to cell size. NMR measurements of pH are unlikely to achieve the anatomical selectivity or spatial resolution of pH microelectrodes (Chapter 6) for years, if they ever do. In the foreseeable future, pH_i determined by NMR methods will be an average of pH in millions of cells. Like the gross domestic product of a nation, the measurement can be valuable for many purposes, as long as it is interpreted with due regard for its limitations. At present, Pi is generally assumed to exist in all brain cells at concentrations of a few millimolar, inferred from Pi content per unit volume of wet tissue. If heterogeneity of Pi distribution among brain cells exists, it is not currently taken into account, and it cannot be until independent measurements of Pi concentrations at the cellular level are made.

Most other factors affecting pH measurement by NMR are technical features of NMR data collection and analysis. Magnetic field strength places a fundamental limit on the sensitivity and resolution of all NMR measurements, which generally are more accurate and precise in direct proportion to field strength. At the 1.5–2 Tesla fields most often used for ^{31}P MRS of human brain, the natural line width of Pi is 4 Hz (0.11–0.16 ppm), equivalent to a spin-spin relaxation time (T_2) of 80 milliseconds (Gadian et al., 1979). These technical features set a limit on how precise pH measurement from Pi can be, since measurement of chemical shift change can rarely be more precise than one-tenth of a linewidth. Low spectral signal-to-noise ratio, line widths broadened by uncorrectable magnetic field inhomogeneity (Gadian, 1982), and low digital resolution all degrade the precision of pH_i measurements from any NMR indicator. In principle, line fitting of NMR signals to Lorentzian or Gaussian models in the frequency domain ought to improve precision, but early achievements have been modest. Some published data based on signal analysis in the time domain retain uncertainties in the range of 0.04–0.17 ppm, corresponding to errors up to 0.14 pH unit (Haselgrove and Elliott, 1991; Joliot et al., 1991; Madden et al., 1991; Morikawa et al., 1993). Under favorable circumstances, improved precision with standard deviations of 0.02–0.06 pH unit is possible (van der Veen et al., 1988; Joliot et al., 1991; Mazzeo and Levy, 1991; Zaim-Wadghiri et al., 1992).

NOVEL FINDINGS FROM NMR STUDIES OF BRAIN pH_i

Most NMR work on brain pH_i done by the mid-1990s belongs to the development and validation phase of a new technology, in which the emphasis is on demonstrating that known phenomena can be detected. However, a few studies have provided new findings with potential to increase understanding of brain pathophysiology:

Divergence of pH_i from [lactate] during increased lactic acid production was demonstrated during experimental status epilepticus (Petroff et al., 1986). In human stroke, a change from acidosis to alkalosis a few days after onset may indicate the outer limit of effectiveness for some kinds of stroke therapy (Levine et al., 1992). Nor has the new capability for measuring pH_i *in vivo* taken long to generate differences that must be resolved by further study: Alkalosis was observed in the epileptic temporal lobes of patients with intractable complex partial seizures by one group of investigators (Hugg et al., 1992) but not by another (Kuzniecky et al., 1992). A more recent study at 4 Tesla agrees with the latter group (Chu et al., 1996).

CONCLUSION

NMR methods are still in an early stage as measures of brain pH_i. Much technological improvement leading to greater accuracy and to better anatomical and time resolution is surely ahead. Even in the forms described in this chapter, which are coarse by the standard of test-tube chemistry, NMR measurement of pH_i from the Pi signal of brain and a number of other tissues is adequate for many research purposes. Its anatomical resolution ranges from a few cubic millimeters in small animals studied in high-field research spectrometers to a few cubic centimeters in the human brain at the usual medical imaging field of 1.5 Tesla. Its time resolution can be under 1 minute for some purposes.

All NMR methods are limited by the requirement that they be made in a magnet; by space and time resolutions that even with plausible improvements are unlikely to reach the cellular or millisecond level; and by assumptions necessary for biological interpretation of the data they generate. Nevertheless, they are the most generally useful means available currently, or in the foreseeable future, for noninvasive measurement of brain pH.

ACKNOWLEDGMENTS

The authors' own work was supported by USPHS grants RO1 NS 27883 and RO1 NS 21708.

REFERENCES

Ackerman JJH, Grove TH, Wong GG, Gadian DG, Radda GK (1980): Mapping of metabolites in whole animals by ^{31}P NMR using surface coils. Nature 283: 167–170.

Adler S, Simplaceanu V, Ho C (1988): Determination of rat brain buffering in vivo by ^{31}P-NMR. J Appl Physiol 64: 1829–1836.

Alger JR, Brunetti A, Nagashima G, Hossmann KA (1989): Assessment of postischemic cerebral energy metabolism in cat by ^{31}P NMR: The cumulative effects of secondary hypoxia and ischemia. J Cereb Blood Flow Metab 9: 506–514.

Andersen BJ, Unterberrg AW, Clarke GD, Marmarou A (1988): Effect of post-traumatic hypoventilation on cerebral energy metabolism. J Neurosurg 68: 601–607.

Argov Z, Bank WJ (1991): Phosphorus magnetic resonance spectroscopy (^{31}P MRS) in neuromuscular disorders. Ann Neurol 30: 90–97.

Arus C, Westler WM, Barany M, Markley JL (1986): Observation of the terminal methyl group in fatty acids of the linolenic series by a new ^{1}H NMR pulse sequence providing spectral editing and solvent suppression. Application to excised frog muscle and rat brain. Biochemistry 25: 3346–3351.

Bailey IA, Williams SR, Radda GK, Gadian DG (1981): Activity of phosphorylase in total global ischaemia in the rat heart: A phosphorus-31 nuclear magnetic resonance study. Biochemical J 196: 171–178.

Bain PG, O'Brien MD, Keevil SF, Porter DA (1992): Familial periodic cerebellar ataxia: A problem of cerebellar intracellular pH homeostasis. Ann Neurol 31: 147–154.

Behar KL (1993): Cerebral metabolic studies in vivo by combined ^{1}H/^{31}P and ^{1}H/^{13}C NMR spectroscopic methods. Acta Neurochir 57: 9–20.

Behar KL, Rothman DL, Fitzpatrick SM, Hetherington HP, Shulman RG (1987): Combined ^{1}H and ^{31}P NMR studies of the rat brain in vivo: Effects of altered intracellular pH on metabolism. Ann NY Acad Sci 508: 81–88.

Behar KL, Rothman DL, Hossman KA (1989): NMR spectroscopic investigation of the recovery of energy and acid-base homeostasis in the cat brain after prolonged ischemia. J Cereb Blood Flow Metab 9: 655–665.

Bernard M, Cozzone PJ (1992): MRS of normal and pathological mycardium. In de Certaines JD, Bovee WMMJ, Podo F (eds): Magnetic Resonance Spectroscopy in Biology and Medicine. New York: Pergamon Press, pp 387–410.

Bickler PE, Litt L, Banville DL, Severinghaus JW (1988): Effects of acetazolamide on cerebral acid-base balance. J Appl Physiol 65: 422–427.

Bloch F, Hansen WW, Packard ME (1946): Nuclear induction. Phys Rev 69: 127.

Brenot F, Aubry L, Martin JB, Satre M, Klein G (1992): Kinetics of endosomal acidification in Dictyostelium discoideum amoebae. ^{31}P-NMR evidence for a very acidic early endosomal compartment. Biochimie 74: 883–895.

Brooks KJ, Porteous R, Bachelard HS (1989): Effects of hypoglycemia and hypoxia on the intracellular pH of cerebral tissue as measured by ^{31}P nuclear magnetic resonance. J Neurochem 52: 604–610.

Brown TR (1992): Practical applications of chemical shift imaging. [Review]. NMR Biomed 5: 238–243.

Burri R, Lazeyras F, Aue WP, Straehl P, Bigler P, Althaus U, Herschkowitz N (1988): Correlation between ^{31}P NMR phosphomonoester and biochemically determined phosphorylethanolamine and phosphatidylethanolamine during development of the rat brain. Dev Neurosci 10: 213–221.

Cady EB, Costello AM, Dawson MJ, Delpy DT, Hope PL, Reynolds EOR, Tofts PS, Wilkie DR (1983): Non-invasive investigation of cerebral metabolism in newborn infants by phosphorus nuclear magnetic resonance spectroscopy. Lancet 1(8333): 1059–1062.

Chance B, Eleff S, Leigh JS (1980): Noninvasive, nondestructive approaches to cell bioenergetics. Proc Natl Acad Sci USA 77: 7430–7434.

Chesler M (1990): The regulation and modulation of pH in the nervous system. Prog Neurobiol 34: 401–427.

Chesler M, Kaila K (1992): Modulation of pH by neuronal activity. Trends Neurosci 15: 396–402.

Chiles C, Sachar-Hill Y, Behar KL (1990): ^{19}F NMR determination of intracerebral pH in rat brain in vivo. Soc Magn Reson Med Abstr 9: 1239.

Chu WJ, Hetherington HP, Kuzniecky RJ, Vaughan JT, Twieg DB, Faught RE, Gilliam FG, Hugg JW, Elgavish GA (1996): Is the intracellular pH different from normal in the epileptic focus of patients with temporal lobe epilepsy? Neurology 47: 756–760.

Civan MM, Lin LE, Peterson-Yantorno K, Taylor J, Deutsch C (1984): Intracellular pH of perfused single frog skin: Combined ^{19}F- and ^{31}P-NMR analysis. Am J Physiol 247: C506–C510.

Clarke K, Stewart LC, Neubauer S, Balschi JA, Smith TW, Ingwall JS (1993): Extracellular volume and transsarcolemmal proton movement during ischemia and reperfusion: A ^{31}P NMR spectroscopic study of the isovolumic rat heart. NMR Biomed 6: 278–286.

Cohen SM, Ogawa S, Rottenberg H, Glynn P, Yamane T, Brown TR, Shulman RG, Williamson JR (1978): ^{31}P nuclear magnetic resonance studies of isolated rat liver cells. Nature 273: 554–556.

Connett RJ (1988): Analysis of metabolic control: New insights using scaled creatine kinase model. Am J Physiol 254: R949–R959.

Corbett RJT, Laptook AR, Hassan A, Nunnally RL (1988): Quantitation of acidosis in neonatal brain tissue using ^{31}P NMR resonance peak of phosphoethanolamine. Magn Reson Med 6: 99–106.

Corbett RJT, Laptook AR, Nunnally RL (1987): The use of the chemical shift of the phosphomonoester P-31 magnetic resonance peak for the determination of intracellular pH in the brains of neonates. Neurology 37: 1771–1779.

Deutsch CJ, Taylor JS (1987): ^{19}F NMR measurements of intracellular pH. In Gupta RK (ed): NMR Spectroscopy of Cells and Organisms. Boca Raton, FL: CRC Press, pp 55–74.

Fishman RA (1992): Cerebrospinal Fluid in Diseases of the Nervous System. Philadelphia: WB Saunders.

Gadian DG (1982): Nuclear Magnetic Resonance and Its Applications to Living Systems. Oxford: Clarendon Press.

Gadian DG, Proctor E, Williams SR, Cady EB, Gardiner RM (1986): Neurometabolic effects of an inborn error of amino acid metabolism demonstrated in vivo by ^{1}H NMR. Magn Reson Med 3: 150–156.

Gadian DG, Radda GK (1981): NMR studies of tissue metabolism. Ann Rev Biochem 50: 69–83.

Gadian DG, Radda GK, Richards RE, Seeley PJ (1979): ^{31}P NMR in living tissue: The road from a promising to an important tool in biology. In Shulman RG (ed): Biological Applications of Magnetic Resonance. New York: Academic Press, pp 463–535.

Gadian DG, Williams SR, Bates TE, Kauppinen RA (1993): NMR Spectroscopy: Current status and future possibilities. Acta Neurochir 57: 1–8.

Gallis JL, Canioni P (1992): MRS of the liver. In de Certaines JD, Bovee WMMJ, Podo F (eds): Magnetic Resonance Spectroscopy in Biology and Medicine. New York: Pergamon Press, pp 345–368.

Gard JK, Ackerman JJH (1983): A ^{31}P NMR external reference for intact biological systems. J Magn Reson 51: 124–127.

Garlick PB, Brown TR, Sullivan RH, Ugurbil K (1983): Observation of a second phosphate pool in the perfused heart by ^{31}P NMR: Is this the mitochondrial phosphate? J Mol Cell Cardiol 15: 855–858.

Garlick PB, Radda GK, Seeley PJ (1979): Studies of acidosis in the ischaemic heart by phosphorus nuclear magnetic resonance. Biochem J 184: 547–554.

Gilboe DD, Kintner D, Anderson ME, Fitzpatrick JH, Emoto SE, Markley JL (1993): Inorganic phosphate compartmentation in the normal isolated canine brain. J Neurochem 60: 2192–2203.

Gillies RJ, Alger JR, den Hollander JA, Shulman RG (1982): Intracellular pH measured by NMR: Methods and Results. In Nucitelli R, Deamer D (eds): Intracellular pH: Its Measurement, Regulation, and Utilization in Cellular Functions. New York: Alan R. Liss, pp 79–104.

Gonzalez-Mendez R, Hahn GM, Wade-Jardetzky NG, Jardetzky O (1988): Comparison of intracellular pH measurements by ^{31}P NMR and weak acid partitioning in Chinese hamster ovary fibroblasts. Magn Reson Med 6: 373–380.

Gyulai L, Bolinger L, Leigh SL, Barlow C, Chance B (1984): Phosphorylethanolamine: The major constituent of the phosphomonoester peak observed by ^{31}P-NMR on developing dog brain. FEBS Lett 178: 137–141.

Halvorson HR, Vande Lind AM, Helpern JA, Welch KMA (1992): Assessment of magnesium concentrations by ^{31}P NMR in vivo. NMR Biomed 5: 53–58.

Haselgrove J, Elliott M (1991): Peak assignment in automatic data analysis. Magn Reson Med 17: 496–508.

Hoehn-Berlage M, Okada Y, Kloiber O, Hossman KA (1989): Imaging of brain tissue pH and metabolites. NMR Biomed 2: 240–245.

Hoult DI, Busby SJW, Gadian DG, Radda GK, Richards RE, Seeley PJ (1974): Observation of tissue metabolites using ^{31}P nuclear magnetic resonance. Nature 252: 285–287.

Hugg JW, Laxer KD, Matson GB, Maudsley AA, Husted CA, Weiner MW (1992): Lateralization of human focal epilepsy by ^{31}P magnetic resonance spectroscopic imaging. Neurology 42: 2011–2018.

Jayasundar R, Hall LD, Bleehan NM (1992): Comparison of pH measurements made using ^{31}P NMR and a fiberoptic pH meter. NMR Biomed 5: 360–363.

Joliot M, Mazoyer BM, Huesman RH (1991): In vivo NMR spectral parameter estimation: A comparison between time and frequency domain methods. Magn Reson Med 18: 358–370.

Kushmerick MJ, Dillon PF, Meyer RA, Brown TR, Krisanda JM, Sweeney HL (1986): ^{31}P NMR spectroscopy, chemical analysis, and free Mg^{2+} of rabbit bladder and smooth muscle. J Biol Chem 262: 14420–14429.

Kuzniecky R, Elgavish GA, Hetherington HP, Evanochko WT, Pohost GM (1992): In vivo ^{31}P nuclear magnetic resonance spectroscopy of human temporal lobe epilepsy. Neurology 42: 1586–1590.

Levine SR, Helpern JA, Welch KM, Vande Linde AM, Sawaya KL, Brown EE, Ramadan NM, Deveshwar RK, Ordidge RJ (1992): Human focal cerebral ischemia: Evaluation of brain pH and energy metabolism with P-31 NMR spectroscopy. Radiology 185: 537–544.

Lien YHH, Zhou HZ, Job C, Barry JA, Gillies RJ (1992): In vivo ^{31}P NMR study of early cellular responses to hyperosmotic shock in cultured glioma cells. Biochimie 74: 931–939.

Litt L, Gonzalez-Mendez R, Severinghaus JW, Hamilton WK, Shuleshko J, Murphy-Boesch J, James TL (1985): Cerebral intracellular changes during supercarbia: An in vivo ^{31}P nuclear magnetic resonance study in rats. J Cereb Blood Flow Metab 5: 537–544.

Lundberg P, Vogel HJ (1987): NMR as a noninvasive tool in meat research. Ann N Y Acad Sci 508: 516–522.

Luyten PR, Groen JP, Vermeulen JW, den Hollander JA (1989): Experimental approaches to image localized human ^{31}P NMR spectroscopy. Magn Reson Med 11: 1–21.

Madden A, Leach MO, Sharp JC, Collins DJ, Easton D (1991): A quantitative analysis of the accuracy of in vivo pH measurements with ^{31}P NMR spectroscopy: Assessment of pH measurement methodology. NMR Biomed 4: 1–11.

Mazzeo AR, Levy GC (1991): An evaluation of new processing protocols for in vivo NMR spectroscopy. Magn Reson Med 17: 483–495.

McIlwain H, Bachelard HS (1985): Biochemistry and the Central Nervous system. New York: Churchill Livingstone.

Merle M, Pianet I, Canioni P, Labouesse J (1992): Comparative ^{31}P and ^{1}H NMR studies on rat astrocytes and C6 glioma cells in culture. Biochimie 74: 919–930.

Moon RB, Richards JH (1973): Determination of intracellular pH by ^{31}P magnetic resonance. J Biol Chem 148: 7276–7278.

Moore GJ, Sillerud LO (1993): The pH dependence of chemical shift and spin-spin coupling for citrate. Proc Soc Magn Reson Med 2: 958.

Morikawa S, Inubushi T, Kito K, Kido C (1993): pH mapping in living tissues: An application of in vivo ^{31}P NMR chemical shift imaging. Magn Reson Med 29: 249–251.

Navon G, Ogawa S, Shulman RG, Yamane T (1977): ^{31}P nuclear magnetic resonance studies of Ehrlich ascites tumor cells. Biochemistry 74: 87–91.

Nucitelli R, Deamer D (ed) (1982): Intracellular pH: Its Measurement, Regulation, and Utilization in Cellular Functions. New York: Alan R. Liss.

Ogawa S, Boens CC, Lee T-M (1981): A ^{31}P nuclear magnetic resonance study of pH gradient and the inorganic phosphate distribution across the membrane in intact rat liver mitochondria. Arch Biochem Biophys 210: 740–747.

Pan JW, Hamm JR, Rothman DL, Shulman RG (1988): Intracellular pH in human skeletal muscle by ^{1}H NMR. Proc Natl Acad Sci USA 85: 7836–7839.

Perry T (1982): Cerebral amino acid pools. In Lathja A (ed): Handbook of Neurochemistry. New York: Plenum Press, pp 151–180.

Petroff OAC (1996): The measurement of in vivo and ex vivo pH using nuclear magnetic resonance spectroscopy. Eur J Lab Med (in press).

Petroff OAC, Prichard JW (1995): Measurement of cytosolic pH by nuclear magnetic resonance spectroscopy. In Kraicer J, Dixon SJ (eds): Methods in Neurosciences. Orlando, FL: Academic Press, pp 233–251.

Petroff OAC, Prichard JW, Behar KL, Alger JR, den Hollander JA, Shulman RG (1985): Cerebral intracellular pH by ^{31}P nuclear magnetic resonance spectroscopy. Neurology 35: 781–788.

Petroff OAC, Prichard JW, Ogino T, Avison M, Alger JR, Shulman RG (1986): Combined ^{1}H and ^{31}P nuclear magnetic resonance spectroscopic studies of bicuculline-induced seizures in vivo. Ann Neurol 20: 185–193.

Portman MA, Lassen NA, Cooper TG, Sills AM, Potchen EJ (1991): Intra- and extracellular pH of the brain in vivo studied ^{31}P-NMR during hyper- and hypocapnia. J Appl Physiol 71: 2168–2172.

Prichard JW (1995): The NMR revolution in basic and clinical neuroscience. Neuroscientist 1: 84–94.

Prichard JW (1996a): Brain MRS of human subjects. In Grant DM, Harris RK, Young IR (eds): Encyclopedia of NMR, Vol. 2. Chichester, John Wiley & Sons, pp 1023–1030.

Prichard JW (1996b): Magnetic resonance spectroscopy of the brain: Prospects for clinical application. In Young IR, Charles HC (eds): Magnetic Resonance Spectroscopy: Clinical Applications and Techniques. London, Martin Dunitz, pp 1–25.

Prichard JW, Alger JR, Behar KL, Petroff OA, Shulman RG (1983): Cerebral metabolic studies in vivo by ^{31}P NMR. Proc Natl Acad Sci USA 80: 2748–2751.

Purcell EM, Torrey HC, Pound RV (1946): Resonance absorption by nuclear magnetic moments in a solid. Phys Rev 69: 37–38.

Roberts JKM, Ray PM, Wade-Jardetzky N, Jardetzky O (1980): Estimation of cytoplasmic and vacuolar pH in higher plant cells by ^{31}P NMR. Nature 283: 870–872.

Roberts JKM, Wade-Jardetzky N, Jardetzky O (1981): Intracellular pH measurements by ^{31}P nuclear magnetic resonance. Influence of factors other than pH on ^{31}P chemical shifts. Biochemistry 20: 3389–3394.

Rosenberg JM, Martin GM, Paradis NA, Nowak RM, Walton D, Appleton TJ, Welch KMA (1989): The effect of CO_2 and non-CO_2-generating buffers on cerebral acidosis after cardiac arrest: A ^{31}P NMR study. Ann Emerg Med 18: 341–347.

Rothman DL, Behar KL, Prichard JW, Petroff OAC (1997): Homocarnosine and the measurement of neuronal pH in patients with epilepsy. Magn Reson Med 32: 924–929.

Salhany JM, Pieper GM, Wu S, Todd GL, Clayton FC, Eliot RS (1979): ^{31}P nuclear magnetic resonance measurement of cardiac pH in perfused guinea-pig hearts. J Mol Cell Cardiol 11: 601–610.

Seo Y, Murakami M, Watari H, Imai Y, Yoshizaki K, Nishikawa H, Morimoto T (1983): Intracellular pH determination by a ^{31}P-NMR Technique. The second dissociation constant of phosphoric acid in a biological system. J Biochem 94: 729–734.

Sessler D, Mills P, Gregory G, Litt L, James T (1987): Effects of bicarbonate on arterial and brain intracellular pH in neonatal rabbits recovering from hypoxic lactic acidosis. J Pediatr 111: 817–823.

Slonczewski JL, Rosen BP, Alger JR, Macnab RM (1981): pH homeostasis in escherichia coli: Measurement by ^{31}P nuclear magnetic resonance of methylphosphonate and phosphate. Proc Natl Acad Sci USA 78: 6271–6275.

Stark DD, Bradley WB (1992): Magnetic Resonance Imaging. St. Louis: Mosby Yearbook.

Stewart IM, Chapman BE, Kirk K, Kuchel PW, Lovric VA, Raftos JE (1986): Intracellular pH in stored erythrocytes. Refinement and further characterization of the ^{31}P-NMR methylphosphonate procedure. Biochim Biophys Acta 885: 23–33.

Stubbs M, Bhujwalla ZM, Tozer GM, Rodrigues LM, Maxwell RJ, Morgan R, Howe FA, Griffiths JR (1992): An assessment of ^{31}P MRS as a method of measuring pH in rat tumours. NMR Biomed 5: 351–359.

Taylor JS, Deutsch C (1983): Fluorinated α-methylamino acids as ^{19}F NMR indicators of intracellular pH. Biophys J 43: 261–267.

Taylor DJ, Bore PJ, Styles P, Gadian DG, Radda GK (1983): Bioenergetics of intact human muscle: A ^{31}P nuclear magnetic resonance study. Mol Biol Med 1: 77–94.

van der Veen JW, de Beer R, Luyten PR, van Ormondt D (1988): Accurate quantification of in vivo ^{31}P NMR signals using the variable projection method and prior knowledge. Magn Reson Med 6: 92–98.

van Rijen PC, Luyten PR, Berkelbach van der Sprenkel JW, Kraaier V, van Huffelen AC, Tulleken CA, den Hollander JA (1989): ^{1}H and ^{31}P NMR measurement of cerebral lactate, high-energy phosphate levels, and pH in humans during voluntary hyperventilation: Associated EEG, capnographic, and Doppler findings. Magn Reson Med 10: 182–193.

Veech RL, Lawson JW, Cornell NW, Krebs HA (1979): Cytosolic phosphorylation potential. J Biol Chem 254: 6538–6547.

Vorstrup S, Jensen KE, Thomsen C, Henriksen O, Lassen NA, Paulson OB (1989): Neuronal pH regulation: Constant normal intracellular pH is maintained in brain during low extracellular pH induced by acetazolamide—^{31}P NMR study. J Cereb Blood Flow Metab 9: 417–421.

Walter L, Callies R, Altenburger R (1992): Studies of plant systems by in vivo NMR spectroscopy. In de Certaines JD, Bovee WMMJ, Podo F (eds): Magnetic Resonance Spectroscopy in Biology and Medicine. New York: Pergamon Press, pp 573–610.

Weiner MW, Moussavi RS, Baker AJ, Boska MD, Miller RG (1990): Constant relationships between force, phosphate concentration, and pH in muscles with differential fatigability. Neurology 40: 1888–1893.

Williams SR, Gadian DG, Proctor E, Sprague DB, Talbot DF, Young IR, Brown FF (1985): Proton NMR studies of muscle metabolites in vivo. J Magn Reson 63: 406–412.

Yoshizaki K, Seo Y, Nishikawa H (1981): High-resolution proton magnetic resonance spectra of muscle. Biochim Biophys Acta 678: 283–291.

Younkin DP, Delivoria-Papadopoulos M, Leonard JC, Subramanian VH, Eleff S, Leigh JS, Chance B (1984): Unique aspects of human newborn cerebral metabolism evaluated with phosphorus nuclear magnetic resonance spectroscopy. Ann Neurol 16: 581–586.

Zaim-Wadghiri Y, Diop A, Graveron-Demilly D, Briguet A (1992): Improving data acquisition parameters of ^{31}P in vivo spectra for signal analysis in the time domain. Biochimie 74: 769–776.

SECTION III

Cellular Mechanisms of pH Regulation

CHAPTER 10

Thermodynamics and Physiology of Cellular pH Regulation

MARK O. BEVENSEE AND WALTER F. BORON
Department of Cellular and Molecular Physiology, Yale University School of Medicine, New Haven, CT 06520

INTRODUCTION

The four chapters of the previous section focus on four powerful techniques for measuring pH both inside (pH_i) and/or outside (pH_o) the cell. Using these techniques, investigators have measured pH_i under both steady-state conditions, and during perturbations designed to assess the cellular mechanisms by which cells regulate pH_i. The results from such studies have revealed that brain cells use a host of acid-base transport proteins that (i) cause pH_i to recover from transient acid or alkali loads, and (ii) thereby contribute to maintaining the steady-state pH_i of a cell. The known acid-base transporters responsible for regulating the pH_i of neurons and glia, in both invertebrates and mammals, will be the subject of the remaining five chapters of this section (Chapters 11–15). The goal of this chapter is to provide the reader with a general overview of the physiology of cellular pH regulation. Although the complement of specific acid-base transporters may be different in neurons and glia than in other cells of the body, the general principles governing the pH_i physiology of brain cells is identical to those governing cells elsewhere in the body. To begin, we will review some basic thermodynamic principles relevant in biological systems. Second, we will examine the thermodynamics of proton movement, that is, the forces that govern the directionality of proton movement (see also Chapter 3). It will become apparent that under most conditions the electrochemical gradient favors proton movement into cells, thereby tending to acidify them. Third, we will discuss the thermodynamics and function of acid-base transporters that are responsible for countering the effects of passive proton movements, and thereby maintaining pH_i well above that predicted by the electrochemical gradient for protons. Finally, we will describe how pH_i is controlled by the balance between processes that tend to raise pH_i (acid extruders) and those that tend to lower pH_i (acid loaders).

pH and Brain Function, Edited by Kai Kaila and Bruce R. Ransom
ISBN 0-471-11838-9

THERMODYNAMICS OF ACID-BASE MOVEMENT

Thermodynamics in Biological Systems

Before discussing the forces that govern proton movement, we will first review some of the thermodynamic principles relevant in biological systems. In biology, most experiments are performed at constant temperature (T) and pressure (P). Under such conditions, the spontaneity of a reaction can be described by the change in the Gibbs free energy, which is defined as:

$$\Delta G = \Delta H - T\Delta S \tag{1}$$

where ΔH and ΔS are the changes in enthalpy and entropy, respectively. If ΔG is negative, that is, if the Gibbs free energy decreases, the reaction can occur spontaneously. If ΔG is positive, the reaction can not occur spontaneously. Finally, if ΔG is zero, the reaction is at equilibrium.

Let us further evaluate the enthalpy and entropy terms in Eq. 1. Imagine that we have a quantity of gas contained in a cylinder bounded by a free-moving piston. The pressure inside the cylinder (the "system") is constant. We now withdraw an amount of heat (q_p) from the surroundings and transfer it to the system, which will have two effects. First, the energy of the gas will increase by an amount ΔE, as manifested by an increase in temperature. Second, the volume of the gas will increase and raise the piston, thereby performing an amount of work, $P\Delta V$. Because energy is conserved,

$$q_p = \Delta E + P\Delta V \tag{2}$$

The enthalpy of the system is defined as

$$H = E + PV \tag{3}$$

Thus, combining Eq. 2 and Eq. 3, we have

$$\Delta H = \Delta E + P\Delta V \tag{4}$$

For most biochemical reactions, except those at pressures of thousands of atmospheres, or those involving gases, the $P\Delta V$ term in Eq. 2 is small compared to ΔE. Therefore, ΔH is approximately the same as ΔE:

$$\Delta H \cong \Delta E \tag{5}$$

Returning now to Eq. 1, consider what happens to the free or "usable" energy when glucose is oxidized to CO_2 and H_2O

$$\text{Glucose} + 6\ O_2 \rightarrow 6\ CO_2 + 6\ H_2O \qquad \text{Rx. 1}$$

In order for the reaction to proceed spontaneously, the overall ΔG must be negative. In other words, the free energy of the CO_2 and H_2O formed must be less than that of the glucose and O_2 consumed, so that the reaction proceeds energetically "downhill." However, if Rx. 1 were performed in a calorimeter, then the change in free energy would manifest itself both as heat transferred to the surroundings (ΔH) and as an increase in entropy, which is related to the randomness of the system. In principle, the heat transferred to the surroundings could be harnessed. The entropy contribution ($T\Delta S$) cannot, and thus represents inefficiency in the system. Only at the unreachable temperature of absolute zero could all the free energy released in a chemical reaction be captured as heat. In our glucose example, both ΔH and $-T\Delta S$ are negative, so that the ΔG strongly favors the reaction as written in Rx. 1. We might have chosen a different reaction with a decrease in entropy. In such a case, the reaction could proceed spontaneously only if the heat released to the surroundings (ΔH) has a value of greater magnitude than $T\Delta S$.

In a living cell, the oxidation of glucose in Rx. 1 is coupled to the synthesis of ATP

$$\text{Glucose} + 6\,O_2 + 36\,\text{ADP} + 36\,\text{Pi} \rightarrow 6\,CO_2 + 42\,H_2O + 36\,\text{ATP} \qquad \text{Rx. 2}$$

Thus, some of the free energy released in the oxidation of glucose (Rx. 1) is captured, so to speak, in the synthesis of ATP, a reaction with a positive ΔG. The cell can later use the free energy stored as ATP for such processes as movement, ion transport, and biosynthesis. Of course, the overall ΔG in Rx. 2 is negative. Some of the free energy released in this overall reaction is liberated as heat, and some as entropy.

Another way of viewing Eq. 1 is to ask what would happen to a chemical system, at constant T and P, when we transfer heat (ΔH) from the surroundings to the system. The energy transferred to the system would manifest itself in two ways. Some of that energy transferred from the surroundings would increase the usable or free energy in the system (ΔG). The rest would increase the entropy ($T\Delta S$), and thus represent inefficiency.

Forces that Govern Proton Movement

Electrochemical free energy of protons. According to the laws of thermodynamics, a proton will tend to move from point A to point B if their free energy decreases in the process. Stated differently, protons tend to move passively from a region of high free energy to one of low free energy. Of course, whether and how fast protons actually move from A to B depends on factors other than thermodynamics, as discussed below (p. 179). As noted in the preceding section, the free energy change associated with the movement of protons depends on the change in both the enthalpy and the entropy of the system. However, as Le Châtelier deduced, the free energy change of a reaction also depends on the concentrations of the reactants and products. In the example of protons, i.e. charged hydrogen atoms, their free energy at any point is comprised of both a chemical and an electrical component. The chemical free energy for a mole of protons is defined as

$$\mu_{\text{chem},H^+} = \mu^{O}_{H^+} + RT \ln a_{H^+} \qquad (6)$$

where $\mu^{O}_{H^+}$ is the standard chemical potential, R is the universal gas constant (8.31 J K^{-1} mole^{-1}), T is temperature (K) and a_{H+} is the activity of H^+. The electrical free energy for a mole of protons is defined as

$$\mu_{\text{elec},H^+} = zFV \qquad (7)$$

where z is the valence of the H^+, F is Faraday's constant (96487 C/mole) and V is voltage (V or J/C). Coulomb (C) is the quantity of charge. The total, or electrochemical, free energy of protons is the sum of both the chemical and electrical free energies:

$$\begin{aligned}\mu_{H+} &= \mu_{chem,H+} + \mu_{elec,H+} \\ &= \mu^{O}_{H+} + RT \ln a_{H+} + zFV\end{aligned} \tag{8}$$

To determine whether thermodynamics favors proton movement into or out of a cell, we must compute the change in the protons' free energy as they cross the cell membrane. Let us take as an example protons moving into a cell:

$$(H^+)_O \rightarrow (H^+)_i \qquad \text{Rx. 3}$$

where o and i represent the outside and inside of the cell, respectively. For this reaction as written, the free energy released or consumed is the difference in the electrochemical potentials ($\Delta\mu_{H+}$) between the products and reactants:

Product: $$\mu_{H+,i} = \mu^{O}_{H+,i} + RT \ln a_{H+,i} + zFV_i \tag{9}$$

Reactant: $$\mu_{H+,o} = \mu^{O}_{H+,o} + RT \ln a_{H+,o} + zFV_o \tag{10}$$

$$\Delta\mu_{H+} = RT \ln \frac{a_{H+,i}}{a_{H+,o}} + zFV_m \tag{11}$$

where we have assumed that the standard electrochemical potential for H^+ inside and outside the cell is the same (i.e., $\mu^{O}_{H+,i} = \mu^{O}_{H+,o}$). Also note that the voltage difference between the inside (V_i) and outside (V_o) of the cell is simply the membrane potential (V_m). Converting the natural logarithm (ln) in Eq. 11 to the logarithm to the base 10 (log).

$$\Delta\mu_{H+} = 2.3\, RT \log \frac{a_{H+,i}}{a_{H+,o}} + zFV_m \tag{12}$$

The pH of a solution is defined as $-\log(a_{H+})$. If, for simplicity, we assume that the activity of H^+ is identical to the concentration of H^+, we can rewrite Eq. 12 as:

$$\Delta\mu_{H+} = 2.3\, RT\, (pH_o - pH_i) + zFV_m \tag{13}$$

For a cell with a V_m of -60 mV, a pH_i of 7.0, and a pH_o of 7.3,

$$\Delta\mu_{H+} = 2.3 \times 8.31 \frac{J}{K\ mole} \times 310\ K \times (7.3 - 7.0) + \left[(+1) \times 96487 \frac{C}{mole} \times \left(-0.060 \frac{J}{C}\right)\right]$$

$$\Delta\mu_{H+} = 1.8\ kJ\ mole^{-1} - 5.8\ kJ\ mole^{-1} = -4.0\ kJ\ mole^{-1}$$

Because $\Delta\mu_{H+}$ is negative, free energy decreases and Rx. 3 may proceed spontaneously as written. In the above example, protons may enter the cell passively.

Equilibrium potential for protons. According to Eq. 13, $\Delta\mu_{H+}$ depends on both the voltage difference (V_m), and the pH difference ($pH_o - pH_i$), or ratio of H^+ activities, across the plasma membrane. As an example of the influence of V_m, imagine performing an experiment in which we measure $\Delta\mu_{H+}$ as we make a step change in V_m. Let us assume that pH_o is 7.3, and pH_i is clamped at 7.0. As computed above, $\Delta\mu_{H+}$ will be -4 kJ/mole at a V_m of -60 mV, and H^+ will tend to enter the cell. However, as the cell is progressively depolarized (more positive V_m), $\Delta\mu_{H+}$ will become less and less negative, so that there will be less and less driving force for the positively charged protons to enter. If V_m is increased sufficiently, $\Delta\mu_{H+}$ will reach zero, and there can be no net movement of H^+. The V_m in which $\Delta\mu_{H+} = 0$ is called the equilibrium potential for H^+ (E_{H+}). At V_m values more positive than E_{H+}, $\Delta\mu_{H+}$ will be positive and the driving force will favor proton exit from the cell. To calculate E_{H+}, we simply set $\Delta\mu_{H+}$ equal to 0 in Eq. 13 and solve for V_m:

$$\Delta\mu_{H+} = 0 = 2.3\,RT\,(pH_o - pH_i) + zFV_m \tag{14}$$

$$zFV_m = -2.3\,RT\,(pH_o - pH_i) \tag{15}$$

$$V_m = E_{H+} = \frac{-2.3\,RT\,(pH_o - pH_i)}{zF} \tag{16}$$

In our example above of a cell clamped at a pH_i of 7.0 and bathed in a solution of pH 7.3:

$$E_{H+} = \frac{-2.3 \times 8.31\ \mathrm{J\,K^{-1}\,mole^{-1}} \times 310\ \mathrm{K} \times (7.3 - 7.0)}{(+1) \times 96487\ \mathrm{C\,mole^{-1}}} \cong -18\ \mathrm{mV}$$

At 37°C, the term $-2.3RT/(zF)$ in Eq. 16 reduces to approximately -61 mV, so that for every pH unit of difference between pH_i and pH_o, E_{H+} changes by ~61 mV.

In our example above, $\Delta\mu_{H+}$ will be 0 at a V_m of -18 mV, and there will be no net driving force for H^+ to enter or leave the cell. At potentials more positive than -18 mV, such as might exist transiently during an action potential for example, the driving force will favor H^+ exit. On the other hand, at a V_m of -60 mV, more typical of resting neurons, the driving force will favor H^+ entry.

Substituting the definition of pH into Eq. 16 yields an expression equivalent to the Nernst equation:

$$E_{H+} = \frac{2.3\,RT}{zF} \log \frac{a_{H+,o}}{a_{H+,i}} \tag{17}$$

Thus, E_{H+} is also called the H^+ Nernst potential.

Equilibrium pH for protons. Let us return to the previous example, but instead clamp V_m to -60 mV and pH_o to 7.3 as we make step changes in pH_i. As computed above, $\Delta\mu_{H+}$ will be -4 kJ/mole at a pH_i of 7.0, and H^+ will tend to enter the cell. However, at progressively lower pH_i values, $\Delta\mu_{H+}$ will become less and less negative, so that the driving force for proton entry will decrease. If pH_i is decreased sufficiently, $\Delta\mu_{H+}$ will reach 0, and there can be no net movement of H^+. This definition of equilibrium pH_i

(pH_{ieq}) is analogous to that for the equilibrium potential; in both cases, $\Delta\mu_{H+} = 0$. At pH_i values less than pH_{ieq}, $\Delta\mu_{H+}$ is positive and the driving force favors proton exit from the cell. To calculate pH_{ieq}, we simply set $\Delta\mu_{H+}$ equal to 0 and solve for pH_i:

$$\Delta\mu_{H+} = 0 = 2.3\,RT\,(pH_o - pH_i) + zFV_m \tag{18}$$

$$pH_i = pH_{i_{eq}} = \frac{zFV_m}{2.3\,RT} + pH_o \tag{19}$$

In our example above of a cell clamped at a V_m of -60 mV and bathed in a solution of pH 7.3

$$pH_i = pH_{i_{eq}} = \frac{(+1) \times 96487\text{ C mole}^{-1} \times (-0.060\text{ J C}^{-1})}{2.3 \times 8.31\text{ J K}^{-1}\text{ mole}^{-1} \times 310\text{ K}} + 7.3 \cong 6.3$$

In general, cells have pH_i values much greater than pH_{ieq}, and therefore the electrochemical driving force favors passive H^+ influx.

Equilibrium Potential for Protonated Weak Acids or Deprotonated Weak Bases

As we will see, Eq. 17 is not only the expression for the equilibrium potential for H^+, but also the expression for the equilibrium potential for protonated weak acids (e.g., NH_4^+) and deprotonated weak bases (e.g., HCO_3^-), provided that their uncharged species can freely equilibrate across the cell membrane. Take as an example a cell exposed to a solution containing NH_3 and NH_4^+. In solution, the weak base NH_3 is in chemical equilibrium with its conjugate weak acid NH_4^+ according to the reaction

$$NH_3 + H^+ \rightleftharpoons NH_4^+ \qquad \text{Rx. 4}$$

The equilibrium constant (K) is defined as

$$K = \frac{[H^+][NH_3]}{[NH_4^+]} \tag{20}$$

Rearranging Eq. 20, we obtain

$$\frac{K}{[NH_3]} = \frac{[H^+]}{[NH_4^+]} \tag{21}$$

Assuming that (i) K is the same both inside (K_i) and outside (K_o) the cell, and (ii) the cell membrane is far more permeable to NH_3 than to NH_4^+, so that $[NH_3]_i$ is the same as $[NH_3]_o$, then

$$\frac{K_o}{[NH_3]_o} = \frac{[H^+]_o}{[NH_4^+]_o} = \frac{[H^+]_i}{[NH_4^+]_i} = \frac{K_i}{[NH_3]_i} \tag{22}$$

Thus,

$$\frac{[H^+]_i}{[H^+]_o} = \frac{[NH_4^+]_i}{[NH_4^+]_o} \tag{23}$$

Assuming that activities are equal to concentrations, and expressing the logarithm to the base 10, we can write an expression for the equilibrium potential governing the movement of NH_4^+

$$E_{NH_4^+} = \frac{2.3\,RT}{zF} \log \frac{[NH_4^+]_o}{[NH_4^+]_i} \tag{24}$$

Substituting $[H^+]_o/[H^+]_i$ for $[NH_4^+]_o/[NH_4^+]_i$ in Eq. 24 above yields

$$E_{NH_4^+} = \frac{2.3\,RT}{zF} \log \frac{[H^+]_o}{[H^+]_i} = E_{H^+} \tag{25}$$

Therefore, the equilibrium potentials for NH_4^+ and H^+ are equivalent. Similarly, when a cell is exposed to a solution containing the weak acid CO_2 in chemical equilibrium with its conjugate weak base HCO_3^-, the equilibrium potential for HCO_3^- ($E_{HCO_3^-}$) will equal E_{H^+}, assuming that the $[CO_2]_i$ is the same as $[CO_2]_o$:

$$E_{HCO_3^-} = \frac{2.3\,RT}{zF} \log \frac{[HCO_3^-]_i}{[HCO_3^-]_o} = E_{H^+} \tag{26}$$

In general, the equilibrium potential for any protonated weak acid (e.g., NH_4^+) or deprotonated weak base (e.g., HCO_3^-) will be equal to E_{H^+} if (i) the equilibrium constant for the weak acid or base is the same both inside and outside the cell, and (ii) the uncharged species is equilibrated across the cell membrane.

Thermodynamics versus Kinetics

Introduction. Our discussion has, thus far, focused on the forces affecting the passive movement of H^+, HCO_3^- and other charged acids and/or bases through a hypothetical conductance pathway (e.g., a channel) in the cell membrane. Thermodynamics allows us to predict whether H^+, for example, would *tend* to move passively into the cell, or whether it would *tend* to move out. However, thermodynamics does not address the issue of how fast that movement might be. That is the domain of kinetics. For example, if the H^+ permeability of the membrane were zero, there would be no passive H^+ flux, even if the H^+ driving force ($\Delta\mu_{H^+}$) were large. On the other hand, if the H^+ permeability were infinite, the passive flux would also be infinite, as long as the driving force were non-zero. Thus, permeability of the membrane to H^+ clearly has a major impact on the passive H^+ flux. The source of this permeability could be a "leakage" pathway for protons through the membrane lipid, or a channel, as discussed in the next section.

How important is permeability? To simplify the analysis, we will assume that V_m is zero. Under these conditions, the net flux of H^+ out of the cell (J_H) may be governed by a form of Fick's first law of diffusion:

$$J_H = P \times ([H^+]_i - [H^+]_o) \tag{27}$$

Note that we have defined the flux as the number of moles of H^+ moving through a unit area of cell membrane, per unit time. By convention, the flux is positive for an efflux. This kinetic equation states that, for a fixed difference in H^+ concentrations, J_H is proportional to the permeability P. Viewed another way, for a fixed permeability, J_H is proportional to $\Delta[H^+]$. Does this mean that J_H is proportional to $\Delta\mu_{H+}$? To answer this question, we will consider two examples. In the first, pH_o is 10 and pH_i is 9. The driving force for H^+ is given by an equation analogous to Eq. 18, except that here $\Delta\mu_{H+}$ will be negative for a passive efflux of H^+:

$$\Delta\mu_{H+} = 2.3\,RT\,(pH_i - pH_o) = 2.3\,RT\,(9 - 10) = -2.3\,RT \tag{28}$$

Thus, thermodynamics predicts that the passive flux of H^+, if any, should be an efflux. The efflux predicted by Eq. 27 is

$$J_H = P \times (10^{-9}\,M - 10^{-10}\,M) = P \times 9 \times 10^{-10}\,M$$

In the second example, pH_o is 2 and pH_i is 1. Thus, thermodynamics predicts that the free energy change for H^+ efflux is

$$\Delta\mu_{H+} = 2.3\,RT\,(pH_i - pH_o) = 2.3\,RT\,(1 - 2) = -2.3\,RT$$

which is exactly the same as in the first example. However, the predicted efflux is

$$J_H = P \times (10^{-1}\,M - 10^{-2}\,M) = P \times 9 \times 10^{-2}\,M$$

which is eight orders of magnitude greater.

From this analysis, it is clear that thermodynamics determines *if* there is a passive flux and, if so, the direction of this flux. Kinetics describes the magnitude of the flux. Equation 27, a form of Fick's first law of diffusion, is the simplest kinetic description of a system. If V_m is *not* zero, then one might describe the passive H^+ flux from first principles using the Goldman-Hodgkin-Katz (GHK) current equation discussed below. As we shall see two sections below, the GHK current equation is far more complex than Fick's law, even though it deals with the simplest distribution of potential in the membrane. In the case of a more complex channel or a transporter, it clearly becomes more difficult to describe the kinetics from first principles, although the GHK current equation will likely yield a good first approximation. In the case of transporters, first principles are discarded altogether, and the H^+ efflux from a cell might be represented by a variant of the familiar Michaelis-Menten equation:

$$\frac{J_H}{J_{max}} = \frac{[H^+]_i}{[H^+]_i + K_m} \tag{29}$$

which is purely descriptive, and in which driving forces are totally unrecognizable.

Obviously, kinetics can never contradict thermodynamics. In the case of a system described by Eq. 27, both Fick's law and thermodynamics predict that the net flux of H^+ should be zero when $[H^+]_i = [H^+]_o$. As $[H^+]_i$ begins to deviate ever so slightly from $[H^+]_o$, the flux is approximately proportional to $\Delta\mu_{H+}$, as described by irreversible thermodynamics, discussed on p. 176.

H^+ and HCO_3^- channels. Performing experiments on neurons, Thomas and Meech have demonstrated that a voltage-dependent H^+ conductance is activated at V_m values more positive than −10 mV (Thomas and Meech, 1982). Because −10 mV is more positive than E_{H+} for this preparation, the driving force favors the passive efflux of H^+. A second possible example of an H^+ conductance through a channel is the hypothesized entry of H^+ through a glutamate-gated cation channel in both the turtle cerebellum (Chesler and Rice, 1991) and the rat hippocampal slice (Jarolimek et al., 1989; Chen and Chesler, 1992a). In these brain regions, the conductance is likely to be activated at V_m values more negative than E_{H+}. Therefore, the driving force would favor the influx of H^+ into cells, thereby causing a simultaneous decrease in pH_i and increase in pH_o. Such an increase in pH_o may account for at least some of the stimulation-induced extracellular alkaline shift observed in many brain regions. More recent data, however, suggest that stimulation-induced alkaline shifts may be predominantly mediated by a Ca^{2+}/H^+ ATPase (Chapters 16 and 17). Schwiening et al. (1993) demonstrated that such a Ca^{2+} pump can extrude intracellular Ca^{2+} in exchange for extracellular H^+ in snail neurons, thereby elevating surface pH (see Chapter 15). In the CA1 region of rat hippocampal slices, a Ca^{2+}/H^+-ATPase appears responsible for the extracellular alkaline shift following the Ca^{2+} influx associated with the activation of ionotropic glutamate receptors (Paalasmaa et al., 1994; Smith et al., 1994; Paalasmaa and Kaila, 1996). This Ca^{2+} pump also appears responsible for the intracellular acidification in CA1 neurons of rat hippocampal slices subjected to a depolarization-induced elevation of intracellular Ca^{2+} (Trapp et al., 1997).

As discussed in the previous section, if $[CO_2]_i = [CO_2]_o$, the equilibrium potential for HCO_3^- is equivalent to that for H^+. Therefore, we would expect HCO_3^- to exit a cell whose membrane was permeable to HCO_3^-. Indeed, the data show that HCO_3^- exits GABA-stimulated Cl^- channels from crayfish muscle (Kaila and Voipio, 1987; Kaila et al., 1990), as well as from cells in both the turtle cerebellum (Chen and Chesler, 1990) and rat hippocampus (Chen and Chesler, 1992b; Kaila et al., 1992). In brain tissue, this HCO_3^- conductance is likely to be activated at potentials more negative than $E_{HCO_3^-}$ and E_{H+}. The resultant HCO_3^- efflux would cause both a decrease in pH_i and an increase in pH_o, similar to that caused by the proposed H^+ influx through glutamate-activated cation channels. A GABA-stimulated HCO_3^- conductance may also contribute to the extracellular alkaline shift observed during neuronal activity (Chapter 17).

Goldman-Hodgkin-Katz equation. A theoretical example of a passive ion flux whose thermodynamics and kinetics are embodied in a single equation derived from first principles is the Goldman-Hodgkin-Katz model (Goldman, 1943; Hodgkin and Katz, 1949). Applied to the H^+ ion, the Goldman-Hodgkin-Katz (GHK) equation becomes

$$J_H = \frac{P_H V_m F}{RT} \times \frac{[H^+]_i - [H^+]_o \times \epsilon}{1 - \epsilon} \tag{30}$$

where P_H is the permeability of the membrane to H^+ and $\epsilon = e^{-V_m F/RT}$. This equation simplifies to a form of Fick's first law of diffusion when V_m is zero.

Because the absolute H^+ concentrations are so small, the flux can be appreciable only if the permeability is extremely high. For a cell with a pH_i of 7.0 ($[H^+]_i = 100$ nM) and a pH_o of 7.3 ($[H^+]_o = 50$ nM), the transmembrane H^+ gradient would be only 50 nM. This value is six orders of magnitude lower than a typical transmembrane K^+ gradient, which might be 125 mM (130 mM − 5 mM). Thus, in order for the predicted H^+ flux to be as great as the K^+ flux, P_H would have to be six orders of magnitude greater than P_K. Nevertheless, P_H may indeed be several orders of magnitude greater than P_K (see review by Deamer and Nichols, 1989), so that passive H^+ fluxes may not be negligible. Of course, the passive flux of H^+ may become important under some physiological conditions, particularly when the pH on one side of the membrane is low.

Irreversible thermodynamics. When H^+ is in electrochemical equilibrium across the membrane, both thermodynamics and simple kinetic descriptions (e.g., Fick's law and the GHK equation) predict that the net H^+ flux should be zero. It seems intuitive that J_H should be more or less proportional to $\Delta\mu_{H+}$ when $\Delta\mu_{H+}$ is very close to zero. To illustrate, consider an example in which V_m is zero, and J_H is described by Fick's first law. We will assume that pH_i is 7, and compute the flux as we sweep pH_o between 2 and 12. As shown in Fig. 1A, J_H is proportional to the difference ($[H^+]_i - [H^+]_o$), where P_H is the proportionality constant. That is, the lower the pH_o (i.e., the higher the $[H^+]_o$), the greater the flux in the negative direction (i.e., influx). In fact, the J_H data would fall along the same line even if the fixed pH_i were 6 or 8. Figure 1B, which shows the data from Fig. 1A replotted as a function of $\Delta\mu_{H+}$ (see Eq. 28), is noteworthy in three regards. First, for a given pH_i, J_H is generally not proportional to $\Delta\mu_{H+}$. Second, for different pH_i values, the J_H values are on distinctly different curves. Third, for a given fixed pH_i, when $\Delta\mu_{H+}$ is sufficiently close to zero, J_H is approximately proportional to $\Delta\mu_{H+}$ (see inset Fig. 1B), which is the fundamental assumption behind irreversible thermodynamics. This example is extremely simple: an H^+ moving through a pore with $V_m = 0$ mV. However, even here, irreversible thermodynamics can accurately predict the flux from $\Delta\mu_{H+}$ only over an extremely narrow pH_o range. One might expect irreversible thermodynamics to have even less applicability for more complex systems, such as H^+ moving through transporters with V_m not equal to zero.

THERMODYNAMICS AND FUNCTION OF ACID-BASE TRANSPORTERS

Acid-Base Transporters

As described above, under most conditions the electrochemical gradients for H^+ and related ions (e.g., NH_4^+ and HCO_3^-) favor fluxes that would tend to acidify cells. We could refer to these passive fluxes as acid-loading processes. However, passive fluxes of ions through channels or other pathways through the membrane represent only one general category of acid loaders. A second such category would include biochemical reactions (e.g., ATP hydrolysis) that generate H^+. A third general category of acid loaders includes acid-base transporters that move acid into, or alkali out of the cell. Figure 2 lists four such acid-loading transporters. Of course, if cells had only acid loaders, pH_i would tend to be rather acidic (e.g., 6.3). In fact, cells generally have a resting pH_i in the range 6.8–7.4.

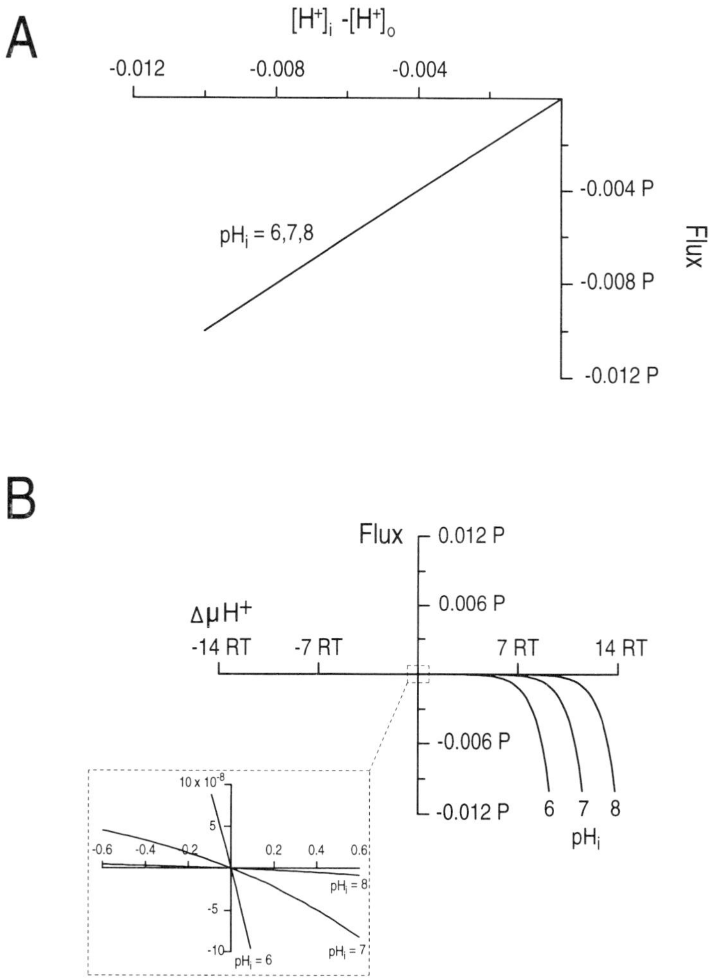

FIGURE 1. Theoretical example describing the relationship between the flux of protons across a cell membrane and either the proton concentration difference ($[H^+]_i - [H^+]_o$) or the chemical free energy change of protons ($\Delta\mu_{H^+}$). The model assumes that the potential across the cell membrane is zero, and can therefore be described by a modified form of Fick's first law of diffusion (see text).
A. Flux *versus* the proton concentration difference. This difference was obtained by varying pH_o from 2 to 12 at a constant pH_i of either 6, 7, or 8. Because of the scaling and the logarithmic nature of the *x*-axis, data for pH_o values greater than the fixed pH_i cannot be read from the graph. For all pH_i values, the plots are both identical and linear, with a slope equal to the proportionality constant P_H. Flux is plotted in units of *P*.
B. Flux *versus* the chemical free energy change of protons. The chemical free energy change was obtained by varying pH_o from 2 to 12 at a constant pH_i of either 6, 7, or 8. For all pH_i values, the plots are nonlinear when $\Delta\mu_{H^+}$ is distant from zero. *Insert.* At $\Delta\mu_{H^+}$ values close to zero, the plots become liner. Flux is plotted in units of *P* and $\Delta\mu_{H^+}$ is plotted in units of *RT*.

Acid loaders

Transporter	Structure	Inhibitors	References
Cl^-/HCO_3^- Exchanger	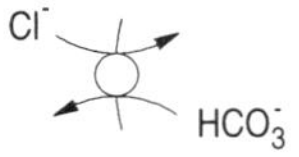	Stilbene derivatives	Reviews by Alper, 1991 Jennings, 1992
1:3 Electrogenic Na^+/HCO_3^- Cotransporter	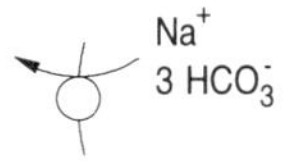	Stilbene derivatives	Reviews by Boron and Boulpaep, 1989
1:1 Electroneutral K^+/HCO_3^- Cotransporter	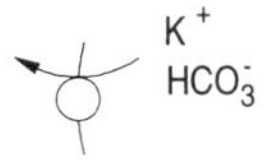	Tetraethylammonium	Hogan et al., 1995a Hogan et al., 1995b
Ca^{2+}/H^+ pump	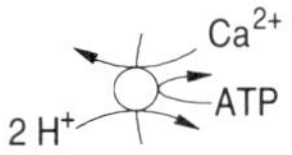	Vanadate	See Chapter 15

FIGURE 2. The major acid loaders in animal cells.

By definition, if acid loaders are present and active, pH_i can be maintained at this level only through the expenditure of energy, that is, by active transport. Because these active-transport mechanisms must either move H^+ out of the cell, or move an alkali such as HCO_3^- into the cell, they are termed acid extruders. The major acid-extruders in animal cells are listed in Fig. 3.

Depending on the energy source used to move H^+ or base against its electrochemical gradient, these acid extruders can be further classified as primary, secondary, and tertiary active transporters. A *primary active transporter,* such as an H^+ pump, derives its energy from ATP hydrolysis. A *secondary active transporter,* such as an Na^+/H^+ exchanger, uses the energy from the electrochemical gradient of another ion (e.g., Na^+). Similarly, a *tertiary active transporter,* such as a monocarboxylate cotransporter, also derives its energy from an ion gradient, albeit one that is established by a secondary active transporter.

In the following two sections, we provide examples of one primary active transporter (i.e., an electrogenic H^+ pump) and one secondary active transporter (i.e., an Na^+/H^+ exchanger).

Thermodynamics of Electrogenic H^+ Pumps

As primary active transporters, H^+ pumps (also referred to as H^+-ATPases) drive protons against their electrochemical gradient by using the energy released from the hydrolysis of ATP to ADP and inorganic phosphate (Pi). In animal cells, three types of H^+ pumps are known (Forgac, 1989). The first is an electroneutral K^+/H^+ pump, which is an E_1E_2 or P-type pump related to the Na^+/K^+ pump. We will not discuss it further. The other two H^+ pumps are both electrogenic, that is, they carry net positive charge. The second of the H^+ pumps is the F_0F_1 pump found in mitochondria. Physiologically, it runs backwards,

Acid extruders

Transporter	Structure	Inhibitors	References
Vacuolar-type H^+pump	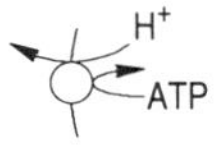	DCCD[1] NBD-Cl[2] NEM[3]	Reviews by Forgac 1989 Gluck and Nelson, 1992
K^+H^+pump	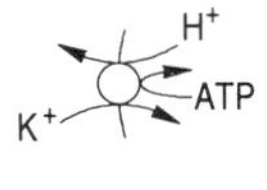	Omeprazole SCH28080	Reviews by Rabon and Reuben, 1990 Wingo and Cain, 1993
Na^+H^+Exchanger	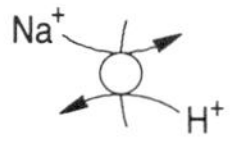	Amiloride Amiloride analogues	Reviews by Yun et al, 1995 Wakabayashi et al., 1997
Na^+-dependent Cl^--HCO^-_3Exchanger	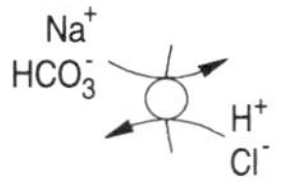	Stilbene derivatives	Reviews by Chesler, 1990 Boron, 1992
1:1 Electroneutral Na^+/HCO^-_3Cotrans-porter	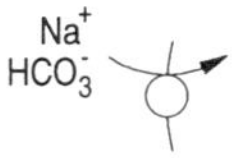	Stilbene derivatives	Dart and Vaughan-Jones, 1992
1:2 Electrogenic Na^+/HCO^-_3Cotrans-porter	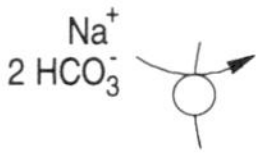	Stilbene derivatives	Review by Deitmer and Rose, 1996
1:1 Electroneutral Na^+/Monocarboxylate Cotransporter	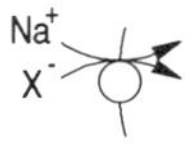		Siebens and Boron, 1987 Nakhoul and Boron, 1988
1:1 Electroneutral H^+/Monocarboxylate Cotransporter		CHC[4]	Review by Poole and Halestrap, 1993

[1]N,N′-dicyclohexylcarbodiimide
[2]7-chloro-4-nitrobenz-2-oxa-1,3-diazole
[3]N-ethylmaleimide
[4]α-cyano-3-hydroxycinnamate

FIGURE 3. The major acid extruders in animal cells. The 1 : 1 electroneutral Na^+/monocarboxylate cotransporter, by itself, does not significantly raise pH_i. However, when the monocarboxylate carried into the cell by the Na^+/monocarboxylate cotransporter is exported together with H^+, via the 1 : 1 electroneutral H^+/monocarboxylate cotransporter, the net effect is the alkalinization of the cell.

driven by a H^+ electrochemical gradient, and synthesizes ATP. The third of the H^+ pumps is the vacuolar or V-type H^+ pump. It acidifies the interior of vacuoles and other organelles, and also extrudes H^+ from certain cells. For an electrogenic H^+ pump operating in the direction of ATP hydrolysis, the reaction is

$$(H^+)_i + ATP \rightarrow (H^+)_o + ADP + Pi \qquad \text{Rx. 5}$$

where the electrochemical free energy change of this reaction ($\Delta\mu_{\text{H-pump}}$) is the sum of the electrochemical free energy change for H^+ moving out of the cell ($\Delta\mu_{H+}$) and the electrochemical free energy change of ATP hydrolysis ($\Delta\mu_{ATP}$). The free energy for H^+ moving out of the cell is given by

$$\Delta\mu_{H+} = 2.3\,RT\,(pH_i - pH_o) - zFV_m \tag{31}$$

which, except for sign, is similar to Eq. 13 (which was written for H^+ entering the cell). The free-energy change for ATP hydrolysis is the chemical free energy of ADP (μ_{ADP}) and Pi (μ_{Pi}), less that of ATP (μ_{ATP}):

$$\begin{aligned}\Delta\mu_{ATP} &= \mu_{ADP} + \mu_{Pi} - \mu_{ATP}\\ &= (\mu^o_{ADP} + RT\ln a_{ADP}) + (\mu^o_{Pi} + RT\ln a_{Pi}) - (\mu^o_{ATP} + RT\ln a_{ATP})\\ &= \mu^o_\alpha + 2.3\,RT\log\frac{a_{ADP}\times a_{Pi}}{a_{ATP}}\end{aligned} \tag{32}$$

where $\mu^o_\alpha = \mu^o_{ADP} + \mu^o_{Pi} - \mu^o_{ATP}$. Note that there are no electrical free energy terms in $\Delta\mu_{ATP}$ because the hydrolysis of ATP does not involve the generation of charge. Therefore,

$$\begin{aligned}\Delta\mu_{\text{H-pump}} &= \Delta\mu_{H+} + \Delta\mu_{ATP}\\ &= 2.3\,RT\,(pH_i - pH_o) - zFV_m + \mu^o_\alpha + 2.3\,RT\log\frac{a_{ADP}\times a_{Pi}}{a_{ATP}}\end{aligned} \tag{33}$$

As an example, for a cell having a V_m of −60 mV, a pH_i of 7.0, and a pH_o of 7.3

$$\begin{aligned}\Delta\mu_{H+} &= 2.3\times 8.31\,\frac{\text{J}}{\text{K mole}}\times 310\text{ K}\times(7.0-7.3) - \\ &\qquad [(+1)\times 96487\,\frac{\text{C}}{\text{mole}}\times(-0.060\,\frac{\text{J}}{\text{C}})]\\ &= 4.0\text{ kJ mole}^{-1}\end{aligned}$$

Assuming that the activities are identical to the concentrations of 2.3, 0.022, and 1.7 mM for ATP, ADP, and Pi, respectively (Nioka et al., 1987),

$$\begin{aligned}\Delta\mu_{ATP} &= -31.0\text{ kJ mole}^{-1} + 2.3\times 8.31\,\frac{\text{J}}{\text{K mole}}\times\\ &\qquad 310\text{ K}\times\log\frac{0.000022\text{ M}\times 0.0017\text{ M}}{0.0023\text{ M}}\\ &= -31.0\text{ kJ mole}^{-1} - 28.3\text{ kJ mole}^{-1} = -59.3\text{ kJ mole}^{-1}\end{aligned}$$

Therefore,

$$\begin{aligned}\Delta\mu_{\text{H-pump}} &= \Delta\mu_{H+} + \Delta\mu_{ATP}\\ &= 4.0\text{ kJ mole}^{-1} - 59.3\text{ kJ mole}^{-1} = -55.3\text{ kJ mole}^{-1}\end{aligned}$$

Because the energy released from the hydrolysis of ATP (−59.3 kJ/mole) in our example is considerably larger than that required to drive H^+ against its electrochemical gradient (4.0 kJ/mole), $\Delta\mu_{\text{H-pump}}$ is negative, and Rx. 5 can proceed spontaneously as written. That is, the enzyme can use the free energy released in the hydrolysis of ATP to fuel the extrusion of H^+ from the cell. The reversal potential of an electrogenic H^+ pump (E_{rev}) can be computed by setting $\Delta\mu_{\text{H-pump}}$ in Eq. 33 equal to 0 and solving for V_m:

$$\begin{aligned}\Delta\mu_{\text{H-pump}} &= 0 = \Delta\mu_{H+} + \Delta\mu_{ATP} \\ &= 2.3\,RT\,(pH_i - pH_o) - zFV_m + \Delta\mu_{ATP}\end{aligned}$$

And,

$$V_m = E_{rev} = \frac{2.3\,RT\,(pH_i - pH_o) + \Delta\mu_{ATP}}{zF}$$

$$E_{rev} = \frac{[2.3 \times 8.31\text{ J K}^{-1}\text{ mole}^{-1} \times 310\text{ K} \times (7.0 - 7.3)] - 59300\text{ J mole}^{-1}}{(+1) \times 96487\text{ C mole}^{-1}} = -633\text{ mV}$$

Therefore, the energy from ATP hydrolysis is sufficient to drive H^+ out of the cell, even at V_m values as negative as ~ -600 mV. Equally impressive is the high equilibrium pH ($pH_{i_{eq}}$), which can similarly be computed by solving for pH_i, after setting $\Delta\mu_{\text{H-pump}}$ in Eq. 33 equal to 0:

$$\begin{aligned}\Delta\mu_{\text{H-pump}} &= 0 = \Delta\mu_{H+} + \Delta\mu_{ATP} \\ &= 2.3\,RT\,(pH_i - pH_o) - zFV_m + \Delta\mu_{ATP}\end{aligned}$$

And,

$$pH_i = pH_{i_{eq}} = \frac{zFV_m - \Delta\mu_{ATP}}{2.3\,RT} + pH_o$$

$$pH_{i_{eq}} = \frac{[(+1) \times 96487\text{ C mole}^{-1} \times (-.060\text{ JC}^{-1})] + 59300\text{ J mole}^{-1}}{2.3 \times 8.31\text{ J K}^{-1}\text{ mole}^{-1} \times 310\text{ K}} + 7.3 = 16.3$$

Thus, at least in principle, a V-type H^+ pump in the cell membrane has sufficient energy available to drive pH_i higher than 14! Viewed differently, with pH_i fixed at 7.3, the electrogenic pump is capable of driving pH_o as low as 0. It is for this reason that cells use ATP-driven H^+ pumps in epithelia (e.g., stomach and distal nephron of kidney) to drive the luminal pH to very low values.

Thermodynamics of Na^+/H^+ Exchangers

In contrast to primary active transporters, secondary active transporters are energized by the downhill movement of an ion whose gradient is, in turn, established by a primary active transporter. For instance, Na^+/H^+ exchangers energize the extrusion of H^+ by linking it to the influx of Na^+. Of course, the out-to-in Na^+ gradient is established by the Na^+/K^+ pump. The reaction of the Na^+/H^+ exchanger is

$$(H^+)_i + (Na^+)_o \rightarrow (H^+)_o + (Na^+)_i \qquad \text{Rx. 6}$$

where the electrochemical free energy change of this reaction ($\Delta\mu_{Na\text{-}H}$) is the sum of the electrochemical free energy change of H^+ moving out of a cell ($\Delta\mu_{H+}$) and the electrochemical free energy change of Na^+ moving into the cell ($\Delta\mu_{Na+}$). The electrochemical energy can be described by a slightly modified form of Eq. 12:

$$\Delta\mu_{H+} = 2.3\,RT\log\frac{[H^+]_o}{[H^+]_i} + zF(V_o - V_i) = 2.3\,RT\log\frac{[H^+]_o}{[H^+]_i} - zFV_m$$

and $\Delta\mu_{Na+}$ is defined as

$$\Delta\mu_{Na+} = 2.3\,RT\log\frac{[Na^+]_i}{[Na^+]_o} + zF(V_i - V_o) = 2.3\,RT\log\frac{[Na^+]_i}{[Na^+]_o} + zFV_m \qquad (34)$$

Therefore,

$$\begin{aligned}\Delta\mu_{Na\text{-}H} &= \Delta\mu_{H+} + \Delta\mu_{Na+}\\ &= 2.3\,RT\log\frac{[H^+]_o}{[H^+]_i} - zFV_m + 2.3\,RT\log\frac{[Na^+]_i}{[Na^+]_o} + zFV_m \qquad (35)\\ &= 2.3\,RT\log\frac{[H^+]_o[Na^+]_i}{[H^+]_i[Na^+]_o}\end{aligned}$$

Note that the voltage terms in this final equation cancel out, because the transporter exchanges 1 H^+ for 1 Na^+ and the process is therefore electroneutral. As an example, let us compute $\Delta\mu_{Na\text{-}H}$ for a cell with a pH_i of 7.0 and a $[Na^+]_i$ of 15 mM, and bathed in a solution with a pH of 7.3 and a $[Na^+] = 140$ mM:

$$\begin{aligned}\Delta\mu_{Na\text{-}H} &= 2.3 \times 8.31\,\frac{J}{K\ mole} \times 310\ K \times \log\frac{(50 \times 10^{-6}\ mM) \times (15\ mM)}{(100 \times 10^{-6}\ mM) \times (140\ mM)}\\ &= -7.5\ kJ\ mole^{-1}\end{aligned}$$

Because $\Delta\mu_{Na\text{-}H}$ is negative in our example, Rx. 6 can proceed spontaneously as written. In other words, the energy released from Na^+ moving down its electrochemical gradient (i.e., into the cell) is sufficient to drive H^+ against its electrochemical gradient (i.e., out of the cell). How high could such a Na^+ gradient drive pH_i? The equilibrium pH (pH_{ieq}) can be computed by solving for $[H^+]_i$ after setting $\Delta\mu_{Na\text{-}H}$ in Eq. 35 equal to 0:

$$\Delta\mu_{Na\text{-}H} = 0 = 2.3\,RT\log\frac{[H^+]_o[Na^+]_i}{[H^+]_i[Na^+]_o}$$

When $\Delta\mu_{Na\text{-}H}$ equals 0, the following must be true:

$$\frac{[H^+]_o[Na^+]_i}{[H^+]_i[Na^+]_o} = 1$$

Therefore,

$$[H^+]_i = [H^+]_o \frac{[Na^+]_i}{[Na^+]_o} = 50 \times 10^{-6}\,\text{mM} \times \left(\frac{15\,\text{mM}}{140\,\text{mM}}\right) = 5.4 \times 10^{-6}\,\text{mM} = 5.4\,\text{nM}$$

A $[H^+]_i$ of 5.4 nM corresponds to a pH_{ieq} of 8.3. Therefore, the energy from the Na^+ gradient in our example is sufficient to drive pH_i to 8.3. However, as discussed above in regard to irreversible thermodynamics, it is difficult to predict how the Na^+/H^+ exchange rate will vary at pH_i values distant from pH_{ieq}. In all cells, the Na^+/H^+ exchanger tends to slow as pH_i increases. However, at least in rat hepatocytes, the Na^+/H^+ exchanger becomes almost totally inactive when pH_i is 7.3, a full pH unit below the pH_{ieq} of 8.3 (Boyarsky et al., 1991). Thus, kinetic constraints inactivate the exchanger, even in the presence of a substantial thermodynamic driving force.

PHYSIOLOGY OF CELLULAR pH REGULATION

Steady-State pH_i

By definition, in the steady state, the rate at which acid is extruded from the cell (J_E) by Na^+/H^+ exchange and other mechanisms must equal the rate at which acid is loaded into the cell (J_L) by mechanisms such as H^+ leak, metabolism and Cl^-/HCO_3^- exchange. It stands to reason that, if J_E exceeds J_L, pH_i will drift upwards. We will see that the rate of the alkaline drift depends on three factors. The rate of alkalinization is obviously proportional to the difference ($J_E - J_L$). The rate also must be inversely proportional to the buffering power (β), which describes the ability of the cell to buffer changes in $[H^+]$. For example, if β were infinite, pH_i could not change at all, regardless of how large ($J_E - J_L$) was. Finally, the rate of alkaline drift will be proportional to the cell's surface-to-volume ratio (ρ). For example, pH_i will change more rapidly in a small cell (with a large ρ) than in a large one. These three statements can be expressed mathematically in an equation that describes how rapidly pH_i changes (dpH_i/dt):

$$\frac{dpH_i}{dt} = \frac{\rho\,(J_E - J_L)}{\beta} \tag{36}$$

The above equation can be regarded as the fundamental law of pH_i regulation. J_E and J_L are fluxes, and typically are given in units of pmole H^+ cm^{-2} s^{-1}, which is only useful for a cell such as a squid giant axon, whose geometry is precisely known. For most preparations it is difficult to determine surface area, and it has become customary to express acid-base movements in units of mM H^+ s^{-1}. In this convention, what might be termed the acid-base "pseudo-flux" (φ) is actually the product of $\rho \times J$. Thus,

$$\frac{dpH_i}{dt} = \frac{(\varphi_E - \varphi_L)}{\beta} \tag{37}$$

Response of a Cell to a Chronic Acid Load

Equation 36 (or Eq. 37) can be particularly useful in interpreting the pH_i changes caused by changes in the activity of acid-base transporters. For example, imagine a situation in which a cell has an initial steady-state pH_i of 7.2 (point *a* in Fig. 4A). If we expose the cell to an agent that simulates an acid-loading mechanism, without affecting any of the acid-extruding mechanisms, J_L will now exceed J_E at the initial steady-state pH_i of 7.2 (point *a*' in Fig. 4A,B). According to Eq. 36, pH_i must begin to decrease (point *a* in Fig. 4C). The pH_i does not decrease indefinitely because, as pH_i decreases, acid extruders will generally become more active, whereas acid loaders will become less active. Therefore, as pH_i decreases in Fig. 4C, J_E and J_L will eventually come into balance (point *b* in Fig. 4A,B) and define a new steady-state pH_i at 7.0, which is lower than the initial pH_i (compare points *b* and *a* in Fig. 4C). According to Eq. 36, the rate, but *not* the magnitude of segment *ab* in Fig. 4C is dependent on β. Therefore, the rate of segment *ab* in Fig. 4C would be slower for a cell with larger β. However, the final pH_i would be the same, regardless of β. An intuitive explanation for why β does not play a role in determining the steady-state pH_i is that β does not appear in Fig. 4A.

In the above example, stimulating an acid-loading mechanism (e.g., a Cl^-/HCO_3^- exchanger) led to a decrease in steady-state pH_i from 7.2 to 7.0. However, we could have achieved a similar decrease in pH_i by inhibiting an acid-extruding mechanism (e.g., a Na^+/H^+ exchanger) instead of stimulating an acid-loading mechanism. This simple analysis leads to the conclusion that changes in steady-state pH_i, in and of themselves, provide no insight into the nature of the acid-base transporter(s) responsible for the shift in pH_i.

Response of a Cell to an Acute Acid Load

In the example of Fig. 4, we subjected a cell to a *chronic* acid load. That is, we imposed a leftward shift in the J_L *versus* pH_i relationship. The result was that pH_i drifted downward and remained there indefinitely. A common experimental maneuver is to load a cell with acid or alkali *transiently*. For example, imagine that we microinjected HCl into a cell for which the relationships J_E *versus* pH_i and J_L *versus* pH_i are as given in Fig. 5A. How far the pH_i initially falls depends on the amount of acid injected, the volume of the cell, and β. For example, if β were infinite, pH_i would not change at all. Imagine that β is physiological, and that the amount of acid injected is sufficient to cause pH_i to fall from 7.2 to 7.0 (point $a \rightarrow b$ in Fig. 5C). Let us assume that this maneuver has no effect on the kinetics of any of the acid-base transporters of the cell, so that the relationships of J_E *versus* pH_i and J_L *versus* pH_i are unaffected. Thus, immediately after the HCl injection, J_E is greater than J_L (points *b* and *b*' in Fig. 5A,B), so that, according to Eq. 36, pH_i must begin to increase (*bc* in Fig. 5C). However, as pH_i increases towards its initial value of 7.2, J_E decreases, and J_L increases, to their initial value (point *c* in Fig. 5A,B). In other words, all the injected HCl will have been extruded by the cell, and the cell will return to exactly its initial steady-state condition. It is interesting to note that, immediately following the injection of HCl (points *b* and *b*' in Fig. 5A,B), dpH_i/dt is dominated by acid-extruding mechanisms. Thus, it is relatively easy to use specific inhibitors or ion substitutions to block specific acid-extrusion mechanisms, and thereby determine their fluxes at a particular pH_i.

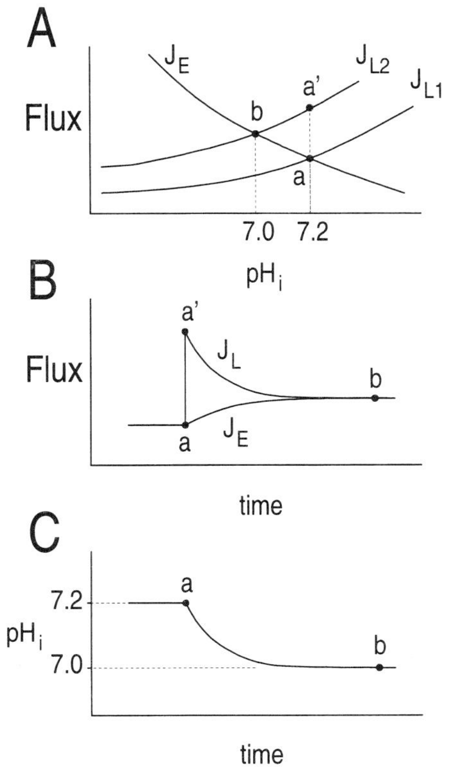

FIGURE 4. The effect on J_E, J_L, and pH_i for a hypothetical cell in which an acid loader has been stimulated.

A. Flux *versus* pH_i. The cell initially has a pH_i of 7.2, and its acid-loading and acid-extrusion rates (J_{L1} and J_E, respectively) are equal at point *a*. Stimulation of an acid loader causes a shift in the acid loading curve from J_{L1} to J_{L2}. Immediately following stimulation, pH_i is still 7.2, but J_{L2} is now greater than J_E at point *a'*. Subsequently, pH_i decreases until J_{L2} equals J_E at a new steady-state pH_i of 7.0 (point *b*).

B. Flux *versus* time. Initially, J_L equals J_E at point *a*. Stimulation of an acid loader causes an instantaneous increase in J_L (point *a'*), but no change in J_E (point *a*). However with time, J_L slowly decreases and J_E slowly increases, until J_L and J_E again balance at point *b*. Note that the flux due to both J_L and J_E is greater at point *b* than just before point *a*.

C. pH_i *versus* time. Initially, the cell pH_i is 7.2 at point *a*. Upon stimulation of an acid loader, the pH_i slowly decreases to a value of 7.0 (point *b*).

SUMMARY AND CONCLUSIONS

As described earlier in this chapter, thermodynamic forces tend to favor the passive movement of acids into, or bases out of cells. To combat the resulting decrease in pH_i, cells utilize acid-extrusion mechanisms to maintain pH_i well above that predicted for the passive distribution of H^+ and related ions. Interestingly, cells also possess acid-loading mechanisms that normally operate to decrease pH_i. In all likelihood, these acid-loading mechanisms respond to transient alkali loads and lower pH_i. In such cases, the passive flux of H^+ into the cells may be too slow, and therefore inadequate to combat the alkali load.

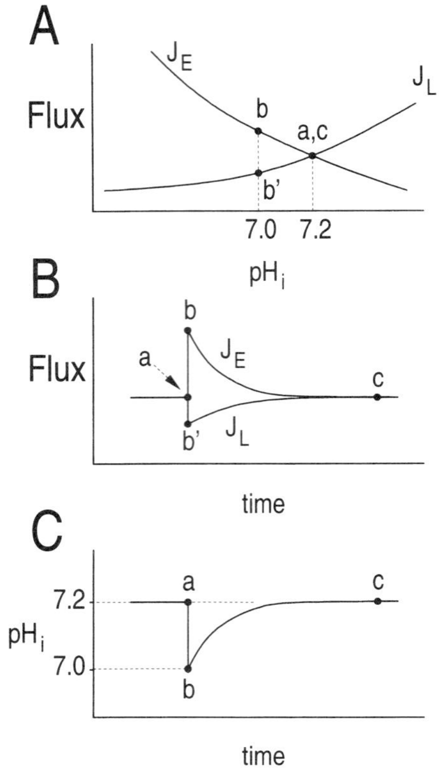

FIGURE 5. The effect on J_E, J_L and pH_i for a hypothetical cell that has been injected with a bolus of HCl.

A. Flux *versus* pH_i. The cell initially has a pH_i of 7.2, and its acid-loading and acid-extrusion rates (J_L and J_E, respectively) are equal at point *a*. Injecting a bolus of HCl into the cell causes an instantaneous decrease in pH_i to 7.0 (point *b* and *b'*). However, at this pH_i, J_E is larger than J_L. Therefore, pH_i begins to increase until J_E and J_L are again equal at the original steady-state pH_i of 7.2 (point *c*). At this point, all of the injected HCl will have been extruded from the cell.

B. Flux *versus* time. Initially, J_L equals J_E at point *a*. The injection of HCl causes an instantaneous increase in J_E (point *b*) and decrease in J_L (point *b'*). However with time, J_E slowly decreases, whereas J_L slowly increases, until J_L and J_E again balance at point *c*.

C. pH_i *versus* time. Initially, the cell pH_i is 7.2 at point *a*. The injection of HCl elicits an instantaneous decrease in pH_i to 7.0 (point *b*). However, because J_E exceeds J_L, pH_i then slowly increases and returns to its initial value of 7.2 (point *c*).

There are two principle advantages for a cell having both acid loaders and acid extruders active in the normal steady state. First, as shown in Fig. 4, changing the activity of an acid loader or an acid extruder can lead to permanent shifts in pH_i. Because many cellular processes are sensitive to changes in pH_i, altering the activity of acid loaders and/or extruders could be a mechanism by which cells change pH_i and thereby modulate cell function in response to a stimulus. The second advantage for a cell having both acid loaders and acid extruders is the ability to control pH_i tightly in response to acute acid-base disturbances. As shown in Fig. 5, when pH_i is perturbed by the introduction of a finite amount of acid or alkali, appropriate changes in J_E and J_L cause pH_i to return to its initial value rapidly. Whether cells use acid loaders and acid extruders to alter pH_i in response to a stimulus, or maintain pH_i in response to acute acid/base disturbances, the goal of the cell is to control pH_i precisely. The price paid by the cell for the exquisite control over pH_i is a large expenditure of energy. This energy expenditure, which is used to generate the ion gradients dissipated by both the acid loaders and acid extruders, is evidently necessary for providing the pH-sensitive cellular processes with a relatively stable acid-base milieu.

ACKNOWLEDGMENTS

We thank Dr. Emile L. Boulpaep for many useful discussions about thermodynamics.

REFERENCES

Alper SL (1991): The band 3-related anion exchanger (AE) gene family. Annu Rev Physiol 53: 549–564.

Boron WF (1992): Control of intracellular pH. In Seldin DW, Giebisch G (eds): The Kidney: Physiology and Pathophysiology. New York: Raven Press, pp 219–263.

Boron WF, Boulpaep EL (1989): The electrogenic Na/HCO_3 cotransporter. Kidney Int 36: 392–402.

Boyarsky G, Rosenthal N, Barrett E, Boron WF (1991): Effect of diabetes on Na^+-H^+ exchange by single isolated hepatocytes. Am J Physiol 260: C167–C175.

Chesler M (1990): The regulation and modulation of pH in the nervous system. Prog Neurobiol 34: 401–427.

Chen JC, Chesler M (1990): A bicarbonate-dependent increase in extracellular pH mediated by $GABA_A$ receptors in turtle cerebellum. Neurosci Lett 116: 130–135.

Chen JCT, Chesler M (1992a): Modulation of extracellular pH by glutamate and GABA in rat hippocampal slices. J Neurophysiol 67: 29–36.

Chen JCT, Chesler M (1992b): Extracellular alkaline shifts in rat hippocampal slice are mediated by NMDA and non-NMDA receptors. J Neurophysiol 68: 342–344.

Chesler M, Rice ME (1991): Extracellular alkaline-acid pH shifts evoked by iontophoresis of glutamate and aspartate in turtle cerebellum. Neuroscience 41: 257–267.

Dart C, Vaughan-Jones RD (1992): Na^+-HCO_3^- symport in the sheep cardiac Purkinje fibre. J Physiol (Lond) 451: 365–385.

Deamer DW, Nichols JW (1989): Proton flux mechanisms in model and biological membranes. J Membr Biol 107: 91–103.

Deitmer JW, Rose CR (1996): pH regulation and proton signalling by glial cells. Prog Neurobiol 48: 73–103.

Forgac M (1989): Structure and function of vacuolar class of ATP-driven proton pumps. Physiol Rev 69: 765–796.

Gluck S, Nelson R (1992): The role of the V-ATPase in renal epithelial H^+ transport. J Exp Biol 172: 205–218.

Goldman DE (1943): Potential, impedance, and rectification in membranes. J Gen Physiol 27: 37–60.

Hodgkin AL, Katz B (1949): The effect of sodium ions on the electrical activity of the giant axon of the squid. J Physiol (Lond) 108: 37–77.

Hogan EM, Cohen MA, Boron WF (1995a): K^+ and HCO_3^--dependent acid-base transport in squid giant axons. I. Base efflux. J Gen Physiol 106: 821–844.

Hogan EM, Cohen MA, Boron WF (1995b): K^+ and HCO_3^--dependent acid-base transport in squid giant axons. II. Base influx. J Gen Physiol 106: 845–862.

Jarolimek W, Misgeld U, Lux HD (1989): Activity dependent alkaline and acid transients in guinea pig hippocampal slices. Brain Res 505: 225–232.

Jennings ML (1992): Anion transport proteins. In Seldin DW, Giebisch G (eds): The Kidney: Physiology and Pathophysiology. New York: Raven Press, pp 503–535.

Kaila K, Voipio J (1987): Postsynaptic fall in intracellular *p*H induced by GABA-activated bicarbonate conductance. Nature 330: 163–165.

Kaila K, Saarikoski J, and Voipio J (1990): Mechanism of action of GABA on intracellular pH and on surface pH in crayfish muscle fibres. J Physiol (Lond) 427: 241–260.

Kaila K, Paalasmaa P, Taira T, Voipio J (1992): pH transients due to monosynaptic activation of GABA-A receptors in rat hippocampal slices. Neuroreport 3: 105–108.

Nakhoul NL, Boron WF (1988): Acetate transport in the S3 segment of the rabbit proximal tubule and its effect on intracellular pH. J Gen Physiol 92: 395–412.

Nioka S, Chance B, Hilberman M, Subramanian HV, Leigh JS Jr, Veech RL, Forster RE (1987): Relationship between intracellular pH and energy metabolism in dog brain as measured by ^{31}P-NMR. J Appl Physiol 62: 2094–2102.

Paalasmaa P, Kaila K (1996): Role of voltage-gated calcium channels in the generation of activity-induced extracellular pH transients in the rat hippocampal slice. J Neurophysiol 75: 2354–2360.

Paalasmaa P, Taira T, Voipio J, Kaila K (1994): Extracellular alkaline transients mediated by glutamate receptors in the rat hippocampal slice are not due to a proton conductance. J Neurophysiol 72: 2031–2033.

Poole RC, Halestrap AP (1993): Transport of lactate and other monocarboxylates across mammalian plasma membranes. Am J Physiol 264: C761–C782.

Rabon EC, Reuben MA (1990): The mechanism and structure of the gastric H,K-ATPase. Annu Rev Physiol 52: 321–344.

Schwiening CJ, Kennedy HJ, Thomas RC (1993): Calcium-hydrogen exchange by the plasma membrane Ca-ATPase of voltage-clamped snail neurones. Proc R Soc Lond 253: 285–289.

Siebens AW, Boron WF (1987): Effect of electroneutral luminal and basolateral lactate transport on intracellular pH in salamander proximal tubules. J Gen Physiol 90: 799–831.

Smith SE, Gottfried JA, Chen JCT, Chesler M (1994): Calcium dependence of glutamate receptor-evoked alkaline shifts in hippocampus. Neuroreport 5: 2441–2445.

Thomas RC, Meech RW (1982): Hydrogen ion currents and intracellular *p*H in depolarized voltage-clamped snail neurones. Nature 299: 826–828.

Trapp S, Lückermann M, Kaila K, Ballanyi K (1997): Acidosis of hippocampal neurones mediated by a plasmalemmal Ca^{2+}/H^+ pump. Neuroreport 7: 2000–2004.

Wakabayashi S, Shigekawa M, Pouysségur J (1997): Molecular physiology of vertebrate Na^+/H^+ exchangers. Physiol Rev 77: 51–74.

Wingo CS, Cain BD (1993): The renal H-K-ATPase: Physiological significance and role in potassium homeostasis. Annu Rev Physiol 55: 323–347.

Yun CHC, Tse C-M, Nath SK, Levine SA, Brant SR, Donowitz M (1995): Mammalian Na^+/H^+ exchanger gene family: structure and function studies. Am J Physiol 269: G1–G11.

CHAPTER 11

Intracellular pH Regulation in Invertebrate Neurons

ROGER C. THOMAS AND CHRISTOF J. SCHWIENING
Department of Physiology, University of Cambridge, Downing Site, Cambridge CB2 3EG, U.K.

INVERTEBRATES AS RESEARCH MODELS IN THE STUDY OF NEURONAL pH REGULATION

Introduction

Invertebrate physiology has long been studied for its own sake, as well as for its relevance to medicine. Most processes of cell physiology seem to be universal across the animal kingdom, as confirmed by biochemistry and molecular biology. The more that we find out about both invertebrate and mammal neurons, the more similar their ion homeostasis mechanisms appear to be. Thus invertebrate neurons have proved to be excellent models for the study of pH regulation. This can be seen from the extensive coverage of invertebrate work in earlier reviews, which should be consulted for more details of early work. We recommend particularly Roos and Boron (1981) and Thomas (1984). In this chapter we will strictly confine our coverage to invertebrate work, although after 1976 there was considerable interaction with work on vertebrate preparations.

Advantages of Using Invertebrate Preparations

The lower cost of the animals and apparatus required, and the lower metabolic rate and greater resistance to hypoxia of invertebrate preparations, have encouraged researchers with low budgets to work on them. But the key advantage offered by the invertebrates to researchers into ion homeostasis is the exceptional size and robustness of some neurons and axons, particularly in molluscs. Before the invention of fluorescent microspectroscopic techniques for recording pH in small cells, the only easy direct way of recording fast pH changes in single cells was to use pH-sensitive microelectrodes. The first pH electrodes used intracellularly had tips over 0.1 mm in diameter and could be used only on very large cells. The largest axon by far is found in cephalopods such as the squid. Such axons were first used in intracellular pH (pH_i) studies with glass mini-electrodes by Caldwell (Caldwell, 1958). His set-up is shown in Fig. 1.

pH and Brain Function, Edited by Kai Kaila and Bruce R. Ransom
ISBN 0-471-11838-9

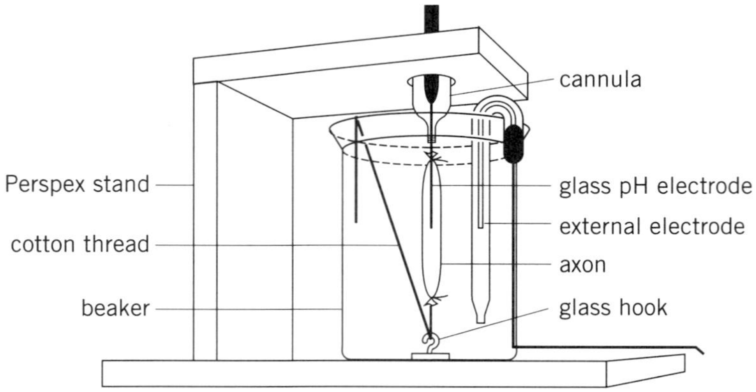

FIGURE 1. Setup used by Caldwell to measure pH in squid giant axons (from Caldwell, 1958).

Caldwell mounted the squid giant axon on a glass cannula and lowered his combined pH and intracellular reference electrode vertically down through the cannula. He found that intracellular pH (pH_i) was about 7 normally, and that it could be rapidly decreased by carbon dioxide, as shown in Fig. 2. This effect of CO_2 on pH_i was well known from theoretical considerations, and had been observed indirectly before, but Caldwell's were the first direct measurements. They confirmed that CO_2 crosses cell membranes very easily.

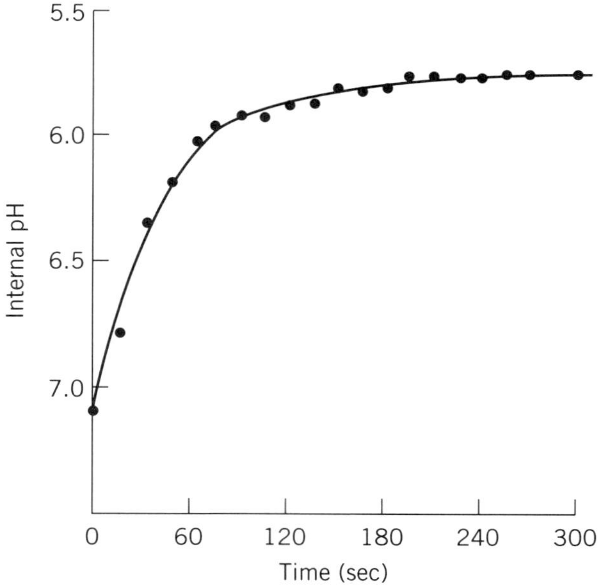

FIGURE 2. Changes in the pH_i of a giant axon from *Loligo forbesi* during the first 300 sec soaking in seawater acidified to pH 5 by saturation with 100% CO_2. Seawater acidified with other (impermeant) acids had only a very slow effect on pH_i (from Caldwell, 1958).

Intracellular pH and Passive Distribution

The internal pH measured by Caldwell was too alkaline to be explained by passive distribution, that is, without a mechanism to extrude H^+ or take up the equivalent OH^- or HCO_3^- ions. For some years this remained controversial, since some measurents with an apparently reliable new design showed a passive distribution. But by 1975 Caldwell's results had been widely confirmed (for references see Roos and Boron, 1981). The first studies of squid axon pH regulation were published soon after this, by Boron and de Weer (1976) and Russell and Boron (1976). They used pH-sensitive electrodes with a long piece of pH-sensitive glass protruding from the tip of an insulating glass micropipette. By this time a new design of pH-sensitive glass microelectrode had been invented (the recessed-tip type, Thomas, 1974a,b) which allowed measurements on other cells, notably the large neurons of the garden snail, *Helix aspersa*. These are so robust that many microelectrodes can be inserted at the same time into one cell. Other invertebrate neurons are smaller and less robust. The first nonmollusc neurons to have their pH_i regulation elucidated were those of the crayfish (Moody, 1981).

ION TRANSPORT MECHANISMS ASSISTING RECOVERY FOLLOWING ACID LOADS

The normal pH_i is only slightly lower than the extracellular pH (pH_o) in all cells so far measured. Since the potential across most cell membranes is about -60mV (negative inside), a passive distribution of H^+ ions would make pH_i much lower than pH_o. There must be some mechanism in the cell membrane that pumps acid equivalents out. Thus most work on pH_i regulation has been directed at the ionic mechanism of pH_i recovery from an imposed acidification (see also Chapter 10).

Methods for Loading a Cell with Acid

Acidifications have been applied to cells in several different ways:

1. Superfusion with a weak acid, usually CO_2. The uncharged molecules easily enter the cell and there dissociate to produce H^+ and anions. For example:

$$CO_2 + H_2O \rightleftharpoons H_2CO_3 \rightleftharpoons H^+ + HCO_3^-$$

   As long as the external pH is maintained by adding bicarbonate to the superfusate, normal pH_i will be restored by the membrane transport mechanisms.
2. Withdrawal of weak base, usually ammonium, after several minutes of exposure. The cell becomes loaded with ammonium ions during the exposure. When external ammonium is removed, the internal ammonium ions mostly leave as ammonia, leaving behind H^+ ions.

$$NH_4^+ \rightleftharpoons NH_3 + H^+$$

3. Injection of a strong acid or ammonium ions, or dialysis with acidified solution.

Figure 3 shows an early example of the effect of both injecting ammonium iontophoretically and superfusion with CO_2 on the pH_i of a snail neuron.

The first injection decreased pH_i by 0.3 units as the injected ammonium ions produced ammonia and H^+. The subsequent recovery took about 20 minutes. Switching to bicarbonate-buffered Ringer's solution equilibrated with 2.5% CO_2 also decreased pH_i by 0.3; however, this resulted from the reaction of CO_2 with intracellular water (later experiments showed that carbonic anhydrase inhibitors slowed pH changes caused by this reaction). The removal of CO_2 produced a much larger pH_i change than was produced by its addition, due to the acid extrusion which occurred during the bicarbonate superfusion.

The second of the above methods, also known as the "ammonium pre-pulse method" of acid loading was first used to study pH_i regulation in the squid giant axon by Boron and De Weer (1976), as shown in Fig. 4.

These methods of acid loading have been used in studies of the ionic mechanism of pH_i regulation in squid axons, and snail, crayfish, leech, and locust neurons. While the molluscan preparations seem to have only one mechanism for acid extrusion, usually described as a Na^+-dependent Cl^-/HCO_3^- exchanger, there are two mechanisms in the other neurons studied, as shown in Fig. 5. One of these is the same as in the mollusc neurons, the other a simple Na^+/H^+ exchanger.

Na^+-Dependent Cl^-/HCO_3^- Exchange: The Main Mechanism

The Na^+-dependent Cl^-/HCO_3^- exchange was first proposed in 1977 by R. C. Thomas. We will consider the evidence for this mechanism for each ion in turn.

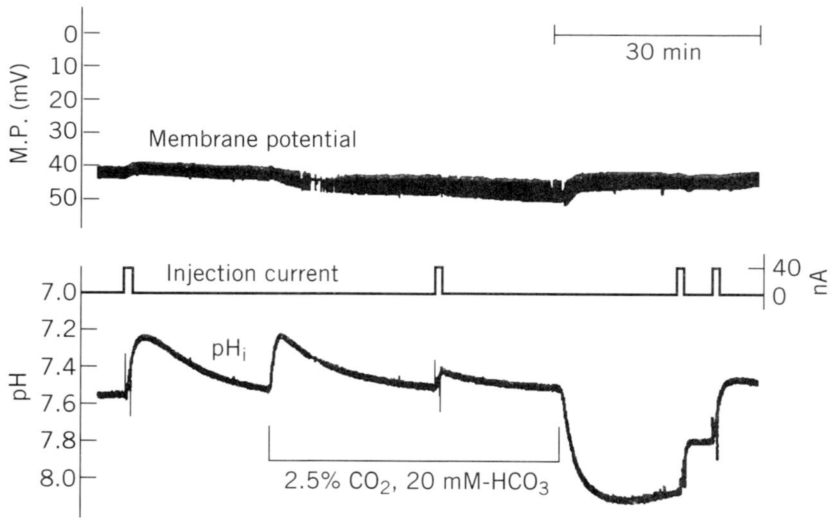

FIGURE 3. Record from an experiment on a snail neuron in an isolated brain in which the cell had been impaled with four microelectrodes: two filled with KCl for potential and current injection, one pH-sensitive, and one filled with NH_4HCO_3 for the injection of NH_4^+ ions (physiologically equivalent to H^+ ions). Except where shown, the preparation was superfused with bicarbonate-free Ringer's solution (from Thomas, 1974a,b).

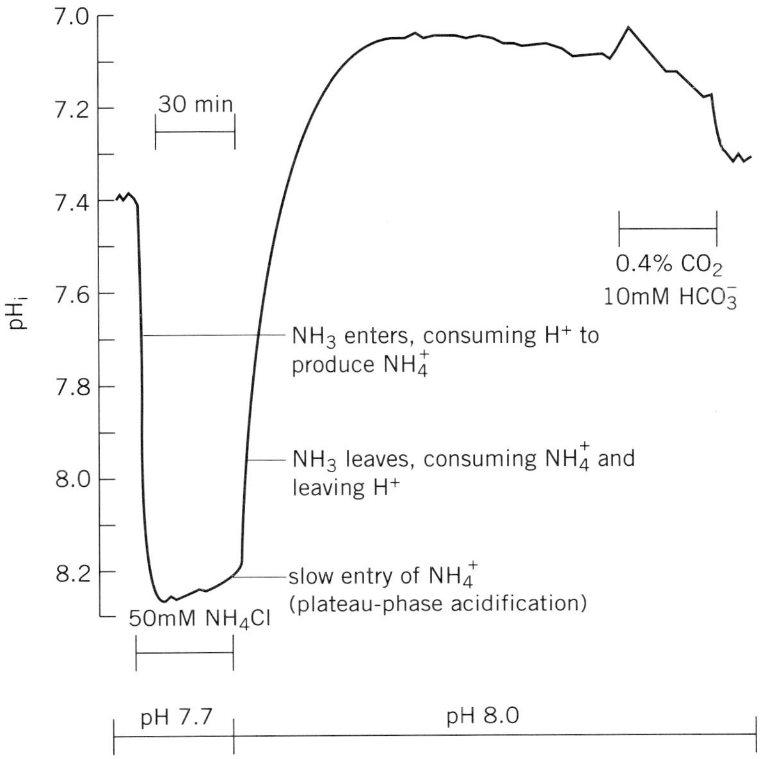

FIGURE 4. The effect of bicarbonate on the squid axon pH_i recovery from an acid load induced by exposure to ammonium. [Reprinted with permission from Nature (Boron and De Weer, 1976) Copyright 1976 Macmillan Magazines Ltd. Description of underlying fluxes added here.]

Bicarbonate. The stimulatory effect of bicarbonate on acid extrusion was first reported by Boron and De Weer in 1976, based on experiments such as that shown above in Fig. 4. The addition of bicarbonate to the superfusate greatly accelerated pH_i recovery from acidification produced by ammonium loading. Similar observations were made on snail neurons (Thomas, 1976a) and since then on many other cells. A problem in understanding the potency of this effect of bicarbonate is that it greatly increases intracellular buffering power. The increased rate of acid efflux will not increase the rate of pH_i recovery in proportion. At the time, the author did not notice the stimulatory effect of bicarbonate on pH_i

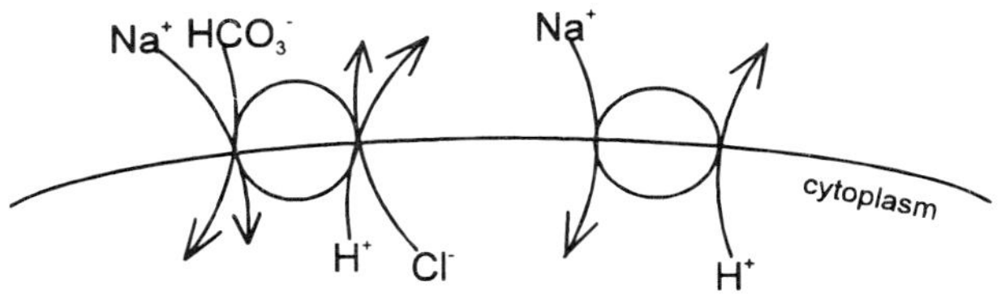

FIGURE 5. Diagram showing the two different models of ion movements during acid extrusion.

recovery. With hindsight he can see in the result shown above as Fig. 3 that acid extrusion was much faster in the presence of bicarbonate. Many workers were, and remain, however, reluctant to use CO_2/HCO_3^- buffered solutions because they require continuous bubbling with gas mixtures and special gas-impermeable tubing. Of course, everyone agrees that bicarbonate is present at relatively high levels *in vivo,* except in some aquatic animals.

A role for bicarbonate was also suggested by the inhibitory effect of SITS (4-acetamido-4′-isothiocyanatostilbene-2,2′-disulfonic acid) on pH_i recovery (Thomas, 1976b). Such stilbene derivatives have since been found to block most bicarbonate-dependent transport mechanisms. The converse is not true—the nominal removal of bicarbonate is not always very effective in blocking all SITS-sensitive acid extrusion. Thus Szatkowski and Thomas (1989) found that SITS decreased the apparent intracellular buffering power of snail neurons by a quarter, presumably by inhibiting residual bicarbonate-dependent pH_i regulation.

Chloride. The importance of intracellular Cl^- was first shown in ingenious experiments on dialyzed squid axons by Russell and Boron (1976). They showed that pH_i did not recover from low values imposed by dialysis unless Cl^- was present internally at the same time that bicarbonate was present externally. Radioactive Cl^- ions were not pumped out of an acid-loaded squid axon without both internal ATP and external bicarbonate. This requirement for ATP and the clear linkage between Cl^- efflux and HCO_3^- uptake (deduced from the effect of external HCO_3^- on pH_i) led Russell and Boron to propose that acid was extruded by an ATP-dependent Cl^-/HCO_3^- exchanger. Thomas (1977) showed later that internal $[Cl^-]$ decreased during pH_i recovery and that Cl^- was, in fact, essential for pH_i recovery in snail neurons that had been acidified by adding CO_2 to the superfusate.

In some cells it is difficult to lower internal $[Cl^-]$ to the point that Cl^--dependent pH_i recovery is blocked effectively. Thus Schwiening and Thomas (1992) found that application of the inhibitory transmitter GABA in Cl^--free solution was needed to deplete locust neurons rapidly. Generally, failure to block pH_i recovery by superfusion with Cl^--free solutions is not convincing evidence against a requirement for internal Cl^-, unless depletion can be proved.

Sodium. The requirement for external Na^+ was established by Thomas (1976c) in snail neurons. His first convincing experiment is reproduced here as Fig. 6. In this experiment Na^+ was replaced by Li^+. When organic substitutes were used, the inhibition was as effective as that seen with SITS (Thomas, 1977). Thus he proposed that pH_i was restored to normal after an acidification by the mechanism shown on the left in Fig. 5. A similar requirement for external Na^+ was revealed later in the squid giant axon (Russell and Boron, 1979) and indeed in most if not all other nerve cells studied.

The stoichiometry of the coupling of Na^+ influx to Cl^- efflux and the extrusion of acid equivalents has been carefully measured in squid axons by Boron and Russell (1983). They measured acid efflux with pH-sensitive microelectrodes, and net Na^+ influx and Cl^- efflux with radioactively labeled isotopes. Under carefully controlled conditions the mean acid extrusion rate was 7.5 $pmol.cm^{-2}.sec^{-1}$. The Na^+ influx was 3.4 and the Cl^- efflux was 3.9. The ratios of these fluxes are very close to 2 H^+ : 1 Na^+ : Cl^-, as required by the scheme in Fig. 5.

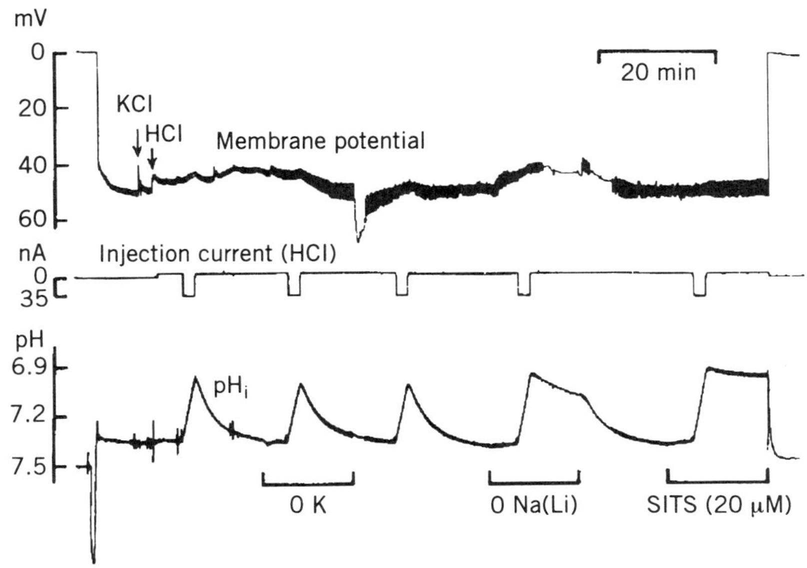

FIGURE 6. The effect of removing external K^+ or Na^+ and of the application of SITS on the pH_i recoveries from the effects of HCl injection (from Thomas, 1976c).

Hydrogen ions. The transport of protons out of the cell by the Na^+-dependent Cl^-/HCO_3^- exchanger was suggested by Thomas (1977) on the grounds of symmetry. To this day there is no evidence which allows a clear choice between proton efflux and the uptake of a second HCO_3^-, or CO_3^{2-}, or indeed a hydroxyl ion. Boron and his group (see Boron and Knakal, 1989, 1992) have thoroughly explored the sensitivity of the squid axon exchanger to ATP analogues, blockers, external pH, and bicarbonate. They have tried to distinguish between several models of acid extrusion, especially the idea that the ion pair Na^+ CO_3^{2-}, a coupling between one Na^+ ion and one CO_3^{2-} ion, might be exchanged for internal Cl^-. Their results are consistent with this scheme, but do not rule out either bicarbonate, hydroxyl, or indeed proton transport.

Role of ATP. In dialyzed squid axons Russell and Boron (1976) had shown that pH_i did not recover if the intracellular fluid contained no ATP. In experiments on snail neurons (see Thomas, 1984) however, metabolic inhibitors clearly failed to block pH_i regulation, although they did block the sodium pump. Boron et al. (1988) therefore investigated the effect of the non metabolized ATP analogue, ATP-γS, on acid extrusion by ATP-depleted axons. They found that this analogue restored acid extrusion, and concluded that activation of the transporter required phosphorylation, but not energy from ATP hydrolysis. Thus both pH_i-regulating transporters are driven by the Na^+ gradient, which itself, of course, depends on ATP hydrolysis.

Na^+/H^+ Exchange

At first it seemed possible that all neurons might have the same mechanism of pH_i regulation, although other cells like mammal muscle had an amiloride-sensitive Na^+/H^+

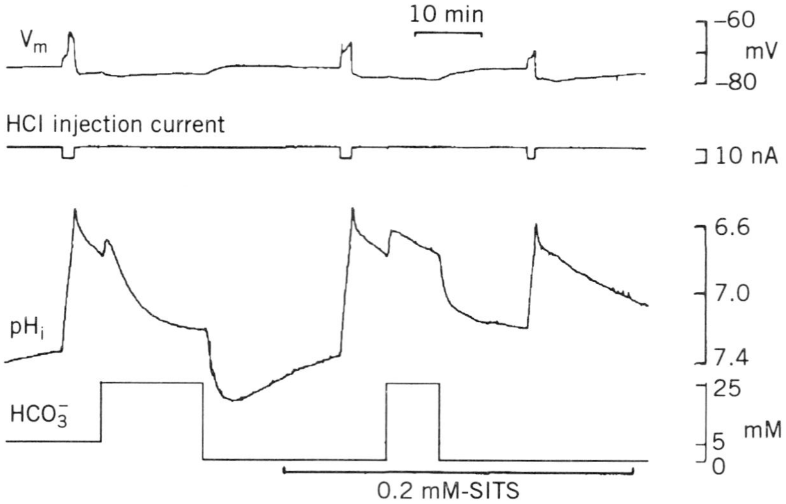

FIGURE 7. Crayfish neurone pH_i recovery from three injections of HCl. The effect of changing external bicarbonate and applying the anion exchange inhibitor SITS (Moody, 1981).

exchanger. But in crayfish neurons, Moody (1981) discovered that pH_i recovery from acidification was not well blocked by either SITS or the removal of bicarbonate, as shown in Fig. 7. In several experiments, he succeeded in decreasing pH_i by iontophoretic injections of HCl. No recovery occurred in the absence of external Na^+. He concluded that both mechanisms shown in Fig. 5 were present, and that each was responsible for about half the acid extrusion when bicarbonate was present. He was unable to block extrusion with amiloride, which inhibits Na^+/H^+ exchange in other preparations, although Gaillard and Rodeau (1987) succeeded with a similar preparation. In the crayfish stretch receptor neuron, Mair et al. (1993) have found that 1 mM amiloride slows pH_i recovery by about 80% in the absence of bicarbonate, as illustrated in Fig. 8. This experiment shows nicely the corresponding changes in internal $[Na^+]$ recorded with a Na^+-sensitive microelectrode, the greater increase in $[Na^+]$ with amiloride probably being due to a direct inhibitory effect of amiloride on the Na^+-pump. Other experiments on crayfish neurons have been described by Moser (1985) and Moser et al. (1988).

A similar dual mechanism has been found in leech neurons (Schlue and Thomas, 1985; Deitmer and Schlue, 1987, 1988), in locust neurons (Schwiening and Thomas, 1992), and possibly in crab neurons (Golowasch and Deitmer, 1993).

Relative Importance

It is not known why there are two mechanisms for acid extrusion from crayfish, leech, and locust neurons, but only one in molluscs. The Na^+-dependent Cl^-/HCO_3^- exchanger is more efficient energetically than simple Na^+/H^+ exchange, since twice as much acid is extruded or neutralized for each Na^+ taken up. Both mechanisms are driven by the Na^+ gradient, although the experiments of Russell and Boron (1976) show that ATP plays a crucial role. Perhaps Na^+/H^+ exchange is primarily a mechanism for volume rather than

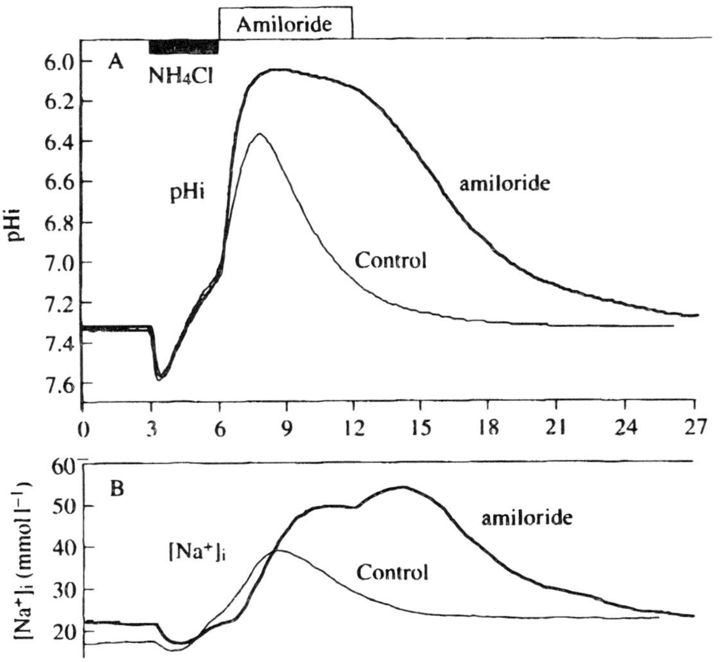

FIGURE 8. The effect of 1 mM amiloride on pH$_i$ and [Na$_i^+$] changes induced by ammonium exposure in crayfish neuron bathed in bicarbonate-free saline (from Mair et al. 1993, with permission of the Company of Biologists).

pH$_i$ regulation. This leaves unanswered the question of how mollusc neurons regulate their volume without any apparent Na$^+$/H$^+$ exchange! Perhaps they don't.

Apparent Electrogenic pH$_i$ Regulation

Both the mechanisms for pH$_i$ regulation described above are electrically neutral. In other cells, particularly glial cells, electrogenic mechanisms have been described. These are covered elsewhere in this book (Chapters 13 and 14), but typically there is a coupled influx of 2 HCO_3^- ions and 1 Na$^+$ ion. Neutral acid extrusion can nevertheless cause large increases in membrane potential (E_m) by an indirect mechanism, a most spectacular example of which is shown in Fig. 9.

In this experiment membrane potential (E_m) increased by 25 mV as pH$_i$ recovered from the first and last acidifications. When pH$_i$ recovery was inhibited by removal of external Na$^+$, there were only small changes in E_m. If the acid extrusion was itself electrogenic, the hyperpolarization should be largest at the peak of the acidification, since acid extrusion will be maximal at this point. But the peak increases in E_m did not occur until much later. Recovery after a period in Na$^+$-free saline, which would lower [Na$^+$]$_i$, occurred with smaller effects on E_m. These points suggest that the E_m increase was caused by the Na$^+$-pump, which is well known to be electrogenic, as it extruded the Na$^+$ entering during acid extrusion. The input resistance of the locust neurons was large enough to convert the Na$^+$-pump current into a large increase in E_m.

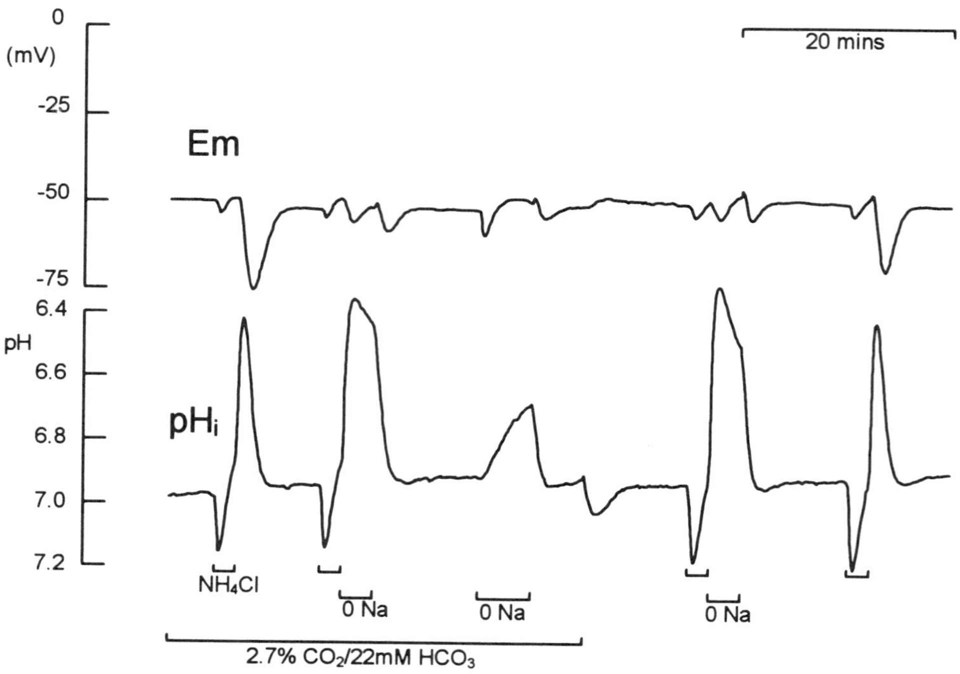

FIGURE 9. Membrane potential and pH_i recorded with a double-barreled combination pH microelectrode from a locust neuron. The animal had been starved for 2 weeks. Acid loads were applied four times, at intervals, by the ammonium pre-pulse method. For the first part the preparation was superfused with CO_2/HCO_3^- buffered saline (from Schwiening, 1990, see also Schwiening and Thomas, 1992).

MECHANISMS ASSISTING RECOVERY FROM ALKALINE LOADS

Since the normal values of pH_o, pH_i, and membrane potential imply a continuous passive entry of H^+ ions and exit of bicarbonate and hydroxyl ions, most work on pH_i has looked at acid extrusion. However, Frey and Schlue (1993) have recently made a careful study of leech Retzius neuron recovery from alkalinization. They raised pH_i by first equilibrating in a normal pH solution with added weak acid (acetate or carbonic) and then removing or reducing the weak acid. They found that 0.5 mM Zn^{2+} slowed the pH_i return to normal, and concluded that proton entry through channels (see next section) contributed to the normal pH_i recovery. They did not discuss the possibility that Zn^{2+} also blocks calcium channels. Some of the normal pH_i recovery might be due to H^+ entry in exchange for Ca^{2+} ions extruded by the Ca^{2+}/H^+ ATPase (Schwiening et al., 1993; see also Chapters 15–17.)

PROTON CHANNELS

In 1979 R. W. Meech and R. C. Thomas discovered independently that the pH_i in snail neurons that had been depolarized by current or by elevated $[K^+]$ recovered from acidifi-

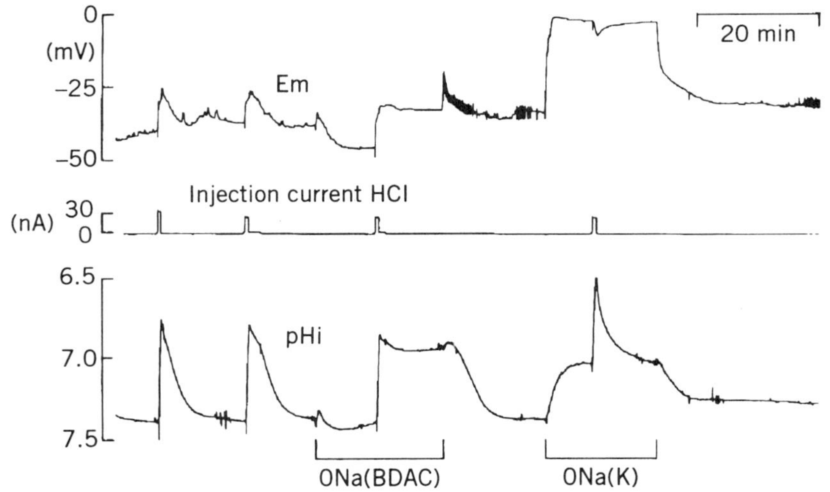

FIGURE 10. The effect on snail neuron pH_i recovery from HCl injections of replacing external Na^+ with BDAC [(bis-(2-hydroxyethyl) dimethylammonium] or K^+. (R. C. Thomas, unpublished 1979 experiment.)

cation much faster than normal. An experiment showing this is illustrated in Fig. 10. When external Na^+ was replaced by an organic substitute, pH_i recovery was blocked; however, when it was replaced by K^+, causing depolarization of the membrane, pH_i recovered from an acid injection about twice as fast as in the controls. SITS had no effect (see also Thomas, 1979).

The fast recovery in depolarized cells was soon found to be caused by a great increase in the membrane permeability to H^+ ions (Thomas and Meech, 1982, Meech and Thomas, 1987). Depolarization appears to open rectifying channels that are highly permeant to protons leaving the cell. Although the free H^+ concentration is only about 100 nM, or about a million times lower than the K^+ concentration, the H^+ current can be almost as large. H^+ currents have been studied in detail in both pond and land snail neurons (Byerly et al., 1984; Byerly and Moody, 1986; Mahaut-Smith, 1989a,b; and Byerly and Suen, 1989), but have not yet been recorded in any other neuronal preparation. Similar currents have been found in other cells, however, notably axolotl oocytes, human neutrophils, and rat alveolar epithelial cells. A detailed description of these results is outside the scope of this chapter, and the topic has been thoroughly reviewed recently (Lukacs et al., 1993; DeCoursey and Cherny, 1994). We will discuss only the results obtained on neurons, concentrating on pH effects. The physiological significance of the currents carried by H^+ ions is uncertain, and the currents are not easy to distinguish from those carried by K^+ or Ca^{2+}. The available channel blockers are neither easily reversible nor very specific, the most widely used being Zn^{2+}. In land snail neurons this was 80 times more potent at blocking H^+ than Ca^{2+} currents (Mahaut-Smith, 1989b). The flow of H^+ ions will cause the extracellular or surface pH to fall, and pH_i to increase. The volume of extracellular space *in vivo* is much smaller than that of the cytoplasm, so the same flow of H^+ may well cause much bigger pH changes outside than inside a depolarized cell.

Surface pH Effects of H^+ Channels

Experiments with pH-sensitive microelectrodes pressed on the surface of snail neurons have revealed surprisingly fast and large acidifications as long as the superfusate is not well buffered. In the experiment shown in Fig. 11, a neuron was depolarized for a few seconds to a range of potentials from −40 to +40 mV. During each depolarization by more than 40 mV, the surface pH fell rapidly. The pH fell much more in 2 mM than in 20 mM HEPES, but nearly as much in 20 mM bicarbonate as in 2 mM HEPES (see also Thomas, 1988). It appears that carbonic anhydrase is absent from the surface of the cell, so that the bicarbonate is unable to act as an effective buffer.

The physiological importance of such extracellular pH changes as might be generated by H^+ flux out of nerve cells is unknown. They would perhaps help to offset the effects of H^+ uptake by the Ca^{2+}/H^+ ATPase that maintains the normal low calcium level (Schwiening et al., 1993; but see also Schwiening, 1994).

Intracellular pH Effects of Proton Channels

Intracellular pH is very well buffered. The quantity of protons required to leave the cell to increase pH_i by 0.1 is at least 1 mmol per liter of cell water. This is about 500 times more than for calcium. In the experiment shown in Fig. 11, the pH_i changes are undetectable, although the surface pH changed by as much as 0.3. In very small cells, it is possible that proton channels might provide a mechanism to help regulate pH_i, but the period during which a nerve cell membrane potential is in the range to open the channels is normally very brief. Nevertheless experimenters who clamp cells at positive potentials for long periods need to be wary of consequent possible changes in pH_i.

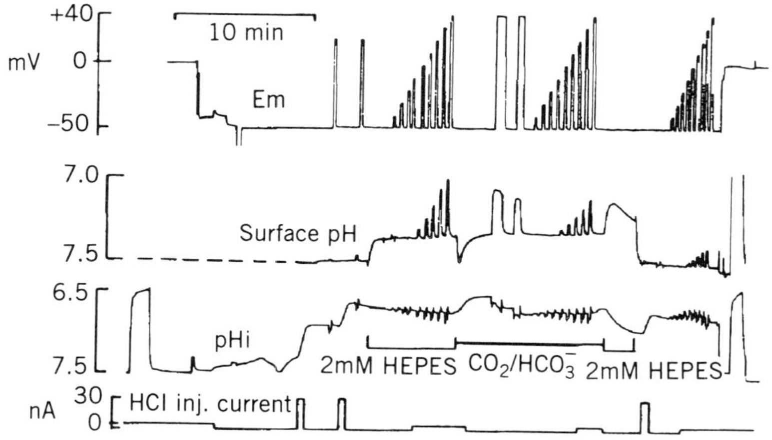

FIGURE 11. Records of membrane potential (E_m), surface pH, pH_i, and injection current from a voltage-clamped snail neuron. The surface and intracellular pH were recorded with two combined reference/pH microelectrodes. Except where indicated, the preparation was superfused with Ringer's solution buffered either by PIPES (pH 6.5, 20 mM) or HEPES (pH 7.5, 20 mM). The normal pH_i regulation was blocked by SITS (from Thomas, 1987).

CONCLUSIONS

Perhaps the most important physiological points to arise from the work described above are (i) that bicarbonate really is a vital ion that is essential for normal H^+ homeostasis, and (ii) that proton channels may exist in other, nonmolluscan neurons. It is hard for us to be objective about the relevance of work on invertebrate neurons, but we think it has made an enormous contribution to knowledge of pH homeostasis. A great many processes were first discovered in the large, tough cells found in these deceptively simple animals, and most have been since found in mammalian cells. Technical developments such as the whole-cell patch clamp and fluorescent indicators have made it easier to work on mammalian cells, so proportionately far less work is now done on mollusc and other invertebrate neurons. This means that new lessons and cross-fertilization of techniques from comparative physiology are less likely to happen and that fewer chances to encounter alternative mechanisms may inhibit interpretation of unexpected findings. Physiology especially should not be a mammocentric subject.

ACKNOWLEDGMENTS

We are grateful to the Medical Research Council for support over many years.

REFERENCES

Boron WF, De Weer P (1976): Active proton transport stimulated by CO_2/HCO_3^-, blocked by cyanide. Nature 259: 240–241.

Boron WF, Knakal RC (1989): Intracellular pH-regulating mechanism of the squid axon. Interaction between DNDS and extracellular Na^+ and HCO_3^-. J Gen Physiol 93: 123–150.

Boron WF, Knakal R (1992): Na^+-dependent Cl-HCO_3 exchange in the squid axon. Dependence on extracellular pH. J Gen Physiol 99: 817–837.

Boron WF, Russell JM (1983): Stoichiometry and ion dependencies of the intracellular-pH-regulating mechanism in squid giant axons. J Gen Physiol 81: 373–399.

Boron WF, Hogan E, Russell JM (1988): pH-sensitive activation of the intracellular-pH regulation system in squid axons by ATP-gamma-S. Nature 332: 262–265.

Byerly L, Moody WJ (1986): Membrane currents of internally perfused neurones of the snail, Lymnaea stagnalis, at low intracellular pH. J Physiol (Lond) 376: 477–491.

Byerly L, Suen Y (1989): Characterisation of proton currents in neurones of the snail, Lymnaea stagnalis. J Physiol (Lond) 413: 75–89.

Byerly L, Meech R, Moody W (1984): Rapidly activating hydrogen ion currents in perfused neurones of the snail, lymnaea stagnalis. J Physiol (Lond) 351: 199–216.

Caldwell PC (1958): Studies on the internal pH of large muscle and nerve fibers. J Physiol (Lond) 142: 22–62.

DeCoursey TE, Cherny VV (1994): Voltage-activated hydrogen ion currents. J Membr Biol 141: 203–223.

Deitmer JW, Schlue WR (1987): The regulation of intracellular pH by identified glial cells and neurones in the central nervous system of the leech. J Physiol (Lond) 388: 261–283.

Deitmer JW, Schlue WR (1988): Intracellular acidosis of identified leech neurones produced by substitution of external sodium. Brain Res 462: 233–241.

Frey G, Schlue WR (1993): pH recovery from intracellular alkalinization in Retzius neurones of the leech central nervous system. J Physiol (Lond) 462: 627–643.

Gaillard S, Rodeau JL (1987): Na^+/H^+ exchange in crayfish neurons: dependence on extracellular sodium and pH. J Comp Physiol B 157: 435–444.

Golowasch J, Deitmer JW (1993): pH regulation in the stomatogastric ganglion of the crab Cancer pagurus. J Comp Physiol [A] 172: 573–581.

Lukacs GL, Kapus A, Nanda A, Romanek R, Grinstein S (1993): Proton conductance of the plasma membrane: properties, regulation, and functional role. Am J Physiol 265: C3–C14.

Mahaut-Smith M (1989a): Separation of hydrogen ion currents in intact molluscan neurones. J Exp Biol 145: 439–454.

Mahaut-Smith M (1989b). The effect of zinc on calcium and hydrogen ion currents in intact snail neurones. J Exp Biol 145: 455–464.

Mair N, Moser H, Fresser F (1993): Contribution of the Na^+/H^+ antiporter to the regulation of intracellular pH in a crayfish stretch-receptor neuron. J Exp Biol 178: 109–124.

Meech RW, Thomas RC (1987): Voltage-dependent intracellular pH in Helix aspersa neurones. J Physiol 390: 433–452.

Moody W (1981): The ionic mechanism of intracellular pH regulation in crayfish neurones. J Physiol 316: 293–308.

Moser H (1985): Intracellular pH regulation in the sensory neurone of the stretch receptor of the crayfish (Astacus fluviatilis). J Physiol (Lond) 362: 23–38.

Moser H, Heinemeyer D, Asal M, Schlue WR (1988): Effects of procaine on intracellular pH and its regulation: Measurements with pH-selective micro-electrodes in Retzius neurones of the leech. Pflügers Arch 412: 589–596.

Roos A, Boron WF (1981): Intracellular pH. Physiol Rev 61: 296–434.

Russell JM, Boron WF (1976): Role of chloride transport in regulation of intracellular pH. Nature 264: 73–74.

Russell JM, Boron WF (1979): Intracellular pH regulation in squid giant axons. Biol Bull 157: 392.

Schlue WR, Thomas RC (1985): A dual mechanism for intracellular pH regulation by leech neurones. J Physiol (Lond) 364: 327–338.

Schwiening CJ (1990): Intracellular pH regulation in locust neurones. PhD Thesis, University of Bristol.

Schwiening CJ (1994): Intracellular pH changes recorded under voltage and current clamp in snail neurones in isolated ganglia. J Physiol 479: 44P.

Schwiening CJ, Thomas RC (1992): Mechanism of pH_i regulation by locust neurons in isolated ganglia—a microelectrode study. J Physiol 447: 693–709.

Schwiening CJ, Kennedy HJ, Thomas RC (1993): Calcium-hydrogen exchange by the plasma membrane Ca-ATPase of voltage-clamped snail neurons. Proc R Soc Lond B 253: 285–289.

Szatkowski MS, Thomas RC (1989): The intrinsic intracellular H^+ buffering power of snail neurones. J Physiol (Lond) 409: 89–101.

Thomas RC (1974a): Intracellular pH of snail neurones measured with a new pH-sensitive glass micro-electrode. J Physiol 238: 159–180.

Thomas RC (1974b): The effect of bicarbonate on the intracellular buffering power of snail neurones. J Physiol (Lond) 241: P103–P104.

Thomas RC (1976a): The effect of carbon dioxide on the intracellular pH and buffering power of snail neurones. J Physiol (Lond) 255: 715–735.

Thomas RC (1976b): Comparison of the Na^+ and H^+ pumps in a snail neurone. J Physiol (Lond) 263: 212P.

Thomas RC (1976c): Ionic mechanism of the H^+ pump in a snail neurone. Nature 262: 54–55.

Thomas RC (1977): The role of bicarbonate, chloride and sodium ions in the regulation of intracellular pH in snail neurones. J Physiol (Lond) 273: 317–338.

Thomas RC (1979): Recovery of pH_i in snail neurones exposed to high external potassium. J Physiol (Lond) 296: 77P.

Thomas RC (1984): Experimental displacement of intracellular pH and the mechanism of its subsequent recovery. J Physiol (Lond) 354: 3P–22P.

Thomas RC (1987): The effect of depolarisation and external buffers on the surface pH of voltage-clamped snail neurones. J Physiol 382: 130P.

Thomas RC (1988): Changes in the surface pH of voltage-clamped snail neurones apparently caused by H^+ fluxes through a channel. J Physiol 398: 313–327.

Thomas RC, Meech RW (1982): Hydrogen ion currents and intracellular pH in depolarised voltage-clamped snail neurones. Nature 299: 826–828.

CHAPTER 12

pH Regulation in Mammalian Neurons

MARK O. BEVENSEE AND WALTER F. BORON
Department of Cellular and Molecular Physiology, Yale University School of Medicine, New Haven, CT 06520

INTRODUCTION

The first modern studies of intracellular pH (pH_i) regulation in neurons were made in the mid-1970s, and focused on invertebrate neurons. As discussed in the previous chapter, the experimental approach in these early studies was to use pH-sensitive microelectrodes with glass sensors. Hinke's exposed-tip microelectrode (Hinke, 1967), with a typical tip diameter of 50–100 μm, could only be used in truly giant cells, such as the giant axon of the squid (Boron and De Weer, 1976a). Even Roger Thomas's recessed-tip microelectrode (Thomas, 1974), which can have tip diameters as low as ~0.5 μm, only could be used easily with relatively large invertebrate preparations, such as neurons from the garden snail *Helix aspersa*. However, two advances in the field have contributed to the growing number of studies on pH_i regulation in small mammalian neurons. First, the development and commercial availability of modern pH-sensitive fluorescent dyes have provided investigators with noninvasive probes for measuring pH_i. The first and perhaps the most widely used of these dyes is 2',7'-biscarboxyethyl-5,6-carboxyfluorescein (BCECF) (Rink et al., 1982). A discussion of fluorescence indicators is presented in Chapter 8 of this book. Second, the ability to either culture (see Goslin and Banker, 1991) or acutely isolate neurons (Kay and Wong, 1986) has allowed investigators to study acid-base transport conveniently in single mammalian neurons. Earlier studies on mammalian neuronal preparations were necessarily restricted to synaptosomes or neuronal cell lines, which may not as closely reflect the physiology of neurons *in vivo*.

In all cells that have been examined, with the exception of mammalian erythrocytes, pH_i is closely regulated, probably because so many cellular processes are sensitive to pH (see Roos and Boron, 1981). For mammalian neurons, the regulation of pH_i is especially important because many ion channels, especially those associated with neuronal activity, are sensitive to changes in either pH_i or extracellular pH (pH_o) (see Chapters 20–25). However, the relationship between pH and neural activity works in both directions. For example, neural activity results in the movement of acid-base equivalents across both neuronal and glial cell membranes. As discussed in Chapter 16, electrical activity can

pH and Brain Function, Edited by Kai Kaila and Bruce R. Ransom
ISBN 0-471-11838-9

elicit a decrease in the pH_i of mammalian neurons due to (i) H^+ influx mediated by the Ca^{2+}/H^+ pump in the plasma membrane and (ii) HCO_3^- efflux mediated by ligand-gated anion channels (e.g., the $GABA_A$ stimulated Cl^- channel). Also, electrical activity may stimulate metabolism and thus elevate the production of CO_2 or lactic acid. To the extent that these products remain inside the neuron, pH_i will decrease. Acid-extrusion mechanisms may have evolved in neurons to combat such activity-induced acid loads.

Another example of an activity-induced pH_i change is the depolarization-induced intracellular alkalinization of astrocytes that is apparently triggered by neuronal activity according to the following scheme (see Chapter 14): Neuronal activity leads to an increase in $[K^+]_o$ that depolarizes the astrocyte, leading to an increase in pH_i and a decrease in pH_o. Thus, it is possible that neuronal activity could lead to opposite pH_i changes in neurons and astrocytes. Of course, a change in pH_i will almost always be accompanied by a change in pH_o (Chapter 17). When the pH_i change is the result of an acid-base flux across the cell membrane (e.g., Ca^+/H^+ pumping), pH_i and pH_o tend to move in opposite directions. When acid is generated *de novo* inside the cell (e.g., lactic acid), and some of this acid is retained by the cell, then pH_i and pH_o both tend to decrease. Finally, any attempt by the cell to regulate pH_i will tend to cause pH_i and pH_o to shift in opposite directions.

Just as neuronal activity can cause changes in pH, changes in pH can cause changes in neuronal activity, by affecting ion channels and other cellular processes. The influence of neuronal activity on pH changes in the brain, and *vice versa,* is accentuated by the close apposition of neuronal and glial cell membranes. Thus, one would expect that it would be important for neurons, and also glia, to have mechanisms for regulating their pH_i, and for these mechanisms to adapt to changes in extracellular fluid. These mechanisms would not only help modulate the pH_i and pH_o changes during neuronal firing, but would also help reestablish the initial pH values following cessation of activity.

In the next section of this chapter, we will first provide an overview of the factors that affect the steady-state pH_i in neurons, and then, in the subsequent section, examine the roles played by specific acid-base transporters in regulating the pH_i of mammalian neurons. The data reviewed is from work on synaptosomes, neurons *in situ,* neuronal cell lines, cultured neurons, and acutely isolated neurons. We will see that the strategies that neurons use to regulate pH_i seem to be a subset of the strategies that cells in general use to accomplish the same task.

STEADY-STATE pH_i

Balance Between Acid Loading and Acid Extrusion

If protons were passively distributed across the plasma membrane of cells with a membrane potential of -60 mV, in a solution of pH 7.3, pH_i would be ~6.3 (see Chapter 10). If, for some reason, pH_i were above ~6.3, H^+ would tend to diffuse into the cell until pH_i fell to ~6.3. Thus, the observation that steady-state pH_i is in fact generally far higher than ~6.3 implies that either cells have no H^+ leak, or if they do, they have some sort of active-transport mechanism for removing acid as rapidly as it accumulates in the cytoplasm. As discussed in Chapter 10, the steady-state pH_i of a cell is determined by the balance between the rates of acid-extrusion (φ_E) and acid loading (φ_L). Acid-extruding processes are those that either move H^+ out of cells or move bases such as HCO_3^- into cells. Conversely, acid-loading processes are those that either move H^+ into cells, or bases such as HCO_3^- out. In the steady state, φ_E and φ_L are equal. However, if φ_E and φ_L are out of balance, pH_i will drift upward or downward according to the equation

$$dpH_i/dt = (\varphi_E - \varphi_L)/\beta$$

where β is the H^+ buffering power of the cell. For example, in a cell initially with a stable pH_i ($\varphi_E = \varphi_L$), applying a growth factor may stimulate an acid extruder such as the Na^+/H^+ exchanger. In this case, φ_E initially will exceed φ_L, and pH_i will increase until a new steady-state pH_i is reached. At this time, φ_E and φ_L once again come into balance. However, at this new, higher steady-state pH_i, both φ_E and φ_L are greater than at the original steady-state pH_i. In general, steady-state pH_i can be increased either by stimulating acid extrusion or inhibiting acid loading.

Some Factors Affecting Steady-State pH_i

Spontaneous differences in steady-state pH_i in the absence of CO_2/HCO_3^-. In studies on mammalian neuronal preparations, the average steady-state pH_i in the absence of CO_2/HCO_3^- varies over a rather wide range. For example, in separate studies, rat synaptosomes had an internal pH between 6.71 and 6.94 (Nachshen and Drapeau, 1988; Sánchez-Armass et al., 1994), whereas rat cerebellar Purkinje cells had a pH of 7.37 (Gaillard and Dupont, 1990). In at least one case, neurons in the absence of CO_2/HCO_3^- had a rather broad range of steady-state pH_i values even within a single study. That is, in the nominal absence of CO_2/HCO_3^-, neurons acutely isolated from the CA1 region of the hippocampus of the mature rat had a pH_i as low as 6.3 or as high as 7.6 (Bevensee et al., 1996). Moreover, as shown in Fig. 1, the distribution of initial pH_i values could be fitted by the sum of two Gaussians, one with a peak at 6.68 (low-pH_i neurons) and the other with a peak at 7.32 (high-pH_i neurons).

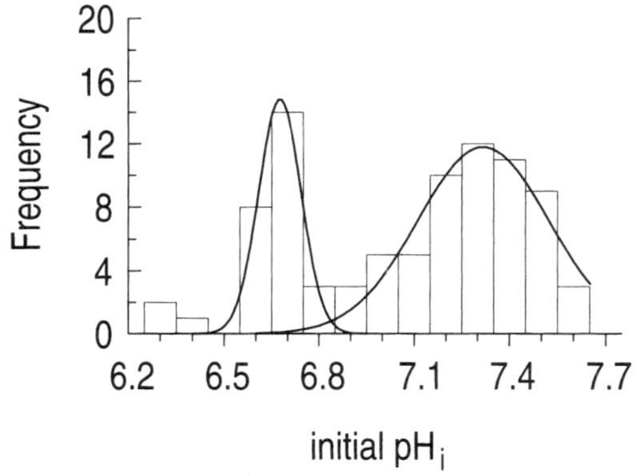

FIGURE 1. Frequency distribution of steady-state pH_i in the nominal absence of CO_2/HCO_3^- of 86 CA1 neurons acutely isolated from the hippocampi of 30 mature rats. The histogram bin width is 0.1 pH unit. The distribution was fitted by the sum of two Gaussians with means of 6.68 and 7.32. Reproduced from Bevensee et al. (1996), with permission from J. Physiol (Lond). Copyright 1996, Cambridge University Press.

Effects of temperature. It is well established that, for poikilotherms, increasing the temperature causes a decrease in the steady-state pH of both the blood and intracellular fluid (see review by Reeves, 1977), with the pH decreasing in parallel with the pH of neutral water. Interestingly, a recent study has demonstrated the opposite temperature dependence for steady-state pH_i in cultured rat hippocampal neurons (Baxter and Church, 1996). At 37°C, the neurons had a steady-state pH_i of 7.23, approximately 0.4 higher than the pH_i of 6.84 at 18°C–22°C. It is perhaps worth noting that, in the nominal absence of CO_2/HCO_3^-, the combined Gaussian distributions of steady-state pH_i at 37°C and 18°C–22°C for these cultured fetal hippocampal neurons (Baxter and Church, 1996) look remarkably similar to the sum of two Gaussian distributions of steady-state pH_i at 37°C for acutely isolated adult rat hippocampal neurons (Bevensee et al., 1996). Thus the pH_i distribution for cultured neurons at 18°C–22°C is similar to that of acutely isolated "low-pH_i" neurons at 37°C, and that of cultured neurons at 37°C is similar to that of acutely isolated "high-pH_i" neurons at 37°C. The similarity extends to rates of pH_i recovery from acid loads: pH_i recovery in the cultured neurons was slow at 18°C–22°C (as for "low-pH_i" neurons), but fast at 37°C (as for "high-pH_i" neurons).

Effects of CO_2/HCO_3^-. Compared to pH_i values in the absence of CO_2/HCO_3^-, the average steady-state pH_i in the presence of CO_2/HCO_3^- may be either higher, the same as, or lower, depending on either the cell type and/or the initial pH_i value. For example, pH_i has been reported to be higher in the presence of CO_2/HCO_3^- by as little as 0.09 in rat cortical neurons (Ou-yang et al., 1993; Mellergård et al., 1994), or as much as 0.29 in acutely isolated rat CA1 neurons with a low initial pH_i (Schwiening and Boron, 1994; Bevensee et al., 1996). On the other hand, in both rat synaptosomes (Nachshen and Drapeau, 1988; Sánchez-Armass et al., 1994) and acutely isolated rat CA1 neurons with a high initial pH_i (Bevensee et al., 1996), adding CO_2/HCO_3^- has no effect on the average steady-state pH_i. In rat cerebellar Purkinje cells, the average steady-state pH_i is actually ~0.3 lower in the presence than in the absence of CO_2/HCO_3^- (Gaillard and Dupont, 1990). Also, the change in average steady-state pH_i when neurons are exposed to CO_2/HCO_3^- may depend on temperature. For example, in cultured hippocampal neurons from rat, CO_2/HCO_3^- elicits an increase in the pH_i of the neurons at 18°C–22°C, but not at 37°C (Baxter and Church, 1996).

Although the varied effects on steady-state pH_i of adding CO_2/HCO_3^- might seem inscrutable, they are probably understandable in terms of the relative effects of CO_2/HCO_3^- on acid extruders and acid loaders. For example, adding CO_2/HCO_3^- may equally stimulate both HCO_3^--dependent acid extruders and acid loaders. In this case, steady-state pH_i would not change. On the other hand, if CO_2/HCO_3^- stimulated HCO_3^--dependent acid extruders more than acid loaders, as might be the case when pH_i is initially low, then steady-state pH_i would increase. The same would be true if the cell had a much more powerful HCO_3^--dependent acid extruder, or if the acid extruder had a much higher HCO_3^- affinity than the acid loader. Finally, if CO_2/HCO_3^- stimulated HCO_3^--dependent acid loaders more than acid extruders, as might occur when pH_i is initially high, then the addition of CO_2/HCO_3^- would lower steady-state pH_i.

Regional pH differences within a cell. At least in principle, the pH_i could be different in different regions of the cell. Caution is advised in interpreting regional differences, however, especially when the regions are rather close together or not separated by a substantial diffusion barrier. For example, differences in binding of the dye to different cellular components could produce different calibration patterns, and thus different apparent pH_i values in nearby regions of a cell. On the other hand, it is not unreasonable to expect differences in widely different parts of the neuron, which may indeed have very different physiologies. For example, it has been reported that, in mouse neuroblastoma cells (N2A) and rat pheochromocytoma cells (PC12), the pH_i in the soma is 0.2–0.3 pH units lower than at the tip of a neurite in the same cell (Dickens et al., 1989).

pH_i Measurements of Neurons *In Situ*

Recently, investigators have used fluorescent intracellular dyes to measure the pH_i of neurons *in situ*. For instance, Ritucci et al. (1996) demonstrated that neuronal cell bodies in brainstem slices from rat preferentially accumulate BCECF when the slices are incubated in the acetoxymethyl ester form of BCECF. Using the perforated patch-clamp technique, the authors demonstrated that these BCECF-loaded neurons had electrophysiological characteristics indicative of healthy cells. For instance, the neurons had resting potentials of ~−60 mV, and could be induced to generate action potentials. When slices are incubated in a room temperature solution containing CO_2/HCO_3^-, the neurons in the dorsal and ventral medullary chemosensitive regions have an average pH_i of 7.32 and 7.38, respectively.

Other investigators have measured the pH_i of neurons in brain slices by dialyzing the neurons with a patch-pipette solution containing BCECF free acid. Using such a technique to dye load neurons in the rat hippocampal slice, Trapp et al. (1996b) reported an average pH_i of 7.24 when neurons were current clamped, and a value of 7.1 when neurons were voltage clamped at a holding potential of −60 mV. Using a similar technique to load BCECF into dorsal vagal neurons from rat brainstem slices, Trapp et al. (1996a) reported a pH_i of 7.27 when neurons were voltage clamped at a holding potential of −50 mV. Under such conditions, however, 40% of the neurons displayed a steady decrease in pH_i of ~0.25 pH units 1/h.

MECHANISMS OF pH_i REGULATION IN MAMMALIAN NEURONS

Acid Extruders That Are Independent of HCO_3^-: The Na^+/H^+ Exchanger

Properties of the transporter. Na^+/H^+ exchangers are acid-extruding transporters present in virtually every mammalian cell, excluding erythrocytes. These transporters not only regulate pH_i, but also may play important roles in transepithelial Na^+ transport and cell-volume regulation, and may also influence cell proliferation, if only in the absence of CO_2/HCO_3^- (see reviews by Yun et al., 1995; Noël and Pouysségur, 1995). Na^+/H^+ exchangers generally are inhibited by amiloride and amiloride analogues (Clark and Limbird, 1991). Given the prevailing Na^+ and H^+ gradients, these transporters normally move external Na^+ into the cell and intracellular H^+ out of the cell.

Isoforms. The first Na^+/H^+ exchanger to be cloned, named NHE1 (Sardet et al., 1989), is the "ubiquitous" Na^+/H^+ exchanger found in most mammalian cells. Exceptions include erythrocytes, the opossum kidney (OK) cell line (Montrose and Murer, 1990), and cells in certain segments of the rabbit proximal tubule (Geibel et al., 1989). Presently, three other isoforms of NHE1 have been cloned (NHE2, NHE3, and NHE4). Although found predominantly in epithelial cells in the kidney, intestine, and stomach, these other isoforms are also present in brain (see reviews by Noël and Pouysségur, 1995; Ma and Haddad, 1997). In the next several paragraphs we will discuss the evidence for the functional expression of Na^+/H^+ exchangers in various neuronal preparations.

pH_i changes caused by Na^+/H^+ exchangers. Early work done on rat brain synaptosomes showed that an influx of $^{22}Na^+$ is coupled to H^+ efflux (measured with the pH-sensitive intracellular dye acridine orange or with the pH-sensitive extracellular dye pyranine) with a 1 : 1 stoichiometry (Sauvaigo et al., 1984; Jean et al., 1985). The $^{22}Na^+$ and H^+ fluxes were inhibited by amiloride, and its more potent derivatives dimethylamiloride (DMA) and ethylisopropylamiloride (EIPA). In another study, in which pH_i was measured using the pH-sensitive dye BCECF, rat brain synaptosomes were acid loaded by a brief pre-exposure to a solution containing NH_3/NH_4^+ (Nachshen and Drapeau, 1988). The technique of briefly exposing cells to a solution containing NH_3/NH_4^+ to acid load cells, termed the NH_4^+-prepulse technique, is discussed in Roos and Boron (1981). Following the acid load, the pH_i recovery of brain synaptosomes in the nominal absence of CO_2/HCO_3^- was substantially slowed in the absence of external Na^+ or in the presence of amiloride. Sánchez-Armass et al. (1994) also provided evidence for Na^+/H^+ exchange in rat brain synaptosomes by demonstrating that removing or reducing external Na^+ elicits a decrease in steady-state pH_i, and also slows the pH_i recovery from an acid load. However, the authors noted that they were unable to assess the amiloride sensitivity of the transporter because amiloride quenched the BCECF. In summary, data from both $^{22}Na^+$-flux and pH_i studies indicate that presynaptic nerve terminals isolated from the rat brain contain an amiloride-sensitive Na^+/H^+ exchanger.

There have also been several studies demonstrating Na^+/H^+ exchange in neuronal cell lines, especially neuroblastoma cells. For example, Moolenaar et al. (1981), working on the mouse neuroblastoma clone N1E, found that Na^+-loaded cells exhibit an amiloride-sensitive H^+ uptake (as measured by an increase in pH_o) when the extracellular Na^+ is removed. When the extracellular Na^+ was returned, the cells exhibited an amiloride-sensitive H^+ release, as calculated from the amount of KOH that had to be added to a cell suspension to maintain pH_o at 7.4. In a later study, in which pH_i was measured with BCECF, Moolenaar et al. (1984) demonstrated that serum-starved neuroblastoma cells, studied in the nominal absence of CO_2/HCO_3^-, alkalinize in response to the tumor promoter 12-O-tetradecanoyl-phorbol-13-acetate (TPA). The alkalinization requires Na^+/H^+ exchange activity because the TPA-induced increase in pH_i does not occur in the absence of external Na^+ or in the presence of amiloride. However, in this study, TPA may not have actually stimulated the Na^+/H^+ exchanger.

In the above studies, the authors examined serum-starved neuroblastoma cells. Work on both mouse neuroblastoma and rat C6 glioma cells indicates that there is no Na^+/H^+ exchange activity in the normal steady state when cells are examined in the presence of serum. Na^+/H^+ exchange activity only becomes apparent after cells are re-exposed to fetal calf serum after having been serum-starved for at least 4 hours (Benos and Sapirstein, 1983).

Another neuronal cell line, rat pheochromocytoma cells (PC12), also has an active Na^+/H^+ exchanger capable of returning pH_i to its initial steady-state level following an acid load imposed by an NH_4^+ prepulse. In PC12 cells, in which pH_i was measured with the pH-sensitive dye 2,3-dicyanohydroquinone (DCH), the pH_i recovery following an NH_4^+-induced acid load was inhibited by removing extracellular Na^+ or adding amiloride (Dickens et al., 1989).

Within the past decade or so, it has become possible to maintain primary cultures of mammalian neurons. Since then, a number of investigators have studied pH_i-regulating mechanisms of mammalian neurons in culture. A Na^+/H^+ exchange mechanism has been documented in nearly all cultured neurons studied so far. In cerebellar Purkinje cells, it is the predominant acid extruder. In BCECF-loaded rat sympathetic neurons (Tolkovsky and Richards, 1987), cerebellar Purkinje cells (Gaillard and Dupont, 1990), cerebellar granule cells (Pocock and Richards, 1992), and hippocampal and neocortical neurons (Amos and Richards, 1996), for example, pH_i recovery from an acid load in the nominal absence of CO_2/HCO_3^- requires external Na^+ and is inhibited by amiloride. In the experiment shown in Fig. 2A, a cerebellar Purkinje cell was acid loaded twice by applying and withdrawing a solution containing 20 mM NH_3/NH_4^+. Whereas the pH_i recovered from the first acid load, the pH_i recovery was reversibly inhibited after the second acid load by removing external Na^+. In a different Purkinje cell (Fig. 2B) acid loaded by an NH_4^+ prepulse, pH_i did not recover when 1.5 mM amiloride was applied to the cell simultaneously with the removal of the 10 mM NH_3/NH_4^+ solution. However, the pH_i increased when the amiloride was removed.

Na^+/H^+ exchange has also been documented in neurons *in situ*. In experiments performed on neurons in slices of the medulla of rat, Ritucci et al. (1997) demonstrated that amiloride-sensitive Na^+/H^+ exchange is responsible for the pH_i recovery from acid loads imposed by raising the partial pressure of CO_2.

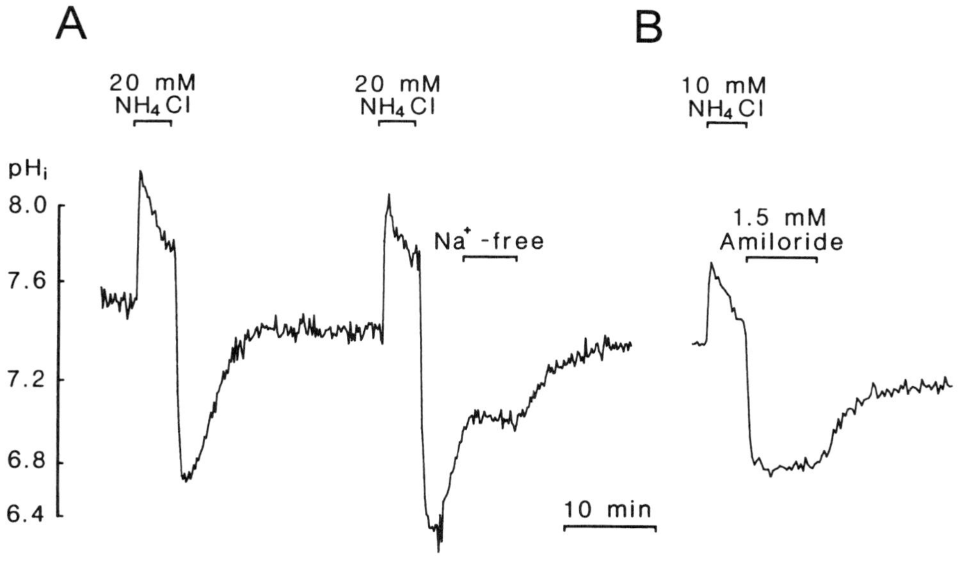

FIGURE 2. Amiloride-sensitive Na^+/H^+ exchange in cultured Purkinje cells from the cerebellum of rat. **A.** A Purkinje cell in the nominal absence of CO_2/HCO_3^- was acid loaded twice by being exposed to a solution containing 20 mM NH_3/NH_4^+. Following the first acid load, the pH_i recovered to near the pH_i at the onset of the experiment. However, following the second acid load, the pH_i recovery was inhibited by removing external Na^+. The pH_i continued to recover when the cell was re-exposed to external Na^+. **B.** In another Purkinje cell acid loaded by a brief exposure to a solution containing NH_3/NH_4^+, the pH_i decreased and did not recover when NH_3/NH_4^+ was removed simultaneously with the addition of 1.5 mM amiloride. Modified from Gaillard and Dupont (1990), with permission from J. Physiol (Lond). Copyright 1990, Cambridge University Press.

Na^+/H^+ exchangers with a low sensitivity to amiloride and its analogues. Although the neuronal preparations described above have Na^+/H^+ exchangers that, like those found in most mammalian cells, are inhibited by amiloride analogues (see Clark and Limbird, 1991), some mammalian neurons have an acid extruder that behaves as if it were a Na^+/H^+ exchanger that is relatively insensitive to amiloride and its derivatives. For example, neurons cultured from rat cerebral cortex predominantly use a HCO_3^--independent mechanism to recover from an acid load (assessed with BCECF) because the acid extrusion is similar regardless of whether the neurons are acid loaded by an NH_4^+ prepulse in the nominal absence of CO_2/HCO_3^-, or by increasing $[CO_2]$ at constant pH_o (Ou-yang et al., 1993). Although the pH_i recovery from a CO_2-induced acid load required Na^+, it was only inhibited about one third by amiloride or 5-N-methyl-N-isobutylamiloride (MIA). A second example of low sensitivity to amiloride analogues is neurons cultured from the hippocampus of the fetal rat. As measured with BCECF, the pH_i recovery from an acid load imposed via an NH_4^+ prepulse is Na^+-dependent, but unaffected by amiloride or a number of its analogues (Raley-Susman et al., 1991; Baxter and Church, 1996). Amiloride analogues and HOE 694 (another inhibitor of Na^+/H^+ exchange) also do not affect the steady-state pH_i of cultured fetal rat hippocampal neurons (Baxter and Church, 1996). A third example of low amiloride sensitivity is seen in neurons acutely isolated from the CA1 region of the hippocampus of the immature rat (Schwiening and Boron, 1994; Bevensee et al., 1996). The pH_i recovery from an acid load in these CA1 neurons requires Na^+ and occurs in the nominal absence of CO_2/HCO_3^-, but is also insensitive to amiloride or hexamethylamiloride (HMA). In fact, the amiloride analogue EIPA actually raises steady-state pH_i and increases the rate of pH_i recovery from an acid load. Presently, there is no good explanation for the increase in pH_i by EIPA[1]. Thus, in all three of these preparations, the Na^+-dependent, HCO_3^--independent acid extruder may be an amiloride-insensitive Na^+/H^+ exchanger, similar in this respect to the Na^+/H^+ exchangers at the apical membranes of certain epithelia, which have a relatively low sensitivity to these compounds (see review by Noël and Pouysségur, 1995). However, it should be emphasized that one cannot yet rule out the possibility that the transporter in question may not be a Na^+/H^+ exchanger at all, but rather a novel, Na^+-dependent transporter that has a superficial resemblance to a Na^+/H^+ exchanger.

Recently, Bookstein et al. (1996) found that NHE4 mRNA is highly expressed in the rat hippocampus, particularly in the pyramidal-cell layer of the CA regions. Moreover, NHE4 expressed in NHE-deficient fibroblasts is relatively insensitive to inhibition by amiloride analogues. It is tempting to speculate that the amiloride-insensitive NHE4 may be responsible for the Na^+/H^+ exchange activity observed in CA1 pyramidal neurons (Schwiening and Boron, 1994; Bevensee et al., 1996).

Effect of harmaline. Two groups have reported that the putative Na^+/H^+ exchangers that are poorly sensitive to amiloride and its analogues are inhibited by harmaline, a hallucinogenic alkaloid with a β-carboline backbone called harmane (Raley-Susman et al., 1991; Ou-yang et al., 1993). However, it should be noted that the absorbance and fluores-

[1]50 μM EIPA also elicits an increase in steady-state pH_i in both rat osteoclasts (Ravesloot et al., 1995) and forebrain astrocytes (Boyarsky et al., 1993; M.O. Bevensee, G. Frey, and W.F. Boron, unpublished observations). The EIPA-induced alkalinization is independent of Na^+ in both of these preparations, and insensitive to the organic anion transport inhibitor probenecid in the astrocytes. In the rat osteoclasts, a similar pH_i increase is observed with HMA, yet is not due to the interference of HMA with BCECF fluorescence.

cence characteristics of this compound markedly interfere with the BCECF fluorescence signal (Ou-yang et al., 1993; Schwiening and Boron, 1994; Baxter and Church, 1996), making it difficult to draw quantitative conclusions about this inhibitor.

Na^+/H^+ exchange activity in high- versus low-pH_i CA1 neurons. We pointed out previously that CA1 neurons acutely isolated from the hippocampus of the mature rat have a bimodal pH_i distribution in the nominal absence of CO_2/HCO_3^-. As might be expected, neurons with a relatively low initial pH_i have a low level of activity of the putative Na^+/H^+ exchanger, whereas those with a relatively high initial pH_i have a much higher activity (Bevensee et al., 1996).

Acid Extruders That Are Independent of Both HCO_3^- and Na^+

Evidence for the possible existence of an H^+ pump. In a preliminary report, Richards and Pocock (1989) noted that, in approximately half of the cultured mammalian CNS neurons they examined, the pH_i recovery from an acid load (imposed by the NH_4^+-prepulse technique) occurs in the absence of both external CO_2/HCO_3^- and Na^+. In a substantially depolarized neuron, where the H^+ electrochemical gradient would favor the efflux of H^+, a Na^+-independent pH_i recovery from an acid load could be due to the activation of an H^+ conductance (Meech and Thomas, 1977; and Chapter 11). For instance, in snail neurons, such an H^+ conductance is evident only at membrane voltages close to 0 mV (Thomas and Meech, 1982). However, if there were a substantial H^+ permeability in a well-polarized neuron, where the H^+ gradient would favor the influx of H^+, then this pathway would mediate a decrease, rather than an increase, in pH_i. Thus, it is tempting to speculate that the Na^+-independent mechanism might be an H^+ pump.

Data on freshly isolated pyramidal neurons from the hippocampal CA1 region also are consistent with the presence of an H^+ pump. In such cells from both immature and mature rats, removing external Na^+ elicits a decrease in pH_i, presumably due to the reversal of the putative Na^+/H^+ exchanger. In some of these neurons, especially those with relatively low initial pH_i values, pH_i recovers in the continued absence of external Na^+. In the experiment shown in Fig. 3, a CA1 neuron in a nominally CO_2/HCO_3^--free solution had an initial pH_i

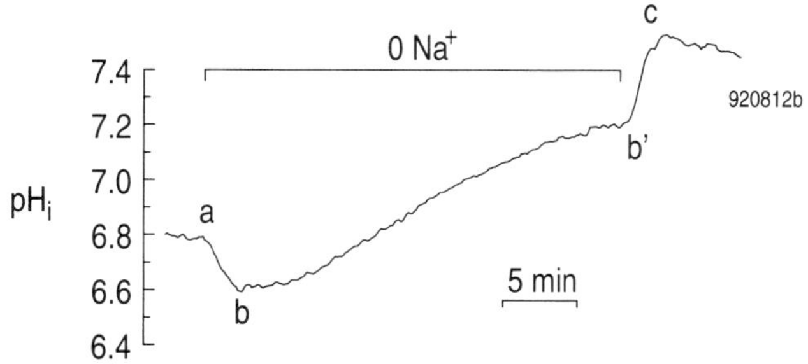

FIGURE 3. Na^+-independent pH_i recovery in a CA1 neuron acutely isolated from the hippocampus of an immature rat. The neuron had an initial pH_i of ~6.8 in the nominal absence of CO_2/HCO_3^- prior to point *a*. When external Na^+ was removed, the pH_i first decreased (*ab*), and then increased (*bb'*) in the continued absence of Na^+. When Na^+ was returned to the bath solution, pH_i increased further (*b'c*). Reproduced from Bevensee et al. (1996), with permission from J. Physiol (Lond). Copyright 1996, Cambridge University Press.

of ~6.8 prior to point *a*. When the neuron was exposed to a Na^+-free solution, the pH_i initially decreased to a pH_i of ~6.6 (*ab*). However, in the continued absence of external Na^+, the pH_i then spontaneously increased to ~7.2 (*bb'*), a value ~0.4 higher than the initial pH_i prior to removing external Na^+. When the neuron was returned to a solution containing Na^+, the pH_i further increased (*b'c*), presumably due to Na^+/H^+ exchange. The segment-*bb'* Na^+-independent pH_i recovery is unaffected by bafilomycin, an inhibitor of the vacuolar-type proton pump (M. O. Bevensee and W. F. Boron, unpublished observations). Because it is not customary to record membrane potential during experiments in which pH_i is measured fluorometrically, we can not exclude the possibility that the neuron in Fig. 3 is considerably depolarized in the absence of Na^+, and that the Na^+-independent pH_i increase is due to activation of a H^+ conductance. It is interesting to recall that the putative H^+ pump in the S3 segment of the rabbit proximal tubule, which also manifests itself by a pH_i increase similar to that in segment *bb'*, is insensitive to N-ethylmaleimide (NEM), an inhibitor the vacuolar-type proton pump (Chen and Boron, 1995). The mechanism in the S3 segment was inhibited by reducing intracellular ATP, accomplished by removing external glucose and amino acids, while at the same time applying high concentrations of the metabolic inhibitors cyanide and iodoacetic acid (Chen and Boron, 1995).

H⁺/Lactate cotransport. The accumulation of lactic acid in the brain is thought to contribute to the brain damage caused by anoxia, ischemia, and hyperglycemia (Plum, 1983; see Chapters 28 and 29). Under such conditions, the lactic acid generated from excessive anaerobic metabolism in astrocytes and neurons is believed to be transported to the extracellular space by an electroneutral H^+/lactate cotransporter. This transporter has been functionally identified in primary cultures of neurons and glial cells. For example, Walz and Mukerji (1988), assaying lactate release by monitoring the formation of NADH during the enzymatic conversion of extracellular lactate to pyruvate, found that both astrocytes and neurons from the cerebral cortex of the rat release lactate into the extracellular medium. In both cell types, lactic acid efflux was inhibited by pyruvate and the nontransportable lactate analogue DL-*p*-hydroxyphenyl-lactate, but not by the monocarboxylic acid transport inhibitor α-cyano-3-hydroxycinnamate (CHC). In another series of experiments, Schneider et al. (1993) used double-barreled pH-sensitive microelectrodes to study H^+/lactate cotransport in isolated rat spinal roots. They found that exposing the roots to organic anions such as L-lactate caused an initial increase in pH_o, followed by a decrease (Fig. 4). (These results imply that the exposure to lactate produced an initial decrease in pH_i.) In contrast, removing L-lactate caused an initial decrease in pH_o, followed by an increase. Making the roots hypoxic led to a decrease in pH_o, suggestive of lactic-acid efflux from the cells. With both lactate addition and/or removal and hypoxia, CHC and the stilbene derivative 4,4'-diisothiocyanatostilbene-2,2'-disulfonic acid (DIDS) substantially inhibited the pH_o changes.

There is also evidence that lactate can be transported *into* neurons from the extracellular environment. As reviewed by Tsacopoulos and Magistretti (1996), the lactate released from glycogen stores in astrocytes could serve as an energy source for neurons that incorporate the lactate. Using a radioactive tracer method, Dringen et al. (1993) demonstrated a CHC-sensitive, lactate uptake mechanism in neuronal cultures from embryonic rat brain. In another group of experiments, Nedergaard and Goldman (1993) focused on the lactic-acid flux into neurons and astrocytes cultured from the forebrain of the rat embryo. Measuring pH_i with BCECF, they found that the rate of intracellular acidification was a saturable function of the amount of externally applied lactic acid, indicative of carrier-

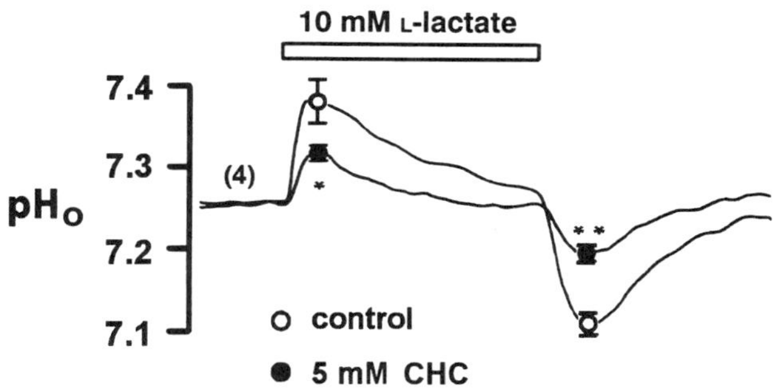

FIGURE 4. H^+/lactate cotransport in isolated spinal roots from rat. When the spinal root was exposed to a solution containing 10 mM L-lactate, the extracellular pH (pH_o), as measured with a double-barreled pH-sensitive microelectrode, initially increased (○) and then subsequently decreased to the pH_o at the onset of the experiment. Removing the L-lactate solution had the opposite effects. Both the increase in pH_o with the addition of L-lactate, as well as the decrease in pH_o with the removal of L-lactate, were significantly reduced in the presence of the H^+/lactate cotransport inhibitor CHC (●). Modified from Schneider et al. (1993), with permission from Neuroscience. Copyright 1993, Elsevier Sciences, Ltd.

mediated transport. However, none of the aforementioned inhibitors, such as CHC and DIDS, slowed the lactic acid influx.

Acid Loaders That Are Independent of Both HCO_3^- and Na^+

Ca^{2+}/H^+ pump. The regulation of $[Ca^{2+}]_i$ is especially important in neurons because elevated levels of $[Ca^{2+}]_i$ can result in neural cell damage, as seen during glutamate-induced neurotoxicity (Choi, 1988, 1995). Because the electrochemical gradient favoring Ca^{2+} influx is large, neurons must extrude Ca^{2+} in order to stabilize $[Ca^{2+}]_i$. Ca^{2+} extrusion is especially important with electrical activity, which causes the Ca^{2+} permeability of neurons to increase markedly. The conventional wisdom is that the major pathway for Ca^{2+} extrusion is an ATP-driven Ca^{2+} pump, with the Na^+/Ca^{2+} exchanger playing an important role at high $[Ca^{2+}]_i$ values. Schwiening et al. (1993) demonstrated in snail neurons, that vanadate blocks a Ca^{2+} pump that exchanges intracellular Ca^{2+} for extracellular H^+ (see also Chapter 15).

In neurons in the rat hippocampal slice, a similar Ca^{2+}-pump is responsible for extruding Ca^{2+} and elevating pH_o in response to stimuli-induced neuronal activity (Paalasmaa et al., 1994; Smith et al., 1994; Paalasmaa and Kaila, 1996). Recently, Trapp et al. (1996b) provided evidence for a Ca^{2+}-pump in hippocampal neurons *in situ* loaded with either BCECF, or the Ca^{2+}-sensitive fluorophore Fura-2, via diffusion from a patch pipette. Depolarization of these neurons to 0 mV under voltage-clamp conditions elicits both an increase in $[Ca^{2+}]_i$ and decrease in pH_i. The $[Ca^{2+}]_i$ increase is due to the increased activity of voltage-gated Ca^{2+} channels (VGCCs) because it is inhibited by either applying Ni^{2+} plus Cd^{2+}, or by removing external Ca^{2+}. The pH_i decrease is due to activation of a plasma membrane Ca^{2+}-pump, which responds to the elevation of $[Ca^{2+}]_i$ by exchanging intracellular Ca^{2+} for extracellular H^+. The intracellular acidification can be inhibited by vanadate and eosin. Inhibiting the VGCCs prevents the depolarization-induced pH_i decrease, whereas inhibiting the Ca^{2+}-pump partially inhibits the return of $[Ca^{2+}]_i$ to baseline following the depolarization-induced $[Ca^{2+}]_i$ increase. A similar Ca^{2+}-pump may also play a role in regulating pH_i of primary cultures of neocortical neurons from rat (OuYang et al., 1994).

Acid Extruders That Transport HCO_3^-: The Na^+-Driven Cl^-/HCO_3^- Exchanger

Properties of the transporter. As discussed in Chapter 11, the Na^+-driven Cl^-/HCO_3^- exchanger is the predominant acid-base transporter in invertebrate preparations, such as the squid giant axon and the snail neuron. The Na^+-driven Cl^-/HCO_3^- exchanger is responsible for the recovery of pH_i from acid loads (Boron and De Weer, 1976a). It has an absolute requirement for external Na^+ (Thomas, 1977) and HCO_3^- (Boron and De Weer, 1976b), and for intracellular Cl^- (Russell and Boron, 1976), and is most active at low pH_i values. In the process of exchange, 1 Na^+ moves into the cell and 1 Cl^- moves out for each pair of protons neutralized inside the cell (Boron and Russell, 1983). The transporter is inhibited by stilbene derivatives such as 4-acetamido-4'-isothiocyanatostilbene-2,2'-disulfonic acid (SITS) and DIDS (Thomas, 1976).

pH_i changes caused by Na^+-driven Cl^-/HCO_3^- exchange. The Na^+-driven Cl^-/HCO_3^- exchanger has been identified in low-pH_i CA1 neurons acutely isolated from the hippocampus of the immature rat (Schwiening and Boron, 1994). When CA1 neurons are switched from a nominally CO_2/HCO_3^--free bath solution to a 5% CO_2/25 mM HCO_3^- solution, pH_i initially decreases, due to the entry of CO_2 and formation of H^+ and HCO_3^- (presumably accelerated by intracellular carbonic anhydrase [see Pasternack et al., 1993]). Subsequently, during the "plateau phase," the pH_i increase is due to active HCO_3^- uptake (Schwiening and Boron, 1994). The pH_i increase does not occur in the absence of external Na^+ or in the presence of 100 μM DIDS, indicating that the transporter probably is either a Na^+/HCO_3^- cotransporter or a Na^+-driven Cl^-/HCO_3^- exchanger. Distinguishing between the two alternatives would require determining whether or not the transporter requires internal Cl^-. Schwiening and Boron (1994) found that the alkalinization during the plateau phase of a CO_2/HCO_3^- exposure requires internal Cl^-. As shown in the experiment in Fig. 5, the plateau-phase alkalinization progressively decreased in rate and magnitude with successive exposures to CO_2/HCO_3^- in the absence of external Cl^- (compare segments *ab, ef,* and *gh*). Presumably, the plateau-phase alkalinization is due to a Na^+-driven Cl^-/HCO_3^- exchanger that depletes the cell of intracellular Cl^- in the absence of external Cl^-. When the neuron was then exposed to a Cl^--containing solution for only about 3 minutes (*ij*), a subsequent exposure to CO_2/HCO_3^- in the absence of external Cl^- elicited a transient acidification, followed by a large and rapid alkalinization (*jk*), similar to the first one at the start of the experiment. Apparently, even a 3-minute exposure to a Cl^--containing solution is enough to increase $[Cl^-]_i$ to a level sufficient to support Na^+-driven Cl^-/HCO_3^- exchange. A K^+/Cl^- cotransporter, such as the one in neocortical pyramidal neurons (Thompson et al., 1988), or an $Na^+/K^+/Cl^-$ cotransporter, could account for at least some of the $[Cl^-]_i$ increase when CA1 neurons are re-exposed to a Cl^--containing solution.

The Na^+-driven Cl^-/HCO_3^- exchanger also appears to make a major contribution to acid extrusion in freshly isolated CA1 neurons with a high initial pH_i (Bevensee et al., 1996). Cultured hippocampal neurons from the fetal rat also have an acid-extrusion mechanism that requires HCO_3^- and Na^+, is at least partially blocked by DIDS (Raley-Susman et al., 1991; Baxter and Church, 1996), and is inhibited by removing external Cl^- (Baxter and Church, 1996).

Baxter and Church (1996) compared pH_i regulation in cultured fetal rat hippocampal neurons both at 18°C–22°C and 37°C. They found that applying DIDS to the neurons at 18°C–22°C, but not at 37°C, elicits a decrease in steady-state pH_i, suggesting that the HCO_3^- transporter contributes to the maintenance of steady-state pH_i only at the lower of

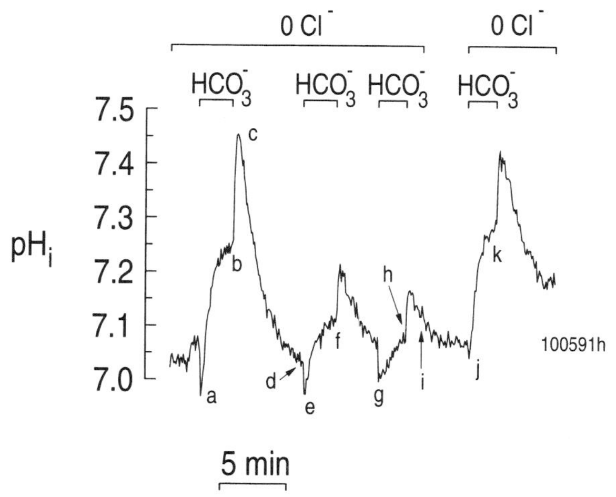

FIGURE 5. $[Cl^-]_i$ dependence of the CO_2/HCO_3^--dependent alkalinization in a CA1 neuron acutely isolated from an immature rat. The neuron had an initial pH_i of ~7.05 in the absence of CO_2/HCO_3^- and external Cl^- at the start of the experiment. When the neuron was exposed to a Cl^--free, CO_2/HCO_3^--buffered solution, the pH_i decreased due to CO_2 influx (point *a*), and then promptly increased (*ab*), due to HCO_3^- uptake. Returning the neuron to a nominally CO_2/HCO_3^--free solution caused pH_i to increase rapidly (*bc*), due to CO_2 efflux, and then to decrease (*cd*), due to acid-loading mechanisms. When the neuron was exposed to the Cl^--free, CO_2/HCO_3^--buffered solution a second and third time, the CO_2/HCO_3^--induced alkalinization progressively decreased (*ef* and *gh*, respectively), presumably reflecting depletion of intracellular Cl^-. For ~3 minutes, the neuron was then exposed to a solution containing Cl^- (*ij*), which presumably reloaded the cytoplasm with Cl^-. When the neuron was then exposed to the Cl^--free, CO_2/HCO_3^--buffered solution, pH_i rapidly recovered (*jk*), as at the beginning of the experiment. Modified from Schwiening and Boron (1994), with permission from J. Physiol (Lond). Copyright 1994, Cambridge University Press.

these two temperatures. These results are consistent with the notion that warming from 18°C–22°C to 37°C increases the activity of the Na^+/H^+ exchanger more than the Na^+-driven Cl^-/HCO_3^- exchanger.

Other cultured mammalian neurons also appear to have a HCO_3^--dependent acid extrusion mechanism, which may in fact be a Na^+-driven Cl^-/HCO_3^- exchanger. For example, sympathetic neurons cultured from the rat have a higher steady-state pH_i in the presence than in the absence of CO_2/HCO_3^- (Tolkovsky and Richards, 1987). When these sympathetic neurons are acid loaded by an NH_4^+ prepulse, the pH_i recovery is enhanced by CO_2/HCO_3^-. On the other hand, the pH_i recovery is unaffected by stilbene derivatives. In cerebellar granule cells cultured from the rat, the pH_i recovery following an acid load requires Na^+ and is enhanced by CO_2/HCO_3^- (Pocock and Richards, 1992). As in sympathetic neurons, SITS failed to inhibit the pH_i recovery (Pocock and Richards, 1992). The latter authors also found that removing extracellular Cl^- failed to block the pH_i recovery, although there is no evidence that this maneuver depleted intracellular Cl^-.

Finally, in some neuronal preparations, HCO_3^--dependent acid-extrusion processes contribute little, if anything, to the pH_i recovery from an acid load. In rat brain synaptosomes acid loaded by an NH_4^+ prepulse, Nachshen and Drapeau (1988) observed that the pH_i recovery rate in CO_2/HCO_3^- is unaffected by SITS or DIDS, or the removal of external Cl^-. Because the rates of pH_i recovery were the same in the absence and the presence of CO_2/HCO_3^-, the authors concluded that a Na^+-driven Cl^-/HCO_3^- exchange mechanism did not contribute to the pH_i recovery in the presence of CO_2/HCO_3^-. However, if the rates of pH_i recovery from an acid load are the same in the presence and nominal absence of CO_2/HCO_3^-, the acid-extrusion rate was probably greater in the presence of CO_2/HCO_3^-, inasmuch as the intracellular buffering power was likely to have been greater in the presence of CO_2/HCO_3^-. Supporting the conclusion that synaptosomes lack significant HCO_3^- transport is the observation (made while measuring pH_i with either pyranine or BCECF) that the pH_i recovery rate from an acid load is unaffected by 100 μM DIDS, or the carbonic anhydrase inhibitor acetazolamide (100 and 500 μM) (Sánchez-Armass et al., 1994). The resting pH_i of these synaptosomes was also unaffected by acetazolamide. Whether rat pheochromocytoma (PC12) cells also lack HCO_3^--dependent acid extruders is unclear. On the one hand, because the pH_i recovery (measured with BCECF) from an acid load was the same in the presence and absence of HCO_3^-, the authors concluded a HCO_3^- transporter was not present (Dickens et al., 1989). On the other hand, because intracellular buffering power was probably greater in the presence of HCO_3^- (and thus the acid-extrusion rate was probably greater in CO_2/HCO_3^-), a HCO_3^--dependent acid extrusion mechanism may actually have contributed to the pH_i recovery in CO_2/HCO_3^-.

Acid Loaders That Transport HCO_3^-: The Cl^-/HCO_3^- Exchanger

Properties of the transporter. The Cl^-/HCO_3^- exchanger was originally identified in erythrocytes, where it appears as the third band on SDS-polyacrylamide gel electrophoresis (see review by Steck, 1978). In nonerythroid cells, the Cl^-/HCO_3^- exchanger normally behaves as an acid loader, moving HCO_3^- out of the cell in exchange for extracellular Cl^-. As might be expected, Cl^-/HCO_3^- exchangers are most active at elevated pH_i (Vaughan-Jones, 1982; Simchowitz and Roos, 1985; Olsnes et al., 1986; Boyarsky et al., 1988). A hallmark of Cl^-/HCO_3^- exchangers is that they are inhibited by disulfonic stilbene derivatives, such as DIDS. Although Cl^-/HCO_3^- exchangers generally transport acid-base equivalents much faster in the presence than in the nominal absence of CO_2/HCO_3^-, this HCO_3^- dependence can sometimes be difficult to demonstrate. For example, the Cl^-/HCO_3^- exchanger in rat osteoclasts can transport HCO_3^- at appreciable rates even when the cells are incubated in a nominally CO_2/HCO_3^--free solution. Only when the solutions are carefully equilibrated with O_2 or N_2 does Cl^-/HCO_3^- exchange activity come to a halt (Siegel et al., 1993).

Isoforms. The first Cl^-/HCO_3^- exchanger to be cloned, named AE1 (Kopito and Lodish, 1985), is the one found in erythrocytes. Subsequently, two other AE isoforms were cloned, AE2 and AE3. AE3, in particular, has been localized to nearly all neurons throughout the mouse and rat brain (Kopito et al., 1989; Kudrycki et al., 1990; Morgans and Kopito, 1993), as well as rat retina (Kobayashi et al., 1994). In adddition, Havenga et al. (1994) have used RT-PCR and mRNA from human frontal cortex to identify a cDNA fragment (AEO) that has a high degree of homology to AE1.

pH_i changes caused by Cl^-/HCO_3^- exchangers. Based on experiments using the pH-sensitive dye DCH, Dickens et al. (1989) have reported that cultured rat pheochromocytoma (PC12) cells have a Cl^-/HCO_3^- exchanger. When cells were exposed to a Cl^--free solution in the presence of HCO_3^-, pH_i increased. This alkalinization was inhibited by nominally removing CO_2/HCO_3^-, or by pretreating the cells with DIDS. Cl^-/HCO_3^- exchangers have been demonstrated in cortical neurons and cerebellar Purkinje cells cultured from the rat. In rat cortical neurons, the Cl^-/HCO_3^- exchanger is the major mechanism by which the cells recover from acute intracellular alkali loads. Monitoring pH_i with BCECF, Ou-yang et al. (1993) used the approach introduced by Ganz et al. (1989) to alkali load cells and monitor the subsequent recovery at a physiological CO_2/HCO_3^-. With the cortical neurons initially incubated in a solution containing 15% CO_2/50 mM HCO_3^- (pH = 7.35), Ou-yang et al. switched the solution to one containing 5% CO_2/20 mM HCO_3^- (similar pH). Following the transient pH_i increase, which is due to the efflux of CO_2 and represents an intracellular alkali load, pH_i decreased to a value similar to that in the presence of 15% CO_2/50 mM HCO_3^- (Fig. 6A). The rate and magnitude of this acidification was reduced by removing external Cl^- (Fig. 6C), or by adding DIDS (Fig. 6D), but not by adding the Na^+/H^+ exchange inhibitor amiloride (Fig. 6B).

The Cl^-/HCO_3^- exchanger has also been found in cultured rat cerebellar Purkinje cells, in which pH_i was also measured with BCECF (Gaillard and Dupont, 1990). With the Purkinje cells exposed to a CO_2/HCO_3^--buffered saline, removing extracellular Cl^- led to an alkalinization that was unaffected by the absence of external Na^+, and yet was inhibited by DIDS.

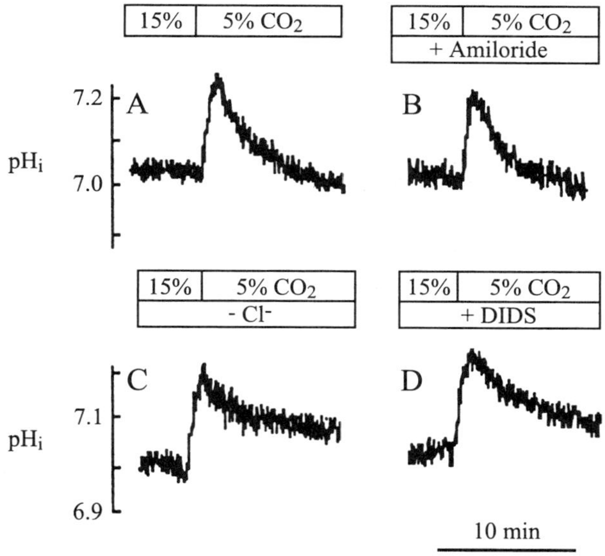

FIGURE 6. Contribution of Cl^-/HCO_3^- exchange to the recovery of pH_i from an alkali load in a neuron cultured from the rat cortex. **A.** When a neuron was switched from a solution containing 15% CO_2 /50 mM HCO_3^- to one containing 5% CO_2 /20 mM HCO_3^- (at constant pH_o), the pH_i initially increased, due to CO_2 efflux, thus alkali loading the cell. Following the alkali load, the pH_i slowly decreased. **B.** Following an alkali load as described in (A), the subsequent pH_i decrease was unaffected by the presence of 1.0 mM amiloride. **C.** In the absence of external Cl^-, the rate and magnitude of the pH_i decrease from the alkali load were smaller than that seen in (A) and (B). **D.** In the presence of 0.5 mM DIDS, the rate and magnitude of the pH_i decrease from the alkali load were smaller than that seen in (A) and (B). Modified from Ou-yang et al. (1993), with permission from J. Cereb. Blood Flow Metab. Copyright 1993, Raven Press, Ltd.

In contrast to the above two preparations, sympathetic and hippocampal neurons cultured from rat do not appear to have substantial Cl^-/HCO_3^- exchange activity. In sympathetic neurons from the superior cervical ganglion, removing external Cl^- in the presence or absence of HCO_3^- had no effect on pH_i, as measured with BCECF, or on the pH_i recovery from an acid load (Tolkovsky and Richards, 1987). Applying DIDS or SITS in either the presence or absence of HCO_3^- also had no effect on either the steady-state pH_i, or on the rate of pH_i recovery from an acid load. Raley-Susman et al. (1991) concluded that hippocampal neurons cultured from the rat fetus also do not appear to have Cl^-/HCO_3^- exchange activity. In experiments in which pH_i was measured with BCECF, removing external Cl^- failed to elicit an increase in pH_i in the presence of CO_2/HCO_3^-.

One would expect a Cl^-/HCO_3^- exchanger to operate in reverse, and thereby function as an acid extruder, under conditions in which the cell is markedly acidic (e.g., pH < 6.5). For example, Baxter and Church (1996), working with cultured fetal rat hippocampal neurons, demonstrated that when these BCECF-loaded neurons were acid loaded by applying and removing NH_4^+, a portion of the subsequent pH_i recovery was dependent both on DIDS and HCO_3^-, yet unaffected by removing external Na^+. As further evidence for a Cl^-/HCO_3^- exchanger operating in reverse at low pH_i, the authors demonstrated that the pH_i recovery could be accelerated by removing external Cl^-.

Tissue localization of Cl^-/HCO_3^- exchangers. Using *in situ* hybridization procedures, Raley-Susman et al. (1993) found AE3 mRNA expression in all neuronal cell fields of the hippocampus in both the adult (Fig. 7A) and fetal (Fig. 7B) rat. However, the hy-

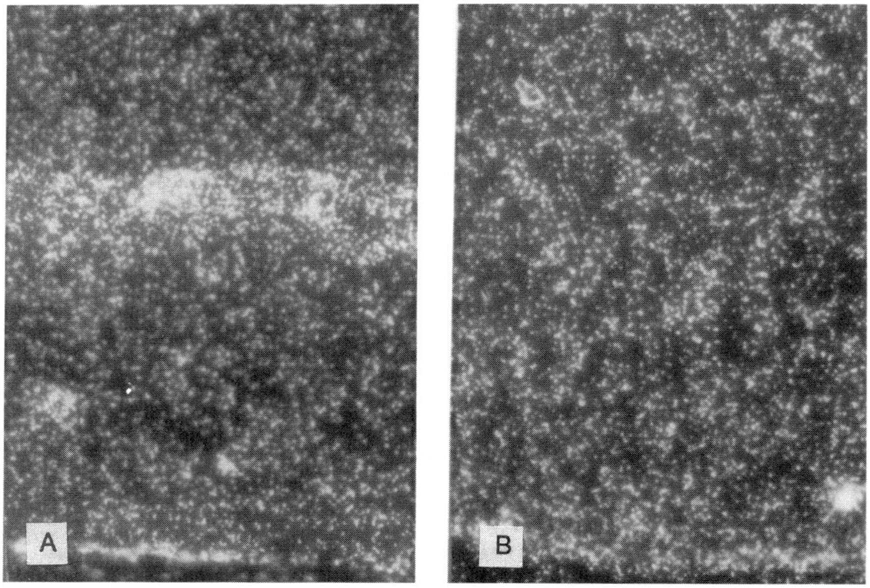

FIGURE 7. The expression of Cl^-/HCO_3^- exchange (AE3) mRNA in the CA1 region of the hippocampus from both the adult (A) and fetal day-18 (B) rat. Rats were perfused with 4% paraformaldehyde and probed with [^{35}S]UTP-labeled antisense mRNA of AE3. The view is a 135× magnification of an autoradiogram, made from a 25-μm section from either the adult or fetal rat brain. The adult preparation has an intense band of activity. Modified from Raley-Susman et al. (1993), with permission from Brain Res. Copyright 1993, Elsevier Sciences, Ltd.

bridization was stronger in the hippocampus of the adult than in the hippocampus of the fetus. The authors found that acutely isolated CA1 neurons from the adult rat hippocampus also have functionally active Cl^-/HCO_3^- exchange that elevates pH_i in response to the removal of external Cl^- in the presence of CO_2/HCO_3^-. In contrast, functional Cl^-/HCO_3^- exchange was not detectable in CA1 neurons from the fetal rat hippocampus.

CONCLUDING REMARKS

As discussed in the previous section, mammalian neurons utilize a variety of acid-base transporters to regulate pH_i. The acid extruders and acid loaders that have been identified are shown in Fig. 8. There are three interesting observations to be made regarding differences in transporter activity observed among different neurons. First, not all mammalian neurons possess the same complement of acid-base transporters. For example, the Cl^-/HCO_3^- exchanger is present in cultured rat cerebellar Purkinje cells, but not in cultured rat sympathetic neurons from the superior cervical ganglion. Second, two different types of neurons may have the same functional transporter, but with different sensitivities to pharmacological agents. For example, amiloride inhibits the Na^+/H^+ exchanger in cultured rat cerebellar Purkinje cells, but not the putative Na^+/H^+ exchanger in CA1 neurons from the rat hippocampus. Another example is that CHC inhibits the H^+/lactate cotransporter in neurons from the rat cerebral cortex, but not the cotransporter in isolated rat spinal roots. The third interesting observation is that transport may be developmentally regulated. For example, the Cl^-/HCO_3^- exchanger in the rat hippocampus appears functionally active in the adult, but not in the fetus. Because the number of studies done on pH_i regulation in mammalian neurons is so small, it is likely that there will turn out to be

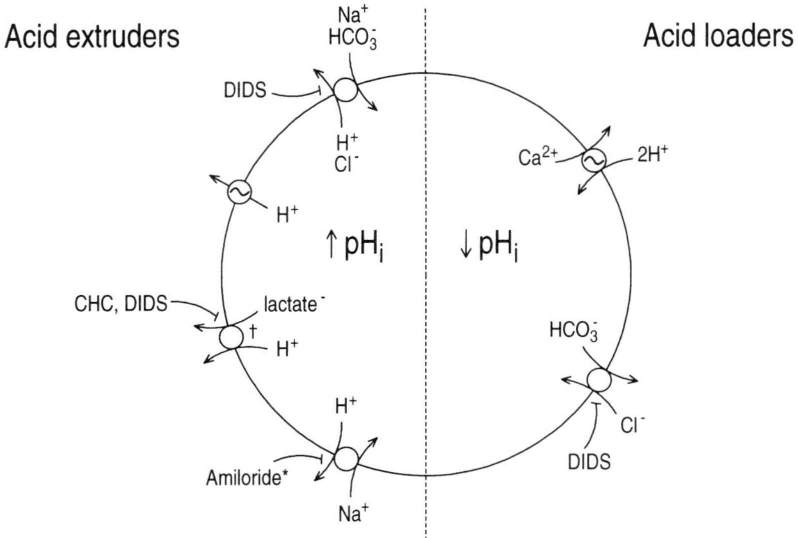

FIGURE 8. Some of the pH_i-regulating mechanisms identified in mammalian neurons. (*) Amiloride and its analogues are poor inhibitors of the putative Na^+/H^+ exchanger in some preparations. (†) Lactate and H^+ can be transported *into* neurons and astrocytes via H^+/lactate cotransport in some preparations.

many more examples of each of these three classes of differences. The question then arises as to why there is such a diversity of acid-base transporters employed by various mammalian neurons. The differences in acid-base transporters may be related to the differences in the electrophysiology of neurons under both physiological and pathophysiological conditions.

REFERENCES

Amos BJ, Richards CD (1996): Intrinsic hydrogen ion buffering in rat CNS neurones maintained in culture. Exp Physiol 81: 261–271.

Baxter KA, Church J (1996): Characterization of acid extrusion mechanisms in cultured fetal rat hippocampal neurones. J Physiol (Lond) 493: 457–470.

Benos DJ, Sapirstein VS (1983): Characteristics of an amiloride-sensitive sodium entry pathway in cultured rodent glial and neuroblastoma cells. J Cell Physiol 116: 213–220.

Bevensee MO, Cummins TR, Haddad GG, Boron WF, Boyarsky G (1996): pH regulation in single CA1 neurons acutely isolated from the hippocampi of immature and mature rats. J Physiol (Lond) 494: 315–328.

Bookstein C, Musch MW, DePaoli A, Xie Y, Rabenau K, Villereal M, Rao MC, Chang EB (1996): Characterization of the rat Na^+/H^+ exchanger isoform NHE4 and localization in rat hippocampus. Am J Physiol 271: C1629–C1638.

Boron WF, De Weer P (1976a): Intracellular pH transients in squid giant axons caused by CO_2, NH_3, and metabolic inhibitors. J Gen Physiol 67: 91–112.

Boron WF, De Weer P (1976b): Active proton transport stimulated by CO_2/HCO_3^-, blocked by cyanide. Nature 259: 240–241.

Boron WF, Russell JM (1983): Stoichiometry and ion dependencies of the intracellular-pH-regulating mechanism in squid giant axons. J Gen Physiol 81: 373–399.

Boyarsky G, Ganz MB, Sterzel RB, Boron WF (1988): pH regulation in single glomerular mesangial cells. II. Na^+-dependent and -independent Cl^--HCO_3^- exchangers. Am J Physiol 255: C857–C869.

Boyarsky G, Ransom B, Schlue W-R, Davis MBE, Boron WF (1993): Intracellular pH regulation in single cultured astrocytes from rat forebrain. Glia 8: 241–248.

Chen LK, Boron WF (1995): Acid extrusion in S3 segment of rabbit proximal tubule II. Effect of basolateral CO_2/HCO_3^-. Am J Physiol 268: F193–F203.

Choi DW (1988): Calcium-mediated neurotoxicity: relationship to specific channel types and role in ischemic damage. Trends Neurosci 11: 465–469.

Choi DW (1995): Calcium: still center-stage in hypoxia-ischemic neuronal death. Trends Neurosci 18: 58–60.

Clark JD, Limbird LE (1991): Na^+-H^+ exchanger subtypes: a predictive review. Am J Physiol 261: C945–C953.

Dickens CJ, Gillespie JI, Greenwell JR (1989): Interactions between intracellular pH and calcium in single mouse neuroblastoma (N2A) and rat pheochromocytoma cells (PC12). Quart J Exp Physiol 74: 671–679.

Dringen R, Wiesinger H, Hamprecht B (1993): Uptake of L-lactate by cultured rat brain neurons. Neurosci Lett 163: 5–7.

Gaillard S, Dupont J (1990): Ionic control of intracellular pH in rat cerebellar Purkinje cells maintained in culture. J Physiol (Lond) 425: 71–83.

Ganz MB, Boyarsky G, Sterzel RB, Boron WF (1989): Arginine vasopressin enhances pH_i regulation in the presence of HCO_3^- by stimulating three acid-base transport systems. Nature 337: 648–651.

Geibel J, Giebisch G, Boron WF (1989): Basolateral sodium-coupled acid-base transport mechanisms of the rabbit proximal tubule. Am J Physiol 257: F790–F797.

Goslin K, Banker G (1991): Rat hippocampal neurons in low-density culture. In Banker G, Goslin K (eds): Culturing Nerve Cells. Cambridge, MA: MIT Press, pp 251–281.

Havenga MJE, Bosman GJCGM, Appelhans H, De Grip WJ (1994): Expression of the anion exchanger (AE) gene family in human brain. Identification of a new AE protein: AEO. Molec Brain Res 25: 97–104.

Hinke JAM (1967): Cation-selective microelectrodes for intracellular use. In Eisenman G (eds): Glass Electrodes for Hydrogen and Other Cations. Principle and Practice. New York: Dekker, pp 464–477.

Jean T, Frelin C, Vigne P, Barbry P, Lazdunski M (1985): Biochemical properties of the Na^+/H^+ exchange system in rat brain synaptosomes. Interdependence of internal and external pH control of the exchange activity. J Biol Chem 260: 9678–9684.

Kay AR, Wong RKS (1986): Isolation of neurons suitable for patch-clamping from adult mammalian central nervous systems. J Neurosci Methods 16: 227–238.

Kobayashi S, Morgans CW, Casey JR, Kopito RR (1994): AE3 anion exchanger isoforms in the vertebrate retina: Developmental regulation and differential expression in neurons and glia. J Neurosci 14: 6266–6279.

Kopito RR, Lodish HF (1985): Primary structure and transmembrane orientation of the murine anion exchange protein. Nature 316: 234–238.

Kopito RR, Lee BS, Simmons DM, Lindsey AE, Morgans CW, Schneider K (1989): Regulation of intracellular pH by a neuronal homolog of the erythrocyte anion exchanger. Cell 59: 927–937.

Kudrycki KE, Newman PR, Shull GE (1990): cDNA cloning and tissue distribution of mRNAs for two proteins that are related to the Band 3 Cl^--HCO_3^- exchanger. J Biol Chem 265: 462–471.

Ma E, Haddad GG (1997): Expression and localization of Na^+/H^+ exchangers in rat central nervous system. Neuroscience 79: 591–603.

Meech RW, Thomas RC (1977): The effect of calcium injection on the intracellular sodium and pH of snail neurones. J Physiol (Lond) 265: 867–879.

Mellergård P, Ouyang Y-B, Siesjö BK (1994): The regulation of intracellular pH is strongly dependent on extracellular pH in cultured rat astrocytes and neurons. Acta Neurochir Suppl 60: 34–37.

Montrose MH, Murer H (1990): Polarity and kinetics of Na^+-H^+ exchange in cultured opossum kidney cells. Am J Physiol 259: C121–C133.

Moolenaar WH, Boonstra J, van der Saag PT, de Laat SW (1981): Sodium/proton exchange in mouse neuroblastoma cells. J Biol Chem 256: 12883–12887.

Moolenaar WH, Tertoolen LGJ, de Laat SW (1984): Phorbol ester and diacylglycerol mimic growth factors in raising cytoplasmic pH. Nature 312: 371–374.

Morgans CW, Kopito RR (1993): Generation of truncated brain AE3 isoforms by alternate mRNA processing. J Cell Sci 106: 1275–1282.

Nachshen DA, Drapeau P (1988): The regulation of cytosolic pH in isolated presynaptic nerve terminals from rat brain. J Gen Physiol 91: 289–303.

Nedergaard M, Goldman SA (1993): Carrier-mediated transport of lactic acid in cultured neurons and astrocytes. Am J Physiol 265: R282–R289.

Noël J, Pouysségur J (1995): Hormonal regulation, pharmacology, and membrane sorting of vertebrate Na^+/H^+ exchanger isoforms. Am J Physiol 268: C283–C296.

Olsnes S, Tønnessen TI, Sandvig K (1986): pH-regulated anion antiport in nucleated mammalian cells. J Cell Biol 102: 967–971.

Ou-yang Y, Mellergård P, Siesjö BK (1993): Regulation of intracellular pH in single rat cortical neurons in vitro: A microspectrofluorometric study. J Cereb Blood Flow Metab 13: 827–840.

OuYang YB, Mellergård P, Kristián T, Kristiánova V, Siesjö BK (1994): Influence of acid-base changes on the intracellular calcium concentration of neurons in primary culture. Exp Brain Res 101: 265–271.

Paalasmaa P, Kaila K (1996): Role of voltage-gated calcium channels in the generation of activity-induced extracellular pH transients in the rat hippocampal slice. J Neurophysiol 75: 2354–2360.

Paalasmaa P, Taira T, Voipio J, Kaila K (1994): Extracellular alkaline transients mediated by glutamate receptors in the rat hippocampal slice are not due to a proton conductance. J Neurophysiol 72: 2031–2033.

Pasternack M, Voipio J, Kaila K (1993): Intracellular carbonic anhydrase activity and its role in GABA-induced acidosis in isolated rat hippocampal pyramidal neurones. Acta Physiol Scand 148: 229–231.

Plum F (1983): What causes infarction in the ischemic brain?: The Robert Wartenberg lecture. Neurology 33: 222–233.

Pocock G, Richards CD (1992): Hydrogen ion regulation in rat cerebellar granule cells studied by single-cell fluorescence microscopy. Eur J Neurosci 4: 136–143.

Raley-Susman KM, Cragoe EJ Jr, Sapolsky RM, Kopito RR (1991): Regulation of intracellular pH in cultured hippocampal neurons by an amiloride-insensitive Na^+/H^+ exchanger. J Biol Chem 266: 2739–2745.

Raley-Susman KM, Sapolsky RM, Kopito RR (1993): Cl^-/HCO_3^- exchange function differs in adult and fetal rat hippocampal neurons. Brain Res 614: 308–314.

Ravesloot JH, Eisen T, Baron R, Boron WF (1995): Role of Na-H exchangers and vacuolar H^+ pumps in intracellular pH regulation in neonatal rat osteoclasts. J Gen Physiol 105: 177–208.

Reeves RB (1977): The interaction of body temperature and acid-base balance in ectothermic vertebrates. Annu Rev Physiol 39: 559–586.

Richards CD, Pocock G (1989): Hydrogen ion regulation in cultured mammalian CNS neurones. Acta Physiol Scand Suppl 582: 36.

Rink TJ, Tsien RY, Pozzan T (1982): Cytoplasmic pH and free Mg^{2+} in lymphocytes. J Cell Biol 95: 189–196.

Ritucci NA, Erlichman JS, Dean JB, Putnam RW (1996): A fluorescence technique to measure intracellular pH of single neurons in brainstem slices. J Neurosci Methods 68: 149–163.

Ritucci NA, Dean JB, Putnam RW (1997): Intracellular pH response to hypercapnia in neurons from chemosensitive areas of the medulla. Am J. Physiol 273: R433–R441.

Roos A, Boron WF (1981): Intracellular pH. Physiol Rev 61: 296–434.

Russell JM, Boron WF (1976): Role of chloride transport in regulation of intracellular *p*H. Nature 264: 73–74.

Sánchez-Armass S, Martínez-Zaguilán R, Martínez GM, Gillies RJ (1994): Regulation of pH in rat brain synaptosomes. I. Role of sodium, bicarbonate, and potassium. J Neurophysiol 71: 2236–2248.

Sardet C, Franchi A, Pouysségur J (1989): Molecular cloning, primary structure, and expression of the human growth factor-activatable Na^+/H^+ antiporter. Cell 56: 271–280.

Sauvaigo S, Vigne P, Frelin C, Lazdunski M (1984): Identification of an amiloride sensitive Na^+/H^+ exchange system in brain synaptosomes. Brain Res 301: 371-374.

Schneider U, Poole RC, Halestrap AP, Grafe P (1993): Lactate-proton co-transport and its contribution to interstitial acidification during hypoxia in isolated rat spinal roots. Neuroscience 53: 1153–1162.

Schwiening CJ, Boron WF (1994): Regulation of intracellular pH in pyramidal neurones from the rat hippocampus by Na^+-dependent Cl^--HCO_3^- exchange. J Physiol (Lond) 475: 59–67.

Schwiening CJ, Kennedy HJ, Thomas RC (1993): Calcium-hydrogen exchange by the plasma membrane Ca-ATPase of voltage-clamped snail neurons. Proc R Soc Lond 253: 285–289.

Siegel W-K, Ravesloot J-H, Baron R, Boron WF (1993): The anion exchanger of neonatal rat osteoclasts. FASEB J 7: A13.

Simchowitz L, Roos A (1985): Regulation of intracellular pH in human neutrophils. J Gen Physiol 85: 443–470.

Smith SE, Gottfried JA, Chen JCT, Chesler M (1994): Calcium dependence of glutamate receptor-evoked alkaline shifts in hippocampus. Neuroreport 5: 2441–2445.

Steck TL (1978): The Band 3 protein of the human red cell membrane: A review. J Supramol Struct 8: 311–324.

Thomas RC (1974): Intracellular pH of snail neurones measured with a new pH-sensitive glass micro-electrode. J Physiol (Lond) 238: 159–180.

Thomas RC (1976): Ionic mechanism of the H^+ pump in a snail neurone. Nature 262: 54–55.

Thomas RC (1977): The role of bicarbonate, chloride and sodium ions in the regulation of intracellular pH in snail neurones. J Physiol (Lond) 273: 317–338.

Thomas RC, Meech RW (1982): Hydrogen ion currents and intracellular *p*H in depolarized voltage-clamped snail neurones. Nature 299: 826–828.

Thompson SM, Deisz RA, Prince DA (1988): Outward chloride/cation co-transport in mammalian cortical neurons. Neurosci Lett 89: 49–54.

Tolkovsky AM, Richards CD (1987): Na^+/H^+ exchange is the major mechanism of pH regulation in cultured sympathetic neurons: Measurements in single cell bodies and neurites using a fluorescent pH indicator. Neuroscience 22: 1093–1102.

Trapp S, Lückermann M, Brooks PA, Ballanyi K (1996a): Acidosis of rat dorsal vagal neurons *in situ* during spontaneous and evoked activity. J Physiol (Lond) 496: 695–710.

Trapp S, Lückermann M, Kaila K, Ballanyi K (1996b): Acidosis of hippocampal neurones mediated by a plasmalemmal Ca^{2+}/H^+ pump. Neuroreport 7: 2000–2004.

Tsacopoulos M, Magistretti PJ (1996): Metabolic coupling between glia and neurons. J Neurosci 16: 877–885.

Vaughan-Jones RD (1982): Chloride-bicarbonate exchange in the sheep cardiac purkinje fibre. In Nuccitelli R, Deamer DW (eds): Intracellular pH: Its Measurement, Regulation and Utilization in Cellular Functions. New York: Liss, pp 239–252.

Walz W, Mukerji S (1988): Lactate release from cultured astrocytes and neurons: A comparison. Glia 1: 366–370.

Yun CHC, Tse C-M, Nath SK, Levine SA, Brant SR, Donowitz M (1995): Mammalian Na^+/H^+ exchanger gene family: structure and function studies. Am J Physiol 269: G1–G11.

CHAPTER 13

pH Regulation in Invertebrate Glia

JOACHIM W. DEITMER
Abteilung für Allgemeine Zoologie, FB Biologie, Universität Kaiserslautern, Postfach 3049, D-67653 Kaiserslautern, Germany

INTRODUCTION

The regulation of pH by glial cells in nervous systems invites three basic questions:

1. What are the mechanisms of intraglial pH regulation?
2. Are these mechanisms involved in controlling the pH in the extracellular spaces?
3. What significance has pH regulation by glial cells on the function of nervous systems?

In order to evaluate the role of glial cells in the overall H^+ homeostasis in the nervous tissue, the production of acid/base equivalents by neurons and glial cells, and the pH changes in the extracellular, interstitial, spaces must be taken into account. The latter generate feedback influencing the performance of neurons and glial cells, and may hence be part of the neuron–glial communication.

All we know about pH regulation by glial cells in invertebrate nervous systems comes from studies on giant glial cells of the leech *Hirudo medicinalis* L. This nervous system was one of the first to be used for studies on the function of glia (Kuffler and Potter, 1964), and appears particularly suitable for investigating properties of single identified glial cells. Some of the glial cells are very large and can easily be impaled by several microelectrodes. The segmental ganglia of this annelid worm contain the neuronal cell bodies in their periphery and a neuropil in the center. The neuronal cell bodies are surrounded by an outer, and the neuropil by an inner ganglion capsule (Fig. 1). In these ganglia, three types of macroglial cells have been described: There are six packet glial cells, which ensheath the neuronal cell bodies, two giant glial cells in the central neuropil, and two connective glial cells in the nerve fiber tracts between the segmental ganglia. In addition, the ventral nerve cord contains microglial cells, which tend to accumulate at the site of injury and/or inflammation.

pH and Brain Function, Edited by Kai Kaila and Bruce R. Ransom
ISBN 0-471-11838-9

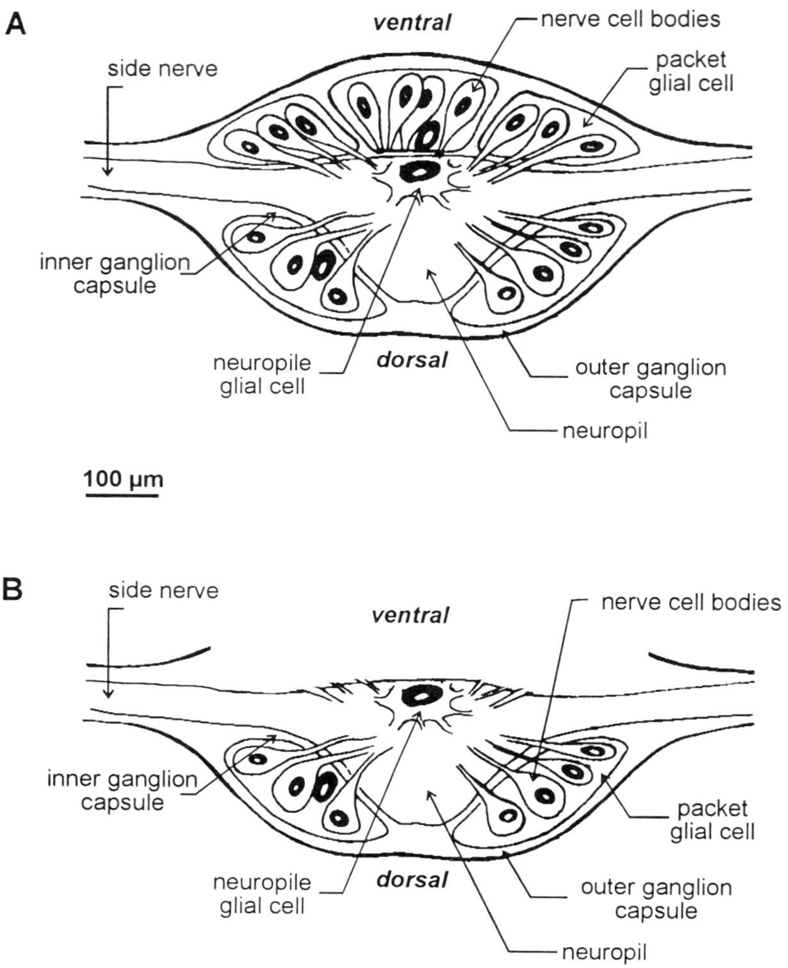

FIGURE 1. Schematic representation of leech segmental ganglia. **A.** Cross-section of an intact ganglion. **B.** Cross-section of an opened ganglion, where the neuropil glial cells are exposed. Calibration bar applies to both (A) and (B). From Nett, 1994, modified after Kuffler and potter, 1964.

The central neuropil of a ganglion contains the neuronal cell processes, dendrites, and axons and is the location where the majority of central synaptic contacts are established. The giant glial cells in the neuropil are hence in the midst of synaptic interaction and information processing. Their cell bodies are about 80 μm in diameter, and the cell extends with numerous branches across the width of the neuropil ($\sim$300 μm), where they come close to numerous neuronal processes, including synapses (Coggeshall and Fawcett, 1964). Most pH measurements have been carried out in these giant neuropil glial cells, and in the extracellular spaces of the neuropil in the vicinity of these glial cells and neurites. In the present chapter, the essentials of pH regulation as investigated extensively in single glial cells of the leech, and their impact on the overall H^+ homeostasis in the ganglionic neuropil will be discussed. Some of our knowledge on pH regulation and proton signalling has been pioneered in these leech glial cells, and later confirmed in vertebrate glial cells (see Chapter 14).

MEASUREMENT OF INTRAGLIAL pH VALUES

Techniques of pH_i Measurements in Leech Glial Cells

Measurements of pH in leech glial cells have predominantly used double-barrelled, pH-sensitive microelectrodes, which contain a small droplet of a neutral carrier as H^+ sensor in its tip (Ammann et al., 1981). These electrodes with a tip diameter around 1 μm are inserted into the cell and provide an *on-line* recording of the intracellular pH (pH_i). Only recently the ion-sensitive fluorescent dye 2',7'-(biscarboxyethyl)-5,6-carboxyfluorescein (BCECF) was used to record pH_i in these cells (Nett, 1994; Nett and Deitmer, 1996). The anion salt of this dye has been injected iontophoretically into the neuropil glial cell. The calibration of the BCECF signal, taking the ratio 440 nm to 495 nm excitation wave lengths, was carried out by H^+-permeabilizing the cell membrane with nigericin in high-[K^+] saline at different external pH values; in addition, the BCECF ratio signal was directly compared to simultaneous pH-sensitive microelectrode recordings (Nett and Deitmer, 1996). Recordings with both techniques were performed for comparison of the steady-state pH_i values and the pH_i changes obtained with each technique. Figure 2A, B shows pH_i recordings in a neuropil glial cell, with either a double-barrelled pH-sensitive microelectrode or BCECF, during acid loading the cell with the NH_4^+ prepulse technique and the subsequent pH_i recovery in the absence and presence of CO_2/HCO_3^-. It is evident that with both techniques very similar pH_i changes are measured. This is confirmed by the *simultaneous* measurement of pH_i with the two techniques (Fig. 2C).

While the use of pH-sensitive microelectrodes is an "invasive" technique, since the cell has to be impaled by the electrode, pH-sensitive fluorescent dyes, such as BCECF, can, in principle, be loaded noninvasively via their membrane-permeable acetomethylester (AM), which is cleaved in the cell by intrinsic diesterases and hence trapped in the cell (see Chapters 6–8). In invertebrate cells, such as the leech glial cells, this type of dye loading does not work as well as in vertebrate cells, and the dye needs to be iontophoretically or pressure-injected into the cell (Pfeiffer, 1992; Munsch and Deitmer, 1992). In order to prevent continuous loss of the dye from the neuropile glial cell, a procedure has been developed by which a small constant current provides a continuous injection of the dye throughout the experiment, thereby maintaining a high dye concentration and fluorecence level (Munsch and Deitmer, 1995; Nett and Deitmer, 1996). Since the neuropil glial cell has a low input resistance, a typical injection current of 2–5 nA hyperpolarizes the cell membrane by only 1–3 mV. Under voltage-clamp conditions, the injection current is directly compensated by the additional current electrode.

Other differences between the two ways of measuring pH_i are that pH-sensitive microelectrodes give point measurements of pH_i and are, due to their very high electrical resistance, slow as compared to fluorescent dyes, which are fast and can provide a spacial resolution of intracellular H^+ distribution (Nett and Deitmer, 1996). With the latter technique, the absolute pH_i value can only be determined by H^+/K^--permeabilizing the cells with nigericin at the end of an experiment. This may be less accurate than determining pH_i with a microelectrode, and reveals the absolute pH_i value only after an experiment. Indeed, in a direct comparison of the fluorescent dye method and pH-sensitive microelectrodes in the leech neuropil glial cell, the calibrated values for pH_i deviated by ~0.1 pH unit for the two techniques (Fig. 2C) (Nett and Deitmer, 1996).

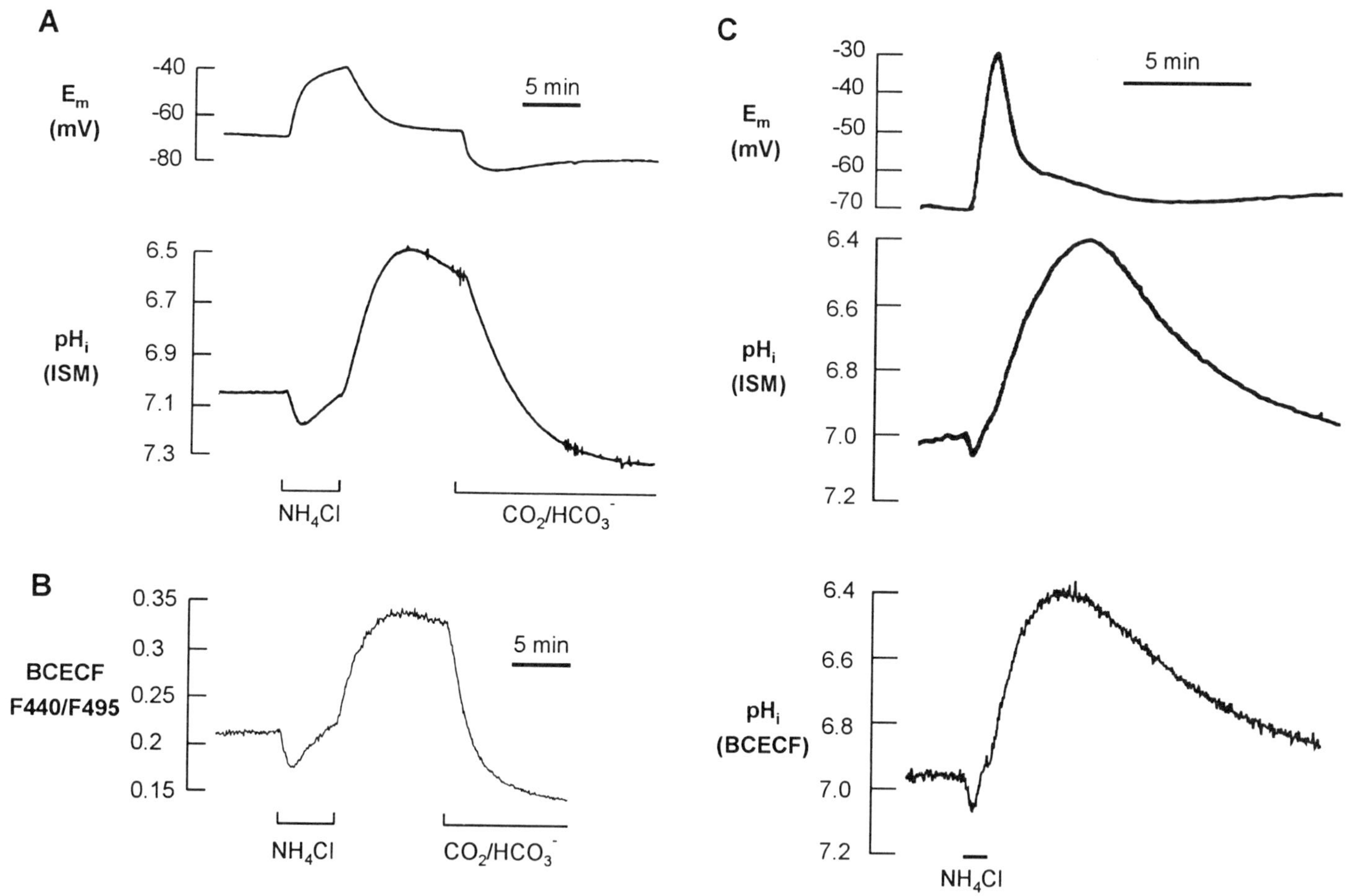
A
Em (mV)
-40
-60
-80
5 min
pHi (ISM)
6.5
6.7
6.9
7.1
7.3
NH4Cl
CO2/HCO3-
B
BCECF F440/F495
0.35
0.3
0.25
0.2
0.15
5 min
NH4Cl
CO2/HCO3-
C
Em (mV)
-30
-40
-50
-60
-70
5 min
pHi (ISM)
6.4
6.6
6.8
7.0
7.2
pHi (BCECF)
6.4
6.6
6.8
7.0
7.2
NH4Cl

Steady-State Intracellular pH Values

The pH_i value measured in neuropil glial cells is dependent on the buffer system used, that is, on the presence or absence of added CO_2/HCO_3^-. In HEPES-buffered, CO_2/HCO_3^--free saline, the pH_i was between 6.87 (Deitmer and Schlue, 1987) and 6.96 (Deitmer, 1992a), while in the presence of CO_2/HCO_3^-, the pH_i was higher by up to 0.3 pH units, its mean values being 7.18 in 2% CO_2/11 mM HCO_3^-, and 7.25 in 5% CO_2/24 mM HCO_3^- (pH of all salines 7.4). All experiments on this cell indicated the importance of CO_2/HCO_3^- for the maintenance of an alkaline steady-state pH_i (Deitmer and Schlue, 1987, 1989; Deitmer, 1991, 1992a; see also Deitmer and Rose, 1996). The pH_i values measured in leech connective glial cells were somewhat more alkaline, being 7.37 in a saline buffered with 2% CO_2/11 mM HCO_3^- (Szatkowski and Schlue, 1992).

In studies, where the ganglion was opened, and the neuropil glial cells exposed (see Fig. 1B; Munsch and Deitmer, 1992), pH_i was more alkaline than in intact ganglia in HEPES-buffered, CO_2-free saline (Deitmer, 1992b). Since in intact ganglia the pH in the extracellular spaces (pH_e) is on average 0.2 pH unit lower (only in HEPES-buffered saline) than the bath saline (pH_o) and the pH of opened ganglia (Deitmer, 1992b), the effective pH in the microenvironment of the neuropil glial cells differs in intact and opened ganglion preparations. When the neuropil was exposed, the neuropil extracellular spaces had a pH close to that in the bath, both in the presence and absence of CO_2/HCO_3^- (i.e., pH_e had the same value as pH_o).

The time course of the pH_i changes as measured with BCECF can be somewhat faster than those measured with the microelectrodes (see Fig. 2A), but usually differed very little. This might be due to measuring the pH_i integral over the large size of the neuropil glial cell body, which would tend to obscure even fast pH changes, if they occur locally in a cell. In contrast, pH recording with a microelectrode gives a point measurement from somewhere in the cytoplasm, and hence depends on the diffusion of H^+/OH^- and/or H^+ buffer to the point of recording. In addition, ion-sensitive microelectrodes are relatively slow due to their high electrical resistance (see Chapter 6).

In leech glial cells, the pH_i values measured were 6.96 and 7.25 in CO_2/HCO_3^--free saline and in 5% CO_2/24 mM HCO_3^--buffered saline, respectively. These values correspond to free H^+ concentrations of 110 nM (CO_2/HCO_3^--free) and 56 nM (5% CO_2/24 mM HCO_3^-). According to the Nernst equation, the H^+ equilibrium potentials, calculated from these values at pH_o 7.4, are −26 mV (CO_2/HCO_3^--free) and −9 mV (5% CO_2/24 mM HCO_3^-), at room temperature. Thus, the presence of CO_2/HCO_3^- produces a difference of 17 mV in the H^+/OH^- equilibrium potential. Since the membrane hyperpolarizes

FIGURE 2. Intracellular pH (pH_i) measurements using ion-sensitive microelectrodes (ISM) and the fluorescent dye BCECF in leech neuropil glial cell. **A.** Recording of pH_i and membrane potential (E_m) with a double-barrelled pH-sensitive microelectrode during and after application of 20 mM ammonium chloride, and after addition of 5% CO_2/24 mM HCO_3^-. Note the difference in the rate of pH_i recovery in the absence and presence of CO_2/HCO_3^-. **B.** Fluorescence recording at excitation wavelength ratio of 440 nm to 495 nm in another cell using the same experimental protocol as in (A). **C.** Simultaneous recording of pH_i using both techniques in the same cell, which was acid loaded by an ammonium pulse in CO_2/HCO_3^--free saline (from J. W. Deitmer, unpublished; (B, C) from W. Nett and J. W. Deitmer, unpublished).

in CO_2/HCO_3^-, the H^+ gradient across the glial cell membrane is around 25 mV larger in CO_2/HCO_3^--buffered saline, where it amounts to about 60 mV, than in HEPES-buffered, CO_2-free saline. The HCO_3^- concentration calculated from the pH_i value of 7.25 according to the modified Henderson-Hasselbalch equation is 17 mM at an external HCO_3^- concentration of 24 mM (Deitmer, 1991, 1992b).

INTRACELLULAR pH BUFFERING AND REGULATION

For their pH_i regulation, invertebrate glial cells seem to use buffering and transport systems that have also been described in other cell types, including neurons and vertebrate glial cells (see Chapters 10–12 and 14). With most cells they share the use of a Na^+/H^+ exchanger and a Cl^-/HCO_3^- exchanger; with most epithelial cells and other glial cells they share a Na^+/HCO_3^- cotransporter, which is apparently absent in neurons and in most nonepithelial cells.

Little is known about the role of second messengers on pH_i regulation in glial cells. The steady-state pH_i and the pH_i changes in leech glial cells were unaffected by intracellular Ca^{2+} transients evoked by membrane depolarization (Deitmer et al., 1993), suggesting that the homeostasis of H^+ and Ca^{2+} is not necessarily interrelated in these glial cells. There is ample evidence that neurotransmitters and hormones may affect intraglial pH; this will be discussed as intracellular H^+ signals.

Intraglial Buffering

The term "buffering" in biological systems often includes all mechanisms which rapidly mop up H^+, processes that are usually too rapid to be followed by pH_i probes. In addition to physicochemical buffering, intracellular sequestration or sometimes even extrusion of acid/base equivalents are considered as part of cellular H^+ buffering. This broad meaning of "buffering" has therefore been termed "H^+ muffling" (Thomas et al., 1991), to differentiate between true physicochemical buffering and all processes, including membrane transport of acid/base equivalents, which lead to a dampening or "muffling" of pH changes. The intrinsic intraglial buffering power (β_i) includes the cytoplasmic physicochemical buffering, as well as the organellar buffering power, and is measured as the "instantaneous" change in pH_i in response to the addition of strong acid or base, in the nominal absence of CO_2/HCO_3^- (see Chapters 2–4 and 10). The total intracellular buffering power is the sum of the intrinsic buffering power and the CO_2/HCO_3^--dependent buffering power. The latter is negligible in the nominal absence of CO_2/HCO_3^-, where the intracellular HCO_3^- concentration, due to diffusion of CO_2 ($\sim$8 μM) from air into the tissue is less than 0.2 mM. However, there is considerable additional intracellular buffer capacity in the presence of 2–5% CO_2/10–24 mM HCO_3^-.

In leech glial cells an intrinsic buffering power of 20–30 mM (on average ~25 mM) has been determined using the NH_4^+/NH_3 and the CO_2/HCO_3^- application techniques (Deitmer and Schlue, 1987). The CO_2/HCO_3^--dependent buffering power can, to a first approximation, be calculated directly from the intracellular HCO_3^- concentration (see Chapter 2). In a typical leech glial cell with a pH_i of 7.25 in a 5% CO_2/24 mM HCO_3^--buffered saline, a CO_2/HCO_3^--dependent buffer capacity of 39 mM is added to the intrinsic buffering power, to give a total mean buffering power of 64 mM. Thus, the CO_2/HCO_3^--dependent buffering power is, on average, more than 50% greater than the intrinsic buffering power under these conditions in leech glial cells. Similar values for the intracellular buffer capacity have been reported for other cells.

The presence of carbonic anhydrase activity in the leech central nervous system makes a substantial contribution to intra- and extracellular buffering (Thomas et al., 1991; Deitmer, 1992b). In particular, the rapid, reversible conversion of CO_2 to H^+ and HCO_3^- catalyzed by this enzyme affected pH buffering in the extracellular spaces. Evidence for the occurrence of this enzyme in leech glial cells came from physiological (Deitmer and Schlue, 1989) and immunohistochemical (Riehl and Schlue, 1990) studies. The pH_e transients evoked by a brief application of NH_4^+ have been used as a relative measure of extracellular buffering (Deitmer, 1992b). As shown in Fig. 3, the CO_2/HCO_3^--dependent increase in extracellular buffering, indicated by the reduction of the NH_4^+-induced pH_e transients in the presence of CO_2/HCO_3^- (Fig. 3A), is reversed by inhibiting the carbonic anhydrase (e.g., by ethoxyzolamide or furosemide; Fig. 3B). A similar effect on pH_e transients was observed with the carbonic anhydrase inhibitor acetazolamide, which is believed to reduce mainly extracellular carbonic anhydrase activity, since it permeates cell membranes less freely than, for example, ethoxyzolamide. It appears that the access of functional carbonic anhydrase activity to the extracellular space (ECS) to buffer pH_e in both invertebrate and vertebrate nervous systems has great impact for the H^+ homeostasis in these tissues (see also Chapter 3 and 16–18).

Acid-Base Transport

In leech glial cells two acid-base transport systems have been identified for active pH_i regulation, an amiloride-sensitive Na^+/H^+ exchanger, and an electrogenic Na^+/HCO_3^- cotransporter. The latter has been characterized in some detail in the neuropil glial cell, including its voltage-dependence, its stoichiometry, and its modulation by some neurotransmitters and second messengers (see also Deitmer, 1995; Deitmer and Rose, 1996). The Cl^--dependent mechanisms of pH_i regulation have also been reported in leech glial cells (Deitmer and Schlue, 1987; Szatkowski and Schlue, 1992), but apart from some evidence for the existence of a Na^+-dependent and a Na^+-independent Cl^-/HCO_3^- exchanger, these have not been studied in any detail in leech or other invertebrate glial cells.

The Na^+/H^+ exchanger. The main evidence in leech glial cells for a Na^+/H^+ exchanger, which drives H^+ out of the cells by using the energy of the Na^+ gradient with a stoichiometry of 1 : 1, are the following:

1. Removal of external Na^+, or addition of external amiloride, an inhibitor of Na^+/H^+ exchange (Benos, 1982), leads to an intracellular acidification.
2. pH_i recovery from an acid load occurs in CO_2/HCO_3^--free saline.
3. pH_i recovery from an acid load is dependent on external Na^+ and a large inwardly directed Na^+ gradient.
4. pH_i recovery from an acid load is independent of anions such as Cl^- and HCO_3^-.
5. pH_i recovery is sensitive to amiloride (Deitmer and Schlue, 1987; Szatkowski and Schlue, 1992).

The pH_i recovery in the absence of CO_2/HCO_3^- (Fig. 2) is assumed to be largely due to the activity of the Na^+/H^+ exchanger.

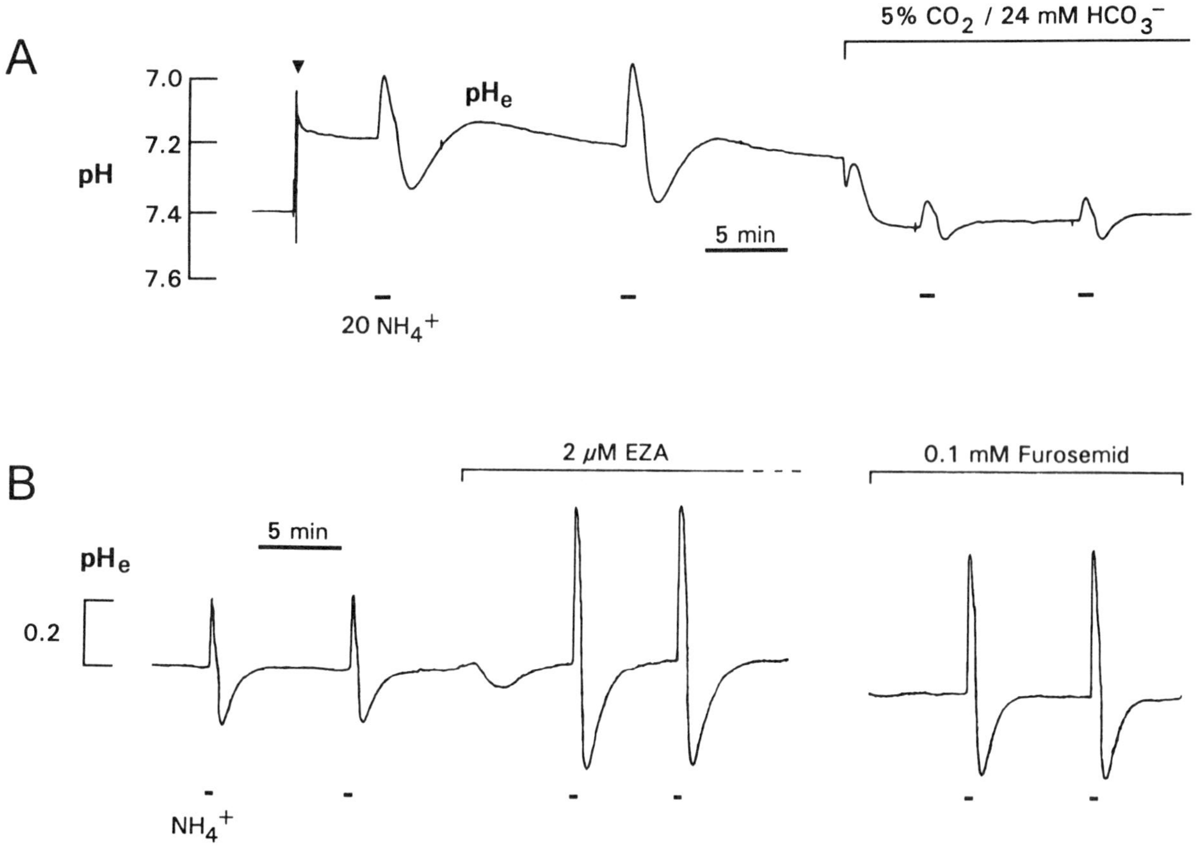

FIGURE 3. Extracellular pH (pH_e) recording with a double-barrelled pH-sensitive microelectrode in the neuropil of a leech segmental ganglion, using brief ammonium pulses, to displace pH_e according to the extracellular buffering capacity. **A.** In the absence and presence of CO_2/HCO_3^-. B. After inhibition of the carbonic anhydrase by ethoxyzolamide (EZA) and furosemide in the presence of 5% CO_2/24 mM HCO_3^-. Field potentials are not shown here (from J. W. Deitmer, unpublished).

As will be discussed, active pH regulation in leech glial cells is not completely inhibited by amiloride, even in the nominal absence of CO_2/HCO_3^-; the remaining HCO_3^- derived from the CO_2 in air in an open system (producing ~150–200 μM HCO_3^-) appears to be sufficient to maintain a pH_i value higher than expected from a H^+ equilibrium. Removal of external Na^+, however, supresses pH_i recovery from an acid load completely, indicated by the approach of pH_i to acid values expected for a passive H^+ distribution across the cell membrane.

Na^+/HCO_3^- cotransporter. The first evidence for a Na^+/HCO_3^- cotransporter in glial cells, after its detection in proximal tubule cells of the salamander kidney (Boron and Boulpaep, 1983), was found in giant neuropil glial cells of the leech (Deitmer and Schlue, 1987, 1989). This cotransporter has now been identified in a variety of glial cells, including astrocytes in the optic nerve of the mudpuppy *Necturus* (Astion and Orkand, 1988), in cultured mouse oligodendrocytes (Kettenmann and Schlue, 1988), in freshly dissociated Müller glial cells of the salamander retina (Newman, 1991), and in cultured mammalian astrocytes (Brune et al., 1994; Shrode and Putnam, 1994; Pappas and Ransom, 1994).

In leech glial cells, the Na^+/HCO_3^- cotransporter was detected during the change from HEPES-buffered, CO_2/HCO_3^--free saline to a solution containing CO_2/HCO_3^-. Addition of CO_2/HCO_3^- resulted in an intracellular alkaline shift of 0.1–0.3 pH unit, a rise in intracellular Na^+, and a membrane hyperpolarization (Deitmer and Schlue, 1989; Deitmer, 1992a). A simultaneous recording of all three parameters with a triple-barrelled microelectrode in a leech neuropil glial cell is shown in Fig. 4. All changes were reversible when CO_2/HCO_3^- was removed again.

The impact of CO_2/HCO_3^- on active pH_i regulation from an acid load in this cell is striking. Using the NH_4^+ prepulse technique to acid load the cell, the pH_i recovery was 6–10 times faster in the presence of CO_2/HCO_3^- than in its absence (see Fig. 2A).

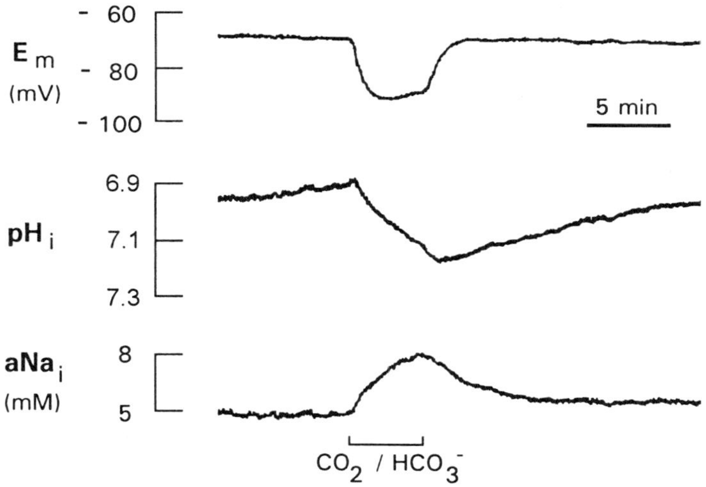

FIGURE 4. Simultaneous measurement of membrane potential (E_m, upper trace), intracellular pH (pH_i, middle trace), and intracellular Na^+ activity (a_{Na_i}, lower trace) in a leech neuropile glial cell during the exposure to 5% CO_2/24 mM HCO_3^-, using a triple-barrelled ion-sensitive microelectrode (from J. W. Deitmer and H. P. Schneider, unpublished).

This would indicate an even larger difference in acid-base flux across the cell membrane due to the increased buffering power in CO_2/HCO_3^--containing saline. This is largely due to the activation of an inwardly directed Na^+/HCO_3^- cotransport. The concomitant hyperpolarization of the glial cell membrane is not due to membrane conductance changes, but indicates the electrogenic nature of this cotransporter, which operates with a stoichiometry of 1 Na^+ and 2 HCO_3^- (Deitmer and Schlue, 1989; Munsch and Deitmer, 1994; Deitmer and Schneider, 1995). In the absence of external Na^+, addition of CO_2/HCO_3^- neither produced an intracellular alkalinization nor a membrane hyperpolarization or outward current in leech glial cells, indicating that there is no detectable HCO_3^- conductance in these cells (Munsch and Deitmer, 1994). These CO_2/HCO_3^--dependent changes in intracellular pH and Na^+ and in membrane potential could be significantly reduced by the anion transport inhibitor diisothiocyanato-stilbene-2,2′-disulfonic acid (DIDS). The DIDS effect observed needs the presence of both external Na^+ and HCO_3^-, suggesting that the blockade of the Na^+/HCO_3^- cotransport is the main effect of DIDS with respect to pH_i regulation in these cells (Deitmer and Schlue, 1989; Deitmer, 1991).

Under voltage-clamp conditions, the addition of CO_2/HCO_3^- produced an outward current; subsequent reduction of the external pH by lowering the HCO_3^- concentration evoked an inward current. This is consistent with an inward electrogenic Na^+/HCO_3^- cotransport upon addition of CO_2/HCO_3^-, and a reversed, outward contransport upon lowering pH and the HCO_3^- concentration, (Deitmer, 1991; Munsch and Deitmer, 1994).

With the intracellular Na^+ and pH (and thus also HCO_3^-) measured, the equilibrium potential (or reversal potential) of the Na^+/HCO_3^- cotransporter, E_{NaHCO_3}, is given by (Boron and Boulpaep, 1983; Deitmer and Schlue, 1989):

$$E_{NaHCO_3} = \frac{RT}{(n-1)F} \ln \frac{[Na^+]_o\,[HCO_3^-]_o^{\,n}}{[Na^+]_i\,[HCO_3^-]_i^{\,n}} \tag{1}$$

where n is the $HCO_3^-:Na^+$ stoichiometry, and R (gas constant), T (temperature), and F (Faraday constant) have their usual meanings. The stoichiometry of this transporter was determined in leech glial cells by measuring intracellular pH (HCO_3^-) and Na^+ (Deitmer and Schlue, 1989; Deitmer, 1992a), by recording the Na^+- and HCO_3^--dependent membrane current in voltage-clamp and its reversal potential (Munsch and Deitmer, 1994), and by a quantitative analysis of the electrical and chemical gradients driving this cotransport (Deitmer and Schneider, 1995). These studies unequivocally indicated a stoichiometry of 2 HCO_3^- : 1 Na^+. In a steady state, where no net transport is mediated by the carrier, the E_{NaHCO_3} was near −75 mV in leech neuropile glial cells, at pH_i of 7.25 and pH_o of 7.4 (in 5% CO_2/24 mM HCO_3^-, 10 mM free intracellular Na^+ and 90 mM external Na^+ concentration; Deitmer, 1992a). This is close to the membrane resting potential of these cells, between −68 and −76 mV (mean −73 mV).

The stoichiometry of 2:1 therefore implies that the cotransporter operates near its equilibrium at the normal resting potential of the giant glial cell. Depolarization stimulates inward, and hyperpolarization stimulates outward Na^+/HCO_3^- cotransport, accompanied by intracellular alkalinization and acidification, and by outward current and inward current, respectively. This renders intraglial pH and acid/base flux across the glia membrane sensitive to membrane potential changes (Deitmer and Szatkowski, 1990; Munsch and Deitmer, 1994; Deitmer and Schneider, 1995). The CO_2/HCO_3^--dependent and membrane potential-dependent changes in pH_i can be directly demonstrated in voltage-clamped glial cells (Fig. 5). A potential step by ±15 mV induced a pH_i shift of

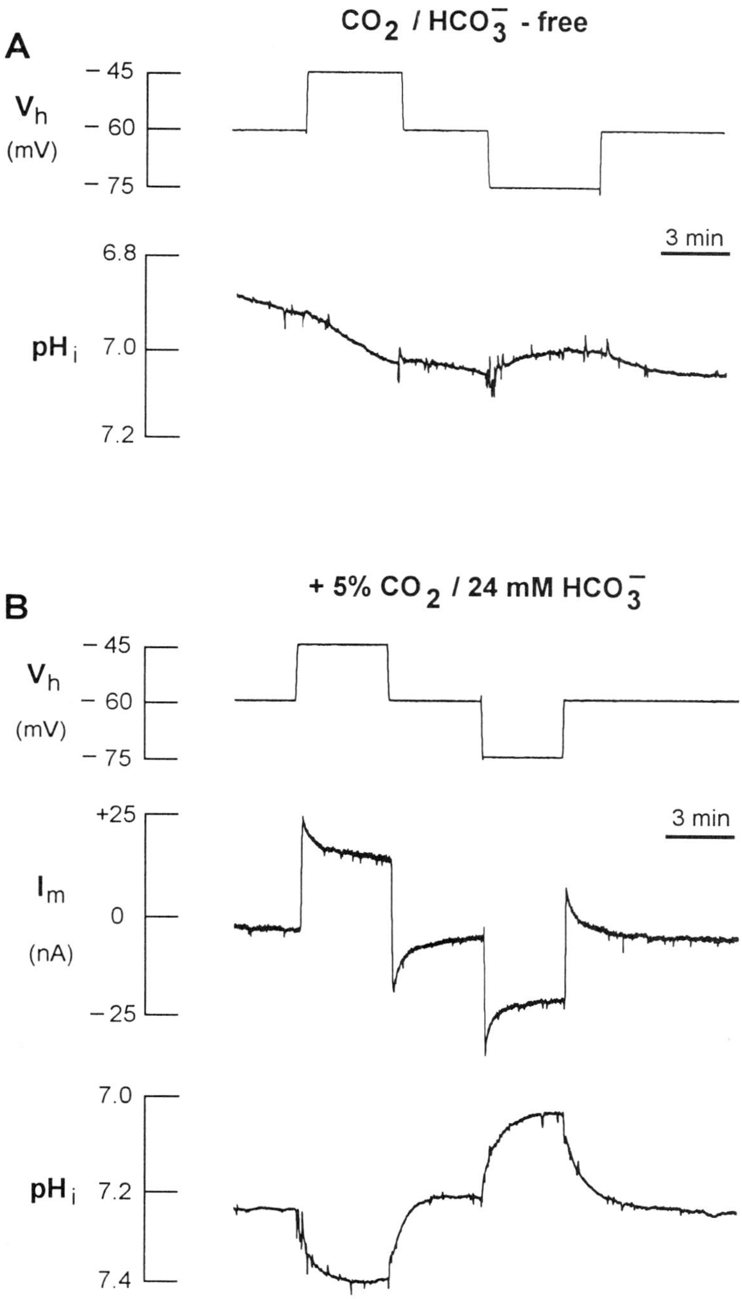

FIGURE 5. Voltage-dependence of the intracellular pH (pH_i) in a voltage-clamped leech neuropil glial cell. **A.** In the absence of CO_2/HCO_3^-. **B.** In the presence of CO_2/HCO_3^-, where the pH_i shifts significantly following voltage steps to ±15 mV from the holding potential (V_h). The membrane current (I_m) is shown in the middle trace in (B) (from J. W. Deitmer and H. P. Schneider, unpublished).

about 0.15 pH units in the presence of CO_2/HCO_3^- (Fig. 5B); in the nominal absence of CO_2/HCO_3^-, however, pH_i changed little with these potential steps (Fig. 5A). The pH_i in leech neuropil glial cells appears tightly voltage-controlled (i.e., voltage-clamped) with a slope of 110 mV/pH_i unit (Deitmer and Schneider, 1995).

Potential steps in voltage-clamped glial cells not only result in pH_i changes, but can also be recorded extracellularly. The acid-base flux across the glial membrane, mediated by the reversible Na^+/ HCO_3^- cotransport (Deitmer, 1991), can be measured at the external surface of the glial cell, or in the narrow extracellular spaces of the neuropil in the immediate vicinity of the giant glial cell (J. W. Deitmer and H. P. Schneider, unpublished). A depolarizing step results in an extracellular acidification, and a hyperpolarizing step in an extracellular alkalinization; the extracellular pH changes thus mirror the intraglial pH changes. This indicates that the membrane potential-dependent Na^+/ HCO_3^- cotransporter across the glial membrane exerts a significant influence also on extracellular pH. This has recently been confirmed also for pH changes evoked by neuronal stimulation (Rose and Deitmer, 1994, 1995a; see next section).

INTRACELLULAR H^+ SIGNALS

Neuronal Activity-Induced pH Transients

Apart from the homeostatic pH control in cells, H^+ may also serve as a *signal,* both intra- and extracellularly (see also Chapter17). Glial cells depolarize during neuronal stimulation due to the increase of the extracellular K^+ concentration (Orkand et al., 1966); cortical stimulation in rat brain depolarized the glial cells, which was accompanied by an intraglial alkalinization (Chesler and Kraig, 1987, 1989). In the leech central nervous system, repetitive stimulation of the nerve root evoked a depolarization of the neuropil glial cells by about 8 mV and a CO_2/HCO_3^--dependent intraglial alkalinization by 0.025 pH unit (Rose and Deitmer, 1995a). This alkalinization was turned into an acidification by all experimental protocols preventing the activation of Na^+/HCO_3^- cotransport, such as (i) removal of either HCO_3^-- or Na^+, (ii) the anion-exchange inhibitor DIDS, and (iii) voltage-clamping the glial cell and thereby preventing the membrane depolarization (Rose and Deitmer, 1994). In fact, if the glial membrane depolarization elicited during nerve root stimulation is suppressed by voltage-clamping the glial cell, the intraglial alkalinization, associated with the glial depolarization, turns into an acidification (Fig. 6A). When the pH_e is recorded in the vicinity of the neuropile glial cell body under these conditions, a distinct alkaline transient preceded an acidification of the ECS only when the glial cell membrane was voltage-clamped to a constant holding potential (Fig. 6B, and Rose and Deitmer, 1994). All these experiments suggest that, in the leech central nervous system, the stimulation-induced intraglial alkalinization, and part of the extracellular pH changes, result from an inward transport of bicarbonate via the electrogenic Na^+/HCO_3^- cotransport activated by the K^+-induced membrane depolarization. Similar results have been reported for rat hippocampal pH transients (Gritchenko and Chesler, 1994). Extracellular pH changes will be considered in more detail in Chapters 17–19.

A rapid and large intracellular acidification caused by low concentrations of ammonium ($\leq$5 mM), which is shuttled between photoreceptors and surrounding glial cells (Tsacopoulos et al. 1997), has recently been monitored in glial cells of the bee retina

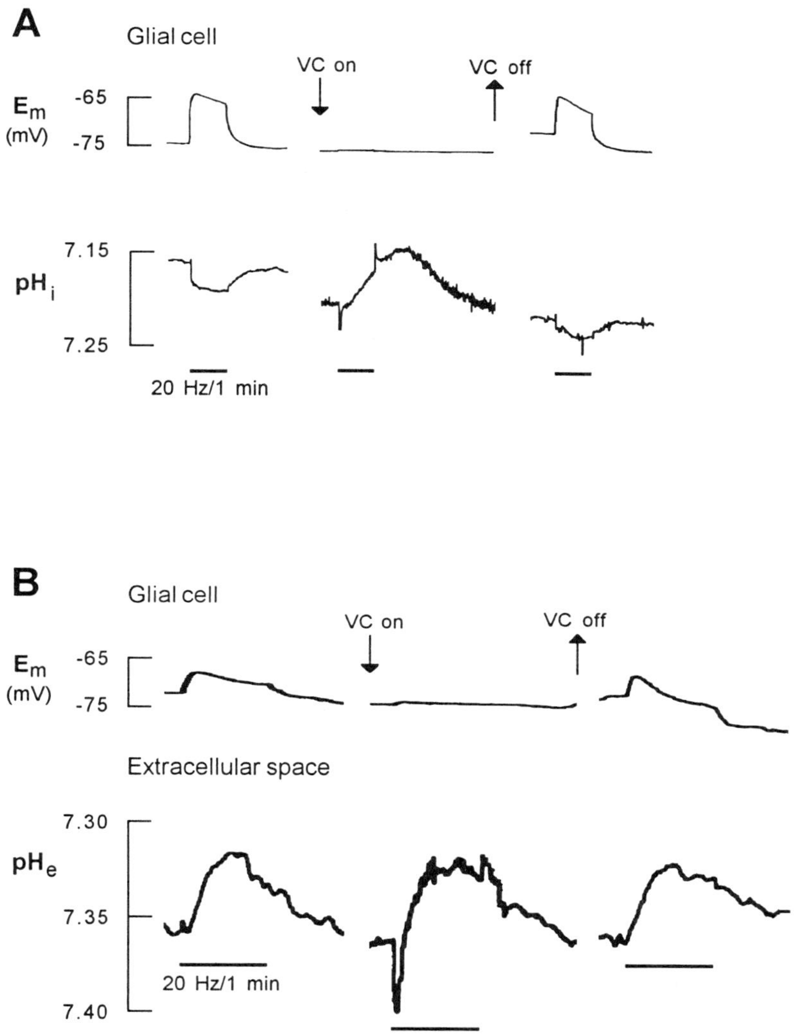

FIGURE 6. A. Intracellular pH (pH_i) and membrane potential (E_m) of a leech neuropil glial cell. **B.** Extracellular pH (pH_e) in the vicinity of the neuropil glial cell and the glial membrane potential during electrical stimulation of a ganglionic side nerve (20 Hz/1min), when the glial cell was depolarized or voltage-clamped (VC on/off). Note the conversion of the intraglial alkalinization into an acidification, and the prominent alkaline transient in the extracellular spaces, when the glial cell depolarization was suppressed by voltage-clamping the cell. (Modified from Rose and Deitmer, 1994).

(Coles et al., 1996a, b). The glial pH changes are believed to depend on the photoreceptor metabolism, which appear to release the ammonium following deamination of alanine. It was concluded that the rapid uptake of ammonium into the glial cells is probably due to an electroneutral NH_4^+-Cl^- cotransport (P Marcaggi, DT Thwaites, JW Deitmer and JA Coles, in preparation).

Action of Neurotransmitters

The activation of different types of neurotransmitter receptors in the membrane of glial cells can induce intra- and extracellular pH transients (see also Chapters 16 and 17, this volume. In nervous systems, the activation of these transmitter receptors can induce pH transients, which may be due to membrane potential changes as a result of H^+ or HCO_3^- flux through receptor channels, or which may be secondary to changes of intracellular [Na^+] and/or [Ca^{2+}]. These topics will be also discussed in Chapters 15–18, and I shall therefore concentrate here on some recent measurements carried out in the leech giant glial cells.

Activation of glutamate receptors of the kainate/AMPA (α-amino-3-hydroxy-5-methylisoxazole-4-propionate) type led to an intracellular acidification in leech neuropil glial cells (Deitmer and Munsch, 1992, 1994). The membrane responses and the intraglial acid transient evoked by kainate in the leech neuropil glial cell are shown in Fig. 7. The pH_i decrease was 0.2–0.3 pH units, irrespective of whether the membrane depolarized by 15–30 mV, or whether the membrane potential was held constant in voltage-clamp. The kainate-induced acidification was blocked by the non-NMDA (N-methyl-D-aspartate) re-

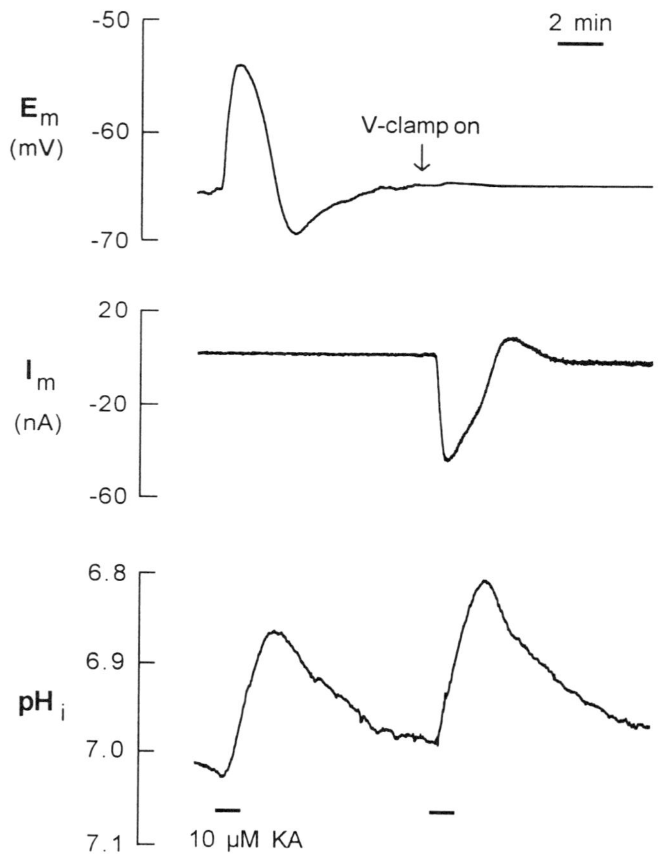

FIGURE 7. Membrane responses, such as membrane potential (E_m, upper trace) and membrane current (I_m, middle trace) in voltage-clamp (V-clamp on) of a neuropil glial cell, and the intraglial pH (pH_i, lower trace) to bath application of the ionotropic glutamate receptor agonist kainate (KA). (Modified from Deitmer and Munsch, 1994).

ceptor antagonist 6-cyano-7-dinitroquinoxaline-2,3-dione (CNQX), and in the presence of Ni^{2+} (Munsch and Deitmer, 1997). The sensitivity to Ni^{2+} suggest that the intraglial acidification might be secondary to the concommitant $[Ca^{2+}]_i$ increase (Munsch et al., 1994), as recently proposed for glutamate-agonist-induced extracellular alkaline transients in vertebrate brain slices (Paalasmaa et al., 1994; Chen and Chesler, 1992; Smith et al., 1994). However, in leech glial cells, the kainate-induced acidification was only partly dependent on the presence of external Ca^{2+}, and there appears to be some evidence for H^+ permeating the kainate/AMPA-gated receptor channel in these cells (Deitmer et al., 1995; J. W. Deitmer and H. P. Schneider, unpublished observations).

When $[Ca^{2+}{}_i]$ transients were elicited by other means (e.g. by 5-hydroxytryptamine [5-HT] or membrane depolarization) they caused no pH_i change or even an alkalinization, due to depolarization-induced Na^+/HCO_3^- cotransport (Deitmer and Szatkowski, 1990). This indicated that there was no direct causal interrelationship between intracellular Ca^{2+} and pH in leech glial cells (Deitmer et al., 1993), as was suggested for snail neurons (Schwiening et al., 1993; and Chapter 15).

Glutamate and aspartate can also acidify leech glial cells by a mechanism unrelated to glutamate receptor activation (Deitmer and Schneider, 1997). It was concluded that this acidification was due to uptake of these amino acids into the glial cells, as has been suggested for salamander retinal Müller glial cells (Bouvier et al., 1992; for a more detailed discussion see Chapter 17 and 19).

In leech glial cells, 5-HT and the cholinergic agonist carbachol caused pH_i changes, which appeared to be secondary to the membrane potential changes elicited by these agonists, and were dependent on the presence of CO_2/HCO_3^- (Deitmer and Szatkowski, 1990; J. W. Deitmer, unpublished). 5-HT hyperpolarized the glial cell membrane by activating a K^+ conductance (Walz and Schlue, 1982; Munsch and Deitmer, 1992), thereby causing an intraglial acidification, while carbachol depolarized the glial membrane, which resulted in an intraglial alkalinization. In the absence of CO_2/HCO_3^-, however, carbachol acidified the glial cytoplasm (Deitmer, 1995), possibly due to H^+ influx through the receptor channel.

The reverse situation was observed with the inhibitory neurotransmitter γ-aminobuyric acid (GABA), which acidified the leech glial cell only in the presence of CO_2/HCO_3^- (Deitmer and Rose, 1996). This was interpreted as being due to some HCO_3^- permeability of the GABA-gated anion channel (for more extensive discussion see Kaila, 1994, and Chapters 16 and 17).

FUNCTIONAL SIGNIFICANCE OF H^+ DYNAMICS

In nervous systems, pH pertubations can induce a variety of changes in cellular functions, either by acting as modulators of biological processes or by acting as messengers, initiating or propagating a messenger cascade, from the modulation of enzyme activities to the induction or inhibition of membrane conductances. They may directly or indirectly influence metabolic pathways, neuron–glial communication, gap junctions and chemical synaptic transmission, cell volume regulation, and neuronal excitability. There is an extensive consideration of the physiological and pathophysiological effects of pH in Chapters 20–33 of this volume. In leech glial cells, intracellular acidification reduced K^+ conductance (Munsch and Deitmer, 1991) and increased the electrical resistance between the two giant neuropil glial cells (T. Munsch and J. W. Deitmer, unpublished).

Based on our substantial, but still incomplete, knowledge of the dynamics of intra- and extracellular pH changes and their mechanisms in nervous systems, we have developed a hypothetical model for the H^+ dynamics in the leech central nervous system (Rose and Deitmer, 1995b; Deitmer and Rose, 1996). Neuronal activity would lead to a local rise in the extracellular K^+ concentration and the release of neurotransmitters. These may result in membrane potential-dependent extra- and intracellular pH changes. The electrogenic Na^+/HCO_3^- cotransporter, activated by glial membrane depolarization would transport HCO_3^- into the cells, and thereby base equivalents out of the ECS. This flux would acidify the ECS or counteract any activity-induced extracellular alkalinization (as may occur following H^+ flux into, or HCO_3^- flux out of the cells). Carbonic anhydrase, which is responsible for the efficacy of bicarbonate-dependent physicochemical buffering in the ECS (Thomas et al.; 1991; Deitmer, 1992b), would mediate the fast delivery of acid-base equivalents from the CO_2/HCO_3^- buffer system. A hypothetical scheme of some of the acid-base transport mechanisms across the leech glial cell membrane is shown in Fig. 8.

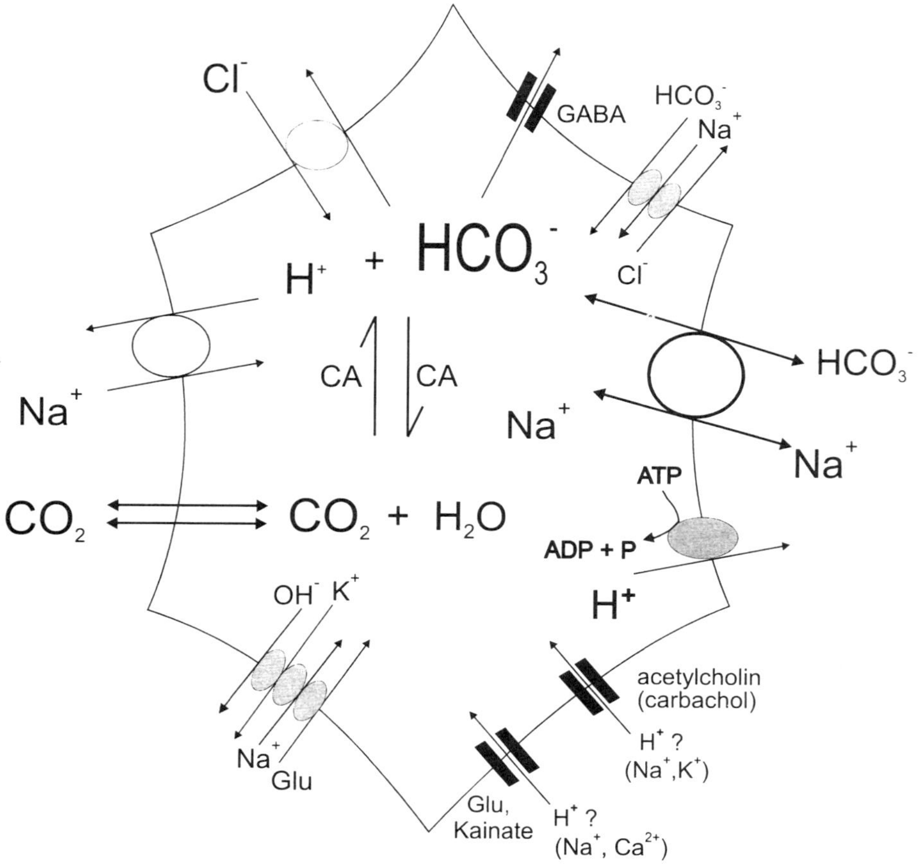

FIGURE 8. Schematic summary of membrane channels and carriers relevant for pH_i in the leech giant glial cell. The carbonic anhydrase (CA) catalyzes a fast equilibrium between $H^+ + HCO_3^-$ and $CO_2 + H_2O$; the cell membrane being permeable to CO_2 and H_2O. An ATP-driven proton pump has been suggested, and the H^+ permeability of glutamate and acetylcholine receptor channels still needs to be shown more directly (hence marked with question mark). The steady-state pH_i, at least in some glial cell types, appears to be dominated by the electrogenic Na^+/HCO_3^- cotransport, which may operate as a voltage-dependent pH_i clamp.

Activity-induced intraneuronal acidification (Rose and Deitmer, 1995a) stimulates neuronal pH_i regulation, which extrudes acid equivalents via the amiloride-sensitive Na^+/H^+ exchanger, and thereby contributes to the stimulus-evoked acidification in the ECS. According to this model, glial cells could mediate the dampening of neuronal excitability by reducing the extracellular alkaline transient, which would tend to raise neuronal activity (Ransom, 1992; Deitmer, 1992b, 1994) and thereby counteract neuronal hyperexcitability, as occurs during epileptiform activity (Aram and Lodge, 1987; see also Chapter 20).

CONCLUSIONS

Glial cells contain an array of different mechanisms by which they control intraglial pH (Fig. 8) and modulate pH changes in the ECS. The leech central nervous system, with its two giant glial cells in the neuropil of each ganglion and the large identifiable neurons, has been a useful preparation to study basic mechanisms of pH regulation. The pH changes measured have been qualitatively very similar to those recorded from vertebrate nervous systems, and both glial and neuronal mechanisms of pH control appear to be similar. In different loci of vertebrate nervous tissue and invertebrate ganglia, however, the pH-controlling mechanisms may differ *quantitatively*. In nervous systems of both invertebrate and vertebrate animal however, dynamic pH changes and the pH control in neurons, glial cells and in the ECS play an important functional role. Glial cells in the leech, and presumably in other species may be H^+ caretakers, which not only regulate their cytoplasmic pH, but also help to control the acid-base content of the neuronal environment. They also shape the often rapid and local pH transients, which may have a signalling function in the information processing in nervous tissue.

ACKNOWLEDGMENTS

I thank all my colleagues who contributed to the work on leech glial cells summarized in this article, and Kai Kaila for helpful comments on an early version of this manuscript. The financial support of the Deutsche Forschungsgemeinschaft to the experimental work is gratefully acknowledged.

REFERENCES

Ammann D, Lantern F, Steiner RA, Schulthess P, Shijo Y, Simon W (1981): Neutral carrier based hydrogen ion selective microelectrode for extra- and intracellular studies. Anal Chem 53: 2267–2269.

Aram JA, Lodge D (1987): Epileptiform activity induced by alkalosis in rat neocortical slices: block by antagonists of N-methyl-D-aspartate. Neurosci Lett 83: 345–350.

Astion ML, Orkand RK (1988): Electrogenic Na^+/HCO_3^- cotransport in neuroglia. Glia 1: 355–357.

Benos DJ (1982): Amiloride: A molecular probe of sodium transport in tissues and cells. Amer J Physiol 242: C131–C145.

Boron WF, Boulpaep EL (1983): Intracellular pH regulation in the renal proximal tubule of the salamander. Basolateral HCO_3^- transport. J Gen Physiol 81: 53–94.

Bouvier M, Szatkowski M, Amato A, Attwell D (1992): The glial cell glutamate uptake carrier countertransports pH-changing anions. Nature 360: 471–474.

Brune T, Fetzer S, Backus KH, Deitmer JW (1994): Evidence for electrogenic sodium-bicarbonate cotransport in cultured rat cerebellar astrocytes. Pflügers Arch 429: 64–71.

Chen JCT, Chesler M (1992): Extracellular alkaline shifts in rat hippocampal slice are mediated by NMDA and non-NMDA receptors. J Neurophysiol 68: 342–344.

Chesler M, Kraig RP (1987): Intracellular pH of astrocytes increases rapidly with cortical stimulation. Am J Physiol 253: R666–R670.

Chesler M, Kraig RP (1989): Intracellular pH transients of mammalian astrocytes. Neuroscience 9: 2011–2019.

Coggeshall RE, Fawcett DW (1964): The fine structure of the central nervous system of the leech, *Hirudo medicinalis*. J Neurophysiol 27: 229–289.

Coles JA, Marcaggi P, Vega C, Cotillon N (1996a): Effects of photoreceptor metabolism on interstitial and glial cell pH in bee retina: evidence for a role for NH_4^+. J Physiol (Lond) 495.2: 305–318.

Coles JA, Thwaites DT, Marcaggi P, Deitmer JW (1996b): Ammonium causes acidification of glial cells isolated from drone retina. J Physiol (Lond) 491.P: 147P.

Deitmer JW (1991): Electrogenic sodium-dependent bicarbonate secretion by glial cells of the leech central nervous system. J Gen Physiol 98: 637–655.

Deitmer JW (1992a): Bicarbonate-dependent changes of intracellular sodium and pH in identified leech glial cells. Pflügers Arch 420: 584–589.

Deitmer JW (1992b): Evidence for glial control of extracellular pH in the leech central nervous system. Glia 5: 43–47.

Deitmer JW (1994): Neuron-glia signal transfer. In Elsner N, Breer H (eds): Proceedings of the 22nd Göttingen Neurobiology Conference 1994, Vol. II. Stuttgart: Thieme Verlag, pp. 164–167.

Deitmer JW (1995): pH regulation. In Kettenmann H, Ransom B (eds): Neuroglial Cells. Oxford University Press: New York, pp 230–245.

Deitmer JW, Munsch T (1992): Kainate/glutamate-induced changes in intracellular calcium and pH in leech glial cells. Neuroreport 3: 693–696.

Deitmer JW, Rose C (1996): pH regulation and proton signalling by glial cells. Progr Neurobiol 48: 73–103.

Deitmer JW, Schlue WR (1987): The regulation of intracellular pH by identified glial cells and neurones in the central nervous system of the leech. J Physiol (Lond) 388: 261–283.

Deitmer JW, Schlue WR (1989): An inwardly directed electrogenic sodium-bicarbonate co-transport in leech glial cells. J Physiol (Lond) 411: 179–194.

Deitmer JW, Schneider HP (1995): Voltage-dependent clamp of intracellular pH of identified leech glial cells. J Physiol (Lond) 485.1: 157–166.

Deitmer JW, Schneider HP (1997): Intracellular acidification in the leech giant glial cell induced by glutamate and aspartate. Glia 19: 111–122.

Deitmer JW, Szatkowski M (1990): Membrane potential dependence of intracellular pH regulation by identified glial cells in the leech central nervous system. J Physiol (Lond) 421: 617–631.

Deitmer JW, Schneider HP, Munsch T (1993): Independent changes of intracellular calcium and pH in identified leech glial cells. Glia 7: 299–306.

Deitmer JW, Schneider HP, Reusch M, Munsch T (1995): Are kainate/AMPA receptor channels H^+ permeable? In Elsner N, Breer H (eds): Proceedings of the 23rd Göttingen Neurobiology Conference 1995, p 699.

Grichtchenko I, Chesler M (1994): Depolarization-induced acid secretion in gliotic hippocampal slices. Neuroscience 62: 1057–1070.

Kaila K (1994): Ionic basis of $GABA_A$ receptor channel function in the nervous system. Prog Neurobiol 42: 489–537.

Kettenmann H, Schlue WR (1988): Intracellular pH regulation in cultured mouse oligodendrocytes. J Physiol (Lond) 406: 147–162.

Kuffler SW, Potter DD (1964): Glia in the leech central nervous system: Physiological properties and neuron-glia relationship. J Neurophysiol 27: 290–320.

Munsch T, Deitmer JW (1991): Intracellular pH modulates the input resistance of neuropil glial cells in the leech central nervous system. In Elsner N, Penzlin H (eds) Proceedings of the 19th Göttingen Neurobiology Conference. Stuttgart: Thieme Verlag, p 385.

Munsch T, Deitmer JW (1992): Calcium transients in identified leech glial cells *in situ* evoked by high potassium concentrations and 5-hydroxytryptamine. J Exp Biol 167: 251–265.

Munsch T, Deitmer JW (1994): Sodium-bicarbonate cotransport current in identified leech glial cells. J Physiol (Lond) 474.1: 43–53.

Munsch T, Deitmer JW (1995): Maintenance of fura-2 fluorescence in identified glial cells and neurones of the leech central nervous system. J Neurosci Methods 57: 195–204.

Munsch T, Deitmer JW (1997): Intracellular Ca^{2+}, Na^{+} and H^{+} transients evoked by kainate in the leech giant glial cell *in situ*. Neurosci Res 27: 45–56.

Munsch T, Nett W, Deitmer JW (1994): Fura-2 signals evoked by kainate in leech glial cells in the presence of different divalent cations. Glia 11: 345–353.

Nett W (1994): Simultane Messung des intrazellulären pH mit pH-sensitiven Mikroelektroden und dem Fluoreszenzfarbstoff BCECF in Neuropilgliazellen des Blutegels *Hirudo medicinalis* L. Diplomarbeit, Universität Kaiserslautern.

Nett W, Deitmer JW (1996): Simultaneous measurements of intracellular pH in the leech giant glial cell using BCECF and ion-sensitive microelectrodes. Biophys J 71: 394–402.

Newman EA (1991): Sodium-bicarbonate cotransport in retinal Müller (glial) cells of the salamander. J Neurosci 11: 3972–3983.

Orkand RK, Nicholls JG, Kuffler SW (1966): Effect of nerve impulses on the membrane potential of glial cells in the central nervous system of amphibia. J Neurophysiol 29: 788–806.

Paalasmaa P, Taira T, Voipio J, Kaila K (1994): Extracellular alkaline transients mediated by glutamate receptors in the rat hippocampal slice are not due to a proton conductance. J Neurophysiol 72: 2031–2033.

Pappas CA, Ransom B, (1994): Depolarization-induced alkalinization (DIA) in rat hippocampal astrocytes. J Neurophysiol 72: 2816–2826.

Pfeiffer F (1992): Farbstoffmarkierungen zur Identifizierung von Nerven- und Gliazellen in Ganglien des Blutegels *Hirudo medicinalis*. Diplomarbeit, Universität Kaiserslautern.

Ransom BR (1992): Glial modulation of neural excitability mediated by extracellular pH: A hypothesis. In Yu ACH, Hertz L, Norenberg MD, Sykova E, Waxman SG (eds): Progress in Brain Research, Vol 94. pp 37–46.

Riehl B, Schlue WR (1990): Localization of carbonic anhydrase in identified glial cells of the leech CNS: Application of histochemical and immunocytochemical methods. J Histochem Cytochem 38: 1173–1178.

Rose CR, Deitmer JW (1994): Evidence that glial cells modulate extracellular pH transients induced by neuronal activity in the leech central nervous system. J Physiol (Lond) 481-1: 1–5.

Rose CR, Deitmer JW (1995a): Stimulus-evoked changes of extra- and intracellular pH in the leech central nervous system. I. Bicarbonate dependence. J Neurophysiol 73: 125–131.

Rose CR, Deitmer JW (1995b): Stimulus-evoked changes of extra- and intracellular pH in the leech central nervous system. II. Mechanisms and maintenance of pH homeostasis. J Neurophysiol 73: 132–140.

Schwiening CJ, Kennedy HJ, Thomas RC (1993): Calcium-hydrogen exchange by plasma membrane Ca-ATPase of voltage-clamped snail neurones. Proc R Soc Lond B 253: 285–289.

Smith SE, Gottfried JA, Chen JCT, Chesler M (1994): Calcium dependence of glutamate receptor-evoked alkaline shifts in hippocampus. Neuroreport 5: 2441–2445.

Shrode LD, Putnam RW (1994): Intracellular pH regulation in primary rat astrocytes and C6 glioma cells. Glia 12: 196–210.

Szatkowski M, Schlue WR (1992): Mechanisms of pH recovery from intracellular acid loads in the leech connective glial cell. Glia 5: 193_200.

Thomas RC, Coles JA, Deitmer JW (1991): Homeostatic muffling. Nature 350: 564.

Tsacopoulos M, Poitry-Yamate CL, Poitry S (1997): Ammonium and gutamate released by neurons are signals regulating the nutritive function of a glial cell. J Neurosci 17: 2383–2390.

Walz W, Schlue WR (1982): External ions and membrane potential of leech neuropile glial cells. Brain Res 239: 119–138.

CHAPTER 14

pH Regulation in Mammalian Glia

CHRISTINE R. ROSE AND BRUCE R. RANSOM
Institute of Physiology, University of Saar Landes, 66421 Hamburg, Germany (C. R. R.)
Department of Neurology, University of Washington, Seattle, WA 98195 (B. R. R.)

INTRODUCTION

Glial cells are the nonneuronal cells of the nervous system. They form a heterogeneous group primarily distinguished by what they lack: axons and action potentials. Within the central nervous system (CNS) there are two main categories of glial cells (or simply glia), macroglia and microglia. Macroglia are the most abundant, if not by number certainly by volume, and are further divided into astrocytes and oligodendrocytes. Macroglia are derived from neuroectoderm like neurons; microglia, on the other hand, are derived, at least in part, from mesoderm (Federoff, 1995). Microglia are closely related to hemopoietic mononuclear cells and may actually arise directly from these cells. In the peripheral nervous system, the main types of glia are Schwann cells, satellite cells of the sympathetic and dorsal root ganglia, and enteric glia.

In the CNS, glial cells outnumber neurons about 10 to 1 and occupy half the volume of the brain. They are located throughout the CNS, but oligodendrocytes predominate in white matter areas, consistent with the fact that they alone provide myelin to axons. The properties of these cells are being defined after years of confusion and neglect. Astrocytes and oligodendrocytes have high resting membrane potentials, multiple types of ion channels, gap junctions that link them with other glial cells, and a wide range of neurotransmitter receptors linked to ion channels or second messenger systems. These cells are not the inert, connective tissue-like elements originally imagined. They share a surprising number of features with neurons and are equipped to send and receive signals from other cells via informational molecules. Finally, it has been a convenient tenet of "glial mythology" that all astrocytes or all oligodendrocytes are basically alike. In actuality, the diversity of regional differences in glial properties is likely to rival that seen with neurons, adding a new dimension of complexity to the brain (Ransom, 1991).

There are two crucial reasons why regulation of intracellular pH (pH_i) in these cells is important for brain function. First, changes in glial pH_i may affect nearby neurons by changing extracellular pH (pH_o). Glia have an extremely intimate anatomic relationship with neurons, and virtually every neuron in the brain has glial neighbors. In vertebrate

pH and Brain Function, Edited by Kai Kaila and Bruce R. Ransom
ISBN 0-471-11838-9

nervous systems the average width of the extracellular space (ECS) is only about 0.02 μm; consequently, the extracellular fluid volume surrounding cells is very small and is quickly altered by transmembrane movements of ions or molecules of any sort (Nicholson, 1995). When glial cells regulate their pH_i, or respond to stimuli with alterations in pH_i (e.g. Ballanyi et al., 1994; Chapter 16), this is usually associated with transmembrane H^+ movements that alter the pH of the common ECS they share with neurons (see Chapter 17). To the extent that neurons are sensitive to changes in pH_o, which they clearly are (see Chapters 20–25, the glial-induced changes in pH_o will influence neuronal function.

In addition to noting the nearness of glia and neurons to each other, it is worth mentioning that astrocytes also bear a special relationship to blood vessels in the brain. Camillo Golgi, more than a century ago, was probably the first to recognize that these cells essentially encase brain capillaries with their end feet. This anatomic feature suggests that astrocytes might influence blood flow by changing pH_o in the immediate vicinity of small resistance vessels in the brain (Newman, 1996).

Second, our increased understanding of glial cell function, which is still primitive in contrast to our knowledge about neurons, indicates that glial cells are importantly involved in nearly everything the brain does. For example, astrocytes play a central role in preventing disruptive accumulation of extracellular K^+ with neural activity (Newman, 1995), they participate in the synthesis and removal of neurotransmitters (Martin, 1995), and they appear to interact with neurons in the context of energy metabolism (Ransom and Fern, 1997). Although the details are currently lacking, changes in pH_i can be confidently predicted to affect all of these glial functions, and changes in brain function would follow. To fix on just a single way that this might occur, consider how glial cells prevent unregulated increases in $[K^+]_o$ during neural activity. Astrocytes, which are thought to be most involved in this process, redistribute extracellular K^+, from sites where neural activity causes it to accumulate, to sites where the extracellular concentration of this ion is normal. The redistribution process utilizes K^+ channels and gap junctions, and both of these are pH sensitive (Ransom, 1995; Tombaugh and Somjen, 1996; Chapter 21, 24).

This review focuses mainly on the regulation of pH_i in astrocytes for the simple reasons that these cells have been the best studied to date.

STEADY-STATE pH_i

The steady-state or baseline pH_i of vertebrate glial cells is known from only a limited number of investigations, where pH_i was determined mainly using three different techniques: pH-sensitive microelectrodes, [^{14}C] 5, 5′-dimethyl-2-4-oxazolidine (DMO) distribution, or fluorescence ratio imaging with the pH-sensitive dye 2′, 7′-bis(carboxyethyl)-5, 6-carboxyfluorescein (BCERCF) (summarized in Table 1). The most common vertebrate glial cell preparations used for studying pH_i regulation were rat or mouse astrocytes, cultured or in the intact brain, and dissociated retinal Müller cells. Only single reports exist about pH_i in oligodendrocytes and Schwann cells, respectively, and these were both tissue culture studies (Table 1).

The baseline pH_i of vertebrate glial cells ranged from pH 6.9 to 7.6, measured in CO_2/HCO_3^- buffer at an extracellular pH (pH_o) of 7.2–7.5 (Table 1). These studies do not reveal a clear trend towards systematic differences in steady-state pH_i between different types of glial cells, nor do they suggest differences related to the pH_i measurement technique, or

TABLE 1 Steady-State pH_i of Vertebrate Glial Cells

Preparation	Technique	pH_i CO_2-free	pH_i CO_2/HCO_3^-	pH_o	Reference
Freshly dissociated salamander retinal Muller cells	BCECF	7.02	7.32	7.4	Newman (1994)
Cultured Schwann cells from rat sciatic nerve	Me2CF	7.17	7.08	7.4	Nakhou et al. (1994)
Cultured oligodendrocytes from mouse spinal cord	Electrodes	7.2	7.6	7.4	Kettenmann and Schlue (1988)
Astrocytes from isolated Necturus optic nerve	Electrodes	7.32	7.39	7.5	Astion et al. (1989, 1991)
Cortical rat astrocytes *in vivo*	Electrodes		7.1	7.3	Chesler and Kraig (1987)
	Electrodes		7.04	7.41	Chesler and Kraig (1989)
	Electrodes		7.31	7.26	Kraig and Chesler (1990)
Glial cells in cat medulla *in vivo*	Electrodes		7.25	7.3–7.5	Ballanyi et al. (1994)
Astrocytes from gliotic rat hippocampal slices	Electrodes		7.03	7.54	Grichtchenko and Chesler (1994a)
Cultured rat astrocytes from cortex	BCECF	6.9	7.0	7.33	Mellergard and Siesjö (1991) Mellergard et al. (1992, 1993)
cortex	BCECF	7.02	7.22	7.4	Kaila et al. (1991)
cortex	BCECF	7.15	7.32	7.4	Shrode and Putnam (1994)
cortex	[^{14}C]DMO		6.93	7.24	Kimelberg et al. (1986)
cerebellum	BCECF	6.96	6.96	7.4	Brune et al. (1994)
forebrain	BCECF	6.89	7.24	7.4	Boyarsky et al. (1993)
hippocampus	BCECF			7.4	Rose and Ransom (1996a)
hippocampus	BCECF	6.88	7.14	7.4	Pappas and Ransom (1993, 1994)
Cultured mouse astrocytes from brain	Electrodes	6.68	7.05	7.35	Wuttke and Walz (1990)
cortex	[^{14}C]DMO	6.97	7.16	7.4	Chow et al. (1991)
cortex	BCECF	7.11		7.4	Brookes and Turner (1994)
C6 glioma	BCECF	7.39	7.5	7.4	Shrode and Putnam (1994)
	Electrodes	7.34		7.4	Erecinska et al. (1993)
	BCECF	7.23		7.3	Jacubovicz et al. (1987)

4′,5′-dimethyl-5 (and 6)-carboxyfluorescein.

related to the nature of the preparation studied (i.e., *in vivo* vs. *in vitro*). Likewise, the temperature under which the experiments were performed (usually either at room temperature, 20°C–22°C, or at ~37°C), did not seem to influence pH_i. One should emphasize, however, that no direct comparison of these variables has been performed in the same study.

In the majority of cells studied, pH_i was 0.1–0.2 pH unit more acidic than pH_o. However, due to the high membrane potential of most glial cells (around −80 mV), glial pH_i was still significantly more alkaline than would be expected from a passive distribution of H^+ across the cell membrane. The relatively alkaline pH_i of glial cells indicates the presence of plasma membrane transporters that extrude intracellular acid from the cells, creating an inwardly directed electrochemical gradient for H^+ (or an outwardly directed gradient for base equivalents). Assuming a representative value of 7.25 for glial pH_i and a pH_o of 7.4, the equilibrium potential for H^+/OH^- ions is ~−9mV (~71mV positive to the typical glial membrane potential).

Glial cell pH_i was generally more alkaline in CO_2/HCO_3^--buffered saline compared to salines buffered with compounds like HEPES in the absence of CO_2/HCO_3^-; nevertheless, the H^+ gradient is inwardly directed in both CO_2/HCO_3^--free and CO_2/HCO_3^--buffered salines (Table 1). As discussed below, the discovery that CO_2/HCO_3^- caused an alkaline shift in baseline pH_i implied that HCO_3^- is used as a substrate for an acid-extruding transporter in glial cells.

In contrast to most other glial cells, the pH_i of cultured Schwann cells was more acid in the presence of CO_2/HCO_3^- (Nakhou et al., 1994). This suggests that in Schwann cells HCO_3^- is being used as a substrate by a pH-regulating transporter that functions as an acid loader.

The importance of maintaining an alkaline pH_i in glial cells, relative to the electrochemical gradient for H^+, is not entirely clear. As in other cells, glia have cytosolic enzymes, like the rate-limiting glycolytic enzyme phosphofructokinase, that are controlled by pH_i. They also have transporters and ion channels, gap junctions are an excellent example (Ransom, 1995), that are pH sensitive (see Chapters 21, 24). If pH_i is to have a meaningful role in controlling the function of these critical molecules it must be "set" at an ideal baseline level to provide for both stability and a dynamic response range for the pH-sensitive molecules.

MECHANISMS OF pH_i REGULATION

Several pH_i-regulating transport mechanisms have been identified in different types of vertebrate glial cells and will be discussed in more detail further on. Plasma membrane transporters which have an alkalinizing influence on pH_i, include a Na^+/H^+ exchanger, a Na^+/HCO_3^- cotransporter, a Na^+-dependent Cl^-/HCO_3^- exchanger, and a H^+-ATPase. A Na^+-independent Cl^-/HCO_3^- exchanger acts as an acid loading transporter. It is likely that different glial cells will have different patterns of transporter expression but our knowledge about this is limited at the present time.

In addition to these pH_i-regulating transporters, a lactic acid carrier, probably an H^+/lactate cotransporter, might contribute to proton movement across the plasma membrane in astrocytes and retinal Müller glial cells (Walz and Mukerji 1988a, 1988b; Staub et al., 1990; Nedergaard and Goldman, 1993; Poitry-Yamate et al., 1995; see Chapter 5). Glutamate uptake is also associated with influx of acid equivalents, as first shown in retinal Müller glial cells (Bouvier et al., 1992). This excitatory amino acid uptake mechanism is robustly expressed in astrocytes and influences pH_i (Bouvier et al., 1992; Amato et al., 1994; Brune and Deitmer, 1995; Rose and Ransom, 1996a; see Chapter 19).

Na^+/H^+ Exchange

The electroneutral Na^+/H^+ exchanger has been found in most mammalian cells where its presence has been sought (Aronson, 1985); certain invertebrate cells, including squid axon, crayfish neurons, and snail neurons, lack Na^+/H^+ exchange (Thomas, 1984). All vertebrate glial cells studied to date exhibit Na^+/H^+ exchange, including Schwann cells, oligodendrocytes, and astrocytes (Table 2). Due to its 1:1 stoichiometry and simple ion requirements, identification of Na^+/H^+ exchange activity in glial cells is often relatively straightforward. In most studies, the Na^+/H^+ exchanger was identified when studying pH_i recovery from an acid load imposed by standard techniques like the NH_3/NH_4^+ prepulse method (Chapter 10) in CO_2/HCO_3^--free saline (which excludes participation of HCO_3^--dependent pH regulating transporters).

An example for such an experimental protocol is given in Fig. 1 for cultured rat oligodendrocytes (Kettenmann and Schlue, 1988). Presence of a plasma membrane Na^+/H^+ exchanger in oligodendrocytes was first suggested by recovery from an NH_4^+-induced acid load in CO_2/HCO_3^--free saline (which means recovery was independent from the presence of HCO_3^-; Fig. 1A). This recovery required extracellular Na^+ (i.e., an inwardly directed Na^+ gradient; Fig. 1C), and was inhibited by amiloride, a commonly used blocker for Na^+/H^+ exchange (Benos, 1982) (Fig. 1B). Moreover, pH_i recovery was not accompanied by significant membrane potential changes, consistent with an electroneutral process. These findings taken together are convincing evidence for the presence of Na^+/H^+ exchange. A fourth criterion for the presence of Na^+/H^+ exchange, which is not shown in this example, is the independence of pH_i recovery from Cl^-. Li^+ is usually able to replace Na^+ for transport (Jean et al., 1986; Shrode and Putnam, 1994), enabling some pH_i recovery from an acid load when Na^+ is replaced by Li^+.

The absolute requirement of a Na^+ gradient for Na^+/H^+ exchange activity suggested that substantial increases in glial intracellular Na^+ concentration ($[Na^+]_i$) might compromise acid secretion, promoting intracellular acidification and/or hindering recovery from an acid load. In cultured hippocampal astrocytes it was recently shown, however, that reduction of the driving force for Na^+/H^+ exchange from 24:1, at steady-state intracellular ion concentrations, to 8:1, during transmitter-induced increases in $[Na^+]_i$, did not cause obligatory pH_i changes in CO_2/HCO_3^--free saline (Rose and Ransom, 1996a). This indicated that astrocytic pH_i regulation is not necessarily jeopardized during short-term, physiological increases in $[Na^+]_i$.

In most glial cells, the Na^+/H^+ exchanger seems to be the predominant mechanism for setting baseline pH_i in CO_2/HCO_3^--free saline, and its activity alone is sufficient to ensure pH_i recovery from an acid load. Still, pH_i in CO_2/HCO_3^--free saline is usually more acidic than in CO_2/HCO_3^--buffered saline (cf. Table 1), indicating that additional, CO_2/HCO_3^--dependent transporters influence baseline pH_i in glial cells.

In addition to having an important role in regulating pH_i, Na^+/H^+ exchange also acts as a major Na^+ entry pathway in many cells and, via Na^+ influx, influences intracellular volume. It has been implicated in a wide range of cellular events, including cell cycle regulation and differentiation (Wakabayashi et al., 1992). Perhaps the most carefully studied of these is the exchanger's role in controlling cell proliferation via signal-induced changes in pH_i (Wakabayashi et al., 1992). The importance of the Na^+/H^+ exchanger in

TABLE 2 pH_i-Regulating Transport Mechanisms in Vertebrate Glial Cells

Preparation	Na^+/H^+ Exchange	Na^+/HCO_3^- Cotransport	$Na^+/Cl^-/HCO_3^-$ Exchange	Cl^-/HCO_3^- Exchange	Reference
Salamander retinal Muller cells		+			Newman and Astion (1991), Newman (1991, 1996)
Cultured rat Schwann cells	+			+	Nakhou et al. (1994)
Cultured mouse oligodendrocytes	+	+			Kettenmann and Schlue (1988)
Astrocytes from gliotic rat hippocampal slices		+			Grichtchenko and Chesler (1994b)
Astrocytes from Necturus optic nerve	+	+			Astion and Orkand (1988), Aston et al. (1991)
Cultured rat astrocytes from cortex	+		+	+	Mellergard et al. (1992, 1993)
cortex	+	+	+	+	Shrode and Putnam (1994)
cortex	+		+		Kimelberg et al. (1979), Kimelberg (1981)
cerebellum	+	+			Brune et al. (1994)
forebrain	+	+			Boyarski et al. (1993)
hippocampus	+	+			Pappas and Ransom (1993, 1994) O'Connor et al. (1994)
Cultured mouse astrocytes from	+	+		+	Chow et al. (1991)
cortex					Chow et al. (1991)
cortex		+			Brookes and Turner (1994)
C6 glioma	+		+		Shrode and Putnam (1994)
	+				Erecinska et al. (1993)
	+				Jacubovitz et al. (1987)
	+				Jean et al. (1986)
	+				Benos and Sapirstein (1983)

cell cycle regulation derives from the fact that it is rapidly activated by a variety of mitogenic and nonmitogenic signals, such as growth factors, oncogenes, neurotransmitters, hormones, chemotactic peptides, phorbol esters, lectins, and osmotic change (Grinstein et al., 1989). None of these secondary roles for the exchanger have yet been explored in glial cells.

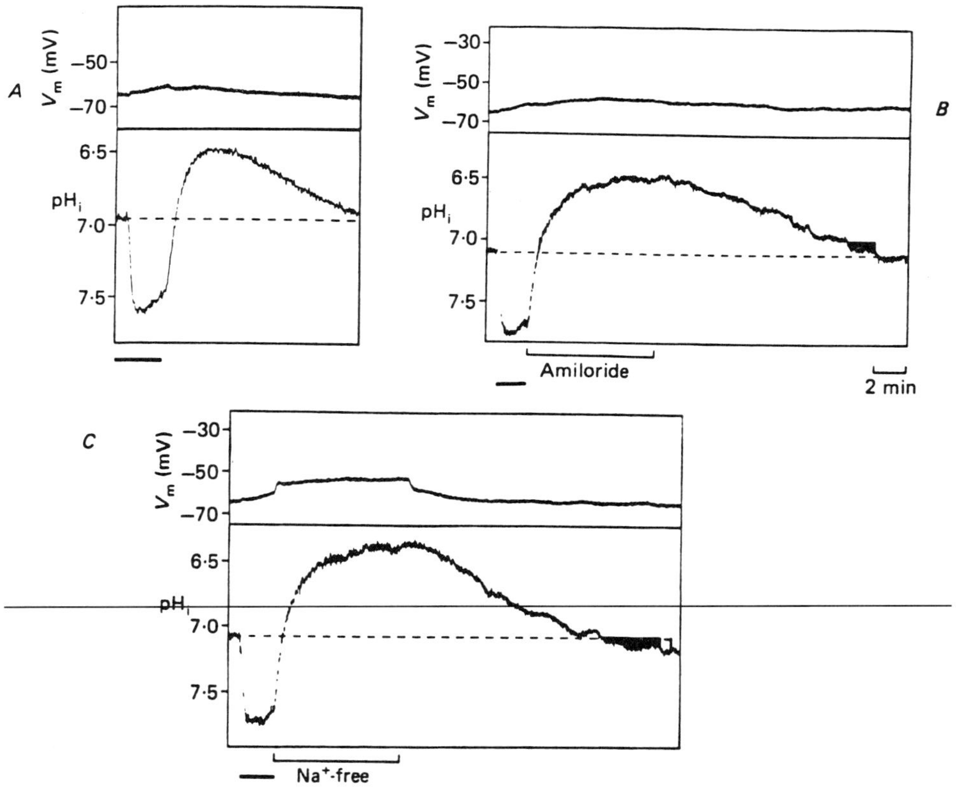

FIGURE 1. pH$_i$ recovery in mouse oligodendrocytes in HCO_3^--free solution. pH$_i$ and membrane potential were simultaneously recorded in an oligodendrocyte using a double-barreled pH-sensitive microelectrode. An acid load was imposed by brief exposure to 10 mM NH_4^+ as indicated by the bars (i.e., the acid load follows a brief alkaline shift). **A.** In the absence of HCO_3^-, the cell rapidly recovers from an acid load. **B.** and **C.** Amiloride (1 mM) or exposure to Na^+-free solution reversibly blocks recovery from an acid load. These results indicate the existence of Na^+/H^+ exchange in oligodendrocytes. From Kettenmann and Schlue (1988), with permission.

Na^+/HCO_3^- Cotransport

After its description in leech glial cells (Deitmer and Schlue, 1987, 1989; see Chapter 13), Na^+/HCO_3^- cotransport has been identified in a variety of different vertebrate glial cell types (Table 2): astrocytes from mudpuppy optic nerve (Astion and Orkand, 1988; Astion et al., 1991), salamander retinal Muller glial cells (Newman, 1991; Newman and Astion, 1991), cultured mouse oligodendrocytes (Kettenmann and Schlue, 1988), and rat and mouse astrocytes (Boyarski et al., 1993; Brune et al., 1994; Brookes and Turner, 1994; Chow et al., Grichtchenko and Chesler, 1994a; O'Connor et al., 1994; Pappas and Ransom, 1994; Shrode and Putnam, 1994). The exceptions among glial cells may be developing Schwann cells (Nakhou et al., 1994) and C6 glioma cells (Shrode and Putnam, 1994), which appear not to express this pH-regulating mechanism. In contrast to its presence in most glial cells, Na^+/HCO_3^- cotransport activity is not found in neurons.

Like Na^+/H^+ exchange, Na^+/HCO_3^- cotransport depends on the inwardly directed Na^+ gradient and is independent of the presence of Cl^-. Obviously, in contrast to Na^+/H^+ exchange activity, Na^+/HCO_3^- cotransport activity requires HCO_3^--and, therefore, can only be robustly expressed in CO_2/HCO_3^--buffered saline (though some residual Na^+/HCO_3^- cotransport activity may be possible in nominally CO_2/HCO_3^--free salines, because they are usually equilibrated with air and contain about 100 μM HCO_3^-). In most preparations, Na^+/HCO_3^- cotransport is inhibited by stilbene derivatives like SITS (4-acetamido-4′-isothiocyanatostilbene-2,2′-disulphonic acid) or DIDS (4,4′-didiothiocyanatostilbene-2,2′-disulphonic acid), which are, however, not specific for this transporter and also block other anion transport mechanisms like Cl^-/HCO_3^- exchange (Cabantchik and Rothstein, 1972; Cabantchik and Greger, 1992). Some studies have reported glial Na^+/HCO_3^- cotransport that is stilbene-insensitive (Kettenmann and Schlue, 1988; Grichtchenko and Chesler, 1994a).

An essential feature of Na^+/HCO_3^- cotransport is the fact that it is electrogenic in most preparations, transporting 1 Na^+ for each 2 or 3 HCO_3^-. This special property is responsible for the fact that activation of Na^+/HCO_3^- cotransport not only changes pH_i, but also generates a current that can lead to membrane potential changes. A reversal potential for Na^+/HCO_3^- cotransport can be easily calculated and depends on the stoichiometry of the Na^+ and HCO_3^- transport (Newman and Astion, 1991). Therefore, the direction of Na^+/HCO_3^- cotransport, and possibly its rate, can be influenced by membrane potential.

Evidence of Na^+/HCO_3^- cotransport activity in glial cells can be obtained by observing pH_i and/or membrane potential changes when switching from a CO_2/HCO_3^--free to a CO_2/HCO_3^--buffered saline. An example of this is shown in Fig. 2, taken from experiments on glial cells in *Necturus* optic nerve (Astion et al., 1991). In switching from a CO_2/HCO_3^--free to a CO_2/HCO_3^--containing solution, there is an initial CO_2-induced acidification, followed by an alkaline shift in baseline pH_i. At the same time, the glial cell

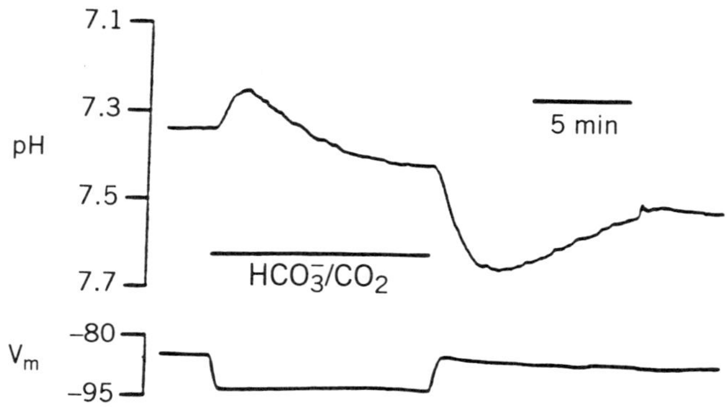

FIGURE 2. The effects of CO_2/HCO_3^- application on pH_i and membrane potential of a glial cell in *Necturus* optic nerve. pH_i and membrane potential were recorded simultaneously with a double-barreled pH-sensitive microelectrode. Switching from a nominally HCO_3^--free solution, to one containing 12.5 mM HCO_3^- and 2.3% CO_2, results in membrane hyperpolarization and a transient intracellular acidification that rapidly converts to an alkaline shift. From Astion et al. (1991), with permission.

hyperpolarized. The HCO_3^--induced hyperpolarization is strongly inhibited by SITS and is dependent on extracellular Na^+ (Astion and Orkand, 1988), indicating activation of inward Na^+/HCO_3^- cotransport in these optic nerve glial cells. In cultured rat hippocampal astrocytes, switching from CO_2/HCO_3^--free to CO_2/HCO_3^--buffered saline results in an increase in intracellular $[Na^+]_i$, probably due to activation of inward Na^+/HCO_3^- cotransport (Rose and Ransom, 1996b).

In salamander retinal glial cells (i.e., Müller cells), raising external HCO_3^--concentration induced an outward, hyperpolarizing current that was Na^+-dependent, Cl^--independent, and blocked by DIDS, indicating the presence of electrogenic Na^+/HCO_3^- cotransport in these cells (Newman, 1991; Newman and Astion, 1991). Several studies reported similar characteristics of HCO_3^--dependent currents and/or pH_i changes from cultured mammalian astrocytes implying Na^+/HCO_3^- cotransport activity (Boyarski et al., 1993; Brune et al., 1994; Brooks and Turner, 1994; Chow et al., 1991; Grichtchenko and Chesler, 1994b; O'Connor et al., 1994; Pappas and Ransom, 1994; Shrode and Putnam, 1994).

Despite common ionic requirements and pharmacological properties, Na^+/HCO_3^- cotransport in astrocytes and Müller glial cells probably differs in one important way. Whereas the stoichiometry of Na^+/HCO_3^- cotransport has been suggested to be 1:2 for cultured rat and mouse astrocytes (Brookes and Turner, 1994; O'Connor et al., 1994), the stoichiometry for Müller glial cells is 1:3 (Newman, 1991). Consequently, the reversal potential for Na^+/HCO_3^- transport in vertebrate astrocytes has been reported to be around -70 to -80 mV, which is very close to the resting membrane potential of these cells (estimates of reversal potential range from -63 mV (Brookes and Turner, 1994), to -86 mV (O'Connor et al., 1994). In contrast, the reversal potential for Na^+/HCO_3^- transport in Müller cells is only about -26 mV (Newman, 1991). This difference in reversal potentials may result in differences in physiological behavior of the transporters. Because the reversal potential of Na^+/HCO_3^- cotransport is relatively close to the resting membrane potential of astrocytes, membrane depolarizations can generate inward Na^+/HCO_3^- cotransport and intracellular alkalinization, whereas membrane hyperpolarizations would promote outward transport and acidification (Pappas and Ransom, 1994). Astrocytic Na^+/HCO_3^- cotransport, therefore, links glial membrane potential with pH_i (or vice versa), enabling the astrocytes to influence extracellular pH during neural activity (see depolarization-induced alkalinization further on). In the case of Müller cells, the more depolarized reversal potential means that this pH-regulating transporter could not become a net acid exporter until the cell was depolarized to >-26 mV. Nevertheless, the Müller cell Na^+/HCO_3^- would be qualitatively influenced by membrane potential in the same manner as its counterpart in astrocytes and may serve, in a similar manner, to link membrane potential to both intracellular and extracellular pH regulation.

Because Müller cells have high membrane potentials like other macroglial cells, the very positive reversal potential of Na^+/HCO_3^- cotransport in these cells ensures that net cotransport is outwardly directed under physiological conditions. An interesting feature of Müller cell Na^+/HCO_3^- cotransport is that it is not equally distributed across the cell's surface, but seems to be concentrated at the cells' endfeet, facing the vitreous humor (Fig. 3) (Newman and Astion, 1991). It was suggested that these properties of Müller cell Na^+/HCO_3^- cotransport enable the cells to serve as "CO_2-sinks," by taking up metabolically produced CO_2 from the photoreceptors and transferring it directly into the vitreous humor (Newman, 1996).

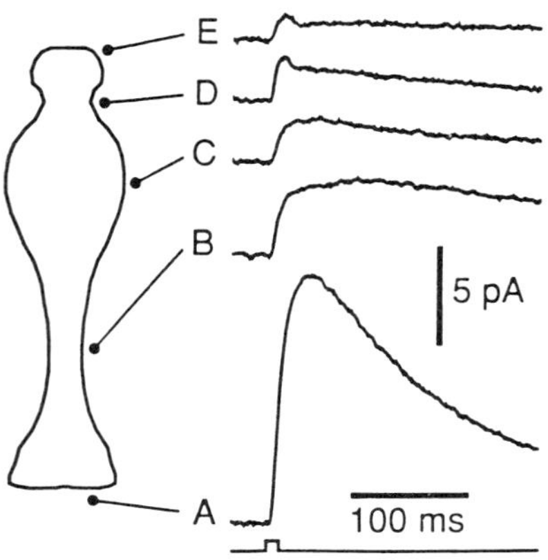

FIGURE 3. Na^+/HCO_3^- cotransport sites are localized at the end feet of retinal glial cells. This is demonstrated here by the fact that extracellular application of HCO_3^--produces the greatest outward current when ejected onto the cell endfoot (point A in the figure) of an isolated salamander Müller cell (drawn on the left). Smaller currents were elicited by ejection of HCO_3^- onto other points of the cell (B–E). Recall that the Müller cell endfoot abuts the vitreous humor of the eye. From Newman and Astion, 1991.

Na^+-dependent Cl^-/HCO_3^- Exchange

Acid extrusion via Cl^-- and HCO_3^--dependent transport in glia was first suggested from studies in cultured rat astrocytes, where a pH-dependent and SITS-inhibitable efflux of $^{36}Cl^-$ was observed (Kimelberg et al., 1979; Kimelberg, 1981). This transport mechanism is coupled to Na^+ movement, and acid extrusion is achieved by using the inwardly directed Na^+ gradient to transport HCO_3^- into cells in exchange for Cl^-. This mechanism, therefore, was called Na^+-dependent Cl^-/HCO_3^- exchange (Thomas, 1984). In cultured rat cortical astrocytes, Mellergard et al. (1993) described a pH_i recovery from an acid load that was not only dependent on Na^+ and HCO_3^-, and inhibited by the stilbene DIDS, but also required the presence of extracellular Cl^-. In agreement with this study, Shrode and Putnam (1994) found that the alkalinization of rat cortical astrocytes and C6 glioma cells upon addition of HCO_3^- was largely blocked by removal of extracellular Cl^-. Both studies, therefore, concluded that Na^+-dependent Cl^-/HCO_3^- exchange was present in these cells (Table 2).

Na^+-dependent Cl^-/HCO_3^- exchange can be distinguished from Na^+/HCO_3^- cotransport mainly by its requirement for intracellular Cl^-. The determination of dependence on intracellular Cl^- can be challenging, however, because removal of extracellular Cl^- for short periods during an experiment might not induce a drop in intracellular $[Cl^-]$ that is large enough to inhibit Na^+-dependent Cl^-/HCO_3^- exchange completely. In addition, both Na^+ -dependent Cl^-/HCO_3^- exchange and Na^+/HCO_3^- cotransport are blocked by stilbenes, making blockade of transport by these agents ineffective as a means of discriminating one transporter from the other.

Cl^-/HCO_3^- Exchange

In sharp contrast to the Na^+-dependent Cl^-/HCO_3^- exchanger, the Cl^-/HCO_3^- exchanger is independent of the transmembrane Na^+ gradient and transports intracellular HCO_3^- out of the cells in exchange for extracellular Cl^- (i.e., it depends on the transmembrane Cl^- gradient; Roos and Boron, 1981). It therefore acts as an acid loader under normal conditions and is thought to play a role during recovery from alkaline loads. This transporter has not been extensively studied in vertebrate glial cells. In cultured Schwann cells, removal of Cl^- in the absence of Na^+ caused pH_i to increase (Nakhou et al., 1994), and in cultured rat astrocytes, a DIDS-inhibitable, Na^+-independent recovery from an alkaline load was observed (Mellergard et al., 1993; Shrode and Putnam, 1994).

The physiological importance of this transporter is not obvious, because Cl^-/HCO_3^- exchange degrades the inwardly directed H^+ gradient whose maintenance is energy-consuming. Moreover, given the resting inwardly directed H^+ gradient and constant metabolic production of intracellular acid (i.e., CO_2 and lactate), an additional acid loader would not seem to be necessary for recovery from an intracellular alkalinization. This transporter may help maintain $[Cl^-]_i$ in addition to furosemide-sensitive $Na^+/K^+/2Cl^-$ cotransport in glial cells (Waltz and Hertz, 1983).

H^+ Pump

All the glial acid extrusion mechanisms described above are energized by the electrochemical Na^+ gradient across the plasma membrane and, therefore, depend ultimately on Na^+/K^+-ATPase activity. In addition to these secondary transport mechanisms, cultured rat hippocampal astrocytes have been reported to possess an acid extrusion mechanism that was independent of Na^+ and HCO_3^- (Pappas and Ransom, 1993).This unusual mechanism enabled a slow recovery from an acid load in CO_2/HCO_3^--free saline, even after removal of extracellular Na^+.

Like Na^+-independent acid extrusion observed in the kidney (Manger and Koeppen, 1992), the Na^+- and HCO_3^--independent acid extrusion in hippocampal astrocytes was blocked by either bafilomycin Al or N-ethylmaleimide, signifying transport by a vacuolar-type H^+-ATPase in these cells. Because the V-type H^+-ATPase is electrogenic, it is theoretically possible, but not necessary, that membrane potential can influence its rate of function, with depolarization tending to enhance transport. For this reason, it could contribute to depolarization-induced alkalinization, or DIA, in astrocytes (Pappas and Ransom, 1993).

Summary of pH_i Regulation in Vertebrate Glial Cells

Several distinct pH_i regulating mechanisms have been identified in vertebrate glial cells. Glial cells differ in the expression of these mechanisms. All vertebrate glial cell types seem to express a Na^+/H^+ exchanger that acts as an acid extruder and can operate independently of membrane potential and in the absence of CO_2/HCO_3^-. In addition to Na^+/H^+ exchange, at least one HCO_3^--dependent acid extrusion mechanism is present in all glial cells, which is, with the exception of Schwann cells (Table 2), Na^+/HCO_3^- cotransport and/or Na^+-dependent Cl^-/HCO_3^- exchange. The small number of studies available on glial pH_i regulation and the growing appreciation of the diversity of these cells (Ransom, 1991) prevent categorical conclusions about their expression of Na^+/HCO_3^- cotransport versus Na^+-dependent Cl^-/HCO_3^- exchange. What is apparent is that in the nervous system, Na^+/HCO_3^- cotransport is a primarily glial transport mechanism that is not found in neurons. Interestingly, however, Na^+/HCO_3^- cotransport is absent from C6 glioma cells (Table 2), indicating that this pH_i regulator may be lost with neoplastic transformation. This could have functional significance given the apparent link between pH_i and glial proliferation (Pappas et al., 1994).

MODULATION OF pH_i

Intraglial pH is not a fixed parameter that is tightly held at a certain steady-state value, but can change, in a variety of different situations such as neural activity, under the influence of hormones and growth factors, or during pathological events. In the following section, factors and mechanisms are discussed that can influence intraglial pH. The subtext is that alterations in pH_i can have physiological significance, and can function as yet another level of neural integration or information transfer within the brain. More detailed descriptions of pH changes observed during activity or pathological events in the brain can be found elsewhere in this book (e.g., Chapters 16–19 and 28–31).

Effects of pH_o

In primary rat astrocyte cultures, a strong correlation between pH_o and pH_i was found, inasmuch as changing pH_o led to changes in pH_i with a slope that varied from 0.3 (Pappas et al., 1996) to 0.76 (Mellergard and Siesjo, 1991). Because of this relatively steep dependence of astrocytic pH_i on pH_o, a phenomenon not usually observed for neurons, it was sometimes questioned if these glial cells fulfill the requirements for active pH_i regulation in terms of their ability to keep pH_i constant and independent of pH changes in the surrounding medium. Still, at any given pH_o, an inwardly directed electrochemcial H^+ gradient was maintained, indicating that while the absolute glial pH_i is not regulated, the H^+ gradient across the membrane is (Mellergard and Siesjo, 1991). As will be discussed below, this characteristic of astrocytic pH_i regulation might contribute to the cells' ability to influence pH_o.

Depolarization-Induced Alkalinization

When astrocytes are depolarized, they exhibit an alkaline shift. This is observed *in vivo* (Chesler and Kraig, 1987, 1989) and in cultured astrocytes (Fig. 4A) (Boyarsky et al., 1993, Brookes and Turner, 1994; Pappas and Ransom, 1994, O'Connor et al., 1994). Oligodendrocytes behave in a similar fashion (Kettenmann and Schlue, 1988). This general phenomenon was first described in kidney epithelial cells and has been referred to as depolarization-induced-alkalinization or DIA (Siebens and Boron, 1989). The term DIA is somewhat misleading because it implicitly suggests that only depolarization affects cellular pH_i; in fact, in astrocytes under physiological conditions, pH_i is a continuous function of membrane potential, such that hyperpolarization causes an acid shift in pH_i (Pappas and Ransom, 1994).

The phenomenon of DIA is interesting for several reasons. First, the mechanism underlying DIA has been challenging to understand. Second, it is a mechanism whereby various depolarizing, or hyperpolarizing, stimuli predictably affect glial pH_i. Certain important pH-sensitive glial properties can therefore be influenced by changes in membrane potential, including gap junction communication (Ransom, 1995), glucose metabolism (see Ransom, 1992), and glutamate metabolism (Brookes and Turner, 1993). Third, changes in glial membrane potential affect glial pH_i by influencing transmembrane H^+ movement, which means movement of H^+ into or out of the extracellular space (Ransom

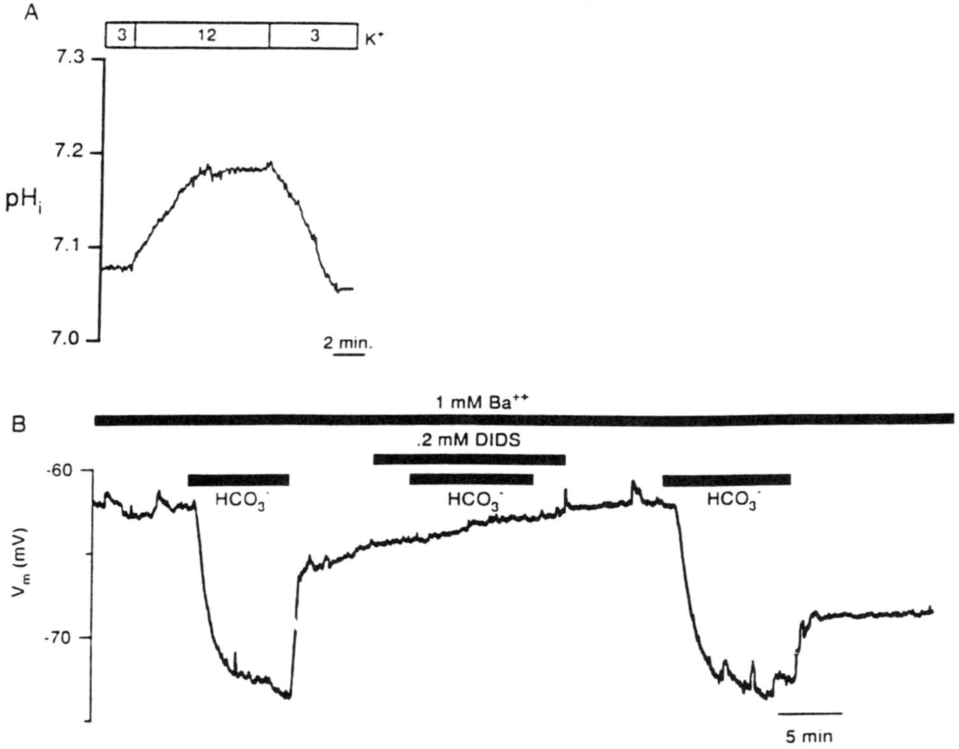

FIGURE 4. Depolarization-induced alkalinization in rat hippocampal astrocytes. **A.** pH_i is monitored in a rat hippocampal astrocyte, using the fluorescent pH indicator BCECF, while transiently depolarizing the cell by the application of 12 mM $[K^+]_o$. The cell undergoes a reversible alkaline shift in CO_2/HCO_3^--buffered solution. From Pappas and Ransom (1994), with permission. B. Astrocytes show a reversible hyperpolarization when switched from a HCO_3^--free solution to one containing 25 mM HCO_3^- at a constant pH_o. This hyperpolarization is reversibly blocked by 0.2 mM DIDS. These whole-cell patch-clamp recordings were made in the presence of 1 mM Ba^{2+}, which blocked resting potassium conductance and enhanced the response magnitude. From O'Connor et al. (1994), with permission.

et al., 1986; Ransom, 1992; Grichtchenko and Chesler, 1994a, Newman, 1996). Glial cells, therefore, may significantly influence brain pH_o in response to depolarizing stimuli such as increased $[K^+]_o$ during neural activity or excitatory transmitter release (Chesler, 1990; Ransom, 1992; Deitmer and Rose, 1996, see also Chapter 13). Due to the multitude of processes in the nervous system that are affected by pH_o, including neuronal excitability, glial influence on pH_o represents a form of glial–neuronal communication (Ransom, 1992; Deitmer and Rose, 1996).

In vertebrate glial cells, DIA could be caused by several mechanisms (Ransom, 1992). One possibility is that nearby neuronal activity might lead glial cells to increase outward transport, and perhaps production, of lactic acid via a H^+/lactic acid carrier, thereby alkalinizing the cell interior (cf. Walz and Mukerji, 1988a, b). Bafilomycin-sensitive H^+ transport, as found in cultured hippocampal astrocytes (Pappas and Ransom, 1993), could be activated by changes in membrane potential. Most evidence, however, supports the idea that DIA is primarily due to depolarization-induced activation of inwardly directed electrogenic Na^+/HCO_3^- cotransport.

In cultured hippocampal astrocytes, both elevation of $[K^+]_o$ and depolarization of the cells by Ba^{2+} leads to a largely CO_2/HCO_3^--dependent, Cl^--independent intraglial alkalinization that is markedly reduced by the inhibitor DIDS (Pappas and Ransom, 1994). The pH event stimulated by depolarization has the trademark features of Na^+/HCO_3^- cotransport, and these cells also exhibit electrophysiological evidence of this transporter (O'Connor et al., 1994). They hyperpolarize when switched from a HCO_3^--free to a HCO_3^--containing medium, and this event is blocked by DIDS (Fig. 4B). Likewise, the K^+-dependent alkalinization in cultured cerebellar astrocytes was CO_2/HCO_3^--dependent, Na^+-dependent, Cl^--independent, and inhibited by DIDS (Brookes and Turner, 1994). For astrocytes in hippocampal slices, DIA is largely accounted for by Na^+/HCO_3^- cotransport, but there is a component of depolarization-induced extracellular acidification that remains in the absence of extracellular Na^+ (Grichtchenko and Chesler, 1994a,b).

NEUROTRANSMITTERS

Glial cells have an astonishing array of neurotransmitter receptors and uptake mechanisms (von Blankenfeld and Kettenmann, 1991; Cull-Candy et al., 1991; see Kettenmann and Ransom, 1995). The best-known, and best-understood, effects of neurotransmitters on glial cells are transmitter-induced changes in glial membrane potential coupled to changes in intracellular Na^+, K^+, and/or Ca^{2+} concentrations (e.g., Rose and Ransom, 1996b). Neurotransmitters, like δ-aminobutyric acid (GABA) or glutamate, however, not only change the concentration of these cations, but can also have significant effects on the H^+ concentration of vertebrate glial cells.

Neurotransmitters can affect pH_i in several ways. It has been demonstrated that $GABA_A$-coupled Cl^- channels possess a significant conductance for HCO_3^- (Kaila and Voipio, 1987). Opening of these channels in astrocytes can, therefore, lead to efflux of HCO_3^- from the cells, and cause an intracellular acidification (Kaila et al., 1991) and extracellular alkalinization (Chesler and Kaila, 1992). Glutamate and aspartate, on the other hand, which also acidify vertebrate glial cells (Bouvier et al., 1992; Brookes and Turner, 1993; Brune and Deitmer, 1995; Rose and Ransom, 1996a), seem to operate via HCO_3^--independent mechanisms (Chapter 16). In Müller glial cells, the glutamate-induced acidification is probably entirely due to uptake of glutamate into the glial cells via a transporter that also moves an acid equivalent into the cells (Bouvier et al., 1992; Chapter 19). In other words, it remains unclear if the intraglial acidification is due to influx of H^+ or efflux of base equivalent. This transporter also accepts aspartate as a substrate. A significant contribution of transport-mediated uptake to the glutamate/aspartate-induced intraglial acidification was also suggested by studies on cultured astrocytes (Brune and Deitmer, 1995; Rose and Ransom, 1996a). D-aspartate is a substrate for the transporter but does not activate the ionotropic receptor. Application of D-aspartate to cultured hippocampal astrocytes caused a pure acidification accompanied by a large increase in $[Na^+]_i$ (Fig. 5). This was not blocked by CNQX, an ionotropic receptor antagonist.

Apart from glutamate-uptake mediated acidification, theoretical but unproven ways that glutamate could influence glial pH are by H^+ influx through unspecific cation chan-

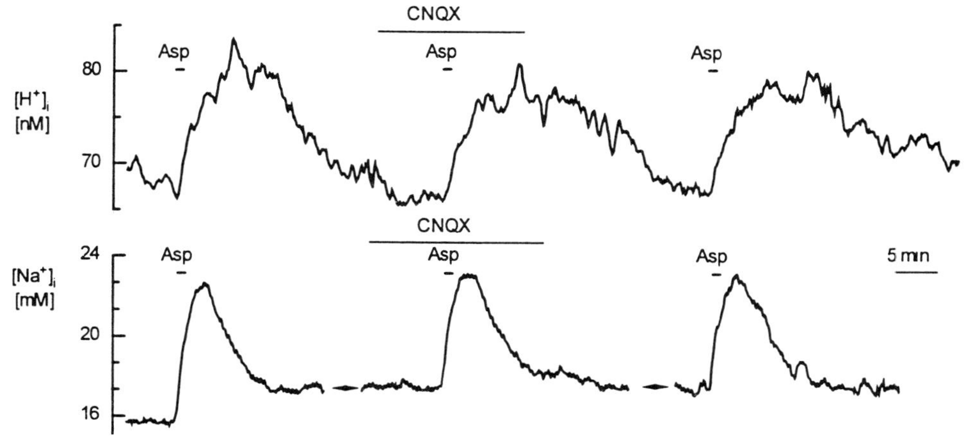

FIGURE 5. Effects of D-aspartate on $[H^+]_i$ and $[Na^+]_i$ transients in rat hippocampal astrocytes. Using fluorescence ratio imaging with BCECF or sodium-binding benzofuran isophthalate (SBFI), changes in $[H^+]_i$ and $[Na^+]_i$, respectively, were measured during brief applications of D-aspartate, which is a substrate for the glutamate transporter but does not activate the ionotropic receptor. D-aspartate produces reversible increases in $[H^+]_i$ and $[Na^+]_i$ that are not blocked by the ionotropic glutamate receptor antagonist CNQX. From Rose and Ransom (1996a), with permission.

nels coupled to glutamate receptors (Chen and Chesler, 1992), or by activation of a Ca^{2+}/H^+-ATPase following glutamate-induced increases in intracellular Ca^{2+} concentration (Schwiening et al., 1993; Paalasmaa et al., 1994; Smith et al., 1994). In a manner that is reminiscent of DIA, glutamate can alkalinize astrocytes by depolarizing them, via coupled cation channels, and activating electrogenic Na^+/HCO_3^- cotransport (Rose and Ransom, 1996a). A strong intracellular alkalinization has been reported following stimulation of astrocyte metabotropic glutamate receptors (Amos, 1996). The mechanism of this pH_i change is not clear.

MITOGENS

Glial cell differentiation, morphology, and other cell functions are modulated by hormones and growth factors (Draves et al., 1986; Arenander et al., 1988; Hösli and Hösli, 1993; Pearce and Wilkin, 1995). Although reports about hormone-induced pH_i changes in glial cells are rare, and the underlying mechanisms not well understood, the effects of these ligands on pH_i-regulating transporters are established in a variety of other cell types. These studies suggest that the transporters known to be responsible for regulation of steady-state pH_i could also take part in transmembrane signaling by hormones in glial cells.

The transporter which has been most intensively studied concerning its regulation and activation by hormones or growth factors is the Na^+/H^+ exchanger (Noel and Pouyssegur, 1995). In addition to being stimulated by intracellular acidification, Na^+/H^+ activity is activated by hormones or growth factors in several cell types (reviewed by Busa and Nuccitelli, 1984; Busa, 1986; Grinstein and Rothstein, 1986, Moolenaar, 1986). This activation, which is likely to be due to intracellular phosphorylation of the Na^+/H^+ exchanger by tyrosine kinases or protein kinase C (Clark and Limbird, 1991), can induce an alkalinization of resting pH_i by several tenths of a pH unit. Studies on regulation of Na^+/H^+ activity in vertebrate glial cells are still sparse. In glioma cells, Na^+/H^+ transport can be stimulated by serum, epidermal growth factor, bradykinin, beta-adrenergic agonists, or phorbol esters (Benos and Sapirstein, 1983; Sapirstein and Benos, 1984; Isom et al., 1987).

It has been speculated that stimulation of Na^+/H^+ exchange, leading to subsequent intracellular alkalinization, might be an intracellular signal necessary for cell proliferation or differentiation (Busa and Nuccitelli, 1984; see also Pappas et al., 1994). In Schwann cells, a cellular alkalinization subsequent to stimulation of Na^+/H^+ exchange was required to potentiate cellular proliferation following mitotic stimulation (Saunders et al., 1989). In general, the physiological consequences of Na^+/H^+ exchange stimulation are quite unclear, especially since the same agents that stimulate Na^+/H^+ exchange can influence HCO_3^--dependent pH_i-regulating transporters at the same time, sometimes resulting in little or no change of pH_i (Ganz et al., 1989; Ganz and Boron, 1994; Geibel et al., 1990; Lagadic-Gossmann et al., 1992). Other studies have suggested that the mitogenic effects of Na^+/H^+ exchanger stimulation are mediated by the increase in intracellular $[Na^+]$ which accompanies Na^+/H^+ activation, rather than by the pH_i increase (Blackshaw and Warner, 1976; Breckenridge and Warner, 1982; Rosoff and Cantley, 1983).

FUNCTIONAL IMPLICATIONS AND CONCLUSIONS

The functional implications of pH_i regulation in mammalian glial cells are not well understood. Some glial properties may be strongly modulated by changes in pH_i. The predominant resting conductance in glial cells is provided by K^+ channels. Cation channels in general, including K^+ channels, are influenced by changes in both external and internal pH (Tombaugh and Somjen, 1996; Chapter 21, this volume). The usual effect of H^+ on K^+ current is to inhibit it. Although the magnitude of this inhibition has been relatively modest in neurons, it could still have important consequences in glial cells because their behavior is strongly influenced by this channel type. The question of pH effects on glial K^+ channels deserves more study.

Glial cells express gap junctions and functional intercellular communication to a greater extent than any other cell in the adult nervous system (Ransom, 1995). These channels permit the free intercellular diffusion of molecules up to a molecular weight of ~1000 and are strongly influenced by changes in pH_i, with acidification causing the channels to close (Chapter 24). The functions of gap junctional communication in glial cells are still actively under investigation, but it is now apparent that these channels help regulate intracellular ion concentrations (Rose and Ransom, 1997). They also mediate a unique form of intercellular communication between astrocytes that consists of propagated waves of increased intracellular Ca^{2+} concentration stimulated by glutamate, or electrical or mechanical perturbation (Charles et al., 1992). Finally, gap junctions may be important for glial regulation of $[K^+]_o$ (Ransom, 1996). It seems very likely that changes in glial pH_i associated with brain activity (e.g., DIA) would affect gap junction communication between these cells (see also Pappas et al., 1996). Recent evidence showing that axonal activity transiently increases glial coupling supports this possibility (Marrero and Orkand, 1996). This activity-dependent modulation of gap junction communication may be the result of the associated increase in $[K^+]_o$, because the effect was mimicked by application of a solution with high [K^+] (Marrero and Orkand, 1996). The glial depolarization that occurs with increased $[K^+]_o$ would induce an intracellular alkaline shift, as discussed above (Fig. 4) (Pappas and Ransom, 1994), and this would tend to open more channels. If astrocyte coupling assists $[K^+]_o$ regulation (Ransom, 1996), rapid increases in coupling during periods of $[K^+]_o$ increase would seem to be a useful adaptation to augment this process.

The expression of robust electrogenic Na^+/HCO_3^- cotransport in astrocytes sets the stage for a fascinating type of glial–neuronal interaction (Ransom, 1992; see also Chesler, 1990). This interaction can be stated as a hypothesis: Neuronal activity increases $[K^+]_o$, causing glia, via enhanced Na^+/HCO_3^- cotransport, to generate an extracellular acidification that serves as a feedback control mechanism to dampen ongoing neural activity and/or lessen general excitability (Fig. 6). The individual steps in this paradigm (numbered 1–4) are supported by substantial experimental evidence. The tendency for neural activity to increase $[K^+]_o$ is well established and the magnitude of this increase is graded with activity (Ransom, 1992). As illustrated in Fig. 6, and discussed earlier, the activity-dependent increase in $[K^+]_o$ depolarizes adjacent glial cells and induces DIA (Fig. 6, Step 2). The transmembrance influx of HCO_3^- associated with DIA is expected to produce an extracellular acid shift (Fig. 6, Step 3), and this has now been directly demonstrated using retinal glial cells (Newman, 1995; see also Ransom et al., 1986).

Although it has long been known that brain acidification depresses neural activity (see discussion in Balestrino and Somjen, 1988 and Chapters 20–25), the details about how this is mediated remain incomplete (Fig. 6, Step 4). These effects may be mediated presynaptically by intracellular acidification (Lee et al., 1996), postsynaptically at the level of neurotransmitter receptors (Tang et al., 1990), or by direct proton-modulation of postsynaptic ion channels (Tombaugh and Somjen, 1996). Neuronal behavior can be extremely sensitive to changes in pH_o (Balestrino and Somjen, 1988; Lee et al., 1993). For example, changes of pH_o of only 0.05 unit can significantly alter complex neural activity in rat hippocampal brain slices (Lee et al., 1993). It appears that pH_o changes of this magnitude may well occur with routine neural activity and much greater changes can be seen with

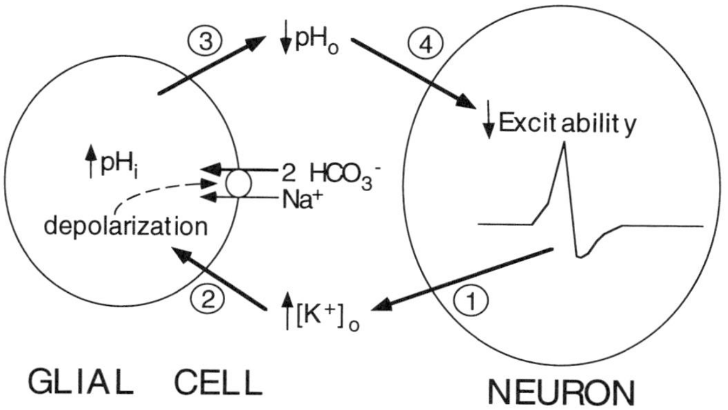

FIGURE 6. Diagram of the hypothesized glial–neuronal interaction mediated by changes in pH_o. Active neurons cause an increase in $[K^+]_o$ (1) that depolarizes nearby glial cells (2) causing depolarization-induced alkalinization (DIA). The mechanism of DIA is an increase in Na^+/HCO_3^- cotransport caused by the depolarization. The transmembrane influx of HCO_3^- also causes an acid shift in pH_o (3). In a feedback-like fashion, the acid shift in pH_o reduces the excitability of the active neurons (4). See text for details.

intense activity (Ransom, 1992; Ransom et al., 1986; Chesler and Kaila, 1992). While the mechanism outlined here is feasible, the extent to which it operates during normal brain activity has yet to be proven.

In summary, the mechanisms of pH_i regulation in mammalian glia are becoming better understood and several general features are now apparent. Like neurons, and most other cells, they actively extrude acid to maintain their pH_i at more alkaline levels than dictated by the H^+ equilibrium potential. They robustly express electrogenic Na^+/HCO_3^- cotransport, and this distinguishes them from neurons, which tend not express this transport mechanism. Glial cells, but not neurons, show DIA. A tantalizing form of pH_i-mediated regulatory interaction between glia and neurons is suggested by these facts, and its further investigation remains a central challenge in the field of pH and brain function.

ACKNOWLEDGMENTS

This work was supported by NIH grant NS 15589.

REFERENCES

Amato A, Ballerini L, Attwell D (1994): Intracellular pH changes produced by glutamate uptake in hippocampal slices. N Neurophysiol 72: 1686–1696.

Amos BI, Mathie A, Richards CD, Chesler M (1996): Astrocyte intracellular pH is increased by activation of metabotropic glutamate receptors. Soc Neurosci Abstr 22: 322.8.

Arenander AT, Lim R, Varnum B, Cole R, Heschmann HR, Derellis J (1988): Astrocyte response to growth factors and hormones—early molecular events. In Reier PJ, Bunge RP, Seil FJ (eds): Current Issues in Neural Regeneration Research, Vol. 48. New York: Alan R. Liss, pp. 257–269.

Aronson PS (1985): Kinetic properties of the plasma membrane Na^+/H^+ exchanger. Ann Rev Physiol 47: 545–560.

Astion ML, Orkand RK (1988): Electrogenic Na^+/HCO_3^- cotransport in neuroglia. Glia 1: 355–357.

Astion ML, Chvatal A, Orkand RK (1989): Na^+/H^+ exchange in glial cells of Necturus optic nerve. Neurosci Lett 107: 167–172.

Astion ML, Chvatal A, Orkand RK (1991): Further studies of electrogenic Na^+/HCO_3^- cotransport in glial cells of Necturus optic nerve: Regulation of pH_i. Glia 4: 461–468.

Balestrino M, Somjen GG (1988): Concentration of carbon dioxide, interstitial pH and synaptic transmission in hippocampal formation of the rat. J Physiol (Lond) 93: 161.

Ballanyi K, Muckenhoff K, Bellingham MC, Okada Y, Scheid P, Richter DW (1994): Activity-related pH changes in respiratory neurones and glial cells of cats. Neuroreport 6: 33–36.

Benos DJ (1982): Amiloride: A molecular probe of sodium transport in tissues and cells. Am J Physiol 242: C131–C145.

Benos DJ, Sapirstein VS (1983): Characteristics of an amiloride-sensitive sodium entry pathway in cultured rodent glial and neuroblastoma cells. J Cell Physiol 116: 213–220.

Blackshaw SE, Warner AE (1976): Alterations in resting membrane properties during neural plate stages of development of the nervous system. J Physiol (Lond) 255: 231–247.

von Blankenfeld G, Kettenmann H (1991): Glutamate and GABA receptors in vertebrate glial cells. Molec Neurobiol 5: 31–43.

Bouvier M, Szatkowski A, Amato A, Attwell D (1992): The glial cell glutamate uptake carrier counter transports pH-changing anions. Nature 360: 471–474.

Boyarski G, Ransom B, Schlue WR, Davis MBE, Boron WF (1993): Intracellular pH regulation in single cultured astrocytes from rat forebrain. Glia 8: 241–248.

Breckenridge LJ, Warner AE (1982): Intracellular sodium and the differentiation of amphibian embryonic neurones. J Physiol (Lond) 332: 392–413.

Brookes N, Turner RJ (1993): Extracellular potassium regulates the glutamine content of astrocytes: Mediation by intracellular pH. Neurosci Lett 160: 73–76.

Brookes N, Turner RJ (1994): K^+-induced alkalinization in mouse cerebral astrocytes is mediated by reversal of electrogenic Na^+/HCO_3^- cotransport. Am J Physiol 36: C1633–C1640.

Brune T, Deitmer JW (1995): Intracellular acidification and Ca^{2+} transients in cultured rat cerebellar astrocytes evoked by glutamate agonists and noradrenaline. Glia 14: 153–161.

Brune T, Fetzer S, Kakus KH, Deitmer JW (1994): Evidence for electrogenic sodium-bicarbonate cotransport in rat cerebellar astrocytes. Pflugers Arch 429: 64–71.

Busa WB (1986): Mechanisms and consequences of pH-mediated cell regulation. Ann Rev Physiol 48: 389–402.

Busa WB, Nuccitelli R (1984): Metabolic regulation via intra-cellular pH. Am J Physiol 246: R409–438.

Cabantchik ZI, Greger R (1992): Chemical probes for anion transporters of mammalian cell membranes. Am J Physiol 262: C803–C827.

Cabantchik ZI, Rothstein A (1972): The nature of the membrane sites controlling anion permeability of human red blood cells as determined by studies with disulfonic stilbene derivatives. J Membr Biol 10: 311–330.

Charles AC, Naus CCG, Zhu D, Kidder GM, Dirksen ER, Sanderson MJ (1992): Intercellular calcium signaling via gap junctions in glioma cells. J Cell Biol 118: 195–201.

Chen JCT, Chesler M (1992): Extracellular alkaline shifts in rat hippocampal slice are mediated by NMDA and non-NMDA receptors. J Neurophysiol 68: 342–344.

Chesler M (1990): The regulation and modulation of pH in the nervous system. Progr Neurobiol 34: 401–427.

Chesler M, Kaila K (1992): Modulation of pH by neuronal activity. Trends Neurosci 15: 396–403.

Chesler M, Kraig RP (1987): Intracellular pH of astrocytes increases rapidly with cortical stimulation. Am J Physiol 253: R666–R670.

Chesler M, Kraig RP (1989): Intracellular pH transients of mammalian astrocytes. Neuroscience 9: 2011–2019.

Chow SY, Yen-Chow YC, White HS, Woodbury DM (1991): pH regulation after acid load in primary cultures of mouse astrocytes. Dev Brain Res 60: 69–78.

Clark JD and Limbird LE (1991): Na^+/HCO_3^- exchanger subtypes—a predictive review. Am J Physiol 261: C945-C953.

Cull-Candy SG, Wyllie JA, Traynelis SI (1991): Excitatory amino acid-gated channel types in mammalian neurones and glia. In: Excitatory Amino Acids and Synaptic Function. New York. Academic Press.

Deitmer JW, Rose CR (1996): pH regulation and proton signaling by glial cells. Prog Neurobiol 48: 73–103.

Deitmer JW, Schlue WR (1987): The regulation of intracellular pH by identified glial cells and neurones in the central nervous system of the leech. J Physiol (Lond) 388: 261–283.

Deitmer JW, Schlue WR (1989): An inwardly directed electrogenic sodium-bicarbonate co-transport in leech glial cells. J Physiol (Lond) 411: 179–194.

Draves DJ, Mauley NB, Timiras PS (1986): Glial hormone receptors: Thyroid hormones and microtubules in gliomas and neuroblastomas. In Federoff S, Vernadakis A (eds): Astrocytes: Cell Biology and Pathology of Astrocytes, Vol. 3. Orlando, FL: Academic Press, pp 183–201.

Erecinska M, Nelson D, Dagani F, Deas J, Silver IA (1993): Relations between intracellular ions and energy metabolism under acidotic conditions: A study with nigericin in synaptosomes, neurons, and C6 glioma cells. J Neurochem 61: 1356–1368.

Fedoroff S (1995): Development of Microglia. In Ransom BR, Kettenmann H (eds): Neuroglia. New York: Oxford University Press, pp 162–181.

Ganz MB, Boron WF (1994): Long-term effects of growth factors on pH and acid-base transport in rat glomular mesengial cells. Am J Physiol 266: F576–F585.

Ganz MB, Boyarski G, Sterzel RB, Brown WF (1989): Arginine vasopressin enhances pH_i regulation in the presence of HCO_3^- by stimulating three acid-base transport systems. Nature 337: 648–651.

Geibel J, Giebisch G, Boron WF (1990): Angiotensin II stimulates both Na^+/H^+ exchange and Na^+/HCO_3^- cotransport in the rabbit proximal tubule. Proc Natl Acad Sci USA 87: 7917–7920.

Grichtchenko II, Chesler M (1994a): Depolarization-induced alkalinization of astrocytes in gliotic hippocampal slices. Neuroscience 62: 1071–1078.

Grichtchenko II, Chesler M (1994b): Depolarization-induced acid secretion in gliotic hippocampal slices. Neuroscience 62: 1057–1070.

Grinstein S, Rothstein A (1986): Mechanisms of regulation of the Na^+/H^+ exchanger. J Membr Biol 90: 1–2.

Grinstein D, Rotin D, Mason MJ (1989): Na^+/H^+ exchange and growth factor-induced cytosolic pH changes. Role in cellular proliferation. Biochim Biophys Acta 988: 73–97.

Hösli E, Hösli S (1993): Receptors for neurotransmitters on astrocytes in the mammalian central nervous system. Progr Neurobiol 40: 477–506.

Isom LL, Cragoe EJ, Limbird LE (1987): (2-adrenergic receptors accelerate Na^+/H^+ exchange in neuroblastoma x glioma cells. J Biol Chem 262: 6750–6757.

Jacubovicz DE, Grinstein S, Klip A (1987): Cell swelling following recovery from acidification in C6 glioma cells: An in vitro model of post-ischemic brain edema. Brain Res 435: 138–146.

Jean T, Frelin C, Vigne P, Lazdunski M (1986): The Na^+/H^+ exchange system in glial cell lines. Properties and activation by an hyperosmotic shock. Eur J Biochem 160: 211–219.

Kaila K, Voipio J (1987): Postsynaptic fall in intracellular pH induced by GABA-activated bicarbonate conductance. Nature 330: 163–165.

Kaila K, Panula P, Karhunen T, Heinonen E (1991): Fall in intracellular pH mediated by $GABA_A$ receptors in cultured rat astrocytes. Neurosci Lett 126: 9–12.

Kettenmann H, Ransom BR (eds) (1995): Neuroglia. New York: Oxford University Press.

Kettenmann H, Schlue WR (1988): Intracellular pH regulation in cultured mouse oligodendrocytes. J Physiol (Lond) 406: 147–162.

Kimelberg HK (1981): Active accumulation and exchange transport of chloride in astroglial cells in culture. Biochim Biophys Acta 646: 179–184.

Kimelberg HK. Biddlecome S, Bourke RS (1979): SITS-inhibitable Cl^- transport and Na^+-dependent H^+ production in primary astroglial cultures. Brain Res 173: 111–124.

Kimelberg HK, Bowman CL, Hirata H (1986): Anion transport in astrocytes. Ann N Y Acad Sci 481: 334–353.

Lagadic-Gossmann D, Vaughan-Jones RD, Buckler KY (1992): Adrenaline and extracellular ATP switch between two modes of acid extrusion in the guinea pig ventricular myocyte. J Physiol (Lond) 458: 385–407.

Lee JH, Vollmer TL, Ransom BR (1993): Extracellular pH and epileptiform activity in rat hippocampal slices. Soc Neurosci Abstr 19: 1870.

Lee JH, Taira T, Pihlaja K, Ransom BR, Kaila K (1996): Effects of CO_2 on excitatory transmission in the rat hippocampal slice. Brain Res 706: 210–216.

Manager TM, Koeppen BM (1992): Characterization of acid base transporters in cultured outer medullary collecting duct cells. Am J Physiol 263: F996–F1003.

Marrero H, Orkand RK (1996): Nerve impulses increase glial intercellular permeability. Glia 16: 285–289.

Martin DH (1995): The role of glia in the inactivation of neurotransmitters. In Kettenmann H, Ransom BR (eds): Neuroglia. New York: Oxford University Press, pp 732–745.

Mellergard PE, Siesjö BK (1991): Astrocytes fail to regulate intracellular pH at moderately reduced extracellular pH. Neuroreport 2: 695–698.

Mellergard PE, Ou-Yang YB, Siesjö BK (1992): The regulation of intracellular pH in cultured astrocytes and neuroblastoma cells, and its dependence on extracellular pH in a HCO_3^- free medium. Can J Physiol Pharmacol 70: S293–S300.

Mellergard PE, Ou-Yang YB, Siesjö BK (1993): Intracellular pH regulation in cultured rat astrocytes maintained in CO_2/HCO_3^--containing media. Exp Brain Res 95: 371–380.

Moolenaar WH (1986): Effects of growth factors on intracellular pH regulation. Ann Rev Physiol 48: 363–376.

Nakhou NL, Abdoulnour-Nakhoul S, Khuri R, Lieberman EM, Hargittai PT (1994): Intracellular pH regulation in rat Schwann cells. Glia 10: 155–164.

Nedergaard M, Goldman SA (1993): Carrier-mediated transport of lactic acid in cultured neurones and astrocytes. Am J Physiol 265: R282–R289.

Newman EA (1991): Sodium-bicarbonate cotransport in retinal Müller (glial) cells of the salamander. J Neurosci 11: 3972–3983.

Newman EA (1994): A physiological measure of carbonic anhydrase in Muller cells. Glia 11: 291–299.

Newman EA (1995): Glial cell regulation of extracellular potassium. In Kettenmann H, Ransom BR (eds): Neuroglia. New York, Oxford University Press, New York, pp 717–731.

Newman EA (1996): Acid efflux from retinal glial cells generated by sodium bicarbonate cotransport. J Neurosci 16: 159–168.

Newman EA, Astion ML (1991): Localization and stoichiometry of electrogenic sodium bicarbonate cotransport in retinal glial cells. Glia 4: 424–428.

Nicholson C (1995): Extracellular space as the pathway for neuron-glial cell interaction. In Kettenmann H, Ransom BR (eds): Neuroglia. New York, Oxford University Press, pp 377–384.

Noel J, Pouyssegur J (1995): Hormonal regulation, pharmacology, and membrane sorting of vertebrate Na^+/H^+ exchanger isoforms. Am J Physiol 268: C283–C296.

O'Connor E, Sontheimer H, Ransom BR (1994): Rat hippocampal astrocytes exhibit electrogenic sodium-bicarbonate co-transport. J Neurophysiol 72: 2580–2589.

Paalasmaa P, Taira T, Voipio J, Kaila K (1994): Extracellular alkaline transients mediated by glutamate receptors in the rat hippocampal slice are not due to a proton conductance. J Neurophysiol 72: 2031–2033.

Pappas CA, Ransom BR (1993): A depolarization-stimulated bafilonycin-inhibitable H^+ pump in hippocampal astrocytes. Glia 9: 280–291.

Pappas CA, Ransom BR (1994): Depolarization-induced alkalinization (DIA) in rat hippocampal astrocytes. J Neurophysiol 72: 2816–2826.

Pappas CA, Ullrich N, Sontheimer H (1994): Reduction of glial proliferation by K^+ channel blockers is mediated by changes in pH_i. Neuroreport 6: 193–196.

Pappas CA, Rioult MG, Ransom BR (1996): Octanol, a gap junction uncoupling agent, changes intracellular $[H^+]$ in rat astrocytes. Glia 16: 7–15.

Pearce B, Wilkin GP (1995): Eicosanoid, purine and hormone receptors. In Kettenmann H, Ransom BR (eds): Neuroglia. New York: Oxford University Press, pp 377–384.

Poitry-Yamate CL, Poitry S, Tsacopoulos M (1995): Lactate released by Muller cells is metabolized by photoreceptors from mammalian retina. J Neurosci 15: 5179–5191.

Ransom BR (1991): Vertebrate glial classification, lineage, and heterogeneity. Ann N Y Acad Sci 633: 19–26.

Ransom BR (1992): Glial modulation of neural excitability mediated by extracellular pH: A hypothesis. In Yu ACH, Hertz L, Norenberg MD, Sykova E, Waxman SG (eds): Progress in Brain Res, Vol 94. pp 37–46.

Ransom BR (1995): Gap junctions. In Kettenmann H, Ransom BR (eds): Neuroglia. New York: Oxford University Press, pp 299–318.

Ransom BR, Fern R (1997): Does astrocytic glycogen benefit axon function and survival in CNS white matter during glucose deprivation? Glia (in press).

Ransom BR, Carlini WG, Connors BW (1986): Brain extracellular space: Developmental studies in the rat optic nerve. Ann N Y Acad Sci 481: 87–105.

Ransom BR (1996): Do glial gap junctions play a role in extracellular ion Homeostasis? In Spray DC, Dermietzel R (eds): Gap Junctions in the Nervous System, Austin: R. G. Landes: PP. 159–173.

Rose CR, Ransom BR (1996a): Mechanisms of H^+ and Na^+ changes induced by glutamate, kainate, and D-aspartate in rat hippocampal astrocytes. J Neurosci 16: 5393–5404.

Rose CR, Ransom BR (1996b): Intracellular sodium homeostasis in rat hippocampal astrocytes. J Physiol (Lond) 491: 291–305.

Rose C, Ransom BR (1997): Gap junctions equalize intracellular Na^+ concentrations in astrocytes. Glia 20: 299–307.

Roos A and Boron WF (1981): Intracellular pH. Physiol Rev 61: 296–433.

Rosoff PM, Cantley LC (1983): Increasing the intracellular Na^+ concentration induces differentiation in a pre-B lymphocyte cell line. Proc Natl Acad Sci USA 80: 7547–7550.

Sapirstein VS, Benos DJ (1984): Influence of hormones on Na^+/H^+ exchanger. J Neurochem 43: 1098–1105.

Saunders RD, Brandon YW, De Vries GH (1989): Role of intracellular pH in the axolemma- and myelin-induced proliferation of Schwann cells. J Neurochem 52: 1576–1581.

Schwiening CJ, Kennedy HJ, Thomas RC (1993): Calcium-hydrogen exchange by plasma membrane Ca-ATPase of voltage-clamped snail neurons. Proc R Soc Lond B 283: 285–289.

Shrode LD, Putnam RW (1994): Intracellular pH regulation in primary rat astrocytes and C6 glioma cells. Glia 12: 196–210.

Siebens AW, Boron WF (1989): Depolarization induced alkalinization in proximal tables. 1. Characteristics and dependence on Na^+. Am J Physiol 25: F342–F353.

Smith SE, Gottfried JA, Chen JCT, Chesler M (1994): Calcium dependence of glutamate receptor-evoked alkaline shifts in hippocampus. Neuroreport 5: 443–470.

Staub F, Baethmann A, Peters J, Weight H, Kempski O (1990): Effects of lactic acidosis on glial cell volume and viability. J Cereb Blood Flow Metab 10: 866–876.

Tang CM, Dichter M, Morad M (1990): Modulation of the *N*-methyl-d-aspartate channel by extracellular H^+. Proc Natl Acad Sci USA, 87: 6445–6449.

Thomas RC (1984): Experimental displacement of intracellular pH and the mechanism of its subsequent recovery. J Physiol (Lond) 354: 3P–22P.

Tombaugh GC, Somjen GG (1996): Effects of extracellular pH on voltage-gated Na^+ and Ca^{2+} currents in isolated rat CAl neurons. J Physiol (Lond) 493: 719–732.

Wakabayoshi S, Sandet C, Fafouymoux P, Counillon L, Meloche S, Pages G, Pouyssegur J (1991): Structure and function of the growth factor-activatable Na^+/H^+ exchanger (NHE l). Rev Physiol Biochem Pharmacol 119: 157–186.

Walz W, Hertz L (1983): Intracellular ion changes of astrocytes in response to extracellular potassium. J Neurosci Res 10: 411–423.

Walz W, Mukerji S (1988a): Lactate release from cultured astrocytes and neurones: A comparison. Glia 1: 366–370.

Walz W, Mukerji S (1988b): Lactate production and release in cultured astrocytes. Neurosci Lett 86: 296–300.

Walz W, Schlue WR (1982): External ions and membrane potential of leech neuropile glial cells. Brain Res 239: 119–138.

Wuttke WA, Walz W (1990): Sodium-independent and bicarbonate-independent regulation of intracellular pH in cultured mouse astrocytes. Neurosci Lett 117: 105–110.

CHAPTER 15

pH Consequences of Calcium Regulation

CHRISTOF J. SCHWIENING AND ROGER C. THOMAS
Department of Physiology, University of Cambridge, Downing Site, Cambridge CB2 3EG, United Kingdom

IMPORTANCE OF INTRACELLULAR CALCIUM

Calcium plays a pivotal physiological role in the regulation of many cellular processes (see review by Tsien and Tsien, 1990). Intracellular calcium transients are associated with secretion, changes in electrical activity, cell proliferation, growth, movement and so on. However, these calcium transients do not occur in isolation from other ions: changes in intracellular pH also frequently occur. Chapter 16–19 of this book deal with activity-dependent changes in pH, some of which are dependent on calcium influx. In this chapter we attempt to describe how calcium and hydrogen ions interact during calcium regulation in neurons. In particular we will concentrate on the plasma membrane calcium pump.

COMPARISON OF CALCIUM AND HYDROGEN IONS

Intracellular hydrogen ion concentration ($[H^+]_i$) and buffering are quite different from that of Na^+, K^+, and Cl^-, but there are some similarities with calcium (see Table 1). Calcium has a low intracellular concentration and is only weakly buffered. The large transmembrane gradient for calcium and poor intracellular buffering allows for large and rapid excursions of intracellular calcium concentration ($[Ca^{2+}]_i$). Intracellular hydrogen ions are well buffered and there is no large transmembrane gradient so most intracellular pH (pH_i) changes are small and slow. Hence pH_i and calcium transients look very different, and any linkage between the two ion species will produce transients that look dissimilar. The mechanisms producing the linkage between calcium and pH have different sensitivities to the two ions so that there is little symmetry in the interactions. A change in intracellular calcium may cause a pH_i transient but a pH_i transient may well have no effect on intracellular calcium.

pH and Brain Function, Edited by Kai Kaila and Bruce R. Ransom
ISBN 0-471-11838-9

TABLE 1 Important Features of Cellular Calcium and pH

Feature	Ca^{2+}	H^+
Usually expressed as	$[Ca^{2+}]$	pH ($=-\log[H^+]$)
Extracellular activity	~5mM (pCa 2.3)	~pH 7.4 (40 nM)
Intracellular activity	~30nM (pCa 7.5)	~pH 7.2 (60 nM)
Equilibrium potential	~+150 mV	~+10 mV
Intrinsic intracellular buffering power at normal intracellular activity	~40 μM Ca^{2+}/pCa	~10mM H^+/pH
Intrinsic intracellular buffering power during changes in ionic activity	Increases dramatically with increasing $[Ca^{2+}]_i$	Fairly independent of pH_i
Plasma membrane channels	Many types of voltage and agonist gated	Only outwardly rectifying voltage gated
Lipid permeability	Negligible	Negligible as H^+, but high as weak acids or bases (e.g., CO_2)
Uptake into stores	Yes; mitochondria and endoplasmic reticulum	Some uptake by vesicles
Release from stores	IP_3 and calcium-induced calcium release	No
Regulation by plasma membrane pumps	ATP-dependent pump, Na^+/Ca^{2+}	Na^+-dependent Cl^-/HCO_3^- exchange, Na^+/H^+
Second messenger	Well-established role in many cells	Possible limited role

CALCIUM REGULATION

The long-term maintenance of low intracellular calcium levels is essential for normal cell functioning. Pioneering work on the squid giant axon (Blaustein and Hodgkin, 1969; DiPolo et al., 1979) demonstrated the presence of two plasma membrane pumps capable of doing this—the Na^+/Ca^{2+} exchanger and the Ca^{2+}-ATPase. The Na^+/Ca^{2+} exchanger was believed to have a low affinity, but a high capacity, for calcium transport, while the Ca^{2+}-ATPase had a high affinity for calcium, but a low capacity. The activity of the Na^+/Ca^{2+} exchanger is low at resting calcium levels and the Ca^{2+}-ATPase is thought to set the resting intracellular calcium level. However, the Na^+/Ca^{2+} exchanger was thought to be the dominant process for extruding calcium loads. More recent work on calcium regulation in neuronal preparations has made the picture less clear. In all studies of calcium regulation in synaptic terminals and axons, the Na^+/Ca^{2+} exchanger has been found to play a significant role in calcium regulation. Most of these studies were of ^{45}Ca fluxes in rat brain synaptosomes (Blaustein and Oborn, 1975; Sanchez-Armass and Blaustein, 1987; Dahan et al., 1991). Some workers have also used fluorescent dyes (Nachshen, 1986; Carvalho et al., 1991). Gill et al. (1981) showed that inverted plasma membrane vesicles derived from guinea pig synaptosomes, which are devoid of internal organelles, also exhibit Na^+/Ca^{2+} exchange. However, many studies of calcium regulation in the neuronal cell body show little or no sign of Na^+/Ca^{2+} exchange: bullfrog sympathetic ganglion cells (Nohmi and Kuba, 1984); *Aplysia* neurons (Levy and Tillotson, 1988); rat dorsal root ganglion cells (Thayer and Miller, 1990); dissociated mouse sensory neurons (Duchen et al., 1990), and cultured rat and chick sensory neurons (Benham et al., 1992; Mironov et al., 1993); embryonic mouse hippocampal neurons (Koch and Barish, 1991); snail neurons (Schwiening et al., 1993;

Kennedy and Thomas, 1995); cultured rat septal neurons (Bleakman et al., 1993). It is therefore possible that Na^+/Ca^{2+} exchange is important only in nerve terminals and not in the cell body (Miller, 1991). The Ca^{2+}-ATPase (calcium pump) would thus be the main mechanism operating within the nerve cell body, and if $[Ca^{2+}]_i$ is to remain low in the long term then all the calcium entering must leave by it.

INTERACTIONS BETWEEN CALCIUM AND HYDROGEN IONS

It has been known for some time that the uptake of cations by isolated mitochondria can occur in exchange for H^+ (Bartley and Amoore, 1958; Chappell et al., 1962; Thomas et al., 1969). In 1977, Meech and Thomas showed that the pressure injection of large amounts of calcium (~4mM Ca^{2+} per liter of cell volume) into snail neurons caused an intracellular acidification, which they concluded was due to calcium being exchanged for H^+ inside the cell. Meech and Thomas (1980) presented further evidence that this acidification was due to the mitochondrial release of hydrogen ions, but they also mentioned a second Ca^{2+}/H^+ exchange mechanism, which they thought was passive binding to calcium buffers. After much further work, it now appears that there may be at least four distinct sites for Ca^{2+}/H^+ interactions in neurons. A rise in intracellular calcium may cause hydrogen ions to be displaced from intracellular calcium buffers, released from mitochondria, exchanged for calcium at the endoplasmic reticulum, and/or exchanged for calcium at the plasma membrane.

Intracellular Buffers

Following reports of intracellular pH changes during electrical activity and intracellular cation injection, various workers suggested that intracellular buffers might bind either Ca^{2+} or H^+ at a single site and that this would cause a reciprocal relationship between $[Ca^{2+}]_i$ and $[H^+]_i$ (Busa and Nuccitelli, 1984). Despite the much lower buffering power for calcium than hydrogen, changes of pH_i seem to produce little or no change in $[Ca^{2+}]_i$. It is possible that such buffers are only present in low concentrations and are of little physiological importance in comparison to the other acidifying mechanisms. Indeed Ou-Yang et al. (1994). who studied the effect of acid-base changes on intracellular calcium in neurons, made no mention of intracellular buffers capable of binding calcium or hydrogen ions.

Mitochondria

That there is exchange or release of H^+ by mitochondria during large calcium loads is beyond doubt; but there is some question as to the physiological relevance of this process. There is a growing body of evidence that mitochondria may be involved in muffling (Thomas et al., 1991) large but physiological Ca^{2+} loads (see Gunter et al., 1994 for a review), as well as preventing large changes in $[Ca^{2+}]_i$ during and following glutamate excitotoxicity (Schinder et al., 1995). Mitochondria may well be the source of the large acid loads (0.5 pH unit) seen during exposure to nonlethal concentrations of N-methyl-D-aspartate (NMDA) in cultured fetal rat hippocampal neurons (Irwin et al., 1994). Werth and Thayer (1994) suggested that mitochondria are involved in sequestering Ca^{2+} loads following more than 25 action potentials in cultured rat dorsal root ganglion cells. Figure 1

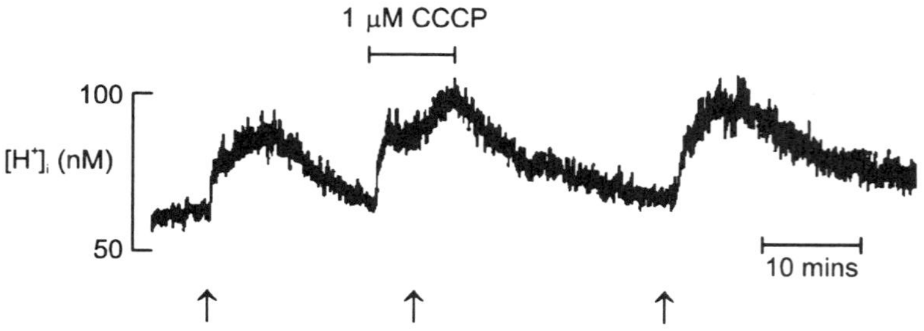

FIGURE 1. Partial inhibition of depolarization-induced acidifications by the mitochondrial uncoupler CCCP. Arrows mark 30-second application of 50 mM K^+ (reproduced from Werth and Thayer, 1994).

shows a representative trace of $[H^+]_i$ (reproduced from Werth and Thayer, 1994) during three 30-sec depolarizations with high potassium solutions. The depolarizations caused a rise in $[H^+]_i$ from ~60 nM (pH 7.22) to ~90 nM (pH 7.5). Mitochondrial uncouplers (such as carbonyl cyanide-m-chlorophenyl hydrazone (CCCP)) and blockers reduced this pH change. Both Werth and Thayer (1994) and Irwin et al. (1994) failed to discuss the possibility that the pH changes that they recorded could have been due to plasma membrane exchange of calcium for hydrogen ions. Also, probably unknown to them, they were recording intracellular pH using a blocker of the calcium pump (see section on calcium pump inhibitors and pH effects of the calcium pump).

Endoplasmic Reticulum

Intracellular Ca^{2+} accumulation by membrane-bound vesicles isolated from muscle cells was first shown by Hasselbach and Makinose in 1961. The calcium uptake consumed ATP and was by nonmitochondrial stores, presumably the sarcoplasmic reticulum. Calcium uptake by neuronal endoplasmic reticulum (ER) was first demonstrated by Otsuka, Ohtsuki, and Ebashi (1965). Henkart, Reese, and Brindley (1978) showed that squid axon ER could sequester Ca^{2+} at physiological levels. There is some evidence that in muscle cells the sarcoplasmic reticulum calcium pump countertransports H^+ (Chiesi and Inesi, 1980), although there is no evidence we know of from neurons. In 1984, Neering and McBurney showed in dorsal root ganglion cells that caffeine, which is known to potentiate calcium-induced calcium release, caused Ca^{2+} to be released from intracellular stores. The ER can therefore both accumulate and release calcium following a calcium load. However, there has been little work on the role of the ER in calcium muffling. Recent work on snail neurons (Orkand and Thomas, 1995) suggests that the rise in calcium during long, small depolarizations can be augmented by calcium-induced calcium release, while that caused by short, large depolarizations is not. In conclusion, although it is clear that interfering with the neuronal ER can influence calcium regulation, there is no information about the pH_i consequences of ER calcium handling.

Plasma Membrane Calcium Pump

Calcium-dependent ATPase activity was first described by Dunham and Glynn (1961) in human red cell ghosts. In 1966, Schatzmann showed that an ATPase pumped calcium out of red blood cells. Since then the Ca^{2+}-ATPase has been found in all cells examined. It has been purified from erythrocyte membrane (Niggli et al., 1979), its sequence has been deduced from rat brain cDNA (Shull and Greeb, 1988), and several isoforms have been described (Shull and Greeb, 1988, 1989). The hydrolysis of ATP liberates hydrogen ions, a process that might have some intracellular pH consequences. However, it is also known that the calcium pump counter-transports hydrogen ions in red blood cells (Carafoli and Zurini, 1982; Niggli, et al., 1982) and across the sarcolemma (Dixon and Haynes, 1989). Researchers working on neuronal pH regulation seemed to be unaware of the possibility of any pH effects of calcium extrusion, although some were aware of the reverse: pH effects on calcium extrusion. In the next two sections we deal with pH effects on, and of, the calcium pump, and we show how they are related to the movement of acid equivalents.

pH effects on the calcium pump. The first description of pH effects on the calcium pump came from Dipolo and Beaugé (1982), who showed that high extracellular pH (pH 8.8) inhibited sodium-independent calcium efflux from dialyzed squid axons. They proposed that high pH would increase the calcium affinity of an external site so as to favor reverse reaction of the pump, thereby causing a decrease in the forward reaction. Work at the same time on red blood cells (see following section) suggested a thermodynamic role for the inhibition of calcium pump activity by high extracellular pH. In 1992, Benham et al. used the inhibition of the plasma membrane calcium pump by high pH to demonstrate its importance in calcium homeostasis in rate sensory neurons. High extracellular pH also raises basal $[Ca^{2+}]_i$ and causes incomplete calcium recovery following a calcium load in goldfish retinal bipolar cells (Kobayashi and Tachibana, 1995) and snail neurons, as shown in Fig. 2. This shows an experiment where a snail neuron was voltage-clamped and injected with two fluorescent dyes. The neuron was depolarized to various potentials at normal and high extracellular pH. The recovery of intracellular calcium during exposure to high extracellular pH was both incomplete and greatly slowed. Unpublished work by C. J. Schwiening, H. J. Kennedy and R. C. Thomas on snail neurons also suggests that

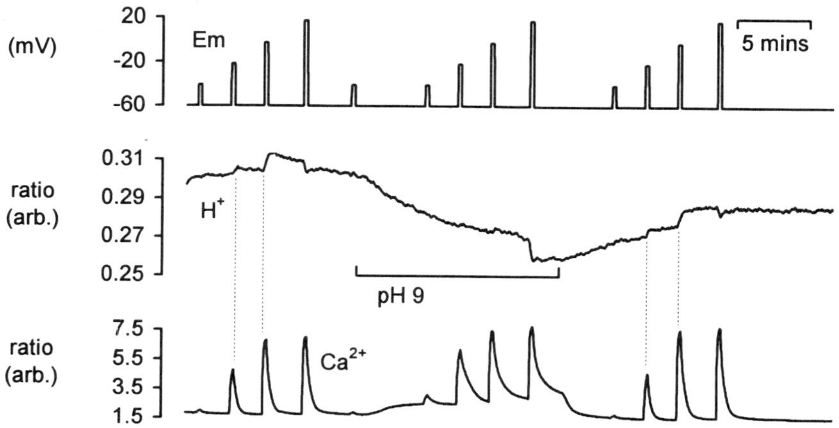

FIGURE 2. The effect of brief depolarization on intracellular $[H^+]$ and $[Ca^{2+}]$ transients at normal and high extracellular pH. Intracellular $[H^+]$ was measured with BCECF-dextran, which was co-injected with Fura-2. Both fluorescence ratios are plotted so that increasing ionic concentration are upwards (C. J. Schwiening, unpublished).

the high pH inhibition of the calcium pump may become less effective with time. Pumping of calcium out of snail neurons is not maximal at normal extracellular pH. Extracellular acidification to pH 6.5 increases the rate of intracellular calcium recovery from a calcium load. In contrast, little is known about the effect of intracellular pH on the plasma membrane calcium pump.

pH effects of the calcium pump. Although work on squid axons showed that calcium extrusion was affected by pH, there was no evidence that calcium extrusion itself changed pH. There are two reasons why such pH changes might not have been seen or ascribed to calcium extrusion. First, in the heyday of the recessed-tip pH-sensitive microelectrode, Meech and Thomas (1977) had reported intracellular pH changes following large calcium injection into snail neurons. These pH changes were believed to be due to mitochondrial release of protons following calcium uptake. Thus mitochondria became the main focus of intracellular interactions between calcium and hydrogen ions. Although, Ahmed and Connor (1980) reported intracellular pH changes induced by calcium influx they did not speculate upon the source of the protons. Secondly, when the fluorescent pH indicators became available workers were unaware that these compounds may also inhibit the calcium pump (see section on calcium pump inhibitors).

Experiments on red blood cells suggested a direct coupling between hydrogen ion and calcium fluxes. Niggli, Siegel, and Carafoli (1982) demonstrated that the purified calcium pump from red blood cells countertransported hydrogen ions. However, Rossi and Schatzmann (1982) were adamant that this was not the case: "The hypothesis of a Ca^{2+}-proton exchange, for which no compelling evidence has been brought forward so far, can therefore be dismissed." In 1983, Smallwood and coworkers provided more evidence for the countertransport of hydrogen ions on the red blood cell calcium pump. Unfortunately, those working on intracellular pH regulation in nerve cells must have either been unaware of the transport of H^+ on the calcium pump (see pH_i reviews by Thomas, 1984; Frelin et al., 1988) or have assumed that the calcium pump was not involved in calcium regulation in their preparation.

The first direct evidence for the countertransport of acid equivalents on the neuronal plasma membrane calcium pump did not appear until much later. Schwiening, Kennedy, and Thomas (1993) discovered that it was possible to use calcium-sensitive microelectrodes to detect calcium being pumped out of snail neurons. Extracellular calcium transients only became visible when extracellular calcium was lowered. They found that the only mechanism by which calcium was extruded out of snail neurons was the calcium pump. They also showed that alkaline extracellular pH transients occurred on snail neurons following increases in intracellular calcium. They raised intracellular calcium by depolarization and by iontophoretic calcium injection. Such alkalinizations following depolarization have been observed previously. They were tentatively ascribed to the inward movement of protons through voltage-gated channels (Thomas, 1988), and later to the exposure of external binding sites for hydrogen ions during calcium influx (Thomas, 1989). Both of these suggestions are incompatible with the experiment shown in Fig. 3. Surface alkalinizations occurred on depolarizing snail neurons beyond about -30 mV. In Fig. 3 it can be seen that depolarization to ~ -20 mV caused a surface alkalinization that continued for some time following repolarization. During this period proton channels were closed, and there was no flux of calcium through voltage-gated channels. Depolarization to more positive potentials opened outwardly rectifying proton channels, causing transient surface acidifications. The increases in extracellular pH were inhibited by intracellu-

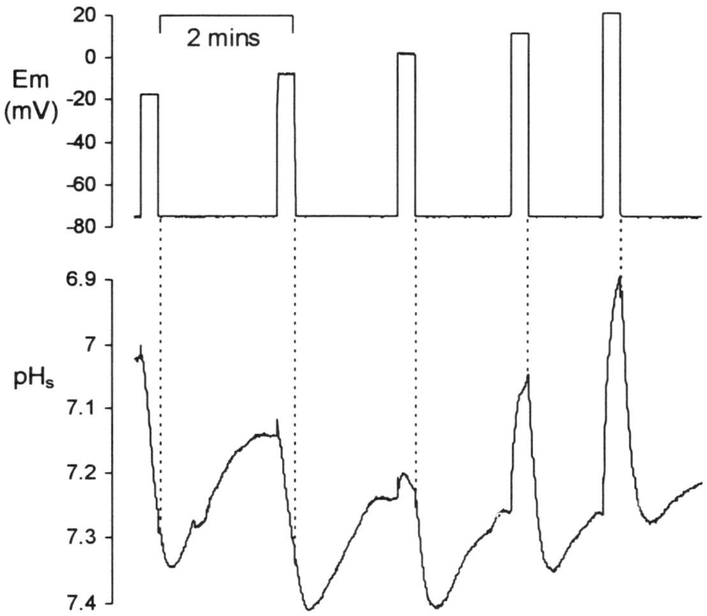

FIGURE 3. Surface pH changes measured with a pH-sensitive microelectrode from a snail neuron in response to depolarizing voltage steps. Surface alkalinizations occurred during depolarization to ~−20 mV and ~−10 mV and after repolarization from all test potentials. Surface acidifications occurred only during depolarization to positive potentials (C. J. Schwiening, unpublished data).

lar vanadate, as was the extrusion of calcium. This led to the conclusion that the calcium pump in snail neurons countertransported acid equivalents, presumably H^+.

The stoichiometry of what is perhaps better referred to as the Ca^{2+}/H^+ pump has not been addressed directly in a neuronal preparation. The electroneutral coupling ratio of 1 Ca^{2+} to 2 H^+ seems most likely, since there have been no currents or voltage changes attributed to the Ca^{2+}/H^+ pump. Such changes would occur with any other simple coupling ratio of calcium and hydrogen ions. It is also possible that the pump might catalyze the cotransport of calcium and two OH^- or base equivalents.

The pumping of calcium out of snail neurons not only causes extracellular pH changes but also pH_i changes (see pH_i transients marked by dotted lines in Fig. 2, and Schwiening, 1994). Although pH_i transients associated with calcium entry are easy to observe experimentally, it is difficult to be sure which of the four processes outlined in this chapter are responsible for generating them. Attempting to show the involvement of the plasma membrane calcium pump by blocking it will raise $[Ca^{2+}]_i$ and increase mitochondrial hydrogen ion release. On the other hand blocking mitochondrial function first could decrease ATP levels and reduce both ER and plasma membrane calcium transport. Recent work by Trapp et al. (1996) and Paalasmaa and Kaila (1996) provides some evidence for intracellular and extracellular pH transients in the rat hippocampus as a result of calcium extrusion on the Ca^{2+}/H^+ pump. These extracellular pH changes may well be rapid enough to modulate NMDA receptor-mediated currents (Gottfried and Chesler, 1996). This raises the exciting possibility that changes in neuronal excitability and synaptic plasticity might be modulated dynamically by protons as well as calcium.

Inhibition of the calcium pump. One of the major problems in studying the calcium pump is the apparent lack of a potent or specific inhibitor. The Ca^{2+} ATPase can of course be inhibited by the reduction of ATP or the addition of vanadate, but these are not specific enough to be experimentally very useful. High extracellular pH will slow the pump (see pH effects on the calcium pump) but also has other consequences. However, in 1993, Gatto and Milanick published evidence that the red blood cell calcium pump can be inhibited fairly selectively by low concentrations of a range of common pH indicators and intracellular markers! For instance 30 μM BCECF (purchased from Calbiochem) produced a 50% inhibition of the calcium pump, while only 40–50 nM eosin was required for a similar effect. Preliminary experiments on snail neurons show that intracellularly injected eosin blocks calcium extrusion from snail neurons at very low levels. Figure 4 shows that even externally applied eosin blocks calcium regulation in a snail neuron. Eosin could therefore be a useful specific inhibitor of the calcium pump. It is unclear what concentration of pH indicators most workers have inside their cells, but many may have been inhibiting a mechanism that would have interested them. There is some evidence that BCECF-dextran does not block the calcium pump. (Fig. 2). More recent BCECF experiments by Trapp et al. (1996) demonstrate that the intracellular dialysis of 100 μM eosin inhibits a portion of the depolarization-induced intracellular acidification.

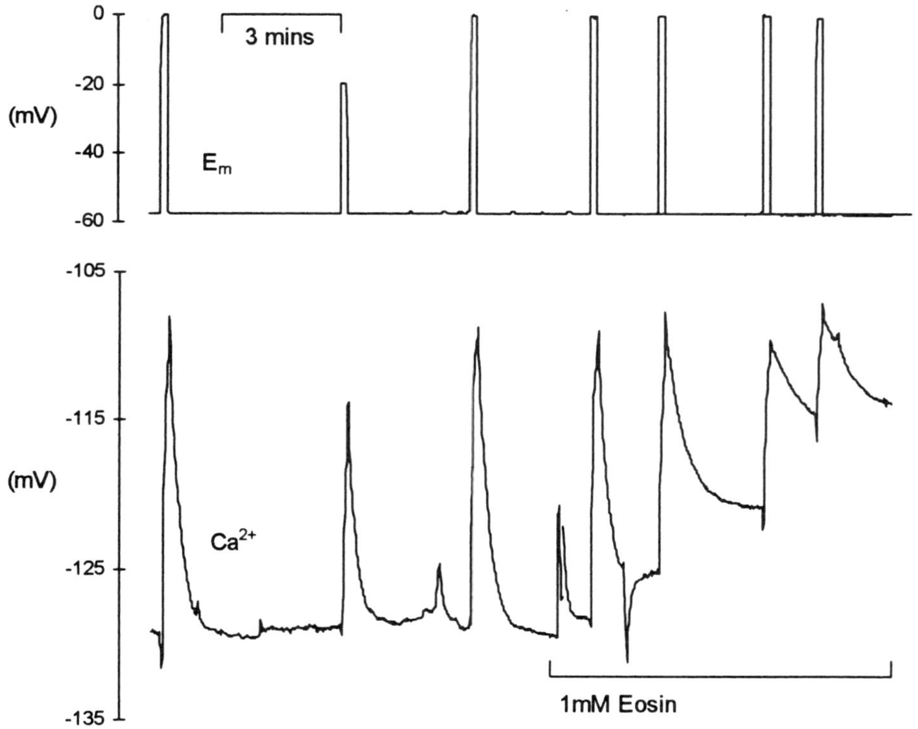

FIGURE 4. Effect of externally applied eosin on intracellular calcium transients. Intracellular calcium measured with an ion-selective microelectrode in a voltage-clamped snail neuron. Depolarization from ~−55 mV to −20 mV and to 0 mV caused a transient increase in $[Ca^{2+}]_i$. Eosin was added to the superfusate at the time shown [R. C. Thomas, unpublished].

CONCLUSION

Given that neurons probably extrude most, if not all, calcium on the Ca^{2+} pump, the countertransport of hydrogen ions will have important consequences for ionic homeostasis. In electrically excitable cells, which experience large but transient calcium loads, the plasma membrane calcium pump is potentially the main intracellular acid loader. As such, those interested in neuronal pH_i regulation, or working in "bicarbonate-free" media, will need to pay attention to the cell membrane potential. It might be especially significant if calcium channels and pumps are spatially localized. Given the slow calcium diffusion through the cytosol and the high concentrations that can occur in microdomains directly under the plasma membrane, it is possible that high rates of Ca^{2+}/H^+ pump activity might occur. This activity could generate transient pH gradients. Of course, those workers using only fluorescent pH_i indicators, which might inhibit the pump and do not provide any indication of the membrane potential, may find the study of this pump difficult.

ACKNOWLEDGMENTS

Supported by the Medical Research Council.

REFERENCES

Ahmed Z, Connor JA (1980): Intracellular pH changes induced by calcium influx during electrical activity in molluscan neurones. J Gen Physiol 75: 403–426.

Bartley W, Amoore JE (1958): The effects of manganese on solute content of rat-liver mitochondria. Biochem J 69: 348–360.

Benham CD, Evans ML, McBain CJ (1992): Ca^{2+} efflux mechanisms following depolarization evoked calcium transients in cultured rat sensory neurones. J Physiol (Lond) 455: 567–583.

Blaustein MP, Hodgkin AL (1969): The effect of cyanide on the efflux of calcium from squid axons. J Physiol (Lond) 200: 497–527.

Blaustein MP, Oborn CJ (1975): The influence of sodium on calcium fluxes in pinched-off nerve terminals in vitro. J Physiol (Lond) 247: 657–686.

Bleakman D, Roback JD, Wainer BH, Miller RJ, Harrison NL (1993): Calcium homeostasis in rat septal neurons in tissue-culture. Brain Res 600: 257–267.

Busa WB, Nuccitelli R (1984): Metabolic regulation via intracellular pH. Am J Physiol 246: R409–R438.

Carafoli E, Zurini M (1982): The Ca^{2+}-pumping ATPase of plasma membranes. Biochim Biophys Acta 683: 279–301.

Carvalho AP, Bandeira-Duarte C, Ferreira IL, Coutinho OP, Carvalho CM (1991): Sodium-calcium exchange in nerve-terminals—influence on internal Ca^{2+} and neurosecretion. Ann N Y Acad Sci 639: 300–311.

Chappell JB, Greville GD, Bicknell KE (1962): Stimulation of respiration of isolated mitochondria by manganese ions. Biochem J 84: 61P.

Chiesi M, Inesi G (1980): Adenosine 5′-triphosphate dependent fluxes of manganese and hydrogen ions in sarcoplasmic reticulum. Biochemistry 19: 2912–2918.

Dahan D, Spanier R, Rahamimoff H (1991): The modulation of rat brain Na-Ca exchange by K^+. J Biol Chem 266: 2067–2075.

DiPolo R, Beaugé L (1982): The effect of pH on Ca^{2+} extrusion mechanisms in dialyzed squid axons. Biochim Biophys Acta 688: 237–245.

DiPolo R, Rojas HR, Beaugé L (1979): Vanadate inhibits uncoupled Ca efflux but not Na-Ca exchange in squid axons. Nature 281; 228–229.

Dixon DA, Haynes DH (1989): Ca^{2+} pumping ATPase of cardiac sarcolemma is insensitive to membrane-potential produced by K^+ and Cl^- gradients but requires a source of counter-transportable H^+. J Membr Biol 112: 169–183.

Duchen MR, Valdeolmillos M, O'Neill SC, Eisner DA (1990): Effects of metabolic blockade on the regulation of intracellular calcium in dissociated mouse sensory neurones. J Physiol (Lond) 424: 411–426.

Dunham ET, Glynn IM (1961): Adenosine triphosphatase activity and the active movements of alkali metal ions. J Physiol (Lond) 156: 274–293.

Frelin C, Vigne P, Ladoux A, Lazdunski M (1988): The regulation of the intracellular pH in cells from vertebrates. Eur J Biochem 174: 3–14.

Gatto C, Milanick MA (1993): Inhibition of the red blood cell calcium pump by eosin and other fluorescein analogues. Am J Physiol 264: C1577–C1586.

Gill DL, Grollman EF, Kohn LD (1981): Calcium transport mechanisms in membrane vesicles from guinea pig brain synaptosomes. J Biol Chem 256: 184–192.

Gottfried JA, Chesler M (1996): Temporal resolution of activity-dependent pH shifts in rat hippocampal slices. J Neurophysiol 76: 2804–2807.

Gunter TE, Gunter KK, Sheu SS, Gavin CE (1994): Mitochondrial calcium transport: Physiological and pathological relevance. Am J Physiol 267: C313–C339.

Hasselbach W, Makinose M (1961): Die calciumpumpe der "Erschlaffungsgrana" des muskels und ihre abhangigkeit von der ATP-spaltung. Biochem Z 333: 518–528.

Henkart MP, Reese TS, Brindley FJ (1978): Endoplasmic reticulum sequesters calcium in the squid giant axon. Science 202: 1300–1302.

Irwin RP, Lin SZ, Long RT, Paul SM (1994): *N*-methyl-D-aspartate induces a rapid, reversible, and calcium-dependent intracellular acidosis in cultured fetal rat hippocampal neurons. J Neurosci 14: 1352–1357.

Kennedy HK, Thomas RC (1995): Intracellular calcium and its sodium-independent regulation in voltage-clamped snail neurones. J Physiol (Lond) 484.3: 533–548.

Kobayashi K, Tachibana M (1995): Ca^{2+} regulation in the presynaptic terminals of goldfish retinal bipolar cells. J Physiol (Lond) 483.1: 79–95.

Koch RA, Barish ME (1991): Sodium-calcium exchange in cultured embryonic mouse hippocampal neurons. J Cell Biol 115: 22a.

Levy S, Tillotson D (1988): Effects of Na^+ and Ca^{2+} gradients on intracellular free calcium in voltage clamped aplysia neurones. Brain Res 474: 333–342.

Meech RW, Thomas RC (1977): The effect of calcium injection on the intracellular sodium and pH of snail neurones. J Physiol (Lond) 265: 867–879.

Meech RW, Thomas RC (1980): Effect of measured calcium chloride injections on the membrane potential and internal pH of snail neurones. J Physiol (Lond) 298: 111–129.

Miller RJ (1991): The control of neuronal Ca^{2+} homeostasis. Prog Neurobiol 37: 255–285.

Mironov S, Usachev YM, Lux HD (1993): Spatial and temporal control of intracellular free Ca^{2+} in chick sensory neurons. Pflugers Arch 424: 183–191.

Nachshen DA, Sanchez-Armass S, Weinstein AM (1986): The regulation of cytosolic calcium in rat brain synaptosomes by sodium-dependent calcium efflux. J Physiol (Lond) 381: 17–28.

Neering IR, McBurney RN (1984): Role for microsomal Ca storage in mammalian neurones? Nature 309: 158–160.

Niggli V, Penniston JT, Carafoli E (1979): Purification of the (Ca^{2+} + Mg^{2+})-ATPase from human erythrocyte membranes using a calmodulin affinity column. J Biol Chem 254: 9955–9958.

Niggli V, Sigel E, Carafoli E (1982): The purified Ca^{2+} pump of human-erythrocyte membranes catalyzes an electroneutral Ca^{2+}-H^+ exchange in reconstituted liposomal systems. J Biol Chem 257: 2350–2356.

Nohmi M, Kuba K (1984): Effects of Na^+ gradient on the intracellular Ca^{2+} oscillation in the sympathetic ganglion cell: Na-Ca exchange in the neurone cell soma. Brain Res 324: 171–174.

Orkand RK, Thomas RC (1995): Effects of low doses of caffeine on $[Ca^{2+}]_i$ in voltage-clamped snail (Helix aspersa) neurones. J Physiol (Lond) (in press).

Otsuka M, Ohtsuki I, Ebashi S (1965): ATP-dependent Ca binding of brain microsomes. J Biochem (Tokyo) 58: 98–104.

Ou-Yang YB, Mellergård P, Kristán T, Kristiánova V, Siesjö BK (1994): Influence of acid-base changes on the intracellular calcium concentration of neurons in primary culture. Exp Brain Res 101: 265–271.

Paalasmaa P, Kaila K (1996): Role of voltage-gated calcium channels in the generation of activity-induced extracellular pH transients in the rat hippocampal slice. J Neurophysiol 75: 2354–2360.

Rossi JPFC, Schatzmann HJ (1982): Is the red cell calcium pump electrogenic? J Physiol (Lond) 327: 1–15.

Sanchez-Armass S, Blaustein MP (1987): Role of sodium-calcium exchange in regulation of intracellular calcium in nerve terminals. Am J Physiol 252: C595–C603.

Schatzmann HJ (1966): ATP-dependent Ca^{2+} extrusion from human cells. Experientia 22: 364–368.

Schinder AF, Olson E, Spitzer NC, Montal M (1995): The role of mitochondria in glutamate-induced neurotoxicity. Biophys J 68: A384.

Schwiening CJ (1994): Intracellular pH changes recorded under voltage and current clamp in snail neurones in isolated ganglia. J Physiol (Lond) 479P: 44P.

Schwiening CJ, Kennedy HJ, Thomas RC (1993): Calcium-hydrogen exchange by the plasma membrane Ca-ATPase of voltage-clamped snail neurons. Proc R Soc Lond B 253: 285–289.

Shull GE, Greeb J (1988): Molecular cloning of two isoforms of the Ca^{2+} transporting ATPase from rat brain. Structural and functional domains exhibit similarity to Na^+, K^+, and other cation transport ATPase. J Biol Chem 263: 8646–8657.

Shull GE, Greeb J (1989): Molecular cloning of a 3rd isoform of the calmodulin-sensitive plasma membrane Ca^{2+}-transporting ATPase that is expressed predominantly in brain and skeletal muscle. J Biol Chem 263: 2195–2202.

Smallwood JI, Waisman DM, Lafreniere D, Rasmussen H (1983): Evidence that the erythrocyte calcium-pump catalyzes a $Ca^{2+}:nH^{2+}$ exchange. J Biol Chemistry 258: 1092–1097.

Thayer SA, Miller RJ (1990): Regulation of the intracellular free calcium concentration in single rat dorsal root ganglion neurones in vitro. J Physiol (Lond) 425: 85–115.

Thomas RC (1984): Experimental displacement of intracellular pH and the mechanism of its subsequent recovery. J Physiol (Lond) 354: 3P–22P.

Thomas RC (1988): Changes in the surface pH of voltage-clamped snail neurones apparently caused by H^+ fluxes through a channel. J Physiol (Lond) 398: 313–327.

Thomas RC (1989): Proton channels in snail neurones—Does calcium entry mimic the effects of proton influx? Ann N Y Acad Sci 574: 287–293.

Thomas RC, Manger JR, Harris EJ (1969): Cation uptake as the basis for production of proton pulses by mitochondria at the anaerobic-aerobic transition. Eur J Biochem 11: 413–618.

Thomas RC, Coles JA, Deitmer JW (1991): Homeostatic muffling. Nature 350: 564.

Trapp S, Lückermann M, Kaila K, Ballanyi K (1996): Acidosis of hippocampal neurones mediated by a plasmalemmal Ca^{2+}/H^{+} pump. Neuroreport 7: 2000–2004.

Tsien RW, Tsien RY (1990): Calcium channels, stores and oscillations. Annu Rev Cell Biol 6: 715–760.

Werth JL, Thayer SA (1994): Mitochondria buffer physiological calcium loads in cultured rat dorsal root ganglion neurons. J Neurosci 14: 348–356.

SECTION IV

Activity-Induced Changes in pH

CHAPTER 16

Activity-Evoked Changes in Intracellular pH

KLAUS BALLANYI AND KAI KAILA
Physiologisches Institut, Universität Göttingen, Humboldtalle 23, Göttingen, Germany D-37073 (K.B.); Division of Animal Physiology, Department of Biosciences, University of Helsinki, FIN-00014 Helsinki, Finland (K.K.)

INTRODUCTION AND BACKGROUND

During intense neuronal activity, transmembrane ion shifts can lead to substantial changes in the composition of the intracellular milieu of both neurons and adjacent glial cells. It is well established that fluxes through voltage- or ligand-gated cation channels can evoke perturbations of the intracellular activities of Na^+ and K^+ by up to several tens of a millimole, whereas intracellular Ca^{2+} increases from the nanomolar into the micromolar range. On the other hand, activation of γ-aminobutyric acid (GABA) or glycine receptor-coupled anion channels produces millimolar changes in the intracellular Cl^- activity. For ion and volume homeostasis, these intracellular changes are reversed by plasmalemmal ion exchange mechanisms and transport ATPases (Grafe and Ballanyi, 1987; Kaila, 1994; Ballanyi 1995a,b).

Under resting conditions, the intracellular pH (pH_i) in neurons and glial cells is maintained at a level significantly more alkaline than would be expected on the basis of a passive transmembrane distribution of H^+. This is achieved by plasmalemmal acid-extruding mechanisms, such as the Na^+/H^+ exchange and the Na^+-dependent Cl^-/HCO_3^- exchange (Chapters 10–14). In glial cells, a further important mechanism for acid extrusion is electrogenic Na^+/HCO_3^- cotransport (Deitmer and Rose, 1996; Chapters 13–14). Traditionally, it has been assumed that a constant pH_i is a pre-condition for normal cell function, and that even small changes of pH_i interfere with the activity of enzymes or membrane-bound proteins. However, it is now becoming increasingly evident that the activity of certain kinds of enzymes and receptors is "tuned" to be responsive to small changes in pH_i (Chapter 1). This has raised the interesting possibility that dynamic changes of intracellular $[H^+]$ have a second-messenger function, similar to that of intracellular $[Ca^{2+}]$, which critically determines neuronal excitability.

pH_i Shifts in the Intact Brain

Prior to reviewing studies on different *in vitro* preparations, the question should be addressed, whether activity-dependent variations of pH_i occur in the brain of intact animals.

pH and Brain Function, Edited by Kai Kaila and Bruce R. Ransom
ISBN 0-471-11838-9

Nuclear magnetic resonance (NMR) spectroscopy has been applied *in vivo* for determination of brain pH under a variety of experimental conditions (Chapter 9). This technique is restricted to measurements of whole-tissue pH_i with a rather low temporal and spatial resolution. Nevertheless, NMR measurements have clearly indicated that pH_i in brain cells undergoes large changes during seizure activity, as well as during metabolic and respiratory disturbances (Chapter 30). In agreement with these findings, microelectrode measurements of neuronal pH_i in rat hippocampal CA1 neurons *in vivo* showed that a progressive hypoxia- or ischemia-evoked depolarization triggered an initial, small intracellular alkalinization, that turned into an acidification of more than 1 pH unit (Silver and Erecinska, 1990).

The only information on intraneuronal pH changes associated with physiological (rather than pathophysiological) activity of neurons *in vivo* stems from experiments done with H^+-selective microelectrodes in expiratory neurons of the ventral respiratory group (VRG), in the lower brainstem of cats (Ballanyi et al., 1994). During periodic inspiration-related hyperpolarizations, caused by summation of GABA- and/or glycinergic inhibitory postsynaptic potentials (IPSPs) for a period of 1–2 seconds, pH_i decreased by about 0.1 pH unit (Fig. 1). In contrast, such rhythmic changes in pH_i were not detected in the axons of the expiratory neurons (Fig. 1). So far, there are no measurements of glial pH_i changes caused by physiological activity *in vivo*. However, the fact that substantial extracellular pH transients, linked to neuronal activity have been recorded in studies *in vivo* (Chesler and Kaila, 1992; Chapter 17, this volume) supports (albeit indirectly) the view that considerable changes in pH_i are evoked by various patterns of neuronal activity in the intact brain.

pH_i TRANSIENTS ASSOCIATED WITH NEURONAL ACTIVITY

A wealth of experiments done on *in vitro* preparations have shown that electrical stimuli, as well as the application of transmitter receptor agonists, can lead to profound changes in neuronal and glial pH_i. Before discussing the effects of various kinds of stimuli, it is useful first to examine the effects of changes in membrane potential (E_m) on pH_i. However, as will be evident later on, the effects on neuronal pH_i of both synaptic and pharmacological activation of excitatory (glutamate-gated) ionotropic receptors might be largely attributable to the concomitant changes in E_m. In contrast to this, activation of GABA- or glycine-gated anion channels has a more direct effect on pH_i, an effect which is due to their HCO_3^- permeability.

Depolarization-Induced Changes in pH_i

The effects of changes in E_m on pH_i have been studied either by direct injection of current into the cell under examination, or by elevating $[K^+]_o$. The latter technique is the only feasible one in a number of experimental approaches, such as those involving fluorescence measurements in cell cultures. However, it should be noted that an elevation of $[K^+]_o$ may have a number of direct and indirect effects on ion-transport mechanisms that are not directly related to the concomitant depolarization.

Changes in neurons. Detailed analysis of the voltage-dependence of neuronal pH_i was started by Thomas and Meech in work on snail neurons (Thomas and Meech, 1982; Meech and Thomas, 1987; see also Chapter 15, this volume). These experiments showed that depolarization from the resting E_m to about -10 mV (whether achieved by current injection or elevated $[K^+]_o$) produced a fall of pH_i that was reduced in the nominal

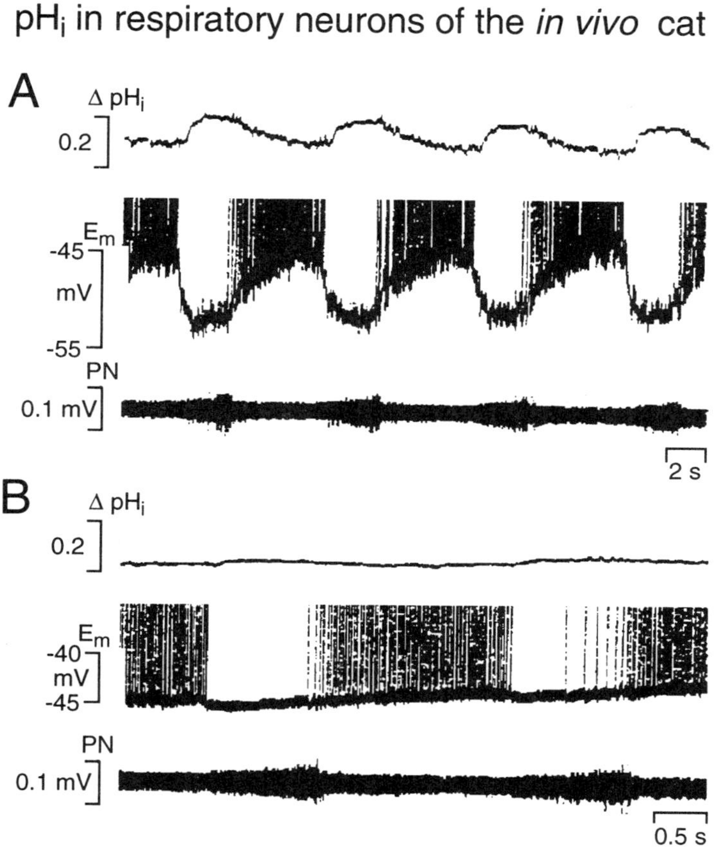

FIGURE 1. A. In a somatic recording from an expiratory neuron of the ventral respiratory group (VRG) with a double-barrelled pH-sensitive microelectrode, rhythmic fluctuations of membrane potential (E_m) were accompanied by decreases of intracellular pH (pH_i); acid shifts are upward deflections. **B.** In contrast, pH_i remained unaffected during rhythmic action potential discharges in an axon of a different expiratory neuron. PN, phrenic nerve activity. (Modified after Ballanyi et al., 1994, with kind permission.)

absence of extracellular Ca^{2+}. However, upon displacement of E_m to positive values, large alkaline pH_i shifts were observed. Under these conditions, pH_i appeared to be determined by a passive transmembrane distribution of H^+, indicative of the activation of a proton conductance. This voltage-dependent H^+ conductance and the associated pH_i transients have been well characterized in molluscan neurons (Byerly et al., 1984; Byerly and Moody, 1986), but a similar conductance mechanism has not been demonstrated in any mammalian neuron as yet. A noteworthy feature of these H^+ channels is the intense rectification of the current that permits only efflux of protons from the cell (see below).

That an increase in $[K^+]_o$ leads to a fall in pH_i has been shown also in lamprey neurons (Chesler, 1990), as well as in cultured rat dorsal root ganglion cells (Werth and Thayer, 1994). Recent work has demonstrated that direct electrically imposed depolarizations produce prominent acid transients in neurons of rat hippocampal (Fig. 2) (Trapp et al., 1996b) and brainstem slices (Lückermann et al., 1997a; Trapp et al., 1996a).

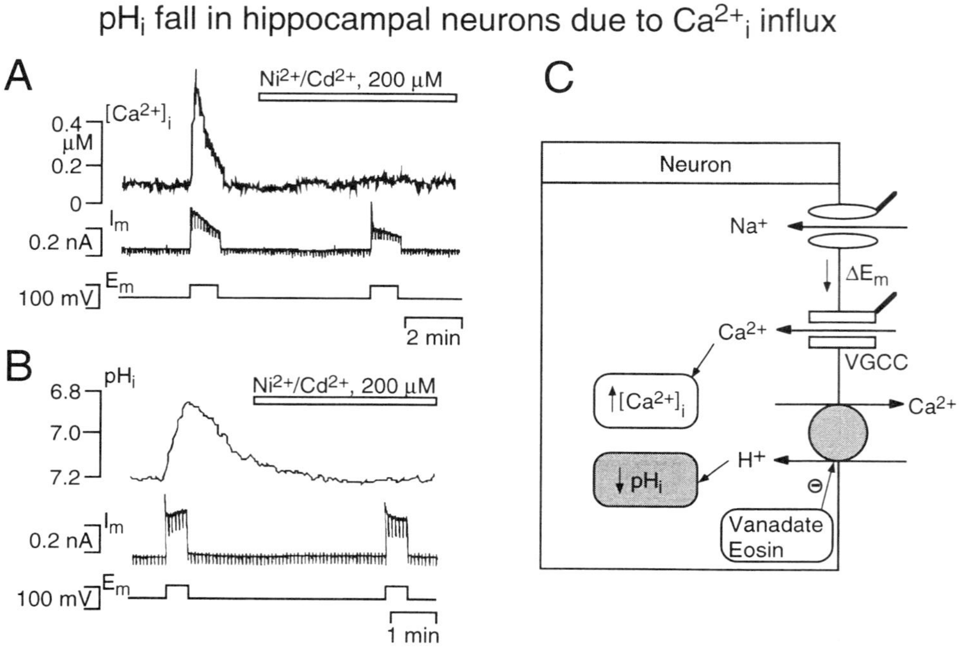

FIGURE 2. **A.** In a CA1 pyramidal neuron of a rat hippocampal slice, dialyzed with Fura-2, depolarization induced an outward current and a prominent increase in intracellular Ca^{2+} concentration, $[Ca^{2+}]_i$. **B.** In a different cell, dialyzed with BCECF, such depolarization evoked a delayed fall of intracellular pH (pH_i). These pH_i and $[Ca^{2+}]_i$ transients were reversibly blocked by a mixture of 200 μm Ni^{2+} and Cd^{2+}. (Modified after Trapp et al., 1996b, with kind permission.) **C.** The schematic drawing illustrates that opening of voltage-gated Ca^{2+} channels (VGCC) by excitation-related membrane potential (E_m) changes leads to a rise of $[Ca^{2+}]_i$. This $[Ca^{2+}]_i$ transient is reversed by activation of a vanadate- and eosin-inhibitable Ca^{2+}/H^+ ATPase in the plasma membrane, which is apparently responsible for the observed fall of pH_i.

Membrane depolarizations due to trains of action potentials lead to a fall in pH_i by up to 0.2 pH units in invertebrate (Ahmed and Connor, 1980) as well as in mammalian (Trapp et al., 1996a,b) neurons. An intracellular acidification was also noted to accompany the depolarizing receptor potential of dark-adapted barnacle photoreceptors during illumination (Brown and Meech, 1979). Taken together, the present data suggest that depolarization of neurons often leads to the generation of a large and sustained intracellular acid load. As will be shown in a later section, a major component of this neuronal acid load is most probably secondary to the entry and subsequent extrusion of Ca^{2+}.

Changes in glia. In the presence of CO_2/HCO_3^-, various kinds of glial cells respond to K^+-induced membrane depolarizations with a pronounced rise in pH_i (Deitmer and Rose, 1996; Chapters 13–14). In neuropil glial cells of the medicinal leech, it was recently shown that similar intracellular alkalinizations occur (at constant extracellular K^+) upon application of depolarizing steps under voltage-clamp conditions whereas a major fall of pH_i results from membrane hyperpolarization (Deitmer and Schneider, 1995). In contrast, pH_i is not affected during hyperpolarization of voltage-clamped neurons (Trapp et al., 1996a).

Synaptic Excitation

Excitation in neurons. In motoneurons of the isolated frog spinal cord as well as in leech lateral pressure neurons, repetitive stimulation of nerve rootlets produced a sustained fall of pH_i of about 0.2 pH units (Fig. 3) (Endres et al., 1986; Rose and Deitmer, 1995). The conclusion that these effects were triggered by excitatory synaptic transmission gained support from observations showing that exposure of the spinal cord to micromolar concentrations of the agonists of various glutamate receptor subtypes (kainate, quisqualate and N-methyl-D-aspartate [NMDA]) produced motoneuronal acidifications, which were similar to those evoked by stimulation (compare Figs. 3 and 4). Application of selective glutamate receptor blockers inhibited both these agonist-evoked depolarizations and pH_i decreases (Endres et al., 1986). Similar neuronal acidifications upon

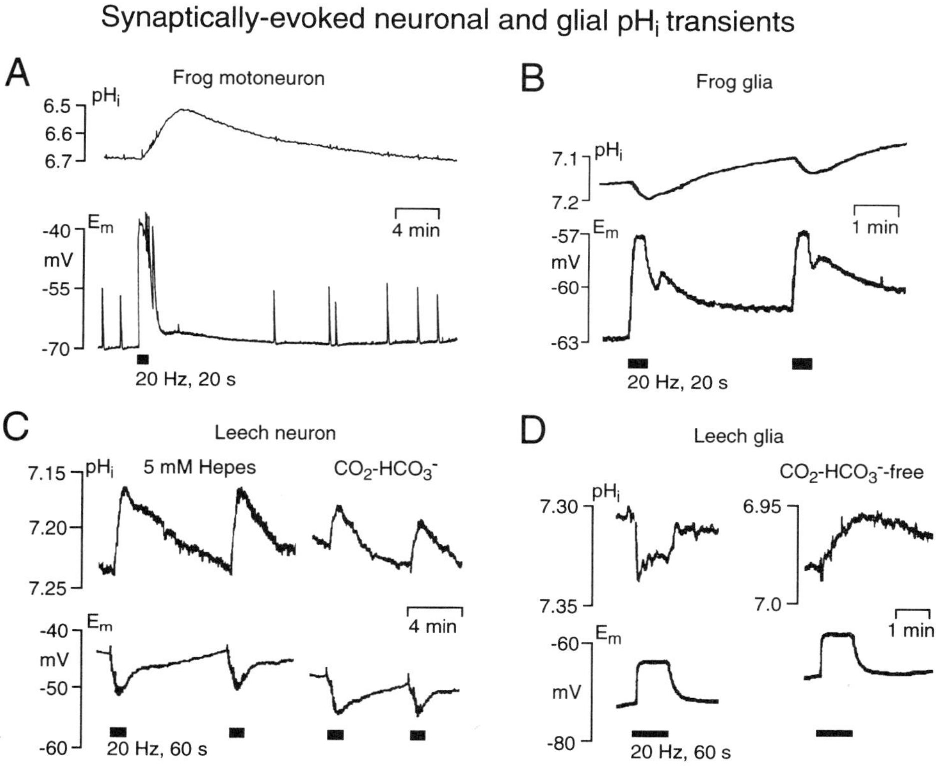

FIGURE 3. A. In a motoneuron of the isolated frog spinal cord, electrical stimulation of dorsal roots led to a depolarization and a decrease of intracellular pH (pH_i). **B.** In a neighboring glial cell, such stimulation produced a smaller depolarization, which was accompanied by an increase in pH_i. **C.** In a lateral pressure neuron of a leech segmental ganglion, nerve stimulation in both a CO_2/HCO_3^--containing and CO_2/HCO_3^--free HEPES-buffered solution produced an intracellular acidosis. (Stimulus-related action potentials are not visible in the membrane potential (E_m) recording because of low pass filtering at 5 Hz.) **D.** In a neuropil glial cell of the latter preparation, a stimulus-related rise of pH_i was observed in CO_2/HCO_3^--containing saline, whereas an intracellular acidosis was revealed in HEPES-buffered solution. (A, B, unpublished recordings by K. Ballanyi, P. Grafe, and W. Endres. C, D modified after Rose and Deitmer, 1994, 1995, with kind permission.) All recordings were done with pH-sensitive microelectrodes.

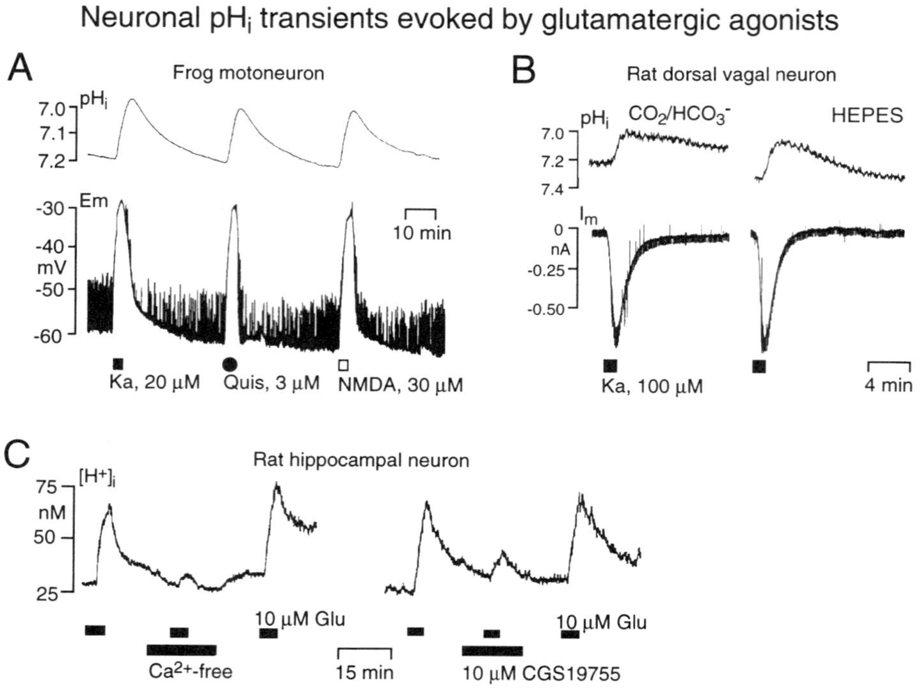

FIGURE 4. **A.** In a motoneuron of the isolated frog spinal cord, application of kainate (Ka), quisqualate (Quis) and N-methyl-D-aspartate (NMDA) to the bath led to a depolarization and a concomitant decrease of intracellular pH (pH_i) as recorded with a double-barrelled, pH-sensitive microelectrode. **B.** In a BCECF-dialyzed dorsal vagal neuron (DVN) of a brainstem slice, application of Ka to the bath led to an inward current and a delayed fall of pH_i, which persisted after removal of CO_2/HCO_3^- from the superfusate. **C.** In cultured hippocampal neurons, ester-loaded with SNARF, glutamate (Glu) evoked a long-lasting fall of pH_i, which was almost completely blocked in the absence of extracellular Ca^{2+} (left trace in B) or after adding the NMDA receptor antagonist CGS19755 (right trace in B). (A modified after Endres et al., 1986, B, from Trapp et al., 1996a, C modified after Wang et al., 1994, with kind permission.)

activation of ionotropic glutamate receptors were recently seen in voltage-clamped dorsal vagal neurons of rat brainstem slices (Fig. 4) (Trapp et al., 1996a), in hippocampal neurons in culture (Fig. 4) (Hartley and Dubinsky, 1993; Wang et al. 1994; Irwin et al., 1994) or *in situ* (Lückermann et al., 1997b), as well as in horizontal cells of the catfish retina (Dixon et al., 1993). In addition to these findings, there is evidence that activation of metabotropic glutamate receptors is associated with a fall of neuronal pH_i, which can be larger than 0.5 pH unit (Phenna et al. 1995).

Excitation in glia. In contrast to the acid pH_i shifts in the motoneurons of frog spinal cord, repetitive stimulation of dorsal roots (Fig. 3) as well as application of glutamatergic agonists (Fig. 5) produced a depolarization and a concomitant intracellular alkalinization in adjacent glial cells (K. Ballanyi, P. Grafe, and W. Endres, unpublished). These depolarizations were apparently due to K^+ release from the motoneurons during excitation (see above). A similar intracellular alkalosis has recently been reported also for leech neuropil glial cells during electrical stimulation of anterior side nerves (Rose and Deitmer, 1994,

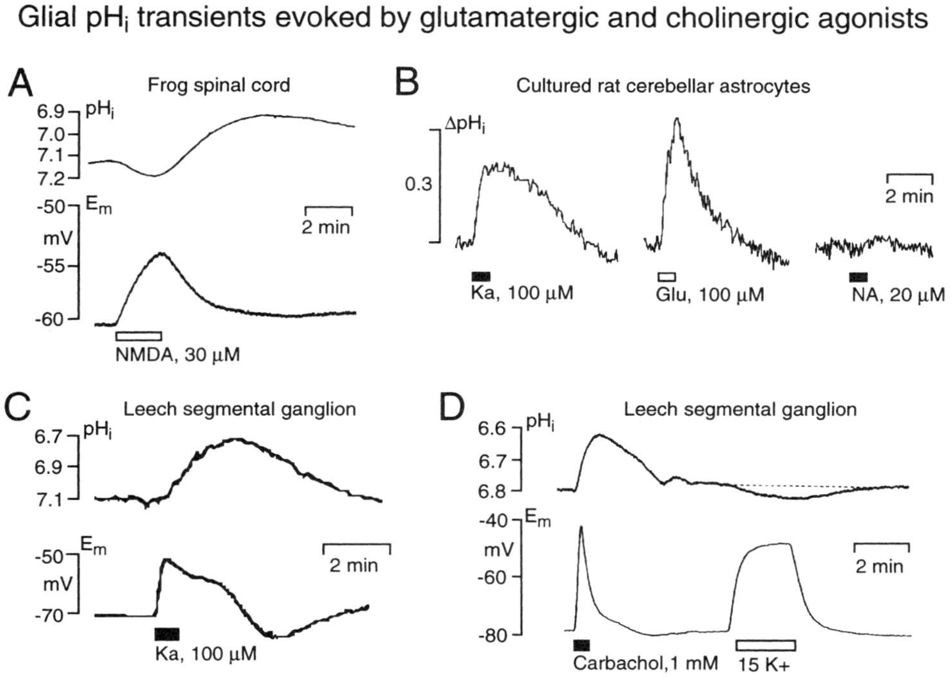

FIGURE 5. A. In a glial cell of the isolated frog spinal cord, N-methyl-D-asparte (NMDA) produced a moderate depolarization and a concomitant increase in intracellular pH (pH_i), followed by a late acidification in CO_2/HCO_3^--containing solution. **B.** In CO_2/HCO_3^--free saline kainate (Ka) and glutamate (Glu) produced an intracellular acidosis in a cultured astrocyte, whereas pH_i was not affected by noradrenaline (NA). **C,D.** In CO_2/HCO_3^--free superfusates an acidosis developed during administration of Ka (in C) or the cholinergic agonist carbachol (D) in a leech neuropil glial cell. Note that elevated extracellular K^+ (in D) led to only a small increase of glial pH_i. Recordings in A,C and D were done with double-barrelled pH-sensitive microelectrodes, cell in B was ester-loaded with BCECF. (A, unpublished recording by K. Ballanyi, P. Grafe, and W. Endres. B, modified from Brune and Deitmer, 1995, C, From Schlue et al., 1991; D, from Ballanyi and Schlue, 1989, with kind permission.)

1995). In these invertebrate glial cells, it was demonstrated that the stimulus-evoked alkalinization turned into a pronounced acidification upon removal of CO_2/HCO_3^- from the bath fluid (Fig. 3) (Rose and Deitmer, 1995). Under these conditions, a substantial intracellular acidosis was also revealed in the neuropil glial cells during administration of agonists of nicotinic acetylcholine (Ballanyi and Schlue, 1989) or ionotropic glutamate receptors (Schlue et al., 1991; Deitmer and Rose, 1996). Similarly, in cultured mammalian glial cells, activation of glutamate receptors in HEPES (N-2-hydroxy-ethylpiperazine-N′-2-ethane sulphonic acid)-buffered saline has been reported to decrease intracellular pH by several tenths of a pH unit (Fig. 5) (Brune and Deitmer, 1995). However, in the presence of CO_2/HCO_3^-, activation of glial acetylcholine (K. Ballanyi, unpublished) or glutamate receptors (Fig. 5) (K. Ballanyi, P. Grafe, and W. Endres, unpublished) evoked an initial alkalinization that turned into a sustained acidosis. Such biphasic pH_i responses were also found in protoplasmic astrocytes *in vivo* during depolarizations which were evoked by either tetanic stimulation or by spreading depression (Chesler and Kraig, 1989). These results indicate that the early glial alkalinization, caused by depolarization-induced activation of the electrogenic Na^+/HCO_3^- cotransport, masks a slowly generating bicarbonate-independent acid load (see below; Deitmer and Rose, 1996; Chapter 13).

Synaptic Inhibition

Inhibition in neurons. In addition to excitatory receptors, activation of inhibitory receptor channels that mediate Cl^--dependent IPSPs is known to produce an intracellular acidosis. Changes in pH_i mediated by $GABA_A$ receptors have been most quantitatively examined in crayfish opener muscle fibers, as well as in crayfish stretch-receptor neurons with the aid of H^+-sensitive microelectrodes (Kaila and Voipio, 1987; Kaila et al., 1990; Voipio et al., 1991). More recently, it was shown on isolated hippocampal pyramidal neurons of rats that a near-saturating concentration of bath-applied GABA evokes a picrotoxin-sensitive fall of about 0.2 pH unit in fluorescence measurements with 2′,7′-bis-(2-carboxyethyl)-5,6-carboxyfluorescein (BCECF). These GABA-induced neuronal acidifications were reversibly suppressed by the carbonic anhydrase (CA) inhibitor acetazolamide (Fig. 6) (Pasternack et al., 1993). In dorsal vagal neurons *in situ,* administration of GABA or glycine caused a similar fall in pH_i, which was comparable in amplitude to that evoked by kainate in these cells (compare Figs. 4 and 6) (Trapp et al., 1996a; see also Lückermann et al., 1997a). As expected, the $GABA_A$ receptor antagonist bicuculline blocked both the GABA-evoked current and the intracellular acidification (Trapp et al., 1996a; Lückermann et al., 1997a).

With respect to their biophysical properties, the anion channels gated by GABA and glycine are very similar (Kaila, 1994). Hence, it is not surprising that activation of glycine-gated anion channels leads to a strychnine-sensitive fall in pH_i, in neurons in the region of the VRG of thin brainstem slices from neonatal rats (Lückermann et al., 1997a). As will be discussed in detail below, the acid load linked to the activation of $GABA_A$ and glycine receptors is caused by a channel-mediated net efflux of HCO_3^-.

Inhibition in glia. Far less is known about the behavior of pH_i during activation of glial $GABA_A$ receptors. According to a recent review by Deitmer and Rose (1996), GABA can lead to a depolarizing or hyperpolarizing membrane response and acidification of neuropil glial cells, which depends on the presence of extracellular CO_2/HCO_3^-. Similar results were obtained in cultured rat astrocytes, in which muscimol-induced activation of $GABA_A$ receptors elicited a concentration- and CO_2/HCO_3^--dependent fall of pH_i (Kaila et al., 1991a).

IONIC AND METABOLIC MECHANISMS OF pH_i TRANSIENTS

The results described so far clearly demonstrate that neuronal activity leads to profound alterations of pH_i in neurons and glial cells. In this section, we discuss mechanisms that are likely to make different quantitative contributions to these pH_i changes during distinct forms of activity:

1. Conductive fluxes of acid-base equivalents
2. Carrier-mediated fluxes of acid-base equivalents
3. Metabolic production or consumption of acid equivalents

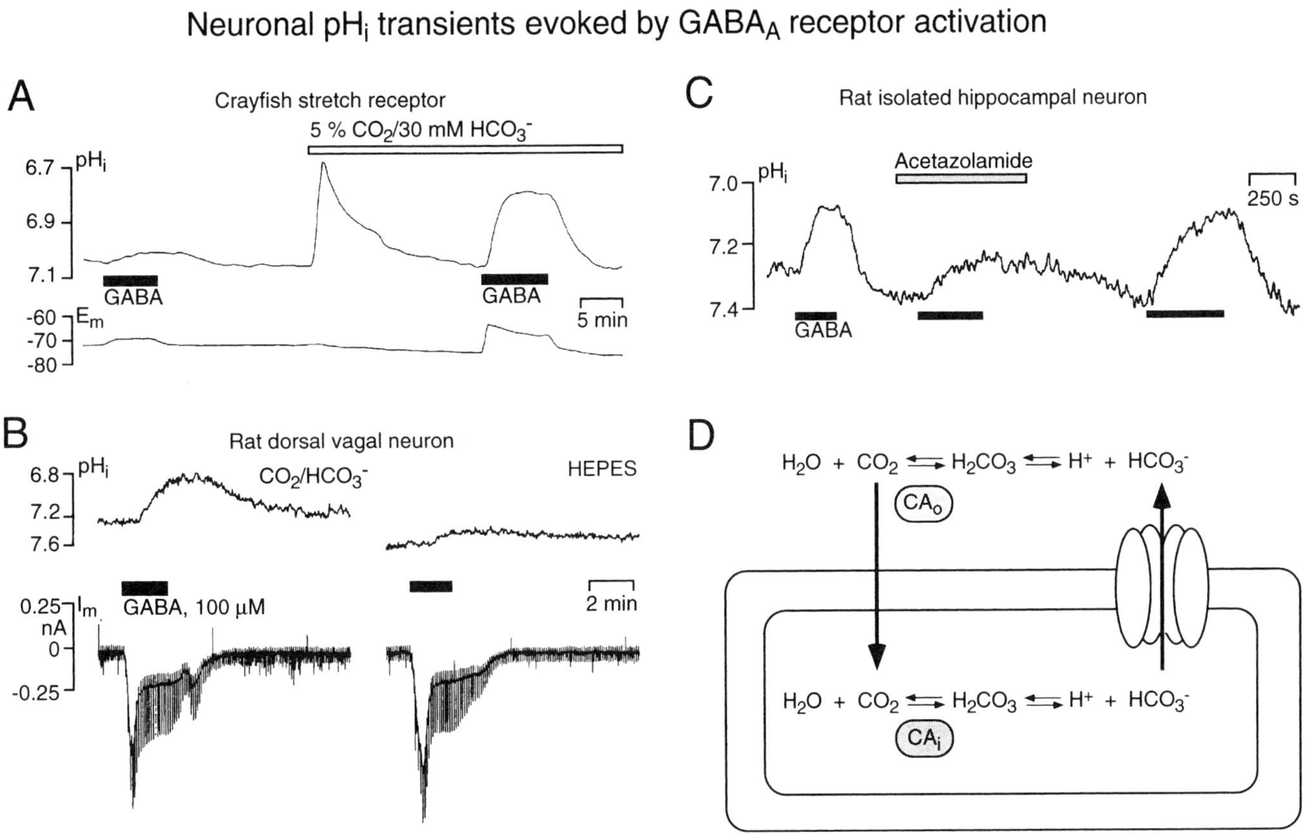

FIGURE 6. **A.** In a crayfish stretch receptor, recorded with a double-barrelled pH-sensitive microelectrode, application of γ-amino butyric acid (GABA) to the bath led to a small depolarization and a concomitant fall in pH$_i$. Both responses were dramatically enhanced after addition of CO_2/HCO_3^- to the superfusate. **B.** A similar CO_2/HCO_3^--dependent pH$_i$ fall was observed in a voltage-clamped, BCECF-dialyzed dorsal vagal neuron. **C.** In an acutely isolated hippocampal pyramidal cell, which was ester-loaded with BCECF, GABA produced an intracellular acidosis, which was reversibly attenuated by addition of 100 µM acetazolamide, an inhibitor of the carbonic anhydrase. (CA$_i$ = intracellular carbonic anhydrase; CA$_o$ = extracellular carbonic anhydrase). **D.** Schematic illustration of the mechanism of neuronal acidification induced by GABA (or glycine). For further information, see text. (A taken from Voipio et al., 1991, B taken from Trapp et al., 1996a; C modified from Pasternack et al., 1993, D, modified from Kaila, 1994, with kind permission).

Acid Load Caused by Activation of Ligand-Gated Anion Channels

The relevant physicochemical factors that must be considered when examining the pH effects of conductive, transmembrane movements of bicarbonate (Fig. 6) are related to HCO_3^- being an anion of a weak acid (carbonic acid, H_2CO_3). If channels with a significant HCO_3^- permeability are activated, the electrochemical HCO_3^- driving force will determine the direction of the conductive net movement. It is easy to show that at a fixed CO_2 partial pressure, the equilibrium potentials of H^+ and HCO_3^- are identical (E_{H+/HCO_3^-}) (see Kaila and Voipio, 1990). Due to active regulation of pH_i, their value is somewhere around -10 to -20 mV in brain cells. This means that activation of an anion channel with a major permeability to bicarbonate in a resting cell will lead to an efflux of HCO_3^-. However, the conjugate acid of bicarbonate, carbonic acid (H_2CO_3) and, in particular, the anhydride CO_2 are highly permeable in cell membranes. Hence, a channel-mediated net efflux of bicarbonate will lead to a net influx of CO_2, which, followed by hydration and dissociation of the latter, produces a fall in pH_i. This mechanism is consistent with the inhibitory action on the $GABA_A$ receptor-mediated acidosis that is seen upon blocking intracellular CA activity (see below; Fig. 6).

To give an illustration of the thermodynamic and physiological factors underlying a pH_i shift attributable to a maintained HCO_3^- conductance, let us assume that pH_i in a brain cell is 7.1, extracellular pH (pH_o) is 7.3, and resting E_m is -75 mV. Under these conditions $E_{HCO_3^-} = E_{H+} = -12$ mV, which gives a HCO_3^- driving force of -63 mV. The cell membrane is equipped with an extruder of acid equivalents which, of course, is a prerequisite for the initial condition assumed (i.e. $E_{HCO_3^-} > E_m$). Activation of an anion channel with a major bicarbonate permeability (e.g., $GABA_A$ or glycine receptors) will initiate a fall in pH_i, but pH_i falls only until active acid extrusion is equal to the net efflux of HCO_3^-. It is noteworthy that in practically all cells studied so far, activation of acid extrusion is steeply dependent on pH_i in a manner such that extrusion is enhanced by a low pH_i (Chapter 10). Hence, even with a maintained bicarbonate conductance, the H^+ and HCO_3^- will not attain a passive distribution.

In crayfish opener muscle fibers, application of a near-saturating concentration (0.5 mM) of GABA in a solution equilibrated with 5% CO_2 in air produced an initial net efflux of HCO_3^- of up to 10 mmol (liter of cell water) min^{-1} (Kaila and Voipio, 1987; Kaila et al., 1990). This is an extremely large flux indeed, and hence the subsequent steady-state fall in pH_i was 0.4–0.5 pH units. The GABA-induced acid load and the active extrusion of H^+ equivalents attained a steady-state level of about 2–3 mmol/(liter of cell water) min^{-1} during the plateau acidosis. The drop of the flux to 20–30% of its initial maximum value was mainly attributable to the decrease in the HCO_3^- driving force caused by both the membrane depolarization and the acidosis itself.

It is generally known that the capacity of an open CO_2/HCO_3^- system to buffer rapid pH shifts in response to the addition (or removal) of H^+ or OH^- is strongly increased by the presence of CA, which catalyzes the hydration of CO_2 (Chapters 2–4). However, as pointed out previously (Kaila et al., 1990; Voipio et al., 1995), CA activity will facilitate pH shifts that are due to a transmembrane CO_2/HCO_3^- shuttle such as that caused by activation of HCO_3^--permeable channels. Indeed, inhibition of intracellular CA activity by acetazolamide resulted in a marked slowing of the GABA-induced acidosis in crayfish muscle fibers and neurons (Kaila et al., 1990; Voipio et al., 1991), as well as in rat hippocampal neurons (Fig. 6) (Pasternack et al., 1993).

The above role of intracellular CA also implies that a modest intracellular acid load produced by $GABA_A$ receptor activation in a given cell type cannot be taken as evidence for the absence of a significant HCO_3^- permeability. Indeed, the rather small acidosis evoked by muscimol in cultured rat astrocytes (Kaila et al., 1991a) or the absence of a major GABA- or glycine-induced acidosis of brainstem neurons *in situ* (Lückermann et al., 1997a; Trapp et al., 1996a) might well reflect a lack (or dialysis-related washout) of intracellular CA.

Mechanisms of Acid Loading Linked To Activation of Ligand- and Voltage-Gated Cation Channels

Channel-mediated H^+ influx? The above findings have clearly shown that HCO_3^- efflux through receptor-coupled anion channels is the primary cause for GABA- and glycine-induced acidifications. Thus the simplest explanation for the observed fall of pH_i in response to activation of ligand- and voltage-gated cation channels would be an influx of H^+ through the channel pore. This might sound plausible, as the dimensions of some channels such as those of glutamate or nicotinic acetylcholine receptors allow permeation of cations that are even larger than the physiologically relevant Na^+, K^+, and Ca^{2+} ions (Hille, 1992). However, it is important to realize here that millimolar fluxes of acid equivalents per liter of cell volume must take place in a short period of time in order to create observable changes in pH_i. Such fluxes would require an enormously high H^+ permeability of conventional ligand- or voltage-gated cation channels, since H^+ ions are present at very low concentrations in biological tissues (10^{-7}–10^{-6} M). In fact, experiments on gramicidin-A channels have shown that other permeant ions like Na^+ block the passage of protons across an open cation channel by interrupting the chain of water molecules, the "water wire" that appears to be necessary to mediate "H^+ hopping" through the membrane (DeCoursey and Cherny, 1994).

In contrast, voltage-sensitive H^+ channels of snail neurons (pp. 292–293) appear to have an almost exclusive permeability for this ion species, and their density is orders of magnitude higher than that of other voltage- or ligand-gated cation channels (Thomas and Meech, 1982). A sound argument against the involvement of such proton channels in activity-induced acid pH_i shifts is that these channels open preferentially at a E_m more positive than E_{H^+/HCO_3^-}, thus mediating only efflux of acid equivalents, which will result in an intracellular alkalinization. The H^+-selective channels may, perhaps, be regarded as a "safety valve," activated in some cells by an excessive intracellular acid load (see DeCoursey and Cherny, 1994).

Ca^{2+}-dependent acid loads. It has frequently been suggested that an increase in cytosolic Ca^{2+} ($[Ca^{2+}]_i$) might produce a fall in pH_i in various kinds of cells (Chapter 15). Calcium can be released from intracellular organelles, for instance, mitochondria (Miller, 1991), but the major routes to increase $[Ca^2]_i$ during neuronal activity are voltage-gated Ca^{2+} channels with minor contributions from Ca^{2+}-permeable glutamate receptor channels (e.g., Miyakawa et al., 1992). At present, it seems that a Ca^{2+}-dependent fall in neuronal pH_i can be mediated by the activation of the plasmalemmal ATP-dependent calcium pump (the Ca^{2+}/H^+-ATPase) and by release of H^+ from mitochondria (Chapter 15, Werth and Thayer, 1994; Wang et al., 1994).

Ca^{2+}/H^+-ATPase. In studies with simultaneous measurements of surface pH and $[Ca^{2+}]_i$ in voltage-clamped snail neurons (Schwiening et al., 1993; Chapter 15, depolarization-induced increases in $[Ca^{2+}]_i$ were accompanied by an increase in surface pH. The

recovery of the $[Ca^{2+}]_i$ transients was impaired at elevated extracellular pH. A similar delay in the recovery of $[Ca^{2+}]_i$ was recently found upon administration of glutamate to cultured cortical neurons (Khodorov et al., 1995). This led to the hypothesis that an ATPase, exchanging intracellular Ca^{2+} for external H^+, might be responsible for Ca^{2+} extrusion in these neurons. In agreement with this, intracellular injection of vanadate, an inhibitor of the plasma membrane Ca^{2+}-ATPase (Carafoli, 1991; DiPolo et al., 1979), led to blockade of Ca^{2+} recovery, as was also the case in cultured rat sensory neurons (Benham et al., 1992). Such increases of surface pH are mirrored by a fall in intracellular pH, as recently simultaneously measured in voltage-clamped CA1 hippocampal pyramidal cells *in situ* (Lückermann et al., 1997b). During depolarization to 0 mV for a period of 1 min, pH_i decreases by up to 0.6 pH units in these cells (Trapp et al., 1996b), as well as in different types of medullary neurons (Trapp et al., 1996a; Lückermann, 1997a). As analyzed in detail in the hippocampal neurons, these acidifications were blocked upon elevation of extracellular pH, during intracellular dialysis with the Ca^{2+} pump inhibitors vanadate or eosin, and also during administration of blockers of voltage-dependent Ca^{2+} channels (Fig. 2) (Trapp et al., 1996b; see also Trapp et al., 1996a; Lückermann et al., 1997b). These results suggest that a significant portion of the activity-induced decreases of neuronal pH_i might be due to stimulation of the Ca^{2+}/H^+-ATPase caused by influx of Ca^{2+} (see also Paalasmaa and Kaila, 1996).

Organellar H^+ release. In addition to a putative direct influx of H^+ or activation of a plasma membrane Ca^{2+}/H^+-ATPase, excitation-induced acidifications could be mediated by Ca^{2+}/H^+ exchange within the cell. As one possibility, Ca^{2+} could displace H^+ from acidic binding sites (Thomas, 1989). A further mechanism that needs to be considered is H^+ production resulting from stimulation of (anaerobic) metabolism. In cultured hippocampal neurons, ester-loaded with the pH-sensitive dye carboxy-SNARF, it was demonstrated that activation of N-methyl-D-aspartate (NMDA) receptors produces a long-lasting fall of pH_i (Wang et al., 1994). These acidifications were blocked after substitution of glucose with the metabolic inhibitor 2-deoxyglucose as well as after inhibition of mitochondrial Ca^{2+} sequestration with ruthenium red, whereas the mitochondrial uncoupler carbonyl cyanide-*p*-trifluoro-methoxyphenyl-hydrazone (FCCP) mimicked the response (Wang et al., 1994). Similar results were obtained in isolated rat dorsal root ganglion cells (Werth and Thayer, 1994). In these cells, brief trains (<25) of action potentials elicited $[Ca^{2+}]_i$ transients that recovered to basal levels by a single exponential process. Longer periods of excitation produced a secondary plateau level (450–550 nM) of intracellular Ca^{2+} that was maintained for several minutes, and paralleled by a concomitant fall of pH_i. This prolonged $[Ca^{2+}]_i$ transient as well as the intracellular acidosis were blocked by either the mitochondrial uncoupler carbonyl cyanide chlorophenyl hydrazone (CCCP) or by the electron chain inhibitor antimycin A1 (Werth and Thayer, 1994). From these observations, the authors concluded that large Ca^{2+} loads are initially buffered by fast mitochondrial sequestration that effectively uncouples electron transport from ATP synthesis, thus leading to a decrease of pH_i perhaps by a CO_2- and lactate-related acidosis (Wang et al., 1994; Werth and Thayer, 1994). However, the possibility remains that mitochondrial buffering of Ca^{2+} leading to a significant acidosis takes place under pathophysiological rather than physiological conditions. One should also note that weak-acid mitochondrial uncouplers such as CCCP can have an effect on pH_i that is due to a direct dissipation of the plasmalemmal proton gradient (Kaila et al., 1989).

Alkaline and acid shifts in glial cells. As described earlier, pH_i in glial cells often undergoes a biphasic alkaline-acid change during nervous activity. Such observations have been made in frog glia (Fig. 5) (K. Ballanyi, P. Grafe, and W. Endres, unpublished) and in rat protoplasmic astrocytes *in vivo* (Chesler and Kraig, 1989). This indicates that the alkalinization, which is primarily caused by depolarization-induced activation of Na^+/HCO_3^- cotransport (Chapters 13–14), masks a slowly generating bicarbonate-independent acid load. This view is supported by the recent finding that stimulus-evoked alkalinizations of leech neuropil glial cells turned into acidifications in nominally CO_2/HCO_3^--free, HEPES pH-buffered solutions (Fig. 3). In contrast to this, stimulus-induced acidifications of leech pressure neurons remained principally unaffected by the absence of CO_2/HCO_3^- (Fig. 3) (Rose and Deitmer, 1994, 1995).

As has previously been suggested (Chesler and Kraig, 1989), the activity-induced acid loads in glial cells could be due to metabolism-related generation of lactate. In addition, transmitter substances, released from neurons, might act on glial receptors to induce an intracellular acidification. In line with this assumption, it was demonstrated that agonists of acetylcholine receptors (Ballanyi and Schlue, 1989) and glutamate receptors (Dörner et al., 1990) directly depolarize leech neuropil glial cells. These depolarizations can have an amplitude of up to 40 mV and are accompanied by a substantial fall of pH_i in CO_2/HCO_3^--free salines (Fig. 5) (Ballanyi and Schlue, 1989; Schlue et al., 1991; Deitmer and Munsch, 1994; Deitmer and Rose, 1996). Similar acidifications were also found during administration of glutamate and kainate in cultured astrocytes of the rat cerebellum (Fig. 5) (Brune and Deitmer, 1995).

The K^+-induced depolarization of leech neuropil glial cells leads to a rise of intracellular Ca^{2+} that is not accompanied by a fall, but rather by a slight increase of pH_i in CO_2/HCO_3^--free solutions (Deitmer et al., 1993; Ballanyi, 1995b). Furthermore, activation of noradrenaline receptors in cultured rat cerebellar astrocytes evokes a substantial rise of intracellular Ca^{2+}, whereas pH_i remains unchanged (Fig. 5) (Brune and Deitmer, 1995). On the basis of these findings it appears that in glial cells the plasma membrane Ca^{2+}/H^+-ATPase and mitochondrial Ca^{2+} buffering are not involved in the modulation of pH_i. Other mechanisms need to be considered for activity-induced acidifications in these brain cells. It was recently hypothesized for cultured astrocytes that decreases of pH_i upon administration of glutamate or aspartate are mainly due to activation of the glutamate uptake system (Brune and Deitmer, 1995). Indeed, it was demonstrated earlier in retinal glial cells from amphibia that activation of electrogenic glutamate uptake produces an intracellular acidification by an outward transport of one intracellular OH^- (or HCO_3^-), which accompanies outflux of one K^+ in exchange for 2 extracellular Na^+ and 1 glutamate anion (Bouvier et al., 1992; Chapter 19). Such neurotransmitter uptake systems might also contribute to activity-induced pH_i transients in neurons.

PHYSIOLOGICAL CONSEQUENCES OF pH_i TRANSIENTS

There is increasing evidence that the properties and functions of a large number of voltage- and ligand-gated channels, as well as both primarily and secondarily active ion-transport mechanisms, are sensitive to slight changes in internal and/or external pH (Chapters 20–27). Hence, pH shifts in nervous tissue can have a dual modulatory role, directed at both intracellular and extracellular targets. This conclusion is underscored by the fact that most of the activity-induced pH_i transients in neuronal and glial cells are due to plasmalemmal mechanisms, which can produce simultaneous and tightly linked shifts in both pH_i and pH_o. Here we will discuss possible physiological consequences of pH_i shifts; the functional roles of extracellular pH transients are dealt with elsewhere (Chesler and Kaila, 1992; Chapter 17).

Nervous Activity and pH_i Homeostasis

Neurons are known to possess powerful acid-extrusion mechanisms that are capable of maintaining pH_i homeostasis in the face of large acid loads, amounting to net millimolar fluxes of acid equivalents per liter of cell volume per minute (Chapters 10–12 and 15). However, as pointed out previously (Kaila et al., 1990), surprisingly little attention has been paid on the teleological question related to the functional roles of these acid-extrusion mechanisms. There are several standard techniques that are frequently used in *in vitro* studies to activate plasmalemmal acid-extrusion mechanisms, such as the ammonium prepulse technique or application of exogenous weak acids (e.g., propionate), but it is hardly likely that neurons are endowed with acid extruders to offset displacements of pH_i generated by these maneuvers!

Acid loads caused by energy metabolism (accumulation of CO_2 and lactate) are self-dissipating (Bountra et al., 1988) in that metabolic acids are capable of leaving the cell in their neutral (protonated form) by simple or facilitated diffusion (e.g., DeHemptinne et al., 1983). Hence, in the long run, the 'proton budget' of a cell is not altered by weak-acid end products of metabolism. However, for long-term pH_i homeostasis, a net influx of H^+ ions or an efflux of HCO_3^- must be ultimately balanced by a quantitatively equal extrusion of H^+ or uptake of HCO_3^-. Thus, it is possible that the main task of neuronal acid extruders is to deal with acid loads associated with electrical and synaptic acitivity. These acid loads appear to fall into two main categories: the influx of H^+ associated with neuronal excitation (which presently appears to be primarily mediated by the plasmalemmal Ca^{2+}/H^+-ATPase), and the efflux of HCO_3^- associated with inhibition (caused by activation of GABA- or glycine-gated anion channels).

pH_i Shifts and Cellular Ion Regulation

Activation of plasmalemmal acid-extrusion mechanisms can lead to substantial changes in the intracellular concentrations of strong-electrolyte ions (such as Na^+, K^+, Cl^-) that are co- or countertransported on ion carriers such as the Na^+/H^+ exchanger or on Na^+-dependent as well as Na^+-independent Cl^-/HCO_3^- exchangers. Such effects are of interest, since changes in the transmembrane distributions of the ions driving the acid-extruders must lead to shifts in the corresponding equilibrium potentials, thereby directly affecting membrane electrogenesis. In addition to this, various ions can have neuromodulatory actions that are mediated by allosteric effects on ion channels and transporters. For instance, the activation of the Na^+/H^+ exchanger leads to an influx of Na^+, which may have an influence on E_m due to the activation of the electrogenic Na^+/K^+ pump (Fig. 9 in Chapter 11; see also Ballanyi et al., 1984). On the other hand, bicarbonate-dependent acid extrusion will produce changes in intracellular Cl^-, which may well have an effect on the efficacy of postsynaptic inhibition mediated by $GABA_A$ or glycine receptors (Kaila and Voipio, 1987). Activation of pH_i regulatory mechanisms probably also leads to changes in cellular volume (Hoffman and Simonsen, 1989), thereby affecting neuronal excitability (Jefferys, 1995).

Ion Channels as Targets of Intracellular H^+ Ions

A generalization that is evident from the data reviewed presently is that activity-induced pH_i shifts in neurons are always displacements in the acid direction. A variety of K^+ channels are inhibited by a fall in internal pH (Chapter 21) and hence neuronal pH_i shifts might be expected to produce an increase in excitability. However, the functional consequences of such a proconvulsant effect may easily be overridden by the inhibitory action that an internal acidosis has on certain types of voltage-gated Ca^{2+} channels (Chapter 21). Furthermore, neuronal pH_i decreases are paralleled by alkaline shifts in extracellular pH, which are thought to enhance excitability at the level of synaptic transmission (Chesler and Kaila, 1992; Chapter 17).

It is worth noting that the two major categories of postsynaptic transmitter receptors, the ionotropic receptors gated by glutamate and GABA, are not significantly affected by moderate changes in pH_i. This absence of sensitivity to internal H^+ (but not to external H^+) is intriguing from an evolutionary point of view, especially as certain kinds of nicotinic acetylcholine receptors show a significant sensitivity to pH_i (Chapter 22).

pH_i Effects on Intracellular Enzymes and Second Messenger Systems

One of the best-documented physiological effects of a change in pH_i is the enhancement of the activity of phosphofructokinase (PFK) in response to an alkalosis (Trivedi and Danforth, 1966). The pH_i sensitivity of PFK may be of particular physiological significance for brain energy metabolism. This is because glial cells undergo a rise in pH_i during neuronal activity and this activity-induced alkalosis may serve as a signal to trigger enhanced lactate delivery from glial cells to neurons (Deitmer and Rose, 1996; Chapter 13, this volume). Indeed, direct measurements of lactate production in rat hippocampal brain slices showed an enhanced rate of glycolysis under various experimental conditions known to produce alkaline shifts in glial cells (S. Lauri, and K. Kaila, unpublished observations).

It has been known for a long time that changes in pH_i can have a profound effect on extrusion, sequestration, and buffering of Ca^{2+} at the cellular level, and recent work suggests that, in neurons, a particularly tight link exists between the regulation of intracellular Ca^{2+} and H^+ (Chapter 15). This raises the obvious possibility that a major role for pH_i shifts in the control of cellular functions is to be found in the Ca^{2+}–H^+ interactions at the level of the regulation of these cations as well as at the level of their target proteins. There are numerous ways in which Ca^{2+} and H^+ can interact within a neuron, and a nice example of this is provided by the work of Dixon et al. (1993) on catfish horizontal cells. In this system, application of glutamate resulted in a fall in pH_i, which in turn produced a decrease in currents mediated by high-voltage-gated Ca^{2+} channels. Since the basic building blocks of this chain of events can be found in most neurons, it is likely that similar effects are widespread throughout the nervous system.

CONCLUSIONS

As is the case with numerous other important physiological variables, pH_i is under the control of powerful homeostatic mechanisms. However, superimposed on this *stasis,* pH_i appears to be subject to a controlled variation that allows H^+ ions to assume an intracellular signalling function. As is evident from the work reviewed in this chapter and elsewhere (for references, see above), there is an abundance of both mechanistic and phenomenological data available on the regulation and modulation of pH_i in brain cells. In contrast to this, and despite the extensive amount of speculation on pH_i modulation of nervous function that is customary for review articles (including the present one!), there is a profound lack of experimental work on the functional consequences of activity-induced pH_i transients in the brain. It is likely that future work in this area will lead to the characterization of novel biophysical and intracellular signalling mechanisms involved in the control of neuronal development excitability and synaptic plasticity.

REFERENCES

Ahmed Z, Connor JA (1980): Intracellular pH changes induced by electrical activity in molluscan neurons. J Gen Physiol 75: 403–426.

Ballanyi K (1995a): Modulation of glial K^+, Na^+ and Cl^- activities by the extracellular milieu. In Kettenmann H, Ransom BR (eds): Neuroglial Cells. London, New York: Oxford University Press, pp 289–298.

Ballanyi K (1995b): Functional role of ion transporters and neurotransmitter receptors in glia. In Vernadakis A, Roots B (eds): Neuron-Glia Interrelationships in Phylogeny. Humana Press, Totowa, NJ, pp 197–219.

Ballanyi K, Schlue, WR (1989): Electrophysiological characterization of a nicotinic acetylcholine receptor on leech neuropil glial cells. Glia 2: 330–345.

Ballanyi K, Grafe P, Reddy MM, ten Bruggencate G (1984): Different types of potassium transport linked to carbachol and γ-aminobutyric acid actions in rat sympathetic neurons. Neuroscience 12: 917–927.

Ballanyi K, Mückenhoff K, Bellingham M, Okada Y, Scheid P, Richter DW (1994): Activity related pH changes in respiratory neurons and glial cells of cats. Neuroreport 6: 33–36.

Benham CD, Evans ML, McBain CJ (1992): Ca^{2+} efflux mechanisms following depolarization evoked calcium transients in cultured rat sensory neurones. J Physiol (Lond) 455: 567–583.

Bountra C, Kaila K, Vaughan-Jones RD (1988): Mechanism of rate-dependent pH changes in the sheep cardiac Purkinje fibre. J Physiol (Lond) 406: 483–501.

Bouvier M, Szatkowski M, Amato A, Attwell D (1992): The glial cell glutamate uptake carrier countertransports pH-changing anions. Nature 360: 471–474.

Brown HM, Meech RW (1979): Light induced changes of internal pH in a barnacle photoreceptor and the effect of internal pH on the receptor potential. J Physiol (Lond) 297: 79–93.

Brune T, Deitmer JW (1995): Intracellular acidification and Ca^{2+} transients in cultured rat cerebellar astrocytes evoked by glutamate agonists and noradrenaline. Glia 14: 153–161.

Byerly L, Moody RW (1986): Membrane currents of internally perfused neurones of the snail, *Lymnea stagnalis,* at low intracellular pH. J Physiol (Lond) 376: 477–491.

Byerly L, Meech RW, Moody WJ (1984): Rapidly activating ion currents in perfused neurones of the snail *Lymnea stagnalis*. J Physiol (Lond) 351: 199–216.

Carafoli E (1991): Calcium pump of the plasma membrane. Phys Rev 71: 129–153.

Chesler M (1990): The regulation and modulation of intracellular pH in the nervous system. Progr Neurobiol 34: 401–427.

Chesler M, Kaila K (1992): Modulation of pH by neuronal activity. Trends Neurosci 15: 396–402.

Chesler M, Kraig RP (1989): Intracellular pH transients of mammalian astrocytes. J Neurosci 9: 2011–2019.

DeCoursey TE, Cherny VV (1994): Voltage-activated hydrogen ion currents J Membr Biol 141: 203–223.

DeHemptinne A, Marranes R, Vanheel B (1983): Influence of organic acids on intracellular pH. Am J Physiol 245: C178–C183.

Deitmer JW, Munsch T (1994): Neuron-glia dialogue in the leech central nervous system: Glial cell responses to glutamate and kainate. Verh Dtsch Zool Ges 87.2: 185–194.

Deitmer JW, Rose C (1996): pH regulation and proton signalling by glial cells. Prog Neurobiol 48: 73–103.

Deitmer JW, Schneider HP (1995): Voltage-dependent clamp of intracellular pH of identified leech glial cells. J Physiol (Lond) 485: 157–166.

Deitmer JW, Schneider HP, Munsch T (1993): Independent changes of intracellular Ca and pH in identified leech glial cells. Glia 7: 299–306.

DiPolo R, Rojas HR, Beauge L (1979): Vanadate inhibits uncoupled Ca efflux but not Na-Ca exchange in squid axons. Nature 281: 228–229.

Dixon DB, Takahashi KI, Copenhagen DR (1993): L-glutamate suppresses HVA calcium current in catfish horizontal cells by raising intracellular proton concentration. Neuron 11: 267–277.

Dörner R, Ballanyi K, Schlue WR (1990): Glutaminergic responses of neuropil glial cells and Retzius neurones in the leech central nervous system. Brain Res 523: 111–116.

Endres W, Ballanyi K, Serve G, Grafe P (1986): Excitatory amino acids and intracellular pH in motoneurons of the isolated frog spinal cord. Neurosci Lett 72: 54–58.

Grafe P, Ballanyi K (1987): Cellular mechanisms of potassium homeostasis in the mammalian nervous system. Can J Physiol Pharmacol 65: 1038–1042.

Hartley Z, Dubinsky JM (1993): Changes in intracellular pH associated with glutamate excitotoxicity. J Neurosci 13: 4690–4699.

Hille B (1992): Ionic Channels of Excitable Membranes. Sunderland, MA: Sinauer Associates.

Hoffmann EK, Simonsen LO (1989): Membrane mechanisms in volume and pH regulation in vertebrate cells. Physiol Rev 69: 315–382.

Irwin RP, Lin SZ, Long RT, Paul SM (1994): N-methyl-D-aspartate induces a rapid, reversible and calcium-dependent intracellular acidosis in cultured fetal rat hippocampal neurons. J Neurosci 14: 1352–1357.

Jefferys JGR (1995): Nonsynaptic modulation of neuronal activity in the brain: Electric currents and extracellular ions. Physiol Rev 75: 689–723.

Kaila K (1994): Ionic basis of $GABA_A$ receptor channel function in the nervous system. Progr Neurobiol 42: 489–537.

Kaila K, Voipio J (1987): Postsynaptic fall of intracellular pH induced by GABA-activated bicarbonate conductance. Nature 330: 163–165.

Kaila K, Voipio J (1990): GABA-activated bicarbonate conductance: Influence on E_{GABA} and on postsynaptic pH regulation. In Alvarez-Leefmans FJ, Russell JM (eds): Chloride Channels and Carriers in Nerve, Muscle and Glial cells. New York, London: Plenum Press, pp. 331–353.

Kaila K, Mattsson K, Voipio J (1989): Fall in intracellular pH and increase in resting tension induced by a mitochondrial uncoupling agent in crayfish muscle. J Physiol (Lond) 408: 271–293.

Kaila K, Saarikoski J, Voipio J (1990): Mechanism of action of GABA on intracellular pH and on surface pH in crayfish muscle fibres. J Physiol (Lond) 427: 241–260.

Kaila K, Panula P, Karhunen T, Heinonen E (1991a): Fall in intracellular pH mediated by $GABA_A$ receptors in cultured rat astrocytes. Neurosci Lett 126: 9–12.

Khodorov B, Pimelis V, Vergun O, Storozhevykh T, Fajuk D, Vinskaya N, Arsenjeva E, Khaspekov L, Lyzin A, Isaev N, Andreeva N, Victorov I (1995): Dramatic effects of external alkalinity on neuronal calcium recovery following short-duration glutamate challenge: The role of the plasma membrane Ca^{2+}/H^+ pump. FEBS Lett 371: 249–252.

Lückermann M, Trapp S, Ballanyi K (1997a). GABA- and glycine-mediated fall of intracellular pH in neurons of brainstem slices from rats. J Neurophysiol 77: 1844–1852.

Lückermann M, Paalasmaa P, Kaila K, Ballanyi K (1997b). Activity-related surface and intracellular pH transients in single pyramidal neurons of rat hippocampal slices. In Elsner N, Wässle H (eds): Neurobiology: From Membrane to Mind. Proceedings of the 23rd Göttingen Neurobiology Conference. Stuttgart, New York: Thieme Verlag p. 813.

Meech and Thomas RC (1987): Voltage-dependent intracellular pH in *Helix aspersa* neurones. J Physiol (Lond) 390: 433–452.

Miller RJ (1991): The control of neuronal Ca^{2+} homeostasis. Progr Neurobiol 37: 255–285.

Miyakawa H, Ross WN, Jaffe D, Callaway JC, Lasser-Ross N, Lissman JE, Johnston D (1992). Synaptically activated increases in Ca^{2+} concentration in hippocampal CA1 pyramidal cells are primarily due to voltage-gated Ca^{2+} channels. Neuron 9: 1163–1173.

Paalasmaa P, Kaila K (1996). Role of voltage-gated calcium channels in the generation of activity-induced extracellular pH tansients in the rat hippocampal slice. J Neurophysiol 75: 2354–2359.

Pasternack M, Voipio J, Kaila K (1993): Intracellular carbonic anhydrase activity and its role in GABA-induced acidosis in isolated rat hippocampal pyramidal neurones. Acta Physiol Scand 148: 229–231.

Phenna S, Jane SD, Chad JE (1995): Increased perinuclear Ca^{2+} activity evoked by metabotropic glutamate receptor activation in rat hippocampal neurones. J Physiol (Lond) 486: 149–161.

Rose CR, Deitmer JW (1994): Evidence that glial cells modulate extracellular pH transients induced by neuronal activity in the leech central nervous system. J Physiol (Lond) 481: 1–5.

Rose CR, Deitmer JW (1995): Stimulus-evoked changes of extra- and intracellular pH in the leech central nervous system. I. Bicarbonate dependence. J Neurophysiol 73: 125–131.

Schlue WR, Dörner R, Rempe L, Briehl B (1991): Glial H^+ transport and control of pH. Ann N Y Acad Sci 633: 287–305.

Schwiening CJ, Kennedy HJ, Thomas RC (1993): Calcium-hydrogen exchange by the plasma membrane Ca-ATPase of voltage-clamped snail neurons. Proc Soc Lond B 253: 285–289.

Silver IA, Erecinska M (1990): Intracellular and extracellular changes of $[Ca^{2+}]$ in hypoxia and ischemia in rat brain *in vivo*. J Gen Physiol 95: 837–866.

Thomas RC (1989): Proton channels in snail neurones. Ann N Y Acad Sci 574: 287–294.

Thomas RC, Meech RW (1982): Hydrogen ion currents and intracellular pH in depolarized voltage-clamped snail neurones. Nature 299: 826–828.

Trapp S, Lückermann M, Brooks PA, Ballanyi K (1996a): Acidosis of rat dorsal vagal neurons *in situ* during spontaneous and evoked activity. J Physiol (Lond) 496: 696–710.

Trapp S, Lückermann M, Kaila K, Ballanyi K (1996b): Acidosis of hippocampal neurones mediated by a plasmalemmal Ca^{2+}/H^+ pump. Neuroreport 7: 2000–2004.

Trivedi B, Danforth WH (1966): Effect of pH on the kinetics of frog muscle phosphofructokinase. J Biol Chem 241: 4110–4114.

Voipio J, Pasternack M, Rydquist B, Kaila K (1991): Effect of γ-amino butyric acid on intracellular pH in the crayfish stretch receptor neurone. J Exp Biol 156: 349–361.

Voipio J, Paalasmaa P, Taira T, Kaila K (1995): Pharmacological characterization of interstitial pH transients evoked by selective synaptic and exogenous activation of AMPA, NMDA and $GABA_A$ receptors in the rat hippocampal slices. J Neurophysiol 74: 633–641.

Wang GJ, Randall RD, Thayer SA (1994): Glutamate-induced intracellular acidification of cultured hippocampal neurons demonstrates altered energy metabolism resulting from Ca^{2+} load. J Neurophysiol 72: 2563–2569.

Werth JL, Thayer SA (1994): Mitochondria buffer physiological calcium loads in cultured rat dorsal root ganglion neurons. J Neurosci 14: 348–356.

CHAPTER 17

Activity-Evoked Changes in Extracellular pH

KAI KAILA AND MITCHELL CHESLER
Department of Biosciences, Division of Animal Physiology, University of Helsinki, FIN-00014 Helsinki Finland (K.K.); Department of Physiology and Neuroscience, New York University Medical Center, New York, NY (M.C.)

BACKGROUND

Why Study Extracellular pH Shifts?

The idea that the brain interstitial space is a dynamic communication channel with a multitude of signalling functions (Nicholson, 1979) is gaining increasing support from recent experimental work. In addition to intercellular communication at specialized synapses, a host of mechanisms appear to serve in both local and wide-range signalling. These include paracrine effects of transmitter substances, as well as changes in the ionic composition of the neuronal microenvironment (Nicholson and Rice, 1991; Jefferys, 1995). In particular, the interstitial fluid undergoes pronounced changes in its volume and in its ionic composition during neuronal activity. In the light of current information, activity-induced shifts in the extracellular concentrations of K^+ and Ca^{2+} appear to have strong modulatory actions on neuronal communication and excitability (Heinemann et al., 1990). Recent findings suggest that extracellular H^+ ions share this ion-modulatory function (Gottfried and Chesler, 1994; Taira et al. 1993; see also Chapters 20–25).

It has been known for a long time that nervous activity leads to large changes in the extracellular pH (pH_o) in brain parenchyma (Chesler, 1990; Chesler and Kaila, 1992). Until recently, there has not been much information on the mechanisms involved in these activity-evoked pH_o shifts. However, as is evident from a number of other chapters in this volume, such information is valuable in that it can shed light on the basic biophysical, ion-regulatory, and energy-metabolic mechanisms underlying neuronal function. In addition, the fact that a number of both voltage- and transmitter-gated channels are highly sensitive to H^+ strongly supports the idea that localized pH shifts play an important neuromodulatory role.

The interstitial space is extremely heterogenous with respect to the localization of the putative molecular targets (receptors, channels) of proton modulation. This spatial heterogeneity is, obviously, also true for the sources and sinks (ion carriers, channels) that mediate movements of H^+, HCO_3^- and other acid-base equivalents from, and into, the

pH and Brain Function, Edited by Kai Kaila and Bruce R. Ransom
ISBN 0-471-11838-9

interstitial space. Hence, in order to gain an understanding of the role of pH_o in the control of neuronal function, the various mechanisms that contribute to the net movements of acid-base species have to be characterized. Indeed, as discussed in more detail below, it appears that neuronal activity leads to distinct types of pH_o transients in spatially circumscribed microdomains. To serve in extracellular communication, it is obvious that the sources of localized pH transients must lie in sufficient proximity to H^+-sensitive target molecules.

Most of the current knowledge on pH_o shifts in nervous tissue comes from work done with ion-selective microelectrodes. As discussed elsewhere in this volume (Chapter 6), the electrode technique permits quantitative or semiquantitative measurements of pH_o transients; but, bearing in mind the extremely narrow dimensions of the clefts that make up the interstitial space (Nicholson and Rice, 1991), it is obvious that these measurements do not provide direct information about the pH_o changes that are sensed by individual membrane receptors, channels, and carriers. At the moment, one has to infer, on the basis of conditional evidence, whether a certain pH shift is likely to be particularly effective in acting on a given H^+-sensitive target protein.

Which Compartment Is Being Regulated?

In the field of cellular pH regulation, the intracellular (cytosolic) compartment is the main focus and pH_i (rather than pH_o) is regarded as the regulated variable. The importance of pH_i regulation stems from the well-known sensitivity of intracellular enzymes and channels to changes in cytosolic H^+ activity. In the nervous system, however, the electrical activity of neurons can also be influenced by changes in pH_o (cf. Chapters 20–27). Of course, due to transmembrane acid fluxes, and the limits of interstitial diffusion, the regulation of pH_i and pH_o *in situ* are neither mutually exclusive nor independent of one another. Indeed, glial cells appear to make use of these constraints, regulating pH_o at the "expense" of their cytosolic pH (Chapters 13–14, 29).

Generation and Buffering of pH_o Shifts: Some Basic Facts

Acid-base equivalents are neither generated *de novo,* nor consumed, within the extracellular space, which implies that all pH_o shifts are caused by net movements of acid or base into or out of the interstitium. When considering short-term changes in pH_o *in vivo,* the blood-brain barrier can be regarded as practically impermeant to all acid-base species except for molecular CO_2. Hence, if transmembrane channels or carriers are directly responsible for a given fast change of pH_o evoked by nervous activity, an equal amount of acid and/or base that appears within the interstitium is taken from the intracellular (neuronal and/or glial) compartment. It follows that alkaline shifts in pH_o are linked to acid shifts in pH_i, and vice versa. However, this rule of null sum is, of course, not applicable to situations where acid or base is being produced or consumed by cellular metabolism. For instance, an enhanced generation of metabolic acids (CO_2, lactate) by cellular energy metabolism will produce an acid shift in pH_i followed by a qualitatively similar shift in pH_o. Furthermore, an intracellular acid load does not necessarily generate an acid load in the nearby extracellular microdomain. This is because glial cells have been shown to be capable of spatial buffering of pH_o (Newman, 1996), akin to their well-known role in the spatial buffering of K^+ (Gardner-Medwin, 1983; Nicholson, 1980).

In quantitative terms, the instantaneous effect of a given transmembrane movement of acid-base equivalents on intracellular versus extracellular pH is affected by the relative volumes of the two compartments and their H^+ buffering capacities. In brain parenchyma, the fractional volume of the interstitial space is about 20% (Nicholson and Rice, 1991), although some areas, such as the pyramidal cell layer in the area CA1, appear to be much more tightly packed, with an estimated volume fraction for interstitial space of only about 10% (McBain et al., 1990; Heinemann et al., 1990). The buffering capacity of the interstitial fluid is thought to be due to the CO_2/HCO_3^- system almost exclusively, whereas buffering in the intracellular compartment has both a pure CO_2/HCO_3^--based component as well as a significant "intrinsic" component, in which amino acids/proteins and phosphates play a major role (see Chapter 2). Since pH_o is typically higher than pH_i, the interstitial $[HCO_3^-]$, and hence the buffering capacity of the interstitial fluid, is often greater than the total intracellular buffering capacity. A caveat in this regard is that rapid interstitial buffering depends upon the presence of sufficient carbonic anhydrase activity in the interstitium. As noted further below, rapid pH_o shifts may not be optimally buffered because of limited interstitial activity of this enzyme.

Consideration of the interstitial buffering capacity allows one to estimate the concentration of acid or base that entered or was removed from the extracellular compartment during a pH_o transient. In their study of stimulus-induced alkaline shifts, Kraig et al. (1983) noted that the generation of the observed pH_o changes (which could be as large as 0.1–0.2 pH unit) would require the addition of strong base to the extracellular compartment in concentrations amounting to few millimolar. Thus, while the resulting fall in the interstitial H^+ activity was only in the order of a few tens of nanomolar, the net removal of acid equivalents must have been greater by five to six orders of magnitude.

Typically, the extracellular buffering capacity has been calculated as $2.3[HCO_3^-]_o$, which assumes a fixed CO_2 partial pressure (P_{CO_2}) and virtually instantaneous equilibration between CO_2 and HCO_3^- (see Chapters 2 and 4). However, if these assumptions are not valid, the buffering capacity would have to be considerably lower. A number of observations now indicate that the capacity of the extracellular space to buffer rapid alkaline transients, while substantial, is far lower than previously assumed. Using simultaneous recordings of CO_3^{2-} and H^+, Chesler and coworkers (1994) determined the rise in $[HCO_3^-]_o$ produced during stimulus-induced alkalinizations in the CA1 region of the rat hippocampal slice. A mean alkalinization of 0.059 pH unit was associated with a rise in $[HCO_3^-]_o$ of roughly 1 mM. Thus, it was concluded that the effective buffering power of 17 mM (i.e., 1 mM/0.059) was just 28% of the theoretical value of 60 mM.

Deviations of extracellular buffering power from the theoretically assumed value may have a number of origins. Factors that have received little attention are those related to differences in the diffusion characteristics of various buffer species (see Chapter 4). Furthermore, measurements of P_{CO_2} transients have demonstrated that interstitial P_{CO_2} is not clamped during stimulus-evoked pH_o shifts, and therefore the CO_2/HCO_3^--based buffering should be less than in a system with a rigidly fixed P_{CO_2} (Voipio and Kaila, 1993). Insufficient extracellular carbonic anhydrase (CA), leading to a departure from equilibrium, is a second factor that may diminish the interstitial buffering capacity. Slow hydration of CO_2 was suggested by the unexpectedly small, and occasionally absent increases in $[HCO_3^-]_o$ during stimulus-evoked alkaline shifts. In further support of this notion, addition of CA to either brain slices or exposed cerebral cortex caused the amplitude of extracellular alkaline transients to decrease by more than 50% (Huang et al., 1995). An additional, often unconsidered factor, is the concentration of CA and thus the kinetics of buffering in the intracellular compartment. Although there is abundant evidence for the presence CA in certain kinds of neurons and glial cells (Chapter 3), the presence of CA within particular subtypes of cells has often been debated. The hippocampal CA1 neurons fail to stain for this enzyme when studied with conventional histochemical and immunocytochemical methods. Yet intracellular CA clearly plays a role in CO_2 hydration in these neurons (Pasternack et al., 1993). Whether equilibrium is assured by the quantities of CA available may not always be certain, however.

All the above caveats notwithstanding, consideration of buffering capacities and relative volumes suggests that when a given amount of acid or base traverses a cell membrane, the local change in pH_o can be expected to be greater than the corresponding change in pH_i. Thus, an extracellular vantage point can have advantages over intracellular recording for the detection and analysis of these processes.

Use of Extracellular Carbonic Anhydrase Inhibitors to Study pH_o Transients

If a pH_o shift is directly generated by a transmembrane movement of H^+, inhibitors of carbonic anhydrase are expected to slow buffering, and thereby increase the magnitude of the response. This behavior applies particularly to a class of alkaline shifts linked to neuronal excitation (see pp. 314–320). By contrast, for a separate category of pH_o shifts, which are linked to inhibitory activity, CA is instrumental in the generation of the alkalosis (Kaila, 1994). These CA-dependent alkaline responses arise from shifts in extracellular $[CO_2]$ and $[HCO_3^-]$, as originally described for crayfish muscle (Kaila and Voipio, 1987; Kaila et al., 1990). The extracellular alkalosis evoked by $GABA_A$ receptors is explained by the underlying ionic mechanism. An efflux of HCO_3^- across the $GABA_A$ anion channel gives rise to a neuronal CO_2 sink and leads to a CA-catalyzed net dehydration of carbonic acid in the interstitial space. Carbonic anhydrase inhibitors slow the extracellular dehydration of CO_2 and thereby suppress the alkalosis (Kaila et al., 1990; Saarikoski and Kaila, 1992).

The distinct role of CA in each mechanism, and the opposite behavior in response to inhibition of this enzyme, provide a useful diagnostic test for these two classes of alkalosis. Where a CO_2/HCO_3^--dependent process predominated, the alkaline shift would be suppressed by a CA inhibitor. If alkalosis were generated mainly by an H^+ sink, it would be enhanced. As explained below, this approach offers a pharmacological means of analyzing the multicomponent pH_o shifts associated with mixed excitatory and inhibitory neural activity.

In experiments on brain slices, a selective inhibition of interstitial CA activity can be achieved by application of dextran-conjugated sulfonamide derivatives, which should be completely impermeant to cell membranes (Kaila et al., 1992; Chen and Chesler, 1992a). Benzolamide, a very acidic sulfonamide that is fully dissociated at physiological pH levels, has also been frequently used to inhibit extracellular CA (e.g., Kaila et al., 1990; Chen and Chesler 1992c; Kaila et al., 1992). It is worth noting that acetazolamide, a somewhat more membrane-permeant CA inhibitor (cf. Saarikoski and Kaila, 1992), was originally shown to enhance activity-dependent alkaline shifts in rat cerebellum (Kraig et al., 1983) and in hippocampal slices (Carlini and Ransom, 1986; Walz, 1989). As these effects have been noted using benzolamide and dextran-conjugated sulfonamides (see references above), they appear to have been mainly due to acetazolamide blocking extracellular rather than intracellular CA. It is worth noting here also that acetazolamide has a rather low membrane permeability (Saarikoski and Kaila, 1992). Hence, in experiments on brain slices where inhibition of intracellular CA is to be achieved, more lipophilic CA inhibitors (Maren, 1984) such as ethoxyzolamide are much more efficient than acetazolamide (e.g., Lamsa and Kaila, 1997; Kaila et al., 1997).

Early Findings

The first attempts to measure activity-evoked pH shifts date from the 1930s. Dusser de Barenne et al. (1937, 1938) used macroelectrodes placed on the brain pial surface to study pH changes induced during seizure activity. Similar attempts were made in subsequent decades (Jasper and Erickson, 1941; Wang and Sonnenschein, 1955). A surface acidification was consistently described in these studies. At the time, it was suggested that this acidification was the result of lactic acid generation (Stone et al., 1945). As discussed below, more recent work indicates that activity-dependent acidification of the extracellular space can have a number of distinct origins. A few early studies also described an extracellular alkalinization at the onset of seizure or cortical spreading depression (Dusser de Barenne et al., 1938; Tschirgi et al., 1957).

In retrospect, it is apparent that the early attempts to measure activity-related pH shifts were complicated by unrecognized technical issues. The large size of the pH sensors (millimeters) and their distance from the origin of the pH changes undoubtedly caused significant temporal distortions. A second factor was the influence of extracellular potential shifts. Neuronal activity in the cortex is typically associated with a negative shift in the extracellular DC potential (Birbaumer et al., 1990). If this component is not subtracted from the voltage on the pH electrode, an artifactual alkalinization will be registered. In more recent studies, measurements have been performed using pH microelectrodes with tip diameters of a few microns, enabling pH_o recordings in discrete regions of the CNS. Using an adjacent reference microelectrode, extracellular potential is measured and subtracted from the voltage on the pH electrode.

The first studies using modern configurations of pH and reference microelectrodes revealed an early extracellular alkalinization associated with stimulated neural activity (Heuser et al., 1977; Urbanics et al., 1978; Kraig et al., 1983) and spreading depression (Kraig et al., 1983; Mutch and Hansen, 1984). The initial alkalosis was often superceded by a slower acid shift that recovered to baseline over several minutes (Fig. 1). This alkaline-acid pattern is common to a wide variety of vertebrate preparations, although some variations upon this theme have been noted (Chesler, 1990; Chesler and Kaila, 1992).

Alkaline-acid transients reminiscent of those above have recently been reported in a central ganglion of the leech (Rose and Deitmer, 1995a,b). In addition, activity-related surface alkalinizations have been described on giant neurons of the land snail (Schwiening et al., 1993). While these similar observations across phyla may indicate conservation of a single process, it is important to bear in mind that both alkaline and acid pH_o shifts arise from a variety of mechanisms, which are discussed in detail below.

EXTRACELLULAR ALKALINE SHIFTS

An alkaline shift of pH_o, irrespective of its origin, must be accompanied by either a rise in $[HCO_3^-]_o$, a fall in P_{CO_2}, or a combination of these changes. In their experiments on cat cortex, Urbanics et al. (1978) noted a hyperemia at the onset of stimulus-evoked pH_o shifts and reasoned that a consequent washout of CO_2 might account for the early alkalosis. This hypothesis appeared unlikely after Kraig and colleagues (1983) noted similar

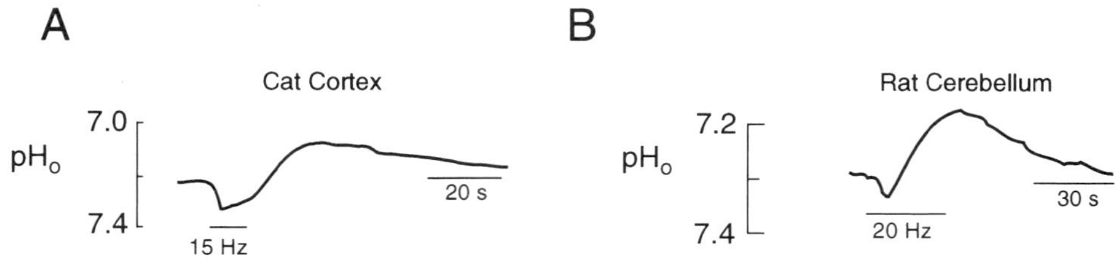

FIGURE 1. Alkaline-acid extracellular pH transients evoked by electrical stimulation. **A.** Stimulation of cat cortex *in vivo* caused an alkaline shift that gave way to a slow acid transient after the termination of the stimulus. **B.** Stimulation of the cerebellar parallel fibers in the rat *in vivo* gave rise to an alkaline shift, followed by an acid transient that rose during the stimulus train. Traces in (A) and (B) are modified from Urbanics et al. (1978) and Kraig et al. (1983), respectively.

alkaline shifts in the *in vitro* cerebellum of the turtle where, of course, blood flow could not be a factor. While more recent work on hippocampal slices has shown that a small fall in local P_{CO_2} may occur at the onset of neural activity (Voipio and Kaila, 1993; Chesler et al., 1994), these changes in P_{CO_2} are small, and do not account for most of the magnitude of the extracellular alkalinization. The subject of activity-induced CO_2 transients is considered in greater detail in Chapter 4.

Kraig et al. (1983) described an early alkaline shift in the molecular layer of the rat cerebellum, evoked by repetitive stimulation of the parallel fibers. Based on earlier studies of stimulus-induced ion shifts (see Nicholson, 1980), these authors noted an unexplained extracellular "anion gap" and suggested a rise in $[HCO_3^-]_o$ as a likely explanation.

Because the extracellular space constricts during repetitive neuronal activity and thereby increases the extracellular concentration of ions (Dietzel and Heinemann, 1986), shrinkage of the interstitial volume was suggested as a possible source for an increase in $[HCO_3^-]_o$ and a consequent alkalosis. However, in subsequent studies on the isolated turtle cerebellum, stimulus-induced volume shifts were shown to occur in the absence of interstitial alkalinization (Chesler and Chan, 1988). Therefore, the stimulus-induced alkaline shifts were most likely due to a transmembrane flux of acid equivalents. Of course, interstial volume shrinkage would still be expected to shape the pH_o changes.

The stimulus-induced alkaline shift in rat cerebellum was abolished when synaptic transmission was blocked by Mn^{2+} (Kraig et al., 1983). Similar findings were later noted in various *in vitro* preparations (Carlini and Ransom, 1986; Chesler and Chan, 1988; Walz, 1989; see also references in Chapter 18). This important observation suggested that the alkaline shift could be a strictly postsynaptic phenomenon. In the next sections, we will review data indicating that stimulation-induced alkaline transients in brain gray matter are, indeed, largely due to postsynaptic events. These fall into two mechanistic categories: (i) a postsynaptic uptake of H^+ which requires external Ca^{2+} and is evoked by neuronal excitation, and (ii) a conductive efflux of HCO_3^- linked to the opening of postsynaptic anion channels associated with inhibitory (typically GABAergic) synaptic transmission.

Neuronal Excitation

Glutamate receptor-mediated alkaline shifts. Direct application of glutamate as well as of specific ionotropic glutamate receptor agonists elicits a transient extracellular alkalosis in CNS gray matter, as was first shown by Endres et al. (1986), in experiments on the isolated frog spinal cord. Simultaneous measurements of motoneuronal pH_i in this work also showed that the extracellular alkaline shift was associated with a fall in pH_i, indicating a transmembrane net movement of acid-base equivalents triggered by activation of glutamate-gated channels. In the isolated turtle cerebellum, a similar synaptically evoked alkaline shift was also clearly linked to activation of postsynaptic glutamate receptors, as the response, studied in bicarbonate-free media, was reversibly abolished by the antagonist kynurenic acid (Chesler and Chan, 1988). In addition, direct iontrophoretic application of glutamate (or aspartate) in cerebellum produced alkaline shifts in HEPES-buffered media that persisted when synaptic transmission was blocked by Mn^{2+} or Cd^{2+} (Chesler and Rice, 1991). Taken together, these data indicated that the glutamate receptor-evoked responses could not be attributed to indirect activation of $GABA_A$ receptors or indirect excitation via local circuits.

A similar effect of ionotropic glutamate receptor activation was also noted in rat hippocampal slices. Direct microdrop application of glutamate produced an immediate extracellular alkaline shift (Jarolimek et al., 1989) similar in appearance to the alkalosis induced by synaptic activation (Carlini and Ransom, 1986). The glutamate-evoked alkalosis was not inhibited by picrotoxin (Chen and Chesler, 1992b) and therefore could not be due to indirect activation of $GABA_A$ receptors. In addition, alkaline shifts were elicited by the more specific ionotropic agonists AMPA (α-amino-3-hydroxy-5-methylisoxazolate-4-propionic acid) and NMDA (N-methyl-D-aspartate) (Smith et al., 1994; Voipio et al., 1995). The responses elicited by glutamate (Chen and Chesler, 1992c) or by AMPA and NMDA (Voipio et al., 1995) were amplified by benzolamide, also differentiating them from GABAergic alkalosis. Finally, the glutamate-evoked alkaloses were not abolished in the presence of tetrodotoxin (TTX) (Chen and Chesler, 1992d; Voipio et al., 1995), which indicates that they were not attributable to spike firing *per se,* although it should be noted that TTX does not block the agonist-induced neuronal depolarization.

A major part of the pH_o recordings in the studies cited above were done in the pyramidal layer of the hippocampal CA1 region. This region has little excitatory synaptic input (Andersen, 1990), and hence it appears that the alkaline shifts were evoked by activation of nonsynaptic glutamate receptors. However, the data available at present indicates that the activation of postsynaptic glutamate receptors is instrumental in generating the alkalosis during afferent activation. It is therefore not surprising that the alkaline shifts evoked by stimulation of the Schaffer collaterals could be blocked by specific antagonists of AMPA and NMDA receptors (Chen and Chesler, 1992b). However, as noted further on, use of these antagonists abolishes direct synaptic activation of the CA1 pyramidal cells, as well as feedforward and recurrent inhibition via GABAergic interneurons. Therefore the contribution of the HCO_3^--dependent GABAergic component to the synaptically-evoked alkalosis could not be assessed by such experiments. The persistence of the Schaffer collateral-evoked alkalosis in HEPES-buffered Ringer's solution (Chesler and Chan, 1988; Gottfried and Chesler, 1994) suggests a substantial non-GABAergic component. However, as discussed below (p. 322), the relative contribution of these mechanisms under normal buffering conditions can vary greatly depending on the pattern of afferent activation.

In addition to the obvious pharmacological distinction between the glutamate receptor-evoked alkalosis and GABAergic responses, several features pointed to important mechanistic differences. The glutamate receptor-evoked alkalosis was maintained (or increased) in HEPES-buffered media (Chesler and Chain, 1988; Chesler and Rice, 1991), whereas HCO_3^--dependent GABAergic responses were blocked under these conditions (Kaila et

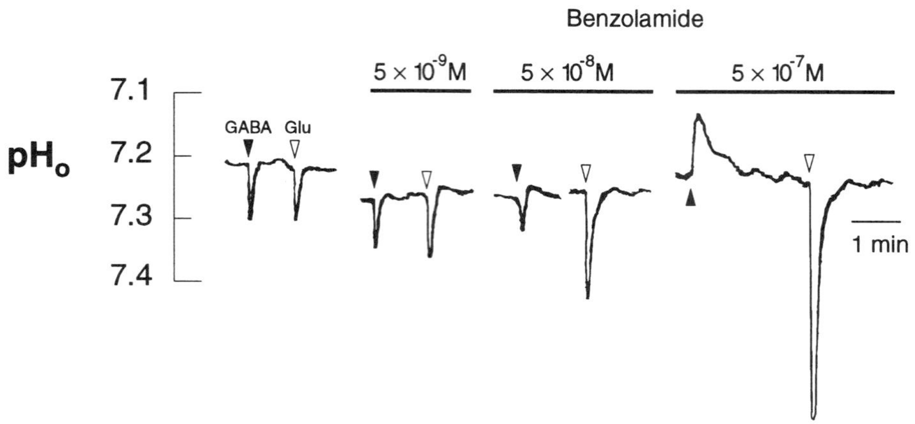

FIGURE 2. Differential effects of a poorly permeant carbonic anhydrase inhibitor, benzolamide, on extracellular alkaline shifts evoked by glutamate versus GABA. Alternate pressure ejection of glutamate (white triangle) or γ-aminobutyric acid (GABA) (black triangle) from a double-barrelled pipette elicited similar alkaline shifts. Inhibition of extracellular carbonic anhydrase with increasing concentrations of benzolamide produced a marked differentiation of these responses. The glutamate-induced alkalosis was amplified several-fold, while the GABA response was diminished, giving way to a brief acid shift. (Modified from Chen and Chesler, 1992c).

al., 1990; Chen and Chesler, 1991). In addition, as illustrated in Fig. 2, the GABAergic alkalosis is blocked by inhibitors of extracellular CA (Kaila et al., 1990; Chen and Chesler, 1992c; Kaila et al., 1992), whereas the glutamate receptor-induced alkaline shifts were enhanced by these agents (Chen and Chesler, 1992c; Voipio and Kaila, 1993; Voipio et al., 1995). These observations are consistent with a mechanism based on a net H^+ sink that is triggered by activation of postsynaptic glutamate receptors (Endres et al., 1986; Chen and Chesler, 1992c). Several of the hypotheses that have been raised to identify this H^+ sink are discussed below.

The cellular H^+ sink. A plausible explanation for synaptically evoked alkaline shifts arose following studies of glutamate transport. Since the uptake of glutamate is coupled to cellular uptake of acid or export of base (Chapter 19), an interstitial alkalosis may be expected. In hippocampal slices, however, the Schaffer collateral-evoked alkalosis was abolished by CNQX and APV (Chen and Chesler, 1992b), agents which block postsynaptic ionotropic glutamate receptors, but not glutamate transport. Moreover, application of the synthetic agonists AMPA and NMDA evoked phenomenologically similar alkaline shifts, although these agents are not transported by the glutamate carrier. It is therefore likely that the contribution of glutamate transport to the interstitial alkaline load is small, and undetectable under normal buffering conditions (Smith et al., 1994).

Given the inward electrochemical gradient for H^+, and a variety of postsynaptic cation channels, a conductive influx of H^+ has often been discussed as a means of generating an H^+ sink (cf. Chesler and Kaila, 1992; Deitmer and Rose, 1996). But while conductive movements of H^+ ions have been observed in certain cells (Thomas and Meech, 1982; DeCoursey and Cherny, 1994; and Chapter 11), there is no direct support for this type of ionic mechanism in the generation of pH shifts in mammalian CNS tissue. One argument against this hypothesis is the conjecture that depolarization would eliminate the driving force for H^+ entry. In addition, there is strong experimental evidence suggesting that the glutamate receptor channels do not serve as an effective H^+-conducting pathway (see Chapter 16). The principal evidence is the dependence of glutamate receptor-evoked alkaline shifts upon external Ca^{2+}. In Ringer's solutions containing Ca^{2+} chelators and no added Ca^{2+}, the alkaline shifts induced by glutamate agonists in hippocampal slices were abolished or strongly reduced (Paalasmaa et al., 1994; Smith et al., 1994). By contrast, concomitant changes in interstitial Na^+ or K^+ were unaffected, indicating normal receptor function in the absence of the H^+ sink. The GABA-induced alkaline shift also remained intact, which indicated that the transmembrane H^+ gradient was not abolished in the absence of Ca^{2+} (Fig. 3). A further relevant observation came from simultaneous measurements of pH_o and extracellular Ca^{2+}, which showed that the initial alkaline shift was coupled to a fast fall in $[Ca^{2+}]_o$ (Paalasmaa et al., 1994).

The Ca^{2+}-dependence of the alkaline shift associated with neuronal excitation has suggested a role for plasmalemmal Ca^{2+}/H^+ exchange. This hypothesis was first proposed by Schwiening et al. (1993; see also Thomas, 1989; and Chapter 15) based on work with snail neurons. Following an influx of Ca^{2+} via voltage-gated Ca^{2+} channels (VGCCs), these investigators noted a surface alkalinization and concomitant intracellular acidification linked to the vanadate-sensitive recovery of intracellular Ca^{2+}. It is notable that this surface alkalosis was generated in response to neuronal depolarization in the absence of synaptic transmission. Recent studies on hippocampal slices have shown that in the absence of synaptic transmission, interstitial alkaline shifts can be induced by direct electrical activation of the CA1 neurons, whether by antidromic stimuli (Paalasmaa and Kaila, 1996; Grichtchenko and Chesler, 1996) or by direct field stimulation (Grichtchenko and Chesler, 1996). These alkaline responses, like those on the surface of snail neurons, were

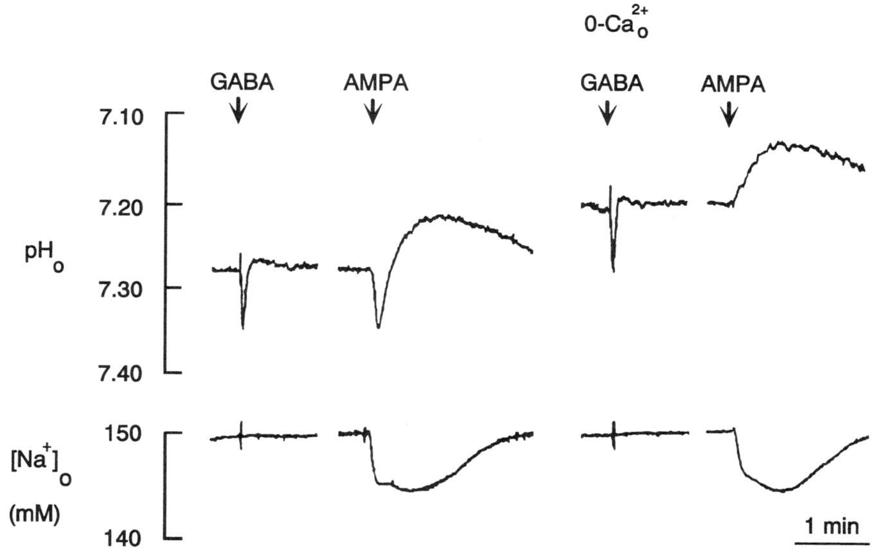

FIGURE 3. Calcium-dependence of alkaline pH_o transients evoked by activation of ionotropic glutamate (AMPA-type) receptors. Removal of extracellular Ca^{2+} selectively abolishes the alkaline pH_o shift evoked by pressure ejection of AMPA, while leaving the GABA-induced alkaline shift, as well as the AMPA-induced fall in extracellular Na^+, unaffected. In all experiments, 1 μM TTX was present. (Modified from Paalasmaa et al., 1994).

abolished in the absence of external Ca^{2+}. The idea that the alkaline pH_o shifts activated by antidromic stimuli are caused by an H^+ sink activated by an influx of Ca^{2+} gained further support from the observations (Paalasmaa and Kaila, 1996) that (i) the alkaline transients were associated with a fall in $[Ca^{2+}]_o$; (ii) both the Ca^{2+} and alkaline shifts were abolished by divalent-cation blockers of VGCCs; and (iii) the alkaline shifts were enhanced by benzolamide.

In CA1 pyramidal neurons, VGCCs are distributed at the soma and throughout the apical dendrites (Spruston et al., 1995). In line with this, during antidromic activation, both the somatic and dendritic fields in area CA1 of hippocampal slices underwent simultaneous extracellular alkaline shifts (Paalasmaa and Kaila, 1996). Since the somata of pyramidal neurons are largely devoid of excitatory synaptic contacts (Andersen, 1990), somatic spiking is likely to be the principal means by which the alkalinization in the cell-body layer is generated, not only by antidromic activation but also by excitatory synaptic transmission (cf. Paalasmaa and Kaila, 1996).

In the dendritic field, the situation is more complicated, since pyramidal-cell dendrites can generate excitatory postsynaptic potentials as well as spikes. During orthodromic synaptic excitation of the CA1 cells, a somatic Na^+ spike has been shown to propagate into the dendrites, activating VGCCs (Spruston et al., 1995). Hence, it was of interest to note that block of the somatic spike by extracellular injection of TTX into the CA1 cell-body layer not only abolished the somatic alkalinization, but also attenuated the alkaline shifts in the dendritic layer, without affecting orthodromic dendritic field potentials (Paalasmaa and Kaila, 1996). These results suggest that in the dendritic field, the alkaline shifts evoked by orthodromic excitation may be triggered by Ca^{2+} entering VGCCs that are activated by back-propagating Na^+ spikes and by excitatory postsynaptic potentials. In addition, Ca^{2+} may enter dendrites via particular subtypes of glutamate receptors, although it is not known whether these influx pathways for Ca^{2+} will activate the plasmalemmal Ca^{2+}/H^+ exchanger. It is important to note here that hyperpolarizing currents sufficient to block somatic Na^+ spikes inhibited 97% of the Ca^{2+} rise in the soma and 90% of the rise in the apical dendrites (Miyakawa et al., 1992). This suggests that Ca^{2+} influx across glutamate receptor-channels does not have a major influence on the H^+ budget of the neuron as a whole.

It is evident from the above studies that the transmembrane movements of H^+ and Ca^{2+} are functionally related. In various preparations, an increase in intracellular free $[Ca^{2+}]$ leads to a fall in pH_i (e.g., Ahmed and Connor, 1980; Meech and Thomas, 1977; Schwiening et al., 1993; Werth and Thayer, 1994; see also Chapters 15 and 16), which has been attributed to a variety of mechanisms in addition to the plasmalemmal Ca^{2+}-ATPase. However, in the case of hippocampal CA1 neurons, recent work provides further evidence for a direct coupling of Ca^{2+} and H^+ movements mediated by the Ca^{2+}/H^+ exchanger. Fluorescence measurements of pH_i and Ca^{2+} in individual cells in hippocampal slices (Trapp et al., 1996) revealed a calcium-linked intracellular acidosis that was inhibited by blockers of VGCCs, as well as by eosin and vanadate, two inhibitors of the transport ATPase (Carafoli, 1991; Gatto and Milanick, 1993; Schwiening et al., 1993).

The data reviewed above are in agreement with the rather simple hypothesis that alkaline pH_o transients associated with neuronal excitation are mainly caused by neuronal depolarizations (usually spiking) large enough to activate VGCCs, which leads to an influx of Ca^{2+} and to an activation of the plasmalemmal Ca^{2+}/H^+-ATPase. This, in turn, would produce a fall in pH_i and a rise in pH_o.

Work related to calcium-dependent cellular processes such as the secretion of synaptic vesicles has shown that, upon the activation of VGCCs, a large increase in the free subplasmalemmal [Ca^{2+}] takes place within less than 1 millisecond (e.g., Llinás et al., 1992). As the plasmalemmal Ca^{2+}/H^{+}-ATPase is also activated by subplasmalemmal Ca^{2+}, the time course of the H^{+} influx on the Ca^{2+}/H^{+}-ATPase might be a prime factor in shaping the temporal properties of the activity-dependent alkaline pH_o shifts associated with neuronal excitation. The work reviewed above is largely based on the use of ion-selective microelectrodes, and a limitation of this technique is its rather poor temporal resolution (Chapter 6). Hence, it is of much interest that recent fluorescence measurements of extracellular alkaline shifts demonstrated a peak of the alkalosis as early as 20 milliseconds after an orthodromic stimulus (Fig. 4) (Gottfried and Chesler, 1996), indicating that pH microelectrode measurements grossly underestimate the speed of these responses. Gaining quantitative insight into the relation of these rapid pH changes with transmembrane Ca^{2+} movements will require high temporal resolution of intracellular Ca^{2+} dynamics with appropriate modeling of the cytoplasmic Ca^{2+} buffering capacity.

Finally, it is prudent to point out here that, while the Ca^{2+}/H^{+}-ATPase is a good candidate as the molecular mechanism underlying alkaline pH_o shifts triggered by neuronal excitation, much more work has to be done to effectively test this hypothesis. Topics of particular concern include the compatibility of the time course of the alkaline transients with the properties of the Ca^{2+}/H^{+}-ATPase, as well as the respective divalent cation (substrate) selectivities. In fact, a recent finding not easily reconciled with the present hypothesis is that Ba^{2+} is able to support electrically evoked alkaline pH_o shifts in hippocampal slices (Grichtchenko and Chesler, 1996; Smith et al., 1997; Tong and Chesler, 1997). However, it should be noted here that the Ba^{2+}-dependent alkaline transients have a rather slow

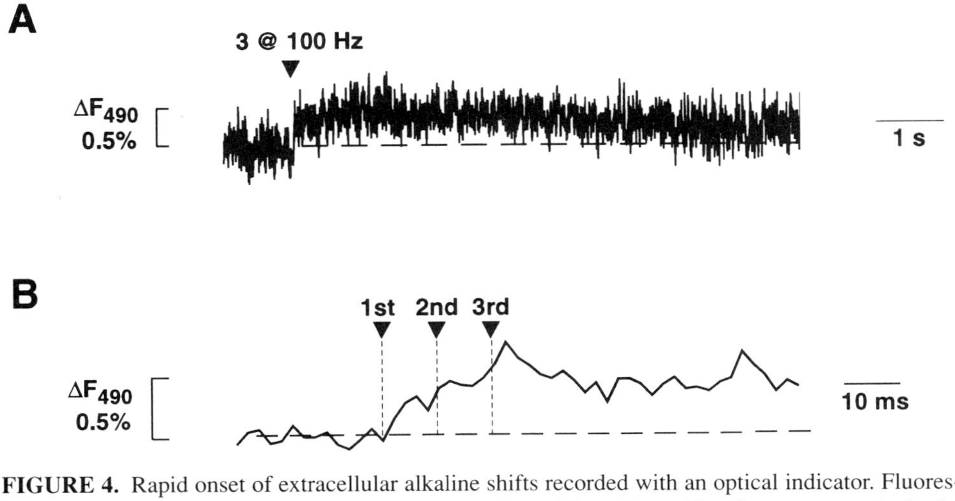

FIGURE 4. Rapid onset of extracellular alkaline shifts recorded with an optical indicator. Fluorescein conjugated to a 2000-kD dextran served as an immobile, pH-sensitive fluorescent probe in the extracellular space of rat hippocampal slices. **A.** When excited by light at 490 nm, rapid stimulation of the Schaffer collaterals (3 shocks at 100 Hz) induced a rapid rise in pH-sensitive emissions at 535 nm (ΔF490) indicative of an alkaline shift. **B.** An expanded timescale indicated a rise of the alkaline shift in less than 20 milliseconds. Records were averaged from 110 traces in two slices. The pH-insensitive emissions elicited by light at 440 nm were unaffected. Arrowheads indicate time of individual stimuli. (Modified from Gottfried and Chesler, 1996).

time course, which may or may not be indicative of yet another ionic mechanism that is involved in the generation of activity-induced pH_o shifts.

Neuronal Inhibition

Whereas there is no evidence for appreciable conductive movements of H^+ linked to excitatory synaptic transmission, the alkaline transients associated with GABAergic inhibition have been shown to arise from a net efflux of HCO_3^- across $GABA_A$ receptor channels. In the light of present information, a significant permeability to HCO_3^- appears to be a universal feature of $GABA_A$ receptors expressed in both vertebrate and invertebrate cells, with the permeability of bicarbonate relative to chloride ranging somewhere between 0.2 and 0.5 (Kaila, 1994). In view of the rather similar permeability properties of the strychnine-sensitive glycine receptor channel (Bormann et al., 1987), it is likely that glycinergic inhibition is also associated with a HCO_3^--dependent rise in pH_o. This prediction gains strength from recent observations in rat medullary neurons where application of both GABA and glycine produces a fall in pH_i (Lückermann et al., 1997).

Exogenous $GABA_A$ receptor. That activation of $GABA_A$ receptors leads to a pH-changing transmembrane flux of bicarbonate was first demonstrated in crayfish muscle fibers (Kaila and Voipio, 1987; Kaila et al., 1990) and neurons (Voipio et al., 1991). The opener muscle fiber of the crayfish proved to be an extremely useful research model in the study of the bicarbonate-dependent actions of GABA, and readily permitted simultaneous measurements of pH_i and of the "surface pH" (pH_s), that is, the pH in the unstirred water layer surrounding the cell. A further advantage of this preparation is the rather small amount of $GABA_A$ receptor desensitization that takes place, even during prolonged (minutes) application of near-saturating agonist concentrations.

Due to acid-extrusion mechanisms that maintain resting steady-state pH_o at a level higher than a passive distribution of HCO_3^- (and H^+) would predict (see Chapters 10–12), the HCO_3^- equilibrium potential is more positive than the resting membrane potential. Therefore, activation of a bicarbonate-permeable conductance pathway, such as the $GABA_A$ receptor, will lead to a net efflux of bicarbonate. This prediction was verified in experiments which showed that exposure of the opener muscle fiber to GABA led to a large (up to 0.5 unit), voltage-sensitive fall in pH_i and a transient rise (by up to 0.1 unit) in pH_s (Kaila and Voipio, 1987; Kaila et al., 1990). Both effects were inhibited in the presence of the $GABA_A$ receptor antagonist, picrotoxin, as well as in solutions devoid of CO_2/HCO_3^-.

Since the unstirred layer is an open compartment for diffusion of acid-base species into the bulk solution, the alkaline shift in pH_s should behave like the time derivative of the change in pH_i (i.e., dpH_i/dt). It was, indeed, evident that the alkaline shift in pH_s peaked at the time when the rate of fall in pH_i was at its maximum (Kaila et al., 1990). On the other hand, when the intracellular acidosis attained its stable maximum value (in the continuous presence of GABA), pH_s relaxed to its original baseline. This is to be expected, since with the intracellular acidosis at its maximum, the channel-mediated net efflux of HCO_3^- must equal the rate of operation of the acid-extruding mechanism, which in the crayfish muscle is mainly a sodium-dependent Cl^-/HCO_3^- exchanger (Galler and Moser, 1986). Hence, at this stage, there can be no net flux of bicarbonate to, or from, the unstirred layer.

In the case of the crayfish opener muscle, it is worth noting that the decay of the pH_s (or pH_o) transient in the continuous presence of the agonist is not due to receptor desensitization, but reflects the attainment of a steady state, where the acid load generated by the channel-mediated net efflux of HCO_3^- equals the rate of plasmalemmal acid extrusion (Kaila et al., 1990). This principle is germane to observations on brain slices, where the alkaline pH_o shift induced by exogenous GABA agonists is also short-lived (cf. Chen and Chesler, 1992c; Kaila et al., 1992). Hence, although desensitization of $GABA_A$ receptors plays a greater role in the mammalian CNS, it is not required for the alkaline shift to have a transient character. In experiments on mammalian brain slices, the major effect of desensitization is probably an attenuation of the peak amplitude of the pH_o response (and on the initial rate of the concurrent change of pH_i), particularly if GABA is applied via the superfusion solution, which must lead to a very slow onset of agonist action.

When HCO_3^- ions leave the cell, intracellular bicarbonate is partly replenished by hydration of CO_2. This, in turn, leads to a CO_2 gradient that drives a net influx of CO_2. Hence, it was interesting to find that application of benzolamide, a poorly permeant CA inhibitor, completely blocked the GABA-induced alkaline pH_s transient (Kaila et al., 1990). This led to the conclusion that the muscle fiber is equipped by an extracellular (probably membrane-bound) CA isoform that has its catalytic site on the external side of the plasma membrane, and plays a role in the regulation of pH_s.

The muscle fiber was also found to have an intracellular CA isoform, since prolonged exposure to acetazolamide had a profound inhibitory action on the GABA-induced fall in pH_i (Kaila et al., 1990). Hence, both intracellular and extracellular CA isoforms assist in the generation of $GABA_A$ receptor-mediated pH_i shifts in the crayfish muscle fiber.

In view of the evolutionary well-conserved permeability properties of $GABA_A$ receptors (Kaila, 1994), it is not surprising that application of $GABA_A$ receptor agonists triggers a transient alkaline pH_o shift in diverse vertebrate CNS tissues, including *in vitro* preparation from the turtle cerebellum (Chen and Chesler, 1990; 1991), and the mammalian hippocampus (Jarolimek et al., 1989; Chen and Chesler, 1992c; Kaila et al., 1992; Voipio et al., 1995) and spinal cord (see Chapter 18). Experiments with inhibitors of CA showed that an extracellular isoform of this enzyme also plays a critical role in the generation of the $GABA_A$ receptor-mediated alkaline pH_o shifts in mammalian brain tissue (Kaila et al., 1992; Chen and Chesler, 1992c; Voipio et al., 1995).

Inhibitory neuronal activity. In addition to $GABA_A$ receptors that are located at postsynaptic sites, neurons are known to have an abundance of extrasynaptic $GABA_A$ receptors. Moreover, glial cells express various kinds of transmitter receptors, including the $GABA_A$ type (e.g., Fraser et al., 1995). Hence, the above demonstrations that application of exogenous agonists leads to the generation of alkaline pH_o transients is not sufficient to show that measurable pH_o shifts are triggered by synaptic activation of $GABA_A$ receptors.

That postsynaptic inhibition mediated by $GABA_A$ receptors does, indeed, lead to significant alkaline shifts was shown in work on rat hippocampal slices. Following complete pharmacological blockade of ionotropic glutamate receptors, activation of a GABAergic pathway evoked an immediate picrotoxin-sensitive alkaline transient in the CA1 cell-body layer (Kaila et al., 1992; Voipio et al., 1995). This alkaline shift was evoked by stimuli applied at 20–200 Hz, but was hardly detectable at lower stimulus frequencies. A characteristic feature of the synaptically evoked $GABA_A$ receptor-mediated alkaline pH_o shift was its short time to peak (1–2 seconds) and brief duration. Hence, it appears that

GABAergic alkaline shifts of the kind described presently are tightly linked to brief high-frequency activation of inhibitory interneurons, such as those that must be triggered by standard stimulation paradigms ("tetanization") used to induce long-term potentiation (Bliss and Collingridge, 1993).

It is also of interest to note that synaptically evoked GABAergic alkaline pH_o shifts occur under experimental conditions where the $GABA_A$ receptor-mediated postsynaptic potential in pyramidal neurons has a pronounced depolarizing component (Voipio et al., 1995; Taira et al., 1995). Both of these features of the postsynaptic response (depolarization and alkaline pH_o transient) are typically seen following high-frequency stimulation of inhibitory interneurons (cf. Grover et al., 1993; Staley et al., 1995) and, at much lower stimulation frequencies in the presence of pentobarbiturate, an allosteric modulator of the $GABA_A$ receptor (Alger and Nicoll, 1982). Recent work (Lamsa and Kaila, 1997; Kaila et al., 1997) suggests that high-frequency stimulation triggers a prolonged activation of a local network of inhibitory interneurons, which is mainly responsible for the pH_o shifts evoked in the presence of antagonists of ionotropic glutamate receptors. The central role of interneuronal network activity in the generation of interstitial ionic shifts in the hippocampus is also evident from recent findings demonstrating that the long-lasting depolarization of pyramidal neurons triggered by high-frequency stimuli is caused by a network-driven, bicarbonate-dependent $[K^+]_o$ transient (Kaila et al., 1997).

Influence of Stimulation Pattern and Neuronal Architecture on Alkaline Shifts

Stimulation of practically any neuronal pathway in mammalian CNS gray matter is bound to lead to the activation of multisynaptic circuits with both excitatory and inhibitory connections. Hence, it is obvious that alkaline shifts evoked at a given recording site are not likely to be attributable to a single cellular or ionic mechanism. This situation is further complicated by the fact that various kinds of neuronal interactions (excitatory/inhibitory, presynaptic/postsynaptic) are differentially affected by the pattern of neuronal activity (e.g., by the frequency and duration of trains of stimuli). An additional factor that is likely to be of importance in the shaping of local pH_o shifts is the fractional volume of neuronal versus glial cells in a given region.

In order to gain an idea about the characteristics of pH shifts that might take place in the interstitial microdomains of the living brain, it is imperative to find out how such responses are affected by local cytoarchitectural features and by various kinds of neuronal activity patterns. Of course, a multisynaptic, mechanistically heterogenous alkaline transient might consist of excitatory "glutamatergic" (H^+-dependent) and inhibitory "GABAergic" (HCO_3^--dependent) parts. Experimentally, it can be difficult to dissect these components. For example, a pharmacological block of glutamatergic excitatory transmission will necessarily abolish inputs that drive inhibitory interneuronal activity. On the other hand, a block of inhibitory transmission (e.g., with picrotoxin) can dramatically increase neuronal firing and thereby augment the "glutamatergic" H^+-dependent alkaline shifts. Interactions between excitatory and inhibitory synaptic inputs may also affect the postsynaptic driving force for HCO_3^- fluxes. Thus, the efflux of HCO_3^- triggered by $GABA_A$ receptor activation may be reduced, or conceivably reversed, if there is a pronounced, simultaneous depolarization due to excitatory inputs.

In view of these considerations, a mechanistically heterogenous alkaline load cannot be considered an algebraic sum of its underlying components, irrespective of how these components are experimentally isolated from one another. Hence, the influence of neuronal architecture and stimulation pattern upon pH_o shifts cannot be discerned by knocking out a single component with a specific postsynaptic receptor antagonist. However, the use of extracellular CA inhibitors has offered an approach to this experimental dilemma (Voipio et al., 1995; Taira et al., 1995). In these experiments on hippocampal slices, the Schaffer collaterals were stimulated with trains with various durations (0.5–20 sec) and frequencies (5–200 Hz), while recording pH_o in either the cell-body or dendritic regions of area CA1. In the dendritic region, benzolamide enhanced the alkaline transients, irrespective of the frequency or duration of the stimulus train. In contrast, the pattern of stimulation had a major influence on pH_o shifts recorded in the cell-body layer (Fig. 5). Here, blocking extracellular CA activity depressed the alkaline transients evoked by brief high-frequency trains of stimuli, whereas the alkaloses induced by lower frequencies of stimulation (5–10 Hz) were enhanced, irrespective of train duration.

The above experiments showed that the alkaline shifts induced by stimulation of Schaffer collaterals are heterogenous in nature, consisting of (i) an H^+-mediated component evoked by neuronal excitation, and enhanced by blocking extracellular CA; and (ii) of a bicarbonate-dependent $GABA_A$ receptor-mediated component that is inhibited by blocking extracellular CA. In particular, the above findings are consistent with the fact that GABAergic synapses are mainly located in the cell-body region, while dendrites receive most of the excitatory inputs (Andersen, 1990). It appears that during continuous high-frequency stimulation of the Schaffer collaterals, the initial alkalosis in the cell-body layer is largely caused by activation of $GABA_A$ receptors; however, during the continuous stimulation this GABAergic component subsides to give way to an alkalosis that is due to excitatory activity (Taira et al., 1995). This transition may be due to a fatigue of inhibition, or to a dissipation of the relevant ionic gradients, or both.

EXTRACELLULAR ACID SHIFTS

Classification of Interstitial Acid Shifts

Acidification of the interstitial space is a common response to intense neuronal activity. Typically, acid shifts have been found to follow the rapid alkalinization described above, appearing and recovering with a time course of minutes (Chesler, 1990; Chesler and Kaila, 1992). There have been notable exceptions where acidification preceded, or occurred in the absence of, an alkaline shift (Syková and Svoboda, 1990; Chvatal et al., 1988), as well as cases where such isolated acid shifts displayed a particularly rapid onset. The variability in the timing of these phenomena may suggest a number of origins for these responses. While the temporal characteristics of these pH shifts may have bearing on their origin, the mechanistic insights offered by consideration of time course alone remain limited.

In evaluating the mechanistic basis for an interstitial acidosis, it is clear that the role of classical acid extrusion mechanisms must be weighed (see Chapters 10–14). Yet, it should be recognized that the plasma membrane is the final common pathway for the appearance of interstitial acid, irrespective of its cellular origin. Thus, while it is important to evaluate the pathway by which acid appears in the extracellular space, plasmalemmal ion transport may or may not represent the primary basis for a given interstitial acidosis.

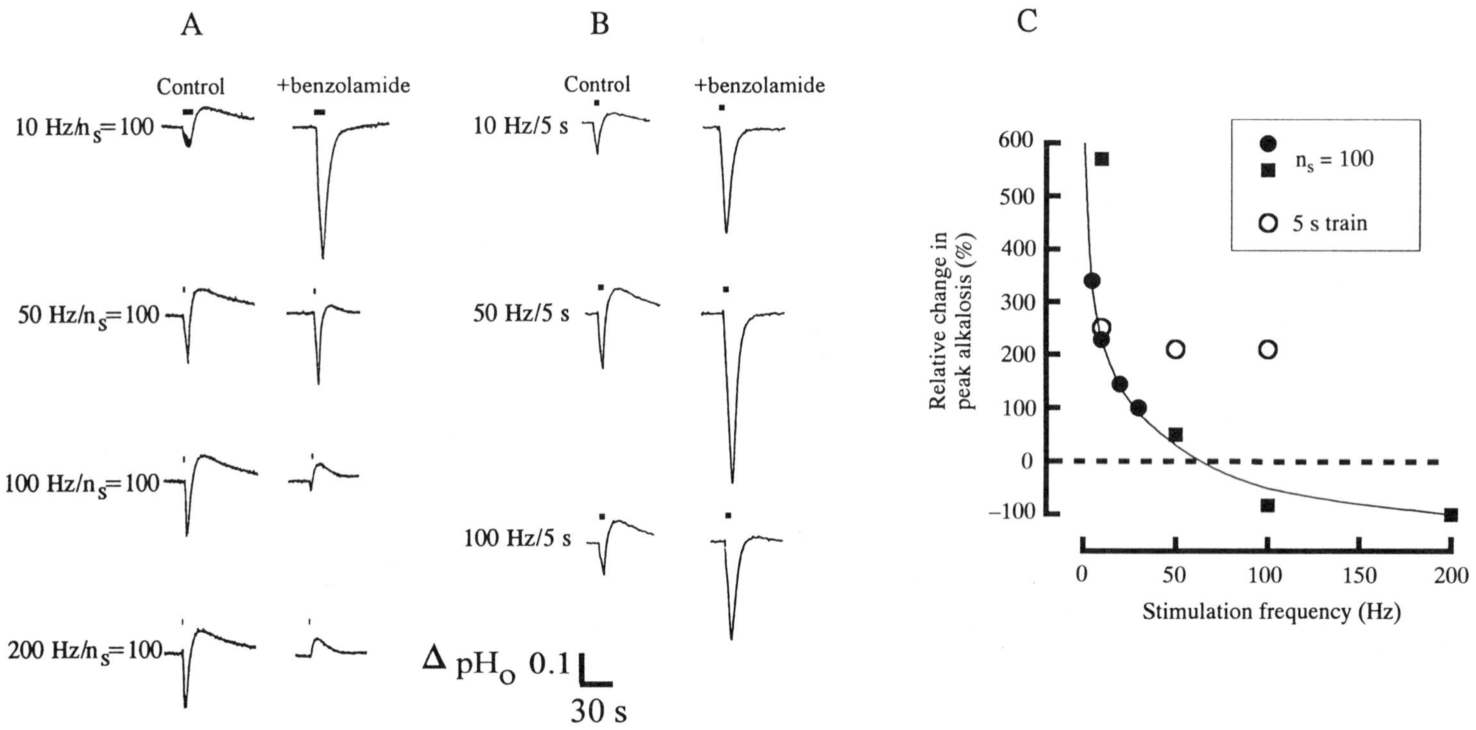

FIGURE 5. Relative contributions of excitatory (glutamatergic) and inhibitory (GABAergic) neuronal activity to alkaline pH_o shifts in the CA1 stratum pyramidale as revealed by differential effects of benzolamide. **A.** A constant number ($n_s = 100$) of pulses evokes an alkaline transient that is enhanced by benzolamide at low stimulation frequencies, but attenuated at high frequencies. **B.** A stimulus train with a constant duration (5 seconds) evokes an alkaline transient that is enhanced by benzolamide irrespective of stimulation frequency. **C.** The peak amplitudes of the alkaline shifts in (A) and (B) and in a third experiment are plotted against stimulation frequency. These data indicate that the alkaline transient evoked by stimulation of Schaffer collaterals is heterogenous, containing a "glutamatergic" (mainly spiking-associated) component, as well as a component directly due to activation of $GABA_A$ receptors. The latter component assumes a dominant role during brief high-frequency trains of stimuli (modified from Taira et al., 1995)

Better distinctions might be made based on the functional origin of an interstitial acidosis. For example, a metabolic basis has been suggested by some investigators (Kraig et al., 1983; Spuler et al., 1987). An extracellular acid shift could be called metabolic if hydrogen ions originate from the cells' respiratory or intermediary metabolism. Clearly, hydrogen ions so generated might cross the plasma membrane by a variety of means. Yet, regardless of the pathway by which acid reaches the extracellular space, the resulting interstitial acidosis could be regarded as metabolic.

The appearance of interstitial acid may also originate from a cytoplasmic acidosis due to organellar transporters. A classic example is the sequestration of Ca^{2+} by mitochondria, which is coupled to the extrusion of H^+ into the cytoplasm (Meech and Thomas, 1980; Werth and Thayer, 1994). Sequestration of Ca^{2+} also occurs by activation of a Ca^{2+}-ATPase in the endoplasmic reticulum. The obligatory exchange of hydrogen ions for Ca^{2+} featured by this class of transporter (see also Chapters 15 and 16) may also contribute to cytoplasmic acidification following neuronal activity. Although entry of Ca^{2+} has most often been associated with organellar exchange for H^+, in a recent study, Ca^{2+} influx into neurons triggered a cytoplasmic acidosis attributed to lactate metabolism (Wang et al., 1994).

An interstitial acid shift may also arise from the activation of plasmalemmal transport without an antecedent fall in cytoplasmic pH. This distinction applies to the voltage-dependent secretion of acid by astrocytes, which has been associated with the electrogenic cotransport of sodium and bicarbonate ions (Chapters 13 and 14). Alternative Na^+-independent mechanisms may also contribute to extrusion of acid by glial cells in the absence of cytoplasmic acidification (Pappas and Ransom, 1993).

In view of these functional distinctions, it appears plausible that a given activity-dependent acidosis could have one or several origins. The relative contribution of metabolic, organellar, and plasmalemmal factors may vary, depending upon the cytoarchitecture of a brain region, its developmental stage, or its functional state.

Metabolic and Organellar Origins of Stimulus-Evoked Acidification

The role of lactic acid. The earliest reports of activity-dependent acidosis suggested a metabolic origin. In response to seizure, decreases in brain surface pH were noted (Dusser de Barenne et al., 1938; Jasper and Erickson, 1941; Wang and Sonnenschein, 1955), and these correlated with increased total tissue lactate (Stone et al., 1945). Similar correlations have been made during the propagation of cortical spreading depression (Scheller et al., 1992). In rat cerebellum, Kraig and colleagues (1983) noted that elevation of $[K^+]_o$ induced an interstitial acidosis that was suppressed by fluoride, a glycolytic inhibitor. Consistent with a role for glycolysis, stimulus-evoked acid shifts have been noted to increase with the glucose concentration of the superfusate (Spuler et al., 1987).

The notion that the tissue generates lactic acid, and thus an acidosis in response to a metabolic challenge, is well established in the literature of brain ischemia (see Chapters 28–30). But the means by which intracellular lactic acidosis results in interstitial acidosis, and how this concept is applicable to activity-dependent acid shifts under normoxic conditions requires further consideration. It is clear that during severe activity, or during the ionic disturbances of spreading depression, lactate generation can increase (Duffy et al., 1975; Stone et al., 1945; Scheller et al., 1992). Since brain cells have plasmalemmal transport systems for lactate (Assaf et al., 1990; Walz and Mukerji, 1988; 1990; Nedergaard and Goldman, 1993), it is reasonable to postulate that the interstitial acidosis arises from the extrusion of lactic acid. Yet, even under hypoxic conditions, lactate transport inhibitors did not effect the resulting interstitial acidification noted in hippocampal slices (Krnjević and Walz, 1990). In addition, the interstitial acidosis induced by the elevation of K^+ in gliotic tissue was also unaffected by lactate transport inhibitors (Grichtchenko and Chesler, 1994b). Thus, it is clear that brain cells have effective means of transporting this metabolite, but there is ample evidence to suggest that activity-dependent interstitial acid shifts can be unrelated to the extrusion or the generation of lactate.

The role of hypercarbia. The implementation of a fast-responding microelectrode for carbon dioxide (Voipio and Kaila, 1993) resulted in a significant advance in the study of activity-dependent acid shifts (see also Chapter 6). Following repetitive synaptic activation of the CA1 pyramidal cells in hippocampal slices, a rise in interstitial P_{CO_2} was noted. This rise in P_{CO_2} accounted for nearly all of the accompanying acid shift. The baseline acidosis of the interstitial space was similarly explained by an elevation of P_{CO_2}. Identical conclusions were reached by Chesler and coworkers, using the complementary technique of interstitial CO_3^{2-} and pH measurements (Chesler and Chen, 1993).

Since oxygen consumption is pronounced in a brain slice (Fujii et al., 1982; Bingmann and Kolde, 1982), it is reasonable to speculate that the generation of interstitial CO_2 is the result of aerobic metabolism. Whereas, *in vivo,* much of the CO_2 due to aerobic respiration can be dispersed by blood flow, the egress of CO_2 across tissue and the unstirred layers of the bath can occur only by diffusion (see Chapter 4). The limited effectiveness of this process is likely to account for the standing CO_2 gradient between the bath and the tissue.

An increase in oxygen consumption, and hence CO_2 production, may similarly explain the generation of slow activity-dependent acid shifts in hippocampal slices (Voipio and Kaila, 1993). The observation of similar acid shifts *in vivo* (Kraig et al., 1983) may suggest a limited rate of CO_2 clearance by blood flow. However, comparable combined studies of pH and P_{CO_2} transients *in vivo* have not been performed.

Alternative Sources of CO_2 and H^+ in the Interstitial Space

It should be pointed out that the titration of intracellular bicarbonate by cytoplasmic H^+ can also generate intracellular CO_2, which would readily cross the plasma membrane (Voipio and Kaila, 1993; Voipio and Ballanyi, 1997; Björkholm et al., 1997). On this basis, both lactic acidosis, and organellar acid transport could indirectly result in the elevation of interstitial P_{CO_2}. However, since the slow acid shifts in hippocampal studies were associated with little or no change in $[HCO_3^-]_o$, it seems unlikely that plasmalemmal acid extruders made a significant contribution. Thus, an early report noting a decrease in the interstitial acid shift in the presence of amiloride (Walz, 1989) might be attributed to this drug's inhibition of low threshold Ca^{2+} channels (Tang et al., 1988) and consequent effects on neuronal excitability. Indeed, the Na^+/H^+ exchanger in hippocampal neurons is notably insensitive to amiloride and its derivatives (Raley-Susman et al., 1991; Schwiening and Boron, 1994). Thus, attempts to manipulate the pH_i of CA1 neurons by blocking Na^+/H^+ exchange with such agents are unfounded.

By contrast, in the leech nervous system, Na^+/H^+ exchange is sensitive to amiloride (Deitmer and Schlue, 1987), and application of this inhibitor diminished interstitial acid shifts induced by electrical activity (Rose and Deitmer, 1995b). In addition, in lamprey brainstem, where neuronal Na^+/H^+ exchange is also amiloride-sensitive (Chesler, 1986), this drug produced an immediate reduction of the baseline interstitial acidosis (Chesler, 1987). Thus, in some preparations, the activation of classic acid extrusion mechanisms by intracellular H^+ may serve as an indirect source of an interstial acidosis.

Direct plasmalemmal sources of H^+. The hallmark of a transient interstitial acidosis generated by plasmalemmal extrusion of H^+ (for instance, on Na^+/H^+ exchange or on lactate/H^+ cotransport) would be a marked fall in $[HCO_3^-]_o$. Since the titration of interstitial bicarbonate would generate CO_2, an apparent "mixed" acidotic picture would be expected. To date, the only direct demonstration of such a response comes from studies of optic nerve, where combined measurements of P_{CO_2} and $[HCO_3^-]_o$ revealed an increase in CO_2 tension and a fall in interstitial bicarbonate following repetitive stimulation (Ransom et al., 1994; see also Voipio and Ballanyi, 1997). However, despite the lack of direct measurements there is reason to speculate that stimulus-evoked acid shifts are of a "mixed" nature in a number of other instances. Insights into this issue are revealed by considering the effects of carbonic anhydrase inhibitors.

When caused by a pure rise in P_{CO_2}, an interstitial acid shift arises due to the hydration of carbon dioxide to carbonic acid (Chapter 4). If the rise in P_{CO_2} is sufficiently rapid, the generation of such an acid shift can be blocked by benzolamide, an inhibitor of extracellular carbonic anhydrase (Voipio and Kaila, 1993). Alternatively, if an acid shift is principally due to net H^+ extrusion, it should be increased by benzolamide, as has been noted in the context of glial acid extrusion (Grichtchenko and Chesler, 1994b; Newman, 1996). Finally, an acid shift may be found insensitive to benzolamide, suggesting a cancellation of effects. This has been noted in rat hippocampal slices during high-frequency stimulation (Taira et al., 1995). In such an instance, both the cellular generation of CO_2 and the net extrusion of acid may have featured prominently.

Glial acid secretion: plasmalemmal transport as a primary source of interstitial acidosis. A decade of evidence has revealed that astrocytes play a key role in the acidification of the extracellular microenvironment during neural activity. A transmembrane flux of acid equivalents was suggested by early recordings of intracellular pH from cortical astrocytes in the rat. During repetitive stimulation of the cortical surface, the cytoplasmic pH of these cells underwent a rapid alkalinization that paralleled the time course of membrane depolarization (Chesler and Kraig, 1987; 1989). These cytoplasmic alkaline shifts arose within seconds, and attained peak magnitudes as great as 0.4 pH unit during repetitive electrical activity, and nearly 1 pH unit during cortical spreading depression. Taking into account the intracellular buffering power, it was reasoned that the observed increase in cytoplasmic pH would require the rapid addition of base equivalents in excess of 10 mM. In view of the magnitude of these changes, it was postulated that a transmembrane acid efflux was the most reasonable basis for these pH shifts.

The astrocytic alkaline shifts appeared related to the membrane potential since their speed and size were correlated with that of the activity-dependent astrocytic depolarization. Support for this idea was gained by preventing the depolarizing response of the astrocytes using Ba^{2+}. With the inward rectifying K^+ channels blocked by Ba^{2+}, cortical astrocytes do not undergo a depolarization during neural activity, despite a rise in external K^+ (Ballanyi et al., 1987). Thus, in the presence of Ba^{2+} both the activity-dependent rise in astrocytic pH_i and the accompanying depolarization were blocked (Chesler and Kraig, 1989). Moreover, correlative measurements of cortical pH_o revealed that Ba^{2+} enhanced an early interstitial alkaline shift. This observation was consistent with the inhibition of depolarization-induced "acid secretion" (i.e., uptake of base) by the astrocytes.

Depolarization-induced alkaline shifts (DIAs) have been noted in a variety of vertebrate glia, including cultured astrocytes (Pappas and Ransom, 1994; Brookes and Turner, 1994), cultured oligodendrocytes (Kettenmann and Schlue, 1988), and retinal Müller cells (Newman, 1991), and have recently been observed in brainstem glial cells *in vivo* (Ballanyi et al., 1994; see also Chapter 14). It has proved especially significant that a similar response can be recorded in the glial cells of the leech, a finding that has suggested phylogenetic conservation of these processes throughout animal evolution. Moreover, the large size of these cells has enabled rigorous study of the mechanism of DIA by Deitmer and colleagues (Deitmer and Rose, 1996). These investigators demonstrated that the DIA in glial cells of the leech was mediated by electrogenic Na^+/HCO_3^- cotransport with a stoichiometry of 1:2 (Deitmer and Schlue, 1989; Deitmer and Szatkowski, 1990; Munsch and Deitmer, 1994 and Chapter 13), indicating that the transporter was close to equilibrium at the resting potential. In this respect, the mechanism was distinct from the 1:3 stoichiometry of Na^+/HCO_3^- cotransport of renal cells, where the gradient for transport was poised in the the outward direction, enabling reabsorption of bicarbonate into the blood (Boron and Boulpaep, 1983).

Subsequent studies of the DIA in cultured astrocytes have indicated a mechanism similar to that of the leech, in which Na^+ and HCO_3^- are transported inward with a 1:2 stoichiometry (Pappas and Ransom, 1994; Brookes and Turner, 1994). In the Müller cells of the retina, however, there exists a 1:3 stochiometry, such that transport is poised outward at the resting potential (Newman, 1991; Newman, 1996). There is recent evidence that the DIA noted in cultured oligodendrocytes (Kettenmann and Schlue, 1988) is also based on a 1:3 coupling ratio (Boussof et al., 1997).

While Na^+/HCO_3^- cotransport appears to be the principal mechanism for the DIA of astrocytes, additional processes may contribute in some systems. An electrogenic, bafilomycin-sensitive H^+ATPase can contribute a small fraction of the alkaline shift in cultured astrocytes (Pappas and Ransom, 1993). In addition, a significant fraction of the DIA in reactive hippocampal astrocytes did not require external Na^+ (Grichtchenko and Chesler, 1994a). Whether this Na^+-independent process represents transmembrane acid extrusion remains to be determined.

Although studies of intracellular pH have demonstrated that DIA is mediated largely, or exclusively by Na^+/HCO_3^- cotransport, there have been fewer studies linking this process to external acidification. A role of glial cells in the generation of activity-dependent acid shifts was suggested by observations in developing optic nerve (Davis et al., 1987) and spinal cord (Jendelova and Syková, 1991). In these neonatal studies, the pure interstitial alkaline shifts noted in the youngest animals gave way to progressively larger acidifications as the tissues matured. The appearance of the acid shifts was correlated with the maturation and proliferation of the neuroglia.

Later work provided direct evidence linking DIAs and interstitial acidification. In an elegant study in the leech, Rose and Deitmer (1994) voltage-clamped the principal ganglionic glial cell, and demonstrated a concomitant loss of extracellular acidification during repetitive stimulation, unmasking an interstitial alkalinization. In vertebrate systems, DIAs have been linked to external acid shifts by studying glial cells in the absence of neurons. In hippocampal astrocytic scars, DIAs have been correlated with acidification of the interstitium. It was notable that the external acid shift induced by elevation of $[K^+]_o$ was diminished in the nominal absence of HCO_3^-, and was abolished in the absence of Na^+ (Grichtchenko and Chesler, 1994b). These depolarization-induced acid shifts were enhanced by the inhibition of interstitial carbonic anhydrase, a result consistent with neither the elevation of CO_2, nor the direct influx of HCO_3^-. This observation led to the hypothesis that the Na^+ and HCO_3^--dependent acid shifts were due to the net transport of Na^+ and CO_3^{2-}. However, since a large component of the DIA persisted in the absence of external Na^+ (Grichtchenko and Chesler, 1994a), it was suggested that a significant fraction of the rise in pH_i was not due to transmembrane acid transport.

The link between DIA and external acidification has recently been studied in single retinal Müller cells from the mudpuppy (Newman, 1996). A major technical advance was achieved by plating the dissociated cells upon a fixed, pH-sensitive fluorophore and imaging pH shifts beneath a single cell. In these studies, depolarization induced a DIA and an external acid shift that was enhanced by benzolamide, as was noted in reactive astrocytic scars (Grichtchenko and Chesler, 1994a,b). This finding in isolated Müller cells represented the first association of external carbonic anhydrase activity with the surface of a single cell from the nervous system.

INTRINSIC PROTON MODULATION OF NEURONAL FUNCTION

Global changes of pH in brain tissue are known to have a pronounced effect on neuronal excitability, with alkaline and acid shifts usually exerting proconvulsant and anticonvulsant actions, respectively (Chapter 20). The presence in neurons of a large number of ligand- and voltage-gated channels sensitive to small changes in pH_o probably plays an important role in effects of the above kind. Perhaps even more interestingly, the abundance of H^+-sensitive target proteins also raises the possibility that, on a much more restricted spatial and temporal scale, activity-induced pH_o shifts may play an important role in the modulation of synaptic transmission and neuronal excitability. Since the H^+ transients underlying activity-induced pH shifts are caused by the workings of cells of the brain parenchyma (neurons and glia), this kind of modulatory effect was originally termed "intrinsic proton modulation" (Taira et al., 1993), to distinguish it from pH effects on nervous function that arise from exogenous sources (e.g., from changes in respiration or from experimental manipulations of ambient pH in experiments *in vitro*). As discussed earlier, pH_o shifts are by necessity paralleled by changes in pH_i. The latter probably have effects of their own on, thereby adding a further dimension to intrinsic H^+ modulation (see Chapter 16).

The functional relevance of depolarization-induced acid secretion from glial cells may lie in its relationship to the extracellular alkaline shifts, as has been discussed in a number of previous reviews (Chesler, 1990; Ransom, 1993; Deitmer and Rose, 1996). It has been suggested that the rapid release of acid can serve to limit, neutralize, or supercede what might otherwise be a large rise in pH_o. Thus, glial acid secretion could serve to modulate the level of excitability of neural ensembles. More intense or widespread activity would give rise to a larger increase in $[K^+]_o$, and correspondingly greater depolarization-induced acid secretion.

At the present, the only direct experimental evidence for intrinsic proton modulation comes from work on the role of alkaline pH_o shifts on hippocampal slices, where the NMDA receptors of pyramidal neurons were found to be subject to the modulatory effect of alkaline transients associated with excitatory transmission (Taira et al., 1993; Gottfried and

Chesler, 1994). Certain NMDA receptors are known to be downmodulated by protons, so that at they are under partial block at physiological pH_o levels (Chapter 22). A key finding in the above experiments was that, in agreement with the facilitatory action that blocking interstitial CA has on the alkaline transients, application of benzolamide led to an enhancement of NMDA receptor-mediated synaptic responses. The recent fluorescence measurements of pH_o shifts (Gottfried and Chesler, 1996) indicate that the extracellular H^+ signals are, indeed, fast enough to modulate synaptic currents.

As such, the above findings would suggest that blocking interstitial CA ought to have a facilitatory effect on nervous excitability. However, at least in hippocampal pyramidal neurons, inhibition of interstitial CA is expected to lead to an enhancement of $GABA_A$ receptor-mediated inhibition as well. There are two factors to be considered here. First, $GABA_A$ receptors with an agonist sensitivity similar to that of synaptic receptors are upmodulated by H^+ ions (Pasternack et al., 1996), which is the opposite of the situation with pyramidal-cell NMDA receptors (Tang et al., 1990). Secondly, the alkaline pH_o shift mediated by $GABA_A$ receptors is suppressed by inhibitors of interstitial CA which is also the opposite of what happens in the case of neuronal excitation. These two qualitatively opposing factors are expected to result in a facilitation of both NMDA and $GABA_A$ receptor-mediated responses upon inhibition of interstitial CA. Indeed, recent work on monosynaptically evoked $GABA_A$ receptor-mediated responses in hippocampal CA1 pyramidal neurons has demonstrated a significant facilitatory action of benzolamide on fast IPSPs and IPSCs (S. Smirnov and K. Kaila, unpublished).

When considering the neuromodulatory actions of pH shifts associated with neuronal excitation and inhibition, it is important to note that the $GABA_A$ receptor channel directly gates transmembrane fluxes of HCO_3^- thereby driving the CO_2/HCO_3^- shuttle; hence the maximum of the alkaline shift is likely to be localized close to the receptor itself. However, as emphasized previously (Paalasmaa et al., 1994), an analogous situation does not hold for the alkaline shifts associated with excitatory transmission. Insofar as these alkaline shifts may be generated by the Ca^{2+}/H^+-ATPase, their localization could be related to the transmembrane pathways that mediate Ca^{2+} movements. Hence, their strongest effects may be exerted on targets that probably include VGCCs in addition to the postsynaptic receptors.

Another factor that might well generate spatial variability of pH_o signals and thereby play a role in proton modulation is the cellular and subcellular localization of the interstitial CA. If this kind of CA activity is due to isoforms residing in glial membranes (see Newman, 1996), then the localization of astrocyte processes would perhaps serve to shape pH_o shifts.

Finally, there is a further point that should be taken into account when judging the effects of blocking interstitial CA activity on neuronal excitability. This factor arises from the spatial and temporal heterogeneity of the activity-induced pH_o shifts. With both excitatory and inhibitory transmission intact, the pH_o transients generated by $GABA_A$ receptors and the Ca^{2+}/H^+-ATPase (or some other H^+ sink) are not likely to sum up in the vicinity of the various H^+-sensitive targets in a manner that could be spatially or temporally resolved by any technique that exists currently for the measurement of pH_o. This conclusion not only underscores the difficulties encountered in experimental studies on intrinsic proton modulation, but it also gives an idea of the versatility of an interstitial communication mechanism that is based on H^+ transients.

CONCLUSIONS

By considering the complex geometry of the brain cell microenvironment and the large number of distinct mechanisms that give rise to transient shifts in interstitial pH, this chapter provides a basis on which a hypothesis of H^+-mediated signalling may be built. Indeed, a number of proteins that are intimately involved in the control of synaptic function and neuronal excitability show significant sensitivity to changes in pH_o within the physiological range. In view of the varied and often localized mechanisms that can rapidly modulate pH_o, a number of neuromodulatory roles can be postulated. In most cases, however, the mechanisms by which pH_o transients are generated, and the targets of these putative H^+ signals, require additional experimental work for a theory of H^+ signalling to gain credence.

It appears likely that a compelling area of emphasis in this regard will be the relationship between Ca^{2+} and H^+ transients, and how these ions interact in the induction of long-term changes in synaptic efficacy and other forms of neuronal plasticity. The relationship between Ca^{2+} and H^+ should also be considered in the context of excitotoxic effects, which are thought to be central mechanisms in the pathophysiology of seizure, and ischemic-traumatic brain injury.

REFERENCES

Ahmed Z, Connor JA (1980): Intracellular pH changes induced by calcium influx during electrical activity in molluscan neurons. J Gen Physiol 75: 403–426.

Alger BE, Nicoll RA (1982): Pharmacological evidence for two kinds of GABA receptor on rat hippocampal pyramidal cells studied in vitro. J Physiol (Lond) 328: 125–141.

Andersen P (1990): Synaptic integration in hippocampal CA1 pyramids. Prog Brain Res 83: 215–222.

Assaf HM, Ricci AJ, Whittingham TS, LaManna JC, Ratcheson RA, Lust WD (1990): Lactate compartmentation in hippocampal slices: Evidence for a transporter. Metab Brain Dis 5: 143–154.

Ballanyi K, Grafe P, ten Bruggencate G (1987): Ion activities and potassium uptake mechanisms of glial cells in guinea-pig olfactory cortex slices. J Physiol (Lond) 382: 159–174.

Ballanyi M, Mückenhoff K, Bellingham MC, Okada Y, Scheid P, Richter DW (1994): Activity-related pH changes in respiratory neurones and glial cells of cats. Neuroreport 6: 33–36.

Bingmann D, Kolde G (1982): Po_2-profiles in hippocampal slices of the guinea pig. Exp Brain Res 48: 89–96.

Birbaumer N, Elbert T, Canavan AGM, Rockstroh B (1990): Slow potentials of the cerebral cortex and behavior. Physiol Rev 70: 1–41.

Björkholm SB, Kaila K, Paalasmaa P, Voipio J (1997): Role of glial Na-dependent bicarbonate uptake in the generation of CO_2 transients and acid extracellular shifts evoked by elevated K^+ or neuronal stimulation in the rat hippocampal slice. Soc Neurosci Abstr 23: 1463.

Bliss TVP, Collingridge GL (1993): A synaptic model or memory: Long-term potentiation in the hippocampus. Nature 361: 31–39.

Bormann J, Hamill OP, Sakmann B (1987): Mechanism of anion permeation through channels gated by glycine and γ-aminobutyric acid in mouse cultured spinal neurones. J Physiol (Lond) 385: 243–286.

Boron WF, Boulpaep EL (1983): Intracellular pH regulation in the renal proximal tubule of the salamander. Basolateral HCO_3^- transport. J Gen Physiol 81: 53–94.

Boussouf A, Lambert RC, Gaillard S (1997): Voltage-dependent Na^+-HCO_3^- cotransporter and Na^+/H^+ exchanger involved in intracellular pH regulation of cultured mature rat cerebellar oligodendrocytes. Glia 19: 74–84.

Brookes N, Turner RJ (1994): K^+-induced alkalinization in mouse cerebral astrocytes mediated by reversal of electrogenic Na^+-HCO_3^- cotransport. Am J Physiol 267: C1633–1640.

Carafoli E (1991): Calcium pump of the plasma membrane. Physiol Rev 71: 129–153.

Carlini WG, Ransom BR (1986): Regional variation in stimulated extracellular pH transients in the mammalian CNS. Soc Neurosci Abstr 12: 452.

Chen JC, Chesler M (1990): A bicarbonate-dependent increase in extracellular pH mediated by $GABA_A$ receptors in turtle cerebellum. Neurosci Lett 116: 130–135.

Chen JCT, Chesler M (1991): Extracellular alkalinization evoked by GABA and its relationship to activity-dependent pH shifts in turtle cerebellum. J Physiol 442: 431–446.

Chen JCT, Chesler M (1992a): pH transients evoked by excitatory synaptic transmission are increased by inhibition of extracellular carbonic anhydrase. Proc Natl Acad Sci USA 89: 7786–7790.

Chen JCT, Chesler M (1992b): Extracellular alkaline shifts in rat hippocampal slice are mediated by NMDA and non-NMDA receptors. J Neurophysiol 68: 342–344.

Chen JCT, Chesler M (1992c): Modulation of extracellular pH by glutamate and GABA in rat hippocampal slices. J Neurophysiol 67: 29–36.

Chen JCT, Chesler M (1992d): NMDA and non-NMDA receptors mediate extracellular alkalinizations in the rat hippocampal slice. Soc Neurosci Abstr 18: 1343.

Chesler M (1986): Regulation of intracellular pH in reticulospinal neurones of the lamprey. *Petromyzon marinus*. J Physiol (Lond) 381: 241–261.

Chesler M (1987): pH regulation in the vertebrate central nervous system: Microelectrode studies in the brain stem of the lamprey. Can J Physiol Pharmacol 65: 986–993.

Chesler M (1990): The regulation and modulation of pH in the nervous system. Prog Neurobiol 34: 401–427.

Chesler M, Chan CY (1988): Stimulus-induced extracellular pH transients in the in vitro turtle cerebellum. Neuroscience 27: 941–948.

Chesler M, Chen JCT (1993): Evoked shifts in extracellular pH, bicarbonate and CO_2 in rat hippocampal slice. Soc Neurosci Abstr 19: 1522.

Chesler M, Kaila K (1992): Modulation of pH by neuronal activity. Trends Neurosci 15: 396–402.

Chesler M, Kraig RP (1987): Intracellular pH of astrocytes increases rapidly with cortical stimulation. Am J Physiol 253: R666–R670.

Chesler M, Kraig RP (1989): Intracellular pH transients of mammalian astrocytes. J Neurosci 9: 2011–2019.

Chesler M, Rice ME (1991): Extracellular alkaline-acid pH shifts evoked by iontophoresis of glutamate and aspartate in turtle cerebellum. Neuroscience 41: 257–267.

Chesler M, Chen JCT, Kraig RP (1994): Determination of extracellular bicarbonate and carbon dioxide concentrations in brain slices using carbonate and pH-selective microelectrodes. J Neurosci Methods 53: 129–136.

Chvatal A, Jendelova P, Kriz N, Syková E (1988): Stimulation-evoked changes in extracellular pH, calcium and potassium activity in the frog spinal cord. Physiologia Bohemoslovaca 37: 203–212.

Davis PK, Carlini WG, Ransom BR, Black JA, Waxman SG (1987): Carbonic anhydrase activity develops postnatally in the rat optic nerve. Dev Brain Res 31: 291–298.

DeCoursey TE, Cherny VV (1994): Voltage-activated hydrogen ion currents. J Membr Biol 141: 203–223.

Deitmer JW, Rose CR (1996): pH regulation and proton signalling by glial cells. Prog Neurobiol 48: 73–103.

Deitmer JW, Schlue WR (1987): The regulation of intracellular pH by identified glial cells and neurones in the central nervous system of the leech. J Physiol (Lond) 388: 261–283.

Deitmer JW, Schlue W-R (1989): An inwardly directed electrogenic sodium bicarbonate co-transport in leech glial cells. J Physiol (Lond) 411: 179–194.

Deitmer WJ, Szatkowski M (1990): Membrane potential dependence of intracellular pH regulation by identified glial cells in the leech central nervous system. J Physiol (Lond) 421: 617–631.

Dietzel I, Heinemann U (1986): Dynamic variations of the brain cell microenvironment in relation to neuronal hyperactivity. Ann NY Acad Sci 481: 72–86.

Duffy TE, Howse DE, Plum F (1975): Cerebral energy metabolism during experimental status epilepticus. J Neurochem 24: 925–934.

Dusser de Barenne JG, McCulloch WS, Nims LF (1937): Functional activity and pH of the cerebral cortex. J Cell Comp Physiol 10: 277–289.

Dusser de Barenne JG, Marshall CS, McCulloch WS, Nims LF (1938): Observations of the pH of arterial blood, the pH and electrical activity of the cerebral cortex. Am J Physiol 124: 631–636.

Endres W, Ballanyi K, Serve G, Grafe P (1986): Excitatory amino acids and intracellular pH in motoneurons of the isolated frog spinal cord. Neurosci Lett 72: 54–58.

Fraser DD, Duffy S, Angelides KJ, Perez-Velazquez JL, Kettenman H, MacVicar BA (1995): $GABA_A$/benzodiazepine receptors in acutely isolated hippocampal astrocytes. J Neurosci 15: 2720–2732.

Fujii T, Baumgärtl H, Lübbers DW (1982): Limiting section thickness of guinea pig olfactory cortical slices studied from tissue pO_2 values and electrical activities. Pflügers Arch 393: 83–87.

Galler S, Moser H (1986): The ionic mechanism of intracellular pH regulation in crayfish muscle fibres. J Physiol (Lond) 374: 137–151.

Gardner-Medwin AR (1983): A study of the mechanism by which potassium moves through brain tissue in the rat. J Physiol (Lond) 335: 353–374.

Gatto C, Milanick MA (1993): Inhibition of the red blood cell calcium pump by eosin and other fluorescein analogues. Am J Physiol 264: C1577–C1586.

Gottfried JA, Chesler M (1994): Endogenous H^+ modulation of NMDA receptor-mediated EPSCs revealed by carbonic anhydrase inhibition in rat hippocampus. J Physiol (Lond) 478: 373–378.

Gottfried JA, Chesler M (1996): Temporal resolution of activity-dependent pH shifts in rat hippocampal slices. J Neurophysiol 76: 2804–2807.

Grichtchenko II, Chesler M (1994a): Depolarization-induced alkalinization of astrocytes in gliotic hippocampal slices. Neuroscience 62: 1071–1078.

Grichtchenko II, Chesler M (1994b): Depolarization-induced acid secretion in gliotic hippocampal slices. Neuroscience 62: 1057–1070.

Grichtchenko II, Chesler M (1996): Calcium- and barium-dependent extracellular alkaline shifts evoked by electrical activity in rat hippocampal slices. Neuroscience 75: 1117–1126.

Grover LM, Lambert NA, Schwartzkroin PA, Teyler TJ (1993): Role of HCO_3^- ions in depolarizing $GABA_A$ receptor-mediated responses in pyramidal cells of rat hippocampus. J Neurophysiol 69: 1541–1555.

Heinemann U, Stabel J, Rausche G (1990): Activity-dependent ionic changes and neuronal plasticity in rat hippocampus. Prog Brain Res 83: 197–214.

Heuser D, Astrup J, Lassen NA, Nilsson B, Nordberg K, Siesjö BK (1977): Are H^+ and K^+ factors for the adjustment of cerebral blood flow to changes in functional state? A microelectrode study.

In Ingvar DH, Lassen NA (eds): Cerebral Function, Metabolism and Circulation. Copenhagen: Munksgaard, pp 216–217.

Huang W, Smith SE, Chesler M (1995): Addition of carbonic anhydrase augments extracellular pH buffering in rat cerebral cortex. J Neurophysiol 74: 1806–1809.

Jarolimek W, Misgeld U, Lux HD (1989): Activity dependent alkaline and acid transients in guinea pig hippocampal slices. Brain Res 505: 225–232.

Jasper H, Erickson TC (1941): Cerebral blood flow and pH in excessive cortical discharge induced by metrazol and electrical stimulation. J Neurophysiol 4: 333–347.

Jefferys JGR (1995): Nonsynaptic modulation of neuronal activity in the brain: Electric currents and extracellular ions. Physiol Rev 75: 689–723.

Jendelova P, Syková E (1991): Role of glia in K^+ and pH homeostasis in the neonatal rat spinal cord. Glia 4: 56–63.

Kaila K (1994): Ionic basis of $GABA_A$ receptor channel function in the nervous system. Prog Neurobiol 42: 489–537.

Kaila K, Voipio J (1987): Postsynaptic fall in intracellular pH induced by GABA-activated bicarbonate conductance. Nature 330: 163–165.

Kaila K, Saarikoski J, Voipio J (1990): Mechanism of action of GABA on intracellular pH and on surface pH in crayfish muscle fibres. J Physiol (Lond) 427: 241–260.

Kaila K, Paalasmaa P, Taira T, Voipio J (1992): pH transients due to monosynaptic activation of $GABA_A$ receptors in rat hippocampal slices. Neuroreport 3: 105–108.

Kaila K, Lamsa K, Smirnov S, Taira T, Voipio J (1997): Long-lasting GABA-mediated depolarization evoked by high-frequency stimulation in pyramidal neurons of rat hippocampal slice is due to a network-driven, bicarbonate-dependent K^+ transient. J Neurosci 17, 7662–7672.

Kettenmann H, Schlue WR (1988): Intracellular pH regulation in cultured mouse oligodendrocytes. J Physiol (Lond) 406:147–162.

Kraig RP, Ferreira-Filho CS, Nicholson C (1983): Alkaline and acid transients in cerebellar microenvironment. J Neurophysiol 49: 831–850.

Krnjević K, Walz W (1990): Acidosis and blockade of orthodromic responses caused by anoxia in rat hippocampal slices, at different temperatures. J Physiol (Lond) 422: 127–144.

Lamsa K, Kaila K (1997): Ionic mechanisms of spontaneous GABAergic events in rat hippocampal slices exposed to 4-aminopyridine. J Neurophysiol 78:2582–2591.

Llinás R, Sugimori M, Silver RB (1992): Microdomains of high calcium concentration in a presynaptic terminal. Science 256: 677–679.

Lückermann M, Trapp S, Ballanyi K (1997): GABA-and glycine-mediated fall of intracellular pH in rat medullary neurons in situ. J Neurophysiol 77: 1844–1852.

Maren TH (1984): The general physiology of reactions catalyzed by carbonic anhydrase and their inhibition by sulfonamides. Ann N Y Acad Sci 429: 568–579.

McBain CJ, Traynelis SF, Dingledine R (1990): Regional variation of extracellular space in the hippocampus. Science 249: 674–677.

Meech D, Thomas RC (1980): Effect of measured calcium chloride injections on the membrane potential and internal pH of snail neurones. J Physiol (Lond) 298: 111–129.

Meech RW, Thomas RC (1977): The effect of calcium injection on the intracellular sodium and pH of snail neurones. J Physiol (Lond) 265: 867–879.

Miyakawa H, Ross WN, Jaffe D, Callaway JC, Lasser-Ross N, Lisman JE, Johnston D (1992): Synaptically activated increases in Ca^{2+} concentration in hippocampal CA1 pyramidal cells are primarily due to voltage-gated Ca^{2+} channels. Neuron 9: 1163–1173.

Munsch T, Deitmer JW (1994): Sodium-bicarbonate cotransport current in identified leech glial cells. J Physiol (Lond) 474: 43–53.

Mutch WAC, Hansen AJ (1984): Extracellular pH changes during spreading depression and cerebral ischemia: Mechanisms of brain pH regulation. J Cerebr Blood Flow Metab 4: 17–27.

Nedergaard M, Goldman SA (1993): Carrier-mediated transport of lactic acid in cultured neurons and astrocytes. Am J Physiol 265: R282–R289.

Newman EA (1991): Sodium-bicarbonate cotransport in retinal Müller (glial) cells of the salamander. J Neurosci 11: 3972–3983.

Newman EA (1996): Acid efflux from retinal glial cells generated by sodium bicarbonate cotransport. J Neurosci 16: 159–168.

Nicholson C (1979): Brain-cell microenvironment as a communication channel. In Schmitt FO, Warden FG (eds): The Neurosciences: Fourth Study Program. Cambridge, MA: MIT Press, pp 457–473.

Nicholson C (1980): Dynamics of the brain cell microenvironment. Neurosci Res Prog Bull 18: 175–322.

Nicholson C, Rice ME (1991): Diffusion of ions and transmitters in the brain cell microenvironment. In Fuxe K, Agnati LF (eds): Volume Transmission in the Brain: Novel Mechanisms for Neural Transmission, Advances in Neuroscience, Vol. 1. New York: Raven Press, pp 279–294.

Paalasmaa P, Kaila K (1996): Role of voltage-gated calcium channels in the generation of activity-induced extracellular pH transients in the rat hippocampal slice. J Neurophysiol 75: 2354–2359.

Paalasmaa P, Taira T, Voipio J, Kaila K (1994): Extracellular alkaline transients mediated by glutamate receptors in the rat hippocampal slice are not due to a proton conductance. J Neurophysiol 72: 2031–2033.

Pappas C, Ransom BR (1993): A depolarization-stimulated, bafilomycin-inhibitable H^+ pump in hippocampal astrocytes. Glia 9: 280–291.

Pappas CA, Ransom BR (1994): Depolarization-induced alkalinization (DIA) in rat hippocampal astrocytes. J Neurophysiol 72: 2816–2826.

Pasternack M, Voipio J, Kaila K (1993): Intracellular carbonic anhydrase activity and its role in GABA-induced acid transients in isolated rat hippocampal pyramidal neurones. Acta Physiol Scand 148: 229–231.

Pasternack P, Smirnov S, Kaila K (1996): Protons modulate $GABA_A$ receptors in a dual manner in acutely isolated rat hippocampal neurones. Neuropharmacology 35: 1279–1288.

Raley-Susman KL, Sapolsky RM and Kopito RR (1991). Regulation of intracellular pH in cultured hippocampal neurons by an amiloride-insensitive Na^+/H^+ exchanger. J Biol Chem 266: 2739–2745.

Ransom BR (1993): Glial modulation of neural excitability mediated by extracellular pH: A hypothesis. Prog Brain Res 94: 37–46.

Ransom BR, Voipio J, Paalasmaa P, Kaila K (1994): Mechanisms of activity-dependent extracellular acid shifts in white matter of the mammalian brain. Soc Neurosci Abstr 20: 575.

Rose CR, Deitmer JW (1994): Evidence that glial cells modulate extracellular pH transients induced by neuronal activity in the leech central nervous system. J Physiol (Lond) 481: 1–5.

Rose CR, Deitmer JW (1995a): Stimulus-evoked changes of extra- and intracellular pH in the leech central nervous system. I. Bicarbonate dependence. J Neurophysiol 73: 125–131.

Rose Y, Deitmer JW (1995b): Stimulus-evoked changes of extra- and intracellular pH in the leech central nervous system. II. Mechanisms and maintenance of pH. J Neurophysiol 73: 132–140.

Saarikoski J, Kaila K (1992): Simultaneous measurement of intracellular and extracellular carbonic anhydrase activity in intact muscle fibres. Pflügers Arch 421: 357–363.

Scheller D, Kolb J, Tegtmeier F (1992): Lactate and pH change in close correlation in the extracellular space of the rat brain during cortical spreading depression. Neurosci Lett 135: 83–86.

Schwiening CJ, Boron WF (1994): Regulation of intracellular pH in pyramidal neurones from the rat hippocampus by Na^+-dependent Cl^--HCO_3^--exchange. J Physiol (Lond) 475: 59–67.

Schwiening CJ, Kennedy HJ, Thomas RC (1993): Calcium-hydrogen exchange by the plasma membrane Ca-ATPase of voltage-clamped snail neurons. Proc R Soc Lond Biol 253: 285–289.

Smith SE, Amos BJ, Chesler M (1997): Dynamics of activity-evoked alkaline shifts and intracellular Ca and Ba transients in hippocampus. Soc Neurosci Abstr 23: 136.

Smith SE, Gottfried JA, Chen JCT, Chesler M (1994): Calcium dependence of glutamate receptor-evoked alkaline shifts in hippocampus. Neuroreport 5: 2441–2445.

Spruston N, Schiller Y, Stuart G, Sakmann B (1995): Activity-dependent action potential invasion and calcium influx into hippocampal CA1 dendrites. Science 268: 297–300.

Spuler A, Endres W, Grafe P (1987): Metabolic origin of activity-related pH changes in mammalian peripheral and central unmyelinated fibre tracts. Pflügers Arch 408: R69.

Staley KN, Soldo BL, Proctor WR (1995): Ionic mechanisms of neuronal excitation by inhibitory $GABA_A$ receptors. Science 269: 977–981.

Stone WE, Webster JE, Gurdjian ES (1945): Chemical changes in the cerebral cortex associated with convulsive activity. J Neurophysiol 8: 233–240.

Syková E, Svoboda J (1990): Extracellular alkaline-acid-alkaline transients in the rat spinal cord evoked by peripheral stimulation. Brain Res 512: 181–189.

Taira T, Smirnov S, Voipio J, Kaila K (1993): Intrinsic proton modulation of excitatory transmission in rat hippocampal slices. Neuroreport 4: 93–96.

Taira T, Paalasmaa P, Voipio J, Kaila K (1995): Relative contributions of excitatory and inhibitory neuronal activity to alkaline transients evoked by stimulation of Schaffer collaterals in the rat hippocampal slice. J Neurophysiol 74: 643–649.

Tang C-M, Presser F, Morad M (1988): Amiloride selectively blocks the low treshold (T) calcium channel. Science 240: 213–215.

Tang C-M, Dichter M, Morad M (1990): Modulation of the N-methyl-D-aspartate channel by extracellular H^+. Proc Natl Acad Sci USA 87: 6445–6449.

Thomas RC (1989): Proton channels in snail neurones. Does calcium entry mimic the effects of proton influx? Ann N Y Acad Sci 574: 287–293.

Thomas RC, Meech RW (1982): Hydrogen ion currents and intracellular pH in depolarized voltage-clamped snail neurones. Nature 299: 826–828.

Tong CK, Chesler M (1997): Activity-related alkaline pH shifts in rat lateral geniculate nucleus. Soc Neurosci Abstr 23: 652.

Trapp S, Lückermann M, Kaila K, Ballanyi K (1996): Acidosis of hippocampal neurons mediated by a plasmalemmal Ca^{2+}/H^+ pump. Neuroreport 7: 2000–2004.

Tschirgi RD, Inanaga K, Taylor JL, Walker RM, Sonnenschein R (1957): Changes in cortical pH and blood flow accompanying spreading depression and convulsion. Am J Physiol 190: 557–562.

Urbanics R, Leniger-Follert E, Lübbers DW (1978): Time course of changes of extracellular H^+ and K^+ activities during and after direct electrical stimulation of the brain cortex. Pflügers Arch 378: 47–53.

Voipio J, Ballanyi K (1997): Interstitial P_{CO_2} and pH, and their role as chemostimulants in the isolated respiratory network of neonatal rats. J Physiol (Lond) 499:527–542.

Voipio J, Kaila K (1993): Interstitial P_{CO_2} and pH in rat hippocampal slices measured by means of a novel fast CO_2/H^+-sensitive microelectrode based on a PVC-gelled membrane. Pflügers Arch 423: 193–201.

Voipio J, Pasternack M, Rydqvist B, Kaila K (1991): Effect of γ-aminobutyric acid on intracellular pH in the crayfish stretch-receptor neurone. J Exp Biol 156: 349–361.

Voipio J, Paalasmaa P, Taira T, Kaila K (1995): Pharmacological characterization of interstitial pH transients evoked by selective synaptic and exogenous activation of AMPA, NMDA and $GABA_A$ receptors in the rat hippocampal slice. J Neurophysiol 74: 633–642.

Walz W (1989): pH shifts evoked by neuronal stimulation in slices of rat hippocampus. Can J Physiol Pharmacol 67: 577–581.

Walz W, Mukerji S (1988): Lactate release from cultured astrocytes and neurons: A comparison. Glia 1: 366–370.

Walz W, Mukerji S (1990): Simulation of aspects of ischemia in cell culture: Changes in lactate compartmentation. Glia 3: 522–528.

Wang RIH, Sonnenschein RR (1955): pH of cerebral cortex during induced convulsions. J Neurophysiol 18: 130–187.

Wang GJ, Randall RD, Thayer SA (1994): Glutamate-induced intracellular acidification of cultured hippocampal neurons demonstrates altered energy metabolism resulting from Ca^{2+} loads. J Neurophysiol 72: 2563–2569.

Werth JL, Thayer SA (1994): Mitochondria buffer physiological calcium loads in cultured rat dorsal root ganglion neurons. J Neurosci 14: 348–356.

CHAPTER 18

Extracellular pH and Ionic Shifts Associated with Electrical Activity and Pathological States in the Spinal Cord

EVA SYKOVÁ
Department of Neuroscience, A Joint Institute of the 2nd Medical Faculty, Charles University and the Institute of Experimental Medicine, Academy of Sciences of the Czech Republic, Prague, Czech Republic

INTRODUCTION

It has become evident that neuronal activity leads to functionally significant changes in extracellular K^+, Ca^{2+}, pH, and cellular volume. Activity-related transient changes in extracellular pH (pH_o) in the CNS had already been described in the late 1980s (for review see Chesler, 1990; Syková, 1992; Chapter 17). Early studies with ion-selective microelectrodes had also revealed pH_o changes in the spinal cord evoked by repetitive electrical stimulation of spinal nerves (Chvátal et al., 1988; Syková et al., 1988; Syková and Svoboda, 1990). It has been found that pH_o changes depend on the intensity, frequency, and duration of the stimulation. In the spinal dorsal horns, the pH_o fluctuates by several tenths of a pH unit during stimulation of afferent input. However, the greatest changes have been found during pathological events (e.g., anoxia; Syková et al., 1994). Moreover, it has become apparent that transmembrane ionic fluxes during neuronal activity and pathological states result in cell swelling and compensatory transient changes in extracellular space (ECS) volume and geometry. This has led to the concept of the ECS as a communication and modulation channel (Nicholson, 1980; Syková, 1983, 1991, 1992, 1997; Nicholson and Syková, in press), whose ionic and chemical composition, including changes in acid-base balance, size, and geometry affect neuronal activity and glial cell function. This chapter, which will focus on transient pH_o changes in the spinal cord, will therefore also describe transient changes in K^+ and Ca^{2+} concentrations, and ECS volume and geometry.

RESTING LEVEL OF THE EXTRACELLULAR pH

As with the brain (see Chapter 2), the extracellular pH (pH_o) resting level or so-called pH_o baseline in the unstimulated spinal cord of the rat is lower than in blood (Syková and

pH and Brain Function, Edited by Kai Kaila and Bruce R. Ransom
ISBN 0-041-11838-9

Svoboda, 1990). The resting values of pH_o were established *in vivo* by comparing the potential of pH-sensitive microelectrodes in cerebrospinal fluid above the cord (pH about 7.35–7.40), in Ringer's solutions of various pH values placed above the cord, and in different regions of the cord (Syková and Svoboda, 1990). The pH_o values in the spinal cord varied between 7.10 and 7.35. In the dorsal horn of the isolated spinal cord, the resting pH values were significantly more alkaline in 3- to 6-day-old (P3-P6) rats (pH_o about 7.2) than in older rats (P10–P14; pH_o about 7.1; Jendelová and Syková, 1991). The extracellular K^+ concentration, $[K^+]_o$, at P3–P6 was about 3.9 mM, while at P10–P14 it was about 4.3 mM. In the ventral horn, the pH_o was usually slightly more alkaline (7.25–7.35) than in the dorsal horn. Similarly, in the isolated frog spinal cord, the pH_o resting level in dorsal horn was always 0.1–0.2 pH unit more acidic than the pH of Ringer's solution (7.35) (Chvátal et al., 1988).

In the unstimulated spinal cord, the pH_o baseline had a depth profile similar to that of the extracellular K^+ concentration (Svoboda et al., 1988). After blockage of synaptic activity by applying Mn^{2+} at a concentration of 2–4 mM onto the spinal cord surface, an alkaline shift in pH_o of 0.05–0.10 pH unit was found. It has been shown previously that an elevated extracellular K^+ concentration found in the dorsal horn is associated with a high level of spontaneous activity of dorsal horn interneurons, due to continuous excitatory input from the periphery (Syková et al., 1983; Svoboda et al., 1988). These observations suggest that the pH_o resting level in the dorsal horn is more acid than in the ventral horn. This may be related to release of H^+, movement of an equivalent ion, or the metabolic generation of CO_2.

ACTIVITY-RELATED TRANSIENT pH_o CHANGES

Neuronal activity results in rapid shifts in extracellular ionic composition. Using ion-selective microelectrodes, $[K^+]$, $[Ca^{2+}]$, and pH changes, along with their origins, time courses, and homeostasis, and their functions in physiological signaling, have been described in great detail (for review see Nicholson, 1980; Syková, 1983; Walz, 1989; Chesler, 1990; Syková, 1992). For example, Fig. 1 demonstrates simultaneous records of transient changes in $[K^+]_o$ and pH in the spinal cord of the rat evoked by repetitive stimulation of the dorsal spinal root. Most frequently, alkaline-acid-alkaline transient changes have been found in adult rats. However, acid-alkaline or pure acid shifts have also been observed after electrical or adequate, e.g. tactile, stimulation, particularly during development (Chvátal et al., 1988; Jendelová and Syková, 1991; Syková and Svoboda, 1990).

Electrical Stimulation

Single electrical pulses applied through a pair of needles inserted intramuscularly into the plantar region of an ipsilateral hind limb and direct stimulation of the sciatic nerve or the spinal dorsal root transiently decrease pH_o in the dorsal horn of the corresponding lumbar segment by about 0.01 pH unit. These small acid shifts are preceded either by no visible alkaline shifts or by scarcely discernible ones. Repetitive electrical stimulation at a frequency of 10–100 Hz elicits in the adult rat either a biphasic acid-alkaline or a triphasic alkaline-acid-alkaline response (Fig. 1A,B). The amplitude and duration of the pH_o changes increase with the intensity, duration, and frequency of stimulation. In adult rats

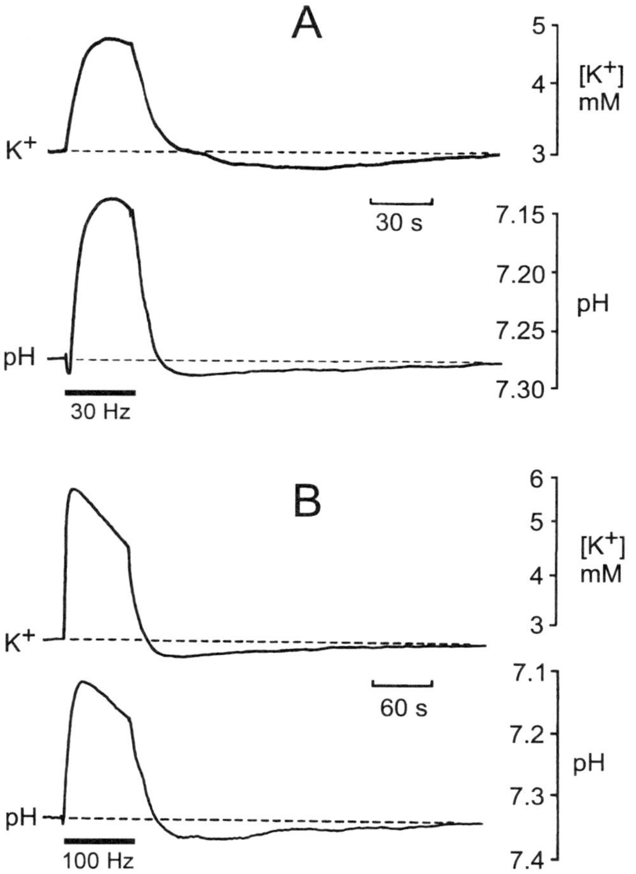

FIGURE 1. $[K^+]_e$ and pH_o changes in dorsal horn of segment L4 of rat spinal cord evoked by repetitive bipolar electrical stimulation with acupuncture needles (**A:** 30 Hz, 30 sec; **B:** 100 Hz, 60 sec) in plantar muscles of the ipsilateral hind paw. Horizontal bar indicates stimulus duration. A K^+- and a pH-sensitive microelectrode were inserted separately from the dorsal spinal surface. Recordings from a depth of 600–700 μm. (Modified from Syková and Svoboda, 1990 with permission).

and frogs, the dominant acid shift of about 0.1–0.2 pH unit is sometimes preceded by a small alkaline shift of about 0.01 pH unit (Fig. 1A). The rising phase of the increase in $[K^+]_o$ corresponds to the initial alkaline shift. The rise time of the activity-related acid shift is always slower. The $[K^+]_o$ peak in the adult rat spinal cord is attained in 5–8 sec, while the peak of the acid shift is reached in 10–20 seconds (Fig. 1B). When stimulation is continued beyond this, a gradual decrease of both transients, $[K^+]_o$ and pH_o, occurs after the peak levels are reached (Fig. 1A and B). Unlike the rises, the recoveries of the $[K^+]_o$ and pH_o changes to the prestimulation values have similar time courses. After returning to the prestimulation $[K^+]_o$ values, a poststimulation K^+ undershoot is accompanied by a poststimulation alkaline "undershoot" (Figs. 1 and 2) (Syková and Svoboda, 1990).

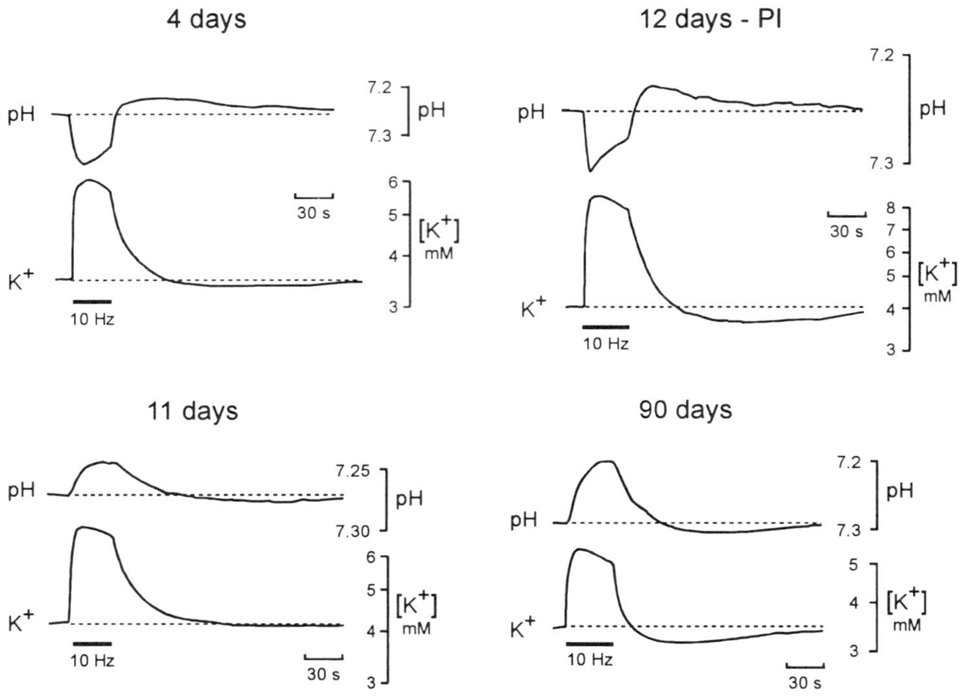

FIGURE 2. Stimulation-evoked pH_o and $[K^+]_o$ changes in the spinal dorsal horn of rats 4, 11, and 90 days old, and in a 12-day-old rat that received a single dose of x-irradiation at lumbar spinal cord at P1 (PI). Note that the stimulation of the dorsal root at a frequency of 10 Hz evoked an alkaline shift in the 4-day-old pup, which was accompanied by an increase in $[K^+]_o$. When stimulation was discontinued, a poststimulation acid shift of smaller amplitude appeared, which was accompanied by a K^+-undershoot. Similar changes were found in 12-day-old animals with postnatal x-irradiation. In the 11-day-old and 90-day-old (adult) rats, the stimulation evoked a monophasic acid shift and a smaller increase in $[K^+]_o$. (From Syková et al., 1992 with permission).

The depth profiles of both the alkaline and acid pH_o shifts in the spinal cord are similar to the depth profile of the stimulation-evoked increase in $[K^+]_o$. Maximal changes in pH_o evoked in the adult rat by electrical stimulation are found in the lower dorsal horn at a depth of about 500–700 μm from the dorsal spinal surface (i.e., in the area of the largest amplitude of field potentials evoked by single electrical pulses of supramaximal intensity, as recorded by a reference-barrel of pH-sensitive microlelectrode). When manipulator micrometer readings are matched with histological data, this area corresponds to Rexed laminae IV–VI.

Adequate Stimulation

Adequate stimulation of the afferent input (e.g., touching or stroking of the skin, squeezing of the toes) also results in transient pH_o changes in the spinal dorsal horn (Syková and Svoboda, 1990). Although these changes are accompanied by a rise in $[K^+]_o$, they are generally of smaller amplitude than those evoked by electrical stimulation. This might explain the fact that initial alkaline shifts are rarely observed. Nevertheless, the acid shifts in pH_o evoked by adequate stimuli always have slower rise times than the $[K^+]_o$ increases. Innocuous stimuli (touching or stroking of the lateral aspect of a hind paw skin with a brush or cotton wick) result in a transient acid shift of about 0.01 pH unit. Acute nociceptive stimuli (squeezing of the toes or application of several drops of hot water), lead to an acid shift of about 0.01–0.05 pH unit, which sometimes persists for 1–2 minutes. Nociceptive stimuli lasting about 1.5–2 hours (e.g., subcutaneous injection of turpentine) result in an acid shift in the spinal dorsal horn of the L4–L5 spinal segment. An acid shift of about 0.05–0.1 pH unit is accompanied by an increase in $[K^+]_o$, which has a similar time course. The $[K^+]_o$ increase and pH_o decrease begin about 2–10 minutes after injection, reach their maximum in about 15 minutes and then persist for 1–2 hours. The maximal changes were found in Rexed laminea III–V, at depths of 400–500 μm from dorsal spinal surface (Syková and Svoboda, 1990) and were accompanied by ECS volume decrease (Svoboda and Syková, 1991). The ionic changes and ECS volume decrease corresponded to the time course of the nociceptive reaction as tested on naive and unrestrained rats by measuring the duration of the licking response (Svoboda et al., 1988). These findings indicate that the ionic and volume changes may be either the cause or the result of the long-term hyperexcitability and synaptic hyperactivity in the dorsal spinal horns.

pH_o Changes During Postnatal Development

Stimulation-evoked changes in $[K^+]_o$ and pH_o differ markedly in early postnatal spinal cord compared with the respective ion shifts in the adult. There has been considerable speculation that these differences derive from developmental changes in the function and number of spinal cord glial cells (Jendelová and Syková, 1991; Syková et al., 1992). Given their extensive membrane surface, the glia are ideally suited to control extracellular ions, and there is considerable evidence for their role in the control of extracellular K^+ homeostasis (for review, see Syková, 1983, 1992). There are also increasing data that suggest a principal role for glia in the regulation of pH_o in the spinal cord.

Activity-related changes in both $[K^+]_o$ and pH_o were studied in spinal cords during early postnatal days, since glial cell proliferation, maturation, and myelination occur postnatally, and more slowly than maturation of neurons. Substantial differences have been found in the first and second postnatal week (Jendelová and Syková, 1991; Syková et al., 1992). In the neonatal rat spinal cord, stimulation-evoked changes in $[K^+]_o$ are much larger than in the adult animal, and pH_o transients differ substantially from the pH_o transients found in adult animals (Jendelová and Syková, 1991). At P3–P6, the $[K^+]_o$ changes evoked in the dorsal horn by a single electrical stimulus are as large as 1.5–2.5 mM. Such changes in $[K^+]_o$ can be evoked in the adult rat spinal cord only with stimulation at a frequency of 10–30 Hz (Svoboda et al., 1988). The K^+ peak level is achieved in neonatal spinal cords *in vitro* at lower frequencies than in adult animals *in vivo* and its level (6.0–6.8 mM) is significantly higher than in adult rats. The K^+ peak level progressively decreases with age to about 4.6–5.2 mM (note that the K^+ peak level in adult rat spinal cords is typically lower than that in cat or frog spinal cord; see Kříž et al., 1975; Chvátal et al., 1988; Syková and Svoboda, 1990).

At P3–P6, single as well as repetitive electrical stimulation evokes a dominant alkaline shift that is followed by a smaller poststimulation acid shift when the stimulation is discontinued. Single pulses evoke alkaline shifts of as much as 0.05 pH unit, and repetitive stimulation at a frequency of 10 Hz alkaline shifts of 0.1–0.15 pH unit (Fig. 2). At P7–P9, the changes during stimulation become biphasic (i.e., alkaline-acid shifts). At P10–P14, the stimulation evokes acid shifts of 0.02–0.10 pH unit, which are sometimes preceded by barely discernible alkaline shifts, as is the case in adult rats. However, the dominant acid shifts in isolated spinal cords from these young animals are still smaller than in adult rats *in vivo* (Syková and Svoboda, 1990). These and other findings suggest that neuronal activity results primarily in alkaline shifts in pH_o. The maturational changes would therefore be consistent with the development of rapid acid efflux from the glial cells. Thus, extracellular alkaline shifts would be depressed when the glial acid extrusion was increased, and enhanced when glial acid extrusion was blocked.

Distinct populations of identified glial cells were recently described in the developing rat spinal cord slices, namely astrocytes, oligodendrocytes, oligodendrocyte precursor cells, and presumably astrocyte or "early" precursor cells. These four types of glial cells could be distinguished in the gray matter of rat spinal cord slices at P1–P19 by their pattern of membrane currents, as revealed with the whole-cell patch-clamp technique, and by their morphological and immunocytochemical features. When randomly accessing glial cells, the number of mature glial cells—astrocytes and oligodendrocytes—increased from 31% at P1–P3 to 76.5% at P13–P15 (Chvátal et al., 1995). These findings support the previous conclusion that control of extracellular ionic homeostasis in spinal cord is closely related to gliogenesis (Jendelová and Syková, 1991).

Activity-related $[K^+]_o$ and pH_o changes in spinal cords were also studied after "early" postnatal x-irradiation (PI)—a procedure that blocks gliogenesis but leaves the neurons intact. Gilmore (1963) showed that PI effects in the neonatal rat spinal cord include inhibition of myelin formation within the spinal cord without an extensive change in myelin formation in the spinal roots, marked decreases in the number of neuroglial nuclei, and hypertrophy of astrocytes without widespread alterations in the spinal neurones. Immunohistochemical staining revealed that the typical development of glial fibrillary acidic protein (GFAP) reaction observed at P3–P14 in nonirradiated (control) pups was disrupted in rats irradiated with a single dose at P1 (Syková et al., 1992). At P10–P14, PI, GFAP staining of radial glia in white matter was disrupted, and GFAP staining of astrocytes in gray matter revealed typical astrogliosis: astrocytes had unusually large soma, and thick, long processes.

At P10–P14, PI, the stimulation-evoked $[K^+]_o$ and pH_o changes resembled those found in control animals during their early postnatal days (i.e., at P3–P6; Fig. 2). The $[K^+]_o$ peak value had not decreased from the neonate level, and the stimulation at a frequency of 10 Hz increased $[K^+]_o$ by about 4.0 mM. In addition, these animals showed persistance of a dominant alkaline shift, typical for animals at P3–P6. Furthermore, measurements of the $[K^+]_o$ and pH_o baselines in unstimulated pups at P10–P14 also revealed values similar to those found in control animals at P3–P6. These results show that postnatal x-irradiation which blocks gliogenesis impairs normal development of extracellular K^+ and pH_o homeostasis.

The alkaline shifts are enhanced in bicarbonate-free, HEPES-buffered saline. At P3–P8, the carbonic anhydrase inhibitor acetazolamide has either no effect on the alkaline shifts or it slightly enhances them. This enhancement is apparently due to blockage of the poststimulation acid shifts (Fig. 3). Acetazolamide decreases by 70–80% the acid shifts at P10–P14.

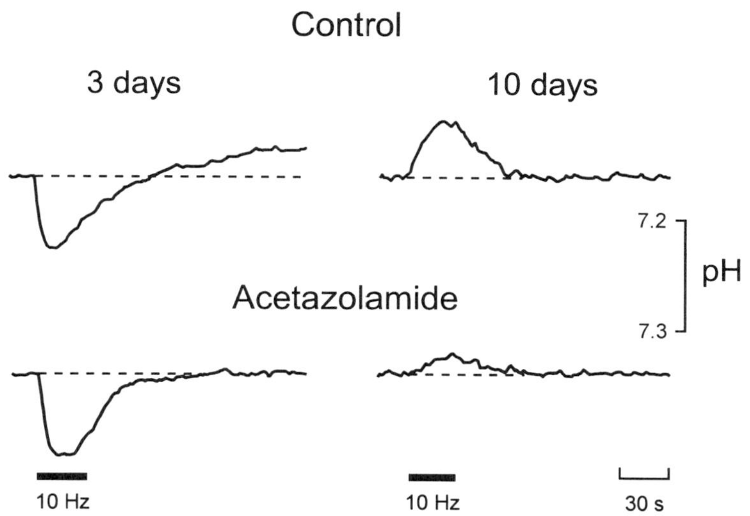

FIGURE 3. Effect of acetazolamide (5 mM) on pH_o changes evoked by stimulation at a frequency of 10 Hz in 3- and 10-day-old rats. The alkaline shift was not changed, while the acid shifts during and after stimulation were substantially decreased. (After Jendelová and Syková, 1991 with permission).

MECHANISMS OF ACTIVITY-RELATED pH CHANGES IN SPINAL CORD

pH homeostasis as well as K^+ homeostasis in the spinal cord is ensured by a variety of mechanisms in both neurons and glial cells (for reviews see Syková, 1983; Walz, 1989; Chesler, 1990; Syková, 1992).

Mechanisms of Alkaline Shifts

As already reported, the alkaline shifts in the spinal cord of adult animals are to a great extent masked by late acidification. However, the alkaline shifts are enhanced in HEPES-buffered saline (Syková, 1989; Syková et al., 1992). This result is probably due to the lower buffering capacity of the HEPES-buffered media, or to the attenuation of the acid shifts (see Chapters 2 and 17). Similar enhancement has also been achieved using CA inhibitors (Fig. 3). The inhibitors of acid extrusion in CNS neurons and glia (amiloride, SITS 4-acetamido-4′-isothiocyanatostilbene-2,2′-disulfonic acid, DIDS 4,4′ diazidostilbene-2,2′-disulfonic acid, partly blocked the stimulation-evoked acid shifts and, therefore, enhanced the alkaline shifts (Syková and Svoboda, 1990). The alkaline shifts were abolished by blockage of synaptic transmission by Mn^{2+} or Mg^{2+}, while the acid shifts were unaffected. These results suggest that alkaline shifts are a consequence of synaptic transmission, that is, that they have postsynaptic and neuronal origins (Jendelová and Syková, 1991; Syková et al., 1992).

Stimulation-evoked alkaline shifts in the isolated rat spinal cord at P3–P6 were also blocked by about 50% with the GABA antagonist picrotoxin at a concentration range of 10^{-6} M to 5×10^{-6} M (Fig. 4A). At this concentration, the K^+ transients evoked

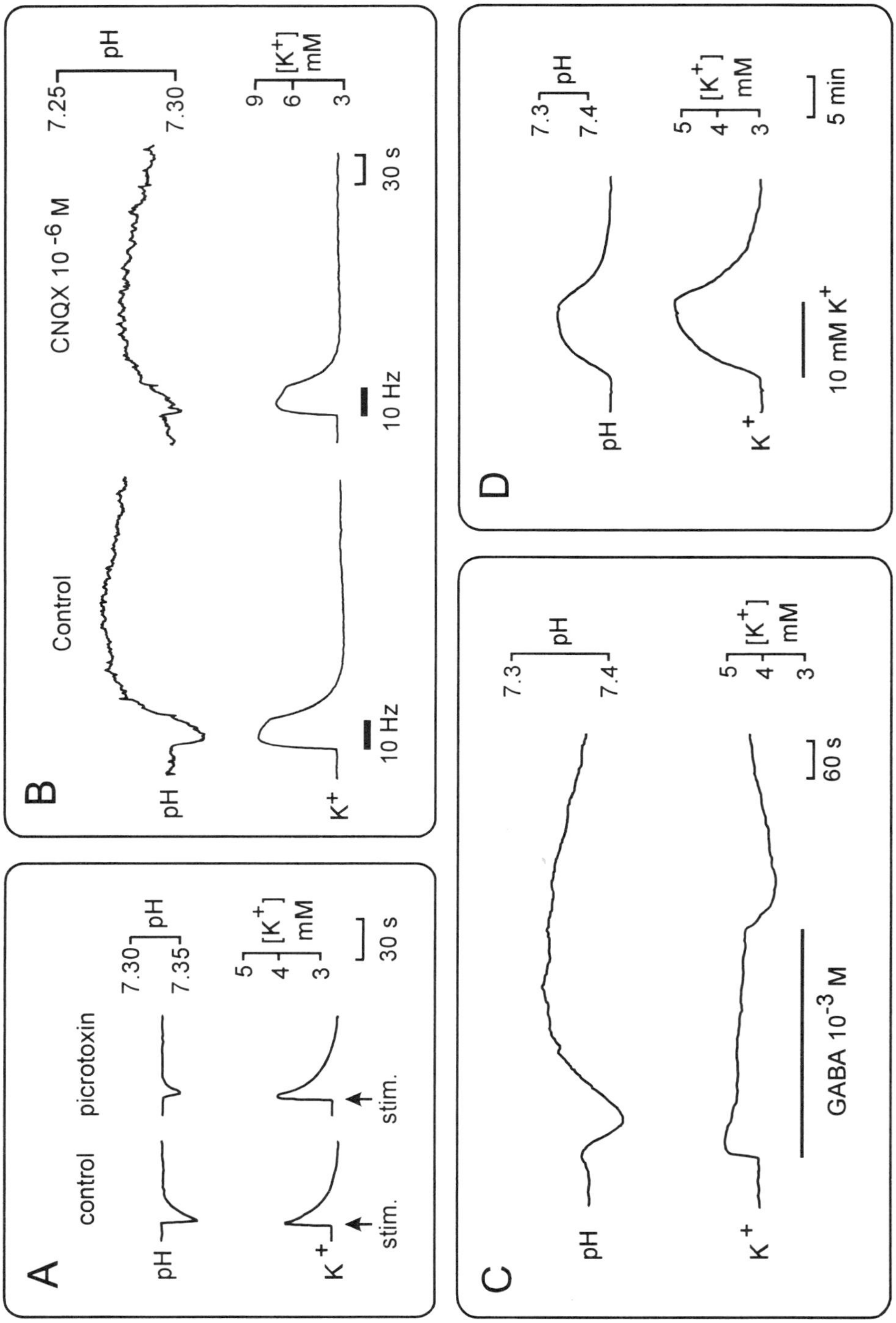

A
control
picrotoxin
pH
K+
stim.
stim.
7.30
7.35
pH
5
4
3
[K+] mM
30 s
B
Control
CNQX 10 -6 M
pH
K+
10 Hz
10 Hz
7.25
7.30
pH
9
6
3
[K+] mM
30 s
C
pH
K+
GABA 10-3 M
60 s
7.3
7.4
pH
5
4
3
[K+] mM
D
pH
K+
10 mM K+
7.3
7.4
pH
5
4
3
[K+] mM
5 min

simultaneously were not affected, while at higher concentrations picrotoxin increased K^+ transients and the K^+ baseline (Syková et al., 1992). Application of various glutamate antagonists also blocked the stimulation-evoked alkaline shifts; however, they also partially blocked the K^+ transients (Fig. 4B) (Jendelová et al., 1994a,b). Subtype selective antagonists revealed that the application of 10^{-5} M MK801 (a noncompetitive N-methyl-D-aspartate [NMDA] receptor antagonist) as well as 10^{-4} M ketamin, resulted in an approximately 50% block of the alkaline shift. CNQX, a competitive AMPA/kainate receptor antagonist, produced a 50% block of the alkaline shifts at a concentration 10^{-6} M (Fig. 4B), a 70–80% block at 10^{-5} M, and an almost complete block at 10^{-4} M.

Superfusion with γ-aminobutyric acid (GABA) (10^{-3} M) of the isolated rat spinal cords of rats at P3–P14 as well as of x-irradiated rats, elicited a transient alkaline shift in the pH_o baseline of about 0.06 pH unit (Fig. 4C), which was blocked by picrotoxin at 5×10^{-6} M (Syková et al., 1992). The alkaline shifts evoked in our experiments by GABA were accompanied by an increase in K^+; however, the superfusion of spinal cord with Ringer's solution containing elevated K^+ produced an acid shift in the pH_o baseline (Fig. 4D). It has been shown that the GABA-gated anion channel has significant permeability to HCO_3^- (Bormann et al., 1987; Kaila and Voipio, 1987). Since the electrochemical gradient for HCO_3^- is outward, release of GABA can evoke a HCO_3^--dependent fall in the intracellular pH of neurons and a concomitant rise in pH_o. In nominally HCO_3^--free solution, buffered with HEPES, GABA evoked no alkaline shift in the rat spinal cord at P3–P14; instead, a slow acid shift was observed. These data may be readily explained by the gating of $GABA_A$ channels giving rise to an HCO_3^- efflux. This mechanism would partially explain the afferent-evoked alkaline shifts of the dorsal horn (see Chapter 17). However, these results also indicate that the observed alkaline shifts can be mediated by other mechanisms, including HCO_3^--independent ones.

In this connection, we have recently found in isolated rat spinal cord at P3–P14 that application of glycine and substance P, in concentrations which produce a very small increase in $[K^+]_o$ (i.e., 10^{-3}M and 10^{-6}M, respectively) also results in alkaline shifts in the dorsal about 0.03 pH unit (Fig. 5), (Jendelová et al., 1994a,b). Application of glutamate receptor agonists, namely L-glutamate, L-aspartate, NMDA, α-amino-3-hydroxy-5-methylisoxazole-4-propionic acid (AMPA), and kainate, results in alkaline shifts of 0.03–0.06 pH unit (Fig. 5). Moreover, Endres et al. (1986) found that NMDA, quisqualate, and kainate produced a concentration-dependent intracellular acidification in motoneurons of the isolated frog spinal cord, and this was accompanied by triphasic pH_o changes (alkaline-acid-alkaline). The alkaline shifts (except for the NMDA-induced ones) were, however, not affected by high $[Mg^{2+}]$. On the other hand, the alkaline shifts were abolished in nominally Ca^{2+}-free solutions, and sometimes even a small acid shift was

FIGURE 4. A: Stimulation of the dorsal root with a single electric pulse evoked an alkaline shift that was accompanied by an increase in $[K^+]_e$. Application of picrotoxin (10^{-6} M) decreased the alkaline shift to about 50%. **B:** This evoked in the dorsal horn of the isolated rat spinal cord of 5-day-old rat by repetitive electrical stimulation (10 Hz) of the dorsal root were partially blocked by the antagonist of the AMPA/kainate receptor CNQX. The alkaline shifts were reduced to about 50%. **C:** The effect of GABA on the pH_o and baselines in the isolated spinal cord of 4-day-old rat. Note the alkaline shift and increase in $[K^+]_o$ during the application of GABA. **D:** Simultaneous pH_o and $[K^+]_o$ records in the dorsal horn during perfusion of the spinal cord with a Ringer's solution containing 10 mM K^+. (A, B and C after Syková et al., 1992 with permission).

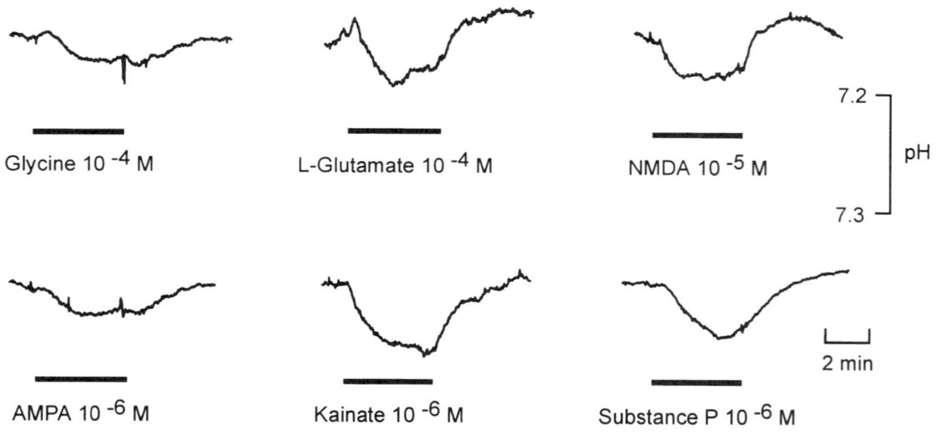

FIGURE 5. Extracellular alkaline shifts evoked in isolated rat spinal cord by application of glycine, glutamate receptor agonists (aspartate, glutamate, NMDA, AMPA, kainate), and substance P. (E. Syková and P. Jendelová, unpublished results)

observed, suggesting that the alkaline shifts in the spinal cord may arise from ATP-dependent Ca^{2+}/H^+ exchange (Schwiening et al., 1993; Paalasmaa et al., 1994; Smith et al., 1994). Moreover, in our studies, the alkaline shifts in pH_o were of the same amplitude in newborn pups (P3–P6), in pups at P10–P14, and in x-irradiated animals, suggesting no substantial role for glial cells.

Mechanisms of Acid Shifts

Stimulation-evoked acid shift. Stimulation-evoked acid shifts are likely to have multiple mechanisms and perhaps several components with variable time courses (see Chapters 2 and 17). Acid shifts in the spinal cord are clearly related to gliogenesis, since they are absent in the early postnatal days, and since this absence persists in the rats after early postnatal x-irradiation, a procedure which blocks gliogenesis. The acid extrusion mechanisms are, therefore, fully developed only in mature glial cells. It has been repeatedly shown that excessive neuronal activity and stimulation result in alkaline shifts in glial pH_i (Chapters 13–14). To determine whether K^+-evoked glial depolarization in the spinal cord is required to trigger the acid shifts, the isolated cord was superfused with Ba^{2+}, a blocker of glial K^+ channels. Indeed, the addition of 0.5 mM Ba^{2+} into isolated spinal cords resulted in dramatic increases of the K^+ peak, on average up to about 6.5 mM, compared to an average control response prior to Ba^{2+} application of about 2.5 mM K^+ (Fig. 6). This suggests that the K^+ buffering by glia was blocked (Syková et al., 1992). The addition of Ba^{2+} increased the K^+ peak value only in rats at P10 and older, while the already high K^+ peak value in rats at P3–P6 and in animals after x-irradiation (i.e., in animals with incomplete or blocked gliogenesis) was either not affected or the evoked increase was not significant. In the spinal cord of pups at P3–P6, application of Ba^{2+} enhanced the stimulation-evoked alkaline shifts by about 100% and depressed the poststimulation acid shifts. At P10–P14, Ba^{2+} decreased the acid shifts by 60% and sometimes even reversed the acid shift to alkaline (Fig. 6). However, in rats x-irradiated at P1 to block gliogenesis, Ba^{2+} had no effect on the stimulation-evoked alkaline shift in pH_o.

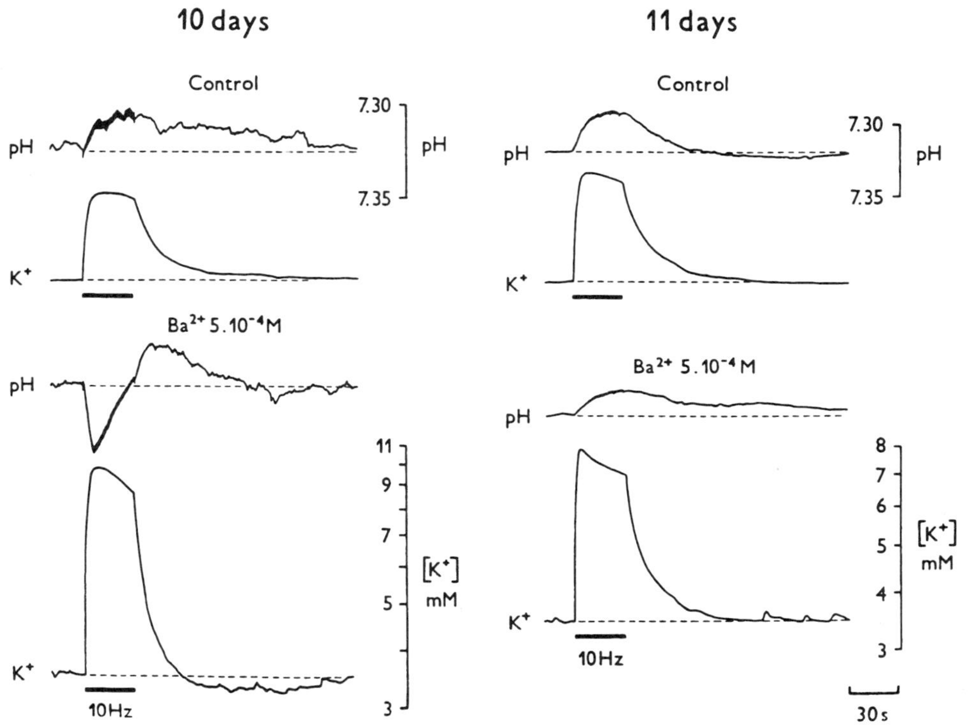

FIGURE 6. Effect of Ba^{2+} (5×10^{-4} M) on pH_o and $[K^+]_o$ changes in the spinal dorsal horn of 10- and 11-day-old rats. pH_o and $[K^+]_o$ changes were evoked by the stimulation of the dorsal root at a frequency of 10 Hz. Note the reversal of acid shift into alkaline shift in the 10-day-old rat and the decrease in acid shift in the 11-day-old rat. Concomitantly, Ba^{2+} enhanced the $[K^+]_o$ ceiling level. (From Syková et al., 1992 with permission).

These results strongly support the idea that activity-evoked acid shifts in pH_o are related to depolarization of glial cell membranes by elevated $[K^+]_o$.

K^+-evoked acid shifts. To examine to what extent activity-related $[K^+]_o$ changes in spinal cord result in acid shifts in pH_o, isolated spinal cords of frogs and rats were superfused with Ringer's solutions containing elevated K^+ (Chvátal et al., 1988; Jendelová and Syková, 1991). Increases in $[K^+]_o$ to 10 mM resulted in concomitant acid shifts of about 0.2 pH unit (see Fig. 4D). The application of K^+ evoked similar changes at P4 and P11; however, the rise in $[K^+]_o$ and the decrease in pH_o were slower at P10–P14, presumably due to the dramatic decrease in the extracellular space volume, which slowed down diffusion of K^+ into the spinal cord (Syková and Chvátal, 1993; Lehmenkühler et al., 1993; Prokopová et al., 1997). It is therefore evident that a rise in $[K^+]_o$ results in an acid shift in pH_o.

Poststimulation Alkaline Undershoot

A poststimulation alkaline shift, called also "poststimulation alkaline undershoot", was clearly demonstrated in the spinal cord of rats as well as frogs (Chvátal et al., 1988; Syková and Svoboda, 1990). During repetitive electrical stimulation of an afferent input, $[K^+]_o$ and

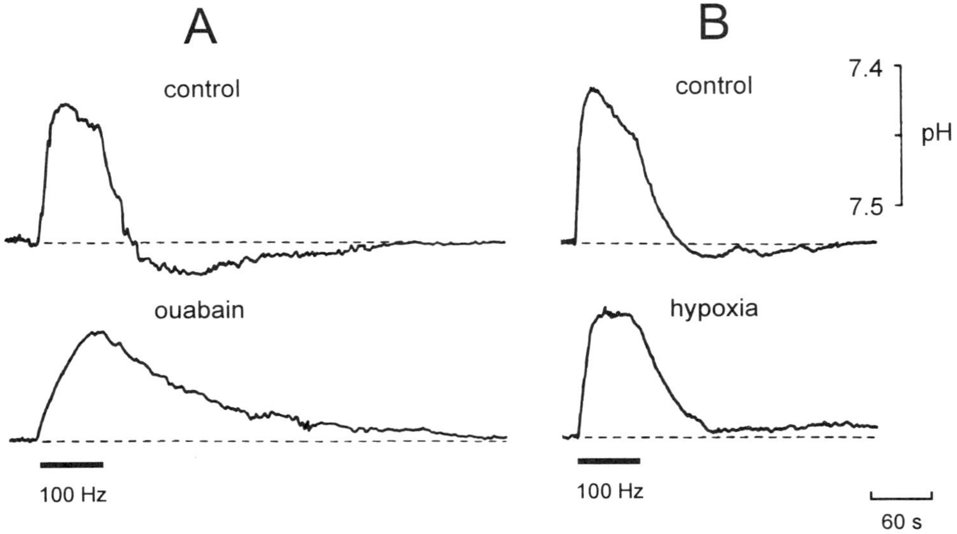

FIGURE 7. Depression of pH_o changes evoked by repetitive electrical stimulation (100 Hz, 60 sec) after application of 10^{-4} M ouabain to the surface of the spinal cord (**A**), and after partial respiratory arrest when blood pressure decreased to 70 mmHg (hypoxia) (**B**). The data are from two animals. (After Syková and Svoboda, 1990 with permission).

pH_o reach their peak. When the stimulation continues, a gradual decrease of both transients, $[K^+]_o$ and pH_o, occurs immediately after the peaks are reached. Recoveries of both the $[K^+]_o$ and pH_o levels to their original baselines take 2–5 minutes and have similar time courses. The poststimulation alkaline undershoot (pH_o decrease below the prestimulation pH_o level) has a very similar time course to the poststimulation K^+ undershoot. Anoxia and ouabain, the cardiac glycoside which inhibits Na^+/K^+-ATPase, have been shown to significantly slow K^+ recovery from the stimulation-evoked increases in $[K^+]_o$ and to block poststimulation K^+ undershoot in the frog and cat, as well as in rat spinal cord (Chvátal et al., 1988; Kříž et al., 1975; Krnjevic and Morris, 1975; Syková et al., 1985). Application of ouabain onto the spinal cord surface of rats *in vivo* decreases the pH_o baseline by about 0.05 pH unit. The rise time of the stimulation-evoked acid shift is prolonged, as is the poststimulation pH_o recovery to the original base line (Syková and Svoboda, 1990). The poststimulation alkaline undershoot is diminished, while the amplitude of the acid shift is not substantially decreased (Fig. 7A). Mild hypoxia has similar effects on pH_o transient changes (Fig. 7B). These findings suggest that the $[K^+]_o$ and pH_o recoveries and "undershoots" may have common underlying mechanisms involving the Na^+/K^+ pump.

pH_O CHANGES IN SPINAL CORD DURING PATHOLOGICAL STATES

Pathological states are accompanied by lack of energy, seizure activity, excessive release of transmitters and neuroactive substances, neuronal death, glial cell loss or proliferation, cell swelling, production of metabolites, etc. (for review see Siesjö et al., 1991; Syková, 1992 see also Chapters 5, 28–31). These changes undoubtedly also lead to changes in extracellular pH

(Kraig et al., 1987), which in turn modulate neuronal and glial cell functions (Chesler, 1990). Described below are two examples of the pH_o changes, occurring in spinal cord during anoxia and during experimental autoimmune encephalomyelitis (EAE), an experimental model of multiple sclerosis.

Anoxia and Ischemia

The pH_o in the spinal cord is altered during anoxia and/or ischemia. In adult rats, within 2 min after respiratory arrest, blood pressure begins to increase and pH_o begins to decrease (by about 0.1 pH unit), while the $[K^+]_o$ is not yet changed. With the subsequent blood pressure decrease, the pH_o decreases by 0.6–0.8 pH unit to an actual pH level of 6.4–6.6 (Syková and Svoboda, 1990; Syková et al., 1994). This increase is accompanied by a

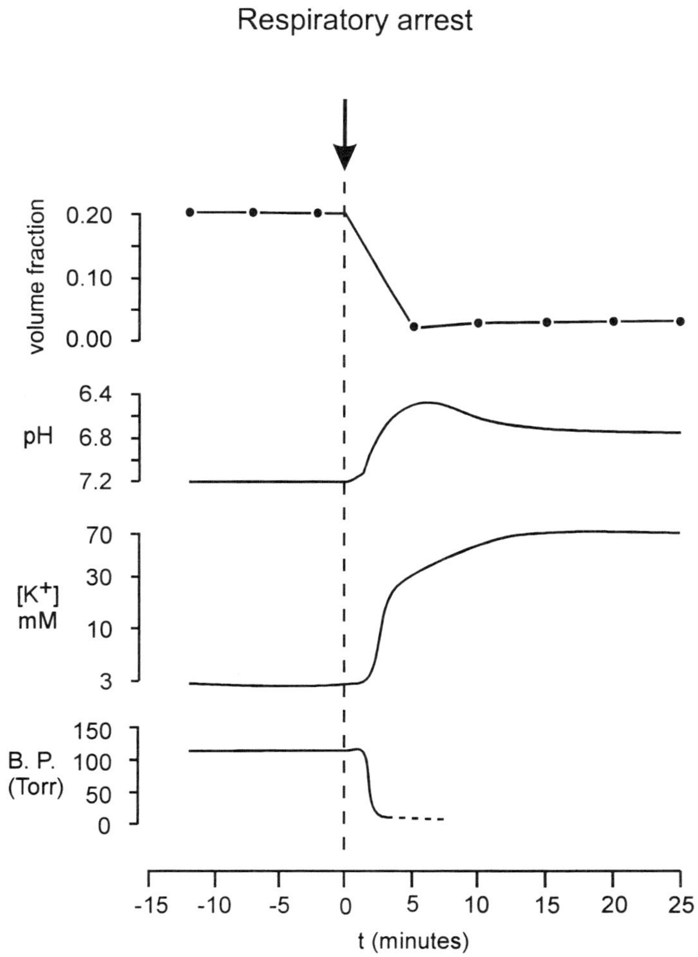

FIGURE 8. Decrease in extracellular space volume fraction, increase of extracellular K^+ concentration, and decrease of extracellular pH in the L4 spinal segment at a depth of 600 μm, as recorded after respiratory arrest. Graph labeled B.P. shows concomitantly recorded changes in blood pressure. (Adapted from Syková et al., 1994 with permission).

steep rise in $[K^+]_o$ to about 50–70 mM and by an extracellular space volume decrease of about 80% (Fig. 8) (Syková et al., 1994). During early postnatal development (P5), the anoxia-evoked $[K^+]_o$ increase, pH_o decrease, and ECS volume decrease, are of about the same magnitude as in adult rats, but these changes develop with a time course about 10 times as long (Vargová et al., 1994; Voříšek and Syková, 1997). This appears consistent with the well-known resistance of immature CNS to anoxia.

Experimental Autoimmune Encephalomyelitis (EAE)

pH shifts and extracellular space volume changes were studied in Lewis rats during EAE. EAE was induced by injection of guinea-pig myelin basic protein (MBP), which resulted in typical morphological changes in the spinal cord, namely an inflammatory reaction, astrogliosis, blood-brain barrier damage, demyelination, and paralysis. We have recently reported that EAE was also accompanied by a statistically significant increase in the ECS volume, and that the increase had a time course similar to the manifestation of clinical signs and edema formation (Šimonová et al., 1996). In addition, we found that during maximal clinical signs, the pH_o resting level, which normally varies in adult spinal cords between 7.10–7.35, is in EAE rats more alkaline, between 7.45–7.65. This resembles the situation in immature animals, with a larger ECS volume, proliferating astrocytes, incomplete myelination, and a pH_o of about 7.6 (Fig. 9). These results show that EAE rats develop impairment of ionic and volume homeostasis in the spinal cord, presumably resulting from glial cell dysfunction.

ECS VOLUME CHANGES AND pH_o HOMEOSTASIS

Dramatic changes in the volume of cells accompanied by changes in the ECS volume may either aggravate or diminish activity-related alkaline and acid shifts (Syková, 1997; Nicholson and Syková, in press). Since the ECS is an important communication pathway between cells, the diffusion parameters of the ECS affect the movement and accumulation of ions and neuroactive substances within the ECS. Chemical signals as well as metabolites may diffuse via the extracellular space to target cells far distant (micrometer to millimeter range) from the release sites. Diffusion in ECS obeys Ficks's law, subject to two important modifications (Nicholson and Phillips, 1981; Chapter 4 and Chapter 14): First, diffusion in ECS is constrained by the restricted volume of the tissue available for diffusing particles, i.e., by the extracellular volume fraction (α). The concentration of a released substance in the ECS is therefore greater than it would be in a free medium. Second, the free diffusion coefficient, D, is reduced by the square of the tortuosity (λ) to an apparent diffusion coefficient $ADC = D/\lambda^2$, due to an increase in path length for diffusion between two points (because the diffusing substance encounters membrane obstructions, glycoproteins, macromolecules of the extracellular matrix, charged molecules, and glial cell processes). These diffusion parameters of the ECS and their dynamic changes can be studied *in vivo* with the real-time iontophoretic method (Nicholson and Phillips, 1981), which uses ion-selective microelectrodes to follow the diffusion of an extracellular marker applied by iontophoresis (e.g., the tetraethylammonium ion, TEA^+).

Changes in ECS diffusion parameters (ECS volume decrease and tortuosity increase) in the spinal cord result from activity-related transmembrane ionic shifts and cell swelling under physiological conditions (e.g., electrical or adequate stimulation; Svoboda and Syková, 1991), as well as under pathological conditions (e.g., hypoxia, terminal anoxia;

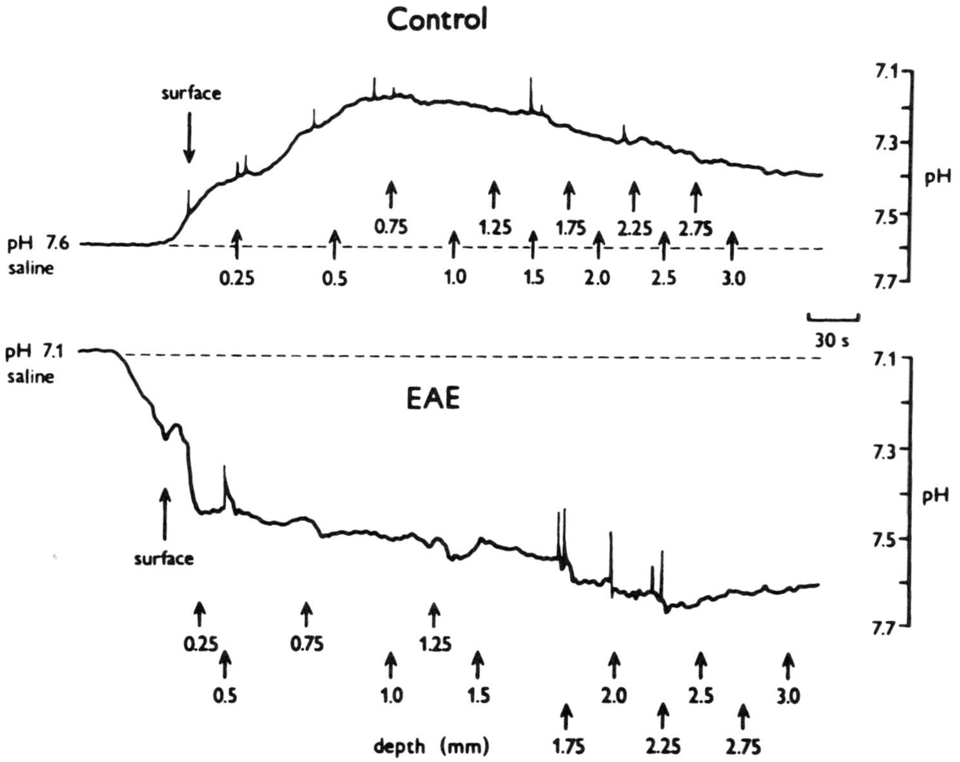

FIGURE 9. Extracellular pH levels at various depths in the spinal cord of the rat. *Control:* The pH-sensitive microelectrode was introduced into the spinal cord of the control animal through saline, the pH of the saline was adjusted to 7.6, and the solution was poured over the dorsal surface of the cord. The arrows indicate the surface and various depths in the spinal cord. *EAE:* A microelectrode was introduced through saline into the spinal cord of experimental autoimmune encephalomyelitis (EAE) animal with maximal signs of paralysis (14 days after injection of myelin basic protein). The pH of the saline was adjusted to 7.1. The microelectrode penetrated the spinal cord until it reached the ventral surface.

Syková et al., 1994), and x-irradiation injury (Syková et al., 1996). It is generally accepted that ECS volume decrease is primarily due to astrocytic swelling, although swelling of neurons, particularly dendrites and fibers also occurs. A number of different mechanisms have been proposed to account for astrocytic swelling. These include uptake of extracellular K^+, acid-base changes, glutamate uptake, and excitatory amino acid-induced swelling (for review see Kimelberg et al., 1993). In the isolated rat spinal cord, application of hypotonic solution, application of Ringer's solution with elevated $[K^+]_o$, and application of glutamate receptor agonists, all result in cell swelling, ECS volume decrease, and ECS tortuosity increase (Vargová et al., 1995, 1997). Measurements on the isolated rat spinal cord also reveal that moderate changes in the pH of Ringer's solution lead to changes in ECS volume, namely an alkaline shift in pH leading to ECS volume increase and an acid shift leading to ECS volume decrease (Fig. 10). Indeed, in the isolated frog spinal cord, repetitive electrical stimulation often results in acid shifts in pH_o that outlast the stimulation by 15–30 minutes (Fig. 11), while the $[K^+]_o$ returns to the original

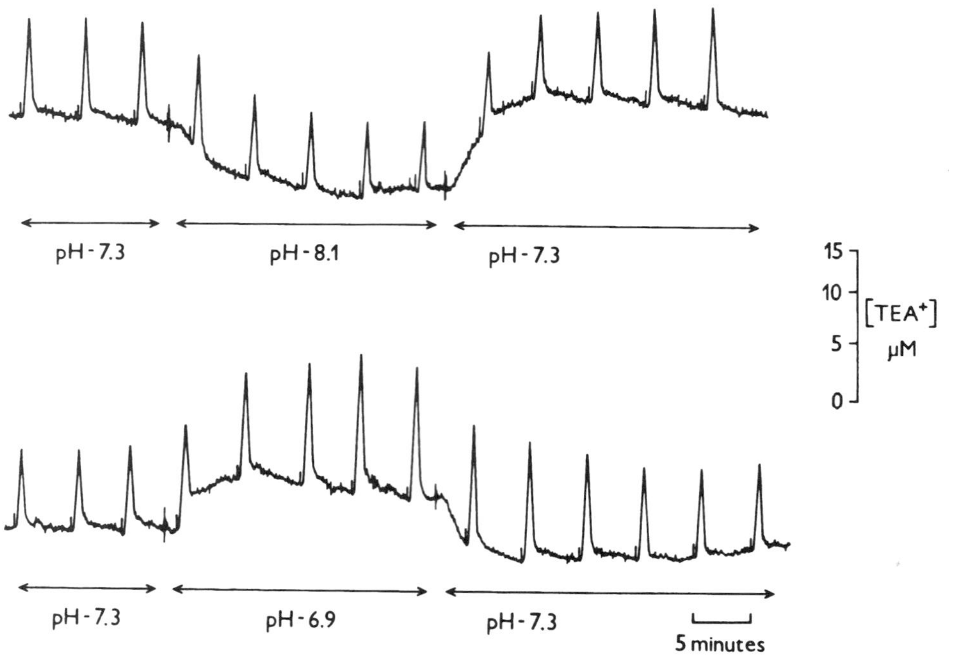

FIGURE 10. The effect of the pH of Ringer's solution on ECS volume in the isolated frog spinal cord. Tetraethylammonium (TEA^+) diffusion curves (fast transient upward deflections) were evoked by iontophoretic application of TEA^+ and recorded with a TEA^+-sensitive microelectrode. The spacing between the TEA^+ iontophoretic pipette and the TEA^+-sensitive microelectrode was 200 μm. *Upper trace:* In alkaline solution (pH = 8.1) the TEA^+ diffusion curves superimposed on the decreasing TEA^+ baseline decrease in amplitude due to ECS volume increase. *Lower trace:* In acidic solution (pH = 6.9), TEA^+ diffusion curves are larger and superimposed on the increasing TEA^+ baseline due to ECS volume decrease. (E. Syková and A. Chvátal, unpublished results).

base line in 3–5 minutes and then decreases to below resting level (K^+ undershoot). This longer-lasting acid shift in pH_o is accompanied by an ECS volume decrease by as much as 50% (Fig. 11). The time course of delayed acidification, unlike the rise in $[K^+]_o$, corresponds to cell swelling that results in ECS volume shrinkage.

A pertinent question also arises in relation to pH_o homeostasis: Are the alkaline or acid shifts aggravated or diminished by changes in the ECS volume and tortuosity? Kraig et al. (1987) noted that, in principle, alkaline shift in pH_o could be a result of stimulation-evoked shrinkage of the ECS. An ECS volume decrease would lead to an increased concentration of HCO_3^-. Because CO_2 readily crosses cell membranes, the selective increase in concentration of HCO_3^- ions would result in an alkaline shift in pH_o.

Our recent studies revealed that the ECS volume fraction in the cortex and the corpus callosum, as well as in the spinal cord in newborn rats, is about twice that in adults (Lehmenkühler et al., 1993; Prokopová et al., 1997; Voříšek and Syková, 1997). We have also found that the percentage of ECS volume shrinkage in high $[K^+]$, hypotonic solutions and during application of excitatory amino acids is much larger in younger rats (Vargová et al., 1995, 1997). In neonates, we also observed dominant stimulation-evoked

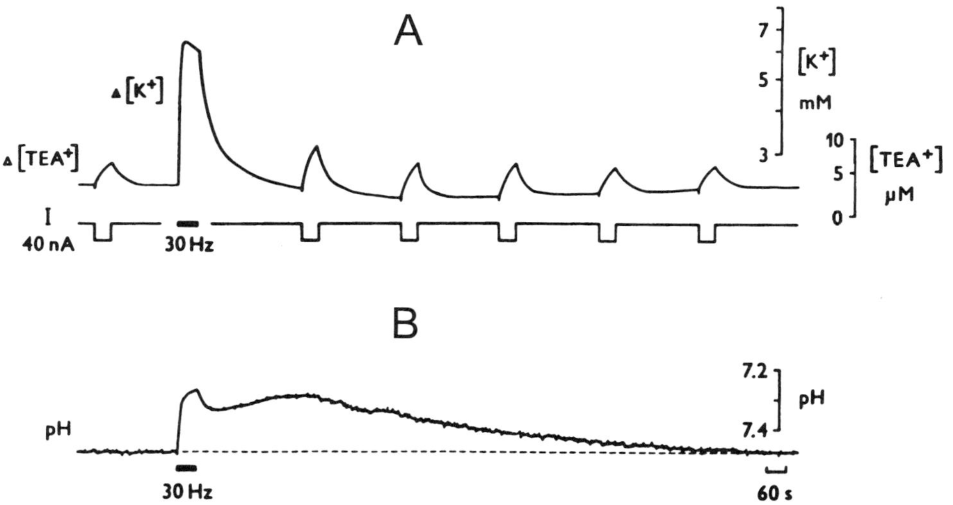

FIGURE 11. Effect of repetitive electrical stimulation of dorsal root on ECS volume and pH in dorsal horn of the isolated frog spinal cord. **A.** *Lower trace:* Current (40 nA, the bottom recordings) was applied with an iontophoretic pipette filled with TEA^+. *Upper trace:* TEA^+ diffusion curve recorded with TEA^+-sensitive microelectrode during iontophoretic applications, and change in $[K^+]_o$ induced by repetitive stimulation (30 Hz, 60 seconds). (TEA^+ electrode is also sensitive to K^+). **B.** Concomitantly recorded pH_o changes with pH-sensitive microelectrode. Three microelectrodes (iontophoretic pipette, TEA^+/ K^+/H^+-selective microelectrodes) were inserted separately from the spinal surface. Recordings from a depth of about 300 μm.

alkaline shifts (Jendelová and Syková, 1991). This suggests that the greater the percentage of the ECS volume change, the greater the alkaline shift. Further experiments on the isolated rat spinal cord revealed that experimentally induced cell swelling and ECS shrinkage evoked either by high $[K^+]$, by hypotonic solutions, or by application of excitatory amino acids, is accompanied by enhancement of the stimulation-evoked alkaline shifts (P. Jendelová, L. Vargová, and E. Syková, unpublished observations). Further experiments need to be done, but the possibility that the ECS volume changes affect the alkaline and acid shifts cannot so far be excluded.

REFERENCES

Bormann J, Hamill OP, Sakmann B (1987): Mechanism of anion permeation through channels gated by glycine and γ-aminobutyric acid in mouse cultured spinal neurons. J Physiol (Lond) 385: 243–286.

Chesler M (1990): The regulation and modulation of pH in the nervous system. Prog Neurobiol 34: 401–427.

Chvátal A, Jendelová P, Kříž N, Syková E (1988): Stimulation-evoked changes in extracellular pH, calcium and potassium activity in the frog spinal cord. Physiologia bohemoslovaca 37: 203–212.

Chvátal A, Pastor A, Mauch M, Syková E, Kettenmann H (1995): Distinct populations of identified glial cells in the developing rat spinal cord slice: Ion channel properties and cell morphology. Eur J Neurosci 7: 129–142.

Endres W, Ballanyi K, Serve G, Grafe P (1986): Excitatory amino acids and intracellular pH in motoneurons of isolated frog spinal cord. Neurosci Lett 72: 54–58.

Gilmore SA (1963): The effect of X-irradiation on the spinal cords of neonatal rats. II. Histological observations. J Neuropathol Exp Neurol 22: 294–301.

Jendelová P, Syková, E (1991): Role of glia in K^+ and pH homeostasis in the neonatal rat spinal cord. Glia 4: 56–63.

Jendelová P, Chvátal A, Šimonová Z, Syková E (1994a): Effect of excitatory and inhibitory amino acids on extracellular pH in isolated rat spinal cord. In Abstracts, 17th Annual Meeting European Neuroscience Association, Vienna, 4-8 September 1994, Oxford University Press, p 199.

Jendelová P, Chvátal A, Syková E (1994b) : Effect of transmitters on extracellular pH baseline in the rat spinal cord. Physiol Res 43: 24P.

Kaila K, Voipio J (1987): Postsynaptic fall in intracellular pH induced by GABA-activated bicarbonate conductance. Nature 330: 163–165.

Kimelberg HK, O'Connor ER, Kettenmann H (1993): Effect of swelling on glial cell function. In Lang J, Häussinger D (eds): Interaction of Cell Volume and Cell Function. Berlin, Heidelberg, New York: Springer-Verlag, pp 157–186.

Kraig RP, Petito CK, Plum F, Pulsinelli WA (1987): Hydrogen ions kill brain at concentrations reached in ischemia. J Cereb Blood Flow Metab 7: 379–386.

Kříž N, Syková E, Vyklický L (1975): Extracellular potassium changes in the spinal cord of the cat and their relation to slow potentials, active transport and impulse activity. J Physiol (Lond) 249: 167–182.

Krnjevic K, Morris ME (1975): Factors determining the decay of K^+ potentials and focal potentials in the central nervous system. Can J Physiol Pharmacol 53: 923–934.

Lehmenkühler A, Syková E, Svoboda J, Zilles K, Nicholson C (1993): Extracellular space parameters in the rat neocortex and subcortical white matter during postnatal development determined by diffusion analysis. Neuroscience 55: 339–351.

Nicholson C (1980): Dynamics of the brain cell microenvironment. Neurosci Res Prog Bull 18: 177–322.

Nicholson C, Phillips JM (1981): Ion diffusion modified by tortuosity and volume fraction in the extracellular microenvironment of the rat cerebellum. J Physiol (Lond) 321: 225–257.

Nicholson C, Syková E (in press) Extracellular space structure revealed by diffusion analysis. Trends Neurosci.

Paalasmaa P, Taira T, Voipio J, Kaila K (1994): Extracellular alkaline transients mediated by glutamate receptors in the rat hippocampal slice are not due to a proton conductance. J Neurophysiol 72: 2031–2033.

Prokopová Š, Vargová L, Syková E (1997): Heterogeneous and anisotropic diffusion in the developing rat spinal cord. Neuroreport 8:3527–3532.

Schwiening CJ, Kennedy HJ, Thomas RC (1993): Calcium-hydrogen exchange by the plasma membrane Ca-ATPase of voltage clamped snail neurons. Proc R Soc Lond Biol Sci 253: 285–289.

Siesjö BK, Ekholm A, Asplund B (1991): Transmitter release, ion and water fluxes, and ischemic brain damage. In Fuxe K, Agnati LF (eds): Volume Transmission in the Brain: Novel Mechanisms for Neural Transmission. New York: Raven Press, pp 539–547.

Šimonová Z, Svoboda J, Orkand P, Bernard CCA, Lassmann H, Syková E (1996): Changes of extracellular space volume and tortuosity in the spinal cord of Lewis rats with experimental autoimmune encephalomyelitis. Physiol Res 45: 11–22.

Smith SE, Gottfried JA, Chen JCT, Chesler M (1994): Calcium dependence on glutamate-evoked alkaline shifts in hippocampus. Neuroreport 5: 2441–2445.

Svoboda J, Motin V, Hájek I, Syková E (1988): Increase in extracellular potassium level in rat spinal dorsal horn induced by noxious stimulation and peripheral injury. Brain Res 458: 97–105.

Svoboda J, Syková E (1991): Extracellular space volume changes in the rat spinal cord produced by nerve stimulation and peripheral injury. Brain Res 560: 216–224.

Syková E (1983): Extracellular K^+ accumulation in the central nervous system. Prog Biophys Molec Biol 42: 135–189.

Syková E (1989): Activity-related extracellular pH transients in spinal cord. Verh Dtsch Zool Ges 82: 153–163.

Syková E (1991): Activity-related ionic and volume changes in neuronal microenvironment. In Fuxe K, Agnati LF (eds): Volume Transmission in the Brain: Novel Mechanisms for Neural Transmission. New York: Raven Press, pp 317–336.

Syková E (1992): Ionic and volume changes in the microenvironment of nerve and receptor cells. In Ottoson D (ed): Progress in Sensory Physiology 13. Heidelberg: Springer-Verlag, pp 1–167.

Syková E (1997): The extracellular space in the CNS: Its regulation, volume and geometry in normal and pathological neuronal function. The Neuroscientist 3: 28–41.

Syková E, Chvátal A (1993): Extracellular ionic and volume changes: The role in glia-neuron interaction. J Chem Neuroanat 6: 247–260.

Syková E, Svoboda J (1990): Extracellular alkaline-acid-alkaline transients in the rat spinal cord evoked by peripheral stimulation. Brain Res 512: 181–189.

Syková E, Hájek I, Chvátal A, Kříž N, Diatchkova G (1985): Changes in extracellular potassium accumulation produced by opioids and naloxone in frog spinal cord: Relation to changes of Na-K pump activity. Neurosci Lett 59: 285–290.

Syková E, Jendelová P, Šimonová Z, Chvátal A (1992): K^+ and pH homeostasis in the developing rat spinal cord is impaired by early postnatal X-irradiation. Brain Res 594: 19–30.

Syková E, Kříž N, Preis P (1983): Elevated extracellular potassium concentration in unstimulated spinal dorsal horn of frogs. Neurosci Lett 43: 293–298.

Syková E, Orkand RK, Chvátal A, Hájek I, Kříž N (1988): Effect of carbon dioxide on extracellular potassium accumulation and volume in isolated frog spinal cord. Pflügers Arch 412: 183–187.

Syková E, Svoboda J, Polák J, Chvátal A (1994): Extracellular volume fraction and diffusion characteristics during progressive ischemia and terminal anoxia in the spinal cord of the rat. J Cereb Blood Flow Metab 14: 301–311.

Syková E, Svoboda J, Šimonová Z, Lehmenkühler A, Lassmann H (1996): X-irradiation-induced changes in the diffusion parameters of the developing rat brain. Neuroscience 70: 597–612.

Vargová L, Škobisová E, Voříšek I, Syková E (1994): Extracellular volume fraction and tortuosity in developing rat brain during anoxia. In Abstracts, 17th Annual Meeting European Neuroscience Association, Vienna, 4-8 September 1994, Oxford University Press, p 16.

Vargová L, Jendelová P, Nicholson C, Syková E (1995): The effect of K^+ and excitatory amino acids on diffusion parameters in isolated rat spinal cord. Physiol Res 44: 5P.

Vargová L, Prokopová Š, Syková E (1997): Developmental changes in extracellular space volume and tortuosity in rat spinal cord gray and white matter. Abstracts, Society for Neuroscience Meeting, New Orleans, p. 1732.

Voříšek I, Syková E (1997): Ischemia-induced changes in the extracellular space diffusion parameters, K^+ and pH in the developing rat cortex and corpus callosum. J Cereb Blood Flow Metab 17: 191–203.

Walz W (1989): Role of glial cells in the regulation of the brain ion microenvironment. Prog Neurobiol 33: 309–333.

CHAPTER 19

pH Changes Associated with Glutamate Uptake

ALESSANDRA AMATO, BRIAN BILLUPS, MURIEL BOUVIER, MAREK SZATKOWSKI, AND DAVID ATTWELL
Department of Physiology, University College London, London, United Kingdom

INTRODUCTION

Glutamate uptake into neurons and glial cells of the central nervous system is important for terminating the synaptic action of neurotransmitter glutamate, and for keeping the extracellular glutamate concentration below neurotoxic levels (Nicholls and Attwell, 1990). To date, four mammalian glutamate uptake carriers have been cloned (Storck et al., 1992; Pines et al., 1992; Kanai and Hediger, 1992; Fairman et al., 1995). They have related structures but differential expression in neurons and glial cells (reviewed by Kanai et al., 1994; Bouvier et al., 1994). Traditionally these transporters have been studied by tracing the movement of radioactive glutamate (reviewed by Hertz, 1979). Such studies characterized the distinctive pharmacology of glutamate uptake, and revealed that uptake only occurs in the presence of a transmembrane sodium gradient, suggesting that cotransport of sodium provides the energy needed for uptake. More recently, uptake has been studied at the single-cell level by whole-cell patch-clamp methods, which make it possible to detect uptake from the membrane current it produces (Brew and Attwell, 1987): the ionic stoichiometry of the uptake process is such that net positive charge enters the cell with each glutamate anion (see Fig. 3). The topic of pH turns out to be an important one for understanding the operation of the glutamate uptake carriers, for three reasons. First, as described below, glutamate uptake generates extra- and intracellular pH changes. Second, changes of pH affect the rate of glutamate uptake, and could thus affect excitatory synaptic transmission in the brain (Takahashi et al., 1995). Third, during brain ischemia, glutamate uptake runs backwards, releasing glutamate into the extracellular space of the brain, where it triggers neuronal death; the acidification occurring in ischemia slows this release of glutamate (Billups and Attwell, 1996).

pH and Brain Function, Edited by Kai Kaila and Bruce R. Ransom
ISBN 0-471-11838-9

ENERGIZING GLUTAMATE UPTAKE

In prokaryotes the membrane transport of substances up their electrochemical gradients is often powered by utilizing energy stored in a transmembrane pH gradient, for example by cotransporting protons with amino acids. In eukaryotes, however, the dominant energy source for membrane carriers that do not get energy directly from ATP is the transmembrane electrochemical gradient for sodium set up by the Na^+/K^+ pump. Understanding the ionic stoichiometry of the glutamate uptake process is important, since the stoichiometry determines how low the extracellular glutamate concentration can (in principle) be pumped by the uptake process. As a design principle, one might expect the uptake carriers to have sufficient accumulative power to lower the extracellular glutamate concentration to a value near the bottom of the dose-response curves for activation of ionotropic and metabotropic glutamate receptors ($< 1\ \mu$M).

Radiotracing studies (reviewed by Bouvier et al., 1994) suggested that 2 Na^+ are transported into the cell with each glutamate anion. In addition, uptake only occurs in the presence of intracellular K^+ (Kanner and Sharon, 1978; Barbour et al., 1988), and the form of the dependence of uptake on the intracellular 12+ concentration, $[K^+]_i$, suggests that 1 K^+ is transported out of the cell on each carrier cycle. However, glutamate uptake generates an inward membrane current (Brew and Attwell, 1987), while a stoichiometry in which 2 Na^+ and 1 glutamate$^-$ enter the cell and 1 K^+ leaves the cell would be electroneutral. Suggestions that a pH-changing ion might be transported on the uptake carrier, and thus might provide the missing charge movement, came from Nelson et al. (1983), who showed that the renal brush border glutamate carrier could be energized by a transmembrane pH gradient, and from Erecinska et al. (1983), who found that glutamate uptake into brain synaptosomes generated an extracellular alkalinization (though they did not exclude the possibility that this was due to Na^+ entry into the cell affecting the activity of Na^+-dependent pH-regulating carriers).

GLUTAMATE UPTAKE IN SALAMANDER RETINAL GLIA GENERATES pH CHANGES

Use of the transmembrane pH gradient to power vertebrate glutamate uptake might be an interesting evolutionary relic of the use by prokaryotes of pH gradients to power transmembrane transport. To investigate this possibility we carried out experiments on the glutamate uptake carrier in salamander retinal glial cells. These cells do not express ionotropic glutamate receptors, but show a large glutamate uptake current (Brew and Attwell, 1987). In these experiments, glutamate uptake was monitored by whole-cell clamping, while measuring pH changes just outside the cell with a pH-sensitive microelectrode, or measuring pH changes inside the cell fluorescently by including the dye 2′,7′-bis(carboxyethyl)carboxyfluorescein (BCECF) in the whole-cell pipette solution. Activating glutamate uptake was found (Bouvier et al., 1992) to result in an extracellular alkalinization (Fig. 1A) and an intracellular acidification (Fig. 1B). The internal acidification evoked by a 20 second exposure to 100 μM glutamate was about 0.2 pH unit, and measurement of the cells' buffering power, volume, and uptake current showed this to be consistent with the movement of roughly one effective proton into the cell per unitary charge of uptake current (Bouvier et al., 1992). The extracellular pH change was much smaller, reflecting the large volume of the bath in which the cell was immersed: *in vivo* the restricted extracellular space would result in a much larger alkaline shift

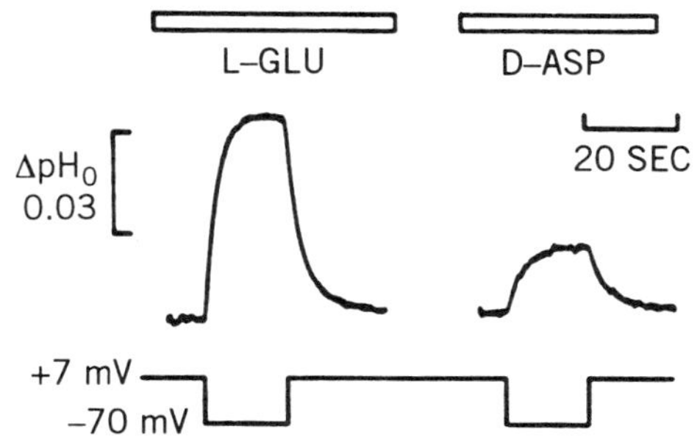

(*a*)

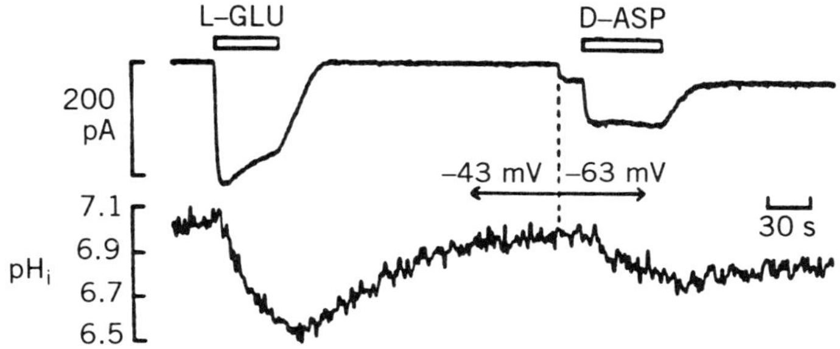

(*b*)

FIGURE 1. pH changes produced by the glutamate uptake carrier in salamander retinal Müller cells (from Bouvier et al., 1992). **A.** Extracellular pH changes (acid downwards) recorded with a pH-sensitive microelectrode positioned just outside the cell, during uptake of 100 μM L-glutamate (L-GLU) and D-aspartate (D-ASP). Uptake was evoked by changing the clamped membrane potential from +7 mV, where uptake occurs at a low rate, to −70 mV (bottom trace), where uptake occurs at a high rate. **B.** Uptake current (top) and intracellular pH change (bottom, measured with BCECF inside the cell) produced by 100 μM L- glutamate (at −43 mV) and D-aspartate (at −63 mV to increase the amount of uptake).

of pH. These pH changes showed the pharmacology of glutamate uptake, being activated by L-glutamate and D-aspartate, but not by the ionotropic or metabotropic receptor agonists kainate, α-amino-3-hydroxy-5-methyl-4-isoxazole propionic acid (AMPA), N-methyl-D-aspartate (NMDA), quisqualate, or 1S,3R-1-aminocyclopentane-1,3-dicarboxylic acid (ACPD). In addition, like uptake, they were abolished by removal of external Na^+, and were reduced by membrane depolarization.

The pH changes produced by glutamate uptake could reflect the transport of a pH-changing ion on the uptake carrier, or an indirect effect caused by Na^+ entry on the uptake carrier raising $[Na^+]_i$ and affecting the Na^+/H^+ exchanger or the Na^+/HCO_3^- cotransporter. To test for the latter possibility, we measured (Bouvier et al., 1992) the intracellular pH changes produced by glutamate uptake in the presence of amiloride (to block Na^+/H^+ exchange) or 4,4′-diisothiocyanatostilhene-2,2′-disulphonic acid (DIDS) (to block Na^+/HCO_3^- cotransport and Na^+-dependent Cl^-/HCO_3^- exchange, although in our nominally HCO_3^--free solutions these carriers should not be very active anyway). Amiloride had little effect on the uptake current or intracellular acidification (see Fig. 2A). DIDS halved the uptake current (Fig. 2B), possibly by binding to an anion transport site on the carrier (see below), but did not affect the intracellular pH change produced per picoamp of uptake current (Fig. 2C). We concluded, therefore, that the pH changes are generated directly by the uptake carrier, and not by pH-regulating carriers. Thus, glutamate uptake is powered by the transmembrane gradients of $[Na^+]$ and of pH.

SOURCE OF THE pH CHANGES: PROTON COTRANSPORT VERSUS COUNTER-TRANSPORT OF HYDROXYL OR BICARBONATE ANIONS

The pH changes seen in Fig. 1 could reflect the cotransport of protons into the cell with glutamate on the uptake carrier, or the counter-transport out of the cell of a pH-changing anion like OH^- or HCO_3^- (Fig. 3). Either of these options would, if occurring with the co-transport of two Na^+ and 1 glutamate$^-$ and the countertransport of a K^+, result in a net inward movement of charge accompanying glutamate uptake.

Bouvier et al. (1992) suggested that it is countertransport of a pH-changing anion, rather than cotransport of a proton, that is responsible for the pH changes, on the basis of the following evidence. First, replacing the Cl^- normally used as an intracellular anion with certain other anions greatly enhanced the glutamate uptake current: with 100 mM intracellular NO_3^-, ClO_4^-, or SCN^- in the recording pipette (and hence in the cell) the uptake current was approximately doubled, tripled, or quadrupled respectively. This result could be explained if these anions competed for transport out of the cell at an anion transport site on the uptake carrier, and were transported at a faster rate than the normally transported anion. Second, although these anions *increased* the uptake current, they *decreased* the size of the extracellular alkalinization occurring per picoamp of uptake current: this was to be expected if they were transported in place of (say) OH^-, because these anions are the salts of strong acids and would not generate a pH change when added to the extracellular solution. Finally, the use of anion-sensitive microelectrodes (nominally Cl^--sensitive, but in fact much more sensitive to ClO_4^- and SCN^-) allowed a direct demonstration of the fact that activating glutamate uptake, with these anions present inside the cell, led to the anions exiting from the cell by a process displaying the pharmacology, and voltage- and sodium-dependence of glutamate uptake.

However, this interpretation of the results put forward by Bouvier et al. (1992) has recently been shown to be invalid, as a result of the discovery that glutamate transporters contain an anion channel in their structure, which is activated during glutamate uptake and which is more permeable to SCN^-, ClO_4^-, and NO_3^- than to Cl^- (Wadiche et al., 1995; Eliasof and Jahr, 1996; Billups et al., 1996). This implies that some of the current generated by uptake carriers is not directly linked to glutamate transport, but is a result of anion flux through the carrier's anion conductance. When NO_3^-, ClO_4^-, or SCN^- are present inside the cell, they exit through the carrier's anion conductance, making the glutamate-evoked inward current larger without producing any extra pH change, thus explaining all the results of Bouvier et al. (1992). Consequently, the results of Bouvier et al. (1992) do *not* allow a distinction to be made between whether the uptake carrier transports an H^+ into the cell or a pH-changing anion out of the cell.

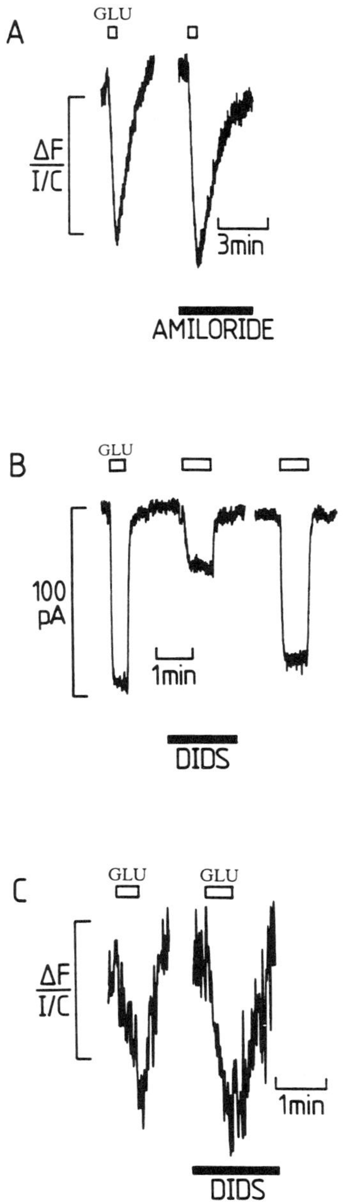

FIGURE 2. Effects of blockers of pH-regulating carriers on the pH changes produced by glutamate uptake (previously unpublished data from the study of Bouvier et al., 1992). **A.** Change of intracellular pH (acid downwards) measured as the fluorescence (F) of BCECF and normalized by the uptake current/cell capacitance *(I/C)*) evoked by 100 μM L-glutamate in control solution (left) and in the presence of 1 mM amiloride. **B.** Uptake currents at −43 mV evoked by 100 μM L-glutamate before, during, and after superfusion of 0.5 mM DIDS. DIDS reduced the uptake current, probably by binding to the OH^- transport site on the uptake carrier. **C.** Change of intracellular pH (fluorescence of BCECF) normalized by uptake current/capacitance (to compensate for the smaller uptake current in DIDS) in control solution (left) and in the presence of 0.5 mM DIDS.

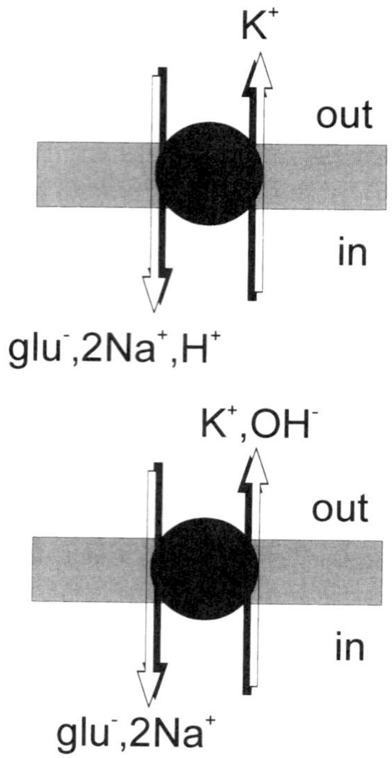

FIGURE 3. Two possible ionic stoichiometries for the glutamate uptake carrier. With each glutamate anion entering the cell, 2 Na^+ are co-transported, 1 K^+ is countertransported, and either 1 OH^- is countertransported or 1 H^+ is cotransported.

It is possible, however, to rule out the notion that the glutamate transporter countertransports HCO_3^- anions. First, the experiments of Bouvier et al. (1992) were performed in nominally bicarbonate-free, HEPES-buffered solution. Measurements of the small amount of HCO_3^- present in the cell, and calculations of how rapidly HCO_3^- could be generated from CO_2 (with carbonic anhydrase blocked; see also Chapter 4), showed that the cell could not provide enough HCO_3^- to account for the observed uptake current (assuming that one HCO_3^- was transported per unitary charge of uptake current). Second, the observed changes of intracellular pH could be accounted for by the movement of about 1 OH^- or H^+ per unitary charge of uptake current, but would require the transport of 9 HCO_3^- ions, an implausibly high number, because the effective pK of 6.1 for HCO_3^- makes this anion a weak base. Finally, inhibiting carbonic anhydrase (a manoeuver that greatly reduces pH changes produced by HCO_3^- movement in other preparations; Kaila et al., 1990; Chapters 16 and 17) had no effect on the uptake-evoked pH changes.

It appears, therefore, that on each carrier cycle 1 glutamate$^-$ and 2 Na^+ enter the cell, while 1 K^+ leaves the cell, and either 1 OH^- is transported out or 1 H^+ is transported in (Fig. 3). In addition some of the current generated by the uptake carrier reflects the flow of Cl^- anions through the carrier's anion conductance. By measuring the intracellular pH change generated by the uptake carrier at different potentials, and showing that application of external glutamate evokes an intracellular acidification both above and below the reversal potential for H^+/OH^-, Billups et al. (1996) showed that the carrier-evoked pH changes are produced by transport of an ion linked to glutamate transport, and not by a passive flow of OH^- ions through the anion conductance.

HIPPOCAMPAL GLUTAMATE UPTAKE: GENERATION OF pH CHANGES

The large size of salamander retinal glial cells greatly facilitated the discovery that the glutamate uptake carrier transports pH-changing ions; but, from a more general perspective, it is obviously desirable to know whether mammalian glutamate uptake behaves similarly. To this end we have studied glutamate-evoked intracellular pH changes in hippocampal slices, by loading the slices with the acetoxymethyl ester of the pH-sensitive fluorescent dye BCECF (Amato et al., 1994; see also Rose and Ransom, 1996).

Applying glutamate to these slices generated an intracellular acidification (Fig. 4). It has been suggested, previously in this book, that similar acidifications can result from proton movement (Chapter 16) or from bicarbcnate movement through GABA- (γ-aminobutyric acid) gated channels. We were surprised to find, therefore, that the glutamate-evoked pH change we observed was not blocked by a combination of 6-cyano-7-nitroquinoxaline-2,3-dione (CNQX) and D-aminophosphonovalerate (APV) (to block glutamate-gated channels) and picrotoxin and strychnine (to block GABA- and glycine-gated anion channels), as shown in Fig. 4. Provided these blockers were present, the pH change showed the pharmacology of glutamate uptake, in that L-glutamate and D-aspartate produced a pH change, while kainate, NMDA, quisqualate, and ACPD did not. (In the absence of these blockers, kainate and quisqualate sometimes produced a pH change, but—unlike the glutamate-evoked pH change—this was blocked by removing calcium, and so could have been due to activation of non-NMDA receptors releasing glutamate, which then generated a pH change by activating glutamate uptake. Alternatively the pH change could be due to calcium entry evoked by activation of glutamate-gated channels raising the intracellular calcium concentration and activating Ca^{2+}/H^+ exchange (Chapter 16). The intracellular acidification was greatly reduced by the uptake blocker

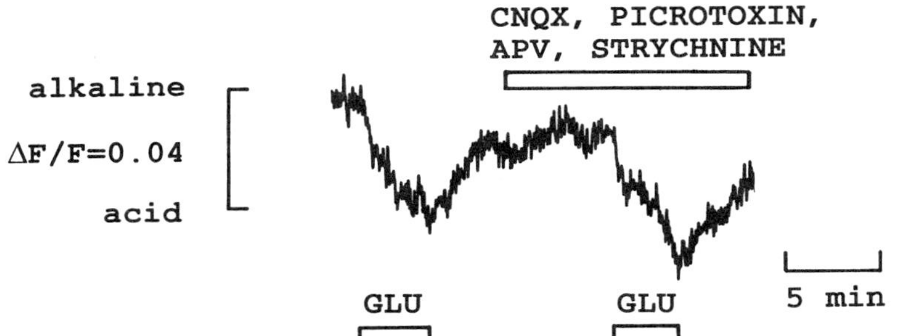

FIGURE 4. Changes in intracellular pH (more acid shown as a downward deflection) evoked by 300 μM L-glutamate in a hippocampal slice; measured as the fluorescence (F) of BCECF (from Amato et al., 1994). The acid shift evoked in the control solution (first application of glutamate) is similar in size to that produced when activation of glutamate-, GABA- and glycine-gated ion channels was blocked with 10 μM CNQX, 40 μM D-APV, 100 μM picrotoxin, and 10 μM strychnine.

L-trans-pyrrolidine-2,4-dicarboxylate (PDC) (Fig. 5). In addition, it showed the ionic dependence of glutamate uptake, being abolished by removing external Na^+ and reduced by raising the external K^+ concentration (Amato et al., 1994).

Blocking Na^+/H^+ exchange with amiloride or blocking Na^+-HCO_3^- cotransport with DIDS had little effect on the intracellular acidification produced by glutamate, which therefore does not appear to be a secondary consequence of Na^+ loading (due to activity of the uptake carrier) affecting these pH-regulating carriers. There are two possible explanations of the glutamate-evoked pH change. One is that, as in salamander retinal glia, an uptake carrier present in the hippocampus counter-transports OH^- ions or cotransports H^+ ions. Alternatively, the pH changes might be produced by metabolic activity caused by the Na^+ loading associated with the activity of the uptake carrier. These two possibilities could be distinguished by measuring the extracellular pH during application of glutamate. If the intracellular acidification is a result of enhanced metabolic activity then the extracellular fluid should also become acid, while if the uptake carrier counter-transports OH^- or cotransports H^+ ions then the extracellular fluid should become alkaline (as for salamander retinal glia, Fig. 1A).

Given the suggestion of a proton flux through glutamate-gated ion channels (Chen and Chesler, 1992a; but see also Paalasmaa et al, 1994 and Chapter 16), and of a bicarbonate efflux through GABA-gated channels (Kaila et al., 1992), which activation of inhibitory interneurons by glutamate might be expected to evoke, why does blocking glutamate- and GABA-gated channels not significantly affect the glutamate-evoked intracellular acidification in hippocampal slices (Fig. 4)? There are several possible reasons for this. First, we used bicarbonate-free (HEPES-buffered) solutions, which may greatly reduce the HCO_3^- flux through GABA-gated channels. Second, desensitization of glutamate receptors produced by the prolonged application of glutamate in our experiments might greatly reduce the pH change produced by any proton influx through the channels linked to receptors, as compared with the situation when the receptors are activated briefly by synaptically released glutamate, as in the experiments of Chen and Chesler (1992a,b). Finally, when a slice is superfused with glutamate-containing solution, the amount of uptake activated throughout the slice may generate fluxes of pH-changing ions which simply outweigh the fluxes that can be produced at the relatively localized groups of ion channels present at synapses.

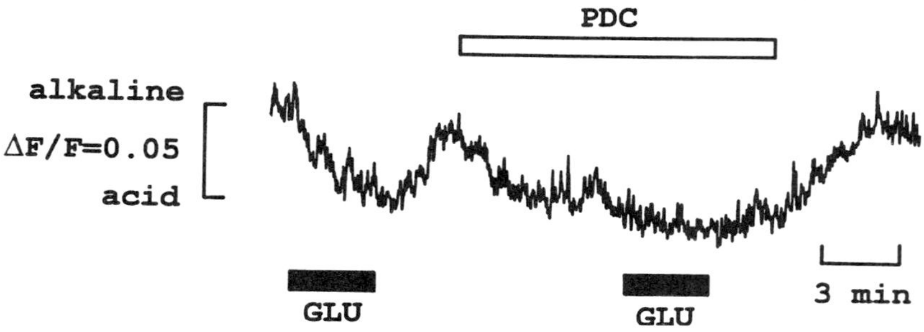

FIGURE 5. The glutamate uptake blocker PDC (200 μM) blocked the acidification produced by 300 μM L-glutamate, and produced an acidification itself, possibly because it is transported on the glutamate uptake carrier (Sarantis et al., 1993). Experiment carried out in the presence of the blockers applied on the right-hand side of Fig. 4. Data from the study of Amato et al. (1994)

EFFECTS OF pH CHANGES ON THE RATE OF GLUTAMATE UPTAKE

Changes of pH might affect glutamate uptake by altering the concentration of transported OH^-/H^+ ions inside and outside the cell, as well as by having direct effects on the carrier protein, and by altering the membrane surface potential and thus altering the effective concentration of charged substrates at the membrane surface (Gilbert, 1971). Understanding the effects of pH is highly relevant to understanding the events in brain ischemia, when the brain pH becomes more acid and when an excessive rise of extracellular glutamate concentration triggers the death of neurons (reviewed by Szatkowski and Attwell, 1994; see also Chapters 28–30).

During glutamate uptake, OH^- ions are transported out of the cell or H^+ ions are transported in. One might expect, therefore, that an acid shift of the intracellular pH could inhibit glutamate uptake by depriving the carrier of substrate OH^- or preventing the carrier from losing H^+ at the inner membrane surface. To investigate this we measured uptake currents in salamander glial cells whole-cell clamped with pipettes filled with highly buffered solutions of different pH (Billups and Attwell, 1996). Decreasing the pipette pH (increasing acidity) reduced the uptake current as predicted. Acidifying the external solution also inhibited the uptake current, apparently independent of any change in the concentration of transported OH^-/H^+, by reducing the carrier's affinity for extracellular Na^+ (Billups and Attwell, 1996).

Thus, during brain ischemia, when the extra- and intracellular pH shift acid by 0.6–1.0 pH unit (Smith et al., 1986), glutamate uptake will be reduced, facilitating the rise of extracellular glutamate concentration to neurotoxic levels.

GLUTAMATE RELEASE IN BRAIN ISCHEMIA

The release of glutamate, which raises the extracellular glutamate concentration to neurotoxic levels during brain ischemia, is thought to occur in two phases. In the first few minutes of ischemia, the fall of ATP levels inhibits the Na^+/K^+ pump, leading to a rise of extracellular $[K^+]$, depolarization, and a burst of action potentials that trigger Ca^{2+}-dependent vesicular release of glutamate. During this first phase, the acidity-induced inhibition of uptake just discussed will potentiate the ongoing rise of extracellular glutamate concentration.

After a few minutes, however, vesicular release is inhibited by the fall of ATP levels that has occurred, and a sudden rise of extracellular K^+ concentration, $[K^+]_o$, occurs to around 60 mM. This triggers the "anoxic depolarization" of cells and has been postulated to result in the glutamate uptake carriers reversing their operation and pumping glutamate out of the cells (Szatkowski et al., 1990; Szatkowski and Attwell, 1994). Release of glutamate by reversed uptake has been shown to be produced by a rise of $[K^+]_o$ (Szatkowski et al., 1990) or as a result of the metabolic inhibition that anoxia produces (Madl and Burgesser, 1993). Release of glutamate will occur until the extracellular glutamate concentration, $[glu]_o$, rises to reach a new equilibrium, where there is no net glutamate flux on the carrier. The stoichiometry of the carrier predicts that the minimum value that $[glu]_o$ can have is (Attwell et al., 1993; see also Zerangue and Kavanaugh, 1996).

$$[glu]_o = [glu]_i \frac{[Na^+]_i^2}{[Na^+]_o^2} \cdot \frac{[K^+]_o}{[K^+]_i} \cdot \frac{[OH^-]_o}{[OH^-]_i} \exp \frac{VF}{RT} \qquad (1)$$

which is about 0.2 μM in normal circumstances, but 260 μM for the ion concentrations pertaining in ischemia (Attwell et al., 1993). The latter value of $[glu]_o$ is sufficient to trigger the death of neurons if maintained for more than a few minutes (Choi et al., 1987). Reversed operation of a carrier with the stoichiometry proposed above would involve the movement of 1 glutamate and 2 Na^+ out of the cell, the movement of 1 K^+ into the cell, and the movement of 1 OH^- into or 1 H^+ out of the cell. By analogy with the inhibitory effect (described on p. 359) of an acid intracellular pH on forward uptake, one might expect the acid extracellular pH occurring in brain ischemia to inhibit reversed uptake by depriving the carrier of extracellular OH^- ions to transport into the cell, or slowing the loss of cotransported H^+ from the carrier at the outer face of the membrane. Indeed, a low external pH does reduce reversed K^+-dependent uptake in salamander glia (Billups and Attwell, 1996). At the pH occurring in ischemia (about 6.1) this slowing is by a factor of about 14. Because the intra- and extracellular pH decrease by about the same amount in ischemia, the *equilibrium level* of glutamate predicted from Eq. 1 is not affected ($[OH^-]_i$ and $[OH^-]_o$ are reduced by about the same factor). However the *rate* at which $[glu]_o$ rises to neurotoxic levels is slowed by a factor of about 14. This slowing will have no practical consequences in cases of prolonged ischemia: eventually $[glu]_o$ will reach a neurotoxic level. For cases of transient ischemia, however, where a blood vessel is only blocked for a period of seconds to minutes, the increase in the time needed for $[glu]_o$ to rise, which is the result of the involvement of OH^- or H^+ in the glutamate transport stoichiometry, could serve a useful neuroprotective role.

SIMILARITIES BETWEEN VARIOUS GLUTAMATE TRANSPORTERS

Data in the literature reveal that many glutamate transporters appear to transport a pH-changing anion or a proton. This is true for the salamander glial cell carrier as described above (Bouvier et al., 1992), for the mammalian neuronal carrier EAAC1 (Kanai et al., 1995), for the carriers in rat forebrain synaptosomes (Erecinska et al., 1983) and in renal brush border cells (Nelson et al., 1983), for the bacterial glutamate transporter *gltP* (de Vrij et al., 1989; Tolner et al., 1992), and for the mitochondrial glu^-/OH^- exchanger (Minn and Gayet, 1977; LaNoue and Schoolwerth, 1979). Since the mammalian glial cell carrier GLT1 cloned by Pines et al. (1992) shows considerable structural homology with *gltP*, it seems likely that countertransport of an OH^- or cotransport of an H^+ is a method of powering glutamate transport that has been well conserved on an evolutionary timescale.

ACKNOWLEDGMENTS

Supported by the Wellcome Trust, Medical Research Council and European Community (contract BMH4-CT95-0571).

REFERENCES

Amato A, Ballerini L, Attwell D (1994): Intracellular pH changes produced by glutamate uptake in rat hippocampal slices. J Gen Physiol 72: 1686–1696.

Attwell D, Barbour B, Szatkowski M (1993): Nonvesicular release of neurotransmitter. Neuron 11: 401–407.

Barbour B, Brew H, Attwell D (1988): Electrogenic glutamate uptake in glial cells is activated by intracellular potassium. Nature 335: 433–435.

Billups B, Attwell D (1996): Modulation of non-vesicular glutamate release by pH. Nature 379: 171–174.

Billups B, Rossi D, Attwell D (1996): Anion channel behaviour of the glutamate uptake carrier in salamander retinal glial cells. J Neurosci 16: 6722–6731.

Bouvier M, Szatkowski M, Amato A, Attwell D (1992): The glial cell glutamate uptake carrier countertransports pH-changing anions. Nature 360: 471–474.

Bouvier M, Amato A, Attwell D (1994): Molecular biology and physiology of glutamate transporters. Sem Neurosci 6: 117–125.

Brew H, Attwell D (1987): Electrogenic glutamate uptake is a major current carrier in the membrane of axolotl retinal glial cells. Nature 327: 707–709.

Chen JCT, Chesler M (1992a): Extracellular alkaline shifts in rat hippocampal slice are mediated by NMDA and non-NMDA receptors. J Neurophysiol 68: 342–344.

Chen JCT, Chesler M (1992b): pH transients evoked by excitatory synaptic transmission are increased by inhibition of extracellular carbonic anhydrase. Proc Natl Acad Sci USA 89: 7786–7790.

Choi DW, Maulucci-Gedde M, Kriegstein AR (1987): Glutamate neurotoxicity in cortical cell culture. J Neurosci 7: 357–368.

De Vrij W, Bulthuis RA, van Iwaarden PR, Konings WN (1989): Mechanism of L-glutamate transport in membrane vesicles from *Bacillus stearothermophilus*. J Bacteriol 171: 1118–1125.

Eliasof S, Jahr CE (1996): Retinal glial cell glutamate transporter is coupled to an anion conductance. Proc Natl Acad Sci USA 93: 4153–4158.

Erecinska M, Wantorsky D, Wilson DF (1983): Aspartate transport in synaptosomes from rat brain. J Biol Chem 258: 9069–9077.

Fairman WA, Vandenberg RJ, Arriza JL, Kavanaugh MP, Amara SG (1995): An excitatory aminoacid transporter with properties of a ligand-gated chloride channel. Nature 375: 599–603.

Gilbert DL (1971): Fixed surface charges. In Adelman WJ Jr (ed): Biophysics and Physiology of Excitable Membranes. New York: Van Nostrand Reinhold, pp 359–378.

Hertz L (1979): Functional interactions between neurons and astrocytes. I. Turnover and metabolism of putative amino acid neurotransmitters. Prog Neurobiol 13: 277–323.

Kaila K, Saariskoski J, Voipio J (1990): Mechanism of action of GABA on intracellular pH and on surface pH in crayfish muscle fibres. J Physiol (Lond) 427: 241–260.

Kaila K, Paalasmaa P, Taira T, Voipio J (1992): pH transients due to monosynaptic activation of $GABA_A$ receptors in rat hippocampal slices. Neuroreport 3: 105–108.

Kanai Y, Hediger MA (1992): Primary structure and functional characterization of a high-affinity glutamate transporter. Nature 360: 467–471.

Kanai Y, Smith CP, Hediger MA (1994): A new family of neurotransmitter transporters: The high affinity glutamate transporters. FASEB J 8: 1450–1459.

Kanai Y, Nussberger S, Romero MF, Boron WF, Hebert SC, Hediger MA (1995): J Biol Chem 270: 16561–16568.

Kanner BI, Sharon I (1978): Active transport of L-glutamate by membrane vesicles from rat brain. Biochemistry 17: 3949–3953.

LaNoue KF, Schoolwerth AC (1979): Metabolite transport in mitochondria. Ann Rev Biochem 48: 871–922.

Madl JE, Burgesser K (1993): Adenosine triphosphate depletion reverses sodium-dependent neuronal uptake of glutamate in rat hippocampal slices. J Neurosci 13: 4429–4444.

Minn A, Gayet J (1977): Kinetic studies of glutamate transport in rat brain mitochondria. J Neurochem 29: 873–881.

Nelson PJ, Dean GE, Aronson PS, Rudnick G (1983): Hydrogen ion cotransport by the renal brush border glutamate transporter. Biochemistry 22: 5459–5463.

Nicholls D, Attwell D (1990): Release and uptake of excitatory amino acids. Trends Pharm Sci 11: 462–468.

Paalasmaa P, Taira T, Voipio J, Kaila K (1994): Extracellular alkaline transients mediated by glutamate receptors in the rat hippocampal slice are not due to a proton conductance. J Neurophysiol 72: 2031–2033.

Pines G, Danbolt NC, Bjørås M, Zhang Y, Bendahan A, Eide L, Koepsell H, Storm-Mathisen J, Seeberg E, Kanner BI (1992): Cloning and expression of a rat brain L-glutamate transporter. Nature 360: 464–467.

Rose CR, Ransom BR (1996): Mechanisms of H^+ and Na^+ changes induced by glutamate, kainate, and D-aspartate in rat hippocampal slices. J Neurosci 16: 5393–5404.

Sarantis M, Ballerini, L, Miller B, Silver RA, Edwards M, Attwell D (1993): Glutamate uptake from the synaptic cleft does not shape the decay of the non-NMDA component of the synaptic current. Neuron 11: 541–549.

Smith M-L, von Hanwehr R, Siesjo BK (1986): Changes in extra- and intracellular pH in the brain during and following ischemia in hyperglycemic and in moderately hypoglycemic rats. J Cereb Blood Flow Metab 6: 574–583.

Storck T, Schulte S, Hofmann K, Stoffel, W (1992): Structure, expression, and functional analysis of a Na^+-dependent glutamate/aspartate transporter from rat brain. Proc Natl Acad Sci USA 89: 10955–10959.

Szatkowski M, Attwell D (1994): Triggering and execution of neuronal death in brain ischemia: Two phases of glutamate release by different mechanisms. Trends Neurosci 17: 359–365.

Szatkowski M, Barbour B, Attwell D (1990): Non-vesicular release of glutamate from glial cells by reversed electrogenic glutamate uptake. Nature 348: 443–446.

Takahashi M, Kovalchuk Y, Attwell D (1995): Pre- and postsynaptic determinants of EPSC waveform at cerebellar climbing fiber and parallel fiber to Purkinje cell synapses. J Neurosci 15: 5693–5702.

Tolner B, Poolman B, Wallace B, Konings WN (1992): Revised nucleotide sequence of the *gltP* gene, which encodes the proton-glutamate-aspartate transport protein of *Eschericia coli* K–12. J Bacteriol 174: 2391–2393.

Wadiche JI, Amara SG, Kavanaugh MP: (1995) Ion fluxes associated with excitatory amino acid transport. Neuron 15: 721–728.

Zerangue N, Kavanaugh MP (1996): Flux coupling in a neuronal glutamate transporter. Nature 383: 634–637.

SECTION V

pH Modulation of Neural Activity

CHAPTER 20

pH Modulation of Neuronal Excitability and Central Nervous System Functions

GEORGE G. SOMJEN AND GEOFFREY C. TOMBAUGH
Department of Cell Biology, Duke University Medical Center, Durham, NC (G.G.S.);
Department of Neuroscience, University of Medicine and Dentistry of New Jersey,
Piscataway, New Jersey (G.C.T.)

THE SCOPE OF THIS CHAPTER

In today's biomedical research we often try to answer questions that were first asked a century or more ago, and so it is in many respects with the significance of hydrogen ions for the functioning of the brain. It has been clear for some time that pH influences numerous biophysical and biochemical functions (Sörensen, 1912), among them the excitability of the central nervous system (Foerster, 1924; Wyke, 1963); yet the mechanisms that underlie these interactions are still only partially understood. This chapter is a review of the evolution over time, as well as the present state, of knowledge concerning the effects of protons and of CO_2 on cerebral function and especially the excitability of central neurons. Specifically, we address the following questions:

- How was the effect of H^+ ions on excitability discovered?
- What are the effects of H^+ on electrical excitability, membrane potential, and synaptic transmission?
- Are the effects of CO_2 explicable in terms of pH?
- Are the effects of H^+ mediated by changes in Ca^{2+} activity?

In Chapter 21, we will continue the discussion of the following specific points:

- How do H^+ ions modulate voltage-dependent ion channel function?
- To what degree can ion channel changes account for the H^+ effects on cerebral function?

PRELUDE

Well before the physical chemistry of acidity was clarified or its significance for bodily functions was appreciated, two persons started two different trails of inquiry that eventually lead

pH and Brain Function, Edited by Kai Kaila and Bruce R. Ransom
ISBN 0-471-11838-9

to the recognition of the power of pH over brain function. One was an obscure country physician in England by the name of Henry Hickmann, the other was one of the giants of nineteenth century French medicine, Jean-Marie Charcot.

According to some historians (Cartwright, 1952), Hickmann was the first to propose that "suspended animation" could be a means to surgery without pain. His ideas and experiments were published in a short pamphlet in 1824 (Hickman, 1824; Wellcome and Malcolm, 1930), some 20 years before the celebrated operations by William Morton and by Horace Wells in the United States that started the age of general anesthesia. Hickmann described methods to render animals unconscious in order to be able to subject them painlessly to surgical procedures. Among the methods he tried and recommended for use in human patients, was the inhalation of "carbonic acid"—meaning, of course, carbon dioxide. Here follows an excerpt from Hickmann's notebook, reproduced by Wellcome and Malcolm (1930), describing such a trial:

> Experiment 4th. A Mouse was confined under a Glass, surrounded by Water. By means of a small tube a foot long I passed carbonic acid Gas very slowly into the glass, respiration ceased in three minutes, I cut all its leg off at the first joint, and plunged it into a basin of cold water, the Animal immediately recovered and ran about the table apparently without pain; the stumps soon healed and I kept it a fortnight, after which I gave it liberty.

While this 19th century experiment may be cruel by today's standards, it should be remembered that its purpose was to alleviate pain in humans as well as animals. Nevertheless, CO_2 did not become a practical anesthetic; diethyl ether, nitrous oxide and chloroform were found to be safer and easier to administer (for additional early references, see Poulsen, 1952; Wyke, 1963). Much later, however, in the 1930s, Meduna advocated the use of CO_2-induced coma to treat psychoneuroses (Meduna, 1950; Gellhorn, 1953a). From Meduna's careful observation of patients treated by CO_2 inhalation emerged a detailed picture of the effect of this gas on humans. In a nutshell, inhalation of 30% CO_2 in 70% O_2 causes at first a feeling of excitation and choking, then involuntary movements, then coma. This sequence of symptoms resembles the well-known phases of the induction of ether anesthesia.

While excess CO_2 depresses the CNS, lack of CO_2 in body fluids can lead to cramps and to convulsions. The content of CO_2 is reduced below normal when breathing exceeds metabolic needs. The term for breathing in excess is *hyperventilation,* to be distinguished from *hyperpnea* which means increased breathing matching increased demand, for example during exercise. CO_2 content can be measured as concentration, or, more commonly and more conveniently, as the partial pressure of the gas, denoted P_{CO_2}.

To trace the story of hyperventilation-induced hyperexcitability (Brown, 1953), we have to begin with the studies of hysterical and epileptic fits, first described in detail at the famous Salpêtrière hospital by Charcot and his colleagues (Owen, 1971). At first, these two conditions, epilepsy and hysteria, were considered to be the extremes of a continuum, with intermediate forms that were called *"les attaques de grande hystéro-épilepsie."* These were described in detail in a doctoral dissertation by Charcot's pupil, Richer (1879).

Richer (1879) defined successive phases in the evolution of hystero-epileptic fits. The process starts with varied psychological prodromals that can last several days. A sensory aura follows and after that carpo-pedal spasms while the patient is still conscious (Fig. 1). The process culminates in an epileptiform attack characterized by loss of consciousness, tonic and then clonic convulsions, followed by relaxation of the musculature and prolonged coma. Unlike the "purely" epileptic patient who gradually emerges from postictal

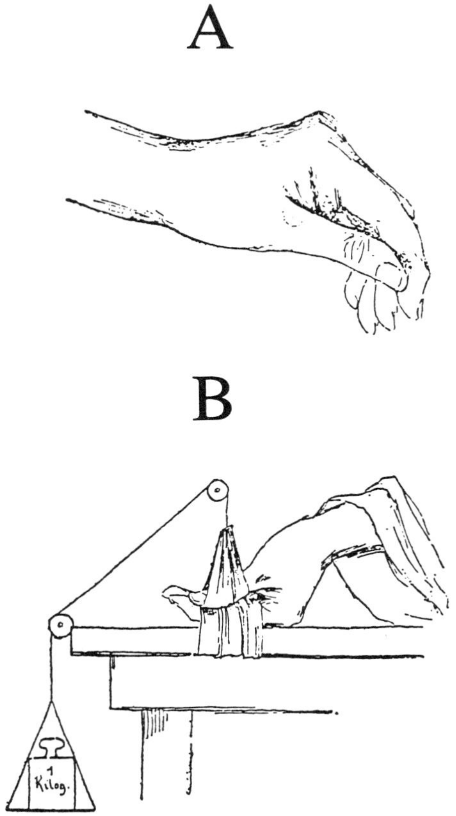

FIGURE 1. Charcot's illustration of hysterical contractures. In reality this is probably carpal spasm due to hyperventilation-induced alkalosis. **A.** The position of the fingers during the spasm. **B.** Demonstration of the strength of the contraction.

depression to a normal state, the hystero-epileptic continues with relapsing fits of bizarre attitudes and various psychomotor signs that again can last for days.

Subsequently Charcot's school sharply distinguished hysteria, considered to be of functional nature, from epilepsy, which is of organic etiology. By this time Charcot was explaining the striking similarity of grand mal seizures and grande hystérie in terms of a "contagion," by which he meant the involuntary imitation of seizures by hysterical patients housed in the same wards as epileptics. Charcot could not know at the time that prolonged hyperventilation, which is common in "grand" hysteria, can cause neurological malfunction that is indistinguishable from "real," (i.e., organic) neurological disease. This insight emerged only several decades after Charcot's death.

HYPERVENTILATION, CO_2, pH, AND Ca^{2+}

It was an early observation that excessive breathing, whether voluntary or forced, is followed by a period of cessation of respiration (Pflüger, 1868). The cause of post-hyperventilation apnea became the subject of debate: Pflüger attributed it to the depletion of CO_2 in blood, while Hering thought it was caused by the accumulation of O_2 (Pflüger, 1868; Ewald, 1873). The matter was settled in favor of Pflüger by Haldane and Priestley (1905) who showed that it is CO_2 that normally drives the respiratory center, and that only when the fraction of O_2 in inspired air falls below 13%, does lack of oxygen also become a stimulus (see also Chapters 26–27).

Hyperventilation not only reduces the CO_2 concentration in blood but it also drives the blood pH into a relatively alkaline range (Collip and Backus, 1920). During experimental voluntary hyperventilation the subjects, who often included the investigators, noticed strange tingling sensations in the face and extremities as well as cramps and tremors in hands and feet (Henderson, 1909; Vernon, 1909). A decade later Grant and Goldman (1920) and Collip and Backus (1920) recognized the similarity between the sensory and motor hyperexcitability associated with hyperventilation, and the clinical condition known as tetany.

Around this period it also became apparent that "spasmophilia" of young children was in fact due to hypocalcemia caused by rickets, entirely similar to the tetany caused by hypoparathyroidism (MacCallum and Voegtlin, 1909; Howland and Marriott, 1918; Freudenberg and György, 1922). As these new insights emerged, it was natural to search for a connection between the tetany of hyperventilation and its other forms. The similarity seemed contradicted in that, in contrast with hypoparathyroidism and rickets, during hyperventilation serum calcium levels were either unchanged or slightly increased (Grant and Goldman, 1920; Brown, 1953). The postulated connection was revived, however, once Rona and Takahashi (1913) proposed that the ratio of ionized to bound calcium in blood plasma depends on pH. Based on this theory, György and Vollmer (1923) attributed the tetany of alkalosis to the reduction of ionized calcium in favor of its bound form. This explanation has been perpetuated in many texts (e.g., Brown, 1953; Gellhorn, 1953a).

As it turned out, this was a classical example of a plausible explanation which eventually was proven wrong. Measured changes of ionized calcium induced by hyperventilation in circulating blood were smaller than expected (Band and Broadway, 1979; Oberleithner et al., 1982). We have measured pH and ionized calcium activity ($[Ca^{2+}]$) with ion-selective electrodes in arterial blood while monitoring neuromuscular function by electromyography in anesthetized cats. $[Ca^{2+}]$ was lowered either by the infusion of sodium citrate, or by hyperventilation. While forced hyperventilation did cause $[Ca^{2+}]$ to decrease, the level was not nearly low enough to account for the tetany (Allen and Somjen, 1983; Allen et al., 1985; Somjen et al., 1987). The culprit must therefore have been the deficit of $[H^+]$, or of CO_2, or of both. The lowered $[Ca^{2+}]$ may have facilitated the effect, but could not have caused it.

Like peripheral tetany, respiratory alkalosis can trigger seizures of central origin, especially in patients suffering from epilepsy. The two classical papers on this topic are by Foerster (1924) and by Rosett (1924). Foerster reported three key observations: First, he found that voluntary hyperventilation precipitated a seizure in 25 out of 45 epileptic patients, and the seizure so induced was identical in its manifestations to those occurring spontaneously in each of the patients. Next he showed that in the state of respiratory alkalosis the electrical threshold of peripheral nerves was reduced, demonstrating the similarity between alkalosis and other forms of tetany. Finally, in two cases that came to surgery under local anesthesia, he stimulated the exposed cerebral cortex and found that the threshold of evoking muscle contractions was reduced when the patients were hyperventilating, thus linking hyperexcitability of the CNS to the hyperexcitability of peripheral nerves.

Rosett (1924) claimed to have performed more than a thousand experiments with hyperventilation in healthy persons as well as various neurological patients. He emphasized the diagnostic utility of the procedure inasmuch as ". . . abnormalities of the central nervous system which are not under ordinary conditions manifested by definite signs and symptoms, are signalled during hyperpnoea in a striking manner . . .".

Lennox, Gibbs, and Gibbs (1936) confirmed that hyperventilation can trigger seizures in epileptic patients, especially those with petit mal pattern in the electroencephalogram (EEG). They also demonstrated that inhalation of CO_2 can suppress seizures. Moreover, in healthy volunteers hyperventilation induced slow, high-amplitude EEG wave patterns, which the authors considered similar to seizure patterns.

Direct and Indirect Central Effects of Hyperventilation

The multiple effects of CO_2 and of its depletion complicate their interpretation. Among them is the powerful control exerted by arterial P_{CO_2} over the cerebral circulation (Gibbs et al., 1947; Brown, 1953). The cerebral vasoconstriction induced by low P_{CO_2} plays an important part in the central disturbances caused by hyperventilation, in addition to the effect of concomitant alkalinization of cerebral pH on neuronal excitability.

Then there is the question of the relative importance of the effect on interstitial calcium ($[Ca^{2+}]_o$) caused by changes of pH, already raised in connection with peripheral tetany. Unlike blood plasma, cerebrospinal and cerebral interstitial fluids contain a negligible amount of protein that binds Ca^{2+}, but bicarbonate, and smaller amounts of carbonate and other anions capable of binding Ca^{2+} in pH-dependent fashion, are present. The increase of ionized calcium activity in the interstitial fluid of a brain tissue slice in vitro caused by acidification of the artificial bathing medium is illustrated in Fig. 2A. Nevertheless, the measured $[Ca^{2+}]_o$ in cerebral neocortex of anesthetized cats changed little during forced hyperventilation and during elevation of inhaled CO_2 (Somjen et al., 1987), and changed even less in the hippocampal formation of rats (Balestrino and Somjen, 1988). This eliminates variation of $[Ca^{2+}]_o$ as a possible mediator of central pH effects. The compensatory regulation of $[Ca^{2+}]_o$ present in intact brain is, however, not available to the hippocampal slice (Fig. 2A).

Finally there is the question whether the CO_2 effects are due to variation of proton activity or to molecular CO_2 itself. While "respiratory" acidosis and alkalosis produce powerful effects on the central nervous system, systemic "metabolic" acidosis and alkalosis have much weaker influence on CNS function. This difference could be taken to support the role of CO_2, not of H^+. This argument lost validity when it was discovered that cerebral pH is regulated independently from systemic pH and that, while CO_2 freely penetrates the blood-brain barrier, H^+ and most other ions do so only to a limited degree (Dusser de Barenne et al., 1938; Swanson et al., 1958; Rapoport, 1964; Fencl et al., 1966; Pappenheimer, 1967; see also Chapters 4, 14 and 32). With this insight the matter remains unresolved: Since CO_2 concentration governs pH, variations of either or both could account for the central effects of respiratory acidosis and alkalosis.

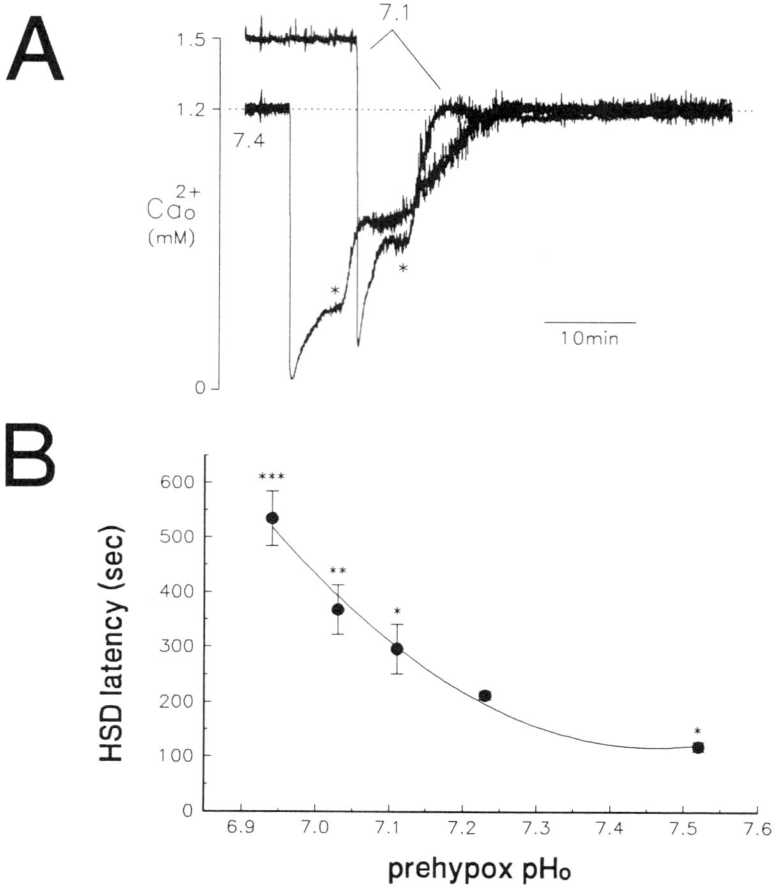

FIGURE 2. Acidosis delays and alkalosis hastens the onset of hypoxic spreading depression-like depolarization (HSD). **A.** Interstitial calcium concentration, $[Ca^{2+}]_o$, recorded with ion-selective microelectrodes in stratum radiatum of CA1 region of two rat hippocampal tissue slices cut from the same hippocampus and recorded simultaneously in a two-well chamber. One slice was in acidified (pH 7.1) medium, the other at control bath pH (7.4). Note different starting levels of $[Ca^{2+}]_o$ at different pH. Oxygen was withdrawn from both slices at the onset of the traces. At the times marked by asterisks, oxygen was restored, and the previously acidotic slice was returned to normal pH. "Resting" $[Ca^{2+}]_o$ is higher in acidosis than at normal pH, due to release of bound calcium. Onset of spreading depression-related decrease of $[Ca^{2+}]_o$ is delayed and its recovery is hastened by acidosis. **B.** Dependence of HSD onset on pre-hypoxic pH_o. "HSD latency" is the time between oxygen withdrawal and the appearance of an SD-like negative potential shift. Data are mean ± SEM. Tissue pH_o was measured with ion-selective microelectrode; normal control pH_o is 7.25, with bath pH of 7.4. P_{CO_2} was constant, and bath pH was altered by titration with HCl or NaOH. Modified from Tombaugh (1994).

CO_2 AND pH EFFECTS ON CORTICAL EXCITABILITY AND CEREBRAL ELECTRICAL ACTIVITY IN EXPERIMENTAL ANIMALS

The literature that appeared until 1963, dealing with the relationship of pH to the excitability and EEG activity of the brain, has been reviewed by Wyke (1963), and only a few highlights will be mentioned here. Among the influential early papers is one by Brody and Dusser de Barenne (1932), who confirmed in anesthetized cats Foerster's (1924) observations on humans, showing that forced hyperventilation lowered the threshold for eliciting movements of contralateral limb muscles by electrical stimulation of the cortical motor area. The effect was much more striking and reproducible when a subliminal dose of strychnine was administered before the hyperventilation.

The reported effects on electrocorticograms (ECoG) of cats and other laboratory animals were rather variable, perhaps because of uncontrolled circulatory changes (emphasized by Ivanov, 1963). Bremer and Thomas (1936) and Gerard et al. (1936) found that a moderately elevated concentration of CO_2 in inhaled air caused increased frequency of ECoG waves, which they interpreted as "increased excitation"; but according to Gerard et al. (1936), high concentrations of CO_2 depressed and eventually abolished the ECoG waves. Pollock (1949) reported increased EEG frequency but depressed amplitude under the influence of 15–30% CO_2 in inspired air.

Dusser de Barenne and coworkers (Dusser de Barenne and McCulloch, 1939; Dusser de Barenne et al., 1938) were the first to measure the pH of the exposed cortex. They found that acidification of the cortex was associated with a depression of both amplitude and frequency of ECoG waves.

At the time of the early EEG studies (Bremer and Thomas, 1936; Gerard et al., 1936; Dusser de Barenne and McCulloch, 1939) the concept of "activation" of the EEG had not yet been introduced. Later Bonvallet et al. (1955) attributed cortical desynchronization to the selective stimulation of the brainstem ascending reticular formation by elevated CO_2. Gellhorn (1953b) suggested that 10% CO_2 in inhaled air depressed the cortex but stimulated the hypothalamus, whereas 30–40% CO_2 caused generalized depression and hence anesthesia. Supporting the concept of excitation emanating from the brainstem, Gellhorn and French (1953) reported that strychnine-induced spiking was augmented by elevated CO_2 in intact brain but was depressed after undercutting the cortex. These authors anticipated the differential effects of pH (or CO_2) on various neuron types, which later became an intensively discussed topic (see pp 382–384). Ivanov (1963), and Gyarfas and Pollock (1952), on the other hand, maintained that cortical and subcortical structures respond similarly to elevated CO_2, both being desynchronized. According to these authors, only the sensitivity of various regions differs, cortex being the most susceptible. More recently, Caspers and Speckmann (1990) emphasized that, at a certain range of elevated CO_2 levels, the EEG shows highly coherent activity frequencies of 60–70 Hz in both hemispheres, suggesting excitation of some subcortical sources, even while the "conventional" EEG frequencies are suppressed.

Seizures and Spreading Depression

The suppression of seizures and of paroxysmal EEG discharges by CO_2 or acid pH has been confirmed many times (Lennox, 1929; Lennox et al., 1936; Pollock, 1949; Esplin and Rosenstein, 1963; Woodbury and Kahler, 1960; Woodbury et al., 1958; Meyer et al., 1961; Balestrino and Somjen, 1988), including recent studies on brain tissue slices (Aram and Lodge, 1987; Jarolimek et al., 1989; Velíšek et al., 1994). There were, however, two exceptions. Gellhorn and Heymans (1948) reported "excitation" (i.e., increased frequency) of both normal and convulsive potentials; and De Jong et al. (1967) found the threshold of lidocaine-induced seizures reduced by elevated arterial P_{CO_2}.

Carbon dioxide also modulates the spreading depression (SD) phenomenon discovered by Leão (1944, 1947). Generally speaking, interventions that increase cerebral excitability or provoke seizures also favor the eruption of SD, while central depressant drugs have the opposite effect. It may then be expected that elevation of P_{CO_2}, or tissue acidification should also inhibit SD, and this has indeed been reported (Laget and de Neverlee, 1962; Gardner-Medwin, 1981; Lehmenkühler et al., 1976; Balestrino and Somjen, 1988; Kristián et al., 1994). Van Harreveld and Stamm (1953) found that inhalation of 7–15% CO_2 converts spreading depression into spreading convulsion (i.e., propagating electrographic seizure). More recently, Tombaugh (1994) has shown that, in hippocampal tissue slices, acidification of the bathing fluid at constant P_{CO_2} postpones the onset of the SD-like depolarization induced by oxygen withdrawal (Fig. 2).

Cerebral DC Potentials

With the introduction of DC-coupled amplifiers, attention turned to the changes of the extracellular "baseline" potential of central gray matter. Dusser de Barenne and McCulloch (1939) associated "facilitation" of cortical electrical activity with a negative "drift" of the DC potential and alkalinization of the cortex; while "extinction" was associated with a positive drift and acidification.

Systemic respiratory and metabolic acidosis causes the cerebral DC potential of rats, rabbits, goats, and dogs to shift in the positive direction, while alkalosis, for example by hyperventilation, causes a negative change (Tschirgi and Taylor, 1958; Goldensohn et al., 1951; Laget and de Neverlee, 1962; Speckmann and Caspers, 1969; Lehmenkühler et al., 1976; Caspers et al., 1987; Balestrino and Somjen, 1988).

Voltage shifts induced by change of blood pH can be recorded between the cerebral ventricles and the blood, or indeed any point in the brain referred to any point on the body outside the central nervous system. These pH-related sustained potential shifts appear to be generated by ion transport across the blood-brain and blood-cerebrospinal fluid (CSF) barriers (Tschirgi and Taylor, 1958; Held et al., 1964), and have been attributed to the regulation of the pH of CSF and of brain tissue. A different view is taken by Caspers et al. (1987; also Caspers, 1993). These authors agree that elevation of systemic P_{CO_2} causes the potential to shift across the entire thickness of the cerebral cortex without the distinctive depth-profile otherwise typical of neuronally generated voltages, yet they emphasize that elevation of P_{CO_2} causes hyperpolarization of most neurons. They consider the neuronal hyperpolarization to be the main generator of the positive potential shift, but acknowledge an additional contribution by the blood-brain barrier. The effects of pH on neuronal membrane potential are discussed in greater detail on pp 382–383.

In cats and monkeys, the polarities are, however, usually reversed: acidosis causes a negative DC shift in these animals (Carpenter et al., 1974; Woody et al., 1970; Caspers et al., 1987). Woody et al. (1970) proposed that the CO_2- or acid-induced positive shift is related to the H^+ gradient across the blood-brain and blood-CSF barriers, while the negative shift has to do with the increased cerebral blood flow caused by CO_2. The two opposite effects would compete, and in different species either the one or the other wins. Caspers et al. (1987) consider the positive shift to be the true effect of CO_2 and they attribute the negative shift to the concomitant brain swelling, which is especially pronounced in cats. More recently Caspers (1993) pointed out that glial cells depolarize, while neurons hyperpolarize, in the cerebral cortex during hypercapnia, and he suggests that a preponderance of glial depolarization could contribute to the negativity in cats.

CO_2 AND pH EFFECTS ON ISOLATED PERIPHERAL NERVES

Whenever an agent has mixed direct and indirect effects on intact animals, usually these effects can best be sorted out by experimentation in vitro, where parameters of the environment can be controlled independently.

Perhaps the first person to test the effect of CO_2 on isolated frog nerve was Waller (1896), who used his own exhaled air as a source of CO_2. His main observation was an increase of the "negative variation," presumably the summed effect of a series of action potentials on the demarcation current, recorded by a moving coil galvanometer and evoked by a burst of tetanic stimulation.

More readily interpreted in terms of present-day neurophysiology are the recordings of compound action potentials that became possible once instruments of sufficiently fast response times were available. The literature of the first half of this century has been briefly summarized by Lorente de Nó (1947) and in greater detail by Wyke (Wyke, 1963). The early papers (Lehmann, 1937; Davis et al., 1928; Necheles and Gerard, 1930; Coraboeuf, 1951; Lorente de Nó, 1947) are in general agreement on the following points: Elevation of CO_2 levels in the environment of an isolated amphibian or mammalian nerve raises the threshold of stimulation, the magnitude of the (indirectly estimated) resting potential, and the amplitude and duration of the action potential. All these actions were considered similar to those caused by "anelectrotonus" (i.e., by hyperpolarization).

CO_2, pH_o, AND pH_i

Once more the question arises whether CO_2 could exert its effects directly, or indirectly by acidifying the medium. In either case, it could act on cell membranes, or, since CO_2 penetrates cell membranes (Waddell, Bates, 1969), within cells. The degree to which CO_2 can acidify intracellular pH (pH_i) depends on the effectiveness of the regulation of cell pH_i by intracellular buffering power, as well as by membrane transport of acid-base equivalents (Siesjö and Messeter, 1971; Chesler, 1990; Ou-Yang et al., 1993; Schwiening and Boron, 1994; reviewed in Chapters 2, 4 and 10–12).

Beginning with the earliest studies, opinions were divided as to whether CO_2 modulates excitability by acidification of the milieu or by an action of the CO_2 molecule itself. For example, Lehmann (1937) considered pH to be the active agent, but Lorente de Nó (1947), Monnier (1955) and Coraboeuf (1951) have argued that it must be CO_2, because acidifying the bath at constant P_{CO_2} had little or no effect various aspects of nerve excitability.

A different result emerged when experimenters learned to remove the connective tissue sheath of nerves, and to isolate single nerve fibers. Once the barrier to diffusion was removed, strong acids or bases proved to affect spike propagation and firing threshold in a manner similar to variation of P_{CO_2} (Meves, 1955; Niedergerke and Stämpfli, 1953). Even Coraboeuf changed his mind, but he insisted that both molecular CO_2 and H^+ ions influence nerve excitability and the two effects are additive (Coraboeuf and Niedergerke, 1953).

It appears therefore that the perineural connective tissue protected the ensheathed axons from penetration by strong acids and bases in a manner somewhat akin to the blood brain barrier. Even so, a direct effect by molecular CO_2, in addition to and independently of its indirect effect exerted through generating H^+, cannot be excluded. As we shall see, the matter is still undecided.

EFFECT OF H^+ AND CO_2 ON NEURONS: INHIBITION AND EXCITATION

Most neurons are inhibited, but some are excited by H^+ and by CO_2. The uniformly depressant effect on peripheral nerves by H^+ ions and by CO_2 contrasts with the mixture of excitation and inhibition caused by elevated CO_2 in the CNS, discovered in the middle decades of this century and reviewed earlier in this chapter. A "net" increase of excitation could, in principle, result either from depression of inhibitory neurons (i.e., disinhibition) or from true stimulation. The concept of disinhibition may be seen as reconciling seemingly contradictory, or at least opposite effects. Yet genuinely opposite responses of different types of central neurons to CO_2 and/or to H^+ cannot *a priori* be excluded. As it turns out, many neurons in the nervous system of molluscs and some in the mammalian CNS, especially in the respiratory control region of the brain stem, can indeed be excited by elevated CO_2. The responses of specialized respiratory neurons to CO_2 are the topic of Chapter 26. Here we review only the general responses of nerve and glial cells to CO_2.

Chalazonitis (1963) was the first to report the effect of CO_2 on *Aplysia* neurons in vitro. Most of these nerve cells became depolarized and, if spontaneously active, increased their firing rate. A few cells were, however, hyperpolarized and inhibited. These findings have been confirmed by others (Brown and Berman, 1970; Walker and Brown, 1970; Carpenter et al., 1974; Caspers and Speckmann, 1990).

Brown and Berman (1970) found that either raising CO_2 concentration, or acidification at constant CO_2 of the seawater bathing the abdominal ganglion of *Aplysia* caused depolarization in about half of the cells and hyperpolarization in the others. Both types of effects were produced, even when synaptic transmission was blocked, thus excluding indirect mediation of excitation by disinhibition. The membrane resistance decreased during both types of responses. At constant interstitial pH [pH_o], however, raising CO_2 concentration was without any effect. They (Brown and Berman, 1970) concluded that CO_2 exerts its effect exclusively by changing pH_o, and that molecular CO_2 and pH_i changes played no part. Walker and Brown (1970) concluded that the effect of acid pH_o was mediated by increasing Cl^- conductance. The direction of the membrane potential change depended on the intracellular Cl^- concentration—more precisely, whether the Cl^- equilibrium potential was negative or positive relative to the prevailing membrane potential. This conclusion calls to mind a report by Meves and Völkner (1958), who also found the polarity of the response to CO_2 of frog skeletal muscle fiber membranes to be dependent on the membrane potential. The reversal potential in frog muscle was, however, -90 mV, which suggests that in these cells K^+ rather than Cl^- was the responsible ion.

Carpenter et al. (1974) confirmed most of the findings of Brown and Berman, (1970) and Walker and Brown (1970), but, unlike the previous authors, claim to have separated the effects caused by CO_2 from those of extracellular acidification. The team of Caspers and Speckmann (1990) reported that by increasing the buffer in the bathing solution of snail ganglia they were able to change a depolarizing response to CO_2 into a hyperpolarization in the same cell. This would suggest that, in these neurons, extracellular acidification causes depolarization while intracellular CO_2 causes hyperpolarization. Acidification at constant P_{CO_2}, which could have confirmed this conclusion, was not tested in this study (Caspers and Speckmann, 1990).

Krnjević et al. (1965) related the activity of cortical and spinal neurons of cats to tissue P_{CO_2}. Raising P_{CO_2} by adding varying concentrations of CO_2 to the inspired air of anesthetized cats resulted in an initial transient increase followed by a depression of spontaneous and of glutamate-evoked firing of cortical neurons. Similar effects were seen in undercut cortex, which excluded excitation mediated by synaptic input from subcortical regions. Intracellular recording showed brief depolarization followed by sustained hyperpolarization during elevation of P_{CO_2}. Earlier, Gill and Kuno (1963) made intracellular recordings from phrenic motoneurons and found an increased threshold to electrical stimulation and an increase of antidromic spike amplitude, probably indicating hyperpolarization. Purely hyperpolarizing responses of cortical neurons of cats to elevated CO_2 levels were reported by Carpenter et al. (1974).

Neurons in the mammalian CNS were studied in detail both *in situ* and *in vitro* by the Münster team (Speckmann and Caspers, 1969, 1974; Caspers et al., 1987; Lehmenkühler et al., 1985, 1989; Caspers and Speckmann, 1990). The majority of cells in forebrain and spinal cord were hyperpolarized; their input resistance and electrical threshold increased and their ability to fire repetitively was blocked when the inspired P_{CO_2} of anesthetized animals was raised. At the same time, intracellularly recorded monosynaptic as well as polysynaptic excitatory post-synaptic potentials (EPSPs) were depressed. A minority of spinal interneurons and many cells in the brainstem depolarized under similar conditions. These authors called the former type I-cells and the latter E-cells. As in molluscan neurons, the response in mammalian spinal dorsal root ganglion and in brain slice preparations *in vitro* could be changed depending on the buffering of the bathing fluid. These authors (Lehmenkühler et al., 1989; Caspers and Speckmann, 1990) attributed CO_2-induced hyperpolarization to intracellular acidification leading to elevation of free intracellular calcium ($[Ca^{2+}]_i$) which, in turn, would cause increase of potassium conductance ($G_{K,Ca}$). Extracellular potassium concentration, $[K^+]_o$, indeed increases during CO_2 elevation (Lehmenkühler et al., 1985; Balestrino and Somjen, 1988), and the excess K^+ could have escaped from cells by the action of $G_{K,Ca}$. An unresolved question relating to the proposed explanation is the degree to which pH_i actually changes under the specific conditions in which the hyperpolarization is observed.

Gruol et al. (1980) applied H^+ ions by microiontophoresis on cultured mouse spinal cord neurons. The application of H^+ transiently depolarized the cells and decreased their input resistance, indicating an inward current (see Chapter 23 for a discussion of proton-induced currents). It also raised the threshold of firing—and while the H^+-induced inward current declined to zero within seconds during prolonged H^+ administration, the threshold remained elevated, indicating independence of the two effects.

Lehmenkühler et al. (1989; also Caspers, 1993) reported that glial cells in rat cortex are depolarized by elevation of CO_2 levels, probably because of the rise in $[K^+]_o$. In cat cortex, Carpenter et al. (1974) found in "unresponsive," presumably glial, cells only small and inconsistent changes of membrane potential.

SYNAPTIC TRANSMISSION IN SPINAL CORD, BRAIN, AND BRAIN SLICES

Spinal reflexes generally have been found to be depressed by inhaled CO_2 (King et al., 1932; Bradley et al., 1956; Esplin and Rosenstein, 1963; Brooks and Eccles, 1947; Kirstein, 1951; Carpenter et al., 1974; Brooks and Bachelard, 1992). Brooks and Eccles (1947) ascribed mixed excitatory-inhibitory effects to hypoxia, but a purely depressant effect to CO_2. Kirstein (1951) found the monosynaptic reflex to be selectively depressed and polysynaptic segmental reflexes to be relatively preserved (see also: Esplin and Rosenstein, 1963; Carpenter et al., 1974) (Fig. 3). This selectivity is unusual, because most generally depressant agent affect multisynaptic pathways more strongly than monosynaptic connections (Bárány, 1947; Megirian et al., 1958). Esplin and Rosenstein (1963) reported a qualitatively similar effect when blocking carbonic anhydrase by acetazolamide and the inhalation of CO_2 (Fig. 3).

Analyzing the depression of monosynaptic transmission, Carpenter et al. (1974) found that the threshold of intraspinal afferent terminals increased during CO_2 inhalation. They interpreted this result to indicate hyperpolarization of presynaptic nerve endings. This finding seemed paradoxical, because presynaptic hyperpolarization is expected to enhance, and not to depress, synaptic transmission (Eccles, 1964). Carpenter et al. (1974) suggested that the CO_2-induced hyperpolarization may have been strong enough to block impulse conduction into the axon terminal region.

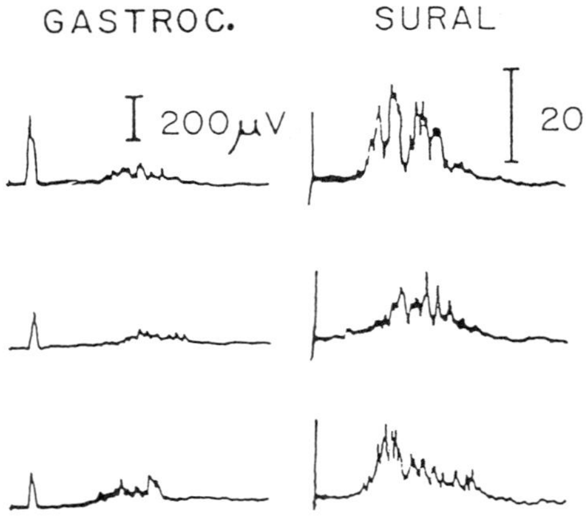

FIGURE 3. Inhalation of CO_2 and administration of acetazolamide depress monosynaptic and spare polysnaptic spinal reflexes. Recordings from a spinal ventral root of spinal cat preparations. *Left column:* Stimulation of a muscle nerve, evoking brief monosynaptic reflex followed by late polysynaptic discharge. *Right column:* Stimulation of a cutaneous nerve, evokes only a polysynaptic reflex. *Top:* Control. *Middle:* During inhalation of 5% CO_2. *Bottom:* After recovery from CO_2, and after administration of acetazolamide, 20 mg/kg body weight intravenously. From Esplin and Rosenstein (1963).

The amplitude of all components of sensory-evoked potentials in cerebral neocortex is also depressed by moderately elevated inspired CO_2 (Krnjević et al., 1965; Caspers and Speckmann, 1990). At levels exceeding 10% inspired CO_2, Krnjević et al. (1965) found, however, that sensory-evoked slow waves become enhanced even though evoked action potentials remain blocked.

Takahashi and Ogawa (1987) found that in slices of cortical tissue the extracellularly recorded presynaptic action potential was strongly depressed when the pH was lowered to 6.2 by reducing bicarbonate concentration of the bathing fluid. This result offers possible support for the idea of blocked propagation into presynaptic terminals, proposed by Carpenter et al. (1974).

Balestrino and Somjen (1988) related the amplitude of monosynaptically transmitted evoked potentials recorded in dentate gyrus of anesthetized rats to tissue pH_o. Potentials and pH were recorded from the same point by double-barreled ion-selective microelectrodes. pH_o was raised by hyperventilation and lowered by adding CO_2 to inspired gas. Transmission proved extremely sensitive to CO_2-induced pH_o change. The focally recorded compound action potential (population spike) amplitude changed on average by 40% when pH_o changed by 0.1 unit. The rate of rise (slope) of the extracellular excitatory postsynaptic potential (fEPSP) changed to a lesser degree (5.3% per 0.1 pH unit) (Fig. 4). These changes proved independent of $[K^+]_o$, $[Ca^{2+}]_o$, and tissue Po_2. The question was: Is the active agent the H^+ ion, or CO_2 itself? To answer this question, evoked potentials were recorded in hippocampal tissue slices *in vitro*. In the CA1 region of this preparation, altering pH_o by varying CO_2 concentration of the bath caused, on average, the orthodromic population spike amplitude to change by 30% when pH_o changed by 0.1 unit. When pH_o was varied by the addition of HCl or NaOH to the bath, a similar but consistently milder effect was seen in the same preparations (mean of 15% per 0.1 unit pH_o). Similar results were obtained by Tombaugh (1994) (Fig. 5). Balestrino and Somjen (1988) concluded that extracellular H^+ ions depress synaptic transmission, but that CO_2 has an additional effect that is independent of pH_o. Moreover, the synaptic blockade by elevated CO_2 level is mainly due to decreased excitability of postsynaptic neurons, while depressed transmission of EPSPs played a lesser but not negligible part.

Krnjević and Walz (1990) acidified hippocampal tissue slices by lowering $[NaHCO_3]$ (which is equivalent to adding H^+ at constant Pco_2), and report depression of the population spike by about 18% per 0.1 unit of pH_o. Lipton and Korol (1981) reduced pH_o by 0.4 units, either by raising CO_2 or by lowering bicarbonate in the bath, and both interventions caused the orthodromic population spike to decrease by 55%. In the CA3 region of hippocampal slices, Jarolimek et al. (1989) found that in the pH range of 7.2–7.5 synaptically transmitted responses were not affected by pH changes, but orthodromic population spikes increased steeply when pH was raised above 7.6 and were strongly depressed at $pH < 7.15$. Recently Lee et al. (1996) have reported, however, that the effect of lowering or raising bath CO_2 concentration on fEPSPs is transient. If bath CO_2 concentration, and with it tissue pH_o, are maintained at a steady abnormal level for an extended period of time, synaptic transmission gradually returns to control level.

The question then is whether CO_2 acts by changing pH_i. Lipton and Korol (1981) thought so, because when they raised the levels of both CO_2 and HCO_3^- so as to leave pH_o unchanged while pH_i decreased by 0.35 unit, the depression of synaptic transmission equalled the depression achieved by raising CO_2 alone. Sick and Balestrino (1987) found, however, that pH_i measured with an intracellular indicator dye changed only slightly when CO_2 concentration (and pH_o) was raised or lowered in a range where synaptic

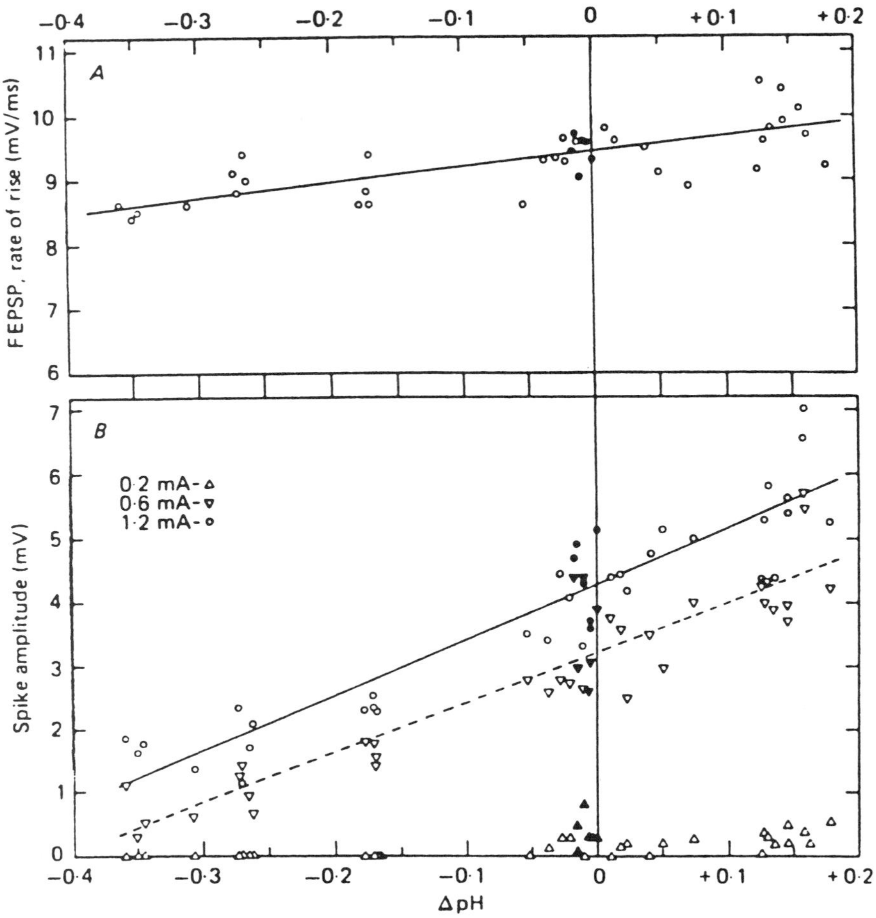

FIGURE 4. The dependence of synaptic transmission on CO_2-induced tissue pH_o changes. pH_o and evoked responses were recorded by double-barreled ion-selective microelectrode in dentate gyrus of a rat under urethane anesthesia. pH_o was raised by forced hyperventilation, and it was lowered by adding CO_2 to inhaled O_2. Key: FEPSP, Extracellularly recorded ("focal") excitatory postsynaptic potential; Spike, Synaptically transmitted extracellular action potential ("population spike"); Filled symbols, Control recordings during spontaneous respiration of room air; Open symbols, During artificial ventilation with O_2 with or without varying CO_2 added. Different symbols for different stimulus intensities (0.2–1.2 mA; see inset). All data from one animal. Each point represents the average of four responses. From Balestrino and Somjen (1988).

transmission was powerfully affected. More recently, Church (1992) changed the bathing solution of hippocampal slices from the usual HCO_3^- /CO_2 buffer to HEPES-buffered solution of the same pH, while recording from CA1 neurons with intracellular electrodes. This transition caused the membrane potential to fall, the threshold for generating action potentials and calcium-spikes to rise, and cell input resistance to decrease. These changes could not be reversed when pH_o was raised by titrating the HEPES-buffered bathing solution, and therefore Church (1992) attributed them to acidification of pH_i. Lee et al. (1996)

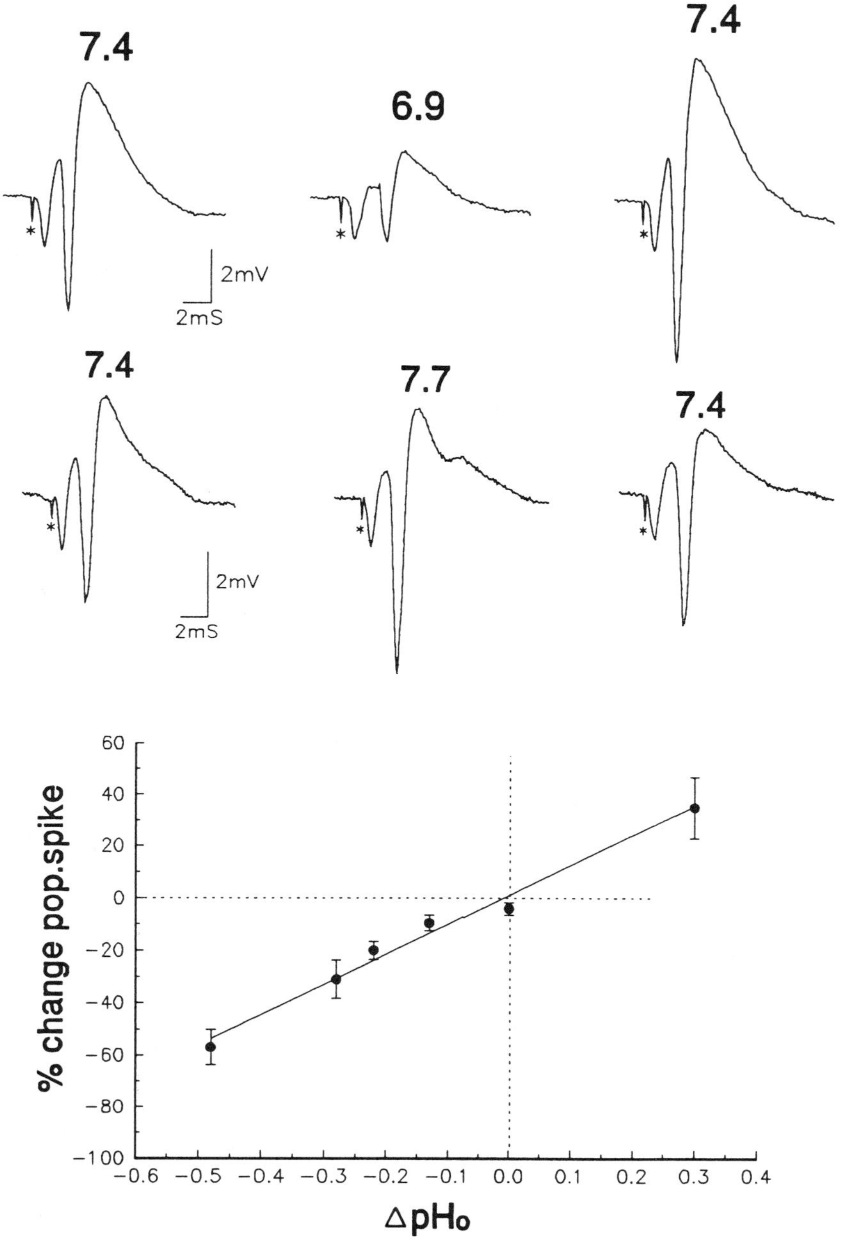

FIGURE 5. The dependence of synaptic transmission on tissue pH_o at constant P_{CO_2}. pH_o and electric responses were recorded by double-barreled ion-selective microelectrode in stratum pyramidale of CA1 region of a hippocampal tissue slice. *Above:* Sample recordings before, during, and after pH_o change. The numbers above the traces indicate bath pH. *Below:* Change of normalized orthodromic population spike amplitude as function of change of pH_o measured with ion-selective microelectrode. From Tombaugh (1994)

also concluded that CO_2 affects synaptic transmission by altering pH_i. They explain the transient nature of the CO_2 effect by the cellular regulation that enables recovery of pH_i to normal levels even while pH_o remains abnormal.

SUMMARY OF pH EFFECTS DISCOVERED "BEFORE VOLTAGE CLAMP"

The research reviewed in this chapter has been conducted without the benefit of the voltage-clamp technique. Chapter 21 reviews pH effects on voltage-gated membrane ion currents, while Chapter 22 deals with agonist-controlled channels. At this point it is appropriate to summarize the most important conclusions derived from investigations "BVC."

1. H^+ ions in the extracellular medium depress the electrical excitability of most nerve cells. This effect is expressed as an increase of resting membrane potential, increase of threshold, and decrease of impulse conduction velocity.
2. CO_2 has a depressant effect which, in some cells at least, cannot fully be accounted for by the pH_o change caused by the dissolved CO_2. The excess depressant effect of CO_2 may be due to acidification of cytosol or perhaps some other intracellular compartment, or to an effect of the CO_2 molecule itself. Molecular CO_2 could act within the membrane, or on intracellular biochemical processes. These issues have not yet been resolved.
3. Unlike most neurons, a number of neurons in the mammalian brain stem and in molluscan nervous systems as well as a few interneurons in the spinal cord are excited by elevated CO_2. Many, but not all, of these cells are believed to have a role in controlling respiration (see Chapters 26 and 27).
4. Even though the signs and symptoms of tetany induced by low $[Ca^{2+}]_o$ and by high pH are similar, and even though pH does influence the ratio of free to bound calcium in physiological fluids, the changes in ionized calcium activity can play only a minor role in causing the increased excitability in alkalosis and decreased excitability in acidosis.
5. Blockade of synaptic transmission by acidosis is due in part to the elevation of the firing threshold of postsynaptic neurons and in part to the depression of synaptic potentials. The relative weight of these two effects may be different in the case of different synapses. The depression of synaptic potentials may be due to reduced transmitter output, to interference with the reaction of transmitters and receptors, or to the blockade of transmitter operated channels (see Chapter 22).

REFERENCES

Allen BW, Somjen GG (1983): The influence of pH on $[Ca^{2+}]$ in circulating blood. Fed Proc 42: 296 (Abstract).

Allen BW, Somjen GG, Sanders DB (1985): Effect of hypocalcemia on neuromuscular function. In Kessler M, Harrison DK, Höper J (eds): Ion Measurements in Physiology and Medicine. Berlin: Springer, pp 243–248.

Aram JA, Lodge D (1987): Epileptiform activity induced by alkalosis in rat neocortical slices: block by antagonists of N-methyl-D-aspartate. Neurosci Lett 83: 345–350.

Balestrino M, Somjen GG (1988): Concentration of carbon dioxide, interstitial pH and synaptic transmission in hippocampal formation of the rat. J Physiol (Lond) 396: 247–266.

Band DM, Broadway JW (1979): Some effects of pH on plasma calcium in the human and cat. J Physiol (Lond) 289: 38P–39P (Abstract).

Bárány EH (1947): A theoretical note concerning the action of drugs on the central nervous system. Arch Int Pharmacodyn 75: 222–226.

Bonvallet M, Hugelin A, Dell P (1955): Sensibilité comparée du système réticulé activateur ascendant et du centre respiratoire aux gaz du sang et a l'adrenaline. J Physiol (Paris) 47: 651–654.

Bradley K, Schlapp W, Spaccarelli G (1956): Effect of carbon dioxide on the spinal reflexes in decapitated cats. J Physiol (Lond) 111: 62P (Abstract).

Bremer F, Thomas J (1936): Action de l'anoxhémie, l'hypercapnie et l'acapnie sur l'activité électrique du cortex. C R Soc Biol (Paris) 123: 1256.

Brody BS, Dusser de Barenne JG (1932): Effect of hyperventilation on the excitability of the motor cortex in cats. Arch Neurol Psychiatry 28: 571–585.

Brooks CMcC, Eccles JC (1947): A study of the effects of anaesthesia and asphyxia on the monosynaptic pathway through the spinal cord. J Neurophysiol 10: 349–360.

Brooks KJ, Bachelard HS (1992): The regulation of intracellular pH studied by ^{31}P- and ^{1}H-NMR spectroscopy in superfused guinea pig cerebral cortex slices. Neurochem Int 21: 375–379.

Brown EB Jr (1953): Physiological effects of hyperventilation. Physiol Rev 33: 445–471.

Brown AM, Berman PR (1970): Mechanism of excitation of Aplysia neurons by carbon dioxide. J Gen Physiol 56: 543–558.

Carpenter DO, Hubbard JH, Humphrey DR, Thompson HK, Marshall W (1974): Carbon dioxide effects on nerve cell function. In Nahas G, Schaefer KE (eds): Carbon Dioxide and Metabolic Regulation. New York: Springer, pp 49–62.

Cartwright FF (1952): English Pioneers of Anaesthesia. (Beddoes, Davy, Hickman). Baltimore: Williams & Wilkins.

Caspers H (1993): DC potentials of the brain. In Haschke W, Speckmann E-J, Roitbak AI (eds): Slow Potential Changes in the Brain. Boston, Birkhäuser, pp 9–20.

Caspers H, Speckmann E-J (1990): Effects of CO_2 on neural function. In Deecke L, Eccles JC, Mountcastle VB (eds): From Neuron to Action. Berlin: Springer, pp 433–441.

Caspers H, Speckmann EJ, Lehmenkühler A (1987): DC potentials of the cerebral cortex. Rev Physiol Biochem Pharmacol 106: 127–178.

Chalazonitis N (1963): Effects of changes in P_{CO_2} and P_{O_2} on rhythmic potentials from giant neurons. Ann N Y Acad Sci 109: 451–479.

Chesler M (1990): The regulation and modulation of the pH in the nervous system. Progr Neurobiol 34: 401–427.

Church J (1992): A change from HCO_3-CO_2 to HEPES-buffered medium modifies membrane properties of rat CA1 pyramidal neurones *in vitro*. J Physiol (Lond) 455: 51–71.

Collip VP, Backus PC (1920): Effect of prolonged hyperpnoea on CO_2 combining power of the plasma, CO_2 tension of alveolar air and excretion of acid and basic phosphate by the kidney. Am J Physiol 51: 568–579.

Coraboeuf E (1951): L'action particulière du CO_2 sur le nerf myélinisé et son indépendence à l'égard du pH. C R Soc Biol 145: 544–547.

Coraboeuf E, Niedergerke R (1953): Kohlensäure und pH-Wirkung an die markhaltigen Einzelfaser des Frosches. Pflügers Arch 258: 103–107.

Davis H, Pascual W, Rice LH (1928): Quantitative studies of the nerve impulse. III. The effect of carbon dioxide on the action current of medullated nerve. Am J Physiol 86: 706–723.

De Jong RH, Wagman IH, Prince DA (1967): Effect of carbon dioxide on the cortical seizure threshold to lidocaine. Exper Neurol 17: 221–232.

Dusser de Barenne JG, McCulloch WS (1939): Factors for facilitation and extinction in the central nervous system. J Neurophysiol 2: 319–355.

Dusser de Barenne JG, Marshall CS, McCulloch WS, Nims LF (1938): Observations of the pH of arterial blood, the pH and electrical activity of the cerebral cortex. Am J Physiol 124: 631–636.

Eccles JC (1964): The Physiology of Synapses. New York: Springer.

Esplin DW, Rosenstein R (1963): Analysis of spinal depressant action of carbon dioxide and acetazolamide. Arch Int Pharmacodyn 143: 498–513.

Ewald A (1873): Zur Kenntniss der Apnoë. Arch ges Physiol (later Pflügers Arch) 7: 575–581.

Fencl V, Miller TB, Pappenheimer JR (1966): The respiratory response to disturbances of acid-base balance, with deductions concerning the ionic composition of cerebral interstitial fluid. In Caveness WF, Walker AE (eds): Head Injury. Philadelphia: JB Lipincott, pp 414–430.

Foerster O (1924): Hyperventilationsepilepsie. Deut Ztschr Nervenheilk 83: 347–356.

Freudenberg E, György P (1922): Die pathogenetische Beziehungen zwischen Tetanie und Rachitis. Münch Med Wchschr 69: 422–423.

Gardner-Medwin AR (1981): Possible roles of vertebrate neuroglia in potassium dynamics, spreading depression and migraine. J Exp Biol 95: 111–127.

Gellhorn E (1953a): Physiological Foundations of Neurology and Psychiatry. Minneapolis: University of Minnesota Press.

Gellhorn E (1953b): On the physiological action of carbon dioxide on cortex and hypothalamus. Electroenceph Clin Neurophysiol 5: 410–413.

Gellhorn E, French LA (1953): Carbon dioxide and cortical spike frequency. Arch Int Pharmacodyn 93: 427–433.

Gellhorn E, Heymans C (1948): Differential actions of anoxia, asphyxia and carbon dioxide on normal and convulsive potentials. J Neurophysiol 11: 261–273.

Gerard RW, Marshall WH, Saul LJ (1936): Electrical activity of the cat's brain. Arch Neurol Psychiatry 36: 675.

Gibbs FA, Maxwell H, Gibbs EL (1947): Volume flow of blood through the human brain. Arch Neurol Psychiatry 57: 137.

Gill PK, Kuno M (1963): Properties of phrenic motoneurones. J Physiol (Lond) 168: 258–273.

Goldensohn ES, Schoenfeld RL, Hoefer PF (1951): The slowly changing voltage of the brain and the electrocorticogram. Electroenceph Clin Neurophysiol 3: 231–236.

Grant SB, Goldman A (1920): A study of forced respiration: Experimental production of tetany. Am J Physiol 52: 209–232.

Gruol DL, Barker JL, Huang L-YM, MacDonald JF, Smith TG (1980): Hydrogen ions have multiple effects on the excitability of cultured mammalian neurons. Brain Res 183: 247–252.

Gyarfas K, Pollock GH (1952): The subcortical effects of carbon dioxide in the cat. Arch Int Pharmacodyn 90: 233.

György P, Vollmer H (1923): Über den Chemismus der Atmungstetanie. Biochem Z 140: 391–400.

Haldane JS, Priestley JG (1905): The regulation of the lung-ventilation. J Physiol (Lond) 32: 225–266.

Held D, Fencl V, Pappenheimer JR (1964): Electrical potential of cerebrospinal fluid. J Neurophysiol 27: 942–959.

Henderson Y (1909): Acapnia and shock. IV. Fatal apnea after excessive respiration. Am J Physiol 25: 310–333.

Hickman HH (1824): A Letter on Suspended Animation, containing Experiments showing that it may be safely employed during Operations on Animals, with the view of ascertaining its probable utility in surgical operations on the Human Subject. Ironbridge: The Author.

Howland J, Marriott WMcK (1918): Observations upon the calcium content of the blood in infantile tetany and upon the treatment by calcium. Quart J Med 11: 289–319.

Ivanov YN (1963): Changes in electrical activity of different brain regions in cats and dogs exposed to carbon dioxide. Fed Proc 22: T13–T18.

Jarolimek W, Misgeld U, Lux HD (1989): Activity-dependent alkaline and acid transients in guinea pig hippocampal slices. Brain Res 505: 225–232.

King CE, Garrey WE, Bryan WR (1932): The effect of carbon dioxide, hyperventilation and anoxemia on the knee jerk. Am J Physiol 102: 305–318.

Kirstein L (1951): Early effects of oxygen lack and carbon dioxide excess on spinal reflexes. Acta Physiol Scand 23(suppl 80): 1–51.

Kristián T, Katsura K-I, Gidö G, Siesjö BK (1994): The influence of pH on cellular calcium influx during ischemia. Brain Res 641: 295–302.

Krnjević K, Walz W (1990): Acidosis and blockade of orthodromic responses caused by anoxia in rat hippocampal slices at different temperatures. J Physiol (Lond) 422: 127–144.

Krnjević K, Randić M, Siesjö BK (1965): Cortical CO_2 tension and neuronal excitability. J Physiol (Lond): 176: 105–122.

Laget P, de Neverlee H (1962): Évolution de la dépression envahissante et du potentiel stable dans l'éveil cortical. J Physiol (Paris) 54: 362–363.

Leão AAP (1944): Spreading depression of activity in the cerebral cortex. J Neurophysiol 7: 359–390.

Leão AAP (1947): Further observations on the spreading depression of activity in the cerebral cortex. J Neurophysiol 10: 409–414.

Lee J, Taira T, Pihlaja P, Ransom BR, Kaila K (1996): Effects of CO_2 on excitatory transmission apparently caused by changes in intracellular pH in the rat hippocampal slice. Brain Res 706: 210–216.

Lehmann JE (1937): The effect of changes in pH on the action of mammalian nerve fibers. Am J Physiol 118: 600–612.

Lehmenkühler A, Speckmann E-J, Caspers H (1976): Cortical spreading depression in relation to potassium activity, oxygen tension, local flow and carbon dioxide tension. In Kessler M, Clark LC, Lübbers DW, Silver IA, Simon W (eds): Ion and Enzyme Electrodes in Biology and Medicine. München: Urban & Scwerzenberg, pp 311–315.

Lehmenkühler A, Caspers H, Kersting U (1985): Relations between DC potentials, extracellular ion activities and extracellular volume fraction in the cerebral cortex with changes in P_{CO_2}. In Kessler M, Harrison DK, Höper J (eds): Ion Measurements in Physiology and Medicine. Berlin: Springer, pp 199–205.

Lehmenkühler A, Bingmann D, Speckmann E-J (1989): Neuronal and glial responses to hypoxia and hypercapnia. Biomed Biochim Acta 48: S155–S160.

Lennox WG (1929): The effect on epileptic seizures of varying the composition of the respired air. J Clin Invest 6: 23–34 (Abstract).

Lennox WG, Gibbs FA, Gibbs EL (1936): Effect on the electroencephalogram of drugs and conditions which influence seizures. Arch Neurol Psychiatry 36: 1236–1245.

Lipton P, Korol D (1981): Evidence that decreases in intracellular pH rapidly inhibit transmission in the guinea pig hippocampal slice. Soc Neurosci Abstr 7: 440.

Lorente de Nó R (1947): Carbon dioxide and nerve function. Studies of the Rockefeller Institute of Medical Research 131: 148–193.

MacCallum W, Voegtlin C (1909): On the relation of tetany to the parathyroid gland and to calcium metabolism. J Exp Med 11: 118–151.

Meduna LJ (1950): Carbon Dioxide Therapy. Springfield: Charles C. Thomas.

Megirian D, Vasey J, Posternak JM (1958): Action différentielle de quelques anesthésiques sur les voies spinales monosynaptiques et polysynaptiques chez le chat. Helv Physiol Acta 16: 241–258.

Meves H (1955): Die Wirkung der Wasserstoffionen und der Kohlensäure aud die Nervenleitungsgeschwindigkeit. Pflügers Arch 261: 249–263.

Meves H, Völkner K-G (1958): Die Wirkung von CO_2 auf das Ruhemembranpotential und die elektrischen Konstanten quergstreiften Muskelfaser. Pflügers Arch 265: 457–476.

Meyer JS, Gotoh F, Tazaki Y (1961): Inhibitory action of carbon dioxide and acetazoleamide in seizure activity. Electroenceph Clin Neurophysiol 13: 762–775.

Monnier AM (1955): Die funktionelle Bedeutung der Dämpfung in der Nervenfaser. Ergebrisse der Physiologic 48: 230–285.

Necheles H, Gerard RW (1930): The effect of carbon dioxide on nerve. Am J Physiol 93: 318–335.

Niedergerke R, Stämpfli R (1953): Die Kohlensäurewirkung an der einzelnen markhaltigen Nervenfaser bei Rheobasebestimmungen. Pflügers Arch 258: 95–102.

Oberleithner H, Greger R, Lang F (1982): The effect of respiratory and metabolic acid-base changes on ionized calcium concentration: In vivo and in vitro experiments in man and rat. Eur J Clin Invest 12: 451–455.

Ou-Yang Y, Mellergård P, Siesjö BK (1993): Regulation of intracellular pH in single rat cortical neurons in vitro: A microspectrofluorometric study. J Cereb Blood Flow Metab 13: 827–840.

Owen ARG (1971): Hysteria, Hypnosis and Healing. The Work of J.-M. Charcot. London: Dobson.

Pappenheimer JR (1967): The ionic composition of cerebral extracellular fluid and its relation to control of breathing. Harvey Lect Ser Academic Press, New York 61: 71–94.

Pflüger E (1868): Ueber die Ursache der Athembewegungen, sowie der Dyspnoë und Apnoë. Arch ges Physiol (later Pflügers Arch) 1: 61–106.

Pollock GH (1949): Central inhibitory effects of carbon dioxide. J Neurophysiol 12: 315–324.

Poulsen T (1952): Investigation into the anaesthetic properties of carbon dioxide. Acta Pharmacol Toxicol 8: 30–46.

Rapoport SI (1964): Cortical pH and the blood-brain barrier. J Physiol (Lond) 170: 238–249.

Richer, P (1879). Étude Descriptive de la Grande Attaque Hystérique ou Attaque Hystéro-épileptique. Thése pour le Doctorat en Médecine. Paris: Adrien Delahaye.

Rona P, Takahashi D (1913): Beitrag zur Frage nach dem Verhalten des calciums im Serum. Biochem Z 49: 370–380.

Rosett J (1924): The experimental production of rigidity, of abnormal involuntary movements and of abnormal states of consciousness in man. Brain 47: 293–336.

Schwiening CJ, Boron WE (1994): Regulation of intracellular pH in pyramidal neurones from the rat hippocampus by Na^+-dependent Cl^--HCO_3^- exchange. J Physiol (Lond) 475: 59–67.

Sick T, Balestrino M (1987): Effects of CO_2, HCO_3^- and HCl on intraxcellular pH in brain slices: Correlation with excitability? Neuroscience 22(suppl): S410 (abstract).

Siesjö BK, Messeter K (1971): Factors determining intracellular pH. In Siesjö BK, Sorensen SC (eds): Ion Homeostasis of the Brain. New York: Academic Press, pp 244–262.

Somjen GG, Allen BW, Balestrino M, Aitken PG (1987): Pathophysiology of pH and Ca^{2+} in bloodstream and brain. Can J Physiol Pharmacol 65: 1078–1085.

Sörensen SPL (1912): Über die Messung und Bedeutung der Wasserstoffionenkonzentration bei biologischen Prozessen. Ergebrisse der Physiologie 12: 393–532.

Speckmann E-J, Caspers H (1969): Verschiebungen des corticalen Bestandspotentials bei veränderungen der Ventilationsgrösse. Pflügers Arch 310: 235–250.

Speckmann E-J, Caspers H (1974): The effect of O_2- and CO_2-tensions in the nervous tissue on neuronal activity and DC potentials. In Rémond A (ed): Handbook of Electroencephalography and Clinical Neurophysiology. Amsterdam: Elsevier, pp 2C: 71–89.

Swanson AG, Stavney LS, Plum F (1958): Effects of blood pH and carbon dioxide on cerebral electrical activity. Neurology 8: 787–792.

Takahashi Y, Ogawa T (1987): Effects of lowered pH of the extracellular solution on non-transsynaptic component of evoked potentials in the rat's in vitro cerebral cortical slice. Tohoku J Exp Med 152: 63–65.

Tombaugh GC (1994): Mild acidosis delays hypoxic spreading depression and improves neuronal recovery in rat hippocampal slices. J Neurosci 14: 5635–5643.

Tschirgi RD, Taylor JL (1958): Slowly changing bioelectrical potential associated with the blood-brain barrier. Am J Physiol 195: 7–22.

Van Harreveld A, Stamm JS (1953): Spreading cortical convulsions and depressions. J Neurophysiol 16: 352–366.

Velíšek L, Dreier JP, Stanton PK, Heinemann U, Moshé SL (1994): Lowering of extracellular pH suppresses low-Mg^{2+}-induced seizures in combined entorhinal cortex-hippocampal slices. Exp Brain Res 101: 44–52.

Vernon HM (1909): The production of prolonged apnoea in man. J Physiol (Lond) 38: xviii–xx (abstract).

Waddell WJ, Bates RG (1969): Intracellular pH. Physiol Rev 49: 285–329.

Walker JL, Brown AM (1970): Unified account of the variable effect of carbon dioxide on nerve cells. Science 167: 1502–1504.

Waller AD (1896): Observations on isolated nerve. Electrical changes a measure of physico-chemical change. Croonian lecture. Proc R Soc Lond 59: 308–312.

Wellcome HS, Malcolm LWG (1930): Henry Hill Hickman Centenary Exhibition at the Wellcome Historical Museum. London: The Wellcome Foundation.

Woodbury DM, Kahler R (1960): The role of carbon dioxide in the nervous system. Anesthesiology 21: 686–703.

Woodbury DM, Rollins LT, Gardner MD, Hirschi WL, Hogan JR, Rallison ML, Tanner GS, Brodie DA (1958): Effects of carbon dioxide on brain excitability and electrolytes. Am J Physiol 192: 79–90.

Woody CD, Marshall WH, Besson JM, Thompson HK, Aleonard P, Albe-Fessard D (1970): Brain potential shift with respiratory acidosis in the cat and monkey. Am J Physiol 218: 275–283.

Wyke B (1963): Brain Function and Metabolic Disorders. The Neurological Effects of Changes in Hydrogen Ion Concentration. London: Butterworths.

CHAPTER 21

pH Modulation of Voltage-Gated Ion Channels

GEOFFREY C. TOMBAUGH AND GEORGE G. SOMJEN
Department of Neuroscience, University of Medicine and Dentistry of New Jersey, Piscataway, NJ (G.C.T.); Department of Cell Biology, Duke University Medical Center, Durham, NC (G.G.S.)

INTRODUCTION

In the preceding chapter, we have presented a historical overview of the research into the relationship between pH and neuronal excitability. Generally, acidosis depresses synaptic transmission and the excitability of nerve cells while alkalosis increases excitability, but there are notable exceptions where cells respond in the opposite way (Chapter 26). In this chapter we examine whether the interaction of protons with voltage-gated ion channels could account for these functional changes or at least contribute to them. As we will see, certain ion channels are quite sensitive to ambient pH changes, supporting the idea that they have evolved not only to serve distinct physiological functions but also as targets for H^+ modulation.

Normal cellular function in the mammalian brain depends on the tight regulation of both extracellular and intracellular pH (pH_o and pH_i). A loss of pH homeostasis can lead to severe, even life-threatening consequences: excessive acidosis in the CNS can lead to coma, while alkalosis can provoke seizures. However, it is now equally clear that interstitial pH is far from static: transient pH_o shifts occur in the vertebrate CNS during spontaneous neuronal activity, respiratory changes, and pathological states (Chesler and Kaila, 1992; also see Chapters 17, 18 and 28). As an example, synchronized neuronal firing is accompanied by an initial extracellular alkalinization followed by a more prolonged acidification. The ionic mechanisms underlying such pH shifts are not fully understood, but it is clear that even small pH changes are capable of influencing synaptic transmission, suggesting that pH transients occurring near somatic and dendritic membranes may act as local feedback signals or in a paracrine fashion at neighboring synapses (Balestrino and Somjen, 1988; Ransom, 1992; Taira et al., 1993, Gottfried and Chesler, 1994). Consistent with this idea, several ligand-gated receptors are sensitive to protons within a physiological pH range (see Chapter 22).

In addition to their synaptic action, pH shifts in brain tissue also exert profound effects on intrinsic neuronal excitability (Caspers and Speckmann, 1972; Chapter 20). Such effects are likely to reflect changes in the properties of voltage-gated membrane channels. On a molecu-

pH and Brain Function, Edited by Kai Kaila and Bruce R. Ransom
ISBN 0-471-11838-9

lar level, these changes may arise from any of several possible mechanisms. Chief among these mechanisms is the neutralization of specific negative charges on the channel proteins. In this way, protons may reduce unitary channel conductance either directly by blocking the pore or indirectly by inducing conformational shifts in the protein subunits, as proposed for the N-methyl-D-aspartate (NMDA) receptor (Traynelis and Cull-Candy, 1991). Alternatively, external protons may reduce channel conductance by binding to negative charges near the mouth of the channel, thereby diluting the local concentration of permeant cations. Protons may also affect channel gating, modulating the time channels spend in an open or closed state. Finally, pH changes may shift the *I–V* relation of a channel by neutralizing negative charges distributed over the surface of the cell membrane, and thereby alter the voltage detected by the channel's voltage sensor (cf. Hille, 1992). In this chapter, nonspecific surface charge effects of H^+ will be referred to as "screening" to distinguish it from the more specific neutralization of negatively charged groups on or near channel proteins.

The pH sensitivity of ion channels in excitable membranes has been of general biophysical interest for many years. In this chapter we focus on such channels in the mammalian CNS; but, with relatively little direct evidence available from mammalian neurons, we quote reports dealing with other systems, limiting our discussion as much as possible to their relevance for understanding pH effects on mammalian ion channels. We will first summarize the reported effects of pH on sodium, potassium, and calcium channels in isolated cell preparations and then discuss the possible physiological consequences of these effects in intact CNS tissue. This synthesis will be largely theoretical, since results obtained from probing channel function *in vitro* with internal or external protons can be very different from those derived from studies *in vivo*.

Na^+ CHANNELS

Effects of Extracellular pH (pH_o) on Macroscopic Currents

In neurons, Na^+ channels underlie the initiation and propagation of action potentials. The conductance of Na^+ channels has long been known to be pH-sensitive. In frog axons, acidification of the external medium restricts Na^+ conductance (G_{Na}); a Hill plot of G_{Na} or I_{Na} as a function of pH_o reveals a pK between 5.0 and 5.5 (Hille, 1968; Drouin and The, 1969; Woodhull, 1973). In the context of channel function, pK is conveniently defined as the midpoint of the titration curve relating ion conductance or current amplitude to pH. Because of its acidic pK, the titration curve for G_{Na} reaches a ceiling on the alkaline side of the usual control pH_o of 7.3 or 7.4. Moreover, G_{Na} is only moderately sensitive to extracellular protons within an acidic pH range considered compatible with mammalian tissue ($7.2 > pH > 6$) (Fig. 1A).

FIGURE 1. Dependence of voltage-gated currents on extracellular pH. **A.** Mean peak Na^+ and K^+ currents recorded from nodes of Ranvier in myelinated frog axons. Current amplitude is normalized to that recorded at pH_o 7.3; pK for $I_{Na} = 5.15$, $I_K = 4.63$ (pK has been defined as the midpoint of the titration curve relating ion conductance to pH). **B.** Single Na^+ channel current amplitudes (pA) from cell-attached patch recordings in guinea pig ventricular myocytes. The data points are mean ± SD; the three curves represent different theoretical models: The continuous curve assumes a single site of H^+ block in the channel, p$K = 5.04$. The dotted curve assumes channel block plus screening of surface charges. The dashed curve assumes only screening of surface charges. **C.** Maximal whole-cell high-threshold Ca^{2+} currents from freshly isolated rat hippocampal CA1 neurons. Data are corrected for "rundown" and normalized to peak amplitude recorded at pH_o 7.4; p$K = 7.1$; slope (Hill coefficient) = 0.6. A: modified from Drouin and The (1969); B: modified from Zheng and Siegelbaum (1991); C: Reprinted with permission from Tombaugh and Somjen, 1996.

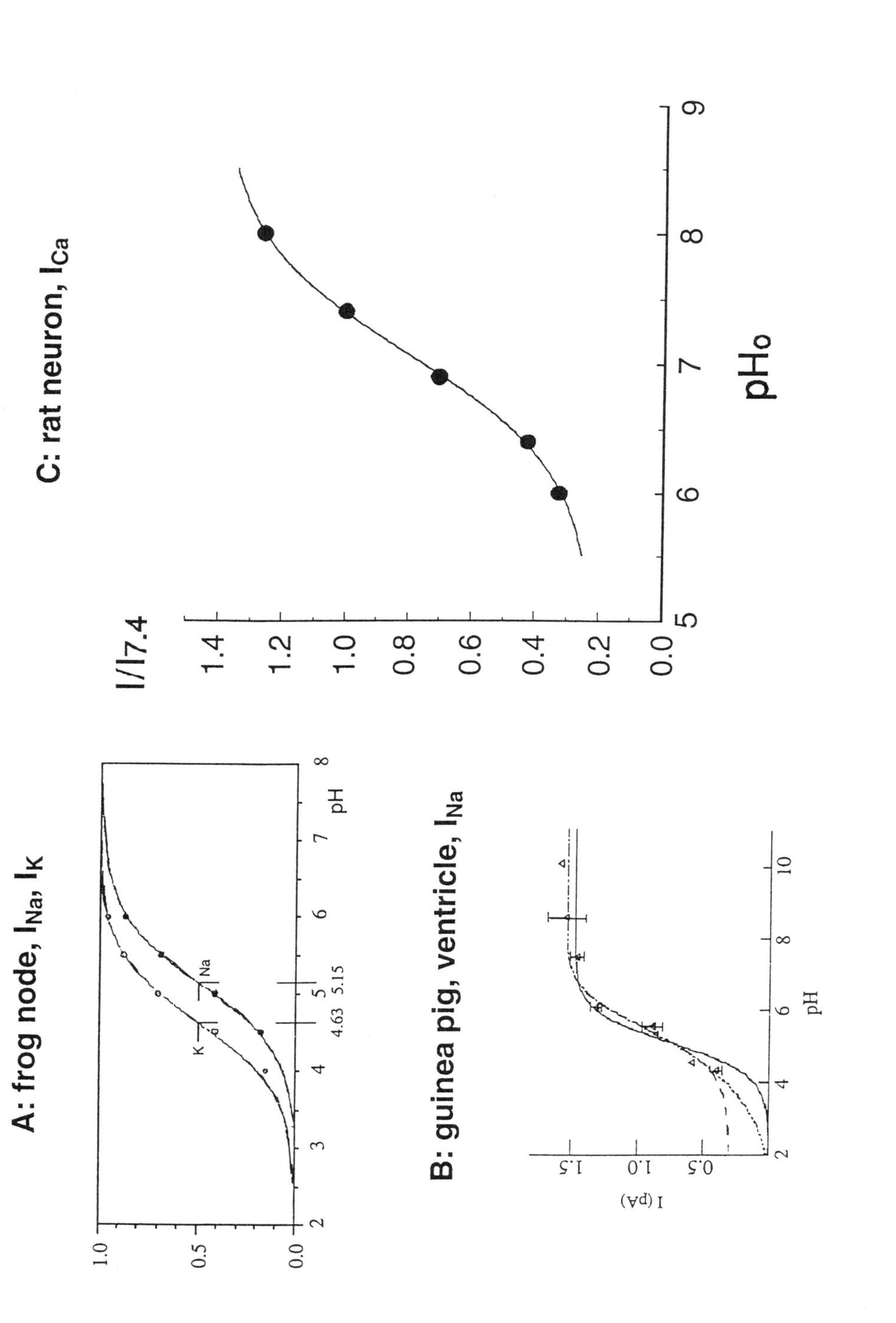
A: frog node, I_{Na}, I_K
1.0
0.5
0.0
2
3
4
5
6
7
8
pH
K
Na
4.63
5.15
B: guinea pig, ventricle, I_{Na}
I (pA)
1.5
1.0
0.5
2
4
6
8
10
pH
C: rat neuron, I_{Ca}
$I/I_{7.4}$
1.4
1.2
1.0
0.8
0.6
0.4
0.2
0.0
5
6
7
8
9
pH_o

Subsequent studies have revealed a pK of 4.8 for *Myxicola* giant axon (Schauf and Davis, 1976); while Carbone et al. (1978) derived a pK of 4.5 for the voltage dependence of the Na^+-activation kinetics of the giant axon of *Loligo vulgaris*. A higher pK of 6.5 had been reported earlier for the maximal Na^+ current evoked by ramp voltage commands in the giant axon of *Loligo pealei* (Stillman et al., 1971). Shrager (1974), however, reported that Na^+ current activation in the crayfish axon was relatively unaffected by external H^+.

Interspecies variation in channel function, however, cannot explain all cases where experimental data disagree. Such is the case for H^+ effects on the steady-state inactivation (h_∞; Hodgkin and Huxley, 1952) of Na^+ channels. Schauf and Davis (1976) reported no significant effect on h_∞ at a pH_o as low as 4.5 while Carbone et al. (1978) derived a pK between 4 and 5 in the same squid axons preparation.

Such divergent results impede reasonable predictions about the pH sensitivity of Na^+ channels in mammalian neurons; it therefore seems surprising that so little empirical data exist for H^+ effects on neuronal Na^+ channels in higher vertebrates. In a recent study of freshly isolated rat hippocampal CA1 neurons, Tombaugh and Somjen (1996) found that moderate acidosis (pH 6.4) reversibly depressed peak Na^+ current amplitude by 15% while alkalosis had the opposite effect (Fig. 2). Published data derived from frog or squid axons cannot be compared directly to mammalian neurons, largely because the physiologically relevant pH range of the former has not been clearly defined. Nevertheless, it is worth noting that the 15% depression of I_{Na} at pH 6.4 in hippocampal neurons is only slightly stronger than that derived by interpolating Hille's (1968) or Drouin and The's (1969) data in the frog axon (Fig. 1A). This suggests that the pH_o sensitivity of Na^+ channel activation may be highly conserved among vertebrates. However, unlike the insensitivity of I_{Na} to elevated pH in the frog node of Ranvier (Hille, 1968), raising pH_o to 8.0 slightly increased I_{Na} amplitude in rat CA1 neurons (Fig. 2B).

It has also been shown in CA1 neurons that neither low nor elevated pH_o (6.4–8.0) had measurable effects on Na^+ current activation kinetics, rate of decay, or steady-state inactivation (Tombaugh and Somjen, 1996). In the frog and squid axons, Na^+ current kinetics are similarly unaffected within or above this pH range, though activation rates are slowed and h_∞ becomes noticeably right-shifted under more acidic conditions ($pH_o < 6$) (Drouin and The, 1969; Carbone et al., 1978).

In addition to these effects of low pH_o on G_{Na} and h_∞, H^+ ions also influence the voltage-dependence of I_{Na} activation. Generally, external acidification shifts the activation voltage to a more depolarized (more positive) level (Drouin and The, 1969). In cells that are not voltage clamped, such a shift would elevate the threshold for triggering an action potential. As we have seen in the previous chapter (Chapter 20), one of the most frequently observed effects of extracellular acidosis in neurons is an increase in the firing threshold.

FIGURE 2. Acidosis reversibly depresses, while alkalosis enhances, Na^+ currents in mammalian central neurons. **A, B.** Rapid, tetrodotoxin-sensitive whole-cell Na^+-currents were evoked by ramp depolarization from acutely isolated rat hippocampal CA1 neurons before, during, and after, exposure either to acidic (pH 6.4) or alkaline (pH 8.0) solutions. Illustrated currents were recorded from two different cells. Holding potential was −70 mV and voltage ramps were applied from −120 mV to +60 mV over a period of 12 milliseconds. The K^+ and Ca^{2+} currents were blocked by including tetraethylammonium-chloride (TEA^+Cl^-) and Cd^{2+} in the bath and Cs^+ in the recording pipette. Reprinted with permission from Tombaugh and Somjen, 1996. **C.** In addition to modulating current amplitude, changes in external pH cause simultaneous shifts in the voltage-dependence of current activation. In the data illustrated, external acidosis shifted the peak current by +3 mV

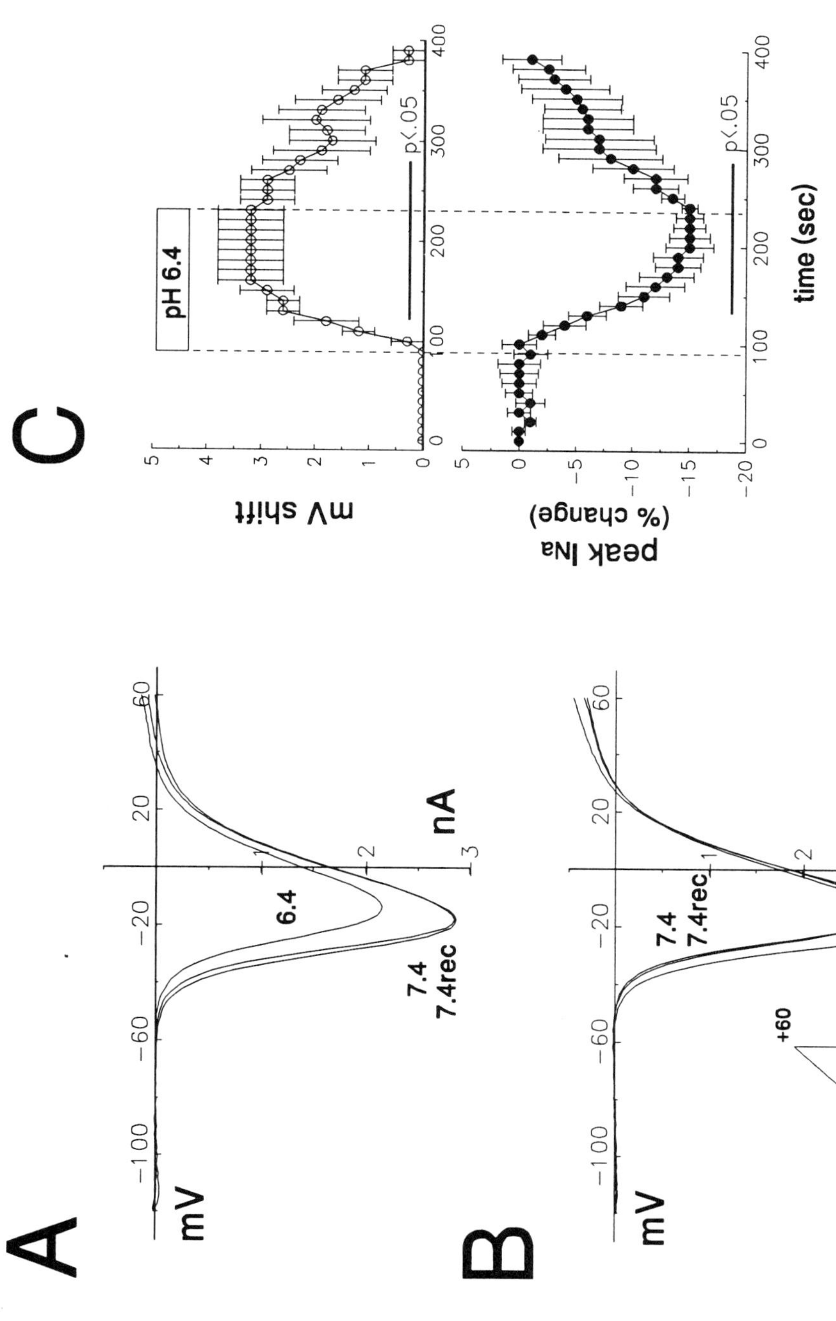
A
B
C
mV
nA
6.4
7.4
7.4rec
8.0
+60
-120
-70
pH 6.4
mV shift
peak INa
(% change)
time (sec)
p<.05

Similar voltage shifts for I_{Na} activation have been observed when the extracellular concentration of other mono- and divalent ions was changed. These apparent nonspecific effects gave rise to the theory that protons not only neutralize fixed negative charges on integral membrane proteins, but that protons also screen the thin layer of negative surface charges distributed over the membrane surface (Hille, 1968; Hille et al., 1975). This screening of negative surface charges would be perceived by a voltage sensor within the channel as a hyperpolarization, resulting in an apparent positive shift of the activation voltage. Data obtained from the squid axon (Schauf and Davis, 1976; Carbone et al., 1978) and more recently from rat CA1 neurons (Fig. 2) (Tombaugh and Somjen, 1996) reveal such a shift of activation voltage during external acidification. Proton screening of surface charges is, however, not the only possible explanation; the titration of specific charged residues on or near the channel's voltage sensor cannot be ruled out. The size of these acid-induced voltage shifts is generally small, on the order of +1 to +3 mV/pH unit above pH 6. Below pH_o 6, larger voltage shifts in the frog axon have been noted (Hille, 1968).

Effects of Intracellular pH (pHi) on Macroscopic Na^+ Currents

As for the effects of intracellular pH on Na^+ channels, Ehrenstein and Fishman (1971) found that Na^+ current amplitude in the perfused squid axon was dramatically reduced at pH_i 6.5 and was more sensitive to pH_i than to pH_o. Qualitatively similar results were obtained in the same preparation by others (Wanke et al. 1980, Carbone et al. 1981) though at a much lower pH_i of <5.5. Earlier, however, Brodwick and Eaton (1978) found that the peak Na^+ current in the squid axon was depressed at the very high pH_i of 11. This suggests that Na^+ channels in some cell types may be inhibited at both acid and alkaline extremes.

Both Brodwick and Eaton (1978) and Carbone (1981) also reported that strong internal alkalosis removed steady-state Na^+ current inactivation. In striking contrast to these findings, as originally reviewed by Moody (1984), the effects of pH_i on h_∞ appear very different when examined in frog skeletal muscle (Nonner et al., 1980), in which inactivation is blocked by reduced pH_i, while current activation remains unaffected. No dramatic shifts in the rapid activation or decay kinetics of Na^+ currents have been observed at elevated pH_i in squid axons (above 6.0), although slight changes are seen at lower pH_i (Carbone et al., 1981). Reports on pH_i effects on whole-cell Na^+ currents of mammalian central neurons are, unfortunately, lacking.

Unitary Analysis of Na^+ Channels

Several mechanisms have been proposed to explain the H^+-mediated effect on G_{Na} (reviewed in Hille, 1992). As mentioned in the Introduction, protons could either alter channel gating or they could reduce unitary channel conductance. The latter could be the consequence of titrating negatively charged groups near the entrance of the channel pore that normally attract Na^+ ions, or of binding to one or more negative groups near the channel pore, directly impeding current flow.

Over the years, arguments have evolved to support such theoretical possibilities. For example, based on the voltage-dependence of the proton block of I_{Na}, Woodhull (1973) suggested that the H^+ ions bound to acid residues within the channel. Begenisich and Danko (1983) supported this conclusion in reporting that internal ions could modify the action of externally applied H^+. Campbell (1982), on the other hand, favored the idea that protons block Na^+ channels by acting on charges on the outside of the membrane. Since then, a few attempts have been made to resolve this debate by the analysis of single Na^+ channel currents.

In heart muscle cells, Zhang and Siegelbaum (1991) found that low pH_o reduced single Na^+ channel conductance with a pK of 5.15 (Fig. 1B), shifted the activation voltage as well as the h_∞ function to more positive levels, and shortened the mean open time of the channels. The authors considered these effects to be consistent with the reduction of negative surface charge by screening a single class of negatively charged sites on the cell surface. More recent studies in skeletal muscle suggest that external H^+ ions modulate Na^+ channels by both blocking the pore and charge screening (Benitah et al., 1997).

The biophysical basis of H^+ action on Na^+ channel permeability in neurons remains for now largely theoretical. As of this writing, only one study of pH effects on the behavior of single Na^+ channels has been reported. Daumas and Anderson (1993) examined individual rat brain Na^+ channels incorporated in artificial neutral lipid bilayers and found that channel conductance was reduced as external pH fell below 6.5, exhibiting a pK of 4.5–5, which is nearly identical to that derived from whole-cell recordings in the frog node of Ranvier. By using 1 M NaCl in the bathing solution, the authors were able to make single-channel measurements under conditions that minimized the relative screening effect of external H^+ on fixed charges near the channel entrance. For this reason, their data argue for a direct blocking action of H^+ on channel permeability and support the original conclusions of Woodhull (1973) that the voltage-dependence of the proton block reflects the binding of H^+ within the channel pore. As for the action of intracellular pH, Daumas and Anderson (1993) found a relatively weak effect of internal acidification on channel conductance (pK = 4.1) compared to that seen with external H^+ in their study of isolated rat brain Na^+ channels.

In addition to acting on Na^+ channel conductance, protons might also influence the open probability or mean open time of the channel via effects on channel gating. At present, neither of these possibilities has been rigorously tested in neuronal membranes, although an effect of external H^+ on Na^+ channel gating appears unlikely from current fluctuation analysis in the frog node of Ranvier (Sigworth, 1980). This negative conclusion is consistent with the observation that a reduction in I_{Na} amplitude at low pH in both isolated CA1 neurons and squid axons is not associated with any change in the kinetics of Na^+ current activation or decay (Schauf and Davis, 1976; Tombaugh and Somjen, 1996).

POTASSIUM CHANNELS

Effects of pH_o

Potassium channels play a dominant role in regulating neuronal excitability by stabilizing membrane potential, as well as by limiting action potential duration and repetitive firing. However, in spite of the diversity of K^+ channels and their importance in brain function, few studies to date have actually examined direct H^+ effects on K^+ channels in vertebrate nervous tissue.

The first voltage-clamp study of pH effects on potassium current in the frog node of Ranvier was reported by Hille (1968). In this study, steady-state K^+ current activation and inactivation (h_∞) were recorded under conditions of varying extracellular pH and were found to be unaffected above pH_o 6.0. These observations were later confirmed by others in both the frog (Drouin and The, 1969) and squid axon (Stillman et al., 1971; Carbone et al., 1978), as well as in intact snail neurons (Alekseev and Zaykin, 1993). However, both the steady-state conductance and activation rate of delayed K^+ currents in frog, squid, and crayfish axons are depressed as the pH_o falls below 5 (Hille, 1968; Shrager, 1974; Schauf and Davis, 1976; Carbone et al., 1978).

The expectation that mammalian K^+ channels might have similar properties was partially borne out in a recent study with isolated rat hippocampal neurons (Tombaugh and Somjen, 1996). In that study, the amplitude and voltage-dependence of both delayed rectifier (I_K) and transient (I_A) K^+ currents were relatively insensitive to external H^+ in a physiological pH range (6.4–8.0). In contrast, steady-state inactivation of both currents was found to be more pH-sensitive than reported in peripheral frog axons: acidosis (pH 6.4) reversibly shifted h_∞ by 8–10 mV to more depolarized voltages (Fig. 3), while alkalosis had no detectable effect.

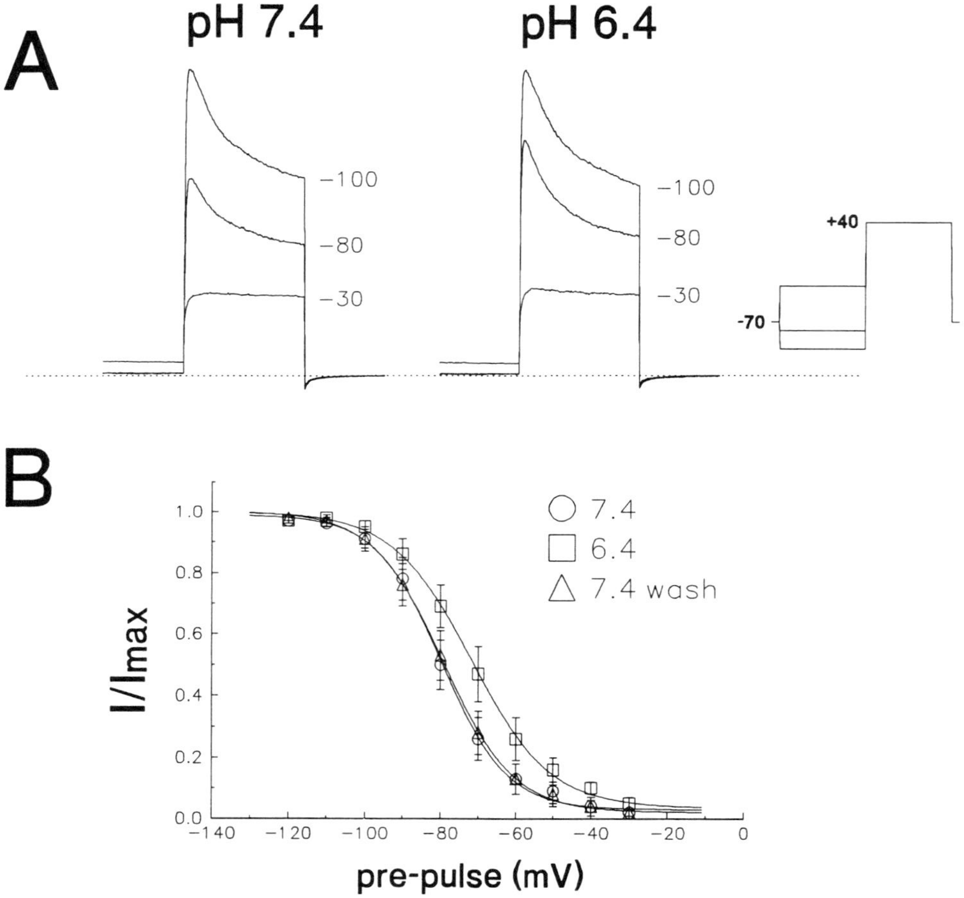

FIGURE 3. Steady state inactivation (h_∞) of the transient potassium current (I_A) is reversibly shifted to more depolarized potentials in isolated rat CA1 neurons. **A.** Representative currents were evoked at +40 mV after a series of increasingly depolarized prepulses in the presence of tetrodotoxin to inhibit Na^+-currents. A subtractive protocol was used to isolate I_A. The voltage protocol is illustrated in the inset. The zero current level is illustrated by a dotted line. **B.** Summary h_∞ plot for I_A ($n = 12$). Reprinted with permission from Tombaugh and Somjen, 1996.

Inwardly rectifying K^+ channels are also present in mammalian central neurons (Penington et al., 1993; Takano et al., 1995). Presumably these channels contribute to G_K at "rest" and therefore help maintain resting membrane potential. External acidification has been shown to depress the inward K^+ current in myocytes (Harvey and Ten Eick, 1989). In a more recent study of HIR and IRK1 inward rectifier channel chimeras in oocytes, Coulter et al. (1995) reported that external H^+ ions selectively reduced unitary conductance of the HIR channel with a pK of 7.4, and they identified several amino acid residues that mediate this effect. Quite surprisingly, a critical histidine residue (h117), identified in their study as essential for pH sensitivity, turned out not to be the titratable site. Instead, their results suggest h117 plays a permissive role by stabilizing a certain tertiary protein structure that exposes other critical amino-acid residues for H^+ binding. In addition to identifying inward rectifiers as potential targets for H^+ modulation, their data have expanded our current thinking of how H^+ may influence channel behavior.

Effects of pH_i

The actions of intracellular pH on K^+ currents have also been examined. Carbone et al. (1978, 1981) and Wanke et al. (1979) reported that perfusion of squid axons with low pH solutions depressed delayed K^+ currents with a p*K* of 6.4–6.9, without changing their activation kinetics. Moreover, they found that the activation and voltage-dependence of I_K exhibited different pH sensitivity (pK) depending on whether pH_o or pH_i was altered. Based on this, the authors concluded that protons act independently on both faces of the cell membrane. In frog axons, internal acidification with weak acids preferentially depressed the "fast", 4-Aminopyridine-sensitive K^+ current (presumably equivalent to I_A) with an apparent p*K* near 7.0 (Schneider et al., 1993); the slower (delayed rectifier) K^+ currents were blunted to a lesser degree. While the general effect of H^+ on K^+ currents appears to be one of inhibition, internal acidosis in perfused snail neurons has been shown to increase the delayed K^+ current (presumably via the contribution of a non-inactivating outward H^+ current) while depressing the transient K^+ current (I_A; Byerly and Moody, 1986). Specific reports of pH_i effects on the steady-state inactivation of K^+ currents are lacking.

In *Helix* neurons, the Ca^{2+}-activated K^+ current ($I_{K\text{-}Ca}$) is reportedly more severely depressed by internal H^+ than is the delayed rectifier current (Meech, 1979), though estimates of the relative p*K*s were not made in that study. It remains uncertain whether this effect on $I_{K\text{-}Ca}$ arose from a direct action of H^+ on the channel, indirectly by blocking Ca^{2+} influx (see p. 407), or via an increase in cytosolic Ca^{2+} buffering (Zucker, 1981). Recordings made from inside-out patches taken from rat hippocampal neurons revealed no change in $I_{K\text{-}Ca}$ channel activity when the "internal" pH was reduced to 6.2 (Wann and Richards, 1994), suggesting that pH-dependent effects on $I_{K\text{-}Ca}$ in intact cells are mediated via one or more cytosolic factors. However, in a study of single high-conductance K^+ channels (presumably $I_{K\text{-}Ca}$) in excised patches of rat synaptosomal membranes, low "internal" pH reduced unitary conductance, increased open probability, and shortened both the open and closed time constants (Habartova et al., 1994). It may be that K^+ channels expressed in different cellular domains exhibit distinct pH sensitivities.

The Ca^{2+}-activated K^+ channels derived from rat skeletal muscle and incorporated into artificial planar bilayers (Laurido et al., 1991) exhibit properties similar to the channels studied by Habartova et al. (1994), in that protons affect the channel only when applied to the internal (i.e., Ca^{2+}-sensitive) side. Acid pH_i reduced the channel's open probability and shifted its activation potential by -23.5 mV/pH unit. The authors concluded that protons weaken Ca^{2+} binding to all conformational states of the channel.

For inward (anomalous) rectifier K^+ channels (IRK), intracellular alkalinization enhanced the whole-cell current in catfish retinal neurons while acidification had little effect (Takahashi and Copenhagen, 1995). This biased sensitivity to internal alkalosis appears to distinguish these neuronal IRK channels from their analogues in other cell types. In rat locus coeruleus neurons, for example, internal alkalinization depressed excitability and activated an outward K^+ current (presumably IRK), while acidification had the opposite effect (Pineda and Aghajanian, 1997). As additional examples, acid pH_i has been shown to inhibit the IRK in frog skeletal muscle, starfish egg cells, and heart muscle cells (Moody and Hariwara, 1982; Blatz, 1984; Ito et al., 1992). Depression by internal H^+, however, does not appear to be a universal property of potassium channels. Low internal pH can enhance the activity of a newly described family of stretch- and fatty acid-activated K channels found in cultured rat neurons (Kim et al., 1995). The contribution of these "new" channels to the resting G_K is not yet clear.

Ca^{2+} CHANNELS

Effects of pH_o

A diverse group of Ca^{2+} channels have been identified in vertebrate neurons, each presumably with a unique density, distribution, and function (Tsien et al., 1988; Takahashi et al., 1989; Bean, 1989; Hess, 1990; Karst et al., 1993). Given the dynamic and often confusing nomenclature for Ca^{2+} channels, we will use a simplified classification to avoid potential confusion between current type and channel type, referring to Ca^{2+} currents as either high-voltage activated (HVA) or low-voltage activated (LVA).

High-threshold (HVA) Ca^{2+} channels, which include L-, N-, P-, Q-, and R-type channels, activate at strongly depolarized potentials, which can arise during high-frequency or synchronized input at multiple excitatory synapses or invasion of the soma and dendrites by back-propagating action potentials (Stuart and Sakmann, 1994). Presynaptic HVA Ca^{2+} channels mediate transmitter release (Miller, 1990), while dendritic and somatic HVA Ca^{2+} channels underlie the prolonged action potential ("complex" spike) in vertebrate neurons (Llinas, 1988). In contrast, low threshold (LVA), or T-type, Ca^{2+} channels exhibit a comparatively small unitary conductance, activate at negative potentials near the threshold for action potentials, inactivate rapidly, and are thought to regulate intrinsic firing behavior in certain cell types (Destexhe et al., 1996). As an example, T-currents have been shown to facilitate spike generation in the inferior olive and superchiasmatic nuclei (Llinas and Yarom, 1981; Akasu et al., 1993).

In general, Ca^{2+} channels appear much more pH-sensitive than either Na^+ or K^+ channels. The effects of external pH on HVA Ca^{2+} currents have been described for a variety of cell types (Ohmori and Yoshii, 1977; Irisawa and Sato, 1986; Krafte and Kass, 1988; Prod'hom et al., 1989; Tytgat et al., 1990; Barnes and Bui, 1991; Klockner and Isenberg, 1994; Tombaugh and Somjen, 1996; Zhou and Jones, 1996). Some investigators have determined the p*K* of the peak HVA Ca^{2+} current in mammalian neurons to be close to the physiological pH_o of 7.2–7.3 (Fig. 1C) (Barnes and Bui, 1991; Tombaugh and Somjen, 1996; cf. Chesler, 1990). This indicates that little more than 50% of maximal HVA channel conductance is available at normal pH_o and that it may be modulated bidirectionally by pH shifts as small as 0.1 pH unit. In line with this reasoning, HVA Ca^{2+} currents in freshly isolated rat hippocampal neurons are reversibly enhanced by external alkalosis

(Fig. 4) (Tombaugh and Somjen, 1996). Low-voltage activated (LVA), or T-type calcium currents have been less extensively studied in this way but appear similarly affected by external H^+, both in cardiac cells and neurons (Tytgat et al., 1990; Tombaugh and Somjen, 1997).

In addition to their effects on macroscopic HVA calcium current amplitude, changes in pH_o shift the voltage-dependence of current activation (Fig. 4). As with Na^+ currents, this shift has been attributed to the charge-screening effect of H^+ (Iijima et al., 1986; Klockner and Isenberg, 1994; Zhou and Jones, 1996). By neutralizing negative charges near the mouth of Ca^{2+} channels and over the membrane surface, protons could theoretically mediate shifts in both voltage dependence and channel conductance via the same charge-screening effect. This point has been challenged for Na^+ channels (Hille, 1968; Woodhull, 1973), and it may not hold for Ca^{2+} channels either. I_{Ca} voltage shifts are, like those for I_{Na}, modest in size (3–8 mV/pH unit), but macroscopic Ca^{2+} currents are nearly 500-fold more sensitive to external H^+ than I_{Na} ($pK = 7.1$ vs. 4.5–5.5). Assuming that a similar surface charge density exists near Na^+ and Ca^{2+} channels, screening of surface charges should affect the voltage sensor of both channel types to a similar degree. Therefore, the data are more consistent with a specific binding of H^+ to one or more titratable acid groups within the Ca^{2+} channel's protein structure. A Hill coefficient of less than 1 derived from whole-cell recordings in both neurons and myocytes suggests that multiple titratable sites on the external face of HVA channels cooperatively mediate the effects of H^+ on Ca^{2+} current amplitude (Barnes and Bui, 1991; Klockner and Isenberg, 1994; Tombaugh and Somjen, 1996a).

It remains unclear whether external H^+ ions influence neuronal Ca^{2+} currents by affecting unitary channel conductance (channel block), gating, or both. Whole-cell HVA currents in hippocampal neurons inactivate more rapidly at low pH_o (Tombaugh and Somjen, 1996), suggesting a shift in channel gating. To date, no single channel data have been reported for external H^+ effects on neuronal Ca^{2+} channels, though three independent studies have confirmed that at $pH_o < 6.5$ protons can reduce unitary L-type Ca^{2+} conductance in cardiac myocytes (Kraft and Kass, 1988; Prod'hom et al., 1987, 1989; Klockner and Isenberg, 1994). Klockner and Isenberg argued that H^+ ions reduce channel conductance by diluting the local Ca^{2+} concentration near the channel mouth. In contrast, Prod'hom et al. (1989) concluded that the protonation site lies on the external aspect of the channel, interacting indirectly with the channel conductance mechanism, and that protonation does not strongly affect the local potential near the channel pore. At physiological $[Ca^{2+}]_o$ and within a pH_o range of 6.5–7.5, however, Prod'hom observed little if any effect of H^+ on macroscopic Ca^{2+} channel conductance, concluding that under normal conditions H^+ depress Ca^{2+} currents primarily through charge screening effects on channel gating.

This idea that H^+ allosterically modulates Ca^{2+} channel conductance has been challenged by Chen et al. (1996), who used site directed mutagenesis of critical pore-forming glutamate residues to modulate or abolish the depressive effects of H^+ on single L-type channels at or above physiological pH_o. The high degree of conservation of these residues among the $\alpha 1$ subunits of different Ca^{2+} channels implies that this type of H^+ modulation exists for all high-threshold Ca^{2+} channels, though whether such a model holds true for neuronal Ca^{2+} channels remains to be seen. In a contemporary report, Zhou and Jones (1996) examined tail currents in frog sympathetic neurons and found that the effects of external H^+ in a near-normal pH_o range arise not from changes in channel conductance but primarily from shifts in the voltage-dependence of channel gating. To this extent, their data are more consistent with the earlier results of Prod'hom et al. (1989).

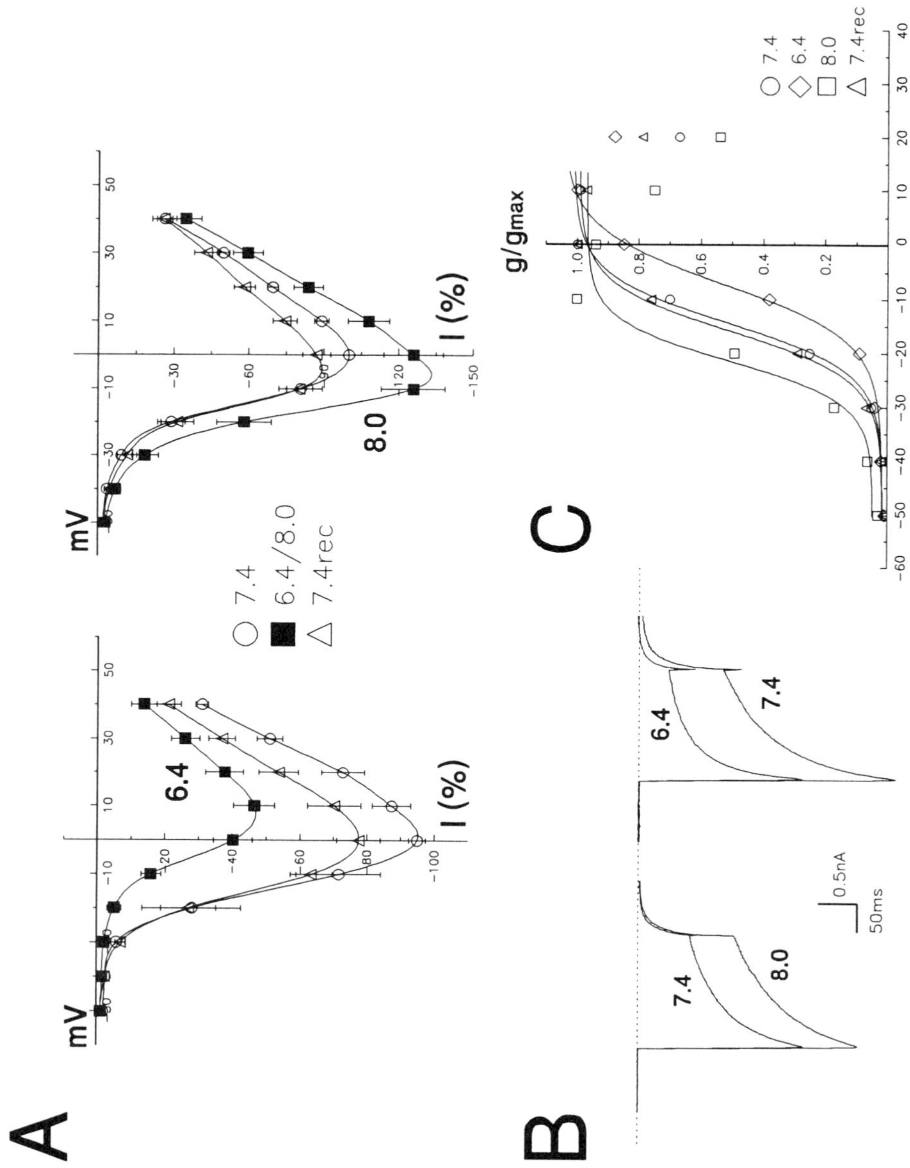
A
mV
I (%)
6.4
8.0
7.4
6.4/8.0
7.4rec
B
7.4
8.0
6.4
7.4
0.5nA
50ms
C
g/gmax
mV
7.4
6.4
8.0
7.4rec

Effects of pH_i

HVA Ca^{2+} channels are also sensitive to cytosolic pH shifts, a finding first demonstrated in *Paramecium* (Umbach, 1982). Similar findings have been reported for other cell types (Irisawa and Sato, 1986; Dixon et al., 1993); however, determining a p*K* and dissecting the underlying mechanism(s) have been hampered by the difficulty in controlling internal pH and by the fact that Ca^{2+} currents exhibit rapid and severe "run-down" in excised patches. In myocytes, using the open-cell-attached patch method to reduce "run-down," Kaibara and Kameyama (1988) reported that H^+ primarily depressed the open-state probability for L-type channels with a p*K* of 6.6. However, pH_i was not measured and it is likely that the actual pH seen by the channel was different from that measured in the bathing solutions. Although single-channel studies in mammalian neurons have not yet been reported, Mironov and Lux (1991) showed that whole-cell HVA currents in chick dorsal root ganglion (DRG) neurons were reversibly enhanced by the weak base NH_4Cl and were depressed by the weak acid sodium acetate. Identical results have been obtained in rat CA1 neurons (Fig. 5) (Tombaugh and Somjen, 1997). The Hill plot derived from DRG neurons suggested a p*K* of 6.5; however, this value is probably also too low, given that pH_o was not held constant and presumably contributed to the internal H^+ effect. Dixon et al. (1993) used the identical method (weak acids) to depress HVA Ca^{2+} currents in catfish retinal cells, but they were able to correlate their measurements with optical recordings of pH_i and reported a p*K* close to 7.2.

As alluded to above, whole-cell HVA Ca^{2+} currents in neurons are likely to reflect a composite of overlapping currents flowing through multiple HVA Ca^{2+} channel types. Because the separation of different HVA Ca^{2+} channels can be difficult, very little is currently known about the relative pH_i sensitivities among HVA channel classes. Recently, a study of isolated rat CA1 neurons exposed to NH_4Cl indicates that N-type currents are more sensitive to internal alkaline shifts than are L-type currents (Tombaugh and Somjen, 1997). Further studies at the single-channel level are needed to address this question more thoroughly.

The pH_i sensitivity of HVA Ca^{2+} channels is likely to turn out to be a feature common to many cell types. This, however, does not appear true of the low-threshold, or T-type Ca^{2+} channels (Fig. 6). In both chick DRG and rat CA1 neurons, as well as chick myocytes, low-threshold currents are sensitive to external pH shifts between 6.4 and 8.0 but are unaffected by exposure to weak acids or bases at constant extracellular pH (Tytgat et al., 1990; Mironov and Lux, 1991; Tombaugh and Somjen, 1997). Though these studies lacked pH_i measurements, they clearly demonstrate that different classes of neuronal calcium channels in a given cell type can exhibit very different pH sensitivities.

FIGURE 4. External pH shifts evoke reversible shifts in the amplitude and voltage-dependence of HVA Ca^{2+} currents in rat CA1 neurons. **A.** HVA Ca^{2+} current *I–V* plots, both during a pH_o change and recovery (rec), are expressed as a percent of the maximum current evoked at pH 7.4. **B.** Representative waveforms, derived from two separate cells, were evoked by depolarization to 0 mV ($V_{hold} = -100$ mV) under either acidic ($pH_o = 6.4$) or alkaline ($pH_o = 8.0$) conditions. **C.** Normalized whole-cell conductance data were derived from the means illustrated in (A) and fitted to separate Boltzman functions. The K^+ and Ca^{2+} currents were blocked by including tetraethylammonium-chloride $(TEA)^+$ and Cd^{2+} in the bath, and Cs^+ in the recording pipette. All data are presented as mean ± SEM. Reprinted with permission from Tombaugh and Somjen, 1996

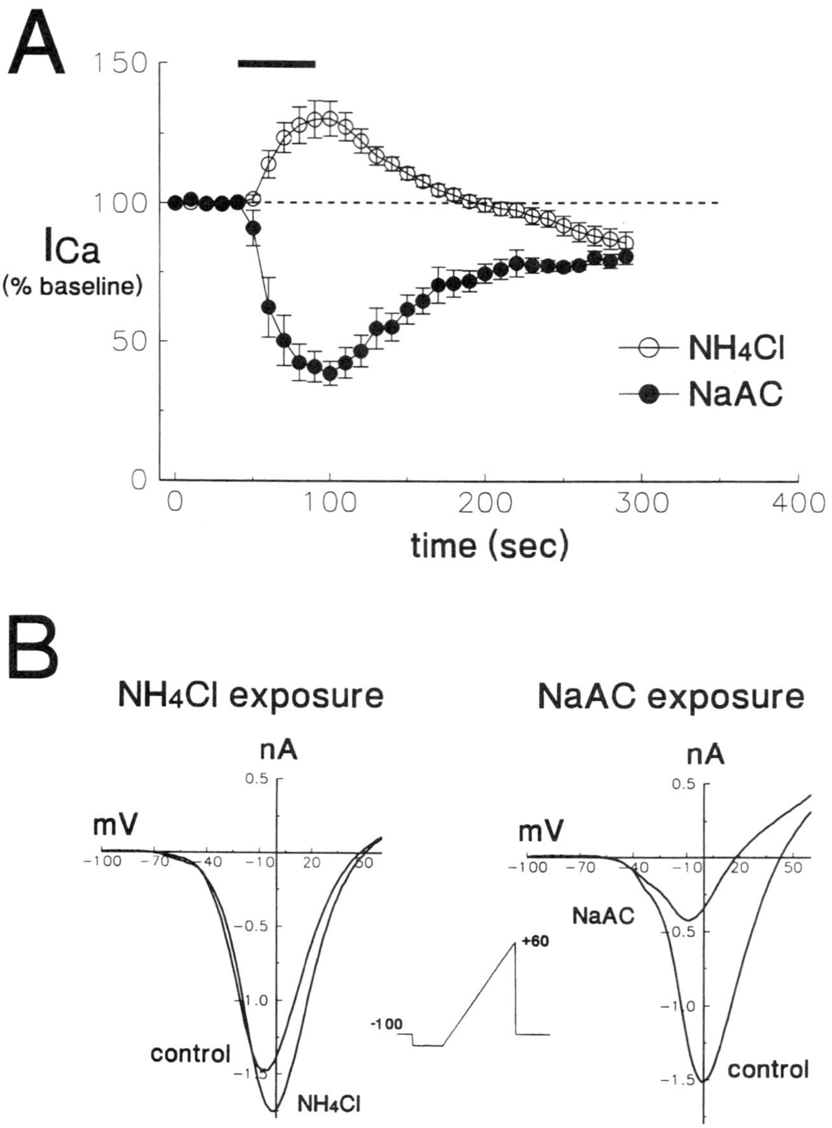

FIGURE 5. Internal acidification depresses while internal alkalinization enhances high-voltage activated (HVA) calcium currents recorded from freshly isolated rat CA1 neurons (patch pipette buffered with 10 mM Bapta/10 mM Hepes). **A.** Bath administration of 20 mM Na-acetate known to cause acidification of pH_i, depresses HVA currents (n = 4). Exposure to 20 mM NH_4Cl, known to alkalinize cells, increases current amplitude (n = 8). External pH was held constant at 7.4; the duration of each exposure is depicted by the heavy bar. Data points (mean ± SEM) represent peak ramp-evoked whole-cell currents expressed as a percent of baseline. **B.** *I–V* profiles of representative currents and the voltage ramp protocol used. Reprinted from Tombaugh and Somjen (1997) with permission by the American Physiological Society.

Like external protons, internal protons also appear to act at discrete sites on Ca^{2+} channel proteins. Mironov and Lux (1991) found that the internal application of agents that selectively modify histidine residues blunted the modulation of HVA Ca^{2+} current amplitude by shifts in pH_i. Similar results were obtained earlier from a study of single L-type channels in myocytes (Kaibara and Kamayema, 1988). In that study, internal acidification decreased the channel's open-state probability and caused a small (15%) depression of unitary conductance.

The similarity between the external and internal p*K*s of HVA Ca^{2+} currents raises the question of whether the reported effects of externally applied H^+ on whole-cell HVA Ca^{2+} currents reflect the indirect action of intracellular H^+. Though this caveat can never be dismissed without measurement of pH_i, three lines of evidence argue against such an indirect effect. First, strong acids like HCl are essentially fully ionized at physiological pH and do not readily cross cell membranes. Second, raising the internal pH buffering in CA1 neurons does not diminish the depression of HVA Ca^{2+} currents by external acidification (Tombaugh and Somjen, 1996), while this maneuver modifies changes in HVA currents during internal acidification (Dixon et al., 1993) or alkalinization (Tombaugh and Somjen, 1997). Third, changing pH_i directly in isolated neurons by internal perfusion re-

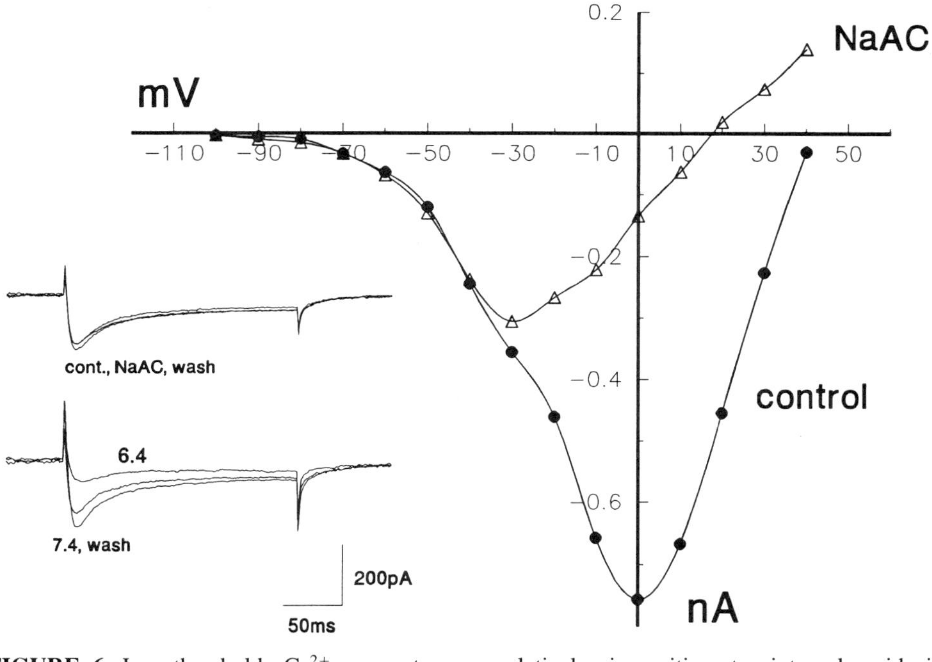

FIGURE 6. Low-threshold Ca^{2+} currents are relatively insensitive to internal acidosis. Current–voltage (*I–V*) curves derived from step-depolarizations in an isolated rat CA1 neuron before, and 2 min after, exposure to 20 mM sodium acetate ($pH_o = 7.4$). Note that the region of the plot from −100 mV to −40 mV is unchanged by sodium acetate, while at more positive voltages, the currents are depressed. Inset shows LVA currents evoked in a different cell by step depolarization to −60 mV ($V_{hold} = -100$ mV), illustrating that LVA currents are depressed by external acidosis ($pH_o = 6.4$) even though they are insensitive to internal acidosis. Unpublished observations, G. C. Tombaugh.

quires strongly buffered solutions (≥100 mM) and long (5–10 min) exchange periods (Byerly and Moody, 1986). Because the measured effects of pH_o on I_{Ca} are relatively rapid (seconds) it seems unlikely that, under the experimental conditions commonly used to measure Ca^{2+} currents, modest changes in pH_o within a physiological range cause any significant pH_i shifts in neurons during whole-cell recording.

GENERAL DISCUSSION AND CONCLUSIONS

In virtually all studies to date, whenever H^+ ions affected voltage-gated ion currents they usually acted to depress them. In many instances, acidification also raised (shifted to a more positive level) the threshold voltage of activation. Having stated these generalizations, it is equally important to point out that the relative pH sensitivities among the diverse ion channels in a given cell can differ greatly, while the pH sensitivity of the same ion current can vary among cell types. These quantitative variations could have important functional consequences; whether a given neuron is depressed or excited by acid pH could depend on whether inward or outward currents are more strongly affected by H^+. This may explain why respiratory neurons (or, more generally, the so-called E-neurons: see Chapters 20, 26, 27) are excited by high CO_2, perhaps reflecting that outward (K^+) currents in these cells are depressed by H^+ while inward Na^+ and Ca^{2+} currents are spared.

Na^+ Channels

It is clear that protons are capable of blunting Na^+ channel conductance in all cells examined, probably via a direct external block upon entering the channel pore, while causing a depolarizing shift in the voltage-dependence of current activation, presumably via a more general charge-screening effect. These changes are noticeable at pH_o 6.5 and are robust below 5.0 but are small or undetectable near a resting physiological pH_o of 7.2. However, it should be emphasized that action potentials are triggered as all-or-none events, so even a slight change in Na^+ channel conductance or open probability at a voltage near threshold would have a comparative large effect on firing probability. Complementary to this is the consistent observation that internal alkalinization above pH 8.0 retards Na^+ channel inactivation. Both of these effects are consonant with the ability of acidosis to depress and alkalosis to enhance cell excitability.

K^+ Channels

Many K^+ channel isoforms and subtypes exist in the vertebrate CNS and have been grouped into a few broadly defined classes: transient (I_A), delayed rectifier (I_K), inward rectifier, and Ca^{2+}-activated channels ($I_{K\text{-}Ca}$). Judging by whole cell voltage-clamp recordings, K^+ currents appear only slightly sensitive to H^+ within a relevant pH range. However, two noteworthy exceptions to this exist. First, the conductance of the HIR inward rectifier is strongly depressed by external H^+ and exhibits a pK near 7.4 (Coulter et al., 1995). Second, low pH_o causes a positive shift in the steady-state inactivation of outwardly rectifying currents in hippocampal neurons, albeit at the low end of the physiological pH range (Fig. 3) (Tombaugh and Somjen, 1996).

As for the effects of H^+, K^+ channels appear generally more sensitive to internal than external pH. The reported internal p*K* of the delayed rectifier conductance in squid axon (6.4–6.9) (Wanke et al., 1979) is sufficiently close to the estimated resting pH_i in mammalian central neurons (7.1–7.2) to support a physiological role for internal H^+, though comparable studies in mammalian neurons have not yet been reported. Because neurons are believed to acidify during cellular activity (Chapter 16), impaired activation of I_K and $I_{K\text{-}Ca}$ channels by low pH_i would tend to relieve the stabilizing influence of I_K on membrane potential, reduce post-spike after-hyperpolarization, and lead to enhanced excitability. Cytosolic acidosis has been implicated in the enhancement of the post-spike depolarization seen in rat axons via its selective depression of fast K^+ currents (Schneider et al., 1993; see also Chapter 31). As mentioned earlier, this mechanism may also play a part in the enhanced spontaneous activity of medullary respiratory neurons at elevated P_{CO_2} (Richter et al., 1993; Chapter 26).

Ca^{2+} Channels

As we have seen, HVA Ca^{2+} channels are especially sensitive to pH. Thus, even modest pH shifts (0.1 pH unit) located near these Ca^{2+} channels could, in principle:

1. Modulate synaptic transmission by acting presynaptically.
2. Affect the temporal summation of excitatory postsynaptic potentials (EPSPs).
3. Influence the time course of membrane depolarization and thus firing threshold and frequency.
4. Indirectly affect the rate of post-spike repolarization by altering $I_{K\text{-}Ca}$.

It is becoming clear that in spite of their general H^+ sensitivity, Ca^{2+} channels exhibit quite different pH_i sensitivities between the two broad classes (HVA and LVA) of channel types. This raises the intriguing possibility that a generalized shift in internal pH might selectively modulate certain aspects of cell function. For example, intradendritic acidosis associated with synaptic activity or burst firing may depress HVA channels without affecting LVA channels, thereby preventing an excessive rise in cytosolic $[Ca^{2+}]$ while preserving LVA channel activity for normal integration and amplification of synaptic input. This scenario is clearly speculative, but it illustrates the capacity of pH transients to act as potent and selective modulators of neuronal function via ion channels.

FINAL REMARKS

One important factor that must be considered in any discussion of proton action on ion channels is the potential interplay between H^+ and other ions, both inside and outside of cells. For example H^+ and Ca^{2+} ions both serve as extracellular charge-screening agents and it is known that Ca^{2+} activity in solution rises as pH falls (Oberleithner et al., 1982; Somjen et al., 1987). In addition, both cations may compete for negative binding sites in the cytosol (Busa and Nuccitelli, 1984). Thus, depending on the direction of a given pH shift, and the relative rates at which H^+ and Ca^{2+} transients are buffered, resulting changes in the local H^+ concentration could either amplify or attenuate direct actions of Ca^{2+} on cell function.

Under normal conditions, the size of pH_o shifts evoked during electrical activity and recorded by ion-selective microelectrodes (ISMs) in brain tissue are relatively small ($\leq$0.2 pH unit; Chesler and Kaila, 1992; Chapter 17). However, as pointed out by Chesler (1990), even the most reliable electrode measurements of pH_o transients may grossly underestimate true local changes in pH_o. Thus, extracellular pH transients recorded *in situ* during cellular activity could reflect much steeper H^+ gradients capable of influencing voltage-gated ion channels locally over a limited range. The same can be said of internal pH transients. Localized pH_i transients, like brief Ca^{2+} transients that can occur in restricted dendritic regions (Eilers et al., 1995), may greatly exceed estimates commonly based on lumped tissue averages or on pH_i measurements recorded from cell bodies. Should such pH_i transients occur near regional "hot spots" for ion channels (e.g., dendritic spine, axon hillock), the resulting modulation of these channels could dramatically influence neuronal function.

The last decade has seen remarkable advances in our understanding of ion channels and the diverse array of factors that modulate their function. Small pH shifts known to occur in the mammalian CNS can exert profound effects on certain ion channels. However, neuronal cells *in situ* appear vastly more sensitive to experimentally imposed pH changes than can readily be accounted for by any single H^+ effect described in this chapter. Thus, it seems likely that if protons play a role in shaping neuronal function via their action on ion channels, they do so through cooperative and probably subtle shifts in the behavior of multiple channel types.

REFERENCES

Akasu T, Shoji S, Hasuo H (1993): Inward rectifier and low-threshold calcium currents contribute to the spontaneous firing mechanism in neurons of the rat superchiasmatic nucleus. Pflügers Arch 425: 109–116.

Alekseev SI, Zaykin AV (1993): Kinetic study of A-type current inactivation in *Lymnaea* neurons. Biochem Biophys Acta 1148: 97–107.

Balestrino M, Somjen GG (1988): Concentration of carbon dioxide, interstitial pH and synaptic transmission in hippocampal formation of the rat. J Physiol (Lond) 396: 247–266.

Barnes S, Bui Q (1991): Modulation of calcium activate chloride current via pH-induced changes in calcium channel properties in cone photoreceptors. J Neurosci 11: 4015–4023.

Bean BP (1989): Classes of calcium channels in vertebrate cells. Ann Rev Physiol 51: 367–384.

Begenisich T, Danko M (1983): Hydrogen ion block of sodium pore in squid giant axons. J Gen Physiol 82: 599–618.

Benitah J, Blaser JR, Marban E, Tomaselli GF (1997): Proton inhibition of sodium channels: mechanism of gating shifts and reduced conductance. J Memb Biol 155: 121–131.

Blatz, AL (1984): Asymmetric proton block of inward rectifier K channels in skeletal muscle. Pflügers Arch 401: 402–407.

Brodwick MS, Eaton DC (1978): Sodium channel inactivation in squid axon is removed by high internal pH or tyrosine specific reagents. Science 200: 1494–1495.

Busa WB, Nuccitelli R (1984): Metabolic regulation via intracellular pH. Am J Physiol 246: R406–R438.

Byerly L, Moody WJ (1986): Membrane currents of internally perfused neurons of the snail, *Lymnaea Stagnalis,* at low intracellular pH. J Physiol (Lond) 376: 477–491.

Campbell DT (1982): Do protons block Na^+ channels by binding to a site outside the pore? Nature 298: 165–167.

Carbone E, Fioravanti R, Prestipino G, Wanke E (1978): Action of extracellular pH on Na^+ and K^+ membrane currents in the giant axon of *Loligo vulgaris*. J Membr Biol 43: 295–315.

Carbone E, Testa PL, Wanke E (1981): Intracellular pH and ionic channels in the *Loligo vulgaris* giant axon. Biophys J 35: 393–413.

Caspers H, Speckmann E-J (1972): Cerebral pO_2, pCO_2 and pH: Changes during convulsive activity and their significance for spontaneous arrest of seizures. Epilepsia 13: 699–725.

Chen X-H, Bezprozvanny I, Tsien RW (1996) Molecular basis of proton block of L-type Ca^{2+} channels. J Gen Physiol. 108:363–374.

Chesler, M (1990): The regulation and modulation of pH in the central nervous system. Prog Neurobiol 34: 401–427.

Chesler M, Kaila K (1992): Modulation of pH by neuronal activity. Trends Neurosci 15: 396–402.

Coulter KL, Perier F, Radeke CM, Vandenberg CA (1995): Identification and molecular localization of a pH-sensing domain for the inward rectifier potassium channel HIR. Neuron 15: 1157–1168.

Daumas P, Anderson OS (1993): Proton block of rat brain sodium channels. Evidence for two proton binding sites and multiple occupancy. J Gen Physiol 101: 27–43.

Destexhe A, Contreras D, Steriade M, Sejnowski TJ, Huguenard JR (1996): *In vivo, in vitro,* and computational analysis of dendritic calcium currents in thalamic reticular neurons. J Neurosci 16: 169–185.

Dixon DB, Takahashi K-I, Copenhagen DR (1993): L-Glutamate suppresses HVA calcium current in catfish horizontal cells by raising intracellular proton concentration. Neuron 11: 267–277.

Drouin H, The R (1969): The effect of reducing extracellular pH on the membrane currents of the Ranvier node. Pflügers Arch 313: 80–88.

Ehrenstein G, Fishman HM (1971): Evidence against hydrogen-calcium competition model for activation of electrically excitable membranes. Nature New Biol 233: 16–17.

Eilers J, Augustine GJ, Konnerth A (1995): Subthreshold synaptic Ca^{2+} signaling in fine dendrites and spines of cerebellar Purkinje neurons. Nature 373: 151–158.

Gottfried JA, Chesler M (1994): Endogenous proton modulation of NMDA receptor-mediated EPSCs revealed by carbonic anhydrase inhibition in rat hippocampus. J Physiol (Lond) 478: 373–378.

Habartova A, Krusek J, Zemmkova H (1994): Sensitivity of high conductance channels in synaptosomal membranes from the rat brain to intracellular pH. Eur Biophys J 23: 71–77.

Harvey RD, Ten Eick RE (1989): On the role of sodium ions in the regulation of the inwardly rectifying potassium conductance in cat ventricular myocytes. J Gen Physiol 94: 329–348.

Hess P (1990): Calcium channels in vertebrate cells. Ann Rev Neurosci 13: 337–356.

Hille B (1968): Charges and potentials at the nerve surface. Divalent ions and pH. J Gen Physiol 5: 221–236.

Hille B (1992): Ion channels of excitable membranes. Sunderland, MA. Sinauer Associates.

Hille B, Woodhull AM, Shapiro BI (1975): Negative surface charge near sodium channels of nerve: Divalent ions, monovalent ions, and pH. Phil Trans R Soc Lond 270: 301–318.

Hodgkin AL, Huxley AF (1952): A quantitative description of membrane current and its application to conduction and excitation in nerve. J Physiol (Lond) 117: 500–544.

Iijima T, Ciani S, Hagiwara S (1986): Effects of the external pH on Ca channels: Experimental studies and theoretical consideration using a two-site, two-ion model. Proc Nat Acad Sci USA 83: 654–658.

Irisawa H, Sato R (1986): Intra- and extracellular actions of proton on the calcium current of isolated guinea pig ventricular cells. Circ Res 59: 348–355.

Ito H, Vereecke J, Carmeliet E (1992): Intracellular protons inhibit inward rectifier K^+ channel of guinea-pig ventricular cells. Pflügers Arch 422: 280–286.

Kaibara M, Kameyama M (1988): Inhibition of the calcium channel by intracellular protons in single ventricular myocytes. J Physiol (Lond) 403: 621–640.

Karst H, Joels M, Wadman WJ (1993): Low threshold calcium current in dendrites of the adult rat hippocampus. Neurosci Lett 164: 154–158.

Kim D, Sladek CD, Aguado-Velasco C, Mathiasen JR (1995): Arachidonic acid activation of a new family of K^+ channels in cultured neuronal cells. J Physiol (Lond) 484: 643–660.

Klockner U, Isenberg G (1994): Calcium channel current of vascular smooth muscle cells: Extracellular protons modulate gating and single channel conductance. J Gen Physiol 103: 665–678.

Krafte DS, Kass RS (1988): Hydrogen ion modulation of Ca channel current in cardiac ventricular cells. J Gen Physiol 91: 641–657.

Laurido C, Candia S, Wolff D, Latorre R (1991): Proton modulation of a Ca^{2+}-activated K^+ channel from rat skeletal muscle incorporated into planar bilayers. J Gen Physiol 98: 1025–1043.

Llinas RR (1988): The intrinsic electrophysiological properties of mammalian neurons: Insights into central nervous system function. Science 242: 1654–1664.

Llinas RR, Yarom Y (1981): Electrophysiology of mammalian inferior olivary neurons in vitro. Different types of voltage-dependent ionic conductances. J Physiol (Lond) 315: 549–567.

Meech RW (1979): Membrane potential oscillations in molluscan "burster" neurons. J Exp Biol 81: 93–112.

Miller RJ (1990): Receptor-mediated regulation of calcium channels and neurotransmitter release. FASEB J 4: 3291–3299.

Mironov SL, Lux HD (1991): Cytoplasmic alkalinization increases high-threshold calcium current in chick dorsal root ganglion neurons. Pflügers Arch 419: 138–143.

Moody W (1984): Effects of intracellular H^+ on the electrical properties of excitable cells. Ann Rev Neurosci 7: 257–278.

Moody WJ, Hagiwara S (1982): Block of inward rectification by intracellular H^+ in immature oocytes of the starfish *Mediaster aequalis*. J Gen Physiol 79: 115–130.

Nonner W, Spalding B, Hille B (1980): Low intracellular pH and chemical agents slow inactivation gating in sodium channels of muscle. Nature 284: 360–363.

Oberleithner H, Greger R, Lang F (1982): The effect of respiratory and metabolic acid-base changes on ionized calcium concentration: In vivo and in vitro experiments in man and rat. Eur J Invest 12: 451–455.

Ohmori H, Yoshii M (1977): Surface potential reflected in both gating and permeation mechanisms of sodium and calcium channels of the tunicate egg cell membrane. J Physiol (Lond) 419: 353–378.

Penington NJ, Kelly JS, Fox AP (1993): Whole-cell recordings of inwardly rectifying K^+ currents activated by 5-HT1A receptors on dorsal raphe neurones of the adult rat. J Physiol (Lond) 469: 387–405.

Pineda J, Aghajanian GK (1997): Carbon dioxide regulates the tonic activity of locus coeruleus neurons by modulating a proton- and polyamine-sensitive inward rectifier potassium current. Neuroscience 77: 723–743.

Prod'hom B, Pietrobon D, Hess P (1987): Direct measurement of proton transfer rates to a group controlling the dihydropyridine-sensitive Ca^{2+} channel. Nature 329:243–246.

Prod'hom B, Pietrobon D, Hess P (1989): Interactions of protons with single open L-type calcium channels. J Gen Physiol 94: 23–42.

Ransom BR (1992): Glial modulation of neural excitability mediated by extracellular pH: A hypothesis. Prog Brain Res 94: 37–46.

Richter DW, Champagnat J, Jacquin T, Benacka R (1993): Calcium currents and calcium-dependent potassium currents in mammalian medullary respiratory neurons. J Physiol (Lond) 470: 23–33.

Schauf CL, Davis FA (1976): Sensitivity of the sodium and potassium channels of *Myxicola* giant axons to changes in external pH. J Gen Physiol 67: 185–195.

Schneider U, Quasthoff S, Mitrovic N, Grafe P (1993): Hyperglycemic hypoxia alters afterpotential and fast K^+ conductance of rat axons by cytoplasmic acidification. J Physiol (Lond) 465: 679–697.

Shrager P (1974): Ionic conductance changes in voltage clamped crayfish axons at low pH. J Gen Physiol 64: 666–690.

Sigworth FJ (1980): The conductance of sodium channels under conditions of reduced current at the node of Ranvier. J Physiol (Lond) 307: 131–142.

Somjen GG, Allen BW, Balestrino M, Aitken PG (1987): Pathophysiology of pH and Ca^{2+} in bloodstream and brain. Can J Physiol Pharmacol 65: 1078–1085.

Stillman IM, Gilbert DL, Lipicky RJ (1971): Effect of external pH upon the voltage dependent currents of the squid giant axon. Biophys J 11: 55a.

Stuart GJ, Sakmann B (1994): Active propagation of somatic action potentials into neocortical pyramidal cell dendrites. Nature 367: 69–72.

Taira T, Smirnov S, Voipio J, Kaila K (1993): Intrinsic proton modulation of excitatory transmission in rat hippocampal slices. Neuroreport 4: 93–96.

Takahashi K, Copenhagen DR (1995): Intracellular alkalinization enhances inward rectifier K^+ current in retinal horizontal cells of catfish. Zoological Sci. 12: 29–34.

Takahashi K, Wakamori M, Akaike N (1989): Hippocampal CA1 pyramidal cells of rats have four voltage-dependent calcium conductances. Neurosci Lett 104: 229–234.

Takano K, Stanfield PR, Nakajima S, Nakajima Y (1995): Protein kinase C-mediated inhibition of an inward rectifier potassium channel by substance P in nucleus basalis. Neuron 14: 999–1008.

Tombaugh GC, Somjen GG (1996): Effects of extracellular pH on voltage-gated Na^+, K^+, and Ca^{2+} currents in isolated rat CA1 neurons. J Physiol (Lond) 493: 719–732.

Tombaugh GC, Somjen GG (1997): Differential sensitivity to intracellular pH among high- and low-threshold Ca^{2+}-currents in isolated rat CA1 neurons. J Neurophysiol 77: 639–653.

Traynelis SF, Cull-Candy SG (1991): Pharmacological properties and H^+ sensitivity of excitatory amino acid receptor channels in rat cerebellar granule cells. J Physiol 433: 727–763.

Tsien RW, Lipscombe D, Madison DV, Bley KR, Fox AP (1988): Multiple types of neuronal calcium channels and their selective modulation. Trends Neurosci 11: 3431–438.

Tytgat J, Nilius B, Carmeliet E (1990): Modulation of the T-type cardiac calcium channel by changes in proton concentration. J Gen Physiol 96: 973–990.

Umbach JA (1982): Changes in intracellular pH affect calcium currents in *Paramecium caudatum*. Proc R Soc Lond [Biol] 261: 209–224.

Wanke E, Carbone C, Testa PL (1979): K^+ channel conductance modified by a titratable group accessible to protons from the intracellular side of the squid axon membrane. Biophys J 26: 319–324.

Wanke E, Carbone E, Testa PL (1980): The sodium channel and intracellular H^+ blockage in squid axon. Nature 287: 62–63.

Wann KT, Richards CD (1994): Properties of single calcium activated potassium channels of large conductance in rat hippocampal neurons in culture. Eur J Neurosci 6: 607–617.

Woodhull AM (1973): Ionic blockage of sodium channels in nerve. J Gen Physiol 61: 687–708.

Zhang J-F, Siegelbaum SA (1991): Effects of external protons on single cardiac sodium channels from guinea-pig ventricular myocytes. J Gen Physiol 98: 1065–1083.

Zhou W, Jones SW (1996): The effects of external pH on calcium channel currents in bullfrog sympathetic neurons. Biophys J, 70: 1326–1334.

Zucker RS (1981): Cytoplasmic alkalinization reduces calcium buffering in molluscan central neurons. Brain Res 225: 155–170.

CHAPTER 22

pH Modulation of Ligand-Gated Ion Channels

STEPHEN F. TRAYNELIS
Department of Pharmacology, Emory University, Atlanta, GA

INTRODUCTION

Although the importance of protons and hydroxide anions is documented throughout the history of biochemistry, the implications of the proton sensitivity of macromolecules have only recently begun to occupy a prominent position in our understanding of neurotransmitter-gated as well as other types of (see Chapters 20, 21, 23, 24) ion channels. This results in part from a curious assortment of barriers. For example, although pH (the negative logarithm of proton activity) is no doubt a useful convention for the synthetic chemist and biologist, the pH range maintained by the mammalian central nervous system (CNS) is relatively narrow; that is, pH 6.8–7.8 could alternatively be discussed in terms of H^+ activity (~16–160 nM). Activity corresponds in a complex manner to the effective concentration and is always less than the nominal concentration by a factor of about 0.8 for protons because of the nonideal behavior of ions in solution (see Chapters 4 and 10). Thus, while the customary "pH" nomenclature can be useful, it can also obscure the fact that protons often act at very low concentrations. In this chapter, we will refer to pH and H^+ concentration whenever possible. In addition, an appreciation of the role of pH in ion channel function also paradoxically suffers from its own importance. Because changes in pH affect so many biochemical processes, sorting out the proton sensitivity of particular ion channels can sometimes be tricky.

In striking contrast to these obstacles, sophisticated measurements performed largely over the past two decades have led to an appreciation of the dynamic nature of extra- and intracellular proton (and corresponding hydroxide ion) concentrations in the brain (Chapters 16–19). Moreover, we also now know that certain pathological conditions (notably seizure and stroke; see Chapters 28–33) produce marked perturbations in interstitial pH, and these predominantly acidic pH changes appear to play important roles in the pathophysiology of these disorders. Against this backdrop, it is noteworthy that almost all ligand-gated ion channels studied thus far are influenced in some way by changes in pH. It therefore seems safe to speculate that the regulation of extra- and intracellular pH holds

pH and Brain Function, Edited by Kai Kaila and Bruce R. Ransom
ISBN 0-471-11838-9

implications for those ion channels that possess functionally important ionizable groups, or are activated or controlled by molecules with pK_a values near physiological pH. Indeed, the closer we examine the proton sensitivity of certain ion channels, the more complex and intriguing it appears.

MECHANISMS OF PROTON REGULATION OF LIGAND-GATED ION CHANNELS

The importance of pH in biology follows from two basic facts: (i) the solvent properties of water are dominated by hydrogen bonding and a high dielectric constant, and (ii) the protein, carbohydrate, and phospholipid components of macromolecules, as well as virtually all neuroactive ligands, contain titratable groups that can control their structure, function, and electrical properties. These two principles define the means by which protons alter receptor function, and the pK_a values of ionizable functional groups dictate the range over which pH changes exert meaningful effects. This chapter surveys the proton regulation of ion channels gated by acetylcholine, γ-aminobutyric acid (GABA), glutamate, intracellular cyclic nucleotides, and intra- and extracellular adenosine-5′-triphosphate (ATP). Embedded in consideration of these receptor systems are two fundamental mechanisms by which alterations in pH can regulate ligand-gated ion channels. Changes in pH can control channel function by acting at a proton sensor within the receptor complex, or can alter the ionization of the ligand, thereby changing the concentration of the active species. Ion channels activated by extracellular glycine and serotonin have been omitted from this discussion, since the pH regulation of their function has not been examined in depth (see Kishimoto et al., 1981; Pinkus et al., 1989; Bolanos et al., 1990; Yakel et al., 1990; Schmieden et al., 1994).

pH Sensitivity of Acetylcholine Receptors

Neuronal nicotinic acetylcholine-gated ion channels are widely distributed throughout the central nervous system, and are thought to mediate both pre- and postsynaptic signals (Sargent, 1993). Whereas the pH sensitivity of neuronal nicotinic receptors has yet to be investigated, the pH sensitivities of nicotinic receptors localized in amphibian and mammalian muscle, as well as in the electric organ of *Torpedo,* have been studied. In general, these peripheral receptors have served as useful models for acetylcholine-activated channels in the CNS, and given the homology between nonneuronal and neuronal acetylcholine receptor subunits (Karlin, 1993), it seems sensible to predict that the effects of protons on nonneuronal nicotinic receptor function (see p 419) will transfer to mammalian neuronal receptors. Consequently, this discussion will focus on the pH sensitivity of acetylcholine-gated ion channels in muscle, electric organ, and the invertebrate nervous system.

Extracellular proton control of acetylcholine receptor function. More is probably known about the endplate nicotinic acetylcholine receptor, including its sensitivity to pH, than about any other ligand-gated channel. Many of the pH experiments performed on nonneuronal acetylcholine-gated receptors have been geared towards unraveling the receptor's structure–function relationship, rather than understanding the biological implications of pH perturbations. Yet, the more we learn about protein structure, the more advisable is a cautious approach to the use of the pH sensitivity alone to identify important amino acid residues on the basis of their pK_a values. Rather, a combination of approaches is necessary. This is because local structure and charge within a protein can alter the pK_a of any ionizable side chain (histidine, cysteine, glutamate, aspartate, arginine, lysine, tyrosine) and changing the pK_a considerably from their free-solution values (Honig and Nicholls, 1995; Kyte, 1995). There are many examples of proteins that contain functionally important residues with altered pK_a values, such as the cyclic nucleotide-gated channel (pK_a Glu_{333} = 7.6; Root and MacKinnon, 1995), lysozyme (pK_a Glu_{35} = 6.1; Inoue et al., 1992), ribonuclease HI (pK_a Asp_{10} = 6.1; Oda et al., 1994), ovomucoid (pK_a Cys_{56} < 2.5; Schaller and Robertson, 1995), and ribonuclease TI (pK_a $His_{40,92}$ = 7.8–7.9; Inagaki et al., 1981). Nevertheless, the pH sensitivity of these receptors remains an important point, given that pH is a dynamic rather than static parameter in both peripheral tissue and the CNS. Indeed, if the pH sensitivity of neuronal nicotinic receptors is similar to that of nonneuronal acetylcholine receptors, then the regulation of extracellular pH under normal (Chapters 16–19) and pathological conditions (Chapters 28–33) holds interesting implications for cholinergic pathways.

In almost every case examined, acidic pH reversibly inhibits acetylcholine receptor function and alkaline pH enhances it (Del Castillo et al., 1962; Scuka et al., 1975; Iljin and Bregestovski, 1977; Landau et al., 1981; Goldberg and Lass, 1983; Palma et al., 1991; Li and McNamee, 1992). Half-maximal inhibiting concentrations (IC_{50}) of protons for native, reconstituted, and recombinant nicotinic receptors in virtually all preparations are in the range of pH 6.5–7.0 (125–395 nM; Fig. 1A), although there is some species variation in the effects on particular receptor subunits. This proton sensitivity suggests that both physiological and pathological changes in extracellular pH are capable of controlling receptor function. However, the full pH inhibition profile appears less steep than expected for a single binding site (Fig. 1A), which suggests that very slight changes in pH may not meaningfully alter receptor function. This shallow inhibition curve is probably the sum of varied effects of pH on different ion channel properties. Unitary channel conductance is reduced in most preparations by low pH values (IC_{50} near pH 5.5 or 4.0 μM; Fig. 1C), a result that appears consistent with the idea that carboxylic acid-containing side chains near the vestibule of the channel control conductance (Imoto et al., 1988), although a reduction in channel conductance by fast proton block cannot be entirely excluded. It will be important to determine whether mutant nicotinic receptors lacking the negatively charged residues near the channel vestibule possess proton-sensitive single-channel conductance values, since chemical modification experiments have yet to conclusively identify the nature of the ionizable groups governing conductance. In contrast to single-channel conductance, the mean channel open time has a biphasic pH sensitivity with a maximum near physiological pH (Fig. 1B) (Landau et al., 1981; Palma et al., 1991; Bouzat et al., 1993). This has been interpreted to suggest at least two titratable groups (species-specific pK_a of roughly 6 and 9) influence channel gating. Although the identity of these groups has not yet been established, chemical modification of presumed extracellular surface ω-carboxylic acid groups with trimethyloxonium removed the pH dependence of single-channel open time (Pappone and Barchfeld, 1990). Slow receptor desensitization also appears to be proton sensitive, being accelerated for *Torpedo* receptors (but not receptors containing the mouse β subunit) at both alkaline and acid pH values (Li and McNamee, 1992). These effects of protons are largely independent of voltage and are not associated with shifts in the reversal potentials for receptor-gated current responses (Ritchie and Fambrough, 1975; Mallart et al., 1976; Trautmann and

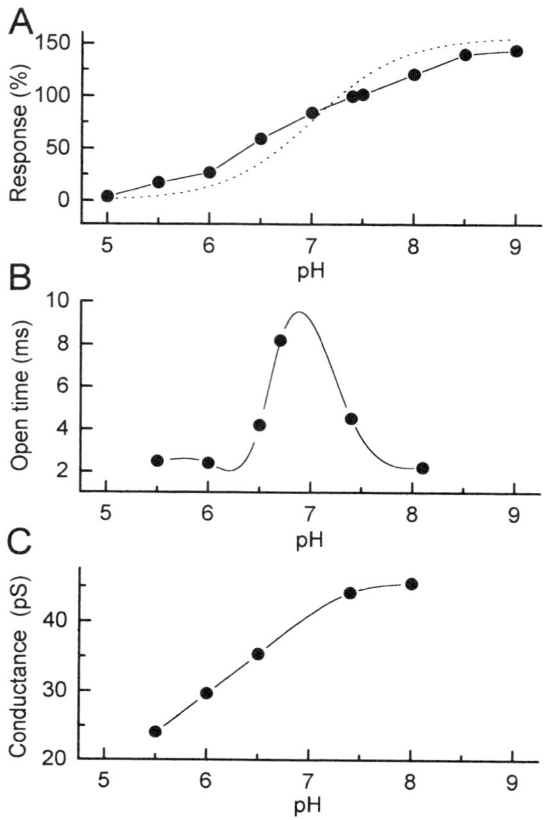

FIGURE 1. A. Reduced extracellular pH inhibits and alkalosis enhances peak acetylcholine-induced (1 μM) currents in voltage-clamped oocytes expressing receptors from *Torpedo* at a holding potential (V_{HOLD}) of -80 mV. Current responses were normalized to the value measured at pH 7.4 (data from Palma et al., 1991 with permission from Springer-Verlag, New York). Broken line shows the expected relationship for a single binding site. **B.** The mean acetylcholine channel open lifetime (BC3H-1 cells) has an optimum near physiological pH ($V_{HOLD} = -100$ mV; minimum resolution 113 μs). **C.** Single-channel conductance of acetylcholine receptors in BC3H-1 cells (determined from ramping V_{HOLD} from the reversal potential to -200 mV) is reduced by acid pH. (B) and (C) are modified from Pappone and Barchfeld (1990). The Journal of General Physiology 96, pp. 1–22 by copyright permission of The Rockefeller University Press.

Zilber-Gachelin, 1976; Landau et al., 1981). However, some investigators have interpreted proton-induced shifts in the relationship between voltage and endplate current decay time to suggest that changes in proton concentration alter the membrane surface charge, thereby altering channel function (Mallart and Molgo, 1978; Dreyer et al., 1979)

Intracellular proton control of acetylcholine receptor function. Although it is well established that muscarinic receptor activation decreases intracellular pH (e.g., Lau et al., 1989; Seo et al., 1994), there are also recent suggestions that acetylcholine-gated ion channels can lower intracellular pH by an undetermined and probably indirect mechanism (Rosario et al., 1991; Grassi et al., 1993; but see Del Castillo et al., 1992). Thus, if activation of acetylcholine-gated channels facilitates the transmembrane movement of a sufficient number of acid equivalents, these receptors may influence the pH within the narrow confines of the extracellular space. In addition, there is also some evidence for inhibition of acetylcholine receptor function by ammonia-induced changes in intracellular pH (Goldberg and Lass, 1983). This observation fits with the anomalous behavior of extracellular NH_4^+ ions noted in permeation studies of acetylcholine receptors in muscle, which might in part reflect intracellular alkalization (Takeda et al., 1980; Adams et al., 1981). These results raise the possibility that intracellular acidification during intense muscle activity may contribute to muscle fatigue by inhibiting acetylcholine receptor function (Pan et al., 1988; Renaud, 1989). Moreover, these findings raise the possibility that a portion of the apparent extracellular actions of pH on acetylcholine receptor function described above might be mediated by intracellular acidification in cells in which extra- and intracellular pH are tightly coupled.

pH sensitivity of acetylcholine receptor ligands and modulators. In contrast to the effect of protons on single-channel properties, agonist potency determined physiologically appears pH-insensitive, even though (as usual) agonist and antagonist binding in isolated membranes show distinct pH optima (Huang et al., 1978; O'Brien and Gibson, 1974; Schmidt and Raftery, 1974; McLaughlin and Hawrot, 1989; also Park and Bursztajn, 1990). However, the acetylcholine receptor in muscle is also selectively regulated by other ionizable compounds, including certain negatively charged lipids and local anesthetics (Kordas, 1970; Neher and Steinbach, 1978; Fong and McNamee, 1986, and references therein). Because the phosphate group in phospholipids and secondary or tertiary amino groups in anesthetics have pK_a values near physiological pH, the concentration of charged species is obviously pH-sensitive. This simple fact has been exploited to gain insight into the mechanisms underlying proton inhibition as well as the actions of these molecules.

Spectroscopic approaches have demonstrated that the association of acetylcholine receptors with phosphatidic acid decreases the pK_{a2} of the phosphate by nearly 2 log units from 8.5 to 6.6, independent of ionic strength (Fig. 2A). Moreover, titration of these specific phospholipids may be responsible for altering the function of the receptor (Fig. 2B), since only receptors reconstituted into membranes that contain phosphatidic acid show pH sensitivity. In addition, phospholipase D can convert inactive receptors in bilayers lacking negatively charged lipids to active receptors, presumably through the production of phosphatidic acid (see Bhushan and McNamee, 1993). There are also hints that other lipids (stearic acid and phosphatidylserine) may play key pH-dependent roles in receptor regulation (Raines and Miller, 1993). Thus, lipid–protein interactions are critical to receptor function, and may constitute a pH sensor that controls acetylcholine receptor function.

Like charged phospholipids, ionizable anesthetics also influence acetylcholine receptor function in a pH-sensitive fashion that appears to reflect their charged state (uncharged species more effective; Horvath et al., 1990). Conversely, local anesthetics lacking an ionizable group do not show pH dependence. These findings are consistent with the pH-dependent potency and rate of action of ionizable local anesthetics acting at voltage-dependent Na^+ channels (Hille, 1977). However, there is a slight twist. In contrast to a presumed internal site of action for local anesthetic block of Na^+ channels, this class of compounds is thought to act from the outside of the nicotinic receptor, suggesting there are important differences in the way voltage- and ligand-gated receptors interact with local anesthetics (discussed in Hille, 1992).

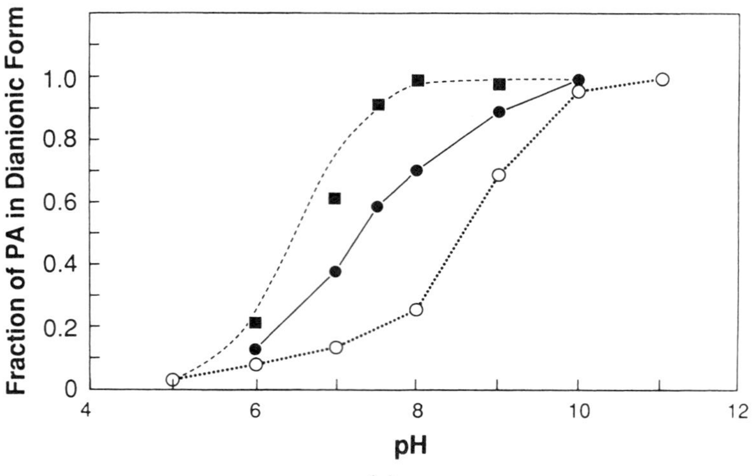

(a)

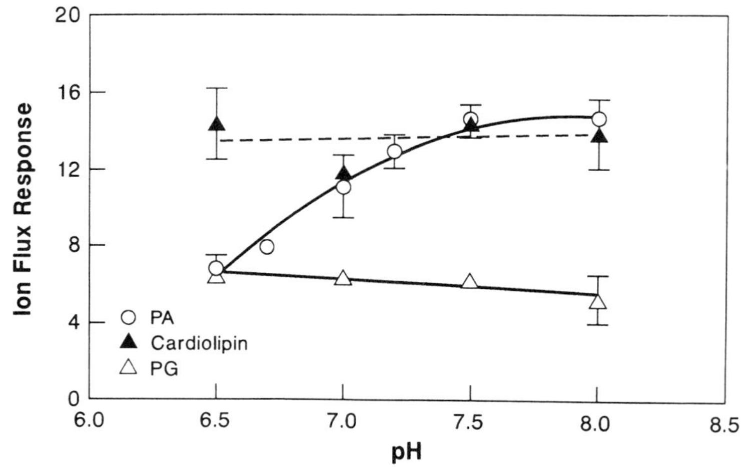

(b)

FIGURE 2. A. Titration of pure phosphatidic acid (PA) (open circles), and PA in reconstituted membrane samples (closed circles) that include acetylcholine receptors purified from *Torpedo* electroplax, as determined by Fourier transform infrared spectroscopy. Filled squares represent receptor-bound phosphatidic acid calculated as a function of pH. **B.** Purified *Torpedo* acetylcholine receptors are only sensitive to pH when reconstituted into membranes containing 20% PA (but not cardiolipin or phosphatidylglycerol). Receptor responses to carbamylcholine where measured using a $^{86}Rb^{+}$ flux assay (reproduced from Bhushan and McNamee, 1993, with permission from the Biophysical Journal).

pH Sensitivity of GABA Receptors

GABA is the dominant fast inhibitory neurotransmitter in the brain, and activates a Cl^- channel that usually mediates a hyperpolarization (reviewed by Kaila, 1994). Receptors responding to GABA are widespread, and are comprised of multiple subunits drawn from a large gene family. $GABA_A$ receptors are bicuculline-sensitive ion channels that have the unique property among ligand-gated channels of directly mediating intra- and extracellular pH changes through the transmembrane ion fluxes they catalyze. In order to fully appreciate the nature of the relationship between $GABA_A$ receptor function and pH, it is important to recognize that an extensive literature exists describing the permeability of mammalian and invertebrate GABA receptors to small weak-acid anions (see Chapter 16–17). Because the neutral form of these molecules are passively distributed across the plasma membrane, extra- and intracellular concentrations of the anionic form will be dictated by the transmembrane pH gradient, which is a function of a variety of active processes. That is, under constant partial pressure of CO_2, the equilibrium potentials for HCO_3^- and H^+ are identical. The transmembrane proton gradient (inside acid—pH 7.1–7.2) translates into an outward driving force for bicarbonate near resting membrane potentials. Bicarbonate can permeate $GABA_A$ channels (P_{HCO3}/P_{Cl-} ~0.2–0.3; see Kaila, 1994), and the efflux of this anion associated with the inwardly-directed current component during channel opening causes an intracellular acidification and extracellular alkalization at a rate governed by the carbonic anhydrase activity. Interestingly, this process is known to occur during normal GABAergic synaptic transmission (Kaila et al., 1992, 1993). Perhaps surprisingly, at membrane potentials near E_{Cl-}, a considerable portion of the GABA-mediated inwardly directed component of current is carried by bicarbonate (Voipio et al., 1991; Kaila et al., 1993). Thus, the unusual ability of $GABA_A$ receptors to control the pH of their microenvironment raises the obvious question of self-regulation, and provides a compelling reason to study in detail the $GABA_A$ receptor's pH sensitivity.

Extracellular proton control of GABA receptor function. A considerable body of evidence suggests that invertebrate GABA-mediated Cl^- conductances are inhibited by alkaline pH and potentiated at acidic pH over the range of pH 5 to 9 (1.3 nM to 13μM H^+; see Fig. 3) (Takeuchi and Takeuchi, 1967; Smart and Constanti, 1982; Kaila 1994), but are insensitive to intracellular pH. This enhancement is independent of the agonist concentration, and appears correlated with a prolongation of synaptic currents. Changes in synaptic kinetics have been suggested to reflect a change in channel gating, although alternative possibilities (e.g., effects on desensitization, and transmitter release and removal) remain to be investigated (Pasternack et al., 1992; Onodera and Takeuchi, 1979). $GABA_A$ receptor function in mammalian neurons also appears to be influenced by extracellular pH (Gruol et al., 1980; Gallagher et al., 1983; Tang et al., 1990; Smart, 1992; Sun & Reis, 1993; Robello et al., 1994; Krishek et al., 1996; see also Enna & Snyder, 1975). In most neurons, low pH enhances GABA receptor function with minimal changes on GABA EC_{50} or desensitization (Robello et al., 1994; Pasternack et al., 1996), although protons increase the EC_{50} of hippocampal receptors with high affinity for GABA (Pasternack et al., 1996).

Like most neurotransmitter receptors, the GABA receptor is comprised of several subunits, and molecular techniques have revealed multiple classes of subunit that can combine to form heteromeric receptors with different pharmacological and biophysical properties (Seighart, 1995). The opposing effects of pH changes observed in different preparations and for different agonist concentrations (Gallagher et al., 1983; Tang et al., 1990; Pasternack et al., 1996) suggest that the nature of pH effects depend on receptor subunit composition. Molecular data addressing the relationship between protons and recombinant $GABA_A$ receptor subunits supports the idea that subunit composition controls receptor pH sensitivity, since different subunit combinations show widely varying pH sensitivity.

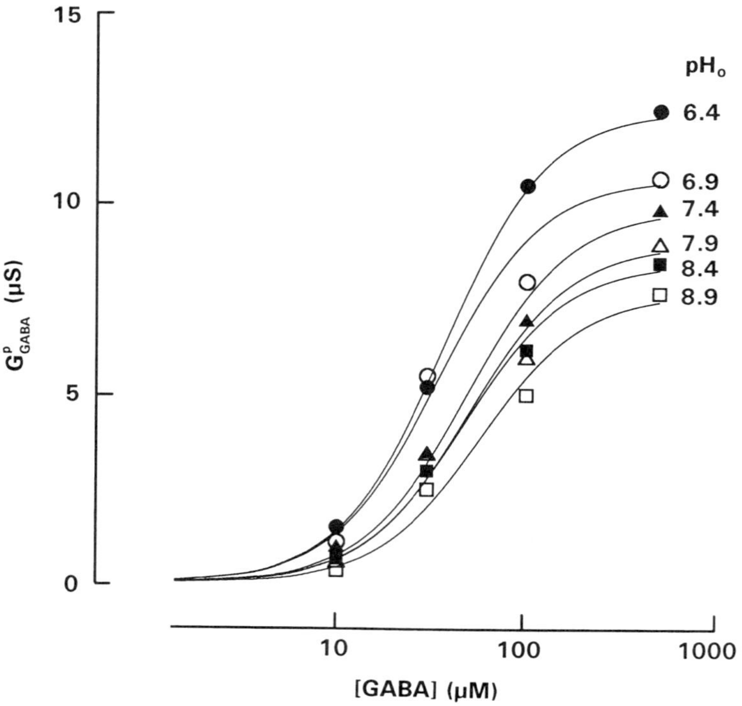

(a)

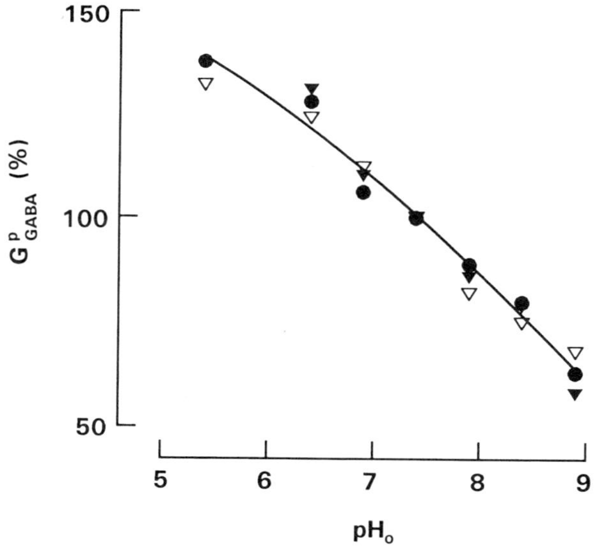

(b)

FIGURE 3. A. Concentration-effect curves are shown for activation of crayfish muscle GABA receptors at varying external pH. **B.** The conductance of crayfish muscle in response to 1 mM GABA is plotted against extracellular pH. Conductance values are expressed as a percentage of those observed at pH 7.4. Data are from three experiments. Reprinted from Pasternack et al., 1992, with kind permission from Elsevier Science Ltd, The Boulevard, Langford Lane, Kidlington OX5 1GB, UK.

Low extracellular pH enhances the function of receptors comprised of $\alpha1\beta1$ (pKa 5.5, 7.5) and $\alpha1\beta1\delta$ (pKa 6.6, 9.4), with titration curves best fit by two sites (Krishek et al., 1996). Thus, the δ subunit increases the pH sensitivity more than an order of magnitude. In contrast, homomeric receptors comprised of $\rho1$ subunit are inhibited at low pH (Wegelius et al., 1996), whereas $\alpha1\beta1\gamma2S$ are pH insensitive (Krishek et al., 1996). Receptors comprised of $\alpha1\beta2\gamma2S$ or $\alpha1\beta1\gamma2S\delta$ exhibit biphasic pH sensitivity, being inhibited at both high and low pH values outside the physiological range (pKa 5.4 and 9.4). These diverse effects of pH on the GABA receptors of differing subunit composition should facilitate molecular characterization of the nature of the proton sensor, which may ultimately provide clues describing how protonation at the extracellular face of this macromolecular structure controls ion channel opening.

What might proton regulation of GABA receptors mean for brain function? The facilitatory effects of increased proton concentration on some $GABA_A$ receptors and the inhibitory effects of protons on glutamate receptors (see p 426) suggest a synergistic relationship between pH and neuronal excitability (Aram and Lodge, 1987; Balestrino and Somjen, 1988; Jarolimek et al., 1989; see also Chapters 17–20). In addition, the $GABA_A$ receptor-mediated alkaline shifts in extracellular pH will tend to potentiate local NMDA receptor responses and potentiate or inhibit GABA receptor function (depending on subunit composition; see Krishele et al., 1996; Wegelius et al., 1996), which emphasizes, to no one's surprise, that a complicated region-specific relationship exists between pH and synaptic excitability. Finally, the acid enhancement of certain GABA receptors could be relevant during pathophysiological conditions, which are known to be associated with pronounced extracellular acidification (see Chapters 28–33).

pH sensitivity of GABA receptor ligands. In contrast to the typical pH optimum for agonist binding to isolated membrane fractions (e.g., Rutherford et al., 1987), the pK_a values for GABA (4 and 10.6) do not dictate any physiologically relevant pH-dependent changes in its ionization (Krishek et al., 1996). However, a variety of antagonists do show distinct pH optima for binding to isolated membranes, which may reflect the ionization of these compounds (Heaulme et al., 1987; Brown et al., 1989). The competitive antagonist bicuculline has a pK_a of 6.48, consistent with the presence of two ionic species (~10% protonated) at physiological pH (Andrews and Johnston, 1973; Smart and Constanti, 1981). The apparent higher potency of the positively charged N-methylbicuculline versus bicuculline in the mammalian CNS primarily reflects the relative amounts of cations at physiological pH. Similarly, pentobarbital (pK_a 8.1) is only about 17% ionized (negative) at physiological pH (7.4), and its actions are enhanced by acidic pH and attenuated at alkaline pH (Robertson, 1989).

Benzodiazepine receptor ligands have long been known to bind in a pH-dependent manner, and these data have been interpreted to suggest than an amino acid residue, such as histidine, is an important determinant of benzodiazepine binding. Consistent with this idea, a variety of selective and nonselective reagents that covalently modify presumed histidine groups can block benzodiazepine binding (e.g., Lambolez and Rossier, 1987; Uusi-Oukari, 1992; Maksay, 1992). A single histidine residue has been identified that appears to control benzodiazepine actions, although it is unclear whether the ionization of this residue accounts for the reported pH sensitivity for benzodiazepine binding (Wieland et al., 1992; Kleingoor et al., 1993).

pH Sensitivity of Glutamate Receptors

Glutamate-gated cation channels mediate excitatory synaptic transmission throughout the central nervous system. These receptors have been classified on the basis of their pharmacology, molecular structure, and biophysical properties into three subtypes. Of these, only the proton-sensitivity of NMDA (N-methyl-D-aspartate) and AMPA (α-amino-3-hydroxy-5-methyl-4-isoxazole propionic acid) receptors has been examined in detail.

Extracellular proton control of neuronal NMDA receptor function. NMDA receptors have attracted widespread interest because of their roles in neuronal development, synaptic transmission and plasticity, learning, and memory (Stevens, 1993; McBain and Mayer, 1994). In addition, NMDA receptor activation is thought to be a causative factor in ischemia-induced neuronal damage secondary to increased levels of extracellular glutamate (Choi, 1992; Attwell et al., 1993; Tombaugh and Sapolsky, 1993; Zorumski and Olney, 1993; see also Chapter 19). Similarly, a link between NMDA receptors and epilepsy has been suspected for more than a decade (Dingledine et al., 1990; Meldrum, 1992). Given the potentially harmful actions of NMDA receptors, one might predict them to be tightly regulated. Indeed, in addition to requiring two distinct agonists, NMDA receptor function is controlled by a host of endogenous ions, including physiological concentrations of H^+.

Although there were hints as early as 1980, 10 years passed before a variety of investigators (all working from different angles) demonstrated that protons directly inhibit NMDA receptors with an IC_{50} value that corresponds to physiological pH, or about 50 nM H^+ (Fig. 4) (Gruol et al., 1980; Giffard et al., 1990; Tang et al., 1990; Traynelis and Cull-Candy, 1990; Vyklicky et al., 1990; Takadera et al., 1992; Rajdev and Reynolds, 1993; Ou-Yang et al., 1994; Yoneda et al., 1994; see also Wu and Christensen, 1996). The ability of nanomolar concentrations of protons to control NMDA receptor function is intriguing for at least three reasons. First, there are more data to suggest that the extracellular H^+ concentration is dynamic during normal brain function than for any of the other allosteric modulators of NMDA receptors (see Chapters 17–19). Because the IC_{50} for proton inhibition is near physiological pH, alterations in pH during synaptic transmission will probably affect NMDA receptor function. It is interesting to note that glutamate receptor activation appears to indirectly mediate a rapid alkaline extracellular pH shift, followed by a slower acid shift (see Chapter 17). Second, because metabolically linked decreases in interstitial pH are a hallmark of ischemia and NMDA receptors are thought to mediate some stroke-induced damage, the H^+ inhibition of receptor function seems uniquely qualified to provide neuroprotective negative feedback (Siesjö, 1985; Obrenovitch et al., 1990; Tombaugh and Sapolsky, 1993; Chapters 28 and 29). Consistent with this idea, mild acidic conditions reduce neuronal death *in vitro* during exposure to glutamate (Kaku et al., 1993; but see O'Donnell and Bickler, 1994). Intense neuronal activity during seizures also leads to the acidification of the interstitial space, which may act as negative feedback to attenuate overactivation of glutamate receptors. Interestingly, experimentally induced decreases in interstitial pH by as little as 0.1–0.2 unit increase seizure threshold (Aram and Lodge, 1987; Balestrino and Somjen, 1988). Third, the proton IC_{50} (60 nM) and association rate suggest that NMDA receptors exist in rapid equilibrium between a fully active (unprotonated) and inactive (protonated) state (Prod'hom et al., 1987; Traynelis and Cull-Candy, 1991).

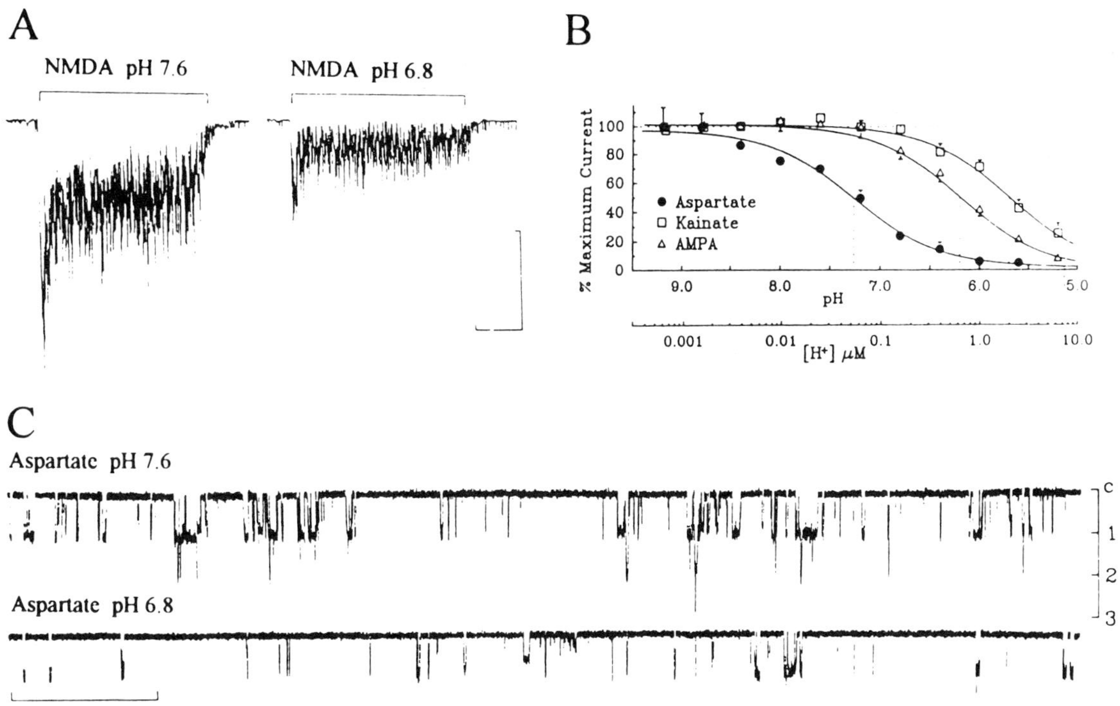

FIGURE 4. **A.** Whole-cell currents activated by NMDA plus glycine in rat cerebellar granule cells. Calibration is 50 pA and 3 seconds ($V_{HOLD} = -80$ mV). **B.** Composite pH inhibition curves are superimposed for whole-cell currents recorded from cerebellar granule cells in response to aspartate plus glycine (IC_{50} pH 7.3), AMPA (IC_{50} pH 6.3), and kainate (IC_{50} pH 5.7); the horizontal axis reflects H^+ activity. **C.** Protons decrease the frequency of aspartate-activated single-channel openings in outside-out membrane patches ($V_{HOLD} = -80$ mV). The number of open channels (1, 2, 3) indicated by vertical bar shown to the right correspond to 4, 8, 12 pA and the time scale is 300 ms; c indicates the channel's closed state. Data are from Traynelis and Cull-Candy, 1991.

How do protons inhibit NMDA receptors? The available data rules out more possible mechanisms than it suggests. The mechanism of proton inhibition of NMDA receptors appears independent of voltage, rapid receptor desensitization, single-channel conductance, and individual channel open time (Giffard et al., 1990; Tang et al., 1990; Traynelis and Cull-Candy, 1990; Vyklicky et al., 1990). Although the co-agonists glutamate and glycine have distinct pH optima for binding in isolated membranes (Head et al., 1980; Monahan and Michel, 1987; McDonald et al., 1990), the minimal effects that protons exert on functionally determined agonist potency cannot account for the inhibition by acid pH. In addition, only extracellular pH changes are capable of inhibiting channel function (Traynelis and Cull-Candy, 1991; Wu and Christensen, 1996), which occurs through a decrease in single-channel opening frequency (Fig. 4C), with only modest changes in the kinetic burst structure of individual receptor openings. These data suggest as a working hypothesis that protonation of a pH sensor (likely candidates being ionizable amino acid side chains, but see Bhushan and McNamee, 1993) on the external face of the receptor inhibits the transition to a nonconducting protein conformation that precedes agonist-induced channel opening.

Extracellular proton control of recombinant NMDA receptor function. One of the most important concepts to emerge from the molecular biology of neurotransmitter receptors is the large degree of receptor diversity. For glutamate receptors, the identification of cDNAs coding for multiple subunits has provided a means to explore the structural, pharmacological, and biophysical differences between receptor subtypes (Nakanishi, 1992; Seeburg, 1993; Hollmann and Heinemann, 1994). Of the various subunits identified, the first subunit discovered (denoted NMDAR1 or NR1) contains the glycine binding site, as well as the pH sensor. Eight functional alternatively spliced NR1 variants have been described (Fig. 5A), and examination of the effect of NR1 splicing on proton inhibition suggests that inclusion of the 21 amino acids encoded by exon5 relieves the tonic inhibition present at physiological pH. Moreover, this effect of exon5 is voltage-independent, which distinguishes it from pH-linked alterations in membrane surface charge (Traynelis et al., 1995). Interestingly, sequence analysis predicts that alternative exon5 exists as a flexible loop that (like the pH sensor) resides at the extracellular surface of the protein (Fig. 5 B,C). The proton sensitivity of heteromeric receptors, comprised of

FIGURE 5. A. The rat NR1 subunit is constructed from 22 exons, 3 of which can be alternately spliced (bold) to form 8 isoforms (arrowheads are stop codons). Exon3 contains a stop codon, and is not shown. Many of the amino acids encoded by exon5 contain pi molecular orbital system ("). Secondary structure was predicted as described by Rost and Sander (1993); similar results were obtained by other prediction algorithms (Chou and Fasman, 1978; Garnier et al., 1978). A weak helix at Asp-200 to Leu-202 was omitted (see Fig. 6). Downward bars represent increased probability of being at the protein surface (Emini et al., 1985). Sequence and intron-exon structure are from Hollmann et al. (1993). **B.** The composite proton inhibition curves for NR1 ± exon5 homomeric receptors expressed in *Xenopus* oocytes have fitted IC_{50} values of 50 and 305 μM H^+ (activity coefficient = 0.8); the physiological range is indicated by a superimposed box, and the arrow shows pH 7.3. (A) and (B) reprinted with permission from Traynelis et al., (1995). Science 268: 873–876. Copyright 1995 American Association for the Advancement of Science). **C.** The pH sensitivity of responses induced by glutamate plus glycine in NR1-exon5/NR2A subunits coexpressed in HEK 293 cells; bars indicate agonist application (1 mM Ca^{2+} and nominally Mg^{2+} free; $V_{HOLD} = -60$ mV; S.F. Traynelis and S.F. Heinemann, unpublished data).

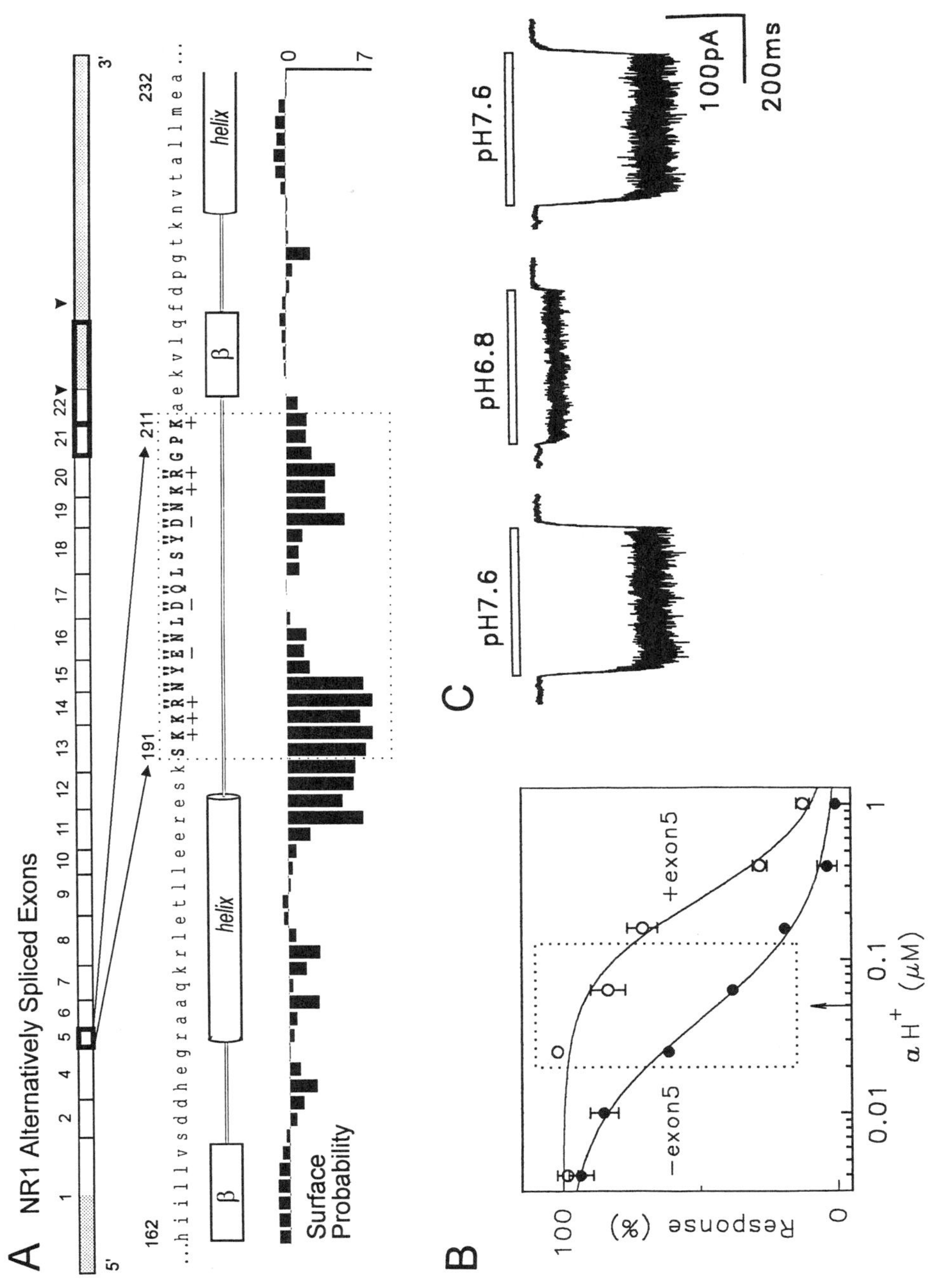
A
NR1 Alternatively Spliced Exons
5'
3'
1
2
4
5
6
7
8
9
10
11
12
13
14
15
16
17
18
19
20
21
22
162
191
211
232
...h i i l l v s d d h e g r a a q k r l e t l l e e r e s k S K K R N Y E N L D Q L S Y D N K R G P K a e k v l q f d p g t k n v t a l l m e a ...
β
helix
β
helix
Surface Probability
0
7
B
100
Response (%)
0
+exon5
−exon5
0.01
0.1
1
a H+ (μM)
C
pH7.6
pH6.8
pH7.6
100pA
200ms

NR1 + NR2 subunits (in the absence of Mg^{2+}), is similar to homomeric NR1, except that the NR2C abolishes all pH sensitivity, adding another level of transcriptional control to the effects of extracellular protons on receptor function.

One way alternative NR1 exon5 could influence the NMDA receptor pH sensitivity would be to introduce residues close enough to the H^+ sensor either to directly shield it or to alter access to it. Interestingly, there is a strong correlation between the relief of proton sensitivity by exon5 and ionic strength, consistent with the shielding hypothesis. In further support of this idea, replacement of a single charged amino acid (Lys-211) with the uncharged residues glutamine or glycine appears to decrease exon5-induced shifts in the proton IC_{50} (Traynelis et al., 1995). It is important to note that the opposite effects of positively charged Lys-211 and protons are compatible with the very distinct nature of these similarly charged ions. Borrowing from the story of K^+ channel inactivation (e.g., Hoshi et al., 1990), it seems reasonable that exon5 could behave as a constitutive tethered modulator of the protein. That is, an allosteric binding site may exist for some portion of the exon5 encoded surface loop, and occupation of this binding site might stabilize the unprotonated state of the pH sensor (Fig. 6). This proposal predicts that compounds should exist which mimic the effects of this loop. Consistent with this prediction, Mg^{2+} can relieve the pH dependence of exon5-lacking (and NR2B-containing) receptors (Paoletti et al., 1995). Similarly, histamine potentiation and voltage- and glycine-independent polyamine potentiation of NMDA responses are known to occur only in proton-sensitive heteromeric receptors that contain the NR2B subunit and lack the NR1 exon5 (Durand et al., 1992, 1993; Bekkers, 1993; Vorobjev et al., 1993; Williams, 1994; Johnson, 1996). Mg^{2+}, polyamine, and histamine potentiation seem to reflect the relief of tonic proton inhibition, perhaps through their association with part of an exon5 extracellular surface recognition site (Johnson, 1996). Moreover, most mutations in the NR1 subunit that alter polyamine sensitivity also alter the receptor's pH sensitivity (Sullivan et al., 1994; Traynelis et al., 1995; Kashiwagi et al., 1996; Williams et al., 1995). These data suggest that competitive polyamine antagonists might serve as the starting point for the development of a new class of compounds that can control the functional consequences of RNA splicing and proton inhibition of the NMDA receptor NR1 subunit.

FIGURE 6. Two hypotheses describing relief of H^+ inhibition by exon5. The far left panel illustrates an assumed pentameric subunit configuration of the NMDA receptor complex, with a cross-sectional area of a single subunit (labeled NR1) highlighted with a box. The effects of NR1 exon5 are shown in two expanded panels: Model I (*top*) and Model II (*bottom*).

Left panels in Models I and II. A small region surrounding the exon5 splice site is assumed to be at the extracellular surface of the protein (Traynelis et al., 1995). This region is predicted to exist within a surface loop, rather than within a helix or sheet. Ser-184 and Glu-192 are shown in all panels; in the presence of exon5, Glu-192 becomes Glu-213 due to the insertion of 21 amino acids. The model predicts that there exists (for simplicity) a single ionizable functional group (denoted X) that influences channel gating. A partial negative charge (δ^-) is shown on X, since it is unknown whether it is neutral or negative in the unprotonated state. K_{acid} is the equilibrium constant for the interaction of unknown group X with protons; in Model II, group X is distant to the exon5 splice junction.

Right Panels in Models I and II. The 5th exon is hypothesized to extend a flexible loop that does not alter the underlying secondary structure. Lys-211 either directly interacts with or shields the proton sensor to stabilize the unprotonated state (*Model I, top*), or indirectly alters the charge distribution surrounding the proton sensor to change its pK_a (*Model II, bottom;* e.g., Coulter et al., 1995). Other NR1 residues known to influence the proton sensor (C744, C798, D342; Sullivan et al., 1994; Williams et al., 1995) are not shown.

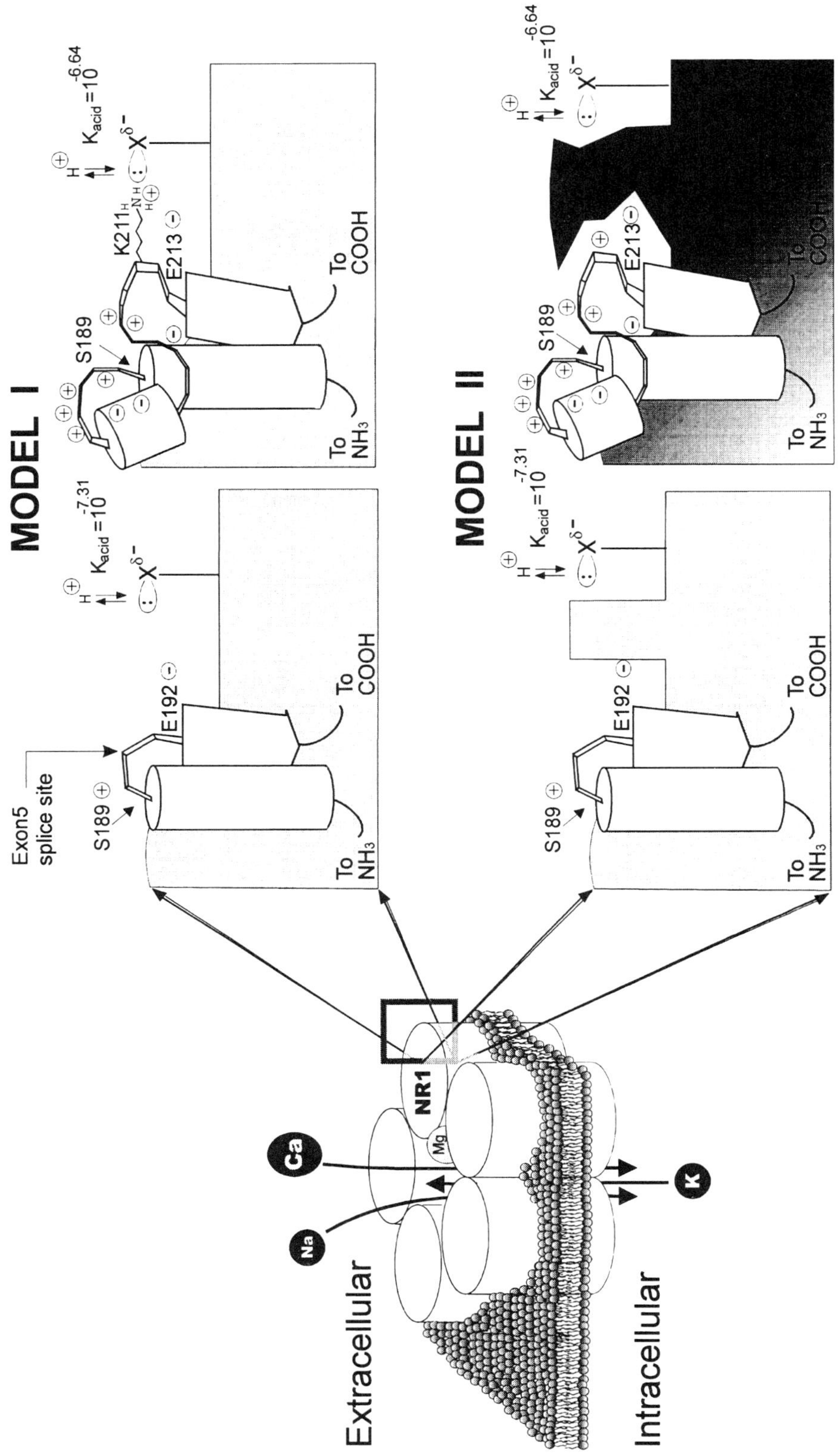

Extracellular
Intracellular
Na
Ca
K
Mg
NR1
Exon5 splice site
S189
E192
To NH3
To COOH
MODEL I
MODEL II
K211
E213
$K_{acid} = 10^{-7.31}$
$K_{acid} = 10^{-6.64}$
$X^{\delta-}$

What implications does the differential proton sensitivity of NMDA receptors hold for neuronal function? Exon5-containing NR1 and NR2 subunit mRNAs are differentially distributed in the CNS throughout development (Standaert et al., 1993; Laurie and Seeburg, 1994). Proton-insensitive NMDA receptors (e.g., NR1+exon5-containing receptors) appear to be fully active at physiological pH, in contrast to receptors that contain NR1-exon5, which are subject to tonic inhibition of about 50% (Sugihara et al., 1992; Hollmann et al., 1993; Traynelis et al., 1995). In addition to having more responsive NMDA receptors, neurons that splice exon5 into the NR1 subunit will also be insensitive to pH changes during normal and pathological functions. One consequence of this differential pH sensitivity might be that proton-insensitive receptors endow their neurons with a greater susceptibility to overactivation (and subsequent damage) during ischemia or seizures, since they are insensitive to negative feedback imposed by the falling pH each condition produces.

pH sensitivity of NMDA receptor ligands. The ionization of the co-agonists glutamate (free solution pK_a values of 2.2, 4.3, and 9.7) and glycine (pK_as of 2.4 and 9.8) is not altered over the pH range under consideration here. However, this is not true for all NMDA receptor ligands, such as ketamine, whose actions are pH-dependent (MacDonald et al., 1991). In addition, the majority of competitive NMDA receptor antagonists are amino acids that contain a negatively charged group positioned four to six atoms away from the α-carbon (Olverman et al., 1988). Active ω-substitutes include $-$COOH and $-PO_3H_2$, and in every case studied, antagonists containing ω-PO_3H_2 are more potent than their ω-COOH counterparts (see Benveniste and Mayer, 1992). The $-$COOH group is planar (sp^2 hybridization) and contains a single negative charge, whereas the $-PO_3H_2$ group is tetrahedral (sp^3 hybridization) and has two negative charges when ionized. Because the antagonist's carboxylic acid and amino groups are ionized at physiological pH (and their charges cancel), the net charge on the molecule will be that of the ω-COOH or ω-PO_3H_2. In contrast to the virtually complete ionization of ω-COOH at physiological pH (yielding a net charge of -1 on the molecule), the pK_a values of, for example, the phosphono group on 2-amino-7-phosphonoheptanoic acid (AP7; pK_{a_1} = 2.6, pK_{a_3} = 7.7; Chenard et al., 1990) indicate that two ionized species of ω-PO_3H_2 will exist at physiological pH, with net molecular charges of -1 and -2. Benveniste and Mayer (1992) examined whether the different ionized forms of ω-PO_3H_2-containing antagonists have different antagonist dissociation constants ($K = K_{off}/K_{on}$; see also Murphy et al., 1987; Fagg and Baud, 1988). Potency of antagonists that differed only by their ω-functional groups was measured at pH values that alter the ratio of doubly to singly charged ω-PO_3H_2 groups but do not affect the ionization of ω-COOH (or the rest of the molecule). They found that increases in pH decreased the K_i values for ω-PO_3H_2 but not ω-COOH antagonists without changing receptor function, suggesting that the doubly charged ω-PO_3H_2 group is the more active species.

These results are important for two reasons. First, the use of compounds whose ionization is not constant over the physiological range is complicated during pathological conditions associated with altered pH values, such as ischemia. This is because the concentration of the active species, and thus the apparent effectiveness, may be reduced. For example, at physiological pH, less than a third of the AP7 molecules contain two negative charges and act with high potency. A modest decrease of 0.5 pH units (to pH 6.8) during experimentally induced ischemia would reduce this percentage threefold to about 10%, a fact that clearly requires consideration in both experimental design and therapeutic strategies. Second, a consideration of the structure of the active species of antagonists should lead to the development of more selective compounds.

Extracellular proton control of AMPA receptor function. Like NMDA receptors, neuronal and recombinant AMPA receptors are inhibited by extracellular protons. However, the reported IC_{50} values for proton inhibition of these receptors range from pH 5.7 to 6.3 (Fig. 4B) (Christensen and Hida, 1990; Traynelis and Cull-Candy, 1991, 1995), making this form of allosteric modulation of more biochemical than physiological interest. It is noteworthy that this inhibition shares many features with that of NMDA receptors, including voltage- and agonist-independence (but see Miralles and Olmos, 1989; Yoneda et al., 1994, for pH-dependent binding). Perhaps the main features of the proton sensor and its actions on channel gating are conserved across glutamate receptor subfamilies.

pH Sensitivity of Nucleotide-Gated Channels

A consideration of the proton regulation of ligand-gated channels would not be complete without a discussion of ion channels gated by intracellular and extracellular nucleotides. One shared feature that links the pH sensitivity of these structurally diverse channel classes is the effect of pH on agonist binding and/or activation, which may reflect the electrostatic nature of the nucleotide-binding pocket.

pH sensitivity of ATP-regulated K^+ channels. There are numerous therapeutically interesting compounds that can either activate or inhibit ATP-sensitive K^+ (K^+-ATP) channels through actions on the associated sulfonylurea receptor (Inagaki et al., 1995; Aguilar-Bryan et al., 1995; Thomas et al., 1995). These drugs act in a pH-sensitive fashion that is linked to their ionization and lipid solubility, as well as the pH-sensitivity of channel function (Findlay, 1992; Forestier et al., 1993; Jahangir et al., 1994; Akiyoshi et al., 1995; Moser et al., 1995). The sulfonylurea ion channel has been found in a variety of tissues, including vertebrate axons, central neurons, and brain microvascular endothelial cells (Ashcroft and Ashcroft, 1990; Ashford et al., 1990; Treherne and Ashford, 1991; Janigro et al., 1993; Trapp and Ballanyi, 1995). In addition to roles in pancreatic β-cells, skeletal muscle, and ventricular myocytes, it is suggested that K^+-ATP channels are involved in neurotransmitter release and the nervous system's response to hypoglycemia, anoxia, mitochondrial deficits, and pain (Murphy and Greenfield, 1991; Jiang et al., 1992; Tromba et al., 1992; Narita et al., 1992; Takano and Noma, 1993; Frey et al., 1993; Haddad and Jiang, 1994; Raffia and Codd, 1994; Roper and Ashcroft, 1995; Trapp and Ballanyi, 1995). At its normal concentration range inside the cell, endogenous ATP is an effective inhibitor for most K^+-ATP channels (IC_{50} values for different subtypes range roughly between 10–1000 μM). Because the ATP levels within many cell types fluctuate modestly, the effect of ATP will often be to exert tonic inhibition on the ion channel. How then does the cell regulate channel function? ATP gating is markedly dependent on a variety of intracellular substances, including intracellular pH, which is highly dynamic and linked to metabolism (see Chapters 5, 10, 11, 16; Busa and Nuccitelli, 1984). Its effects on K^+-ATP channels probably reflect (to differing degrees in different preparations) the pH dependence of the concentration of various ionization states of ATP ($pK_{a_4} = 6.8$), the ionization of the protein, and the intracellular channel surface charge. As is the case for acetylcholine receptors, the H^+ regulation of ATP-controlled K^+-channels is only now being examined in neurons (e.g., Trapp and Ballanyi, 1995), and is better understood in peripheral than in central receptors. Consequently, this discussion will include information about the relationship between ATP-regulated K^+ channel activity and intracellular pH from skeletal muscle, ventricular myocytes, and pancreatic β-cells.

It is important to recognize that ATP-regulated K^+ channels are functionally heterogeneous. Moreover, not all receptors possess the same pH sensitivity, and some have been suggested to be altogether proton-insensitive (Kunzelmann et al., 1989). Unlike acetylcholine-, GABA-, and glutamate-activated receptors, extracellular pH has little effect on ATP-sensitive K^+ channels. By contrast, acidic intracellular pH inhibits channel activity in pancreatic β-cells and myocytes in the absence of ATP with a half-maximal H^+ concentration in the range of 0.04–1.13 μm (pH 6.0–7.5) and a Hill slope of about 2 (Cuevas et al., 1991; Proks et al., 1994; Fan and Makielski, 1993; Frey et al., 1993; Allard et al., 1995). Inhibition of receptor function by acidic pH is not fully reversible, suggesting low pH may facilitate channel rundown (see Fan et al., 1994). Proton inhibition of ATP-regulated K^+ channels occurs with either no change or an increase in the mean open time, and a modest decrease in single-channel unitary conductance without a change in single-channel noise (e.g., Davies et al., 1992; Allard et al., 1995). This latter effect can be described by titration (pK_a 6.25) of the surface charge at the intracellular channel entrance (Vivaudou and Forestier, 1995). In addition, at least one report has described acid-induced subconductance levels at extreme pH values (Fan et al., 1994).

The effects of reduced pH are more physiologically interesting in the presence of ATP. Decreased intracellular pH reduces the extent of ATP inhibition of the channel and increases channel activity (Fig. 7) (Davies, 1990; Cuevas et al., 1991; Standen et al., 1992; Koyano et al., 1993; Proks et al., 1994; Allard et al., 1995; Vivaudou and Forestier, 1995), although the effect of experimental intracellular solutions (often high in Cl^-) is not clear. This may occur by a small change in the sensitivity to the receptor for ATP, which likely could reflect a change in the concentration of ATP^{4-}. In skeletal muscle, decreases in pH have larger effects on the ATP sensitivity, and it has been suggested that protons alter the binding of ATP, perhaps by competing with an ATP molecule for the inhibitory site on the channel protein. This is consistent with the reciprocal relationship that exists between the pH and ATP sensitivity, with ATP being a less effective inhibitor at low pH and low pH having more modest effects at higher ATP concentrations (e.g., Vivaudou and Forestier, 1995).

The pathological and physiological relevance of proton relief of tonic K^+-ATP channel inhibition is complex. Whereas K^+-ATP channels appear to be involved in stimulus secretion coupling in pancreatic β-cells (Ashcroft et al., 1993, 1994) and low intracellular pH relieves inhibition by ATP, the role of intracellular pH in the coupling of glucose metabolism to K^+-ATP channels is more controversial than that for the ADP/ATP ratio (Fan et al., 1994). Similarly, intracellular acidification does not appear linked to K^+-ATP channel function in metabolically inhibited rat uterus or in anoxic dorsal vagal neurons (Heaton et al., 1993; Trapp et al., 1996). By contrast, the intracellular pH will probably be reduced to levels that relieve ATP-induced inhibition of K^+ channels in muscle during extreme exercise or ischemia. This pH reduction can lead to electrical inexcitability, significant increases in extracellular K^+, and vasodilation, with important implications for cardiovascular disease and skeletal muscle physiology (Ashcroft and Ashcroft, 1990; Allard et al., 1995).

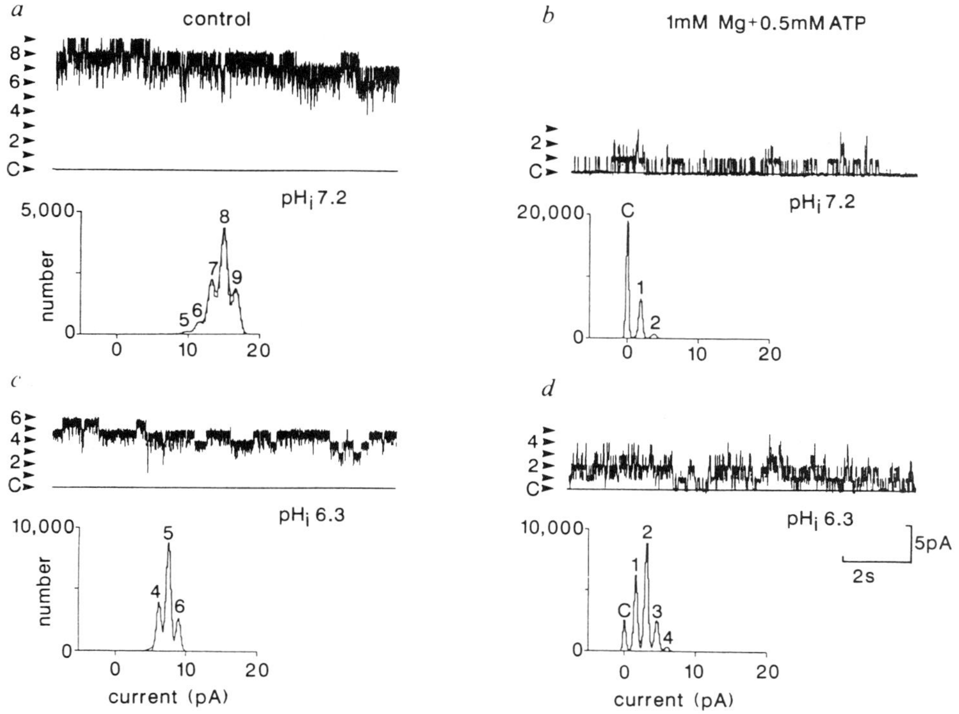

FIGURE 7. Effect of intracellular pH (pH$_i$ in figure) on the activity of ATP-sensitive K^+ channels in inside-out membrane patches excised from skeletal muscle. **A.** Channel activity in the absence of Mg^{2+} and ATP. The closed state is marked by "C," and the number of open channels is shown on both the raw current trace and the event amplitude histogram. **B.** 1 mM Mg^{2+} and 0.5 mM ATP strongly inhibited channel activity in this patch. **C.** Acidic pH also reduced channel activity in the absence of Mg^{2+}/ATP (compare with *A*), but, as shown in (**D**), relieved inhibition by Mg^{2+}/ATP (compare with *C*). Data from Davies, 1990, with permission from Nature.

pH sensitivity of channels gated by intracellular cyclic nucleotide channels. Ion channels gated by cyclic nucleotides are low-conductance cation-selective channels that are blocked by Ca^{2+}, which is a weakly permeant ion (Yau and Baylor, 1989; Colamartino et al., 1991; Zufall and Firestein, 1993). Several effects of pH changes on channel function have been described (Liebman et al., 1984; Tanaka, 1993), prominent among them being an apparent effect at the agonist binding site. For example, Gordon et al. (1996) have shown that protons potentiate rod cyclic nucleotide channels in an agonist-dependent manner. They suggest the amplification of cAMP responses results from protonation of the open state, although the data do not rule out effects on binding or additional states. Mutagenesis studies implicate an acidic residue (Asp-604) as one pH sensor. This negatively charged residue is thought to interact favorably with the highly efficacious agonist cGMP through hydrogen bonding, but unfavorably with a similarly positioned cAMP, unless pH is reduced and Asp-604 protonated (see Fig. 5 in Gordon et al., 1996). Consistent with this idea, Gordon et al. (1996) find that replacement of this aspartate with asparagine greatly reduces the effects of pH, also allowing cAMP to act as a full agonist.

Perhaps not unexpectedly this single residue does not account for all the effects of pH; additional mutagenesis studies have identified another residue (His-468) that also controls pH sensitivity.

Root and MacKinnon (1994) have exploited the kinetic isotope effect of deuterium to study proton transfer within the pore of catfish olfactory cyclic nucleotide-gated channels. Their experiments provide an elegant example of the regulation of single-channel conductance by ionization of presumed intrapore residues (see also Coulter et al., 1995). Cyclic nucleotide channels from olfactory epithelium contain an acidic residue (Glu_{333}) that is thought to be positioned within the channel in a fashion similar to that of the divalent ion binding P-region of Ca^{2+} channels; mutation of the corresponding residue (Asp-447) in K^+ channels alters that receptor's selectivity for K^+ and sensitivity to Ca^{2+} block (Tsien et al., 1987; Heginbotham et al., 1992; Root and MacKinnon, 1993). In ^2H_2O, cyclic nucleotide-gated channels possess three kinetically distinguishable subconductance levels, the relative proportions of which are a function of pH. By measuring the exit rate from the various open states as a function of pH, identical pK_a values (7.6) are obtained for two ionizable residues (presumed intrapore glutamate residues) that appear to control the conductance of the channel. These results are consistent with reports that the ion permeation properties of channels in photoreceptors are altered at acidic pH (Tanaka, 1993). They are intriguing because they suggest that ionization of only two of the four glutamate residues contributed by the respective subunits influences ion permeation. Moreover, the pK_a values for the two glutamate residues are independent, an observation that suggests intrapore shielding by permeating ions imparts important properties to ion channel permeation (see Chen and Eisenberg, 1993, for a discussion of intrapore shielding).

pH sensitivity of channels gated by extracellular ATP. Extracellular ATP is now understood to be an important excitatory neurotransmitter in the central and peripheral nervous systems. The P_{2X} class of receptors are ion channels gated by extracellular ATP and modulated by endogenous ions such as Ca^{2+}, Zn^{2+}, and protons. Physiological changes in pH have meaningful effects on receptor function in rat nodose ganglion neurons, dorsal root ganglion neurons, sensory neurons, and recombinant receptors expressed in *Xenopus* oocytes. Protons potentiate agonist-activated currents, which can reach values that are double those seen at physiological pH; inhibition of receptor function at alkaline pH is nearly complete, with a predicted pK_a for the inhibitory effect between 7.2 and 7.6 (30–80 nM H^+; Hill slope of 1.1) depending on the preparation (King et al., 1996; Li et al., 1996a,b,c). Unlike the K^+-ATP channels described above, the available data do not support the idea that the altered dose–response curve reflects a change in the concentration of different active ionic species (Li et al., 1996b). Rather, it appears that potentiation of ATP receptor responses results from a decrease in the EC_{50} for receptor activation without a change in the maximal achievable current. Proton potentiation appears to be independent of voltage, receptor desensitization, and intracellular acidification, although potentiating levels of H^+ lengthen the deactivation time course (Li et al., 1996c). Interestingly, the potentiating effects of Zn^{2+} on ATP receptors are enhanced at acidic pH levels through a decrease in Zn^{2+} EC_{50}, although the additive effects of pH and Zn^{2+} suggest independent sites of action. The relationship between pH and ATP receptor function may have physiological relevance for even modest changes in interstitial pH in the brain. In the periphery, the proton-mediated potentiation of ATP receptor function on nociceptive neurons may contribute to the transmission of sensory information in conditions of injury or inflammation accompanied by extracellular acidification (Li et al., 1996a,b).

CONCLUSIONS

There are a variety of means by which normal or pathological changes in pH can affect ligand-gated ion channel function. Protons can act at either the extracellular or intracellular face of the receptor to alter the receptor responsiveness to agonists. Changes in pH can also influence ion permeation and the effective ligand concentration, a point that is particularly relevant during pathological conditions associated with marked pH changes. The pH sensitivity of ligand-gated ion channels holds implications for hypersynchronous neuronal activity, neurodegeneration, nociception, and rational drug design. In addition, the presently available data support the generalization (with a few exceptions) that acidic extracellular pH tends to reduce neuronal excitability in the brain, whereas elevated pH appears to enhance it. Thus, the pH sensitivity of ligand-gated ion channels should be a standard part of their characterization. It is obvious that the pH regulation of ion channels holds potentially exciting implications for our understanding of surprisingly diverse phenomena. In addition, future molecular studies of pH regulation of ion channel function will almost certainly provide key insight into the mechanisms of channel activation and regulation.

ACKNOWLEDGMENTS

I thank Drs. K. Ballanyi, R. Dingledine, B. Hille, and G. Somjen for their valuable input. This work was supported by the NIH (NS33777, MH54759), the John Merck Fund, and the Epilepsy Foundation of America.

REFERENCES

Adams DJ, Nonner W, Dwyer TM, Hille B (1981): Block of endplate channels by permeant cations in frog skeletal muscle. J Gen Physiol 78: 593–615.

Aguilar-Bryan L, Nichols CG, Wechsler SW, Clement JP, Boyd AE, Gonzalez G, Herrera-Sosa H, Nguy K, Bryan J, Nelson DA (1995): Cloning of the β cell high-affinity sulfonylurea receptor: A regulator of insulin secretion. Science 268: 423–426.

Akiyoshi M, Kakei M, Nakazaki M, Tanaka H (1995): A new hypoglycemic agent, A-4166, inhibits ATP-sensitive potassium channels in rat pancreatic β-cells. Am J Physiol 268 E185–E193.

Allard B, Lazdunski M, Rougier O (1995): Activation of ATP-dependent K^+ channels by metabolic poisoning in adult mouse skeletal muscle: Role of intracellular Mg^{2+} and pH. J Physiol (Lond) 485: 283–296.

Andrews PR, Johnston GAR (1973): Molecular orbital and proton magnetic resonance studies of bicuculline. Nature New Biology 243: 29–30.

Aram JA, Lodge D (1987): Epileptiform activity induced by alkalosis in rat neocortical slices: Block by antagonists of N-methyl-D-aspartate. Neurosci Lett 83: 345–350.

Ashcroft SJH, Ashcroft FM (1990): Properties and functions of ATP-sensitive K-channels. Cellular Signalling 2: 197–214.

Ashcroft SJH, Niki I, Kenna S, Weng L, Skeer J, Coles B, Ashcroft FM (1993): The β cell sulfonylurea receptor. Ostenson C-G, Efendic S, Vranici M (eds) New Concepts in the Pathogenesis of NIDDM.: New York: Plenum Press.

Ashcroft FM, Proks P, Smith PA, Ammala C, Bokvist K, Rorsman P. (1994): Stimulus-secretion coupling in pancreatic β cells. J Cell Biochem 55S: 54–65.

Ashford MLJ, Boden PR, Treherne JM (1990): Glucose-induced excitation of hypothalamic neurones is mediated by ATP-sensitive K^+ channels. Pflügers Arch 415: 479–483.

Attwell D, Barbour B, Szatkowski M (1993): Nonvesicular release of neurotransmitter. Neuron 11: 401–407.

Balestrino M, Somjen GG (1988): Concentration of carbon dioxide, interstitial pH, and synaptic transmission in hippocampal formation of the rat. J Physiol (Lond) 396: 247–266.

Bekkers JM (1993): Enhancement by histamine of NMDA-mediated synaptic transmission in the hippocampus. Science 261: 104–106.

Benveniste M, Mayer ML (1992): Effect of extracellular pH on the potency of N-methyl-D-aspartic acid receptor competitive antagonists. Mol Pharmacol 42: 679–686.

Bhushan A, McNamee MG (1993): Correlation of phospholipid structure with functional effects on the nicotinic acetylcholine receptor. Biophys J 64: 716–723.

Bolanos FJ, Schechter LE, Miquel MC, Emerit MB, Rumigny JF, Hamon M, Gozlan H (1990): Common pharmacological and physicochemical properties of 5-HT_3 binding sites in the rat cerebral cortex and NG108-15 clonal cells. Biochem Pharmacol 40: 1541–1550.

Bouzat CB, Lacorazza HD, Bonino MB de J, Barrantes FJ (1993): Effect of chemical modification of extracellular histidyl residues on the channel properties of the nicotinic acetylcholine receptor. Pflügers Arch 423: 365–371.

Brown MCS, Lunt GC, Stapleton A (1989): Further characterisation of the [^{35}S]-TBPS binding site of the GABA receptor complex in locust (*Schistocerca gregaria*) ganglia membranes. Comp Biochem Physiol 92C: 9–13.

Busa WB, Nuccitelli RD (1984): Metabolic regulation via intracellular pH. Am J Physiol 246: R409–R438.

Chen D, Eisenberg R (1993): Charges, currents, and potentials in ionic channels of one conformation. Biophys J 64: 1405–1421.

Chenard BL, Lipinski CA, Dominy BW, Mena EE, Ronau RT, Butterfield GC, Marinovic LC, Pagnozzi M, Butler TW, Tsang TY (1990): A unified approach to systematic isosteric substitution for acidic groups and application to NMDA antagonists related to 2-amino-7-phosphonoheptanoate. J Med Chem 33: 1077–1083.

Choi DW (1992): Excitotoxic cell death. J Neurobiol 23: 1261–1276.

Chou PY, Fasman GD (1978): Prediction of the secondary structure of proteins from their amino acid sequence. Methods Enzymol 47: 45–148.

Christensen BN, Hida E (1990): Protonation of histidine groups inhibits gating of the quisqualate/kainate channel protein in isolated catfish cone horizontal cells. Neuron 5: 471–478.

Colamartino G, Menini A, Torre V (1991): Blockage and permeation of divalent cations through the cyclic GMP-activated channel from tiger salamander retinal rods. J Physiol (Lond) 440: 189–206.

Coulter KL, Perier F, Radeke CM, Vandenberg CA (1995): Identification and molecular localization of a pH-sensing domain for the inward rectifier potassium channel HIR. Neuron 15: 1157–1168.

Cuevas J, Bassett AL, Cameron JS, Furukawa T, Myerburg RJ, Kimura S (1991): Effect of H^+ on ATP-regulated K^+ channels in feline ventricular myocytes. Am J Physiol 261: H755–H761.

Davies NW (1990): Modulation of ATP-sensitive K^+ channels in skeletal muscle by intracellular protons. Nature 343: 375–377.

Davies NW, Standen NB, Stanfield PR (1992): The effect of intracellular pH on ATP-dependent potassium channels of frog skeletal muscle. J Physiol (Lond) 445: 549–568.

Del Castillo J, Nelson TE, Sanchez V (1962): Mechanism of the increased acetylcholine sensitivity of skeletal muscle in low pH solutions. J Cell Comp Physiol 59: 35–44.

Del Castillo AR, Vitale ML, Trifaró J-M (1992): Ca^{2+} and pH determine the interaction of chromaffin cell scinderin with phosphatidylserine and phosphatidylinositol 4,5-biphosphate and its cellular distribution during nicotinic-receptor stimulation and protein kinase C activation. J Cell Biol 119: 797–810.

Dingledine R, McBain CJ, McNamara JO (1990): Excitatory amino acid receptors in epilepsy. Trends Pharmacol Sci 11: 334–338.

Dreyer F, Peper K, Sterz R, Bradley RJ, Muller K-D (1979): Drug-receptor interaction at the frog neuromuscular junction. Prog Brain Res 49: 213–223.

Durand GM, Gregor P, Zheng X, Bennett MVL, Uhl GR, Zukin RS (1992): Cloning of an apparent splice variant of the rat N-methyl-D-aspartate receptor NMDAR1 with altered sensitivity to polyamines and activators of protein kinase C. Proc Natl Acad Sci USA 89: 9359–9363.

Durand GM, Bennett MVL, Zukin RS (1993): Splice variants of the N-methyl-D-aspartate receptor NR1 identify domains involved in regulation by polyamines and protein kinase C. Proc Natl Acad Sci USA 90: 6731–6735.

Emini EA, Hughes JV, PerlowDS, Boger J (1985): Induction of hepatitis A virus-neutralizing antibody by a virus specific synthetic peptide. J Virol 55: 836–839.

Enna SJ, Snyder SH (1975): Properties of γ-aminobutyric acid (GABA) receptor binding in rat brain synaptic membrane fractions. Brain Res 100: 81–97.

Fagg GE, Baud J (1988): Characterization of NMDA receptor-ionophore complexes in the brain. In Lodge D (ed): Excitatory Amino Acids in Health and Disease. Chichester: John Wiley & Sons, pp 63–90.

Fan Z, Makielski JC (1993): Intracellular H^+ and Ca^{2+} modulation of trypsin-modified ATP-sensitive K^+ channels in rabbit ventricular myocytes. Circ Res 72: 715–722.

Fan Z, Furukawa T, Sawanobori T, Makielski JC, Hiraoka M (1993): Cytoplasmic acidosis induces multiple conductance states in ATP-sensitive potassium channels of cardiac myocytes. J Membr Biol 136: 169–179.

Fan Z, Tokuyama Y, Makielski JC (1994): Modulation of ATP-sensitive K^+ channels by internal acidification in insulin-secreting cells. Am J Physiol 267: C1036–C1044.

Findlay I (1992): Effects of pH upon the inhibition by sulphonylurea drugs of ATP-sensitive K^+ channels in cardiac muscle. J Pharmacol Exp Ther 262: 71–79.

Fong TM, McNamee MG (1986): Correlation between acetylcholine receptor function and structural properties of membranes. Biochemistry 25: 830–840.

Forestier C, Depresle Y, Vivaudou M (1993): Intracellular protons control the affinity of skeletal muscle ATP-sensitive K^+ channels for potassium-channel-openers. FEBS Lett 325: 276–280.

Frey G, Hanke W, Schlue W-R (1993): ATP-inhibited and Ca^{2+}-dependent K^+ channels in the soma membrane of cultured leech retzius neurons. J Membr Biol 134: 131–142.

Gallagher JP, Nakamura J, Shinnick-Gallagher P (1983): The effects of temperature, pH and Cl^- pump inhibitors on GABA responses recorded from cat dorsal root ganglia. Brain Res 267: 249–259.

Garnier J, Osguthorpe DJ, Robson B (1978): Analysis of the accuracy and implications of simple methods for predicting secondary structure of globular proteins. J Mol Biol 120: 97–120.

Giffard RG, Monyer H, Christine CW, Choi DW (1990): Acidosis reduces NMDA receptor activation, glutamate neurotoxicity, and oxygen-glucose deprivation neuronal injury in cortical cultures. Brain Res 506: 339–342.

Goldberg G, Lass Y (1983): Evidence for acetylcholine receptor blockade by intracellular hydrogen ions in cultured chick myoballs. J Physiol (Lond) 343: 429–437.

Gordon SE, Oakley JC, Varnum MD and Zagotta WN (1996): Altered ligand specificity by protonation in the ligand binding domain of cyclic nucleotide-gated channels. Biochemistry 35: 3994–4001.

Grassi F, Giovannelli A, Fucile S, Eusebi F (1993): Activation of the nicotinic acetylcholine receptor mobilizes calcium from caffeine-insensitive stores in C2C12 mouse myotubes. Pflügers Arch 422: 591–598.

Gruol DL, Barker JL, Huang L-YM, MacDonald JF, Smith TG (1980): Hydrogen ions have multiple effects on the excitability of cultured mammalian neurons. Brain Res 183: 247–252.

Haddad GG, Jiang C (1994): Mechanisms of neuronal survival during hypoxia: ATP-sensitive K^+ channels. Biol Neonate 65: 160–165.

Head RA, Tunnicliff G, Matheson GK (1980): Glutamate receptor binding to cat central, nervous system membranes. Can J Biochem 58: 534–538.

Heaton RC, Wray S, Eisner DA (1993): Effects of metabolic inhibition and changes of intracellular pH on potassium permeability and contraction of rat uterus. J Physiol (Lond) 465: 43–56.

Heaulme M, Chambon J-P, Leyris R, Wermuth CG, Biziere K (1987): Characterization of the binding of [^{3}H]SR 95521, a $GABA_A$ antagonist, to rat brain membranes. J Neurochem 48: 1677–1686.

Heginbotham L, Abramson T, MacKinnon R (1992): A functional connection between the pores of distantly related ion channels as revealed by mutant K^+ channels. Science 258: 1152–1155.

Hille B (1977): The pH-dependent rate of action of local anesthetics on the node of Ranvier. J Gen Physiol 69: 475–496.

Hille B (1992): Mechanisms of block. In Ionic Channels of Excitable Membranes. Sunderland, MA: Sinauer, pp 390–422.

Hollmann M, Heinemann SF (1994): Cloned glutamate receptors. Annu Rev Neurosci 17: 31–108.

Hollmann M, Boulter J, Maron C, Beasley L, Sullivan J, Pecht G, Heinemann SF (1993): Zinc potentiates agonist-induced currents at certain splice variants of the NMDA receptor. Neuron 10: 943–954.

Honig B, Nicholls A (1995): Classical electrostatics in biology and chemistry. Science 268: 1144–1149.

Horvath LI, Arias HR, Hankovszky HO, Hideg K, Barrantes FJ, Marsh D (1990): Association of spin-labeled local anesthetics at the hydrophobic surface of acetylcholine receptor in native membranes from *Torpedo marmorata*. Biochemistry 29: 8707–8713.

Hoshi T, Zagotta WN, Aldrich RW (1990): Biophysical and molecular mechanisms of shaker potassium channel inactivation. Science 250: 533–538.

Huang L-YM, Catterall WA, Ehrenstein G (1978): Selectivity of cations and nonelectrolytes for acetylcholine-activated channels in cultured muscle cells. J Gen Physiol 71: 397–410.

Iljin VI, Bregestovski PD (1977): Chemical nature of functional cholinoreceptor groups of Lymnaea Stagnalis neurons. Acta Physiologica Academiae Scientiarum Hungaricae 49: 221–230.

Imoto K, Busch C, Sakmann B, Mishina M, Konno T, Nakai J, Bujo H, Mori Y, Fukuda K, Numa S (1988): Rings of negatively charged amino acids determine the acetylcholine receptor channel conductance. Nature 335: 645–648.

Inagaki N, Gonoi T, Clement JP, Namba N, Inazawa J, Gonzalez G, Aguilar-Bryan L, Seino S, Bryan J (1995): Reconstitution of I_{KATP}: An inward rectifier subunit plus the sulfonylurea receptor. Science 270: 1166–1170.

Inagaki F, Kawano Y, Shimada I, Takahashi K, Miyazawa T (1981): Nuclear magnetic resonance study on the microenvironments of histidine residues of Ribonuclease T1 and carboxymethylated ribonuclease T1. J Biochem 89: 1185–1195.

Inoue M, Yamada H, Hashimoto Y, Yasukochi T, Hamaguchi K, Miki T, Horiuchi T, Imoto T (1992): Stabilization of a protein by removal of unfavorable abnormal pKa: Substitution of undissociable residue for glutamic acid-35 in chicken lysozyme. Biochemistry 31: 8816–8821.

Jahangir A, Terzic A, Kurachi Y (1994): Intracellular acidification and ADP enhance nicorandil induction of ATP sensitive potassium channel current in cardiomyocytes. Cardiovascular Res 28: 831–835.

Janigro D, West GA, Gordon EL, Winn HR (1993): ATP-sensitive K^+ channels in rat aorta and brain microvascular endothelial cells. Am J Physiol 265: C812–C821.

Jarolimek W, Misgeld U, Lux HD (1989): Activity dependent alkaline and acid transients in guinea pig hippocampal slices. Brain Res 505: 225–232.

Jiang C, Xia Y, Haddad GG (1992): Role of ATP-sensitive K^+ channels during anoxia: Major differences between rat (newborn and adult) and turtle neurons. J Physiol (Lond) 448: 599–612.

Johnson TD (1966): Modulation of channel function by polyamines. Trends Pharmacol Sci 17: 22–27.

Kaila K (1994): Ionic basis of $GABA_A$ receptor channel function in the nervous system. Prog Neurobiol 42: 489–537.

Kaila K, Paalasmaa P, Tairi T, Voipio J (1992): pH transients due to monosynaptic activation of $GABA_A$ receptors in rat hippocampal slices. Neuroreport 3: 105–108.

Kaila K, Voipio J, Paalasmaa P, Pasternack M, Deisz RA (1993): The role of bicarbonate in $GABA_A$ receptor-mediated IPSPs of rat neocortical neurones. J Physiol (Lond) 464: 273–289.

Kaku DA, Giffard RG, Choi DW (1993): Neuroprotective effects of glutamate antagonists and extracellular acidity. Science 260: 1516–1518.

Karlin A (1993): Structure of nicotinic acetylcholine receptors. Curr Opin Neurobiol 3: 299–309.

Kashiwagi K, Fukuchi J-I, Chao J, Igarashi K, Williams K (1996): An aspartate residue in the extracellular loop of the N-methyl-D-aspartate receptor controls sensitivity to spermine and protons. Mol Pharmacol 49: 1131–1141.

King BF, Ziganshina LE, Pintor J, Burnstock G (1996): Full sensitivity of P_{2X} purinoceptor to ATP revealed by changing extracellular pH. Br J Pharmacol 117: 1371–1373.

Kishimoto H, Simon JR, Aprison MH (1981): Determination of the equilibrium dissociation constants and number of glycine binding sites in several areas of the rat central nervous system, using a sodium-independent system. J Neurochem 37: 1015–1024.

Kleingoor C, Wieland HA, Korpi ER, Seeburg PH, Kettenmann H (1993): Current potentiation by diazepam but not GABA sensitivity is determined by a single histidine residue. Neuroreport 4: 187–190.

Kordas M (1970): The effect of procaine on neuromuscular transmission. J Physiol (Lond) 209: 689–699.

Koyano T, Kakei M, Nakashima H, Yoshinaga M, Matsuoka T, Tanaka H (1993): ATP-regulated K^+ channels are modulated by intracellular H^+ in guinea-pig ventricular cells. J Physiol (Lond) 463: 747–766.

Krishek BJ, Amato A, Connolly CN, Moss SJ, Smart TG (1996): Proton sensitivity of the $GABA_A$ receptor is associated with the receptor subunit composition. J Physiol (Lond) 492: 431–443.

Kunzelmann K, Pavenstadt H, Greger R (1989): Characterization of potassium channels in respiratory cells. Pflügers Arch 414: 297–303.

Kyte J (1995): Structure in protein chemistry. New York: Garland.

Lambolez B, Rossier J (1987): Benzodiazepine agonists protect a histidine residue from modification by diethyl pyrocarbonate whereas propyl β-carboline does not. FEBS Lett 219: 301–305.

Landau EM, Gavish B, Nachshen DA, Lotan I (1981): pH dependence of the acetylcholine receptor channel. J Gen Physiol 77: 647–666.

Lau KR, Elliott AC, Brown PD (1989): Acetylcholine-induced intracellular acidosis in rabbit salivary gland acinar cells. Am J Physiol 256: C288–C295.

Laurie DJ, Seeburg P (1994): Regional and developmental heterogeneity in splicing of the rat brain NMDAR1 mRNA. J Neurosci 14: 3180–3194.

Li L, McNamee MG (1992): Modulation of nicotinic acetylcholine receptor channel by pH: A difference in pH sensitivity of *Torpedo* and mouse receptors expressed in *Xenopus* oocytes. Cell Molec Neurobiol 12: 83–93.

Li C, Peoples RW, Weight FF (1996a): Acid pH augments excitatory action of ATP on a dissociated mammalian sensory neuron. Neuroreport 7: 2151–2154.

Li C, Peoples RW, Weight FF (1996b): Proton potentiation of ATP-gated ion channel responses to ATP and Zn^{2+} in rat nodose ganglion neurons. J Neurophysiol 76: 3048-3058.

Li C, Peoples RW, Weight FF (1997): Enhancement of ATP-activated current by protons in dorsal root ganglion neurons. Pflugers Arch 433: 446–54.

Liebman PA, Mueller P, Pugh EN (1984): Protons suppress the dark current of frog retinal rods. J Physiol (Lond) 347: 85–110.

McBain CJ, Mayer ML (1994): N-methyl-D-aspartic acid receptor structure and function. Physiol Rev 74: 723–760.

MacDonald JF, Bartlett MC, Mody I, Pahapill P, Reynolds JN, Salter MW, Scheiderman JH, Pennefather PS (1991): Actions of ketamine, phencyclidine and MK-801 on NMDA receptor currents in cultured mouse hippocampal neurones. J Physiol (Lond) 432: 483–508.

McDonald JW, Penney JB, Johnston MV, Young AB (1990): Characterization and regional distribution of strychnine-insensitive [^{3}H]glycine binding sites in rat brain by quantitative receptor autoradiography. Neuroscience 35: 653–668.

McLaughlin JT, Hawrot E (1989): Structural characterization of α-bungarotoxin-binding proteins from *Aplysia californica.* Mol Pharmacol 35: 593–598.

Maksay G (1992): Modification of benzodiazepine receptors supports the distinctive role of histidine residues. Eur J Pharmacol 227: 57–62.

Mallart A, Molgó J (1978): The effects of pH and curare on the time course of end-plate currents at the neuromuscular junction of the frog. J Physiol (Lond) 276: 343–352.

Mallart A, Dreyer F, Peper K (1976): Current-voltage relation and reversal potential at junctional and extrajunctional ACh-receptors of the frog neuromuscular junction. Pflügers Arch 362: 43–47.

Meldrum BS (1992): Excitatory amino acids in epilepsy and potential novel therapies. Epilepsy Res 12: 189–196.

Miralles A, Olmos G (1989): [^{3}H]Kainic acid binding sites in chick cerebellar membranes. Comp Biochem Physiol C: Comp Pharmacol and Toxicol 93: 321–325.

Monahan JB, Michel J (1987): Identification and characterization of an N-methyl-D-aspartate-specific L[^{3}H]glutamate recognition site in synaptic plasma membranes. J Neurochem 48: 1699–1708.

Moser C, Hehl S, Neumcke B (1995): Inhibition of ATP-sensitive K^+ channels of mouse skeletal muscle by disopyramide. Eur J Pharmacol 284: 35–41.

Murphy KPSJ, Greenfield SA (1991): ATP-sensitive potassium channels counteract anoxia in neurones of the substantia nigra. Exp Brain Res 84: 355–358.

Murphy DE, Schneider J, Boehm C, Lehmann J, Williams M (1987): Binding of [^{3}H]3-(2-carboxypiperazin-4-yl)propyl-1-phosphonic acid to rat brain membranes: A selective, high affinity ligand for N-methyl-D-aspartate receptors. J Pharmacol Exp Ther 240: 778–784.

Nakanishi S (1992): Molecular diversity of glutamate receptors and implications for brain function. Science 258: 597–603.

Narita M, Suzuki T, Misawa M, Nagase H, Nabeshima A, Ashizawa T, Ozawa H, Saito T, Takahata N (1992): Role of central ATP-sensitive potassium channels in the analgesic effect and spinal noradrenaline turnover-enhancing effect of intracerebroventricularly injected morphine in mice. Brain Res 596: 209–214.

Neher E, Steinbach JH (1978): Local anesthetics transiently block currents through single acetylcholine-receptor channels. J Physiol (Lond) 277: 153–176.

Obrenovitch TP, Scheller D, Matsumoto T, Tegtmeier F, Höller M, Symon L (1990): A rapid redistribution of hydrogen ions is associated with depolarization and repolarization subsequent to cerebral ischemia reperfusion. J Neurophysiol 64: 1125–1133.

O'Brien RD, Gibson RE (1974): Two binding sites in acetylcholine receptor from *Torpedo marmorata* electroplax. Arch Biochem Biophys 165: 681–690.

Oda Y, Yamazaki T, Nagayama K, Kanaya S, Kuroda Y, Nakamura H (1994): Individual ionization constants of all the carboxyl groups in ribonuclease HI for *Escheria coli* determined by NMR. Biochemistry 33: 5275–5284.

O'Donnell BR, Bickler PE (1994): Influence of pH on calcium influx during hypoxia in rat cortical brain slices. Stroke 25: 171–177.

Olverman HJ, Jones AW, Watkins JC (1988): [^{3}H]D-2-amino-5-phosphonopentanoate as a ligand for N-methyl-D-aspartate receptors in the mammalian central nervous system. Neuroscience 26: 1–15.

Onodera K, Takeuchi A (1979): An analysis of the inhibitory post-synaptic current in the voltage-clamped crayfish muscle. J Physiol (Lond) 286: 265–282.

Ou-Yang Y, Kristián T, Mellergård P, Siesjö BK (1994): The influence of pH on glutamate- and depolarization-induced increases of intracellular calcium concentration in cortical neurons in primary culture. Brain Res 646: 65–72.

Palma A, Li L, Chen X, Pappone P, McNamee M (1991): Effects of pH on acetylcholine receptor function. J Membr Biol 120: 67–73.

Pan JW, Hamm JR, Rothman DL, Shulman RG (1988): Intracellular pH in human skeletal muscle by ^{1}H NMR. Proc Natl Acad Sci USA 85: 7836–7839.

Paoletti P, Neyton J, Asher P (1995): Glycine-independent and subunit-specific potentiation of NMDA responses by extracellular Mg^{2+}. Neuron 15: 1109–1120.

Pappone PA, Barchfeld GL (1990): Modifications of single acetylcholine-activated channels in BC3H-1 cells. J Gen Physiol 96: 1–22.

Park J, Bursztajn S (1990): The role of pH in the intracellular acetylcholine receptor ligand processing. Biol Cell 70: 177–188.

Pasternack M, Bountra C, Voipio J, Kaila K (1992): Influence of extracellular and intracellular pH on GABA-gated chloride conductance in crayfish muscle fibres. Neuroscience 47: 921–929.

Pasternack M, Smirnov S, Kaila K (1996): Proton modulation of functionally distinct $GABA_A$ receptors in acutely isolated pyramidal neurons of rat hippocampus. Neuropharmacol 35: 1279–1288.

Pinkus LM, Sarbin NS, Barefoot DS, Gordon JC (1989): Association of [^{3}H]zacopride with $5\text{-}HT_3$ binding sites. Eur J Pharmacol 168: 355–362.

Prod'hom B, Pietrobon D, Hess P (1987): Direct measurement of proton transfer rates to a group controlling the dihydropyridine-sensitive Ca^{2+} receptor. Nature 329: 243–246.

Proks P, Takano M, Ashcroft FM (1994): Effects of intracellular pH on ATP-sensitive K^+ channels in mouse pancreatic β-cells. J Physiol (Lond) 475: 33–44.

Raffa RB, Codd EE (1994): Lack of glibenclamide or TEA affinity for opiod receptors: Further evidence for in vivo modulation of antinociceptive at K^+ channels. Brain Res 650: 146–148.

Raines DE, Miller KW (1993): The role of charge in lipid selectivity for the nicotinic acetylcholine receptor. Biophys J 64: 632–641.

Rajdev S, Reynolds IJ (1993): Effects of pH on the actions of dizocilpine at the N-methyl-D-aspartate receptor complex. Br J Pharmacol 109: 107–112.

Renaud JM (1989): The effect of lactate on intracellular pH and force recovery of fatigued sartorius muscles of the frog *Rana Pipiens*. J Physiol (Lond) 416: 31–47.

Ritchie AK, Fambrough DM (1975): Ionic properties of the acetylcholine receptor in cultured rat myotubes. J Gen Physiol 65: 751–767.

Robello M, Baldelli P, Cupello A (1994): Modulation by extracellular pH of the activity of $GABA_A$ receptors on rat cerebellum granule cells. Neuroscience 61: 833–837.

Robertson B (1989): Actions of anaesthetics and avermectin on $GABA_A$ chloride channels in mammalian dorsal root ganglion neurones. Br J Pharmacol 98: 167–176.

Root MJ, MacKinnon R (1993): Identification of an external divalent cation-binding site in the pore of a cGMP-activated channel. Neuron 11: 459–466.

Root MJ, MacKinnon R (1994): Two identical noninteracting sites in an ion channel revealed by proton transfer. Science 265: 1852–1856.

Roper J, Ashcroft FM (1995): Metabolic inhibition and low internal ATP activate K-ATP channels in rat dopaminergic substantia nigra neurones. Pflügers Arch 430: 44–54.

Rosario LM, Stutzin A, Cragoe EJ, Pollard HB (1991): Modulation of intracellular pH by secretagogues and the Na^{+}/H^{+} antiporter in cultured bovine chromaffin cells. Neuroscience 41: 269–276.

Rost B, Sander C (1993): Prediction of protein secondary structure at better than 70% accuracy. J Mol Biol 232: 584–599.

Rutherford DM, Jeffery D, Lunt GC, Weitzman PDJ (1987): GABA binding to receptor sites in locust supraoesophageal ganglia. Neurosci Lett 79: 337–340.

Sargent PB (1993): The diversity of neuronal nicotinic acetylcholine receptors. Annu Rev Neurosci 16: 403–443.

Schaller W, Robertson AD (1995): pH, ionic strength, and temperature dependence of ionization equilibria for the carboxyl groups in turkey ovomucoid third domain. Biochemistry 34: 4714–4723.

Schmidt J, Raftery MA (1974): The cation sensitivity of the acetylcholine receptor from *Torpedo californica.* J Neurochem 23: 617–623.

Schmieden V, Kuhse J, Betz H (1994): The pH dependence of glycine receptor activation: Charged amino groups are essential for ligand-receptor interaction. In Eisner N, Breer B (eds): Proceedings of the 22nd Göttingen Neurobiology Conference, Vol. II p 738. Georg Thieme Verlag, Stuttgart

Scuka M (1975): The amplitude and the time course of the end-plate current at various pH levels in the frog sartorius muscle. J Physiol (Lond) 249: 183–195.

Seeburg PH (1993): The molecular biology of mammalian glutamate receptor channels. Trends Neurosci 16: 359–365.

Seo JR, Steward MC, Larcombe-McDouall JB, Cook LJ, Case RM (1994): Continuous fluorometric mesurement of intracellular pH and Ca^{2+} in perfused salivary gland and pancreas. Pflügers Arch 426: 75–82.

Sieghart W (1995): Structure and pharmacology of γ-aminobutyric $acid_A$ receptor subtypes. Pharmacol Rev 47: 181–234.

Siesjö BK (1985): Acid-base homeostasis in the brain: Physiology, chemistry, and neurochemical pathology. Prog Brain Res 63: 121–153.

Smart TG (1992): A novel modulatory binding site for zinc on the $GABA_A$ receptor complex in cultured rat neurones. J Physiol (Lond) 447: 587–625.

Smart TG, Constanti A (1981): A re-examination of the GABA-inhibitory action of bicuculline on lobster muscle. Eur J Pharmacol 70: 25–33.

Smart TG, Constanti A (1982): A novel effect of zinc on the lobster muscle GABA receptor. Proc R Soc Lond [Biol] 215: 327–341.

Standaert DG, Testa CM, Penney JB, Young AB (1993): Alternatively spliced isoforms of the NMDAR1 glutamate receptor subunit: differential expression in the basal ganglia of the rat. Neurosci Lett 152: 161–164.

Standen NB, Pettit AI, Davies NW, Stanfield PR (1992): Activation of ATP-dependent K^{+} currents in intact skeletal muscle fibres by reduced intracellular pH. Proc R Soc Lond [Biol] 247: 195–198.

Stevens CF (1993): Quantal release of neurotransmitter and long-term potentiation. Cell 72: 55–63.

Sugihara H, Moriyoshi K, Ishii T, Masu M, Nakanishi S (1992): Structures and properties of seven isoforms of the NMDA receptor generated by alternative splicing. Biophys Biochem Res Commun 185: 826–832.

Sullivan JM, Traynelis SF, Chen H-SV, Escobar W, Heinemann SF, Lipton SA (1994): Identification of two cysteine residues that are required for redox modulation of the NMDA subtype of glutamate receptor. Neuron 13: 929–936.

Sun M-K, Reis D (1993): Extracellular H^+ iontophoresis modifies responses to γ-aminobutyric acid and cyanide of reticulospinal vasomotor neurons in rats. Eur J Pharmacol 236: 305–313.

Takadera T, Shimada Y, Mohri T (1992): Extracellular pH modulates N-methyl-D-aspartate receptor-mediated neurotoxicity and calcium accumulation in rat cortical cultures. Brain Res 572: 126–131.

Takano M, Noma A (1993): The ATP-sensitive K^+ channel. Prog Neurobiol 41: 21–30.

Takeda K, Barry PH, Gage PW (1980): Effects of ammonium ions on endplate channels. J Gen Physiol 75: 589–613.

Takeuchi A, Takeuchi N (1967): Anion permeability of the inhibitory post-synaptic membrane of the crayfish neuromuscular junction. J Physiol (Lond) 191: 575–590.

Tanaka JC (1993): The effects of protons on 3′5′-cGMP-activated currents in photoreceptor patches. Biophys J 65: 2517–2523.

Tang C-M, Dichter M, Morad M (1990): Modulation of the N-methyl-D-aspartate channel by extracellular H^+. Proc Natl Acad Sci USA 87: 6445–6449.

Tombaugh GC, Sapolsky RM (1993): Evolving concepts about the role of acidosis in ischemic neuropathology. J Neurochem 61: 793–803.

Thomas PM, Lote GJ, Wohllk N, Haddad B, Mathew PM, Rabl W, Aguilar-Bryan L, Gagel RF, Bryan J (1995): Mutations in the sulfonylurea receptor gene in familial persistent hyperinsulinemic hypoglycemia of infancy. Science 268: 426–429.

Trapp S, Ballanyi K (1995). K_{ATP} channel mediation of anoxia-induced outward current in rat dorsal vagal neurons in vitro. J Physiol (Lond) 487: 37–50.

Trapp S, Luckermann M, Brooks PA, Ballanyi K (1996): Acidosis of rat dorsal vagal neurons in situ during spontaneous and evoked activity. J Physiol 496: 695–710.

Trautmann A, Zilber-Gachelin NF (1976): Further investigations on the effect of denervation and pH on the conductance change at the neuromuscular junction of the frog. Pflügers Arch 364: 53–58.

Traynelis SF, Cull-Candy SG (1990): Proton inhibition of N-methyl-D-aspartate receptors in cerebellar neurons. Nature 345: 347–350.

Traynelis SF, Cull-Candy SF (1991): Pharmacological properties and H^+ sensitivity of excitatory amino acid receptor channels in rat cerebellar granule neurones. J Physiol (Lond) 433: 727–763.

Traynelis SF, Hartley M, Heinemann SF (1995): Control of proton sensitivity of the NMDA receptor by RNA splicing and polyamines. Science 268: 873–876.

Treherne JM, Ashford MLJ (1991): The regional distribution of sulphonylurea binding sites in rat brain. Neuroscience 40: 523–531.

Tromba C, Salvaggio A, Racagni G, Volterra A (1992): Hypoglycemia-activated K^+ channels in hippocampal neurons. Neurosci Lett 143: 185–189.

Tsien RW, Hess P, McCleskey EW, Rosenberg RL (1987): Calcium channels: Mechanisms of selectivity, permeation, and block. Annu Rev Biophys Biophys Chem 16: 265–290.

Uusi-Oukari M (1992): Characterization of two cerebellar binding sites of [^{3}H]Ro 15-4512. J Neurochem 59: 568–574.

Vivaudou M, Forestier C (1995): Modification by protons of frog skeletal muscle K_{ATP} channels: Effects on ion conduction and nucleotide inhibition. J Physiol (Lond) 486: 629–645.

Voipio J, Pasternack M, Rydqvist B, Kaila K (1991): Effect of γ-aminobutyric acid on intracellular pH in the crayfish stretch-receptor neurone. J Exp Biol 156: 349–361.

Vorobjev VS, Sharonova IN, Walsh IB, Haas HL (1993): Histamine potentiates N-methyl-D-aspartate responses in acutely isolated hippocampal neurons. Neuron 11: 837–844.

Vyklicky L, Vlachova V, Krusek J (1990): The effect of external pH changes on responses to excitatory amino acids in mouse hippocampal neurones. J Physiol (Lond) 430: 497–517.

Wegelius K, Reeben M, Rivera C, Kaila K, Saarma M, Pasternack M (1996): The $\rho 1$ GABA receptor cloned from rat retina is down-modulated by protons. Neuroreport 7: 2005–2009.

Wieland HA, Lüddens H, Seeburg PH (1992): Single histidine in $GABA_A$ receptors is essential for benzodiazepine agonist binding. J Biol Chem 267: 1426–1429.

Williams K (1994): Subunit-specific potentiation of recombinant N-methyl-D-aspartate receptors by histamine. Mol Pharmacol 46: 531–541.

Williams K, Kashiwagi K, Fukuchi J-I, Igarashi K (1995): An acidic amino acid in the N-methyl-D-aspartate receptor that is important for spermine stimulation. Mol Pharmacol 48: 1087–1098.

Wu X, Christensen BN (1996): Proton inhibition of the NMDA-gated channel in isolated catfish cone horizontal cells. Vision Res 36: 1521–1528.

Yakel JL, Shao XM, Jackson MB (1990): The selectivity of the channel coupled to the $5\text{-}HT_3$ receptor. Brain Res 533: 46–52.

Yau K-W, Baylor DA (1989): Cyclic GMP-activated conductance of retinal photoreceptor cells. Annu Rev Neurosci 12: 289–327.

Yoneda Y, Enomoto R, Ogita K (1994): Supporting evidence for negative modulation by protons of an ion channel associated with the N-methyl-D-aspartate receptor complex in rat brain using ligand binding techniques. Brain Res 636: 298–307.

Zorumski CF, Olney JW (1993): Excitotoxic neuronal damage and neuropsychiatric disorders. Pharmacol Ther 59: 145–162.

Zufall F, Firestein S (1993): Divalent cations block cyclic nucleotide-gated channel of olfactory receptor neurons. J Neurophysiol 69: 1758–1768.

CHAPTER 23

Proton-Gated Ion Channels in Neurons

STUART BEVAN
Novartis Institute for Medical Sciences, London, United Kingdom

INTRODUCTION

The electrical activity of neurons can be influenced by a wide range of "classical" neurotransmitters (e.g., acetylcholine, γ-aminobutyric acid) and neuromodulators (e.g., substance P) as well as by agents such as the prostaglandins and nitric oxide. Accounts of intercellular signalling usually focus on the ability of these substances to activate receptor-operated ion channels and second messenger systems and to modulate the activity of voltage gated ion channels. In addition, changes in extracellular pH (pH_o) can influence neural activity by modifying synaptic transmission (Balestrino and Somjen, 1988), the activity of excitatory amino acid receptors (Tang et al., 1990; Traynelis and Cull-Candy, 1990; Vyklický et al., 1990), and the normal function of voltage-gated ion channels (see for example Hille 1968, 1973; Iijima et al., 1986; and Chapters 20–22). Acidification of the external medium can activate ion channels in neurons in both the central nervous system (CNS) and the peripheral nervous system (PNS). It is, therefore, likely that changes in the pH in the vicinity of nerve cells have important modulatory effects on neuronal activity (Chapter 17).

This chapter will focus on the ways in which changes in pH_o activate ion channels in neuronal cells. Other effects of changes in pH will be discussed elsewhere in this volume.

TYPES OF PROTON-EVOKED CURRENTS

Most of the characteristics of proton-activated ion channels have been studied in voltage clamp experiments in vertebrate neurons, where the membrane potential has been kept constant and the current evoked by a change in pH_o has been measured. In almost every reported study the response to a challenge with an acidic solution is associated with an increase in membrane conductance due to the opening of ion channels. The evoked currents in vertebrate neurons can be divided into two groups on the basis of their time courses,

pH and Brain Function, Edited by Kai Kaila and Bruce R. Ransom
ISBN 0-471-11838-9

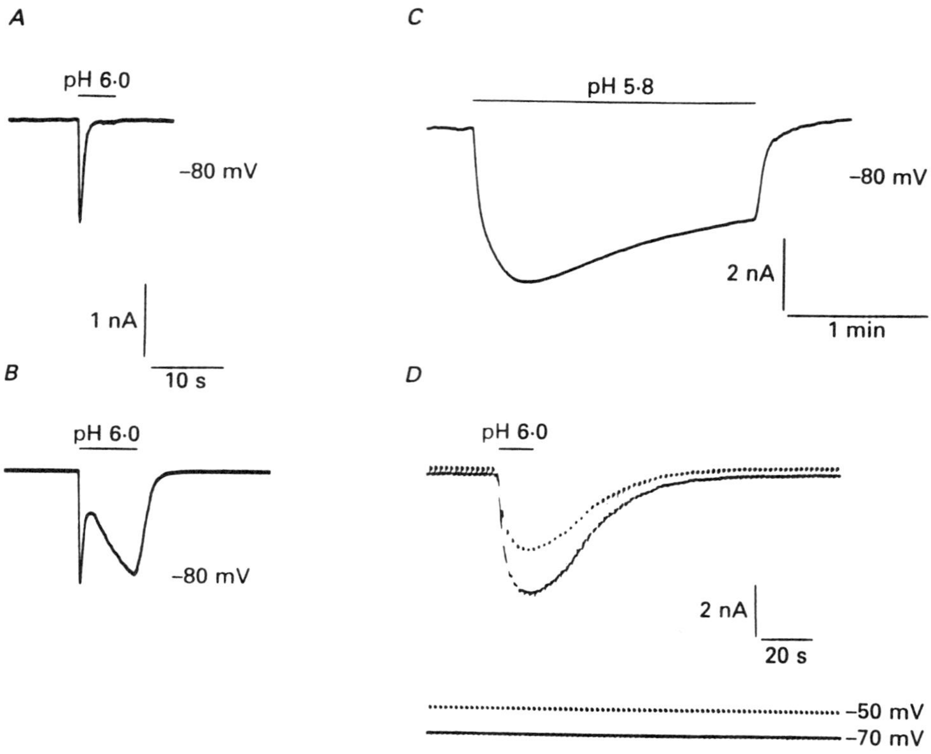

FIGURE 1. Examples of transient and sustained proton-gated currents in rat DRG neurons. **A.** Transient current ($I_{Na(H)}$). **B.** $I_{Na(H)}$ and a sustained current recorded in another DRG neuron. **C.** Prolonged sustained current to 2-minute exposure to acidic medium. **D.** Voltage steps of 1-sec duration, between −50 and −70 mV, show the increase in conductance associated with the sustained current. All holding potentials as shown. Duration of application indicated by horizontal bars. Reprinted from Bevan and Yeats (1991).

pharmacology, and the ion selectivity of the activated channels. One type of current is *transient* and has a duration of only a few seconds in the continued presence of acid solutions, while the other current can be *sustained* for minutes when pH_o is decreased (Bevan and Yeats, 1991). Examples of the two types of current recorded in rat dorsal root ganglion (DRG) neurons can be seen in Fig. 1. The membrane potential changes evoked by the two types of current are slowed slightly by the membrane properties of the neurons but show essentially the same time course.

TRANSIENT CATION CURRENTS

Currents activated by a rapid decrease in pH_o were first described in a subpopulation of rat trigeminal and dorsal root ganglion (DRG) neurons by Krishtal and Pidoplichko (1980). Their presence in small-diameter sensory neurons, many of which are likely to be nociceptive neurons, led to the suggestion that these play a role in nociception (Krishtal and Pidoplichko 1981). While this remains a valid suggestion for these neurons, subsequent investigations have demonstrated that similar transient proton-gated currents can be found in a range of CNS and PNS neurons as well as in some, but not all, glial cells. To date transient currents have been described in DRG neurons of both chick (Konnerth et al., 1987) and rat (Krishtal and Pidoplichko, 1980), chick motoneurones (Hatt et al., 1995), rat trigeminal neurons (Krishtal and Pidoplichko, 1981), superior collicular neurons (Grantyn and Lux, 1988), ventromedial hypothalamic neurons (Ueno et al., 1992) and cortical neurons (Davies et al., 1987), frog DRG neurons (Akaike et al., 1990; Philippi et al., 1995) and parasympathetic neurons (Kim et al., 1990), mouse neuroblastoma cells (Krishtal and Pidoplichko, 1980), and bovine chromaffin cells (Morad and Callewaert, 1988), as well as rat oligodendrocytes and their precursor cells (Sontheimer et al., 1989). Hatt et al. (1995) also reported that they had recorded transient currents from neurons isolated from the hippocampus, spinal cord, and olfactory bulb of the rat, and Davies et al. (1988) cited unpublished accounts of transient currents in chick ciliary ganglion and sympathetic ganglion neurons, and rat neuroblastoma cells. The expression of a transient proton-gated current in a wide range of cells suggests that its role is not restricted to nociception but that the current has a more general function in cell signalling. As the currents have a high permeability to Na^+ they have been termed $I_{Na(H)}$, even though the conductance has a significant permeability to other cations (see below).

Dependence of Current on pH_o

A rapid reduction in pH_o from physiological levels is required to evoke $I_{Na(H)}$ and the current is not seen if the pH is decreased slowly (Konnerth et al., 1987). The amplitude of the current increases with lower final pH_o values. As shown in Fig. 2, the shape of the pH-current amplitude plots can be described by a simple sigmoidal curve of the form

$$I = \frac{I_{max}}{1 + K/[H^+]_o^n}$$

where I_{max} is the maximum amplitude of the evoked current, n is the Hill coefficient or cooperativity factor, K is the dissociation constant for protons and $[H^+]_o$ is the external proton concentration ($pH_o = -\log_{10}[H^+]_o$).

The precise pH-dependence varies between cell types, and the pH_o for half-maximal activation of the current ranges from pH 7.1 in frog parasympathetic neurons (Kim et al., 1990) to pH 5.0 in rat ventromedial hypothalamic neurons (Ueno et al., 1992). The Hill coefficient is greater than 1 for all preparations studied, and published estimates vary between 1.5 (Ueno et al., 1992) and about 4 (Konnerth et al., 1987, see Fig. 2) as documented in Table 1. These values suggest that the opening of each membrane channel requires the binding of at least 2, and perhaps as many as 4, protons.

Ionic Basis of Current

All the published data indicate that Na^+ is the major charge carrier of the current in physiological conditions and estimates of the reversal potential (E_{rev}) for the current agree closely with the Na^+ equilibrium potential (E_{Na}). The activated ion channels show a significant permeability for other monovalent cations (see Table 1) and the original experiments of Krishtal and Pidoplichko (1981) on rat trigeminal ganglion neurons indicated a relative permeability for K^+ of $P_K/P_{Na} = {\sim}0.1$. Other relative permeabilities, calculated from

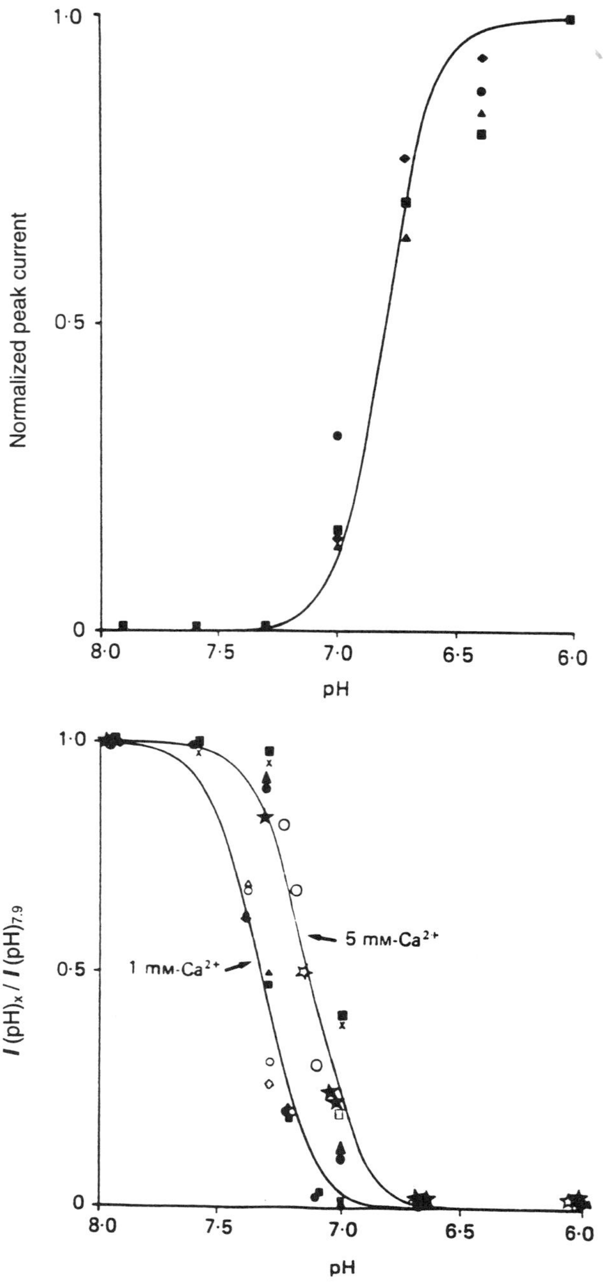

FIGURE 2. *Top*. Relationship between amplitude of $I_{Na(H)}$ and pH of the test solution for chick DRG neurons. The pH_o was stepped from pH 7.9 to levels shown on the abscissa. Amplitudes were normalized to the amplitude evoked by pH 6 medium. The fitted curve shows half-maximum activation at pH 6.83. *Bottom*. Steady-state inactivation of $I_{Na(H)}$. The pH_o was stepped from the levels indicated on the abscissa to pH 6, and amplitudes were normalized to the response evoked by a step change from pH 7.9 to pH 6. Half-maximum inactivation at pH 7.11 (1 mM Ca^{2+}) and pH 7.32 (5 mM Ca^{2+}). All lines are drawn assuming a stoichiometry of 4 H^+: 1 "binding site." Reprinted from Konnerth et al. (1987).

measurements of E_{rev} in chick DRG neurons (Konnerth et al., 1987) and rat ventromedial hypothalamic neurons (Ueno et al., 1992), are $P_{Li}/P_{Na} = 0.8$ and 1.1; $P_K/P_{Na} = 0.18$ and 0.08; $P_{Rb}/P_{Na} = 0.14$ and 0.07, and $P_{Cs}/P_{Na} = 0.04$ and 0.05. Pidoplichko (1992) also reported that ammonium ions were permeable and calculated that P_{NH_4}/P_{Na} was about 0.5 in rat trigeminal ganglion neurons.

Although the channels were originally thought to be impermeable to divalent cations, it has more recently been demonstrated that they have a significant permeability to Ca^{2+}, as well as to some other divalent cations. Inward currents carried by Ca^{2+} can be recorded in the absence of external Na^+ in rat trigeminal ganglion and ventromedial hypothalamic neurons (Kovalchuk et al., 1990; Ueno et al., 1992). The relative permeabilities for various divalent cations have been calculated from measurements of E_{rev} to yield values of $P_{Ca}/P_{Na} = 0.31$, $P_{Ba}/P_{Na} = 0.07$, and $P_{Sr}/P_{Na} = 0.2$ (Kovalchuk et al., 1990).

Activation Kinetics

$I_{Na(H)}$ activates rapidly and usually reaches a peak amplitude in whole cells within a few hundred milliseconds after making a step reduction in pH_o (Table 1). The rising phase of the current in rat ventromedial hypothalamic neurons and frog parasympathetic neurons can be fitted with a single exponential function with time constants (τ_a) of about 5–20 milliseconds (Kim et al., 1990; Ueno et al., 1992). In these preparations, τ_a depends on the final pH_o value and decreases with lower pH_o values. A similar increase in the rate of activation with step changes to lower pH_o values was originally noted by Davies et al. (1988), who reported a 20-fold decrease in the half-time to reach peak amplitude for ensemble currents in outside-out membrane patches when pH_o was stepped from pH 7.9 to either pH 7.2 ($t_{1/2} = 10$–40 ms) or pH 5.7 ($t_{1/2} = 0.4$–0.7 ms). Transient currents in frog DRG neurons activate more slowly than in other preparations ($\tau_a = 1$–4 seconds; Akaike et al., 1990); nevertheless, their rate of activation also increases in more acidic conditions. The rate of activation is independent of membrane potential in all preparations studied. τ_a is influenced by the external Ca^{2+} concentration in at least one preparation, rat ventromedial hypothalamic neurons, where the rate increases when $[Ca^{2+}]_o$ is lowered from 5 mM to 0.01 mM (Ueno et al., 1992). This calcium dependence may vary between preparations as the results of Akaike et al. (1990) and Davies et al. (1988) on frog and chick DRG neurons showed that the rate of activation was not influenced significantly by the external Ca^{2+} concentration.

Inactivation of $I_{Na(H)}$

Evidence for different types of current. The rate and degree of inactivation of $I_{Na(H)}$ varies, and Krishtal and Pidoplichko (1981) described three types of inactivation in rat trigeminal neurons (see Fig. 3). In the majority of neurons (25/32) the current inactivated exponentially to zero with a time constant (τ_i) of 1.5–2 sec. In a few neurons, $I_{Na(H)}$ inactivated completely but at a slower rate (τ_i 3–4 sec), while in 5 of the 32 neurons the current showed an even slower, partial inactivation. Akaike et al. (1990) also reported three types of currents with distinct inactivation rates in DRG neurons of the frog, although they restricted their investigations to one type of current. Three kinetically distinct components of $I_{Na(H)}$, with fast, medium, and slow rates of inactivation, were reported in rat trigeminal ganglion neurons by Pidoplichko (1992), who showed that the medium component could be selectively activated by external ammonium ions (see p. 459).

TABLE 1 Biophysical Properties of $I_{Na(H)}$ in Various Neurons

Preparation	Permeant Ions	EC_{50}	Activation Rate: Whole Cell (w/c) or Outside-Out Patch) o/o		Coperativity Factor	Inactivation Rate (τ)	Reference
Chick DRG neurons	$Na^+ > Li^+ \geq K^+ > Rb^+ > Cs^+$	pH 6.83	200–300 ms	(w/c, time to peak)	~2	300 ms	1,2
			~2 ms	(o/o, time to peak)			
Chick motoneurons	Na^+	pH 6.76	6 ms	(o/o, time to peak, pH 7)	2.67	90–200 ms	3
			0.35 ms	(o/o time to peak pH 6)			
Frog DRG neurons	Na^+	pH 6	$\tau \sim 1$ s	(w/c)	1.97	2–15 s	4
						0.5–1 s	
						> 1 min	
Frog DRG neurons						~0.5 s	5
Rat superior colliculus neurons	Na^+	pH 8	< 1s	(w/c, time to peak)		~1.9 s	6
Rat trigeminal neurons	$Na^+ \geq K^+$	pH 6.2				~0.5 s	7
						~0.5–1 s	
						3–4 s	
						~10 s	8
						(~0.1–0.2 s)	
	$Na^+ > NH^+_4$					~0.5 s	9
						(~3 s)	
	$Na^+ > Ca^{2+} > Sr^{2+} > Ba^{2+}$						10

Rat ventromedial hypothalamic neurons	Na^+	pH 5.0	τ 5–15 ms	(w/c)	1.5	0.15–0.2 s 1–2 s	11
Mouse neuroblastoma cells						3.1 s	7.8
Frog parasympathetic neurons	Na^+	pH 7.1	τ 5 ms	(w/c)	3.2	100 ms	12

References:
1. Konnerth et al. (1987).
2. Davies et al. (1988).
3. Hatt et al. (1995).
4. Akaike et al. (1990).
5. Philippi et al. (1995).
6. Grantyn and Lux (1988).
7. Krishtal and Pidoplichko (1980).
8. Krishtal and Pidoplichko (1981).
9. Pidoplichko (1982).
10. Kovalchuk et al. (1990).
11. Ueno et al. (1992).
12. Kim et al. (1990).

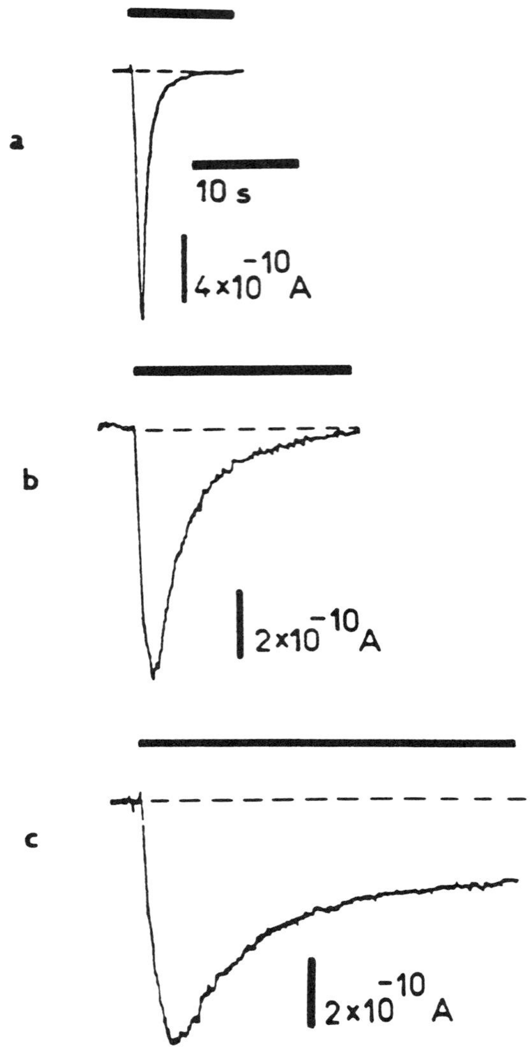

FIGURE 3. Three different types of $I_{Na(H)}$ in chick DRG neurons. Reprinted from Krishtal and Pidoplichko (1981)

Furthermore, this study showed that the fast and slow components could be recorded in isolation by acidifying the external medium after a prior challenge with an alkaline ammonium chloride solution. Evidence for the existence of multiple transient currents is also suggested by the observation that the inactivation phase in rat ventromedial hypothalamic neurons shows two components (Ueno et al., 1992). One component decays exponentially with a τ_i of 150–200 milliseconds, while the other decays with a τ_i of 1–2 seconds. It is unknown whether the kinetically distinct components of $I_{Na(H)}$ represent the activity of separate membrane molecules or different behaviours of a single type of channel.

Effect of membrane potential. The inactivation rate of $I_{Na(H)}$ is not found to be affected greatly by the membrane potential in most preparations (Davies et al., 1988; Akaike et al., 1990; Kim et al., 1990; Ueno et al., 1992). In contrast, Pidoplichko (1992) reported that one component of the transient current (the medium component) showed a voltage-dependent decay phase that inactivated more rapidly at depolarized membrane potentials, with an *e*-fold change in rate for a 25 mV change in potential. Whether this reflects a difference between $I_{Na(H)}$ in different types of neurons remains to be investigated.

Influence of pH$_o$ and [Ca^{2+}]$_o$. The rate of inactivation is influenced by pH_o, and in most studies τ_i has been shown to decrease with lower final pH_o values (Akaike et al., 1990; Kim et al., 1990; Hatt et al., 1995). An exception is the observation by Ueno et al. (1992) that τ_i in rat ventromedial hypothalamic neurons showed little or no pH dependence over the pH range between 5.5 and 4.7, but increased when pH_o was reduced to levels below pH 4.7. Other investigators have not studied the effects of such acid solutions, and it is possible that the difference in behavior is a result of the lower pH_o challenges required to activate the transient current in hypothalamic neurons.

The concentration of Ca^{2+} also affects the rate of inactivation in some, but not all, preparations. Raising the external [Ca^{2+}] decreases τ_i in chick DRG neurons (Davies et al., 1988) and rat ventromedial hypothalamic neurons (Ueno et al., 1992), but has no effect on τ_i in frog DRG neurons (Akaike et al., 1990). The effect of Ca^{2+} on τ_i in chick DRG neurons is reflected to some degree in the behavior of the single-channel currents where decreasing the Ca^{2+} concentration below 0.1 mM causes a small increase in the mean open time (Davies et al., 1988).

Steady-state inactivation. The steady-state inactivation of $I_{Na(H)}$ has been studied by stepping the pH_o to a constant test level from different resting values (Konnerth et al., 1987; Davies et al., 1988; Akaike et al., 1990; Kim et al., 1990; Ueno et al., 1992; Hatt et al., 1995). The relationship between the normalized evoked current and the initial pH_o is sigmoidal and can be fitted by an equation similar to that used to describe the activation of the current (see Fig. 2). Where reported, the slope factor is similar to the slope factor for activation. The pH for 50% inactivation of the transient current (IC_{50}) is higher than that required for 50% activation of the current (EC_{50}). The differences vary from 0.3 to 0.5 pH unit in chick DRG neurons (Konnerth et al., 1987), to 1.2 to 1.6 pH unit in frog DRG and parasympathetic and rat ventromedial hypothalamic neurons (Akaike et al., 1990; Kim et al., 1990; Ueno et al., 1992). Both activation (Akaike et al., 1990) and inactivation (Konnerth et al., 1987) curves are influenced by the concentration of extracellular Ca^{2+} and are shifted in a parallel manner to more acid pH values as [Ca^{2+}]$_o$ is increased (see Fig. 2). A parallel shift in the activation curve without any decrease in the amplitude of the current suggests that Ca^{2+} and protons compete competitively for a binding site on the "receptor/ion channel."

Mechanism of inactivation. It seems unlikely that the ion channels have to be activated in order to enter the inactivated state. Whole cell studies have shown that small changes in pH_o, which do not activate the current significantly, can cause a large degree of inactivation (Konnerth et al., 1987). This is reflected in the plots showing the pH dependence of activation and steady-state inactivation, which demonstrate that inactivation occurs at pH levels that evoke little or no current (Fig. 2). It is feasible that small decreases in pH_o evoke temporally dispersed channel openings that can be overlooked in whole-cell recordings. The results of single-channel experiments are, however, consistent with the conclusions from the whole-cell studies. Inactivation can occur after a conditioning acidification step that fails to produce any channel openings (Davies et al., 1988). These findings suggest that the inactivated state can be entered "directly" from the closed state and does not require prior channel activation.

Recovery from inactivation. $I_{Na(H)}$ recovers from inactivation, once the pH_o returns to more alkaline levels, over a period between 1 and 100 sec (Konnerth et al., 1987; Davies et al., 1988; Ueno et al., 1992; Hatt et al., 1995). The factors that determine the rate of recovery have not been investigated in depth, although the available data suggest that the amplitude of the pH change between the resting and test conditions is an important determinant.

Single-Channel Properties

The properties of single proton-activated channels have been studied in outside-out and inside-out membrane patches from chick DRG neurons (Konnerth et al., 1987; Davies et al., 1988) and motoneurons (Hatt et al., 1995). Application of acidic medium to the internal surface of the plasma membrane in inside-out membrane patches failed to evoke any currents, whereas single-channel currents were evoked when pH_o was decreased in outside-out patches (Davies et al., 1988). In addition no single-channel currents were evoked when the external pH was decreased during cell-attached recordings. Activation of the channels was not influenced by the inclusion or absence of ATP or the intracellular second messenger, cyclic AMP in the patch pipette. Similarly $[Ca^{2+}]_i$, which also can act as a second messenger, could be varied over a wide range (10^{-9} to 10^{-3} M) without affecting the activation or inactivation of $I_{Na(H)}$ (Davies et al., 1988). These findings indicate that the protons act at the outside and not on the inside of the plasma membrane, and that the action is likely to be direct and not involve cyclic AMP or cytosolic Ca^{2+} as intracellular second messengers.

The single channel conductance in the two types of chick neurons is similar: 20 pS in DRG neurons (Konnerth et al., 1988) and 22 pS in motoneurons (Hatt et al., 1995), with NaCl-based solutions on the outside and KCl-based solutions bathing the internal surface of the membrane. The single-channel current amplitudes in DRG neurons are unaffected by external proton concentrations over the range pH 5.7 to pH 7.3. The distribution of channel open times can be fitted by a single exponential function in both cell types and mean open times are similar: 1.4 milliseconds in motoneurons and 0.6–1.5 milliseconds in DRG neurons.

Pharmacological Properties

The majority of the pharmacological studies on the transient proton-evoked current have examined the actions of agents that are known to act on other types of ion channels, particularly voltage-gated Ca^{2+} and Na^+ channels. Some of the agents tested are listed in Table 2.

Inorganic cations. Divalent cations, which inhibit the passage of Ca^{2+} through calcium channels, have been tested for effects on $I_{Na(H)}$. In sodium-containing solutions without Ca^{2+}, addition of divalent cations reduces the amplitude of the current in frog DRG

TABLE 2 Pharmacology of $I_{Na(H)}$

Agent	Effect	Cell Type	Reference
Cd^{2+}	Blocked	Chick DRG	1,2
		Frog DRG	4
		Frog parasympathetic neurons	12
Ni^{2+}	Blocked	Chick DRG	1,2
Increased Ca^{2+}	Reduced	Chick DRG	1,2
		Frog DRG	4
Ba^{2+}	Reduced	Frog DRG	4
Sr^{2+}	Reduced	Frog DRG	4
Tetrodotoxin ($1 > \mu M$)	No effect	Chick DRG	1
		Rat trigeminal neurons	7
		Mouse spinal neurons	3
		Frog parasympathetic neurons	2
		Rat hypothalamic neurons	5
	Blocked	Frog DRG	4
	Blocked with increased inactivation rate	Rat ventromedial hypothalamus (VMH) neurons	11
Amiloride	Blocked	Rat trigeminal neurons	10
	Suppression at negative E_m	Chick DRG	6
ω-conotoxin ($1 > \mu M$)	No effect	Chick DRG	8, 13
		Frog parasympathetic neurons	12
		Rat hypothalamic neurons	5
	Blocked	Frog DRG	4
Diltiazem	Blocked	Chick DRG	1,2
		Frog DRG	4
		Frog parasympathetic neurons	12
		Rat hypothalamic neurons	5
Verapamil	Blocked	Chick DRG	2
	No effect	Rat trigeminal neurons (M type current)	9
Nifedipine	No effect	Chick DRG	2
D 600	Blocked	Frog DRG	4
		Frog parasympathetic neurons	12
		Rat hypothalamic neurons	5
Bay K8644	No effect	Chick DRG	2

References:

1. Konnerth et al. (1987).
2. Davies et al. (1988).
3. Gruol et al. (1980).
4. Akaike et al. (1990).
5. Akaike and Tokutomi (cited in ref. 4).
6. Tang et al. (1988).
7. Krishtal and Pidoplichko (1980).
8. Morad and Callewaert (1988).
9. Pidoplicho (1992).
10. Kovalchuk et al. (1990).
11. Ueno et al. (1992).
12. Kim et al. (1990).
13. Carbone and Lux (1988)

and rat ventromedial hypothalamic neurons with a blocking potency consistent with the following rank order: $Mn^{2+} > Co^{2+} > Ca^{2+} > Sr^{2+} > Ba^{2+} > Mg^{2+}$ (Akaike et al., 1990; Ueno et al., 1992). Cd^{2+}, which is a potent blocker of the voltage-gated calcium current, also inhibits the proton-evoked current by 40–100% at 1 mM (Konnerth et al., 1987; Davies et al., 1988; Akaike et al., 1990; Kim et al., 1990), and Ni^{2+} reduces the current amplitude with a half maximal block at 1–1.5 mM (Davies et al., 1988; Hatt et al., 1995). For some ions, such as Ca^{2+}, Sr^{2+}, and Ba^{2+}, the observed reduction in amplitude undoubtedly results from their ability to permeate the channel (see p. 726). In these cases, the divalent cations, which have lower permeabilities than Na^+ and are present at lower concentrations, compete with Na^+ as a charge carrier. Whether Mn^{2+} and Co^{2+} act in a similar manner or function as nonpermeant channel blockers is unknown.

Organic blockers. Several organic blockers of voltage-gated Ca^{2+} currents (I_{Ca}) inhibit $I_{Na(H)}$ to some degree, although the effect often occurs at concentrations that are much higher than those used to block I_{Ca}. The effects of some of these agents also vary between preparations. For example, ω-conotoxin (3–20 μM) has no effect in chick DRG, frog parasympathetic, and rat hypothalamic neurons (Carbone and Lux, 1988; Kim et al., 1990; Akaike and Tokutomi, unpublished observations cited in Akaike et al., 1990), but does produce a concentration-dependent (10^{-7} to 10^{-5} M) inhibition of $I_{Na(H)}$ in frog DRG neurons (Akaike et al., 1990). Both diltiazem (1 mM) and D600 (0.1–1 mM) block $I_{Na(H)}$ by 30–60% (Konnerth et al., 1987; Akaike et al., 1990; Kim et al., 1990; Akaike and Tokutumi, unpublished observations cited in Akaike et al., 1990), although at these concentrations both drugs show nonspecific effects and inhibit the voltage-gated Na^+ current (Kim et al., 1990). Neither nifedipine (50 μM) nor Bay K8644 has any effect on $I_{Na(H)}$ in chick DRG neurons (Davies et al., 1988).

Amiloride is among the most potent inhibitors of $I_{Na(H)}$ and, at ≥10 μM, produces a strong voltage-dependent block in chick DRG and rat trigeminal ganglion neurons (Korkushko and Krishtal, 1984; Tang et al., 1988; Kovalchuk et al., 1990). For example, 10 μM amiloride reduces the current in the chick neurons by 20% at +20 mV but produces an 80% block at −120 mV (Tang et al., 1988).

Na^+ channel-blocking drugs. The voltage-gated Na^+ channel-blocking drug tetrodotoxin (TTX) has also been tested for its effects $I_{Na(H)}$. High (micromolar) concentrations have no effect in almost all preparations studied (see Table 2). A small increase in the rate of inactivation has been noted in rat ventromedial hypothalamic neurons, although the amplitude of the current is not significantly altered (Ueno et al., 1992). The only preparation, to date, where a clear blocking effect has been noted is frog DRG neurons, where Akaike et al. (1990) found that TTX (0.1–10 μM) produces a concentration-dependent reduction in the amplitude of the current associated with an acceleration in the rate of inactivation. At high concentrations of TTX, the decaying or inactivating phase of the current is biphasic, with a rapid component and a slowed late phase. These biphasic kinetics are suggestive of a channel-blocking mode of inhibition for this drug.

The effects of a toxin from the scorpion (*Leiurus quinquestriatus*), which acts to slow the inactivation of voltage-gated Na^+ channels, were investigated by Ueno et al. (1992) on rat ventromedial hypothalamic neurons. The toxin (1 μg/ml) increases the amplitude of $I_{Na(H)}$ and slows the inactivation rate of a slowly inactivating component of the total current in these cells. In contrast there is no evidence that the rapidly inactivating component of $I_{Na(H)}$ is altered by the presence of the toxin. Veratridine (100 μM), which shifts the activation curve for sodium channels to more hyperpolarized potentials had no effect on either the amplitude or the kinetics of $I_{Na(H)}$ current in the same cell type (Ueno et al., 1992).

Molecular Nature of the Channel

Molecular cloning experiments have resolved the question of the molecular basis of $I_{Na(H)}$. Earlier suggestions were that $I_{Na(H)}$ resulted from activation of either voltage gated calcium channels, transformed from a voltage-gated, Ca^{2+} permeable state to a proton-gated Na^+ selective state (Konnerth et al., 1987; Davies et al., 1988) or modified voltage-activated Na^+ channels (Akaike et al., 1990; Ueno et al., 1992). However, Waldmann et al. (1997) have cloned and characterized a proton gated cation channel (ASIC, acid sensing ionic channel) from a rat brain cDNA library which shows essentially identical properties to the channel responsible for $I_{Na(H)}$ in neurons. Structurally this channel belongs to the amiloride-sensitive Na^+ channel/degenerin family of ion channels (Driscoll and Chalfie, 1991; Canessa et al., 1993; Lingueglia et al., 1993; Waldmann et al., 1996), which have two putative transmembrane spanning regions. *In situ* hybridization experiments have shown that the channel is highly expressed in small diameter DRG neurons and also in the CNS where it is expressed at high levels in the olfactory bulb, cerebral cortex, hippocampus, habenula and cerebellum. *Xenopus* oocytes injected with cRNA encoding the ASIC channel are proton-sensitive and an $I_{Na(H)}$-like current can be evoked by reducing the extracellular pH from 7.4 to less than 7.0 (Waldmann et al., 1997). The biophysical properties of the ASIC current and $I_{Na(H)}$ are similar. The activation of the ASIC current shows a steep dependence on pH_o and the current activates rapidly. The expressed channel discriminates relatively poorly between cations and is permeable to Na^+, Li^+, K^+ and H^+ as well as Ca^{2+}. The calculated relative permeabilities $P_{Li}/P_{Na} = 0.77$, $P_K/P_{Na} = 0.08$ and $P_{Ca}/P_{Na} =$ a 0.4 are very similar to those for $I_{Na(H)}$ in chick DRG neurons (Konnerth et al., 1987) and rat ventromedial hypothalamic neurons. (Ueno et al. 1992). Furthermore, the amplitude of the ASIC $I_{Na(H)}$ current is reduced by raising the external Ca^{2+} concentration in the same manner as $I_{Na(H)}$ in neurons (*cf.* Davis et al., 1988). The single channel conductance estimate of 15 pS for the ASIC channel with NaCl outside is similar to the estimates for $I_{Na(H)}$ in chick DRG (20 pS, Konnerth et al., 1988) and motoneurons (22 pS, Hatt et al., 1995). There are also pharmacological similarities and the ASIC current is blocked by amiloride and its derivatives EIPA and benzamil with IC_{50} values of about 10 μM, similar to the values reported for $I_{Na(H)}$ in neurons (*cf.* Korkushko and Krishtal, 1984; Tang et al., 1988; Grantyn et al., 1989).

The available evidence overwhelmingly suggests that the cloned ASIC channel is responsible for at least some types of $I_{Na(H)}$ in neurons. It remains to be seen whether this single type of channel is responsible for the different kinetic forms of phases of $I_{Na(H)}$ or whether there are related channels expressed in different neurons.

A Transient Cation Current Evoked by Alkalinization

There is good evidence from rat trigeminal ganglion neurons that one of the three components of $I_{Na(H)}$ noted above can be activated and desensitized by a rapid step change in the external medium to a solution containing NH_4Cl (Pidoplichko, 1992). A subsequent reduction in pH_o activates the fast and slow components of $I_{Na(H)}$ selectively (Fig. 4). The NH_4Cl-induced current, which is often the dominant component of the composite proton

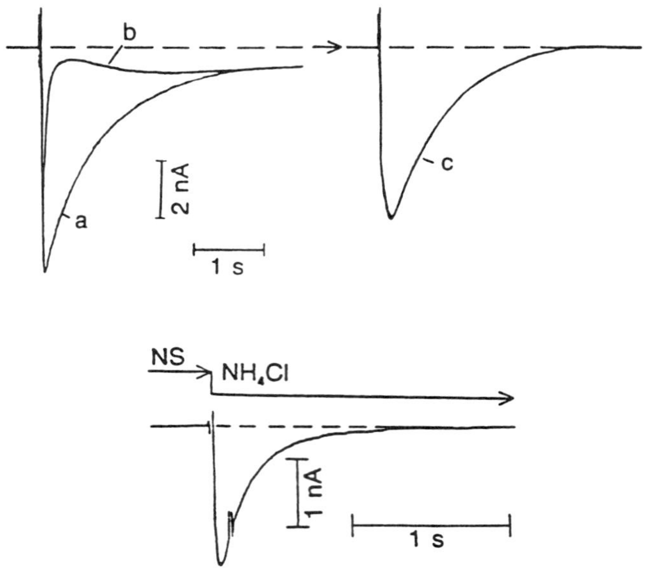

FIGURE 4. *Top*. Trace *a* shows total $I_{Na(H)}$ evoked by pH 6.2 medium in chick trigeminal ganglion neuron (sum of fast, medium, and slow components). In trace *b,* only fast and slow components are evoked when the solution changed from 150 mM NH_4Cl to a pH 6.2 sodium-based solution. Trace *c*. Derived by subtraction of trace *b* from trace *a*. The trace shows the time course of transient current desensitized by NH_4Cl. *Bottom*. Transient current (I_{NH}) evoked by changing from a sodium-based (pH 7.4) solution to 150 mM NH_4Cl (pH 7.4). Reprinted from Pidoplichko (1992).

activated current, appears to be dependent on the concentration of extracellular NH_3 that forms from the added NH_4Cl. The amplitude of the current is related to the concentration of external NH_4Cl at a constant pH_o, with an EC_{50} that corresponds to an external NH_3 concentration of 10^{-4} M. The response evoked by a given concentration of NH_4Cl is also increased by alkalinization of the extracellular medium, which will increase $[NH_3]_o$. The conductance pathway responsible for the current has a high permeability to both Na^+ and NH_4^+ with an estimated relative permeability P_{Na}/P_{NH_4} of about 0.5. Pidoplichko (1992) discussed this current in terms of a sensor for external NH_3, but the data can also be interpreted in other ways. For example, it is well known that a step increase in the concentration of NH_4Cl leads to a rapid increase in internal pH (pH_i) (see Chapters 10–11). It is, therefore, possible that this transient current is evoked not only by external acidification but also by internal alkalinization. Furthermore, the influx of NH_3, which is responsible for the intracellular alkalinization, may lead to a transient acidification of the medium immediately adjacent to the plasma membrane (Chapter 17) as a result of the equilibrium between NH_3 and NH_4^+. The NH_4Cl-evoked current could, therefore, be due to a local acidification, although in this case it is unclear why protons would selectively activate and inactivate only one component of the total proton-gated transient current. Further experimentation will be required to determine the precise mode of activation by NH_3.

Addition of ammonium ions has been reported to depolarize other types of neurons, such as the squid giant axon (Boron and de Weer, 1976), rat hippocampal neurons (Alger and Nicoll, 1983), and frog spinal cord neurons (Nicoll, 1978). In most experiments, the change in membrane potential was not associated with any measurable alteration in membrane resistance, which suggests little or no activation of membrane currents. The depolarizing effect of ammonium ions is likely to be complex and involve several mechanisms (see Szerb and Butterworth, 1992), and it remains possible that alkalinization opens ion channels in other types of neurons. Experiments with rapid changes in ammonium ion concentration may reveal that this is a widespread phenomenon, especially in the light of the observations that many types of neurons express $I_{Na(H)}$ and that a component of this current in trigeminal neurons can be activated by addition of NH_4Cl.

SUSTAINED PROTON-GATED CATION CURRENTS

A reduction in the extracellular pH activates another type of current in a subpopulation of mammalian DRG neurons (Bevan and Yeats, 1991). This current differs in many respects from $I_{Na(H)}$. The time course of the current is slow compared to the transient current. The current usually takes 10–25 seconds to reach a peak amplitude after a sudden maintained reduction in pH_o and persists for many minutes with only a modest reduction in amplitude (see Fig. 1). The expression of the sustained current has not been investigated in depth and the published reports are restricted to sensory neurons. Here fewer neurons express the sustained current than $I_{Na(H)}$. For example, in rat DRG neurons the transient current was found in 53–67% of the cells, whereas the sustained current appeared in 47–49% of cells (Bevan and Winter, 1995). Thus some neurons show only the transient proton-activated current. It remains to be seen whether other types of neurons from the CNS or PNS, or sensory neurons from other species, respond with a sustained current when pH_o is reduced.

Ionic Basis of the Current

The biophysical properties of the sustained current are different from those of $I_{Na(H)}$ (see Bevan and Yeats, 1991), which underlines the separate molecular basis for the two responses. The sustained current is generated by ion flow through channels that discriminate poorly between Na^+, K^+, and Cs^+. Reversal potential experiments revealed relative permeability ratios of $P_K/P_{Na} = 1.32$ and $P_{Cs}/P_{Na} = 1.12$ and ion flux experiments showed that both Rb^+ and guanidinium ions are also permeable (Fig. 5). The channel also has appreciable permeability to Ca^{2+} with a relative permeability ratio P_{Ca}/P_{Cs} of 0.41–0.81 (Zeilhofer et al., 1996, 1997). A large reduction in pH_o is also required to generate the sustained current. The results of electrophysiological and ion flux experiments indicated little or no sustained response to acidification until pH_o is reduced below 6.6. The relationship between pH_o and $^{86}Rb^+$ efflux rate could be fitted by a simple sigmoidal curve of the form noted for $I_{Na(H)}$ with a half-maximal efflux rate at pH 5.8 and a cooperativity factor of about 2 (Fig. 5).

Relationship to Capsaicin-Operated Ion Channels

It has been proposed that protons activate the same ion channels as capsaicin, the pungent ingredient in *Capsicum* peppers that selectively stimulates small-diameter nociceptive sensory neurons (Bevan and Yeats, 1991; Bevan et al., 1993; Bevan and Winter, 1995; Bevan and Docherty, 1993, 1996). This hypothesis is supported by the similar properties of the sustained proton-evoked current and capsaicin-evoked current in rat sensory neurons. The

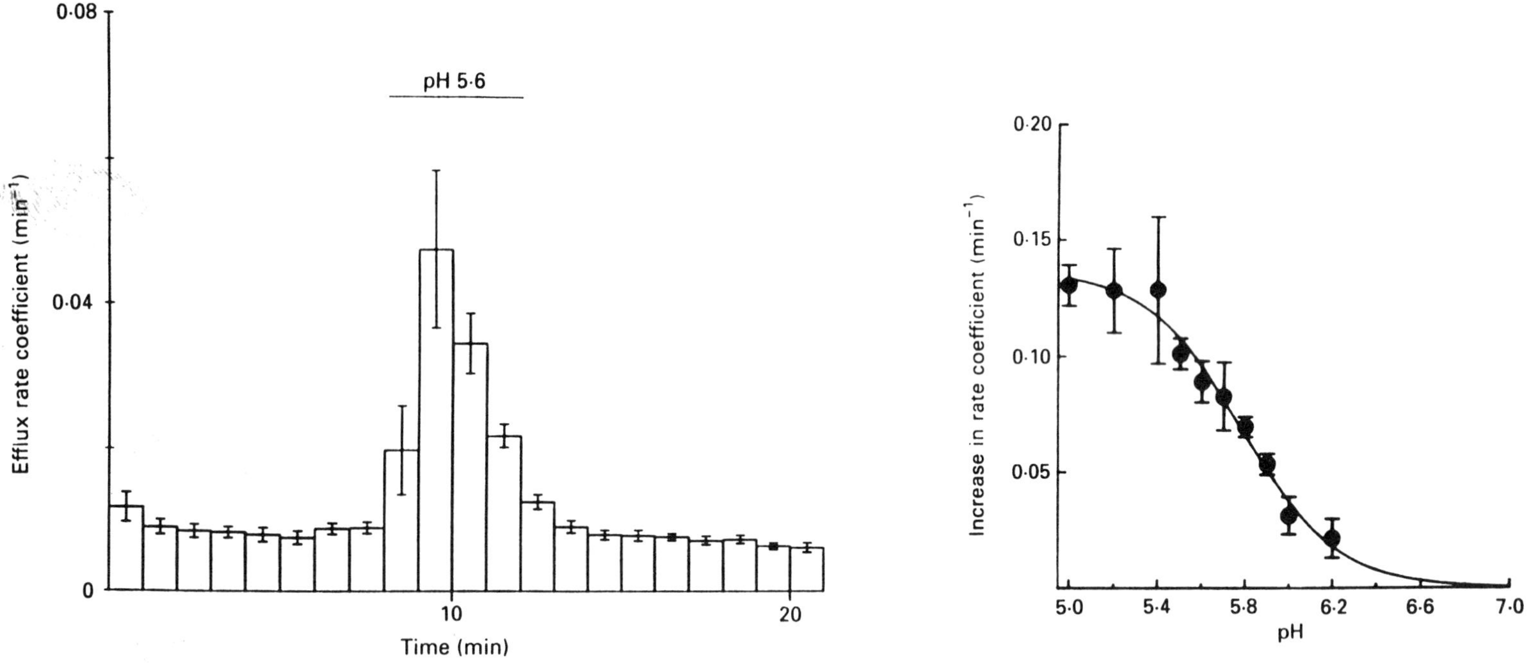

FIGURE 5. *Left*. Efflux of $^{86}Rb^+$ from adult rat DRG neurons evoked by pH 5.6 medium. Mean ± SD for 6 cultures. *Right*. Dose–response curve for $^{86}Rb^+$ efflux from neonatal rat DRG cultures in a NaCl-based solution. Fitted curve with half-maximal rate at pH 5.8 and cooperativity factor of 1.97. Reprinted from Bevan and Yeats (1991).

proton-evoked current is only found in cells that were sensitive to capsaicin (Bevan and Docherty, 1993; Liu and Simon, 1994), and the currents evoked by either stimulus have a similar time course (see e.g., Bevan and Winter, 1995). Preliminary studies on the ion selectivity of the capsaicin-activated conductance also showed that the relative permeabilities to K^+ (P_K/P_{Na}) and Cs^+ (P_{Cs}/P_{Na}) are identical to those given above for the proton-activated conductance (Bevan et al., 1993). Furthermore, the amplitudes of the two current responses are also well correlated and the expression of capsaicin-evoked and sustained proton-evoked currents are regulated to the same degree in culture by the presence or absence of nerve growth factor (Bevan and Winter, 1995). These findings are consistent with the hypothesis that protons and capsaicin open the same ion channels in DRG neurons and suggest that protons are endogenous activators of capsaicin-sensitive ion channels.

Several pieces of evidence challenge this view of a single type of ion channel. First although the capsaicin operated ion channels are Ca^{2+}-permeable (Wood et al., 1988; Bevan and Docherty 1993, 1996) and their activation stimulates the entry of Ca^{2+} from the extracellular medium to raise the concentration of free intracellular Ca^{2+} to high levels (Bleakman et al., 1990; Cholewinski et al., 1993), their calculated Ca^{2+} permeability differs from that of proton-activated currents in DRG neurons (Zeilhoffer et al., 1997). When compared with monovalent cations (M), of assumed equal permeability, the P_{Ca}/P_M ratio is higher for capsaicin activated channels ($P_{Ca}/P_M = 1.38$–1.58) than for proton-gated channels ($P_{Ca}/P_M = 0.55$–0.84). Zeilhofer et al. (1997) have argued that it is unlikely that this difference can be explained by a reduction in Ca^{2+} permeability at lower pH_o, as has been noted for voltage-gated Ca^{2+} channels (Iijima et al., 1986), as higher relative permeability ratios (P_{Ca}/P_M) were noted for responses evoked by capsaicin applied at pH 5.1 than for pH 5.1 medium alone.

Second, Gracía-Hirschfeld et al. (1995) investigated the $[Ca^{2+}]_i$ responses of rabbit trigeminal neurons to rapid reductions in pH_o and found a proton-evoked rise in $[Ca^{2+}]_i$, caused by Ca^{2+} entry from the external medium, in 86% of neurons, whereas they found that only 40% of the proton-sensitive neurons responded to capsaicin. This finding led them to conclude that capsaicin and protons raised $[Ca^{2+}]_i$ by different mechanisms. Third, Oh et al. (1996) found that some rat DRG neurons, and outside-out membrane patches from these neurons, responded to capsaicin but not to acidic external solutions (pH 5.8–6.0). The results of these experiments are clearly at variance with those of Bevan and colleagues noted above, but the reason for the discrepancies are, at present, unclear.

Single-Channel Properties

Little has been reported on the properties of the single-channel currents associated with the sustained response to acidification. A preliminary account noted that the single-channel currents evoked by protons and capsaicin showed very similar properties (Bevan et al., 1993). The amplitudes of the currents evoked by either agent at any given membrane potential and the reversal potentials are very similar. The current–voltage relationships for the currents evoked by both agents showed pronounced outward rectification, so that the currents at positive ($>+40$ mV) membrane potentials are larger than at the equivalent negative (<-40 mV) membrane potentials, even with identical solutions on either side of the membrane. An additional similarity is that the probability of channel opening evoked by either protons or capsaicin is voltage-sensitive and increases when the membrane potential is positive. Given the differences in relative permeabilities for proton- and capsaicin-activated conductances (Zeilhofer et al., 1997), a more detailed comparison of single-channel properties, including those of capsaicin-operated channels at low pH, is warranted.

Relationship to Nerve Activation

In sensory nerve preparations, acidification of the external medium leads to sustained responses such as nerve depolarization (Rang and Ritchie, 1988; Bevan et al., 1992b), action potential firing (Belmonte et al., 1991; Steen et al., 1992), and release of neuropeptides (see e.g., Geppetti et al., 1991; Lou and Lundberg, 1992). Low-pH solutions also evoke the sensation of pain in human volunteers (Lindahl, 1962; Steen and Reeh, 1993). It seems likely that these responses to acidic solutions are initiated, at least in part, by the activation of the sustained proton-activated current, although it is also possible that some nerves can be depolarized and activated by protons in other ways; for example, by decreasing a resting K^+ conductance (Gruol et al., 1980).

CO_2 and Sensory Nerve Activation

Some sensory nerves are activated by a reduction in pH produced by an elevation of the CO_2 concentration (Anton et al., 1992; Steen et al., 1992; Komai and Bryant, 1993; see also Chapters 26–27). This effect can be potentiated by amiloride (Steen et al., 1995a,b), an inhibitor of the Na^+/H^+ exchange mechanism that is a major membrane regulator of pH_i in DRG neurons (S. Bevan and CD Richards, unpublished observations), and suppressed by acetazolamide, a membrane permeant inhibitor of the enzyme carbonic anhydrase (Steen et al., 1992; Komai and Bryant, 1993; Chapter 3). The rate of nerve activation by CO_2 is faster than that seen with phosphate-buffered solutions, which suggests that the site of action of the acid generated by the CO_2 is intracellular (Steen et al., 1992). Carbonic anhydrase is found in sensory neurons (Riley et al., 1988), where it is associated with small-, medium-, and large-diameter neurons (Prabhakar and Lawson, 1995). A working hypothesis is that the CO_2 diffuses across the plasma membrane and forms carbonic acid intracellularly as well as extracellularly. Although the reaction will proceed to equilibrium in the absence of functional carbonic anhydrase, the enzyme will increase the rate of production. Any slow change in pH_i may be reduced by the actions of the membrane regulators of pH_i. Inhibition by acetazolamide can, therefore, result in slower, and perhaps smaller, shifts in pH_i in response to elevated CO_2 levels (e.g., Saarikoski and Kaila, 1992).

The results with CO_2 and acetazolamide raise the possibility that the sustained proton-gated current in DRG neurons is triggered by protons acting on the inside of the plasma membrane. Acidification of the external medium with buffers such as 2-[N-morpholino]ethanesulfonic acid (MES) lead to a slow decrease in pH_i (Bevan and Richards, 1996). An intracellular site of action could therefore explain the slow onset of the sustained proton-gated current after a rapid step change in pH_o.

Pharmacological Properties

There have been few direct studies of the pharmacological properties of the sustained current. Most of the available information has been inferred from the effects of agents on the sustained responses of sensory nerve preparations noted above. Interpretation of the effects of a drug on proton-evoked nerve activity, therefore, requires careful consideration, as any inhibition could be due to an effect on voltage-gated ion channels rather than a block of proton-gated channels.

Ruthenium red. One agent that has been used to investigate proton-evoked response is ruthenium red. This dye was originally described as a noncompetitive inhibitor of capsaicin. It acts at the capsaicin-activated ion channel (Dray et al., 1990; Bevan et al., 1992a; Maggi et al., 1993) to inhibit ion flow through the channels (Wood et al., 1988). In this way, ruthenium red blocks both the release of calcitonin gene-related peptide (CGRP) from sensory nerves and the sensory neuropeptide-mediated muscle contractions evoked by capsaicin (Del Bianco et al., 1991; Franco-Cereceda et al., 1991; Geppetti et al., 1991; Lou et al., 1991; Santicioli et al., 1992). Micromolar concentrations of ruthenium red also block the effects of protons, as measured by a variety of responses in a range of preparations: depolarization of rat vagus nerve (Bevan et al., 1992b), contractions of bronchial smooth muscle (Satoh et al., 1993), nasal irritation (Lou and Lundberg, 1992); and CGRP release from spinal cord (Del Bianco et al., 1991), as well as smooth muscle (Geppetti et al., 1991), cardiac muscle (Franco-Cereceda and Lundberg, 1992; Franco-Cereceda et al., 1994), and skeletal muscle (Santicioli et al., 1993). Ruthenium red appears to be equipotent at blocking the effects of protons and capsaicin and is effective at low micromolar concentrations against these stimuli. At these levels, it shows some selectivity in its actions and does not antagonize the effects of exogenous neuropeptides (CGRP, neurokinin A and substance P), histamine, or nicotine (Amman and Maggi, 1991; Belvisi et al., 1992; Lou and Lundberg, 1992; Satoh et al., 1992; Maggi et al., 1993), whereas at slightly higher concentrations it can inhibit the responses to stimuli such as potassium-induced depolarization and electrical field stimulation (see Amman and Maggi, 1991). These "nonspecific" effects of ruthenium red limit its utility as an antagonist; nevertheless, the available data on the effects of the compound are consistent with the hypothesis that protons activate capsaicin-sensitive sensory nerves by acting on a ruthenium red-sensitive common pathway.

Capsazepine. This compound, which is a structural analogue of capsaicin, is a competitive capsaicin antagonist with submicromolar affinity for its receptor site (see Bevan et al., 1992a; Maggi et al., 1993). Capsazepine shows some selectivity as a capsaicin antagonist as it does not antagonize the responses of DRG neurons to either GABA or ATP (Bevan et al., 1992a), and has little or no effect on potassium-evoked CGRP release at low micromolar concentrations (Santicioli et al., 1993). It seems likely that capsazepine acts at the capsaicin receptor/ion channel and has, therefore, been used as a tool to investigate whether protons and capsaicin act at the same site. Unfortunately the published data do not provide a coherent picture. Micromolar concentrations of capsazepine do not inhibit the proton-evoked efflux of [^{14}C]-guanidinium ions or $^{86}Rb^+$ from cultured DRG neurons (Bevan et al., 1992a), or the acid-induced depolarization of vagus nerves from rat or guinea pig (Bevan et al., 1992b; Fox et al., 1995). In addition, the low pH-induced activation of sensory nerves in a neonatal rat tail-spinal cord preparation is unaffected by capsazepine (Dray et al., 1992). The reported effects on ion currents in sensory neurons are conflicting. Bevan et al. (1992b) noted no obvious, consistent effect on either the sustained or transient ($I_{Na(H)}$) proton-gated currents in rat DRG neurons, whereas Liu and Simon (1994) reported that capsazepine blocks both types of currents in trigeminal ganglion neurons from the same species. Low concentrations of capsazepine do inhibit proton-evoked firing of single afferent fibers innervating the airway of guinea pig (Fox et al., 1995) and block evoked peptide release from sensory nerves in several preparations (Franco-Cereceda and Lundberg, 1992; Lou and Lundberg, 1992; Santicioli et al., 1993), although the mechanism of inhibition has not been elucidated. One possibility is that capsazepine blocks the proton-gated currents in these systems. Alternatively, protons could stimulate the release of a capsaicin-like molecule which interacts with sensory nerves in a capsazepine-sensitive manner as suggested by Lou and Lundberg (1992) and Santicioli et al. (1993). Capsazepine inhibits the ability of prostaglandin I_2 and arachidonic acid to release CGRP from guinea pig heart (Franco-Cereceda et al., 1994), which raises the possibility that protons can act, in part, by liberating a lipid mediator. It has also been suggested that there are multiple types of capsaicin receptor (Lui and Simon, 1994; Szallasi, 1994; Colquhoun et al., 1995), and protons may act in a capsazepine-sensitive manner at some, but not all, receptor subtypes. Clearly the way in which capsazepine acts to inhibit the effects of protons in some, but not all, tissues needs to be investigated further. In summary, the available antagonism data are not consistent with a simple model in which protons and capsaicin always act identically at a single binding site to open an ion channel, but rather suggest that they act at different binding sites on the molecule.

Ca^{2+} channel blockers. The Ca^{2+} channel blocker diltiazem can reduce or abolish the firing of cat corneal and rat skin nociceptors evoked by topical acidic solutions (Pozo et al., 1992; Wegner et al., 1996). The relatively prolonged time courses of the responses to acid in these preparations suggests that the activation is via the sustained proton-activated current. Diltiazem (1 mM) blocks the responses to heat but not to mechanical stimuli in the corneal preparation, a result which suggests some selectivity in the mode of inhibition (Pozo et al., 1992) and would be consistent with an inhibition of the sustained proton-gated current. Experiments on the rat skin-nerve preparation confirmed the inhibitory effect of diltiazem on acid stimuli and showed that the drug acts, with an IC_{50} value ranging from 10^4 to 2×10^{-5} M (Wegner et al., 1996). In this study, however, diltiazem inhibited the nerve firing evoked by mechanical and heat stimuli as well as by electrical stimulation and the authors concluded that diltiazem acts not as a selective inhibitor of proton activated firing, but rather as a use-dependent blocker of action potentials. A direct investigation of the effects of diltiazem on proton-gated channels is warranted in view of the conflicting results on intact nerve preparations.

Non-steroidal anti-inflammatory drugs (NSAIDs). Recent studies have shown that the NSAIDs, salicylic acid, acetylsalicylic acid, ibuprofen, and indomethacin are able to suppress selectively the sensation of pain induced by constant infusion of acidic (pH 5.2) medium into the skin of human volunteers (Steen et al., 1995a, 1996). These drugs inhibit the acid-induced pain when applied topically to the skin, but the effect can be reversed by increasing the infusion rate of the acid solution. The mechanisms that underlie sensory neuron activation in these experiments have not been elucidated, but it is probable that the phenomenon is analogous to the long-lasting proton-induced activation of peripheral nerve terminals seen in rat skin-nerve preparations *in vitro* (Steen et al., 1992; Steen and Reeh, 1993). Such an interpretation is strengthened by a preliminary report that the salicylates cause a concentration-dependent reduction of acid-induced nociceptor firing in the skin-nerve preparation (Stefanidis et al., 1994). The excitation of sensory nerves in human skin is, therefore, likely to involve the sustained proton activated current in some way. One suggestion is that a tonic production of prostanoids sensitizes the sensory nerves to low-pH activation and that the cyclooxygenase inhibitors block prostanoid production (Steen et al., 1996). Although this is a plausible explanation for the effects of the NSAIDs, the possibility that the drugs can also influence the sustained current directly has not been ruled out.

PROTON-EVOKED Cl^- CURRENTS

Proton-gated anion currents have been reported in invertebrate neurons. Brief (10–20 ms) applications of saline at pH 5.5 or lower to neurons of the motor nerve net of the scyphomedusan jellyfish *Cyanea capillata* evoke depolarizations, lasting up to 100 milliseconds, associated with a increase in membrane conductance (Anderson and McKay, 1985). The intrinsic kinetic properties of the response have not been studied, as only very brief applications of acidic solutions were tested. Ion substitution experiments showed that the depolarization is unaffected by changes in the extracellular concentration of either Na^+ or K^+, and by the addition of 5 mM Cd^{2+}, but is profoundly altered by changes in the Cl^- concentration. The quantitative relationship between E_{rev} and the Cl^- concentration suggest that the depolarization is largely dependent on the movement of Cl^-. This suggestion is supported by the finding that the response is blocked by 2 mM 4-acetamido-4′-isothiocyanostilbene-2,2′disulfonic acid (SITS), which can act either as an inhibitor of Cl^- channels or as a blocker of anion transport mechanisms that may alter the distribution of Cl^- ions (Chapters 11–12).

Another study on embryonic neurons of the medicinal leech (*Hirudo medicinalis*) used voltage-clamp methods to examine the current elicited by low-pH (pH 6.7) solutions (Gottmann et al., 1989). This current activates fully within about 500 msec and shows only a small decline (<50%) in amplitude during a 5-second application of the test solution. The data indicate that the response is due to a Cl^- current. Current amplitudes are unchanged if Na^+ is replaced with the bulky choline cation or if Cd^{2+} is added to the medium to block Ca^{2+} currents. Furthermore, the current is abolished if external Cl^- is substituted with either SO_4^{2-} or acetate ions and the reversal potential for the evoked current is close to E_{Cl}.

At present, it is unclear whether anionic proton-gated currents exist in vertebrate neurons. Gruol et al. (1980) noted that mouse spinal cord neurons displayed a proton-evoked transient depolarization, which was probably due to the activation of a transient Na^+ current, when potassium acetate microelectrodes were used. Additional effects, notably a late depolarization associated with an increase in conductance, were noted in some cells when potassium chloride microelectrodes were used. This finding raises the possibility that a chloride-mediated current can be activated by the acid challenge. Further, more detailed, studies are required to investigate the presence of proton-gated anionic currents in these and other vertebrate neurons.

POSSIBLE FUNCTIONS OF PROTON-GATED ION CURRENTS

A clear role for proton-gated currents has not yet been elucidated, although several possible functions have been suggested. Investigations of the physiological relevance of these currents has been hampered by the lack of good, selective pharmacological tools. Nevertheless, there is good circumstantial evidence that these currents are relevant for normal physiological or pathophysiological processes.

Changes in pH_o, as well as pH_i, of both neurons and glia, occur in the CNS in response to neural activity (Kraig et al., 1983; Krishtal et al., 1987; Chesler 1990; Syková and Svoboda 1990; Chapters 16–18), and in response to ischemia and anoxia (Harris and Symon, 1984; Mutch and Hansen, 1984; Kraig et al., 1985; Chesler, 1990; Ransom and Phibin, 1992). Shifts of pH_o in both the alkaline and acid direction have been recorded, and acidification by up to 1 pH unit has been noted in anoxic conditions (Kraig et al., 1985), although the recorded changes are usually smaller. Almost all measurements in neural tissue *in situ* have used pH-sensitive microelectrodes to monitor changes in pH. The relatively slow response of the microelectrodes necessarily limits the temporal resolution of the recorded pH changes (Chapter 6). Nevertheless, significant changes in pH_o can be recorded with this method a few seconds after inducing nerve activity (see also Syková and Svoboda, 1990). pH_o can fall within seconds after stimulation of neural G protein-coupled and tyrosine kinase-linked receptors, and similar results have been noted for ligand-gated and voltage-gated ion channels in isolated cells. These observations suggest that such a response could be a common event in neural tissue (see McConnell et al., 1992). Ion-sensitive electrodes will, in general, monitor changes in bulk extracellular pH, and it is likely that cells in the CNS will be exposed to local, pericellular pH_o changes that are larger than the bulk pH_o changes especially in the areas between tightly apposed cells. There is also one report where a pH sensitive dye was used to show rapid pH_o changes, occurring with a time course of tens of milliseconds, in stimulated brain slices (Krishtal et al., 1987). The available evidence, therefore, suggests that CNS neurons can be exposed to relatively large and rapid changes in pH_o (Chapter 17).

Sensory nerves and other nerves in the PNS can experience a fall in pH_o. This acidification is encountered in normal physiological situations, such as intense muscle activity (Victor et al., 1988), as well as in the pathophysiological conditions of inflammation (McCarty et al., 1966; Stevens et al., 1991) and ischemia (Conn et al., 1959; Jacobus et al., 1977; Corbe and Poole-Wilson, 1980; Knopf et al., 1988; Stahl and Longhurst, 1992). The decrease in pH_o can be large, and acidification to pH 6 has been recorded. In general, acidification in peripheral tissues is relatively slow and the reported changes occur over many seconds or even minutes (see above references; and Chapter 37), although it is possible that peripheral nerves are exposed to local shifts in pH that occur more rapidly than those monitored in the bulk extracellular fluid.

Transient Currents

The discovery of $I_{Na(H)}$ in small-diameter DRG neurons led Krishtal and Pidoplichko (1980, 1981) to propose that the current has a role in nociception. Although this remains possible, the inactivation properties of the current would limit the nociceptive response to conditions in which there were sudden brief changes in pH_o. The relatively widespread occurrence of $I_{Na(H)}$ also suggests other roles. The contents of synaptic vesicles can be acidic with a pH in the range of 5.2–5.5 (see for example Parsons et al., 1993), and it therefore seems likely that the pericellular pH will decrease transiently when exocytosis occurs (see Morad and Callewaert, 1988). Despite this prospect, there is no evidence that protons gate any channels during synaptic transmission, and Morad and Callewaert

(1988) suggested that the main effect of acidification would be to decrease I_{Ca} and so reduce subsequent transmitter release.

New information will be required before we can begin to gain further insights into the roles of $I_{Na(H)}$. For example, optical imaging of local changes in pH_o with ion-sensitive dyes may reveal rapid changes associated with particular stimuli or conditions and suggest possible roles for this current.

Sustained Current

A better case can be made for a nociceptive role of the sustained proton current in sensory neurons, where the available evidence indicates that its expression is restricted to nociceptive C- and Aδ-nerve fibers. Lowering pH_o is a sufficient stimulus to excite nociceptive peripheral neurons. Furthermore, similar decreases in pH_o are required to stimulate neurons in experimental preparations, to generate the sensation of pain in human psychophysical experiments, and to activate the sustained current in isolated sensory neurons. Given that pH_o is known to fall in peripheral tissues to levels that would activate this current (see p. 468), it seems likely that protons do activate nociceptive neurons. This direct effect may be augmented by the ability of protons to sensitize nociceptors to inflammatory mediators such as bradykinin, 5-hydroxytryptamine, histamine and prostaglandin E_2 (Steen et al., 1995b).

A reduction in pH_o, in conditions such as ischemia, will activate sensory nerves and elicit reflex cardiovascular and respiratory changes that increase oxygen supply to the tissues. Sensory nerve activation will also cause a local release of the neuropeptides substance P and CGRP from the nerve terminals in the periphery. The released neuropeptides are thought to play a major protective role and evoke responses such as local vasodilatation, which increases blood flow to the affected area, and plasma extravasation, which is a key part of the inflammatory response. The prolonged time course of sensory nerve activation when pH_o falls suggests that the sustained proton-gated current plays a major role in these adaptive responses.

REFERENCES

Akaike N, Krishtal OA, Maruyama T (1990): Proton-induced sodium current in frog isolated dorsal root ganglion cells. J Neurophysiol 63: 805–813.

Akaike N, Ueno S (1994): Proton-induced currents in neuronal cells. Progress in Neurobiol 43: 73–83.

Akopian AN, Sivilotti L, Wood JN (1996): A tetrodotoxin-resistant voltage-gated sodium channel expressed by sensory neurons. Nature 379: 257–262.

Alger BE, Nicoll RA (1983): Ammonia does not selectively block IPSPs in rat hippocampal cells. J Neurophysiol 59: 1381–1392.

Almers W, McCleskey EW, Palade PT (1984): A non-selective cation conductance in frog muscle membrane blocked by micromolar external calcium ions. J Physiol 353: 565–583.

Amman R, Maggi CA (1991): Ruthenium red as a capsaicin antagonist. Life Sci 49: 849–856.

Anderson PAV, McKay MC (1985): Evidence for a proton-activated chloride current in coelenterate neurons. Biol Bull 169: 652–660.

Anton F, Euchner I, Handwerker HO (1992): Psychophysical examination of pain induced by defined CO_2 pulses applied to the nasal mucosa. Pain 49: 53–60.

Arbuckle J, Docherty RJ (1995): Expression of tetrodotoxin-resistant sodium channels in capsaicin-sensitive dorsal root ganglion neurons of adult rats. Neurosci Lett 185: 70–73.

Balestrino M, Somjen G (1988): Concentration of carbon dioxide, interstitial pH and synaptic transmission in hippocampal formation of the rat. J Physiol (Lond) 396: 247–266.

Belmonte C, Gallar J, Pozo MA, Rebollo I (1991): Excitation by irritant chemical substances of sensory afferent units in the cat's cornea. J Physiol (Lond) 437: 709–725.

Belvisi MG, Miura M, Stretton D, Barnes PJ (1992): Capsazepine as a selective antagonist of capsaicin-induced activation of C-fibres in guinea-pig bronchi. Eur J Pharmacol 215: 341–344.

Bevan S, Docherty, RJ (1993): Cellular mechanisms of the action of capsaicin. In Wood JN (ed): Capsaicin in the Study of Pain. London: Academic Press, pp 27–44.

Bevan S, Docherty, RJ (1996): The ionic basis of capsaicin-evoked responses. In Geppetti P, Holzer P (eds): Neurogenic Inflammation. Boca Raton, FL: CRC Press, pp 55–69.

Bevan S, Forbes CA, Winter J (1993): Protons and capsaicin activate the same ion channels in rat isolated dorsal root ganglion neurones. J Physiol (Lond) 459: 401P.

Bevan S, Geppetti P (1994): Protons: small stimulants of capsaicin-sensitive nerves. Trends Neurosci 17: 509–512.

Bevan S, Hothi S Hughes G, James IF, Rang HP, Shah K, Walpole CS, Yeats JC (1992a): Capsazepine: A competitive antagonist of the sensory neurone excitant capsaicin. Br J Pharmacol 107: 544–552.

Bevan S, Rang HP, Shah K (1992b): Capsazepine does not block the proton-evoked activation of rat sensory neurons. Br J Pharmacol 107: 235P.

Bevan S, Richards CD (1996): Changes in intracellular pH and Ca^{2+} concentration evoked by protons in adult rat isolated dorsal root ganglion neurons. J Physiol (Lond) 494: 76P.

Bevan S. and Szolcsányi J (1990): Sensory neuron-specific actions of capsaicin: mechanisms and applications. Trends Pharmacol Sci 11: 330–333.

Bevan S, Winter J (1995) Nerve growth factor (NGF) differentially regulates the chemosensitivity of adult rat cultured sensory neurons. J Neurosci 15: 4918–4926.

Bevan S, Yeats J (1991): Protons activate a cation conductance in a sub-population of rat dorsal root ganglion neurones. J Physiol (Lond) 433: 145–161.

Bleakman D, Brorson JR, Miller RJ (1990): The effect of capsaicin on voltage-gated calcium currents and calcium signals in cultured dorsal root ganglion cells. Br J Pharmacol 101: 423–431.

Boron WF, De Weer P (1976): Intracellular pH transients in squid axons caused by CO_2, NH_3 and metabolic inhibitors. J Gen Physiol 67: 91–112.

Canessa CM, Horisberger JD, Rossier BC (1993):Epithelial sodium channel related to proteins involved in neurodegeneration. Nature 361: 467–470.

Carbone E, Lux HD (1988): Omega-conotoxin blockade distinguishes Ca from Na permeable states in neuronal calcium channels. Pflügers Arch 413: 14–22.

Chesler M (1990): The regulation and modulation of pH in the nervous system. Prog Neurobiol 34: 401–427.

Cholewinski A, Burgess GM, Bevan S (1993): The role of calcium in capsaicin-induced desensitization in rat cultured dorsal root ganglion neurons. Neuroscience 55: 1015–1023.

Colquhoun EQ, Eldershaw TPD, Bennet KL, Hall JL, Dora KA, Clark MG (1995): Functional and metabolic evidence for two different vanilloid (VN_1 and VN_2) receptors in perfused rat hindlimb. Life Sci 57: 91–102.

Conn HL Jr, Wood JC, Morales GS (1959): Rate of change in myocardial glycogen and lactic acid following onset of coronary occlusion. Circ Res 7: 721–729.

Corbe SM, Poole-Wilson PA (1980): The time of onset and severity of acidosis in myocardial ischaemia. J Mol Cell Cardiol 12: 745–760.

Davies NW, Lux HD, Morad M (1988): Site and mechanism of activation of proton-induced sodium current in chick dorsal root ganglion neurones. J Physiol (Lond) 400: 159–187.

Del Bianco E, Santicioli P, Tramontana M, Maggi CA, Cecconi R, Geppetti P (1991): Different pathways by which extracellular Ca^{2+} promotes calcitonin gene-related peptide release from central terminals of capsaicin-sensitive afferents of guinea-pigs: Effects of capsaicin, high K^+ and low pH media. Brain Res 566: 46–53.

Dray A, Forbes CA, Burgess GM (1990): Ruthenium red blocks the capsaicin-induced increase in intracellular calcium and activation of membrane currents in sensory neurons as well as the activation of peripheral nociceptors *in vitro*. Neurosci Lett 110: 52–59.

Dray A, Patel I, Naeem S, Rueff A, Urban L (1992): Studies with capsazepine on peripheral nociceptor activation by capsaicin and low pH: Evidence for a dual effect of capsaicin. Br J Pharmacol 107: 236P

Driscoll M, Chafie M (1991): The *mec-4* gene is a member of a family of *Caenorhabditis elegans* genes that can mutate to induce neuronal degeneration. Nature 349: 588–593.

Fox AJ, Urban L, Barnes PJ, Dray A (1995) Effects of capsazepine against capsaicin- and proton-evoked excitation of single airway C-fibres and vagus nerve from the guinea pig. Neuroscience 67: 741–752.

Franco-Cereceda A, Lundberg JM (1992): Capsazepine inhibits low pH- and lactic acid-evoked release of calcitonin gene-related peptide from sensory nerves in guinea pig heart. Eur J Pharmacol 221: 183–184.

Franco-Cereceda A, Lou Y-P, Lundberg JM (1991): Ruthenium red inhibits CGRP release by capsaicin and resiniferatoxin but not by ouabain, bradykinin, or nicotine in guinea pig heart: Correlation with effects on cardiac contractillity. Br J Pharmacol 104: 305-310.

Franco-Cereceda A, Källner G, Lundberg JM (1994): Cyclo-oxygenase products released by low pH have capsaicin-like actions on sensory nerves in guinea-pig heart. Cardiovasc Res 28: 365–371

Fukushima Y, Hagiwara S (1985): Currents carried by monovalent cations through calcium channels in mouse neoplastic B lymphocytes. J Physiol 358: 255–284.

García-Hirschfeld J, López-Briones LG, Belmonte C, Valdeolmillos M (1995): Intracellular free calcium responses to protons and capsaicin in cultured trigeminal neurones Neuroscience 67: 235–243.

Geppetti P, Del Bianco E, Patacchini R, Santicioli P, Maggi CA, Tramontana M (1991): Low pH-induced release of calcitonin gene-related peptide from capsaicin-sensitive sensory nerves: Mechanism of action and biological response. Neuroscience 41: 295–301

Gottmann K, Dietzel ID, Lux HD (1988): Proton-induced chloride current and voltage-activated Na^+ and Ca^{2+} currents in embryonic neurons from the medicinal leech (*Hirudo medicinalis*). Neurosci Lett 96: 173–178.

Gottmann K, Dietzel ID, Lux HD, Ruedel C (1989): Proton-induced Na^+ current develops prior to voltage-independent Na^+ and Ca^{2+} currents in neuronal precursor cells from chick dorsal root ganglion. Neurosci Lett 99: 90–94.

Grantyn R, Lux HD (1988): Similarity and mutual exclusion of NMDA- and proton-activated transient Na^+-currents in rat tectal neurons. Neurosci Lett 89: 198–203.

Grantyn R, Perouansky M, Rodriguez-Tébar A, Lux HD (1989): Neuronal precursor cells from the rat brain express proton-activated sodium currents prior to voltage- and transmitter-activated cationic inward currents. Dev Brain Res 49: 150–155.

Gruol DA, Barker JL, Huang L-YM, MacDonald JF, Smith TG Jr (1980): Hydrogen ions have multiple effects on the excitability of cultured mammalian neurons. Brain Res 183: 247–252.

Hablitz JJ, Heinemann U, Lux HD (1986): Step reductions in extracellular Ca^{2+} activate a transient inward current in chick dorsal root ganglion cells. Biophys J 50: 753–757.

Harris RJ, Symon L (1984): Extracellular pH, potassium and calcium activities in progressive ischemia of rat cortex. J Cereb Blood Flow Metab 4: 178–186.

Hatt H, Rosenheimer JL, Smith DO (1995): Proton-activated currents in chick spinal motoneurons. J Comp Physiol A 177: 503–510.

Hess P, Lansmann JF, Tsien RW (1986): Calcium channel selectivity for divalent and monovalent cations: voltage and concentration dependence of single channel current in guinea pig ventricular heart cells. J Gen Physiol 88: 293–319.

Hille B (1968): Charges and potentials at the nerve surface: divalent ions and pH. J Gen Physiol 51: 221–236.

Hille B (1973): Potassium channels in myelinated nerve. Selective permeability to small cations. J Gen Physiol 61: 669–686.

Iijima T, Ciani S, Hagiwara S (1986): Effects of the external pH on Ca channels: Experimental studies and theoretical considerations using a two-site, two-ion model. Proc Natl Acad Sci USA 83: 654–658.

Jacobus WE, Taylor GJ, Hollis DP, Nunally RL (1977): Phosphorus magnetic resonance of perfused working rat hearts. Nature 265: 756–758.

Kettenmann H, Blakenfeld GV, Trotter J (1991): Physiological properties of oligodendrocytes during development. Ann N Y Acad Sci 633: 64–77.

Kim D-K, Tateishi N, Akaike N (1990): Proton-gated sodium current in parasympathetic ganglion cells of frog heart. J Neurophysiol 63: 1060–1067.

Knopf H, Theising R, Hirche H (1988): The effect of desimipramine on ischemia-induced changes in extracellular K^+, Na^+ and H^+ concentrations and noradrenaline release in the isolated rat heart during global isochemia. J Cardiovasc Pharmacol 12 (suppl 1): S8–S14.

Komai M, Bryant BP (1988): Acetazolamide specifically inhibits lingual trigeminal nerve responses to carbon dioxide. Brain Res 612: 122–129.

Konnerth A, Lux HD, Morad M (1987): Proton-induced transformation of calcium channels in chick dorsal root ganglion cells. J Physiol (Lond) 386: 603–633.

Korkushko AO, Krishtal OA (1984): Blocking of proton-activated sodium permeability of the membranes of trigeminal ganglion neurons in the rat by organic cations. Neurofiziologia 16: 557–561.

Kovalchuk YN, Krishtal OA, Nowycky MC (1990): The proton-activated inward current of rat sensory neurons includes a calcium component. Neurosci Lett 115: 237–242.

Kraig RP, Ferreira-Filho CR, Nicholson C (1983): Alkaline and acid transients in cerebellar microenvironment. J Neurophysiol 49: 831–850.

Kraig RP, Pulsinelli WA, Plum F (1985): Hydrogen ion buffering during complete brain ischemia. Brain Res 342: 181–190.

Krishtal OA, Pidoplichko VI (1980): A receptor for protons in the nerve cell membrane of sensory neurons. Neuroscience 5: 2325–2327.

Krishtal OA, Pidoplichko VI (1981): A receptor for protons in the membrane of neurons may participate in nociception. Neuroscience 6: 2599–2601.

Krishtal OA, Osipchuk YV, Shelest TN, Smirnoff SV (1987): Rapid extracellular pH transients related to synaptic transmission in rat hippocampal slices. Brain Res 436: 352–356.

Lindahl O (1962): Pain: A chemical explanation. Acta Rheum Scand 8: 161–169.

Lingueglia E, Voilley N, Waldmann R, Lazdunski M, Barby P (1993): Expression cloning of an epithelial amiloride-sensitive Na^+ channel. A new channel type with homologies to *Caenorhabditis elegans* degenerins. FEBS Lett 318: 95–99.

Liu L, Simon SA (1994): A rapid capsaicin-activated current in rat trigeminal ganglion neurons. Proc Natl Acad Sci USA 91: 738–741.

Lou Y-P, Lundberg JM (1992): Inhibition of low pH evoked activation of airway sensory nerves by capsazepine, a novel capsaicin-receptor antagonist. Biochem Biophys Res Comm 189: 537–544.

Lou Y-P, Karlsson JA, Franco-Cereceda A, Lundberg JM (1991): Selectivity of ruthenium red in inhibiting bronchoconstriction and CGRP release induced by afferent C-fibre activation in the guinea-pig. Acta Physiol Scand 142: 191–199.

McCarty DJ Jr, Phelps P, Pyenson J (1966): Crystal-induced inflammation in canine joints. I. An experimental model with quantification of the host response. J Exp Med 124: 99–114.

McConnell HM, Owicki JC, Parce JW, Miller DL, Baxter GT, Wada HG, Pitchford S (1992): The cytosensor microphysiometer: Biological applications of silicon technology. Science 257: 1906–1912.

Maggi CA, Bevan S, Walpole CSJ, Rang HP, Guiliani S (1993): A comparison of capsazepine and ruthenium red as capsaicin antagonists in the rat isolated urinary bladder and vas deferens. Br J Pharmacol 108: 801–805.

Morad M, Callewaert G (1988): Proton-induced transformation of Ca^{2+} channel: Possible mechanism and physiological role. In Morad M, Nayler W, Kazda S, Schramm M (eds): The Calcium Channel: Structure, Function and Implications. Berlin: Springer-Verlag, pp 71–80.

Mutch WAC, Hansen AJ (1984): Extracellular pH changes during spreading depression and cerebral ischemia: Mechanisms of brain pH regulation. J Cereb Blood Flow Metab 4: 17–27.

Nicoll RA (1978): The blockade of GABA-mediated responses in the frog spinal cord by ammonium ions and furosemide. J Physiol (Lond) 283: 121–132.

Oh U, Hwang SW, Kim D (1996): Capsaicin activates a nonselective cation channel in cultured neonatal rat dorsal root ganglion neurons. J Neurosci 16: 1659–1667.

Parsons SM, Prior C, Marshall IG (1993): Acetylcholine transport, storage, and release. Int Rev Neurobiol 35: 279–390.

Pidoplichko VI (1992): Ammonia and proton gated channel populations in trigeminal ganglion neurons. Gen Physiol Biophys 11: 39–48.

Philippi M, Vyklický L, Kuffler DP, Orkand RK (1995): Serotonin- and proton-induced and modified ionic currents in frog sensory neurons. J Neurosci Res 40: 387–395.

Pozo MA, Gallego R, Gallar J, Belmonte C (1992): Blockade by calcium antagonists of chemical excitation and sensitization of polymodal nociceptors in the cat's cornea. J Physiol (Lond) 450: 179–189.

Prabhakar E, Lawson SN (1995): The electrophysiological properties of rat primary afferent neurones with carbonic anhydrase activity. J Physiol (Lond) 482: (pt 3) 609–622.

Prod'hom B, Pitrobon D, Hess P (1987): Direct measurement of proton transfer rates to a group controlling the dihydropyridine-sensitive Ca^{2+} channel. Nature 329: 243–246.

Rang HP, Ritchie JM (1988): Depolarization of nonmyelinated fibers of the rat vagus nerve produced by activation of protein kinase C. J Neurosci 8: 2606–2617.

Ransom BR, Philbin DM Jr (1992): Anoxia-induced extracellular ionic changes in CNS white matter: The role of glial cells. Can J Physiol Pharmacol 70: S181–S189.

Riley DA, Sanger JR, Matloub HS, Yousif NJ, Bain JL, Moore GH (1988): Identifying motor and sensory myelinated axons in rabbit peripheral nerves by histochemical staining for carbonic anhydrase and cholinesterase activity. Brain Res 453: 79–88.

Saarikoski J, Kaila K (1992): Simultaneous measurement of intracellular and extracellular carbonic anhydrase activity in intact muscle fibres. Pflügers Arch 421: 357–363.

Santicioli P, Del Bianco E, Geppetti P, Maggi CA (1992): Release of calcitonin gene-related peptide-like (CGRP-LI) immunoreactivity from rat isolated soleus muscle by low pH, capsaicin and potassium. Neurosci Lett 143: 19–22.

Santicioli P, Del Bianco E, Figini M, Bevan S, Maggi CA (1993): Effect of capsazepine on the release of calcitonin gene-related peptide-like immunoreactivity (CGRP-LI) induced by low pH, capsaicin and potassium in rat soleus muscle. Br J Pharmacol 110: 609–612.

Satoh H, Lou Y-P, Lee L-Y, Lundberg JM (1992): Inhibitory effects of capsazepine and the NK2 antagonist SR 48968 on bronchoconstriction evoked by sensory nerve stimulation in guinea-pigs. Acta Physiol Scand 146: 535–536.

Satoh H, Lou Y-P, Lundberg JM (1993): Inhibitory effects of capsazepine and SR 48968 on citric acid-induced bronchoconstriction in guinea-pigs. Eur J Pharmacol 236: 367–372.

Sontheimer H, Perouansky M, Hoppe D, Lux HD, Grantyn R, Kettenmann H (1989): Glial cells of the oligodendrocyte lineage express proton-activated Na^+ channels. J Neurosci Res 24: 496–500.

Stahl GL, Longhurst JC (1992): Ischemically sensitive visceral afferents; importance of H^+ derived from lactic acid and hypercapnia. Am J Physiol 262: H748–H753.

Steen KH, Reeh PW (1993): Sustained graded pain and hyperalgesia from harmless experimental tissue acidosis in human skin. Neurosci Lett 154: 113–116

Steen, KH, Reeh PW, Anton F, Handwerker HO (1992): Protons selectively induce lasting excitation and sensitization to mechanical stimuli of nociceptors in rat skin, *in vivo*. J Neurosci 12: 86–95.

Steen, KH, Reeh PW, Kreysel HW (1995a): Topical acetylsalicylic, salicylic acid and indomethacin suppress pain from experimental tissue acidosis in human skin. Pain 62: 339–347.

Steen, KH, Steen AE, Reeh PW (1995b): A dominant role of acid pH in inflammatory excitation and sensitization of nociceptors in rat skin, *in vitro*. J Neurosci 15: 3982–3989.

Steen, KH, Reeh PW, Kreysel HW (1996): Dose-dependent competitive block by topical acetylsalicylic and salicylic acid of low pH-induced cutaneous pain. Pain 64: 71–82.

Stefandis D, Reeh PW, Kreysel HW, Steen KH (1994): Acetylsalicylic and salicylic acid suppress pH induced excitation of rat nociceptors *in vivo*. Soc Neurosci Abstr 20: 15.

Stevens CR, Williams RB, Farrell AJ, Blake DR (1991): Hypoxia and inflammatory synovitis: Observations and speculation. Ann Rheum Dis 50: 124–132.

Syková E, Svoboda J (1990): Extracellular alkaline-acid-alkaline transients in the rat spinal cord evoked by peripheral stimulation. Brain Res 512: 181–189

Szallasi A (1994): The vanilloid (capsaicin) receptor: Receptor types and species differences. Gen Pharmacol 25: 223–243.

Szerb JC, Butterworth RF (1992): Effect of ammonium ions on synaptic transmission in the mammalian central nervous system. Prog Neurobiol 39: 135–153.

Tang CM, Presser F, Morad M (1988): Amiloride selectively blocks the low threshold (T) calcium channel. Science 240: 213–215.

Tang CM, Dichter M, Morad M (1990): Modulation of the N-methyl-D-aspartate channel by extracellular H^+. Proc Natl Acad Sci USA 87: 6445–6449.

Traynelis SF, Cull-Candy SG (1990): Proton inhibition of N-methyl-D-aspartate receptors in cerebellar neurons. Nature 345: 347–350.

Ueno S, Nakaye T, Akaike N (1992): Proton-induced sodium currents in freshly dissociated hypothalamic neurones of the rat. J Physiol (Lond) 447: 309–327.

Vyklický L Jr, Vlachovà V, Krusek J (1990): The effect of external pH changes on responses to excitatory amino acids in mouse hippocampal neurones. J Physiol (Lond) 430: 497–517.

Waldmann R, Champigny G, Voilley N, Lauritzen I, Lazdunski M (1996): The mammalian degenerin MDEG, an amiloride-sensitive cation channel activated by mutations causing neurodegeneration in *Caenorhabititis elegans*. J Biol Chem 271: 10433–10434.

Waldmann R, Champigny G, Bassilana F, Heurteaux C, Lazdunski M (1997): A proton-gated cation channel involved in acid-sensing. Nature 386: 173–177.

Waxman SG, Kocsis JD, Black JA (1995): Type III sodium channel mRNA is expressed in embryonic but not adult spinal sensory neurons, and is re-expressed following axotomy. J Neurophysiol 72: 466–470.

Wegner H, Reeh PW, Brehm S, Kreysel HW, Steen KH (1996): Diltiazem blocks the pH-induced excitation of rat nociceptors together with their mechanical and electrical excitability in vitro. J Neurophysiol 75: 1–10

Wood JN, Winter J, James IF, Rang HP, Yeats J, Bevan S (1988): Capsaicin-induced ion fluxes in dorsal root ganglion cells in culture. J Neurosci 8: 3208–3220.

Zeilhofer HU, Kress M, Swandulla D (1997): Fractional Ca^{2+} currents through capsaicin- and proton-activated ion channels in rat dorsal root ganglion neurones. J Physiol (Lond) 503: 67–78.

CHAPTER 24

Effects of Intracellular pH (and Ca^{2+}) on Gap Junction Channels

DAVID C. SPRAY AND ELIANA SCEMES
Department of Neuroscience, Albert Einstein College of Medicine, Bronx, New York, NY (D.C.S.); Department of Physiology, University of Sao Paulo, Sao Paulo, Brazil (E.S.)

INTRODUCTION

There have been no more controversial or protracted issues in the field of gap junction research than whether H^+ or Ca^{2+} close gap junctional channels and how the channel closure by one or both of these ions is accomplished. Nevertheless, there is general agreement that one or both of these ions provide a critical control mechanism by which gap junction channels are gated, and that this gating may lead to profound changes in intercellular communication in response to tissue acidosis, such as occurs in glia or neurons in response to ischemic brain injury (Chapters 28–30). As a result of their interconnections through gap junction channels, glia form a communication compartment that extends throughout the brain. This syncytium of coupled cells is believed to provide a virtually limitless buffering capacity for dilution of K^+ ions (and probably neurotransmitters and their metabolites as well) and to distribute solutes and water among glial cells, minimizing local volume changes. By contrast, neuronal coupling is more restricted in scope and weaker in strength, and its function is more likely related to electrotonic interactions.

For more than 15 years, our laboratory has been interested in the issues of both how second messengers flow from cell to cell and whether these messengers themselves control junctional permeability. The intent of this chapter is to summarize progress during this time, highlighting the areas that we feel are now fertile for resolution of the remaining questions.

EFFECTS OF Ca^{2+} ON JUNCTIONAL CONDUCTANCE

Numerous experimental studies have been interpreted as providing evidence that intercellular coupling is reduced by elevated cytoplasmic concentrations of both monovalent and divalent cations, chiefly H^+ and Ca^{2+}. The earliest of these studies was performed more

pH and Brain Function, Edited by Kai Kaila and Bruce R. Ransom
ISBN 0-471-11838-9

than a century ago by Engelmann, who first described the phenomenon of "healing over" in cardiac muscle (Engelmann, 1877). The critical finding of this milestone paper was that when a small region of the myocardium was irreversibly injured in physiological saline solution, the tissue immediately adjacent to the injury was transiently rendered quiescent, but then this region bordering the insult recovered its responsiveness. Subsequent studies have shown that the recovery of adjacent cells did not occur as readily when tissue was bathed in solution lacking Ca^{2+} (DeMello et al., 1969; Deleze, 1970), although healing could occur even in reduced Ca^{2+} solution if pH was low enough (DeMello, 1983). The most commonly accepted interpretation of this result is that injury depolarizes the damaged cell, and that the polarization spreading electrotonically through gap junctions among the adjacent cells renders this region quiescent. However, as Ca^{2+} enters the damaged cell, gap junction channel closure will allow the adjacent cells to repolarize because the leaky cell is electrically removed from the circuit. That cardiac cells can be uncoupled by intracellularly injected Ca^{2+} ions has been directly demonstrated (e.g., DeMello, 1975), although the levels of Ca^{2+} necessary to achieve uncoupling in cardiac tissue is higher than that obtained following damage sufficient to cause irreversible contracture (see White et al., 1985).

Abundant support for the uncoupling action of divalent cations has been provided by Loewenstein's group, beginning more than 30 years ago (for reviews, see Loewenstein, 1981; Rose and Rick, 1978). The most detailed of these studies were performed on giant larval salivary gland cells of the midge *Chironomus*. Although it has recently been suggested that arthropod junctions may be formed of proteins that are not homologous to the connexin gene family that encodes gap junctions of vertebrates (Starich et al., 1996), the data obtained on these junctions have largely determined the view that gap junction channels are quite responsive to elevated intracellular Ca^{2+} levels. Several of these papers are particularly noteworthy, both for their novel experimental designs and for the mechanistic insights that they provided in understanding the role of Ca^{2+} ions in uncoupling. One of these is the series of experiments performed by Gilberto Oliveira-Castro (Oliveira-Castro and Loewenstein, 1971), who measured the time course and extent of uncoupling of salivary gland cells in response to membrane damage while the cells were bathed in solutions containing different concentrations of divalent cations. Although changes in coupling were detected at submicromolar levels of extracellular Ca^{2+}, the substantial and reversible uncoupling illustrated in the paper was obtained when Ca^{2+} concentration in the bath was on the order of 30 μM. Another noteworthy series of experiments from this group was the elegant aequorin injections into salivary gland cells performed by Birgit Rose (Rose and Loewenstein, 1975, 1976). In these experiments, Ca^{2+} injections into single cells loaded with this Ca^{2+} indicator resulted in localized luminescence corresponding to elevations in intracellular Ca^{2+}. These were observed to spread rapidly within the cell but then stopped abruptly at margins with neighboring cells. Simultaneous electrical measurements correlated the lack of invasion of Ca^{2+} into the adjacent cells with cell uncoupling, suggesting that Ca^{2+} was not only gap junction impermeant but was also a potent channel blocker. The final intriguing experiment, by Socolar and Politoff (1971), showed that cells previously uncoupled by treatments that increased cytoplasmic Ca^{2+} concentrations could be recoupled by membrane hyperpolarization.

Although each of these experiments has been interpreted as providing evidence that uncoupling of Ca^{2+} ions was both sensitive and universal, each can also be viewed as illustrating either the relative insensitivity of junctional membranes to direct block by Ca^{2+} ions or differences in responses of arthropod junctions compared with those of mammalian cells. For example, the Ca^{2+} concentration that was used in most of the experiments by Oliveira-Castro is now known to be quite high, perhaps three orders of magnitude above the free Ca^{2+} level in the cytoplasm of most cell types. Nevertheless, these values are within the same range as those obtained in our studies of internally perfused teleost cleavage stage embryonic cells (Spray et al., 1982), which we have interpreted as indicating Ca^{2+} insensitivity of junctions except under extreme conditions of membrane breakdown. As is considered in more detail on pp. 483–484, gap junctions in mammalian cells are not blocked by Ca^{2+} ions injected into one cell (Saez et al., 1989; see also Finkbeiner, 1992; Boitano et al., 1992), and Ca^{2+} elevations in photoreceptors of the hydrozoan *Obelia* resulting from illumination spread to the adjacent coupled cells (Dunlap et al., 1987). Finally, reversal of the uncoupling action of Ca^{2+} ions by hyperpolarization now seems explicable in terms of the strong dependence of arthropod junctional conductance on membrane potential, whereby cells uncouple when the membrane is depolarized (see Chanson et al., 1994). This effect is the opposite of that reported in astrocytes, where depolarization was reported to increase coupling, offsetting the uncoupling effect of high $[Ca^{2+}]$ (Enkvist and McCarthy, 1994).

EFFECT OF pH ON JUNCTIONAL CONDUCTANCE

Since the exchange of H^+ for Ca^{2+} ions is a general feature of biological buffering systems (Chapter 15), it is possible that Ca^{2+} ions may act through acidification. In 1977, Luca Turin and Ann Warner showed that blastomeres of early amphibian embryos were electrically uncoupled by exposure to solutions equilibrated with CO_2, and that this uncoupling correlated with intracellular acidification (Turin and Warner, 1977). We extended this study to pairs of cleavage-stage fish and amphibian embryonic cells and found that the relation between normalized junctional conductance (G_j) and intracellular pH gave a good fit with a form of the Hill equation $G_j = K^n/(K^n + [H^+]^n$, where n, the Hill coefficient, was between 4 and 5 and the apparent pK_a of the process was about 7.3 (Fig. 1) (Spray et al., 1981). Our interpretation of these data was that H^+ ions might directly interact with titratable amino acid residues of the proteins forming gap junction channels, and that the Hill coefficient (n) might represent the effective number of highly cooperative binding sites in the junctional channels. If binding sites were on the hemichannels that pair to form complete gap junction channels, $G_j = K^{2n}(K^n + [H^+]^n)^2$; from fits to this equation, the pK_a would be about 7.2 and n between 3.5 and 4.

Several aspects of the effects of intracellular pH on junctional conductance that were observed in this study are noteworthy in the context of studies that followed. First, when we slowly varied intracellular pH, we found that junctional conductance at each pH value was similar, no matter whether pH was increasing or decreasing. This G_j–pH_i relation appeared to indicate that the action of low pH_i on the junctional channel was direct, rather than involving an intermediate process. Second, in teleost and amphibian embryonic cells, where resting intracellular pH (pH_i) was found to be 7.6–7.8, junctional channels were maximally open at resting pH_i. This now appears to be generally the case for mammalian gap junctions as well, so that at least moderate cytoplasmic alkalinization does not markedly affect cell coupling. Third, G_j–pH_i relations with similar Hill coefficients and apparent pK_as were obtained when pH_i was reduced by applying any of the weak acids tested (acetate, propionate, lactate, or CO_2-equilibrated saline; see Chapters 10–12), so that the sensitivity may have been due to acidification *per se* and not to the bathing ions. And, finally, the gating of junc-

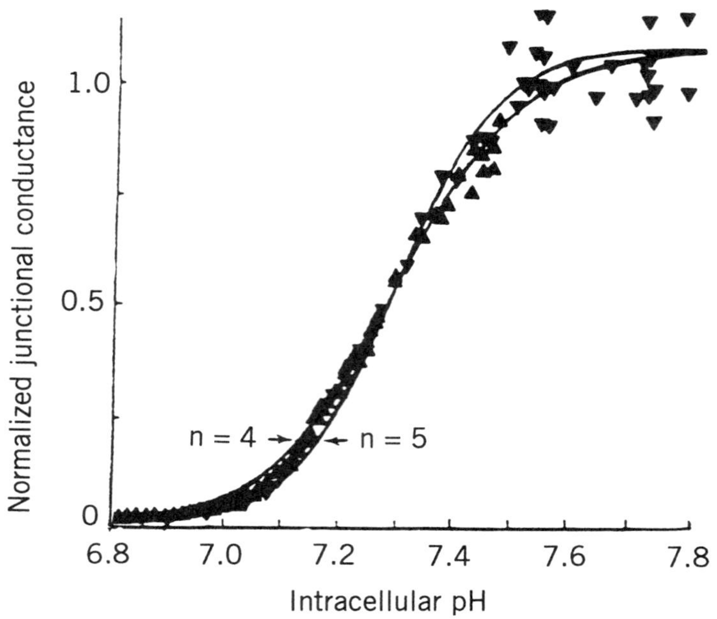

FIGURE 1. Junctional conductance (G_j)-intracellular pH (pH_i) relation for a pair of early blastomeres from amphibian (axolotl) embryos. pH_i was measured with a Thomas-type recessed-tip pH-sensitive glass microelectrode; G_j was measured under current-clamp conditions and is normalized to maximal values at highest pH_i. Downward pointing triangles represent points obtained when pH_i was decreased by exposure to CO_2-equilibrated saline; upward-pointing triangles represent data during recovery from acidification. Solid lines indicate fits to data using a form of the Hill equation [$G_j = K^n/(K^{n} + (H^+)^n)$, where $n = 4$ or 5 and apparent pK_a is 7.3]. Modified from Spray et al., 1981.

tional channels by pH_i was distinct from the effect of transjunctional voltage (V_j) in that acidification was found to reduce conductance to zero, whereas at even large V_j, a substantial residual conductance remained (Spray et al., 1979). This residual voltage-insensitive conductance is now attributed to a conductance substate of junctional channels (Moreno et al., 1994), and the complete closure of junctional channels by pH_i indicates that the gating mechanisms for these two treatments are at least partially distinct.

Each of these observations has had important implications for subsequent studies and for interpretation of the mechanisms by which H^+ ions interact with the junctional channel proteins.

MECHANISM OF ACTION OF pH ON G_j

Subsequent studies by ourselves and others have quantified the action of low pH_i on junctional conductance in the crayfish septate axon (Campos de Carvalho et al., 1984) and in mammalian hepatocytes (Spray et al., 1986b) and cardiac myocytes (Spray and Bennett, 1985; White et al., 1990; Liu et al., 1993). For each preparation the G_j–pH_i relations have shown little or no hysteresis (see, however, hysteresis found in experiments by Peracchia, 1990, and Pressler, 1989) and data gave a good fit with Hill equations, although the apparent pK_as and Hill coefficients differed. In the two mammalian cell types examined, Hill coefficients were markedly different, although the apparent pK_a values were similar, about 6.5. Thus, for channels formed of these gap junctions proteins, pH_i must be lowered by about 0.7 pH unit in order to decrease G_j by 50%.

Vertebrate gap junctions are formed of connexin proteins, which are members of a multigene family. In rodents, thirteen connexins have now been identified and the family is subdivided into Group I (β) or Group II (or α) on the basis of sequence homologies (Bennett et al., 1991). In cardiac myocytes (and in astrocytes; Dermietzel et al., 1989, 1991), the Group II (α) connexin (connexin43; Cx43) is the predominant gap junction protein expressed, whereas in rat hepatocytes (and in oligodendrocytes; Dermietzel et al., 1989, 1997) the major gap junction protein is the Group I (β) connexin, Cx32. All connexin proteins are believed to exhibit similar membrane topology (Fig. 2) (see Bennett et al., 1991), with four membrane spanning domains (M1–M4; M3 is amphiphilic and is believed to line the channel), two extracellular loops (E1, E2, by which the hemichannels or connexins dock across extracellular space), and three cytoplasmic domains, (NT, the amino terminal tail; CL, the cytoplasmic loop or hinge region connecting M2 and M3; and CT, the carboxyl terminal tail of the protein). The regions of these molecules exhibiting the least sequence homology are the cytoplasmic domains, and particularly CL and CT.

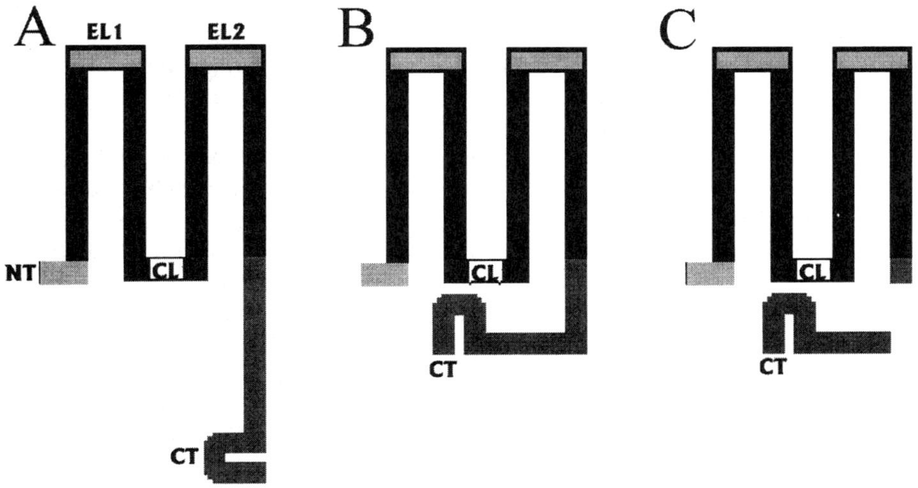

FIGURE 2. "Ball and chain" docking model of Cx43 channel closure by acidification. **A.** Gap junctions are formed of two hemichannels or connexons, each of which contains six connexin molecules, one of which is illustrated here. The extracellular loops (EL1 and EL2) are at the top of the diagram and the amino terminus (NT) and cytoplasmic loop (CL) are on the inside of the cell. **B.** On the basis of site-directed mutagenesis studies, a domain within the CT region of the protein is hypothesized to bind to a region in the cytoplasmic loop (CL) or hinge portion of the molecule. This binding is hypothesized to either effectively plug the channel lumen or to cause a conformational change that blocks junctional conductance. **C.** Coexpression of sequences corresponding to the membrane-spanning regions of Cx43 together with a sequence corresponding to the cytoplasmic tail also provides pH-induced channel closure, presumably through interactions between CL and CT. (B) and (C) drawn to illustrate conclusions from mutagenesis studies (Ek et al., 1994; Morley et al., 1996).

The similar, and nearly neutral apparent pK_as of the Hill relations for hepatocyte and cardiocyte junctions suggested to us that histidine residues might provide the titratable sites responsible for pH gating. In examining the sequence differences between Cx43 and Cx32, we noted that the number of histidine residues in the CL regions of these molecules coincided closely with the Hill coefficients for the G_j–pH_i curves of cardiac myocytes and hepatocytes (Spray and Burt, 1990). We therefore hypothesized that the Hill coefficient might represent cooperative binding sites on a single connexin molecule rather than cooperatively among the six connexin subunits comprising the connexon or channel complex, as had been our original hypothesis.

Mario Delmar's group has evaluated this hypothesis, using site-directed mutagenesis to exchange histidine residues within CL and also to alter regions within the CT domains of Cx43 (Liu et al., 1993; Ek et al., 1994; Morley et al., 1996). These mutant proteins were then expressed in pairs of *Xenopus* oocytes, and sensitivity to intracellular pH has been examined either as the latency to a threshold of uncoupling or by direct measurements of the G_j–pH_i relations of the mutants. These studies revealed that the first histidine in the CL sequence, his-95, plays a key role in pH gating of Cx43, but the mechanism appears to involve interaction with a region of CT rather than direct titration of the histidine residue. Remarkably, injection of mRNA encoding the potent CT region restores pH sensitivity to Cx43 truncation mutants in which the CT domain is absent (Ek-Vitorin et al., 1996; Morley et al., 1996). Thus, this interaction appears to be that of binding of a gating particle (Fig. 2), much like the ball and chain model proposed by Clay Armstrong to account for tetraalkylammonium ion interactions with K^+ channels (Armstrong, 1971). Martin Rook and his colleagues have shown that the other histidine in the CL region of the Cx43 molecule is also important in this gating, perhaps participating in binding of the CT gating particle to the CL channel mouth (Hermans et al., 1995). Recent studies from Camillo Peracchia's laboratory on pH sensitivity of Cx32 channels expressed in *Xenopus* oocytes indicate that interactions between CL and CT residues proximal to M4 are not instrumental in the uncoupling, but interfere with pH gating (Wang and Peracchia, 1997). Further studies are certainly necessary to resolve these mechanistic issues.

This localization of the pH sensitivity to CL and CT reinforces the notion that the gating by pH may be mechanistically different from that of transjunctional voltage. Although our experiments on embryonic cells had indicated that glutaraldehyde and retinoic acid could reduce pH sensitivity without affecting voltage sensitivity to a substantial extent, and that our measurements of voltage sensitivity were similar when performed after moderate reductions in intracellular pH (Spray et al., 1986a), experiments by Vyto Verselis on the same preparation subsequently detected moderate shifts in voltage sensitivity at acidic pH values (Bennett et al., 1988). Verselis et al. (1994) have now emphasized the roles of highly charged residues in E1 and NT in providing polarity to the voltage sensitivity of junctional channels; because of the differences that appear to be present in the mechanisms of these gating changes at the single-channel level (gating to a substate vs. gating to zero conductance), and because different residues appear to be involved, it is now important to evaluate the voltage sensitivity in the Cx43 mutants that affect pH sensitivity.

INTERACTION OF pH AND Ca^{2+} IONS IN THE CONTROL OF JUNCTIONAL CONDUCTANCE

As mentioned above, exchange of H^+ for Ca^{2+} ions is a common mechanism for buffering each ion within the cytoplasm (and is the basis for Ca^{2+} chelation by the most commonly used Ca^{2+} buffer, ethylene glycol-bis (b-aminoethyl ether) N,N,N′,N′tetraacetic acid (EGTA), raising the question of whether Ca^{2+} and H^+ ions might compete for the same sites on the junctional channels. There are several lines of experimentation indicating that such competition may occur, although we believe that the issue is far from rigorously resolved, as is evidenced by the range of values obtained in quantitative studies by different laboratories.

Jan Burt noted that cardiac myocytes uncoupled when treated with a combination of agents at concentrations that led to significant rises in intracellular Ca^{2+} or H^+ levels, but did not uncouple when Ca^{2+} or H^+ were varied individually (Burt, 1987). Several authors have now quantitatively compared sensitivity of gap junctions to intracellular concentrations of both Ca^{2+} and H^+, and the results have differed. Noma and Tsuboi (1987) evaluated junctional conductance between cardiac myocytes following rupture of one cell's membrane in order to expose the junction to unilateral changes in pH and/or Ca^{2+}. In these studies, pH sensitivity (pK_a=6.1, $n = 2.4$) was unaffected by Ca^{2+} concentrations (pKa > 8, 7, or 6.3), while Ca^{2+} sensitivity was reduced moderately at acidic pH values (pH < 6.5). In pairs of cardiac cells, White et al. (1990) have reported that intracellular acidification was ineffective in closing junctional channels when Ca^{2+} was maintained at low levels. This finding is in partial agreement with the report that intracellular buffering with the pH-independent Ca^{2+} chelator 2-bis(2-aminophenoxy ethane-N,N,N,N-tetraacetic acid (BAPTA) protects hepatoma cells from pH-induced uncoupling (Lazrak and Peracchia, 1993), but differs from the recent report by Firek and Weingart (1995) that neonatal myocyte junctions were relatively insensitive to Ca^{2+} (pK_{Ca}=3.5), although effects were additive with acidification. It might be expected that this issue would be resolvable using internal perfusion techniques, through reconstitution of junctional membranes in lipid bilayers, or using recordings from junctional hemichannels, all methods where the intracellular compartment would be accessible to solution exchanges. In internal perfusion experiments on fish embryonic cells, we found similar pH sensitivity to that reported for unperfused cell pairs (Spray et al., 1982), whereas Fidel Ramon and colleagues reported an absence of either pH or Ca^{2+} sensitivity unless the putative intermediary molecule calmodulin was added to the perfusate (Arellano et al., 1986, 1988, 1990). In experiments on bilayers doped with liver junctional membranes, we found little or no Ca^{2+} sensitivity and somewhat lower sensitivity to acidification than in hepatocytes (Campos de Carvalho et al., 1992); however, Young et al. (1987) reported that each of these ions shifted the voltage-dependence of junctional conductance.

Another view concerning the issue of pH and Ca^{2+} as agonists for uncoupling has been championed by Camillo Peracchia, who has argued that Ca^{2+} effects are direct whereas pH effects are indirect (and possibly mediated through Ca^{2+} ions). In support of this argument are his publications showing hysteresis in pH uncoupling effects and direct correlations between Ca^{2+} levels and junctional conductance in crayfish axons (Peracchia, 1990). Peracchia's group has also widely promoted the notion that both pH and Ca^{2+} effects are mediated through the intermediary molecule calmodulin (CaM), based on experiments in which CaM antagonists blocked pH-induced uncoupling (Peracchia and Girsch, 1985; Peracchia, 1987) and, most recently in experiments in which oligonucleotides antisense to the sequence of *Xenopus* CaM were shown to completely block pH sensitivity of the junctional membranes (Peracchia et al., 1996). As mentioned above, this concept is supported by internal perfusion studies showing restoration of Ca^{2+} sensitivity by CaM in crayfish axons (Arellano et al., 1988). However, our experiments with CaM antagonists on crayfish axons and amphibian blastomeres have shown that these agents may increase junctional conductance "per se," and that effects on coupling may be due to nonjunctional conductance decrease. We have not detected a shift in pH sensitivity following application of CaM antagonists (A. C. Campos de Carvalho, V. Verselis, D. C. Spray, unpublished).

IMPLICATIONS OF Ca^{2+} AND pH SENSITIVITY FOR FUNCTIONS OF GLIA AND NEURONS

The most straightforward evidence for an absence of exquisite sensitivity of gap junction channels to elevated cytoplasmic Ca^{2+} levels seems to be the observation that intracellularly injected Ca^{2+} leads to elevated Ca^{2+} concentration in cells that are coupled to it. This phenomenon was first shown in hydrozoan *Obelia* for physiological stimuli (Dunlap et al., 1987) and was then directly demonstrated in rat hepatocytes, in which we showed that Ca^{2+} spread is not dependent on extracellular Ca^{2+} and is blocked by the gap junction inhibitor heptanol (Saez et al., 1989). Subsequent studies have shown intercellular spread of Ca^{2+} waves to be elicited by a variety of stimuli in epithelial, smooth muscle, and numerous other cell types. Spread of Ca^{2+} waves throughout glia has received special attention because of the possibility that these waves might provide long-range signalling throughout the brain and might be responsible for pathological phenomena such as spreading depression.

We have recently examined glial Ca^{2+} waves in astrocytes obtained from mice in which the major astrocyte gap junction protein, Cx43, was deleted through homologous recombination (Reaume et al., 1995). Although these animals die at birth (due to profound right ventricular hyperplasia and pulmonary stenosis), astrocytes were harvested from homozygous Cx43 (−/−) animals and compared with wild-type littermates. In wild-type animals, astrocytes respond to mechanical stimulation with a rise in intracellular [Ca^{2+}] that spreads throughout the stimulated cell within a second, and spreads to adjacent cells at a uniform rate of approximately 20–50 μm/s. At room temperature, this response is gradually damped with distance. Successive and repeated stimulations of one cell lead to changes in [Ca^{2+}] propagated to neighboring cells even when the level of Ca^{2+} in the stimulated cells is still elevated. These results therefore indicate that Ca^{2+} overloaded cells are still in communication with their neighbors. (Although this finding might be interpreted as contradicting the uncoupling action of intracellularly injected Ca^{2+} ions reported by Enkvist and McCarthy (1994), we have not simultaneously recorded membrane potential to evaluate whether membrane depolarization could abolish the uncoupling due to Ca^{2+} ions as proposed by these authors). However, in Cx43 (−/−) astrocytes the conduction velocity, magnitude of Ca^{2+} spread in responding cells, and number of cells participating in the response are significantly reduced (Scemes et al., 1998). The Ca^{2+} waves in both wild-type (Fig. 3A) and Cx43 (−/−) astrocytes are blocked by heptanol and (although attenuated somewhat) are only partially blocked by the purinergic receptor antagonist suramin, so that, although adenosine nucleotide release may contribute an extracellular route for Ca^{2+} signalling, it does not totally account for the spread between mouse astrocytes.

By contrast, regenerative pH signalling does not appear to occur readily between astrocytes from wild-type mice (Fig. 3B). This is presumably due in part to the absence of mechanisms for highly sensitive regenerative pH changes analogous to Ca^{2+}-induced or inositol tris-phosphate (IP_3)-stimulated Ca^{2+} release from intracellular stores, but absence of propagated pH changes also be due to the uncoupling action of acidification in these cells. Although the sensitivity of junctional conductance to pH_i has not yet been quanti-

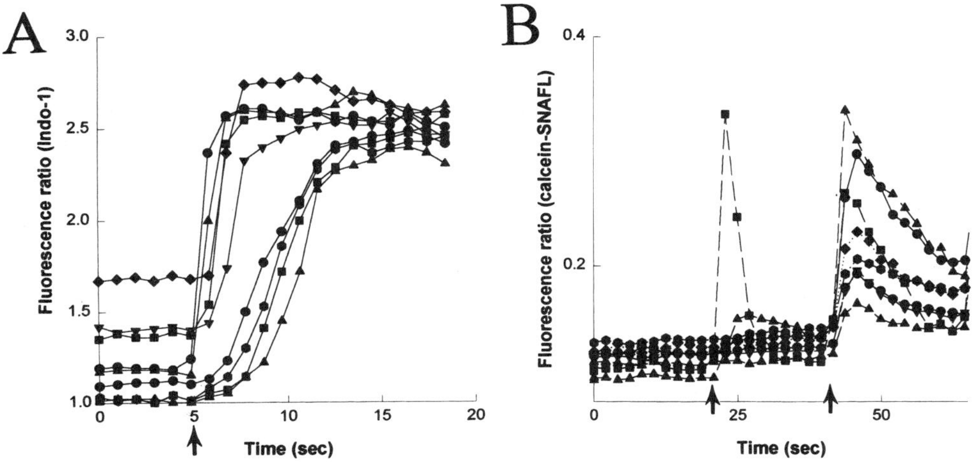

FIGURE 3. Effects of intracellular [Ca^{2+}] and [H^+] rise on intercellular communication. Ca^{2+} waves are shown to spread between cultured wild-type mouse astrocytes (A), whereas local intracellular acidification is shown to be confined to the acid-injected cell (B). **A.** Results from Ca^{2+} imaging experiments performed on a Nikon RCM 8000 real-time confocal microscope using astrocytes loaded with 5 μM Indo1-AM in extracellular medium containing 2.4 mM Ca^{2+}. At the arrow, one cell was mechanically stimulated, generating a rapidly progressing elevation in intracellular Ca^{2+} in other contiguous cells, as indicated by increase in the fluorescence ratios. **B.** Results from pH_i imaging experiments on astrocytes loaded with the pH indicator carboxy-SNAFL (5-[and 6-]-carboxyseminaphthofluorescein-1). At the first arrow, one cell is injected with strongly acidic solution (pH 5); pH_i abruptly decreases and then recovers, but H^+ acidification does not spread to adjacent cells. At the second arrow, the low pH solution is applied to all cells in the field, causing their simultaneous acidification.

fied in astrocytes, these cells do uncouple when exposed to 100% CO_2 (Dermietzel et al., 1991), and acidification by a few tenths of a pH unit may be significant to reduce coupling strength (Pappas et al., 1996). However, the changes in pH and Ca^{2+} that accompany moderate hyposmotic shock do not reduce junctional conductance (Scemes et al., 1997; Scemes and Spray, 1998). The presence of gap junction-mediated Ca^{2+} waves in astrocytes and the absence of long-range pH changes are consistent with the reported independence of mechanisms of extrusion and buffering of these ions in astrocytes (Deitmer et al., 1993, Chapters 13–14).

Coupling between neurons is also reduced when cytoplasmic pH is lowered, and has been reported to increase when pH is raised (Bleasel and Pettigrew, 1994; Cemerikic et al., 1992; Church and Baimbridge, 1991; Bodmer et al., 1988; Giaume et al., 1980; Connors et al., 1984; Campos de Carvalho et al., 1984). Although the impact of increased neuronal coupling on synchronized epileptiform activity has been emphasized, the equally plausible possibility that glial uncoupling by acidification would lead to neuronal synchronization through localized K^+ accumulation should also be explored more thoroughly in slice preparations. Moreover, although some neurons express Cx32 (Dermietzel et al., 1989), this is apparently not the type of gap junction protein present between the most strongly coupled neurons in the CNS. In no vertebrate neuronal system has the sensitivity of junctional conductance to pH_i been quantified, and the lack of identification of the type of connexin in these cells limits inferences about pH sensitivity compared to that of gap junctions in glia.

CONCLUSIONS

Although the details of whether and how pH and Ca^{2+} interact to close gap junction channels remain to be resolved, substantial progress has been made in understanding how changes in intracellular pH may affect gap junction channels. Junctional channels formed of Cx43, as in astrocytes, or of Cx32, as in oligodendrocytes and in some neurons, are expected to be maximally open at normal intracellular pH values. The probability of these channels being open will be reduced by cytoplasmic acidification, resulting in a 50% decrease in junctional conductance at pH_i values of 6.5 or less. Such strong intracellular acidity may occur at least transiently under conditions of ischemic brain injury and would be predicted to have a profound effect on such spatial regulatory functions as K^+ buffering and volume regulation. From the perspective of channel biophysics, the pH gating mechanism appears to involve residues on two distinct cytoplasmic protein domains, thus offering the possibility of developing molecules that might interfere with the docking and render junctional channels pH insensitive.

ACKNOWLEDGMENTS

Current work of our laboratory on glial cells is supported by NIH grants NS 07512 and NS 34931 and a grant from the Muscular Dystrophy Association.

REFERENCES

Arellano RO, Rivera A, Ramon F (1990): Protein phosphorylation and hydrogen ions modulate calcium-induced closure of gap junction channels. Biophys J 57: 363–367.

Arellano RO, Ramon F, Rivera A, Zampighi GA (1986): Lowering of pH does not directly affect junctional resistance of crayfish lateral axons. J Membr Biol 94: 293–299.

Arellano RO, Ramon F, Rivera A, Zampighi GA (1988): Calmodulin acts as an intermediary for the effects of calcium on gap junctions from crayfish lateral axons. J Membr Biol 101: 119–131.

Armstrong CM (1971): Interaction of tetraethylammonium ion derivatives with the potassium channels of giant axons. J Gen Physiol 58: 413–437.

Bennett MVL, Verselis V, White RL, Spray DC (1988): Gap junctional conductance: Gating. In Hertzberg EL, Johnson RG (eds): Gap Junctions. New York: AR Liss, pp 287–304.

Bennett MVL, Barrio L, Bargiello TA, Spray DC, Hertzberg EL, Saez JC (1991): Gap junctions: New tools, new answers, new questions. Neuron 6: 305–320.

Bleasel AF, Pettigrew AG (1994): The effect of bicarbonate-free artificial cerebrospinal fluid on spontaneous oscillations of the membrane potential in inferior olivary neurons of the rat. Brain Res 639: 8–20.

Bodmer R, Verselis V, Levitan IB, Spray DC (1988): Electronic synapses between Aplysia neurons in situ and in culture: Aspects of regulation and measurements of permeability. J Neurosci 8: 1656–1670.

Boitano S, Dirksen ER, Sanderson MJ (1992): Intercellular propagation of calcium waves mediated by inositol triphosphate. Science 258: 292–295.

Burt JM (1987): Block of intercellular communication: Interaction of intracellular H^+ and Ca^{2+}. Am J Physiol 253: C607–C612.

Campos de Carvalho AC, Spray DC, Bennett MVL (1984): pH dependence of transmission at electronic synapses of the crayfish septate axon. Brain Res 321: 279–286.

Campos de Carvalho AC, Eiras LA, Waltzmann M, Hertzberg EL, Spray DC (1992): Properties of channels from rat liver gap junction membrane fractions incorporated into planar lipid bilayers. Braz J Med Biol Res 25: 81–92.

Cemerikic D, Nedeljkov V, Galogaza V, Beleslin BB (1992): Effects of acid Ca^{2+} Ringer on passive electrical properties and intracellular ion activities in leech Retzius neuron. Comp Biochem Physiol 101: 433–451.

Chanson M, Roy C, Spray DC (1994): Voltage-dependent gap junctional conductance in hepatopancreatic cells of *Procambarus clarkii*. Am J Physiol 266: C569–577.

Church J, Baimbridge KG (1991): Exposure to high-pH medium increases the incidence and extent of dye coupling between rat hippocampal CA1 pyramidal neurons in vitro. J Neurosc 11: 3289–3295.

Connors BW, Benardo LS, Prince DA (1984): Carbon dioxide sensitivity of dye coupling among glia and neurons of the neocortex. J Neurosci 4: 1324–1330.

Deitmer JW, Schneider HP, Munsch T (1993): Independent changes of intracellular calcium and pH in identified leech glial cells. GLIA 7: 299–306.

Deleze J (1970): The recovery of resting potential and input resistance in sheep heart injured by knife or laser. J Physiol (Lond) 208: 547–562.

DeMello WC (1983): The influence of pH on the healing over process of mammalian cardiac muscle. J Physiol (Lond) 339: 299–307.

DeMello WC (1975): Effect of intracellular injection of calcium and strontium on cell communication in heart. J Physiol (Lond) 250: 231–245.

DeMello WC, Motta G, Chapean M (1969): A study on the healing over of myocardial cells of toads. Circ Res 24: 475–487.

Dermietzel R, Farooq M, Kessler JA, Althaus H, Hertzberg EL, Spray DC (1997): Oligodendrocytes express gap junction proteins connexin32 and connexin45. Glia 20: 101–114.

Dermietzel R, Traub O, Hwang TK, Beyer E, Bennett MVL, Spray DC, Willecke K (1989): Differential expression of three gap junction proteins in developing and mature brain tissues. Proc Natl Acad Sci USA 86: 10148–10152.

Dermietzel R, Kessler JA, Hertzberg EL, Spray DC (1991): Gap junctions between cultured astrocytes: Immunocytochemical, molecular and electrophysiological analysis. J Neurosoc 11: 1421–1432.

Dunlap K, Takeda K, Brehm P (1987): Activation of a calcium-dependent photoprotein by chemical signalling through gap junctions. Nature 325: 60–62.

Engelmann TW (1877): Vergleichende Untersuchungen zur Lehre von der Muskel- und Nervenelectricitat. Pflügers Arch 15: 116–148.

Ek JF, Delmar M, Perzova R, Taffet SM (1994): Role of histidine 95 on pH gating of the cardiac gap junction protein connexin 43. Circ Res 74: 1058–1064.

Ek-Vitorin JF, Calero G, Morley GE, Coombs W, Taffet SM, Delmar M (1996): pH regulation of connexin-43: Molecular analysis of the gating particle. Biophys J 71: 1273–1284.

Enkvist MO, McCarthy KD (1994): Astroglial gap junction communication is increased by treatment with either glutamate or high K^+ concentration. J Neurochem 62: 489–495.

Finkbeiner S (1992): Calcium waves in astrocytes: Filling in the gaps. Neuron 8: 1101–1108.

Firek L, Weingart R (1995): Modification of gap junction conductance by divalent cations and protons in neonatal rat heart cells. J Mol Cell Cardiol 27: 1633–1643.

Giaume C, Spira ME, Korn H (1980): Uncoupling of invertebrate electronic synapses by carbon dioxide. Neurosci Lett 17: 197–202.

Hermans MM, Kortekaas P, Jongsma HJ, Rook MB (1995): pH sensitivity of the cardiac gap junction proteins, connexin 45 and 43. Pflügers Arch 43: 138–140.

Lazrak A, Peracchia C (1993): Gap junction gating sensitivity to physiological internal calcium regardless of pH in Novikoff hepatoma cells. Biophys J 65: 2002–2012.

Liu S, Taffet S, Stoner L, Delmar M, Vallano ML, Jalife J (1993): A structural basis for the unequal sensitivity of the major cardiac and liver gap junctions to intracellular acidification: The carboxyl tail length. Biophys J 64: 1422–1433.

Loewenstein WR (1981): Junctional intercellular communication: The cell-to-cell membrane channel. Physiol Rev 61: 829–913.

Moreno AP, Rook MB, Fishman GJ, Spray DC (1994): Gap junction channels: Distinct voltage-sensitive and insensitive conductance states. Biophys J 67: 113–119.

Morely GE, Taffet SM, Delmar M (1996): Intramolecular interactions mediate pH regulation of connexin43 channels. Biophys J 1294–1302.

Noma A, Tsuboi N (1987): Dependence of junctional conductance on proton, calcium and magnesium ions in cardiac paired cells of guinea pig. J Physiol (Lond) 382: 193–211.

Oliveira-Castro GM, Loewenstein WR (1971): Junctional membrane permeability: Effects of divalent cations. J Membr Biol 5: 51–77.

Pappas CA, Rioult MG, Ransom BR (1996): Octanol, a gap junction uncoupling agent, changes intracellular $[H^+]$ in rat astrocytes. Glia 16: 7–15.

Peracchia C (1987): Calmodulin-like proteins and communicating junctions. Electrical uncoupling of crayfish septate axons is inhibited by the calmodulin inhibitor W7 and is not affected by cyclic nucleotides. Pflügers Arch 408: 379–385.

Peracchia C (1990): Increase in gap junction resistance with acidification in crayfish septate axons is closely related to changes in intracellular calcium but not hydrogen ion concentration. J Membr Biol 113: 95–92.

Peracchia C, Girsch SJ (1985): Functional modulation of cell coupling: evidence for a calmodulin-driven channel gate. Am J Physiol 248: H765–H782.

Peracchia C, Wang W, Li L, Peracchia LL (1996): Inhibition of calmodulin expression prevents low pH induced gap junction uncoupling in Xenopus oocytes. Pflügers Archiv 431: 379–387.

Pressler ML (1989): Intracellular pH and cell-to-cell transmission in sheep Purkinje fibers. Biophys J 55: 53–65.

Reaume AG, deSousa PA, Kulkarni S, Langille BL, Zhu D, Davies TC, Juneja SC, Kidder GM, Roussant J (1995): Cardiac malformation in neonatal mice lacking connexin43. Science 267: 1173.

Rose B, Loewenstein WR (1975): Permeability of cell junction depends on local cytoplasmic calcium activity. Nature 254: 250–252.

Rose B, Loewenstein WR (1976): Permeability of a cell junction and the local cytoplasmic free ionized calcium concentration: A study with Aequorin. J Membr Biol 28: 87–119.

Rose B, Rick R (1978): Intracellular pH, intracellular free Ca^{2+}, and junctional cell-cell coupling. J Membr Biol 44: 377–415.

Saez JC, Connor JA, Spray DC, Bennett MVL (1989): Hepatocyte gap junctions are permeable to the second messenger, inositol 1,4,5-triphosphate, and to calcium ions. Proc Natl Acad Sci USA 86: 2708–2712.

Scemes E, Spray DC (1998): Hyposmotic stimuli increase junctional conductance and Ca^{2+} wave propagation in astrocytes. Glia, in press.

Scemes E, Rook MB, Spray DC (1997): Hyposmotic shock increases junctional conductance in cultured astrocytes. Intl. Mtg on Gap Junctions, Key Largo, FL, Abstr p 89.

Scemes E, Dermietzel R, Spray DC (1998): Calcium waves between astrocytes from Cx43 KO mice. Glia, in press.

Socolar SJ, Politoff AL (1971): Uncoupling cell junctions in a glandular epithelium by depolarizing current. Science 172: 492–494.

Spray DC, Bennett MVL (1985): Physiology and pharmacology of gap junctions. Annu Rev Physiol 47: 281–303.

Spray DC, Burt JM (1990): Structure-activity relations of the cardiac gap junction channel. Amer. J. Physiol. 258: C195–C205.

Spray DC, Harris AL, Bennett MVL (1979): Voltage dependence of junctional conductance in early amphibian embryos. Science 204: 432–434.

Spray DC, Harris AL, Bennett MVL (1981): Gap junctional conductance is a simple and sensitive function of intracellular pH. Science 211: 712–715.

Spray DC, Stern JH, Harris AL, Bennett MVL (1982): Control of communication via gap junctions: In Hilfer SR, Sheffield JB (eds): Cellular Communication During Ocular Development. Springer-Verlag, NY pp 57–84.

Spray DC, Campos de Carvalho A, Bennett MVL (1986a): Sensitivity of gap junction conductance to H^+ ions in amphibian embryonic cells is independent of voltage sensitivity. Proc Natl Acad Sci USA 83: 3533–3536.

Spray DC, Ginzberg RD, Morales EA, Gartmaitan Z, Arias IM (1986b): Electrophysiological properties of gap junctions between dissociated pairs of rat hepatocytes. J Cell Biol 103: 135–144.

Starich TA, Lee RYN, Panzarella C, Avery L, Shaw JE (1996): eat-5 and unc-7 represent a multigene family in Caenorhabditis elegans involved in cell-cell coupling. J Cell Biol 134: 537–548.

Turin L, Warner A (1977): Carbon dioxide reversibly abolishes ionic communication between cells of early amphibian embryo. Nature 270: 56–57.

Verselis VK, Ginter CS, Bargiello TA (1994): Opposite voltage gating polarities of two closely related connexins. Nature 368: 348–351.

Wang XG, Peracchia C (1997): Positive charges of the initial C-terminus domain of Cx32 inhibit gap junction gating sensitivity to CO_2. Biophys J 73: 798–806.

White RL, Spray DC, Campos de Carvalho AC, Wittenberg BA, Bennett MVL (1985): Some electrical and pharmacological properties of gap junctions between adult ventricular myocytes. Am J Physiol 247: C447–C455.

White RL, Doeller JE, Verselis VK, Wittenberg BA (1990): Gap junctional conductance between pairs of ventricular myocytes is modulated synergistically by H^+ and Ca^{2+}. J Gen Physiol 95: 1061–1075.

Young JD, Cohn ZA, Gilula NB (1987): Functional assembly of gap junction conductance in lipid bilayers: Demonstration that the major 27 kd protein forms the junctional channel. Cell 48: 733–743.

CHAPTER 25

Modulation of Vertebrate Retinal Function by pH

STEVEN BARNES
Neuroscience Research Group, University of Calgary, Faculty of Medicine, Calgary, Alberta, Canada

INTRODUCTION

Many retinal synapses use graded changes in presynaptic membrane potential, not action potentials, to trigger neurotransmitter release. It is this fact that leads to the high degree of sensitivity of retinal electrical activity to external pH. While the same principles apply at numerous other local neural processing networks that utilize graded potentials, the retina of lower vertebrates is particularly well characterized in terms of the interactions between endogenous and experimentally-induced pH changes and neural activity.

Four decades ago the first intracellular recordings of electrical activity in the second-order horizontal cells of the retina revealed that light hyperpolarized these cells in a manner graded with light intensity and established that action potentials were not generated at this early stage in the visual system. It is now generally understood how the neurons of the outer retina (photoreceptors, horizontal cells, and bipolar cells) signal via graded membrane potential changes, while the neurons of the inner retina (amacrine cells, interplexiform cells, and ganglion cells) produce action potentials. Our detailed understanding of how photoreceptors generate their characteristic light responses is rewarding: In darkness, a depolarizing, nonspecific cation conductance in the outer segment balances a hyperpolarizing K^+ conductance in the inner segment to produce a membrane potential near −40 mV. When light activates the visual pigment, an enzymatic cascade reduces the cation conductance, and membrane potential plunges towards the K^+ equilibrium potential. This reduces the release of the photoreceptor neurotransmitter glutamate, which is maximal in the dark, a result that is surprising from the perspective of energy economy but consistent with the neuronal doctrine that depolarization evokes release. Responding to the modulated release of glutamate, second-order horizontal and "OFF" bipolar cells follow the photoreceptor voltage waveform, while a third cell type, the "ON" bipolar, inverts the waveform to depolarize in response to light. This flow of information through the retina is mainly unidirectional, with some feedback provided from the inner to the

pH and Brain Function, Edited by Kai Kaila and Bruce R. Ransom
ISBN 0-471-11838-9

outer retina via interplexiform cells; however, massive amounts of lateral signal transmission occur, mediated by horizontal cells in the outer retina and amacrine cells in the inner retina.

Activity-dependent changes in external pH have also been described in the retinas of lower vertebrates and mammals. These have been characterized with considerable resolution in terms of which retinal cell types are responsible, which may be most affected, and how long the changes take to develop. Taken together, activity-dependent pH changes and the modulation of synaptic activity that has been shown to depend on pH suggest a significant role for endogenous pH changes in modulating retinal activity. The retina can thus serve as a valuable model system for investigating the role of pH in modulating neural activity under both physiological and pathophysiological conditions.

pH CHANGES IN THE RETINA

Direct measurement of extracellular pH in the retina has revealed that the interstitial space is a spatially and temporally dynamic environment in which pH is regulated as a function of retinal depth and the intensity and duration of illumination (Chapter 17). Indeed, careful measurement has revealed that the interstitial space in the retina sees changes in the concentrations of several ionic species, particularly Ca^{2+} and K^{+}, and these changes have been related to known features of specific cellular responses.

The gradient of extracellular pH through the retina and light-evoked changes in proton concentration have been studied in the retinas of lower vertebrates and mammals. In frog retinal eyecups and isolated retinas (Borgula et al., 1989), the largest light-evoked pH changes were recorded with pH-sensitive microelectrodes in the inner plexiform layer, where action potential-producing amacrine and ganglion cells make synapses. These changes were alkalinizations of about 0.03 pH unit and developed fully following 30 sec of steady illumination. Alkalinizations were enhanced during application of flickering lights. Changes in the outer retina, where only graded potentials are recorded, were just under 0.02 pH units and occurred with a similar time course. Inhibition of carbonic anhydrase with acetazolamide and methazolamide increased the light-evoked alkalinizations to about 0.04 pH units.

A similar picture emerged from studies in isolated toad retina (Oakley and Wen, 1989), with the exception that in the inner retina, prominent acidifications were evoked during maintained illuminations, but were again followed by alkalinizations when illuminated ceased. A gradient of pH exists across the retina, with the outer retina (photoreceptor side) being relatively alkaline and the inner retina (ganglion cell side) being relatively acidic. Amiloride, a blocker of Na^{+}/H^{+} exchange, caused a sustained alkalinization in the extracellular pH and reduced the changes in pH associated with light application. The disulfonic stilbene derivative, SITS, an inhibitor of Cl^{-}/HCO_3^{-} exchange (but which also antagonizes certain anion channels, including some $GABA_A$ receptors), also produced small alkalinizations. Inhibition of retinal activity downstream of the photoreceptors blocked the light-evoked inner retinal acidification, but failed to alter the light-evoked alkalinization measured in the outer retina, suggesting that a significant component of this pH increase resulted from acid extrusion from the photoreceptors themselves. This study emphasized the importance of the bicarbonate buffering system in the retina by comparing pH changes evoked by light in bicarbonate-buffered and phosphate-buffered salines, which were much greater in the latter case.

In mammals, the changes in pH caused by light and the retinal pH gradient are generally similar to the descriptions in lower vertebrates. Light stimulation produces outer retinal alkalinizations, and these changes are superimposed upon a gradient of pH which increases with the traverse from the ganglion cells to the photoreceptors. In pig retina (Tsacopoulos and Levy, 1976), pH in the vitreous humor near the retina was measured at 7.04. In the avascular rabbit retina (Murray et al., 1991), pH near the inner limiting membrane, the boundary between the vitreous humor and the ganglion cell nerve fibers, was measured at 7.36 in the dark; light produced alkalinizations of about 0.05 pH units at this position, reflecting the net value of both inner and outer retinal pH changes.

By far the most thorough study was made in cat retina *in situ* (Yamamoto et al., 1992). Here, the pH near the vitreous–retinal boundary, assessed with both large commercially available pH electrodes and double-barreled pH-sensitive microelectrodes, was found to be 7.17 in the dark; this value increased to 7.22 under intense light. Across the depth of the retina in darkness, pH first decreased to a minimum value of 7.03 at a level corresponding to the photoreceptor cell bodies, then increased abruptly, reaching a value close to 7.39 at the level of the photoreceptor outer segments and the pigment epithelium (Fig. 1). At this depth, retinal pH approached that of the blood plasma, owing perhaps to the proximity of the choriocapillaris. Bear in mind, however, that the permeability and transport properties of the blood-brain barrier and pigment epithelium, which has its own pH regulating capabilities (Kenyon et al., 1997), must yet be traversed, are relevant here. The change in extracellular pH during stimulation with light was maximal near the photoreceptor cell bodies, with an average alkalinization of 0.2 pH units. The light-evoked changes occurred with a

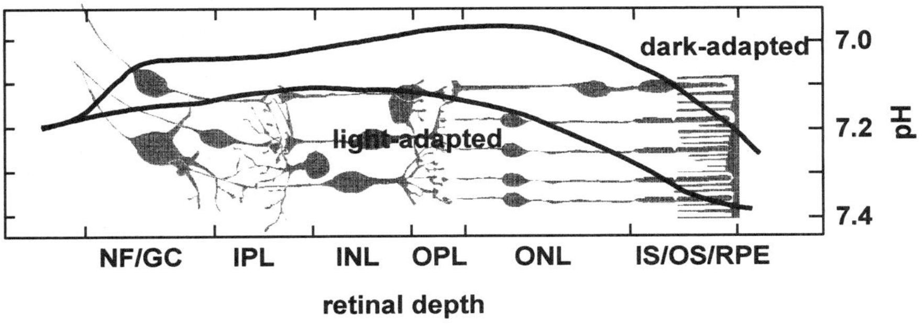

FIGURE 1. Profiles of extracellular pH across the retina in light and dark. Measurements of extracellular pH were obtained with a pH microelectrode as it was advanced through the depth of the retina in an anesthetized cat. During this traverse, the electrode encounters (i) the ganglion cells (GC) and their nerve fibers (NF); (ii) a dense synaptic plexus, the inner plexiform layer (IPL); (iii) the somata of amacrine, bipolar and horizontal cells in the inner nuclear layer (INL); (iv) a thinner synaptic plexus, the outer plexiform layer (OPL); (v) the somata of rod and cone photoreceptors in the outer nuclear layer (ONL); (vi) the inner and outer segments (IS and OS) of the photoreceptors; and finally (vii) the retinal pigment epithelium (RPE). Glial Müller cells traverse the region between ganglion cells and the outer nuclear layer. The depth profile was obtained first after complete dark adaptation, then again after applying rod-saturating diffuse light. The most acidic pH was encountered in darkness in the outer retina, at the level of the photoreceptor cell bodies, which are sites of ATP production by mitochondria and glycolysis in the cytoplasm. This is the same region for which the strongest alkalinizations were observed during prolonged light adaptation. Figure modified from Yamamoto et al., 1992.

time constant of about 1 min, while the recovery of retinal pH upon termination of illumination occurred biphasically with fast (85-sec time constant) and slow (10-min time constant) components.

Anoxia also produces strong acidifications in the vicinity of the photoreceptors (Oakley and Wen, 1989; Yamamoto and Steinberg, 1992) and consequently alters retinal electrical activity by eliminating transmission at the first synapse (photoreceptors to horizontal and/or bipolar cells; Negishi and Svaetichin, 1966; Negishi and Sugawara, 1973; Hankins and Ikeda, 1993).

Several lines of evidence suggest that Müller cells play the central role in buffering and removing protons from the retinal environment. First, inhibition of carbonic anhydrase enhances the light-evoked alkalinizations in the retina and increases the gradient of the retinal pH by raising inner retinal acidity (Oakley and Wen, 1989). Similarly, carbonic anhydrase inhibition increased retinal sensitivity, but not by altering photocurrent responses, which is consistent with the view that Müller cells remove acid from the region near the photoreceptor cell bodies (Donner et al., 1990). Isoforms of carbonic anhydrase are found ubiquitously in retinal neurons and the extracellular space (see Chapter 3), but there is a concentration of the enzyme in Müller cells (Musser and Rosen, 1973; Linser and Moscona, 1984; Moscona, 1985). It follows that the glial cells play a prominent role in the altered spatial and temporal pH changes during inhibition of the enzyme (Newman, 1994). Second, Müller cells show pronounced polarization with regard to the distribution of Cl^-/HCO_3^- exchangers, Na^+/HCO_3^- cotransporters, K^+ channels, and possibly Na^+/H^+ exchangers, on their membranes (Newman, 1991; Newman and Astion, 1991; Kobayashi et al., 1994; Newman, 1996). The distribution of these membrane proteins suggests that a spatial pumping action could occur, wherein CO_2 is moved from the sites of maximal production near the photoreceptors to be dumped at the Müller cell endfeet lying adjacent to the vitreous humor (Donner et al., 1990; Newman, 1991; Newman and Astion, 1991). Some ion channels in Müller cells are themselves sensitive to pH (Bringmann et al., 1997).

Taken together, a picture of exceptional metabolic expense emerges, with a focus on the photoreceptors and their high and changing energy requirements for light and dark operation. In particular, rod photoreceptors have very high energy requirements in darkness, during which time they are maximally depolarized, their inner segment Na^+/K^+-ATPase is most active to support the flow of dark current, and the release of glutamate at their synapses is highest. Glycolysis, which is suppressed during illumination when the photoreceptors are hyperpolarized, lies at the heart of the alkalinization recorded in the extracellular space during light (Winkler, 1981). Following this, the flow of acid away from the photoreceptors across the retina to the vitreous humor and toward the choroidal blood supply, and the buffering capabilities within the retinal environment, in particular those provided by the Müller cells, complete the picture. Thus, the retina provides an excellent example of a nervous system tissue in which pH is a spatially and temporally dynamic variable.

pH SENSITIVITY OF SYNAPTIC TRANSMISSION OF GRADED POTENTIALS

The synaptic transmission of graded potentials is especially sensitive to pH. The cells of the outer retina (e.g., rod and cone photoreceptors, horizontal cells, and bipolar cells) do not generate action potentials and instead signal via graded changes in their membrane potentials. For photoreceptors, the range of membrane potential excursions is limited to values from around −40 mV in darkness, to −60 or −70 mV in bright light. For rod photoreceptors, clipping of hyperpolarizing signals greater than about 5–10 mV occurs at the synapse (Attwell et al., 1987), while for cones, evidence suggests a greater dynamic range for the synapse (Baylor et al., 1971). For both types of photoreceptors, voltage-gated Ca^{2+} channels mediate the lion's share of what is primarily Ca^{2+}-dependent synaptic release of glutamate (Rieke and Schwartz, 1994). Postsynaptic responses in horizontal cells and hyperpolarizing ("OFF") bipolar cells are mediated by a diversity of glutamate-gated channels and are quite variable among species, but can range from −20 or −40 mV in darkness to −60 or −80 mV in the light. Light-evoked photoreceptor hyperpolarizations become depolarizations in ("ON") bipolar cells, the inversion being mediated by a metabotropic glutamate receptor (the APB-receptor), which affects membrane potential via cGMP-phosphodiesterase and cGMP-gated channels (Nawy and Jahr, 1990, 1991).

The brief synopsis of events at the photoreceptor output synapse provided above, sets the stage for understanding the striking sensitivity of synaptic transmission to external pH at this site in the visual system. In particular, a relatively narrow range of presynaptic Ca^{2+} channel activation is utilized for synaptic release, and this requires high gain by the processes wherein membrane potential changes are transduced into glutamate release. As a consequence, a steep exponential relationship between the presynaptic and postsynaptic potentials arises (Attwell et al., 1987).

In cases where extracellular pH can be directly adjusted to known values, a strong dependence of the postsynaptic light response in horizontal and/or bipolar cells has been observed in lower vertebrate retinas (Kleinschmidt, 1991; Barnes et al., 1993; Harsanyi and Mangel, 1993). Figure 2 shows an example of how the light response of a second order retinal neuron (a horizontal cell) is sensitive to pH. As the pH is raised, the response increases exponentially. This effect has been attributed almost entirely to alterations in presynaptic Ca^{2+} influx (Barnes et al., 1993; Harsanyi and Mangel, 1993) and is described below.

To determine the extent to which the photoreceptor light response itself is changed by pH, it would be advantageous to directly control presynaptic membrane potential (in photoreceptors) while simultaneously recording postsynaptic membrane potential (in horizontal or bipolar cells). While this approach has not proven possible, several lines of evidence argue against the likelihood that pH sensitivity of the light response itself is an explanation for the synaptic modulation shown in Fig. 2. The effects of pH on several aspects of the photoreceptor response have been characterized and, over the range of pH where synaptic transmission is most sensitive, there are not large changes in the photoreceptor response. Generally, pH effects on photoreceptors occur mostly in the range below pH 7.0, having much less effect in the range of pH 7.0–8.0, where the most striking alterations of synaptic transmission occur. The light-sensitive current of rods, carried in cGMP-gated channels, is generally suppressed and slowed at extracellular pH 6.5 and below (Gedney and Ostroy, 1978; Leibman et al., 1984; Meyertholen et al., 1986). Intracellular acidification in this same range has complex effects on the light-sensitive current of photoreceptors (Tanaka, 1993), and intracellular pH depends strongly on the ability of photoreceptors to regulate intracellular pH via Na^+/H^+ exchange and HCO_3^- transport (Donner et al., 1990; Kalamkarov et al., 1996; Saarikoski et al., 1997). Other contributors to the photoreceptor response, such as Na^+-Ca^{2+}-K^+ exchange (Hodgkin and Nunn,

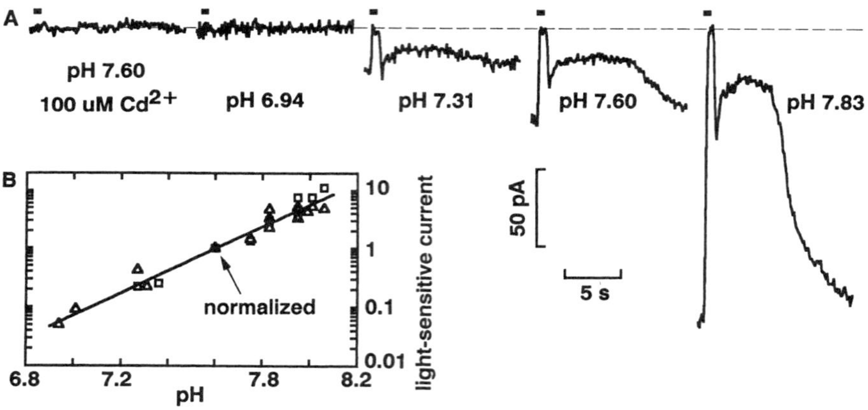

FIGURE 2. Sensitivity of the horizontal-cell light response to extracellular pH. **A.** Postsynaptic currents evoked by half-second steps of bright light (indicated by the bar above each trace) in a voltage-clamped horizontal cell in a tiger salamander retinal slice. Light suppresses a standing, glutamate-mediated inward postsynaptic current that flows in the dark and which maintains the horizontal cells in a depolarized state. Light hyperpolarizes the photoreceptors, reducing glutamate release, and thus the postsynaptic inward current is diminished. Since salamander horizontal cells receive rod and cone input (Marshall and Werblin, 1978), components of their response seen here reflect the activity of both photoreceptor types. In a solution with pH 7.60, inward postsynaptic current is suppressed by light. Alkalinization increased the inward postsynaptic current that had been suppressed by light. Acidification reduced or eliminated the light-suppressible inward current, as did Cd^{2+}, a well-known calcium channel blocker. **B.** Exponential relation between pH and postsynaptic response. The semi-logarithmic plot summarizes synaptic regulation by pH in horizontal cells. *Triangles:* Synaptic current suppression during light application when there is rod and cone input. *Squares:* Postsynaptic current suppression measured during persistent rod hyperpolarization following light application. Reproduced from Barnes et al., 1993 with permission. Copyright National Academy of Sciences.

1987) and phosphodiesterase activity (Miki et al., 1975), are only mildly effected in the pH range between 7 and 8. Finally, the receptor potential component of the electroretinogram, which can be isolated by treating the retina with aspartate to block synaptic transmission of the light response, is slowly affected by extracellular pH, most markedly between pH 6.5 and 7.5 (Sillman et al., 1972; Koskelainen et al., 1993). The increase in photoreceptor response amplitude that accompanies alkalinization is considered to stem from regulatory processes that, in turn, give rise to intracellular alkalinization (Koskelainen et al., 1993). This action appears to be much slower (occurring on a time scale of minutes) than the fast, direct action of extracellular pH on voltage-gated Ca^{2+} channels, which occurs essentially as rapidly as the pH change can be effected at the cell membrane.

The consequences of changing pH on postsynaptic function must also be considered in analyzing the example of synaptic modulation shown in Fig. 2. Since the horizontal cell response was voltage-clamped, there should be no contribution from voltage-gated ion channels in the membrane, and little from Ca^{2+}-gated channels as a consequence. However, we need to consider carefully the contributions to the responses shown in Fig. 2 from pH modulation of the gap junction channels (see Chapter 24) that mediate coupling between horizontal cells (Skrzypek and Werblin, 1983; Borges and Wilson, 1987), the ligand-gated channels (glutamate receptors) that transduce the photoreceptor transmitter

(Marshall and Werblin, 1978; Yang and Wu, 1991; Christensen and Hida, 1990), and on the factors that control the duration of glutamate presence in the synaptic cleft (Bouvier et al., 1992; Chapter 19).

As described elsewhere in this volume (Chapter 24) pH has potent effects on gap junction channels. Classically, acidification is considered to be an inhibitor of gap junction permeability (Spray et al., 1981), and in the retina, horizontal cell coupling via gap junctions is dependent on extracellular pH in this same manner (DeVries and Schwartz, 1989; Qian et al., 1993; Hampson et al., 1994; Dixon et al., 1996). This poses a question about the responses shown in Fig. 2 for tiger salamander: To what degree could the pH-modulated responses arise from current flow through pH-sensitive gap junction channels? Broad-field illumination, which polarizes cells equally, as used in the experiment shown, could produce current changes qualitatively similar to those observed when one horizontal cell in the network is voltage-clamped. As the membrane potential of unclamped neighboring cells became closer to or farther away from its holding potential, the flow of lateral current would change. For example, if one horizontal cell was clamped at -60 mV and if the membrane potential of coupled neighboring cells were -30 mV, current would flow laterally through gap junctions into the clamped cell. During the light response, the neighbor's membrane potential would hyperpolarize and diminish the potential difference between the cells, reducing inward current flow. At the same time, if gap junction permeability were reduced by acidification, less current would flow into the clamped cell, and the lateral current change during a light response would be smaller. Running counter to this possibility, is the fact that horizontal cell bodies in salamander do not form a tightly coupled syncytium (Marshall and Werblin, 1978; Skrzypek and Werblin, 1983), as do the cell bodies in fish and some other species (Dowling, 1987). This circumstance is reflected in the recordings made for Fig. 2, where input resistances are as high as those found in isolated cells. Also, Lucifer yellow coupling of horizontal cells was not observed. Finally, similar observations about the pH sensitivity of synaptic transmission to horizontal cells have been made in non-voltage clamped cells, where no lateral current would flow. This rules out pH modulation of gap junctions as a major contributor to the responses in Fig. 2.

Glutamate-gated channels are also pH-sensitive (Chapter 22). To what extent could this be a source of the pH modulation of synaptic transmission to horizontal cells? The glutamate-gated channels exhibiting the highest degree of pH-sensitivity are the N-methyl-D-aspartate (NMDA) receptors, which are not found on horizontal cells in the species exemplified in Fig. 2 (Yang and Wu, 1991). And while NMDA receptors are pH-sensitive over the same range as is the synaptic response (Tang et al., 1990; Traynellis and Cull-Candy, 1990), the types of non-NMDA glutamate receptors that do mediate the horizontal cell response are not (Tang et al., 1990; Traynellis and Cull-Candy, 1990; Christensen and Hida, 1990).

As described elsewhere in this volume (Chapter 19), pH influences the function of amino acid uptake carriers, and in particular, this has been shown for the glial cells in the retina (Müller cells), which play a role in terminating the action of the glutamate released by photoreceptors (Billups and Attwell, 1996). Glutamate uptake into Müller cells is a complex process, wherein each cycle of glutamate uptake is accompanied by uptake of 2 Na^+ and release of 1 K^+ and 1 effective base equivalent (Bouvier et al., 1992). While this mechanism has been considered in terms of pH-changing potential for the retinal environment, it is evident that uptake of glutamate will also be sensitive to environmental pH (Billups and Attwell, 1996). Acidification of the extracellular space would reduce glutamate uptake, giving rise to an increased lifetime for glutamate in the synaptic cleft, an increased concentration of glutamate present during the sustained release from photoreceptors in darkness, and thus larger postsynaptic responses. However, modulation of glutamate uptake by pH, if occurring by this mechanism, would not produce the supralinear relation observed between extracellular pH and postsynaptic response.

The strongly pH-sensitive population of presynaptic Ca^{2+} channels (see also Chapter 21) form the critical element at the synapse between photoreceptors and second order cells. Descriptions of Ca^{2+} channel pH sensitivity in several preparations (Ohmori and Yoshii, 1977; Iijima et al., 1986), perhaps most notably in cardiac myocytes (Prod'hom et al., 1989; Krafte and Kass, 1988; Kwan and Kass, 1993), provide confirmation of the reports in vertebrate rod and cone photoreceptors (Barnes and Bui, 1991; Barnes et al., 1993). Figure 3 summarizes important features of the pH sensitivity of Ca^{2+} channels in photoreceptors. Ca^{2+} channel currents can be seen to increase with increasing alkalinization, the result of a leftward (negative) shift of the Ca^{2+} channel activation curve (so that more channels are activated at a given

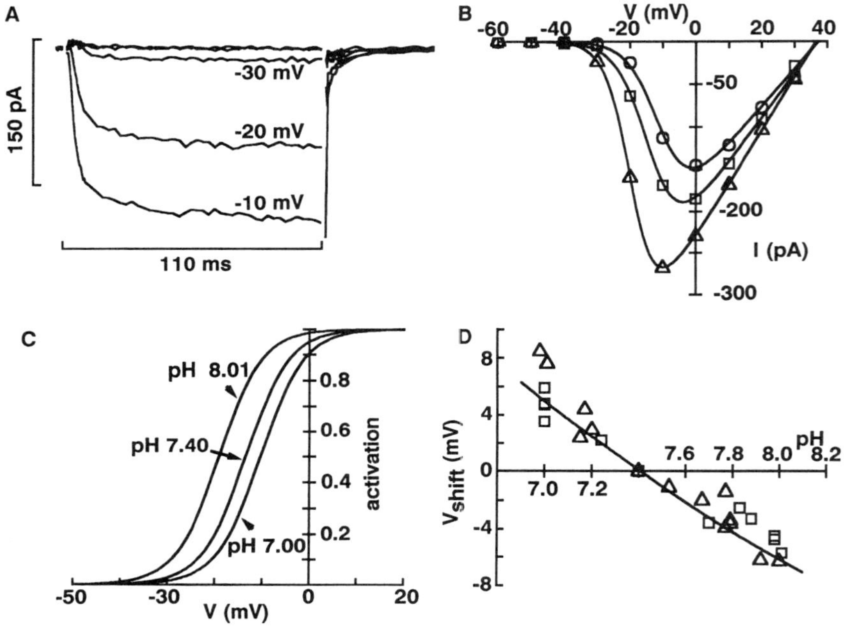

FIGURE 3. Ca^{2+} channel gating and permeation in photoreceptors is modified by extracellular pH. **A.** Ca^{2+} channel currents recorded under voltage-clamp in an isolated tiger salamander rod. **B.** Current–voltage relations generated from the same rod, as extracellular pH was varied: pH 7.00 (*circles*); 7.40 (*squares*); 8.01 (*triangles*). Inward currents were elicited at more negative potentials and had larger maximum amplitudes under alkaline conditions. (Note that activation of these channels occurs at artificially positive potentials at all pH values, due to the high concentration of divalent cations used to amplify Ca^{2+} channel currents, e.g., 20 mM Ba^{2+}). **C.** Boltzmann-fit activation curves derived from the data in (B), with midpoints of −10.4 mV (pH 7.00), −13.5 mV (pH 7.40), and −19.2 mV (pH 8.01). **D.** Summary of activation curve midpoint shifts as a function of pH for rods (*squares*) and cones (*triangles*). Reproduced from Barnes et al., 1993 with permission. Copyright National Academy of Sciences.

membrane potential, Fig. 3C), and an increase in channel conductance or the number of active channels (seen as a steeper *I–V* relation in the fully activated region, Fig. 3B). In other preparations, protons are considered to induce gating and permeation changes in Ca^{2+} channels by binding on or near the channel protein and affecting the electric field seen by the channel voltage sensor, by reducing local external Ca^{2+} concentration or by blocking the channel (Ohmori and Yoshii, 1977; Iijima et al., 1986; Prod'hom et al., 1989; Krafte and Kass, 1988; Kwan and Kass, 1993). For the following analysis, the most germane feature of Ca^{2+} channel pH-sensitivity in the context of pH modulation of synaptic transmission is the lateral shifting of the activation curve (Fig. 3D).

With the focus on presynaptic Ca^{2+} channels as an explanation for pH modulation of synaptic transmission, the following argument emerges: Changes in extracellular pH shift the Ca^{2+} channel activation curve laterally on the voltage axis, and this translates into an equivalent shift along the voltage axis of the synaptic transfer function, a graph describing the relation between presynaptic potential and postsynaptic potential. (The Ca^{2+} influx relation, or Ca^{2+} *I–V* relation, can be related to the synaptic transfer function via a third-power relation). For example, in Fig. 4, a synaptic transfer function for photoreceptor-to-horizontal cell transmission has been created to relate presynaptic potential to postsynaptic current. To assist in the interpretation of the synaptic responses shown in Fig. 2, predictions of postsynaptic current response were made for the same pH values. These estimates were arrived at by shifting laterally the synaptic transfer function by amounts prescribed by the Ca^{2+} channel activation curve shifts shown in Fig. 3D.

The photoreceptor-to-horizontal cell synapse operates on a different synaptic transfer curve at each pH. In darkness at pH 7.6, with the photoreceptor membrane potential near −40 mV (V_{dark} on the presynaptic axis), the horizontal cell generates inward current as shown on the vertical, postsynaptic axis (~60 pA). Strong illumination hyperpolarizes the photoreceptor to V_{light} or beyond, and the horizontal cell response on the vertical axis declines to near zero. The net change of postsynaptic current is described by the vertical line drawn along the right side of Fig. 4, and is labelled 7.60. The changes in postsynaptic current due to strong illumination applied at each pH are also illustrated as vertical lines. These are summarized in the inset to Fig. 4, and can be seen to follow closely the line derived in Fig. 2 for the pH sensitivity of postsynaptic current. This simple model supports the view that pH-induced shifts of presynaptic photoreceptor Ca^{2+} channel activation could account for most of the modulation of postsynaptic horizontal cell responses.

The analysis suggests that endogenous pH changes on the order of 0.1 pH unit will have actions on synaptic transmission at this type of graded potential synapse, being translated to synaptic transfer function shifts at approximately 10 mV/pH unit. This might constitute a form of negative feedback on photoreceptor output. As photoreceptors hyperpolarize in response to light and reduce transmitter release, light-induced alkalinization of extracellular pH would act to increase transmitter release, in effect resetting the synaptic transfer function over a new range of potentials. Slow, time-averaged pH changes, decreasing in darkness and increasing in the light, could function to adapt the synapse to the ambient level of illumination.

For a synapse that operates over a narrow range (in this case, about 5 mV), even the small modulations of Ca^{2+} channel activation that occur with extracellular pH changes will have marked ramifications on the strength of signal transmission. These considerations apply for photoreceptor and bipolar cell synapses in the retina, as well as several other ribbon and/or graded potential synapses in other parts of the nervous system. The functional implications for endogenous pH regulation of synaptic strength are still emerging.

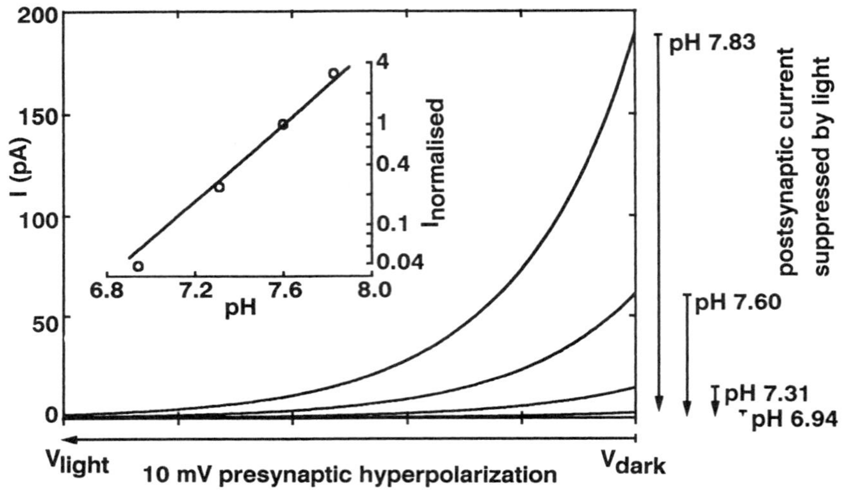

FIGURE 4. Lateral shifting of the photoreceptor-to-horizontal cell synaptic transfer function accounts for the synaptic modulation by pH. A transfer function for the photoreceptor output synapse for tiger salamander retina (Attwell et al., 1987) modified to approximate postsynaptic current in a horizontal cell, $I_{post} = A\ [\exp\ (Vpre/k) - 1]$, where $A = 61$ pA, $k = 2.1$ mV, and V_{pre} is the light-induced change in presynaptic membrane potential of photoreceptors at pH 7.6. The exponential relation was then shifted laterally in the positive direction by 3.3 mV and 7.9 mV, for pH 7.31 and 6.94, respectively, and negative by 2.4 mV for pH 7.83, which are the values given by the data curve describing Ca^{2+} channel gating shifts in Fig. 3. Photoreceptors respond to bright light by hyperpolarizing from their dark potential (V_{dark}, near -40 mV) to potentials at least 10 mV more negative (V_{light}, -50 mV or beyond). The expected changes in postsynaptic current in response to bright light at the different pH values were estimated (vertical lines at right), normalized, and then plotted (inset). They show good agreement with the exponential function found in Fig. 2. Reproduced from Barnes et al., 1993 with permission. Copyright National Academy of Sciences.

MODULATION OF OTHER RETINAL CELL FUNCTIONS BY pH

In the analysis of synaptic output from photoreceptors described in pp. 494–500, several targets for endogenous pH modulation of retinal function were described. These included presynaptic Ca^{2+} channels in photoreceptors, gap junctions between horizontal cells, neurotransmitter (glutamate) receptors on horizontal cells, and a host of enzymatic activities within the photoreceptors. Such consideration is due the numerous similar targets of endogenous pH changes that exist throughout the retina. Here we shall consider a few of the best-known examples.

In addition to the Ca^{2+} channels described in the previous section, there are several additional ion channel targets within photoreceptors themselves. These include Ca^{2+}-gated channels, such as Ca^{2+}-activated Cl^- channels, for which strong pH sensitivity has been described (Barnes and Bui, 1991). Ca^{2+}-activated K^+ channels are also found in photoreceptors (Bader et al., 1982; Barnes and Hille, 1989) and could have roles in pH modulation of response. Voltage-gated channels, such as Kx-channels, are pH-sensitive (Kurenny and Barnes, 1994), as are hyperpolarization activated, nonselective cation "h" channels of rods and cones (Malcolm et al., 1995). Since these channels are generally credited with shaping the photoreceptor voltage waveform, it follows that endogenous changes in pH could have modest effects on the voltage signal generated in response to light. The pH-sensitivity of the voltage-gated Ca^{2+}, K^+, and h channels is generally similar (e.g., roughly 10 mV positive shift of activation per unit pH decrease, mild proton block), while that of the Ca^{2+}-activated channels is much higher. The fact that pH modulation of synaptic transmission from photoreceptors could be explained simply via modulation of presynaptic Ca^{2+} channels argues for the potent role of Ca^{2+} in transmission, with the third-order power relation between Ca^{2+} influx and postsynaptic response reflecting the high gain for this aspect of the synaptic transmission process.

Amacrine cells, which comprise the laterally inhibiting network of cells in the inner retina and which mediate responses to sudden increases or decreases in illumination with the generation of action potentials, have been shown to be electrically and tracer coupled to varying extents in different species (Dowling, 1987). However, unlike the classic uncoupling of horizontal cells that occurs when extracellular pH drops below 7.4 in rabbit (Hampson et al., 1994), tracer-coupling of amacrine cells in this species is not affected by extracellular pH (Hampson et al., 1992).

Whereas the endogenous changes in extracellular pH recorded in the retina presumably arise from primary changes in intracellular pH, and since experimental manipulations of extracellular pH exert their effects from outside and, secondary to regulatory processes inside of cells, it is critical to consider ramifications of internal pH. For example, horizontal-cell Ca^{2+} and K_{IR} channels are sensitive to intracellular pH (Takahashi et al., 1993; Takahashi and Copenhagen, 1995; 1996), and, while these cells conventionally respond to glutamate via ligand-gated channels, it has been shown that, in fact, glutamate increases intracellular acidity in these cells which, in turn, suppresses the voltage-gated Ca^{2+} channels (Dixon et al., 1993; Takahashi and Copenhagen, 1996).

CONCLUSIONS

Endogenous extracellular pH changes of up to 0.2 pH units occur in the retina as a function of light stimulation, reflecting the metabolic requirements and pH-buffering capabilities of various retinal cells under different conditions of illumination. Modulation of the activity of retinal cells by pH has also been measured, emphasizing the high degree of sensitivity of the graded potential synapses in the retina to changes in extracellular pH. One lesson learned from these observations is that the choice an experimenter makes regarding the use of a particular pH buffer (for example HCO_3^- vs. HEPES) or a particular pH value (for example pH 7.4 vs. 7.8), will have profound effects on the recorded activity of retinal cells. The important physiological lesson learned is that pH in the retina is a dynamic variable with modulatory roles in retinal function.

ACKNOWLEDGMENTS

The author is the Roy Allen Investigator in Visual Sciences and is supported by the Medical Research Council of Canada, the Alberta Heritage Foundation for Medical Research, the Johannsen Research Fund of the Fight for Sight research division of Prevent Blindness America and University of Calgary Faculty of Medicine Endowment Awards (Ruth Rannie, Edwin and John L. Gustus).

REFERENCES

Attwell D, Borges S, Wu SM, Wilson M (1987): Signal clipping by the rod output synapse. Nature 328: 522–524.

Bader CR, Bertrand D, Schwartz EA (1982): Voltage-activated and calcium-activated currents studied in solitary rod inner segments from the salamander retina. J Physiol (Lond) 331: 253–284.

Barnes S, Bui Q (1991): Modulation of calcium-activated chloride current via pH-induced changes of calcium channel properties in cone photoreceptors. J Neurosci 11: 4015–4023.

Barnes S, Hille B (1989): Ionic channels of the inner segment of tiger salamander cone photoreceptors. J Gen Physiol 94: 719–743.

Barnes S, Merchant V, Mahmud F (1993): Modulation of transmission gain by protons at the photoreceptor output synapse. Proc Natl Acad Sci USA 90: 10081–10085.

Baylor DA, Fuortes MGF, O'Bryan PM (1971): Receptive fields of cones in the retina of the turtle. J Physiol (Lond) 214: 265–294.

Billups B, Attwell D (1996): Modulation of non-vesicular glutamate release by pH. Nature 379: 171–174.

Borges S, Wilson M (1987): Structure of the receptive fields of bipolar cells in the salamander retina. J Neurophysiol 58: 1275–1291.

Borgula GA, Karwoski CJ, Steinberg RH (1989): Light-evoked changes in extracellular pH in frog retina. Vision Res 29: 1069–1077.

Bouvier M, Szatkowski M, Amato A, Attwell D (1992): The glial cell glutamate uptake carrier counter transports pH-changing anions. Nature 360: 471–474.

Bringmann A, Faude F, Reichenbach A (1997): Mammalian retinal glial (Müller) cells express large-conductance Ca(2+)-activated K+ channels that are modulated by Mg2+ and pH and activated by protein kinase A. Glia 19: 311–323.

Christensen BN, Hida E (1990): Protonation of histidine groups inhibits gating of the quisqualate/kainate channel protein in isolated catfish cone horizontal cells. Neuron 5: 471–478.

DeVries SH, Schwartz EA (1989): Modulation of an electrical synapse between solitary pairs of catfish horizontal cells by dopamine and second messengers. J Physiol (Lond) 414: 351–375.

Dixon DB, Takahashi K, Bieda M, Copenhagen DR (1996): Quinine, intracellular pH and modulation of hemi-gap junctions in catfish horizontal cells. Vision Research. 36: 3925–31.

Dixon DB, Takahashi K-I, Copenhagen DR (1993): L-Glutamate suppresses HVA calcium current in catfish horizontal cells by raising intracellular proton concentration. Neuron 11: 267–277.

Donner K, Hemila S, Kalamkarov G, Koskelainen A, Shevchenko T (1990): Rod phototransduction modulated by bicarbonate in the frog retina: Roles of carbonic anhydrase and bicarbonate exchange. J Physiol (Lond) 426: 297–316.

Dowling JE (1987): The Retina: An Approachable Part of the Brain. Cambridge, MA: Belknap Press.

Gedney C, Ostroy SE (1978): Hydrogen ion effects of the vertebrate photoreceptor. The pK's of ionizable groups affecting cell permeability. Arch Biochem Biophys 188: 105–113.

Hampson EC, Vaney DI, Weiler R (1992): Dopaminergic modulation of gap junction permeability between amacrine cells in mammalian retina. J Neurosci 12: 4911–4922.

Hampson EC, Weiler R, Vaney DI (1994): pH-gated dopaminergic modulation of horizontal cell gap junctions in mammalian retina. Proc R Soc Lond [Biol] 255: 67–72.

Hankins MW, Ikeda H (1993): Consequences of transient retinal hypoxia on rod input to horizontal cells in the rat retina. Vision Res 33: 429–436.

Harsanyi K, Mangel SC (1993): Modulation of cone to horizontal cell transmission by calcium and pH in the fish retina. Vis Neurosci 10: 81–91.

Hodgkin AL, Nunn BJ (1987): The effect of ions on sodium-calcium exchange in salamander rods. J Physiol (Lond) 391: 371–398.

Iijima T, Ciani S, Hagiwara S (1986): Effects of external pH on Ca channels: Experimental studies and theoretical considerations using a two-site two ion model. Proc Natl Acad Sci USA 83: 654–658.

Kalamkarov G, Pogozheva I, Shevchenko T, Koskelainen A, Hemila S, Donner K (1996): pH changes in frog rods upon manipulation of putative pH-regulating transport mechanisms. Vis Res 36: 3029–3036.

Kenyon E, Maminishkis A, Joseph DP, Miller SS (1997): Apical and basolateral membrane mechanisms that regulate pHi in bovine retinal pigment epithelium. Am J Physiol 273: C456–72.

Kleinschmidt J (1991): Signal transmission at the photoreceptor synapse. Role of calcium ions and protons. Ann NY Acad Sci 635: 468–470.

Kobayashi S, Morgans CW, Casey JR, Kopito RR (1994): AE3 anion exchanger isoforms in the vertebrate retina: developmental regulation and differential expression in neurons and glia. J Neurosci 14: 6266–6279.

Koskelainen A, Donner K, Lerber T, Hemila S (1993): pH regulation in frog cones studied by mass receptor photoresponses from the isolated retina. Vision Res 33: 2181–2188.

Krafte DS, Kass RS (1988): Hydrogen ion modulation of Ca channel current in cardiac ventricular cells. J Gen Physiol 91: 641–657.

Kurenny DE, Barnes S (1994): Proton modulation of M-like potassium current in rod photoreceptors. Neurosci Lett 170: 225–228.

Kwan YW, Kass RS (1993): Interactions between H^+ and Ca^{2+} near cardiac L-type calcium channels: Evidence for independent channel-associated binding sites. Biophys J 65: 1188–1195.

Liebman PA, Mueller P, Pugh EN (1984): Protons suppress the dark current of frog retinal rods. J Physiol (Lond) 347: 85–110.

Linser P, Moscona AA (1984): Variable CA II compartmentalization in vertebrate retina. Biology and chemistry of the carbonic anhydrases. Ann NY Acad Sci 429: 430–446.

Malcolm AT, Kurenny DE, Barnes S (1995): Modification to gating of the hyperpolarization-activated cation current, I_h, by pH and calcium in rod photoreceptors. Soc Neurosci Abstr 21: 390.

Marshall LM, Werblin FS (1978): Synaptic transmission to the horizontal cells in the retina of the larval tiger salamander. J Physiol (Lond) 279: 321–346.

Meyertholen EP, Wilson MJ, Ostroy SE (1986): The effects of HEPES, bicarbonate and calcium on the cGMP and calcium. Vision Res 26: 521–533.

Miki N, Baraban JM, Keirns JJ, Boyce JJ, Bitensky MW (1975): Purification and properties of the light-activated cyclic nucleotide phosphodiesterase of rod outer segments. J Biol Chem 250: 6320–6327.

Moscona AA (1985): On glutamine synthetase, carbonic anhydrase and Müller glia in the retina. In Osborne N, Chader G (eds): Progress in Retina Research, Vol 2. Oxford: Pergamon Press, pp 111–135.

Murray DL, Feke GT, Weiter JJ (1991): Preretinal pH changes in the rabbit under conditions of light and dark. Exp Eye R 53: 717–722.

Musser GL, Rosen S (1973): Localization of carbonic anhydrase activity in the vertebrate retina. Exp Eye Res 15: 105–119.

Nawy S, Jahr CE (1990): Suppression by glutamate of cGMP-activated conductance in retinal bipolar cells. Nature 346: 269–271.

Nawy S, Jahr CE (1991): cGMP-gated conductance in retinal bipolar cells is suppressed by the photoreceptor transmitter. Neuron 7: 677–683.

Negishi K, Sugawara K (1973): Evidence for the anoxia sensitivity of the synaptic region at the outer plexiform layer in the fish retina. Vision Res 13: 983–987.

Negishi K, Svaetichen G (1966): Effects of anoxia, CO_2 and NH_3 on S-potential producing cells and on neurons. Pflügers Arch 292: 177–205.

Newman EA (1991): Sodium-bicarbonate cotransport in retinal Müller (glial) cells of the salamander. J Neurosci 11: 3972–3983.

Newman EA (1994): A physiological measure of carbonic anhydrase in Müller cells. Glia 11: 291–299.

Newman EA (1996): Acid efflux from retinal glial cells generated by sodium bicarbonate cotransport. J Neurosci 16: 159–168.

Newman EA, Astion ML (1991): Localization and stoichiometry of electrogenic sodium bicarbonate cotransport in retinal glial cells. Glia 4: 424–428.

Oakley II B, Wen R (1989): Extracellular pH in the isolated retina of the toad in darkness and during illumination. J Physiol (Lond) 419: 353–378.

Ohmori H, Yoshii M (1977): Surface potential reflected in both gating and permeation mechanisms in sodium and calcium channels of the tunicate egg cell membrane. J Physiol (Lond) 267: 429–463.

Prod'hom B, Pietrobon D, Hess P (1989): Interactions of protons with single open L-type calcium channels: Location of protonation site and dependence of proton-induced current fluctuations on concentration and species of permeant ion. J Gen Physiol 91: 23–42.

Qian H, Malchow RP, Ripps H (1993): Gap-junctional properties of electrically coupled skate horizontal cells in culture. Vis Neurosci 10: 287–295.

Rieke F, Schwartz EA (1994): A cGMP-gated current can control exocytosis at cone synapses. Neuron 13: 863–873.

Saarikoski J, Ruusuvuori E, Koskelainen A, Donner K (1997): Regulation of intracellular pH in salamander retinal rods. J Physiol 498: 61–72.

Sillman AJ, Owen WG, Fernandez HR (1972): The generation of the late receptor potential: An excitation-inhibition phenomenon. Vis Res 12, 1519-1531.

Skrzypek J, Werblin FS (1983): Lateral interactions in absence of feedback to cones. J Neurophysiol 49: 1007–1016.

Spray DC, Harris AL, Bennett MVL (1981): Gap junctional conductance is a simple and sensitive function of intracellular pH. Science 211: 712–715.

Takahashi K, Copenhagen DR (1995): Intracellular alkalinization enhances inward rectifier K^+ current in retinal horizontal cells of catfish. Zoological Science 12: 29–34.

Takahashi KI, Copenhagen DR (1996): Modulation of neuronal function by intracellular pH. Neurosci Res 24: 109–116.

Takahashi K, Dixon DB, Copenhagen DR (1993): Modulation of a sustained calcium current by intracellular pH in horizontal cells of fish retina. J Gen Physiol 101: 695–714.

Tanaka JC (1993): The effects of protons on 3′, 5′-cGMP-activated currents in photoreceptor patches. Biophys J 65: 2517–2523.

Tang CM, Dichter M, Morad M (1990): Modulation of the *N*-methyl-D-aspartate channel by extracellular H^+. Proc Natl Acad Sci USA 87: 6445–6449.

Traynelis SF, Cull-Candy SG (1990): Proton inhibition of N-methyl-D-aspartate receptors in cerebellar neurons. Nature 345: 347–350.

Tsacopoulos M, Levy S (1976): Intraretinal acid-base studies using pH glass microelectrodes: Effect of respiratory and metabolic acidosis and alkalosis on inner-retinal pH. Exp Eye Res 23: 495–504.

Winkler B (1981). Glycolytic and oxidative metabolism in relation to retinal function. J Gen Physiol 77: 667–692.

Yamamoto F, Borgula GA, Steinberg RH (1992): Effects of light and darkness on pH outside rod photoreceptors in the cat retina. Exp Eye Res 54: 689–697.

Yamamoto F, Steinberg RH (1992): Effects of systemic hypoxia on pH outside rod photoreceptors in the cat retina. Exp Eye Res 54: 699–709.

Yang XL, Wu SM (1991): Coexistence and function of glutamate receptor subtypes in the horizontal cells of the tiger salamander retina. Vis Neurosci 7: 377–382.

SECTION VI

pH Control of Respiration

CHAPTER 26

Cellular Mechanisms of Sensitivity to pH in the Mammalian Respiratory System

GEORGE B. RICHERSON
Veterans Affairs Medical Center, West Haven, CT; Yale University School of Medicine, New Haven, CT

INTRODUCTION

Of all the functions of the mammalian central nervous system (CNS), respiratory control is one of the most sensitive to pH; and it is one of the primary regulatory mechanisms for preventing acid-base disturbances from inducing dysfunction of other parts of the brain (Chapter 20). The intact respiratory control system responds vigorously to changes in pH of less than 0.1 units. This sensitivity of respiratory output to pH *in vivo* has led to the hypothesis that there exist neurons that are highly sensitive to acid-base disturbances, the central respiratory chemoreceptors. The leading candidate for the location of the chemoreceptors has been the ventrolateral medulla (VLM), but cellular data from neurons in this region have been difficult to obtain. Intracellular recordings from neurons with sensitivity to pH were first obtained from other regions associated with respiration, fueling a longstanding controversy that respiratory chemoreception is located outside the VLM.

Chemosensitivity at the cellular level has now been described in a variety of nuclei linked to respiration. However, it remains unclear which, if any, of these neurons are central respiratory chemoreceptors, and none of these candidate chemoreceptors has yet been studied in sufficient detail to define the cellular mechanisms of chemotransduction. This chapter will review data on the effect of pH on cellular properties of neurons in nuclei of the mammalian brain linked to respiratory control. Understanding the cellular mechanisms of chemosensory transduction in these neurons may provide clues to the location and mechanisms of respiratory chemoreception.

THE NATURE OF CENTRAL CHEMORECEPTORS

In response to an increase in blood CO_2 level, breathing is stimulated (Cunningham et al., 1986; Feldman, 1986; Eldridge et al., 1981; St. John, 1977) (also see Chapter 27). The

pH and Brain Function, Edited by Kai Kaila and Bruce R. Ransom
ISBN 0-471-11838-9

magnitude of the increase in minute ventilation depends on species and experimental conditions, but has been reported to increase from two-fold (St. John, 1977) to nearly six-fold (Cunningham et al., 1986), with respiratory acidosis induced by an increase in P_{CO_2} of about 10 mmHg ($CO_2 \approx 1.3\%$, $\Delta pH \approx -0.08$). The relative contribution of the central chemoreceptors to the hypercapnic response depends on species and blood oxygen levels. It has been estimated that the central contribution is 70–80% of the total response to hypercapnia in humans during normoxia, and that the peripheral contribution is small or negligible, in humans during hyperoxia, and in some animal species at normoxia (O'Regan and Majcherczyk, 1982). The high sensitivity to pH and/or CO_2 level of the central respiratory control system has led to the hypothesis that central respiratory chemoreceptors are a specialized group of neurons (i) each of which individually has a uniquely high stimulatory response to respiratory acidosis, and (ii) whose only function is to monitor the acid-base status of the body and modulate respiration. However, this hypothesis remains speculative. It has not yet been demonstrated at the cellular level that central neurons exist with uniquely high sensitivity to respiratory acid-base disturbances and modulatory effects on respiratory output. There is also no evidence that respiratory chemoreceptors do not play a second role in respiration, such as generation of the rhythm, or shaping of the respiratory output. Thus, it is possible that chemosensitivity is a widely distributed function of the respiratory network. It is also possible that chemosensitive central neurons are not unique to respiratory nuclei, but are also found in other regions where they modulate nonrespiratory CNS functions sensitive to acidosis, including cardiovascular output (O'Regan and Majcherczyk, 1982; Millhorn and Eldridge, 1986), cerebral blood flow (Madden, 1993), pain sensation (Gronroos and Pertovaara, 1994), arousal (Forslid, 1987; Gronroos and Pertovaara, 1994), or seizure threshold (De Jong et al. 1967; Caspers and Speckmann, 1971).

LOCATION OF CENTRAL CHEMORECEPTORS

It is widely accepted that some central respiratory chemoreceptors are located close to the ventrolateral medullary surface, and can respond to changes in pH of cerebrospinal fluid (CSF) (Mitchell et al., 1963; Schlaefke et al., 1970; Schlaefke, 1981; Feldman, 1986). There is strong evidence, using a variety of techniques *in vivo,* that chemoreceptors are present in the VLM. This has been demonstrated with local application of hypercarbic or acidic solution, electrical stimulation, cold block, anesthetic block, and application of kainic acid (Mitchell et al., 1963; Schlaefke et al., 1970; Schlaefke, 1981; Millhorn and Eldridge, 1986; Feldman, 1986; Nattie et al., 1988; Cherniack et al., 1979; Chonan et al., 1990). However, the hypothesis that the VLM is the exclusive site for respiratory chemoreceptors has been controversial (Libscomb and Boyarsky, 1972; O'Regan and Majcherczyk, 1982; Bruce and Cherniack, 1987; Coates et al. 1993; Richerson, 1995). Rather, it has been postulated that changes in pH at the surface of the VLM also affect the pH deeper within the medullary parenchyma, as a result of the blood vessels penetrating the medulla at this site (Libscomb and Boyarsky, 1972; Cragg et al., 1977). Numerous groups have attempted to record from CO_2-sensitive neurons in the VLM, but data was limited to extracellular recordings from neurons (see p 514), and intracellular recordings from chemosensitive glia (Fukuda and Honda, 1975; Fukuda et al., 1978). The extracellular recordings obtained by one group suggested that the cellular responses to acidosis of neurons in the VLM were no different than that from other medullary regions (Miles,

1983). Intracellular recordings from neurons stimulated by acidosis were reported first from other regions, including the *nucleus tractus solitarius* (NTS) (Dean et al., 1990), and hypothalamus (Dillon and Waldrop, 1992). These results fueled a controversy regarding the role of the VLM in chemoreception. An inability to obtain intracellular recordings from chemosensitive neurons in the VLM contributed significantly to this debate, even leading to suggestions that the VLM is not involved in respiratory chemoreception. Recently, intracellular recordings were reported from neurons in the VLM with high sensitivity to respiratory acid-base disturbances (Richerson, 1995), providing cellular data in support of a role for the VLM in chemoreception. However, new evidence suggests that chemosensitivity of neurons outside the VLM may also play a role in respiratory control.

NONSPECIFIC INHIBITION BY INCREASED CO_2

It is often stated that the intrinsic response of most CNS neurons to respiratory acidosis is a "nonspecific" inhibition (cf. Chapter 20). This concept was introduced by Carpenter et al. (1974), who studied neurons from *Aplysia,* neurons in the cat sensorimotor cortex, and the monosynaptic reflex in the cat spinal cord. They found evidence for inhibition by acidosis in each of these systems. This evidence generally agreed with previous work which was primarily obtained by using indirect measures of neuron function such as cortical potentials. From this data, and a review of previous literature, they made two conclusions which have played a major role in guiding interpretation of subsequent work. First, they concluded that "if a neuron is affected by CO_2 at all, it is hyperpolarized and its discharge is decreased. The exceptions to this generalization are specialized neuronal structures concerned with respiration. . ." Second, they concluded that "except for specific CO_2 receptor neurons the responses to CO_2 are not specific and are not conveying information useful to the nervous system." These conclusions have had a major impact on the field of respiratory chemoreception, and are partly responsible for the belief that any neuron that is stimulated by acidosis is involved in respiratory chemoreception, and that neurons must be stimulated by acidosis to be a chemoreceptor. As will be discussed below, even at the present time there are probably not sufficient data to know whether these conclusions are true.

In vivo Recordings of Responses to Acidosis

Acidosis has been described as producing an intrinsic inhibition of respiratory motoneurons. Early studies on respiratory neurons were performed *in vivo,* and were mainly done on large neurons (e.g., phrenic motoneurons) from which stable recordings could be obtained. Gill and Kuno (1963) stimulated descending axons in the upper cervical cord of spinalized animals, and examined the response of phrenic motoneurons before and after inhalation of 6% CO_2. They observed a >50% depression of the phrenic response within 3 minutes, and interpreted this as being due to direct membrane effects. They did not report any change in membrane potential or input resistance.

Microelectrode recordings were made *in vivo* from cat phrenic motoneurons after transection of the spinal cord at the high spinal level to eliminate phasic synaptic drive (Jodkowski and Lipski, 1986). Recordings from three neurons demonstrated a hyperpolarization by 1 mV when the end-tidal CO_2 level was increased from 3.3% to 5.3%. Other monitors of excitability were also found to be inhibited, including orthodromic responses in the phrenic nerve, antidromic field potentials, and spontaneous nerve activity.

During hypercapnia-induced hyperventilation *in vivo,* inspiratory neurons in the nucleus ambiguus increase their activity during inspiration, but hyperpolarize slightly during the expiratory phase by 0.05 mV / mmHg arterial Pco_2 (i.e., 1 mV for a change of 20 mmHg, or ≈ 2.5% change in CO_2) (Mitchell and Herbert, 1974). This was attributed to an intrinsic inhibition of these neurons by CO_2, on the assumption that there were no active currents during this phase of the respiratory cycle. However, it is now known that there is enhancement of after-hyperpolarization by Ca^{2+}-activated potassium current (I_{KCa}) during expiration, resulting from the increase in influx of Ca^{2+} during inspiration, and there is also greater synaptic inhibition during expiration due to an increase in respiratory drive.

Krnjevic et al. (1965) isolated the cerebral cortex of cats from brainstem input, while maintaining normal blood supply, by undercutting the cortical mantle. They studied the effect of inhalation of CO_2 on the response of cortical neurons to iontophoresis of glutamate. In contrast to the purely inhibitory effects described by Carpenter et al. (1974), it was found that when cats inhaled air mixed with 2–30% CO_2, the response of neurons to glutamate was always inhibited at high levels of CO_2, but at lower levels of CO_2 there could be an increase in the response.

In vitro Recordings of Responses to Acidosis

During *in vivo* experiments, it is not clear whether all observations are due to intrinsic membrane effects or whether some responses result from synaptic or nonsynaptic effects from neighboring cells. This has led to the use of *in vitro* preparations to define the intrinsic cellular response of neurons to respiratory acidosis. The effect of hypercapnia on neuronal and glial membrane potential was examined in slices and cultures from the cortex and peripheral sensory neurons (Lehmenkuhler et al., 1989). In response to respiratory acidosis induced by 20% CO_2, cultured sensory neurons depolarize. In brain slices, 20% CO_2 caused cortical neurons on the surface to depolarize, while those in the middle of the slice hyperpolarized. During apnea following ventilation with 100% O_2 *in vivo* (which results in hypercarbia without hypoxia), cortical neurons slowly hyperpolarize when arterial Pco_2 increases over 260 mmHg (CO_2 ≈ 34%). In combination, these results were interpreted to mean that alteration of the extracellular environment by glia leads to reversal of an intrinsic depolarizing response of all neurons to acidosis. Consistent with this hypothesis, cortical glia *in vivo* were found to depolarize substantially in response to apnea following hyperoxia.

The response to metabolic acidosis of neurons in a variety of respiratory and nonrespiratory medullary nuclei (including the ventral medullary surface, ventrolateral NTS, medial cuneate nucleus, dorsal motor nucleus, and medial vestibular nucleus) was examined using extracellular recordings in 300-μm brain slices by changing pH from 7.4 to 7.0 (Miles, 1983). The VLM was not unique, since every region studied contained neurons that were stimulated, other neurons that were inhibited, and still other neurons that were unresponsive to acidosis (Fig. 1). These experiments were performed without blockade of synaptic transmission; however, taken together with the results of Krnjevic et al. (1965) and Lehmenkuhler et al. (1989), these results complicate interpretation of depolarizing responses seen in respiratory nuclei, since depolarizing responses may occur in neurons outside of respiratory regions when they are exposed to acidosis. Thus, they emphasize the point that a depolarizing response, in and of itself, is not proof of a role in respiratory chemoreception. Rather, a high degree of intrinsic sensitivity to physiological levels of CO_2, and connections to respiratory neurons, are more likely to define neurons as respiratory chemoreceptors.

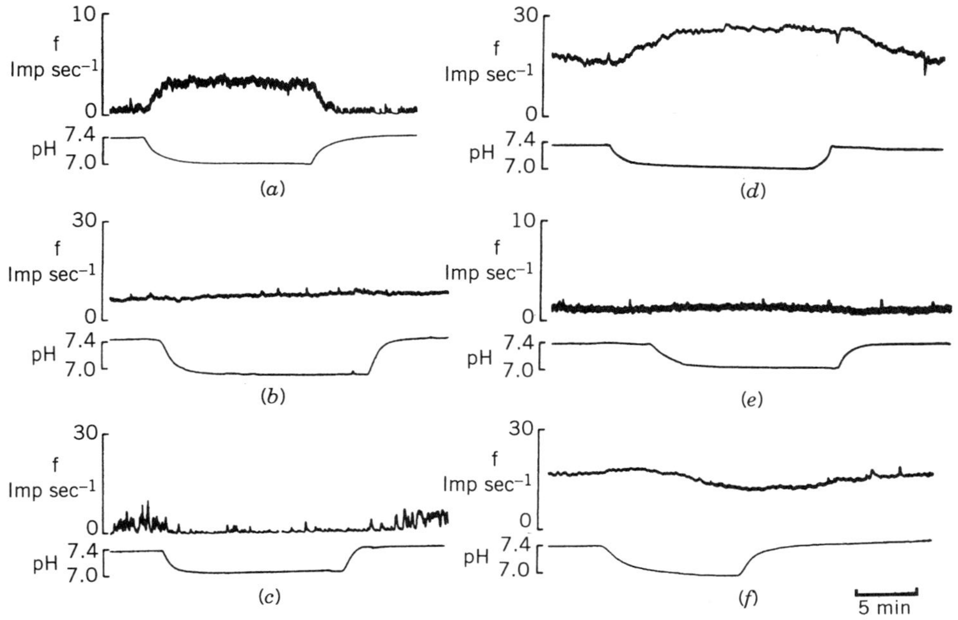

FIGURE 1. Effect of low pH on the firing rate of six neurons measured using extracellular recordings from different regions of medullary slices. In each case the upper trace is firing frequency (f) and the lower trace is pH measured close to the slice. **A, B** and **C.** Neurons in the ventral surface that were respectively excited, not affected, and inhibited by low pH. **D.** Excitatory response to low pH of a neuron recorded ventrolateral to the solitary tract. **E.** Activity of a dorsal vagal neuron that was not affected by low pH. **F.** Low pH inhibited a medial vestibular neuron. Reprinted from Brain Research, 271: 349–353, Miles R. Does low pH stimulate central chemoreceptors located near the ventral medullary surface? With kind permission of Elsevier Science.

Inhibition is considered to be a nonspecific response to respiratory acidosis of CNS neurons. However, when neurons have been studied *in vitro,* the degree of inhibition observed has typically been small. For example, in the NTS of rat brain slices, respiratory acidosis induced by an increase in CO_2 level from 5% to 15% results in a decrease in firing rate from 6.5 Hz to 5.5 Hz in a subset of neurons (Dean et al., 1990). In the medullary raphe and VLM in rat brain slices, the majority of neurons (70%) do not change their firing rate when exposed to respiratory acid-base disturbances over a range of CO_2 levels between 3% and 9% (Richerson, 1995). Thus, although the number of regions studied in isolation has been limited, in most isolated CNS neurons, acidosis either does not cause inhibition, or causes only a small degree of inhibition.

CELLULAR EVIDENCE FOR CHEMORECEPTION IN THE VLM

Lack of evidence at the cellular level for neurons in the VLM with a high degree of intrinsic sensitivity to respiratory acid-base disturbances has been used as an argument against a role for the VLM in chemoreception. There were early reports of glia in the VLM with chemosensitivity (Fukuda and Honda, 1975; Fukuda et al., 1978), but until recently

(Richerson, 1995) cellular data supporting the existence of highly chemosensitive neurons in the VLM was based solely on extracellular recordings.

Extracellular Recordings from the VLM

There is abundant evidence from *in vivo* recordings that neurons in the VLM increase their firing rate in response to systemic respiratory acidosis, but this is true of neurons in many respiratory and nonrespiratory regions of the medulla (Feldman, 1986; Cohen, 1968). Most of these responses are likely mediated by synaptic connections within an intact network.

A number of investigators have used *in vitro* approaches to study the response of neurons in the VLM to acidosis. Neurons stimulated by metabolic acidosis have been recorded with extracellular electrodes in rat brain slices in the arcuate nucleus (parapyramidal region) of the rostral VLM, immediately lateral to the pyramids (Fukuda and Loeschcke, 1977; Fukuda et al. 1980), the same approximate region defined as the rostral chemosensitive zone (Mitchell et al., 1963; Schlaefke, 1981; Feldman, 1986). With normal [Ca^{2+}] and [Mg^{2+}], neurons of the ventral medulla were more likely to be stimulated, and neurons of the dorsal medulla were more likely to be inhibited by acidosis, whereas in 0.35 mM Ca^{2+} and 1.8 mM Mg^{2+} these responses were reversed (Fukuda and Loeschcke, 1977). The authors interpreted this to mean that stimulation by acidosis of ventral neurons was dependent on synaptic input, whereas in dorsal neurons it was intrinsic. However, these experiments were performed in slices as thick as 800 μm, and the ratio of [Mg^{2+}] to [Ca^{2+}] was only 5 : 1. Using 300–600 μm thick horizontal slices of the ventral medulla that excluded all dorsal tissue, the same group later found that 51% of neurons in the rostral chemosensitive zone were stimulated by acidosis, and 32% were inhibited (Fukuda et al., 1980) (Fig. 2). Interestingly, they found that neurons in the intermediate chemosensitive zone were only stimulated by acidosis when that region was connected to the rostral chemosensitive zone. These results provide strong evidence that the rostral ventral medulla contains all the cellular elements required for chemosensitivity. They also illustrate the importance of synaptic projections in mediating responses within brain slices.

Extracellular recordings from neurons in the rostral chemosensitive zone in 300–400 μm rat brain slices have demonstrated neurons stimulated by metabolic acidosis (Jarolimek et al., 1990). These experiments used very slow changes in pH (at constant CO_2), from 7.4 to 7.2, occurring over more than 5 min. In contrast to the results of Fukuda et al. (1980), most neurons studied showed only transient responses to these slow changes in pH (Fig. 3). These results suggest that adaptation could be an important property of some central chemoreceptors. Chemosensitivity was present both in normal Ringer's solution, as well as Ringer's solution with high [Mg^{2+}] (3.8 mM) and zero [Ca^{2+}]. This group also examined the effects of respiratory acidosis using 25% and 50% CO_2, and observed less than a two-fold increase in firing rate in response to 25% CO_2. It is unclear why these neurons respond to metabolic acidosis, but are insensitive to respiratory acidosis.

Neurons from the ventrolateral medulla have been studied in explant culture with extracellular recordings (Neubauer et al., 1991). Of 27 ventral medullary neurons studied, 10 were stimulated by respiratory acidosis (Fig. 4A), with an increase in firing rate of 40–100% when pH was changed by 0.4 unit. In 15 neurons, there was a decrease in firing rate in response to a decrease in pH of 0.4–0.5 unit (Fig. 4B). These neurons showed no response to metabolic acidosis. Thus, the differential sensitivity to respiratory as compared to metabolic acidosis was opposite to that seen by Jarolimek et al. (1990).

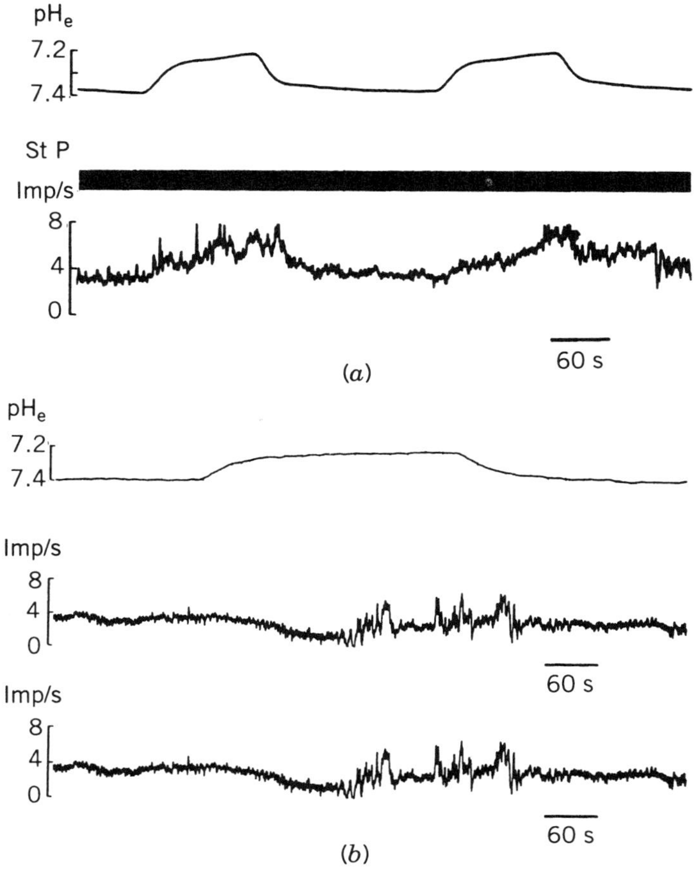

FIGURE 2. Effect of H^+ on tonic activity of VLM neurons recorded extracellularly from brain slices. **A.** Activity was increased by H^+. **B.** H^+ elicited an initial depression of tonic activity, which was followed by appearance of phasic burst on the basic activity. From Fukuda et al., 1980.

Intracellular Recordings from the VLM

The first chemosensitive cells in the VLM from which intracellular recordings were made were nonspiking, did not have synaptic input, and were probably glia (Fukuda and Honda, 1975; Fukuda et al., 1978). These recordings from slices of the rat medulla demonstrated large changes in membrane potential (15–20 mV/pH unit) in response to metabolic acid/base disturbances. The responses were clearly time-locked to the stimulus, were reproducible, and were elicited with changes in pH between 7.0 and 7.8 (Fig. 5). Some cells depolarized, while others hyperpolarized in response to acidosis. Cells with a depolarizing response were more common in the VLM, but were also present in the dorsal medulla. None of the cells tested responded in the presence of 0.35 mM Ca^{2+}/1.8 mM Mg^{2+}.

The first intracellular recordings from CO_2-sensitive neurons in the VLM were recently obtained using perforated patch recordings from thin (100 μm) rat brain slices (Richerson, 1992, 1995). One chemosensitive cell in the VLM had processes that extended in both directions from the cell body along the superficial surface of the VLM (Richerson, 1995), suggest-

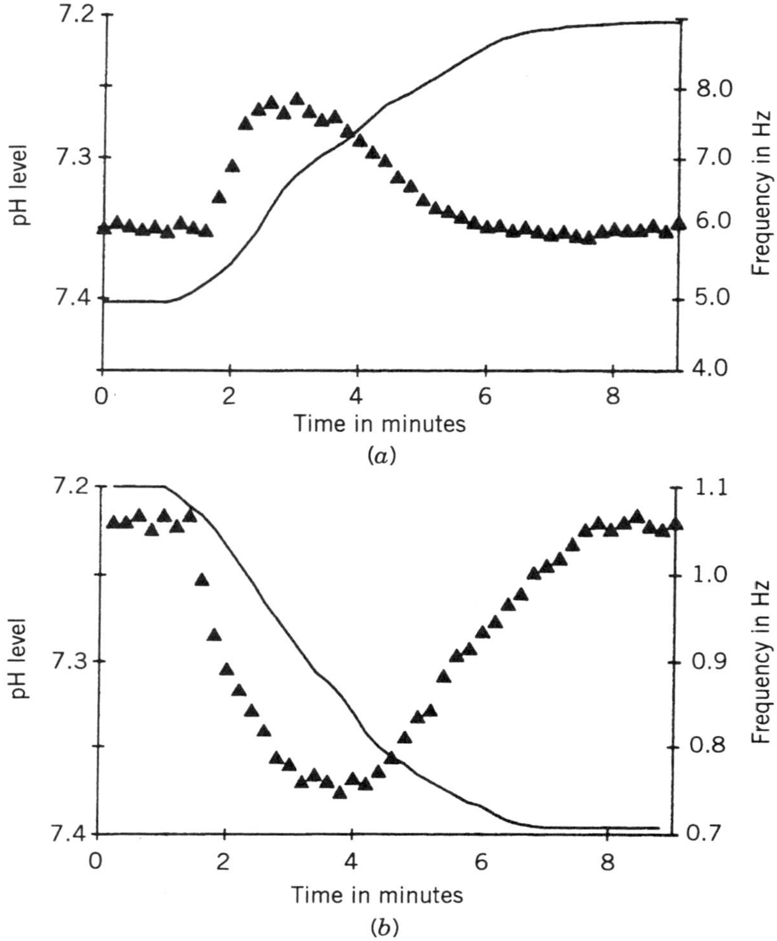

FIGURE 3. Effects of changing pH on the discharge frequency of pH-sensitive neurons in the VLM *in vitro* either by **(A)** reducing $[HCO_3^-]$ from 17 mM (pH 7.40) to 10 mM (pH 7.20) or (**B**) increasing $[HCO_3^-]$ from 10 mM to 17 mM. Extracellular recording. The effects of pH were transient, although pH was continuously changing and finally reached a new steady state. (A) and (B) are recordings from different neurons. pH |—|; neuronal discharge rate ▲. From Jarolimek et al., 1990: Neurons sensitive to pH in slices of the rat ventral medulla oblongata. Pflüegers Arch 416:247–253.

ing that these processes sense changes in pH or CO_2 level on the ventrolateral medullary surface (Schlaefke, 1981), or alternatively from the large number of blood vessels located there.

Using the *in vitro* brainstem and spinal cord preparation of neonatal rats, intracellular recordings have recently been reported from neurons in the ventral medulla that respond to an increase in CO_2 of the bath solution from 2% to 8% (Kawai et al., 1996). This increase in CO_2 of the bath solution causes a decrease in tissue pH at the depth of recording from ≈ 6.7 to near 6.2 (Okada et al., 1993). Responsive neurons were found within the nucleus ambiguus, retrofacial nucleus, and other respiratory regions distant from the ventrolateral medullary surface, which was previously considered to be the sole site of chemosensitive neurons. Many of these neurons fired in phase with respiratory output, in-

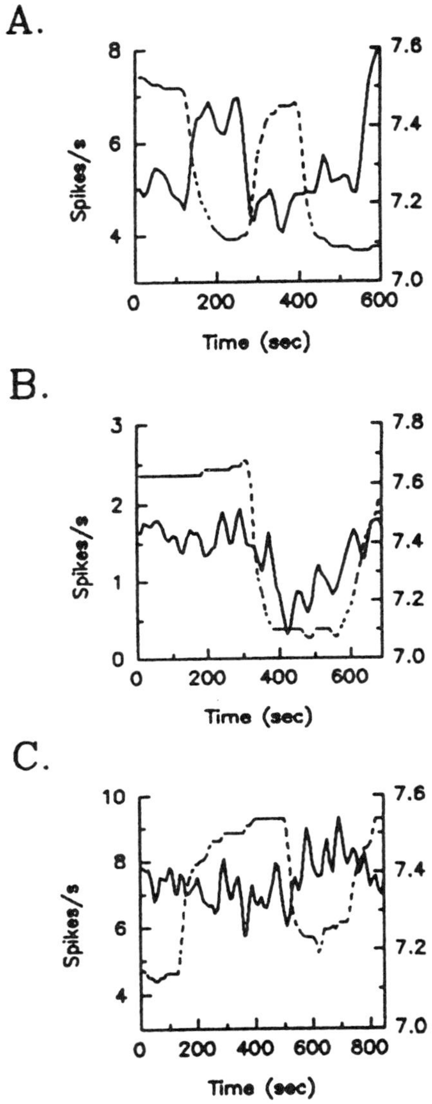

FIGURE 4. CO_2 responses of spontaneously active ventral medullary neurons recorded extracellularly from explant cultures. Panels depict plots of both spike activity (continuous line) and superfusate pH (dashed line) with time. **A.** A neuron excited by hypercapnic acidosis. **B.** A neuron depressed by hypercapnic acidosis. **C.** A neuron whose spontaneous activity was unaffected by hypercapnic acidosis. Reprinted from Neuroscience 45(3), Neubauer et al. Chemosensitivity of medullary neurons in explant tissue cultures, p. 701–708, 1991, with kind permission from Elsevier Science Ltd.

dicating that they were involved in respiratory control. Application of high [Mg^{2+}] (5 mM) / low [Ca^{2+}] (0.2 mM) solution, Cd^{2+} (0.2 mM), or tetrodotoxin (TTX; 0.3–1.5 μm) to the bulk solution disrupted the rhythmic output of the preparation, but did not block the response to acidosis of neurons. Most of the neurons that were responsive had somata that were deep within the tissue, but had dendrites that projected close to the ventral

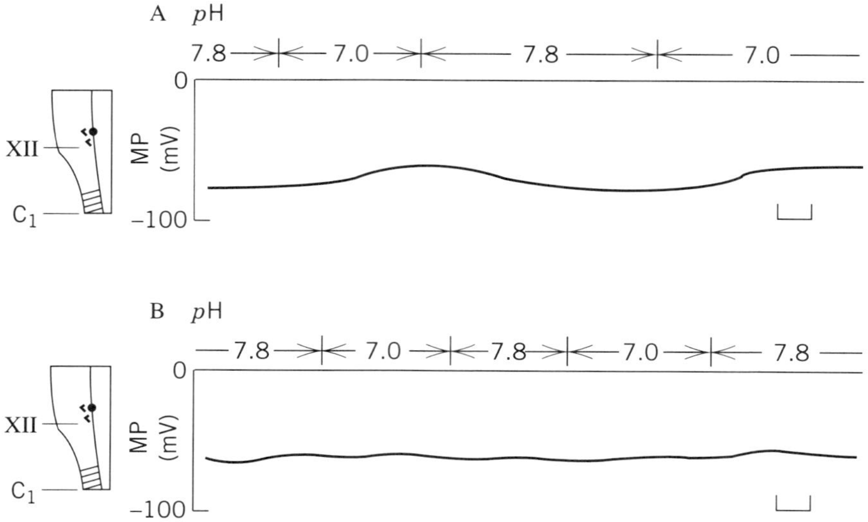

FIGURE 5. Effect of changing the pH of perfusing solution on membrane potential of cells at the surface layer of the rat ventral medulla *in vitro*. **A.** Intracellular recording from a single cell at the area medial to the XIIth cranial nerve (Loeschcke's area). Note the significant changes in potential with changes in pH. **B.** Intracellular recording from a single cell 1 mm cranial to the root of the XIIth cranial nerve and 1 mm lateral to the pyramid. In both (A) and (B) pH was changed with constant P_{CO_2} (42 mmHg). MP = transmembrane potential; XII = XIIth cranial nerve; C_1 = Ist cervical nerve. Asterisk in the schematic representation of the ventral medulla represents the site at which potential was recorded. Bar = 1 min. With permission from Fukuda Y, See WR, and Honda Y: (1980) H^+-sensitivity and pattern of discharge of neurons in the chemosensitive areas of the ventral medulla oblongata of rats in vitro. Pflüegers Arch. 388:53–61.

medullary surface. This was interpreted as indicating that these neurons are intrinsically chemosensitive, and that chemosensitive ion channels are concentrated on the dendrites. Although these results point to a substrate for chemosensitivity on the ventrolateral medullary surface, it is surprising that respiratory neurons with soma distant from the VLM were chemosensitive. It will be important to verify that these responses were intrinsic to these neurons, and determine whether they respond to changes in pH over a more physiologic range. However, these results support the hypothesis that chemosensitivity is a distributed function of the respiratory network.

EVIDENCE FOR CHEMORECEPTION OUTSIDE THE VLM

After exposure of cats or rats to hypercarbia *in vivo*, an increase in c-*fos* immunoreactivity occurs in neurons of the VLM and dorsal medulla (Sato et al., 1992), in the locus coeruleus (Haxiu et al., 1992), and in the medullary raphe (Larnicol et al., 1994). These findings have been interpreted to suggest that chemoreceptors are present in each of these areas and have helped to support the hypothesis that the VLM is not the sole site for chemoreception.

Application of acetazolamide onto the VLM of rats *in vivo* leads to stimulation of respiration (Coates et al., 1991), presumably secondary to development of tissue acidosis (see Chapters 27 and 32). Microinjection of acetazolamide into the brain parenchyma is thought to produce a focal region of acidosis, and stimulates breathing of rats *in vivo* when placed in the VLM, as well as the locus coeruleus, dorsal medulla (Coates et al., 1993), and medullary raphe (Bernard et al., 1996). These results have also provided additional support that chemoreceptors are not restricted to the VLM.

As a result of the difficulty, until recently, of confirming that the VLM contains chemosensitive neurons, and findings consistent with chemosensitivity in other respiratory regions, a number of groups have studied the response to pH of neurons in respiratory regions outside the VLM. Marino and Lamb (1975) recorded from 53 phasic medullary respiratory neurons in cats *in vivo,* and found that every one increased its firing rate in response to inhalation of CO_2. They then used iontophoresis of HCl solution to expose recorded neurons to local acidosis. Of 53 respiratory neurons, none responded to acidosis. Of 74 nonphasic neurons, 3 increased firing during CO_2 inhalation, and 2 of these were also stimulated by iontophoresis of HCl. Neuronal location was determined only by stereotactic coordinates, but the 2 neurons were probably located in the lateral tegmental field. The degree of sensitivity of these neurons is difficult to determine, since iontophoresis was used and extracellular pH (pH_o) was not monitored.

Several lines of evidence have now confirmed early findings (Fukuda and Loeschcke, 1977; Miles, 1983) that neurons of the dorsal medulla are stimulated by acidosis. Kogo and Arita (1990) used local pressure microejection of acidic mock CSF (pH = 6.87) into the medulla of cats *in vivo,* while simultaneously recording extracellular single unit activity from "nonphasic" neurons. They found 5 neurons in the region of the NTS and 5 in the lateral reticular nucleus deep in the VLM that responded with an increase in firing rate to both local extracellular acidosis and injection of a minimal volume of solution with CO_2 of 100% into the vertebral artery, but not to pressure microejection of control mock CSF (pH = 7.43).

Neurons in the NTS have been examined using intracellular and extracellular recordings in transverse slices of the rat medulla in which the ventral half was removed (Dean et al., 1990). Some neurons in the NTS were stimulated by an increase in CO_2 from 5% to 7% (Fig. 6), and others were stimulated by a CO_2 level of 10–15%, even in a solution with 11.4 mM $MgCl_2$ and 0.2 mM $CaCl_2$.

The response to pressure microejection of solution with P_{CO_2} of 296 mmHg ($CO_2 \approx 39\%$) has also been examined in neurons from the rat NTS and nucleus ambiguus in dissociated tissue culture (Rigatto et al., 1992). Many neurons from both regions were stimulated, while neurons from the hippocampus were not responsive. These results are also consistent with the proposal that the nucleus ambiguus and NTS contain chemosensitivity.

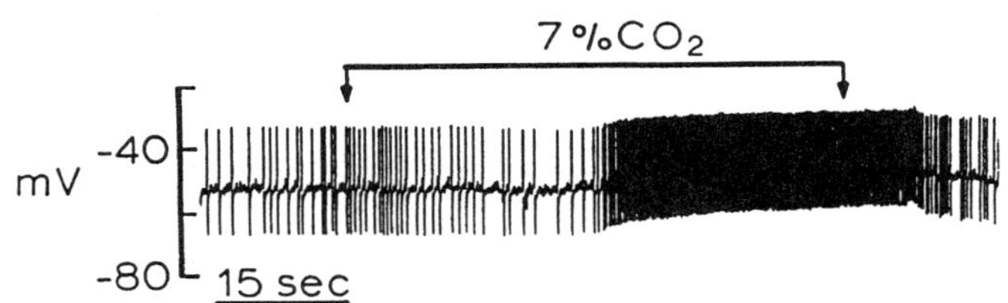

FIGURE 6. Effect of CO_2 on neurons in the NTS in slices of the dorsal medulla. Membrane potential (truncated action potentials) upon changing CO_2 from 5% to 7% (arrows). Resting membrane potential was approximately −50 mV. A holding DC current of −0.06 nA was used to hyperpolarize the neuron and slow down its firing rate. Reprinted from Neuroscience 36(1), Dean et al. Depolarization and stimulation of neurons in nucleus tractus solitarii by carbon dioxide does not require chemical synaptic input. 207–216, 1990 with kind permission from Elsevier Science Ltd.

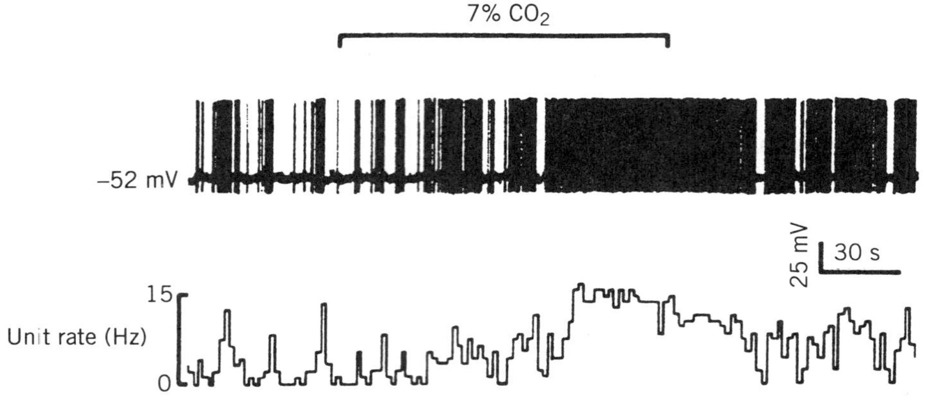

FIGURE 7. Whole-cell patch recording of a caudal hypothalamic neuron. This neuron was excited by hypercapnia. From Dillon and Waldrop, 1992.

The specific role of the hypothalamus in respiration is unknown, but recent recordings from brain slices suggest that neurons from the caudal hypothalamus are sensitive to physiological levels of hypercapnia (Dillon and Waldrop, 1992). Extracellular recordings were made from rat brain slices in an interface chamber. Of 54 neurons tested, 12 increased their firing rate from a mean of 2.8 Hz to 5.2 Hz when the CO_2 level of the gas above the chamber (but not in the perfusate) was increased from 5% to 7%; 6 of 54 cells were inhibited. Using whole-cell patch-clamp recordings, hypercarbia stimulated 7 of 20 neurons, and inhibited 2 of 20 neurons (Fig. 7). These results are interesting, since the hypothalamus is involved in autonomic function, and can modulate respiratory output. Although it remains possible that the hypothalamus is a site for respiratory chemoreception, this is not a widely held view. Therefore, these results suggest that stimulation in response to respiratory acidosis may not be a property limited to respiratory chemoreceptors.

The above data provide evidence that a stimulatory response to acidosis is not a property restricted to the VLM. Until recently, a stimulatory response to acidosis was accepted as sufficient evidence that a neuron is a respiratory chemoreceptor. Thus, *in vitro* experiments have focused on searching for neurons stimulated by acidosis. Acidosis-stimulated neurons have now been detected in a number of regions outside the VLM. Since some of these neurons could be involved in control of other pH-sensitive CNS functions, it will be important to establish criteria for defining a chemosensitive neuron as a respiratory chemoreceptor beyond simply possessing a stimulatory response to acidosis.

CHEMOSENSITIVITY OF NEURONS IN THE MEDULLARY RAPHE

The role of the medullary raphe is unclear, and has been ascribed to a variety of CNS functions, including cardiorespiratory control (Jacobs and Azmitia, 1992). Recently, the medullary raphe of the rat has been proposed to be a site containing chemoreceptors (Richerson, 1992, 1995). In the cat, neurons in the medullary raphe are homologous to a subset of neurons in the parapyramidal region of the VLM (Jacobs and Azmitia, 1992; Filiano et al., 1990). Some neurons of the VLM contain serotonin, as do raphe neurons (Dean et al., 1993; Jacobs and Azmitia, 1992), and neurons in both regions contain thyrotropin-releasing hormone and substance P (Dean et al., 1993; Bayliss et al., 1994). These three neurotransmitters can each stimulate respiratory output (Moss and Inman, 1989; Lalley, 1986; Richerson, 1995; Dekin et al., 1985). Neurons in both the VLM and medullary raphe project widely to cardiorespiratory regions of the CNS, including the NTS, nucleus ambiguus, phrenic motor nucleus, and intermediolateral cell column (Jacobs and Azmitia, 1992; Smith et al., 1989; Connelly et al., 1989). The medullary raphe is located near the ventral medullary surface, deep to the basilar artery, and is highly perfused with large penetrating arteries. This high degree of vascularity would be an advantage in sensing changes in CO_2 of blood or the adjacent extracellular space (Libscomb and Boyarsky, 1972; Cragg et al., 1977).

Perforated patch-clamp recordings were recently made from visually identified neurons of the medullary raphe and VLM in thin (100 μm) rat brain slices (Richerson, 1995). Of 105 neurons tested, 74 did not change their firing rate in response to respiratory acidosis, 16 were stimulated by respiratory acidosis, and 15 were inhibited (Richerson, 1995) (Fig. 8). Respiratory alkalosis resulted in opposite effects. Stimulated and inhibited neurons were found in both the VLM and the medullary raphe, with a high concentration of stimulated neurons within the medullary raphe.

The degree of sensitivity of medullary raphe neurons was high. In neurons stimulated by respiratory acidosis, a change in CO_2 level from 5% to 7% (pH drop of 0.12) resulted in as much as a 65% increase in firing rate, and a decrease in CO_2 level from 5% to 3% (pH increase of 0.12) resulted in as much as a 50% decrease in firing. In contrast to the relatively small nonspecific inhibitory effects of acidosis described in other regions of the nervous system, some neurons of the VLM and medullary raphe were strongly inhibited by respiratory acidosis. For example, an increase in CO_2 from 5% to 6% at constant [$NaHCO_3$] led to as much as a 50% reduction in the firing rate of raphe neurons (Fig. 9). A decrease in CO_2 level from 5% to 4% resulted in a 100% increase in firing rate. This magnitude of inhibition by a physiological degree of respiratory acidosis has not been seen outside the ventral medulla. It has been proposed that neurons inhibited by acidosis are functionally involved in chemoreception, and that a decrease in their firing rate leads to either disinhibition of inspiratory neurons, or a decrease in excitation of expiratory neurons (Richerson, 1995). Neurons stimulated by CO_2 are usually described as "responsive," while those that are inhibited are sometimes called "unresponsive," since inhibition is thought to be nonspecific. However, since it remains unclear whether inhibition of neurons by acidosis plays a functional role, neurons with a highly sensitive inhibitory response to respiratory or metabolic acidosis would be better termed "acidosis-inhibited," and those with a highly sensitive stimulatory response "acidosis-stimulated."

There is *in vivo* evidence to suggest that neurons of the medullary raphe play a functional role in respiratory chemoreception. Electrical stimulation of the medullary raphe of cats *in vivo* causes large changes in minute ventilation (Lalley, 1986). Acetazolamide microinjections into the medullary raphe also stimulate breathing in rats *in vivo* (Bernard et al., 1996). Inhalation of CO_2 results in an increase in firing rate of neurons in the medullary raphe of cats *in vivo* (Veasey et al., 1995) of a magnitude similar to that seen in raphe neurons in brain slices (Richerson, 1995). In addition, hypercarbia induces c-*fos* activation in the raphe of cats *in vivo* (Larnicol et al., 1994).

Abnormalities of homologous neurons in humans have been associated with respiratory control disorders. Some cases of sudden infant death syndrome (SIDS) are thought to be due to a defect in cardiorespiratory control (Richerson, 1997; Hunt, 1992). The arcuate nucleus, the human homologue of the cat and rat medullary raphe and parapyramidal

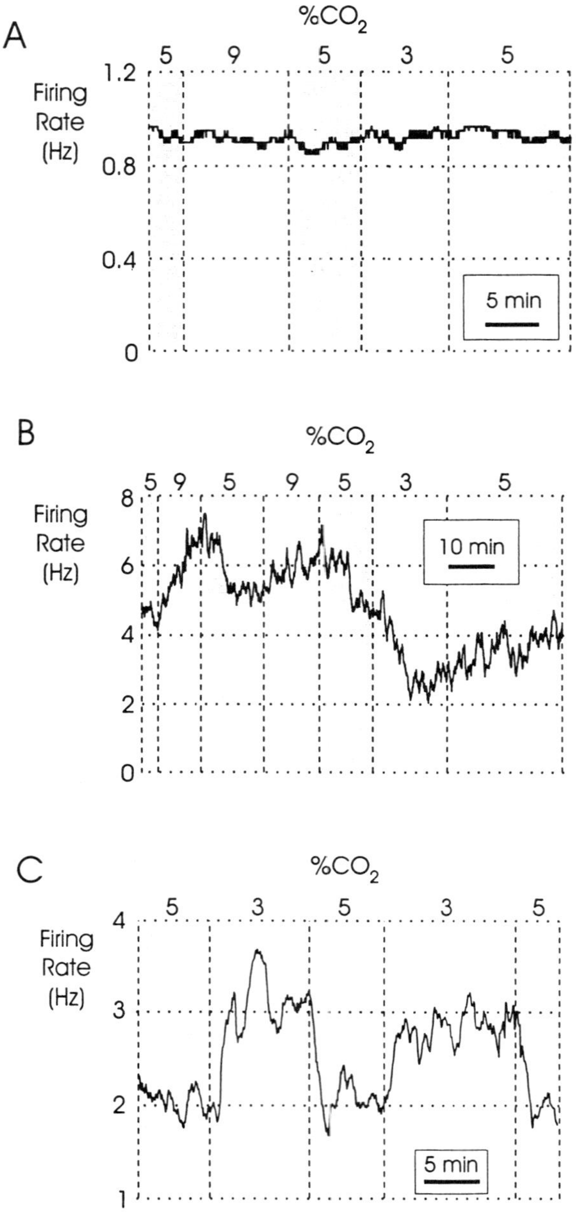

FIGURE 8. Three types of response to respiratory acidosis in medullary raphe neurons recorded using the perforated patch technique. **A.** Firing rate of a neuron with no response, typical of the majority of neurons studied. **B.** Firing rate of a neuron stimulated by an increase in CO_2. **C.** Firing rate of a neuron inhibited by an increase in CO_2. In each panel, firing rate versus time is plotted as a moving average as CO_2 is changed. From Richerson, 1995.

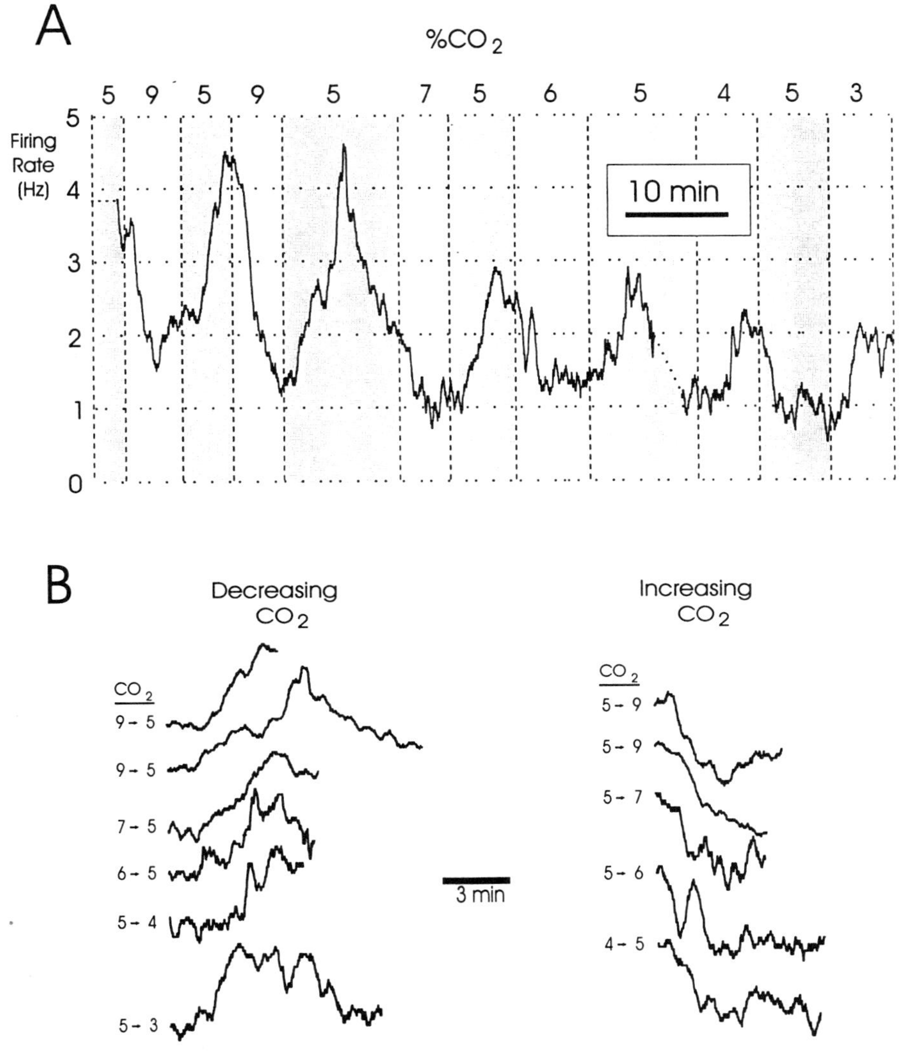

FIGURE 9. Firing rate of a CO_2-inhibited neuron with high sensitivity from the medullary raphe. **A.** Firing rate versus time for entire recording. Note the slow time course of the response, and the adaptation to a maintained CO_2 level during the 3rd, 4th, and 5th periods at 5% CO^2. From Richerson, 1995. **B.** Separation of each segment of recording shown in (A), in order to better illustrate the consistency in time course, the delay in onset, and the adaptation of the response. For each trace in (B), the amplitude has been normalized to the same height to better compare the time courses.

region (Filiano et al., 1990), is decreased in size in some SIDS cases (Filiano and Kinney, 1992), and has been reported to have a decrease in muscarinic receptor binding (Kinney et al., 1995). Abnormalities of the human arcuate nucleus have also been linked to congenital central hypoventilation syndrome (Ondine's curse) (Folgering et al., 1979), a syndrome linked to a defect in central chemoreception. Haddad syndrome (combined Ondine's Curse and Hirschsprung's disease) (Haddad et al., 1978; Verloes et al., 1993), has been proposed to be a result of a defect in a stem serotonergic cell, which would also implicate involvement of serotonergic raphe neurons in chemoreception.

The evidence described in this section supports a role for chemosensitive neurons of the raphe in respiratory control. Further experiments *in vivo* or using semi-intact preparations are needed to define their specific role.

LESSONS FROM PERIPHERAL CHEMORECEPTORS

The cellular mechanisms of chemotransduction have been well characterized for the peripheral chemoreceptors. Although it is tempting to speculate about the mechanisms of central chemoreception based on these data, there are reasons to believe that this could lead to erroneous conclusions. The peripheral chemoreceptors are primarily sensitive to hypoxia, and although hypercarbia potentiates the response to hypoxia, there is little effect of changes in acid-base status in some species during normoxia (O'Regan and Majcherczyk, 1982). In contrast, the central chemoreceptors are highly sensitive to hypercarbia, while the respiratory output of the peripherally chemodenervated CNS is unresponsive or inhibited by hypoxia during eupnea (Millhorn and Eldridge, 1986; Cunningham et al., 1986). However, it may help to consider data from the carotid bodies, since it is possible that some mechanisms are shared.

The carotid bodies contain two cell types, the Type I (glomus) cells, which are thought to be the chemoreceptors, and the Type II cells, which are thought to be supporting (glia-like) cells. The glomus cells make bidirectional synaptic contact with nerve endings of axons that travel in the carotid sinus nerve. Stimulation of the carotid body with metabolic or respiratory acidosis leads to an increase in discharge of the carotid sinus nerve and release of catecholamines, presumably from Type I cells (Rigual et al., 1991).

The effect of acidosis on dissociated Type I cells from the rat has been studied. Buckler et al. (1991) used imaging of intracellular pH and found that these cells have a steep relationship between intracellular and extracellular pH (pH_i and pH_o). A large change in pH_i would enhance chemosensitivity if the site of chemotransduction was on the inside of the cell membrane. They also found that pH_i response of isolated Type I cells to metabolic acidosis, respiratory acidosis, and changes in CO_2 level at constant pH_o, are similar in time course to previously published data on the change in carotid sinus nerve discharge in response to the same stimuli. They concluded that the primary stimulus for chemosensitivity is probably pH_i.

Using calcium imaging and patch-clamp recording from the same preparation (Buckler and Vaughan-Jones, 1994), it was found that hypercapnic acidosis (10–20% CO_2) induces a membrane depolarization and an increase in $[Ca^{2+}]_i$. The increase in $[Ca^{2+}]_i$ is dependent on extracellular Ca^{2+}, and is blocked when membrane depolarization is prevented using voltage clamp. The increase in $[Ca^{2+}]_i$ is also attenuated by nicardipine and Ni^{2+}, suggesting that voltage-dependent Ca^{2+} channels are involved. Furthermore, it is blocked by replacement of Na^+ by N-methyl-D-glucamine, but not by Li^+. It was concluded that membrane depolarization is the primary event, which then leads to a rise in $[Ca^{2+}]_i$ via activation of voltage-gated Ca^{2+} channels. It was also felt that the Na^+/Ca^{2+} exchanger plays little or no role in Ca^{2+} influx. Using depolarizing ramp potentials, evidence was provided that depolarization is a result of a decrease in slope of the evoked current, particularly at positive voltages. This was interpreted as being consistent with a decrease in a potassium current.

Peers and Green (1991) also studied isolated Type I cells of the rat with whole-cell voltage-clamp recordings. They found that when cells are exposed to propionate or acetate (10–20 mM) to induce an intracellular metabolic acidosis without a change in pH_o there is a decrease in K^+ current amplitude. This effect was absent when calcium currents were blocked, but there was no direct effect on Ca^{2+} currents. These results were interpreted as consistent with acidosis acting to inhibit a Ca^{2+}-activated K^+ current from an intracellular site.

In rabbit carotid body Type I cells, Rocher et al. (1991) have proposed a different mechanism. They also observe an increase in $[Ca^{2+}]_i$, but propose that intracellular acidosis leads to a rise in $[Na^+]_i$ via Na^+-dependent acid extrusion, which then causes an increase in $[Ca^{2+}]_i$ by reversal of the Na^+/Ca^{2+} exchanger. The rise in $[Ca^{2+}]_i$ is not blocked by dihydropyridines, but is blocked by ethylisopropylamiloride (EIPA), an antagonist of Na^+/Ca^{2+} exchange, and is blocked by replacement of Na^+ influx with N-methyl-D-glucamine. Thus, in both the rat and rabbit intracellular acidosis has been proposed to lead to a rise in $[Ca^{2+}]_i$, but via two distinct mechanisms. It remains controversial whether these results indicate true species differences, or whether both mechanisms may exist to differing degrees in both species. Buckler and Vaughan-Jones (1994) have suggested that a different interpretation of the results from the rabbit could be consistent with the mechanism proposed in the rat.

Work in the rat carotid body points to depolarization induced by a decrease in pH_i as the primary chemotransduction event, and a rise in $[Ca^{2+}]_i$ as the regulator of downstream events such as neurotransmitter release. If similar mechanisms exist in central chemoreceptors, then the approach of studying the effects of acidosis on membrane potential, firing rate, and ion channel function is likely to be fruitful for elucidation of the primary chemotransduction event.

PRIMARY STIMULUS FOR CENTRAL CHEMOSENSITIVITY

There is strong evidence that the primary stimulus for peripheral chemoreceptors is pH_i. However, evidence for the unique stimulus for central chemoreception is indirect, and less definitive. Data from whole animals (reviewed in Chapters 27, 32) favors a direct effect of pH, but recordings from isolated cells from the medulla have given conflicting results. In slices, neurons from the NTS (Dean et al., 1990) and glia from the VLM (Fukuda and Loeschcke, 1977) that respond to respiratory acidosis also respond to metabolic acidosis. In contrast, neurons in culture from the VLM that are sensitive to respiratory acidosis do not respond to metabolic acidosis (Neubauer et al., 1991). Some neurons in the VLM in slices sensitive to metabolic acidosis are insensitive to respiratory acidosis (Jarolimek et al., 1990). There has been no report of more detailed studies of these individual chemosensitive neurons to determine whether there are cellular effects from changes in CO_2 levels independent of changes in extracellular pH, changes in pH in the absence of CO_2 and $[HCO_3^-]$, or independent changes in pH_i and pH_o. The methods that have been applied to the peripheral chemoreceptors have not yet been used to study chemosensitive CNS neurons, but this is likely to be an active area of study in the future.

SPECULATION ON MECHANISMS OF CENTRAL CHEMOTRANSDUCTION

In the case of the peripheral chemoreceptors, the site of chemoreception has been clearly defined to be a group of cells in the carotid and aortic bodies. In contrast, although a number of CNS regions have been shown to have cells with varying degrees of chemosensitivity, it is not clear which, if any, of these cells are the central respiratory chemoreceptors. Recent data using acetazolamide microinjections *in vivo* suggest that respiratory chemosensitivity may be more widespread than previously believed (Coates et al., 1993; Bernard et al., 1996), and there is overlap between areas found to contain chemosensitivity with this technique and areas found to contain acidosis-stimulated cells *in vitro*. If each of these areas do contain respiratory chemoreceptors, it is possible that the threshold for their response differs widely in each of the different regions. None of the CNS cells described to have chemosensitivity at the cellular level have yet been studied in sufficient detail to define the primary stimulus, the ion channels that are involved, or their projections, neurotransmitter content, or effects on downstream neurons. Thus, there is no clear evidence for the cellular mechanisms of central respiratory chemoreception.

Cellular Properties of Chemoreceptors

Recordings from isolated cells promise to give us detailed information about the mechanisms of chemosensitivity at the cellular level. However, it is important to decide how to determine whether a cell with intrinsic chemosensitivity is or is not a likely candidate for a respiratory chemoreceptor.

In order for central chemoreceptors to carry out their function, it is likely that they have the following cellular properties:

1. *Chemosensitivity.* They should be sensitive to respiratory acid-base disturbances within the range encountered under physiological conditions (pH $\approx$ 7.2 to 7.6). It is likely, but not essential, that neurons specialized to respond to respiratory acidosis have one of the highest degrees of sensitivity of neurons within the nervous system.
2. *Connections with other respiratory nuclei.* In order to influence respiratory output, chemoreceptor neurons should project to one or more of the key respiratory nuclei, such as the NTS, nucleus ambiguus, pre-Bötzinger complex, phrenic motor nucleus, etc. If chemosensitivity is a distributed function of respiratory neurons, then the projections would be inherent in network connectivity.
3. *Appropriate effects on downstream respiratory neurons.* Chemoreceptors should contain neurotransmitters that influence respiratory output.

Chemoreceptors could probably serve their function regardless of the specific characteristics of the following properties:

1. *Amplitude of response.* The respiratory system as a whole is highly chemosensitive. It is therefore sometimes assumed that each individual chemoreceptor neuron has a similar sensitivity. However, it is possible that a smaller degree of chemosensitivity of individual neurons could be amplified via network interactions. Thus, each individual chemoreceptive element does not necessarily have the gain of the entire feedback system.

2. *Uniqueness.* Although the respiratory system as a whole may have a uniquely high sensitivity to CO_2 level and/or pH, there is no evidence that chemosensitivity of the respiratory chemoreceptors is unique. Rather, it is possible that neurons with sensitivity to CO_2 and/or pH as great as the respiratory chemoreceptors could modulate one of the many pH-sensitive nonrespiratory CNS functions.
3. *Sign of response.* It is commonly assumed that all chemoreceptor neurons are stimulated by respiratory acidosis (and inhibited by respiratory alkalosis), and that they in turn stimulate respiratory neurons. However, some chemoreceptors could theoretically be inhibited by respiratory acidosis, and then release a tonic inhibition of inspiratory neurons, or decrease stimulation of expiratory neurons. There is precedence for antagonistic physiological sensors in the thermoregulatory system, with cold-sensitive and warm-sensitive neurons in the hypothalamus (Boulant and Dean, 1986). In an analogous way, there could be two classes of chemoreceptor: excitatory and inhibitory.
4. *Primary stimulus.* It is unclear whether CO_2, pH_o, pH_i, and/or $[HCO_3^-]$ are the primary stimuli for detection of respiratory acidosis. It is possible that more than one of these stimuli interact with the chemotransduction element.
5. *Compartment sensed.* It has been hypothesized that chemoreceptors in the VLM sense pH or CO_2 levels of CSF within the cisterna magna. Although this was the stimulus used in early experimental work, there is no evidence that chemoreceptors specifically sense pH or the CO_2 level of bulk CSF. An alternative is that chemoreceptors detect pH and/or CO_2 levels of blood or adjacent extracellular space, analogous to peripheral chemoreceptors. The ventral medulla is a site with a high concentration of blood vessels (Cragg et al., 1977; Libscomb and Boyarsky, 1972).
6. *Chemoreceptor cell type.* Although chemoreceptors are usually considered to be neurons, glia could be primary chemosensitive elements, or could contribute to the response. The first intracellular recordings from pH-sensitive cells in the VLM were from chemosensitive glia (Fukuda and Honda, 1975; Fukuda et al., 1978). If glia were the primary chemoreceptors, specialized "second order" chemoreceptor neurons must be intimately associated, in order to transduce glial signals into changes in firing rate of respiratory neurons.
7. *Segregation of chemoreceptor function.* The function of respiratory chemoreceptors is not necessarily restricted only to chemoreception. Some chemoreceptors may also contribute to other aspects of respiratory control, such as pattern generation or shaping of the motor output.

Potential Mechanisms of Chemoreceptor Function

Hypotheses about the cellular mechanisms of transduction of acid-base disturbances for respiratory chemoreception will remain speculative until the chemoreceptors have been identified. However, some clues are available from current-clamp recordings from chemosensitive central neurons in respiratory regions. This approach is a necessary first step in identifying the location of chemosensitive neurons, and provides clues about potential mechanisms. However, voltage-clamp recordings will be required to more fully define the ionic mechanisms.

Sharp microelectrode recordings from NTS neurons in slices have demonstrated changes in membrane potential in response to respiratory acidosis (Dean et al., 1990).

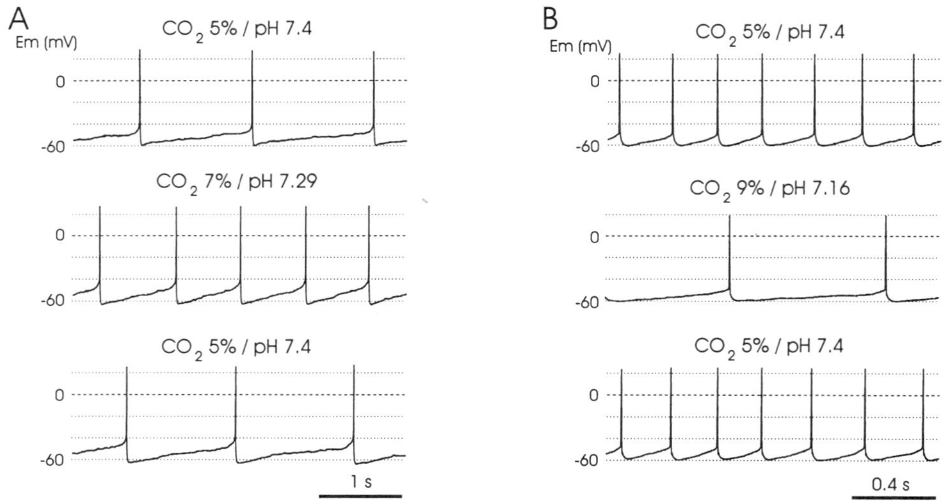

FIGURE 10. Changes in firing rate were accompanied by a change in slope of the ramp depolarization. **A.** Membrane potential changes from a CO_2-stimulated neuron at 5% CO_2 (top), 7% CO_2 (middle), and 5% CO_2 (bottom). **B.** Membrane potential changes from a CO_2-inhibited neuron at 5% CO_2 (top), 9% CO_2 (middle), and 5% CO_2 (bottom). From Richerson, 1995.

Hypercarbia leads to either stimulation, inhibition, or no effect on neurons in this region. Acidosis results in a decrease in the level of after-hyperpolarization in stimulated neurons, and a more negative after-hyperpolarization in inhibited neurons, with little effect on spike threshold in either type of neuron. Several currents that could be responsible for this effect have been described in these neurons (Dekin, 1993; Dekin and Getting, 1987).

In the medullary raphe, beating pacemaker neurons fire action potentials with a high degree of regularity *in vivo* (Jacobs and Azmitia, 1992), and in slices (Fig. 10). Each spike is followed by an after-hyperpolarization, and a linear ramp depolarization to threshold for the next spike. In acidosis-stimulated and acidosis-inhibited neurons in the raphe, acidosis induces opposite changes in slope of the ramp depolarization between spikes (Fig. 10). In most cases there is little change in the level of after-hyperpolarization, the spike threshold, or the mean membrane potential during the ramp depolarization. In the dorsal raphe, currents likely to modulate the membrane potential during the linear ramp depolarization include A-current (I_A) (Aghajanian, 1985; Segal, 1985), Ca^{2+}-activated K^+ current (I_{KCa}), Ca^{2+} currents, and a depolarization-activated cation current (Penington and Kelly, 1993). In the medullary raphe, there is evidence for each of these currents, as well as both a hyperpolarization-activated cation current (I_h) (McCormick and Pape, 1990) and an inwardly rectifying potassium current (I_{kir}) (unpublished observations). One or more of these currents are likely candidates for modulation by pH. Interestingly, both pacemaker activity and chemosensitivity of medullary raphe neurons are rapidly lost during whole-cell patch-clamp recording, presumably due to dialysis of molecules required for chemotransduction. These could potentially be Ca^{2+} or pH buffers, intracellular second messengers involved in chemotransduction, or constitutive factors responsible for maintaining baseline activity of an ion channel that is pH-sensitive. Perforated patch recordings prevent the rundown of the response. Many of the currents listed above are known to be modulated by factors dialyzed during whole-cell recording.

From existing data, some possibilities emerge for the cellular mechanisms of chemoreception. If the primary event in central chemoreception is modulation of an ion channel (as in the rat carotid body), rather than an event with less effect on membrane potential (e.g., H^+-induced changes in membrane transport leading to an increase in $[Ca^{2+}]_i$, e.g., proposed for the rabbit carotid body), then an increase in conductance of an inward current, or a decrease in conductance of an outward current is required to induce an increase in membrane potential or firing rate in response to acidosis in acidosis-stimulated neurons (or alkalosis in acidosis-inhibited neurons). If this is the case, this could result either from one specific ion channel with exquisitely high pH sensitivity, or alternatively, from a complement of ion channels, each with a lesser degree of pH sensitivity, which together lead to a highly sensitive response at the whole-cell level. There are a number of ion channels with various degrees of pH sensitivity (reviewed in Section V). There is evidence that neurons in the respiratory region of the NTS (Dekin and Getting, 1987) possess I_{KCa}, the ion channel linked to chemoreception in the rat carotid body, but the chemosensitivity of I_{KCa} in chemosensitive CNS neurons has not been studied. Two inwardly rectifying K^+ channels described in the kidney, RACTK1 and ROMK1 (Wang et al., 1990; Suzuki et al., 1994; Ho et al., 1993) have a high degree of sensitivity to pH, with an open channel probability that can change from 0.9 at pH 7.4 to 0.1 at pH 7.1. This steep relationship between pH and conductance, centered over the physiological range, could explain the sensitivity at the whole-cell level of acidosis-stimulated neurons. A homologous channel would be an excellent candidate for pH sensitivity in central chemoreceptors, but there is no evidence that these ion channels exist in the medulla.

The specific ion channels responsible for chemosensitivity have not been defined in the chemosensitive neurons studied to date in CNS regions linked to respiratory control. To elucidate the mechanisms of these chemosensitive CNS neurons will require an approach like that which has been successful in the carotid body. Now that several different groups of chemosensitive CNS neurons have been identified, this should be possible.

CONCLUSIONS

There is no clear proof available yet for the location or cellular mechanisms of central respiratory chemoreception. Neurons from a number of CNS regions linked to cardiorespiratory control have been reported to respond to acidosis with either inhibition or stimulation, but it remains to be determined which, if any, of these neurons are involved in respiration. Until recordings are made from neurons in a larger number of brain regions, it will be important to avoid making assumptions about the uniqueness and cellular properties of respiratory chemoreceptors. After the cellular properties of chemosensitive neurons are defined in isolated cells, *in vivo* experiments or experiments in semi-reduced preparations will be essential to determine if chemosensitive neurons play a role in respiratory control, rather than control of other pH-sensitive CNS functions. A combination of systems level, cellular, and molecular approaches will be needed before a comprehensive definition of the cellular mechanisms of respiratory chemoreception is possible.

REFERENCES

Aghajanian GK (1985): Modulation of a transient outward current in serotonergic neurons by 1-adrenoceptors. Nature 315: 501–503.

Bayliss DA, Viana F, Kanter RK, Szymeczek-Seay CL, Berger AJ, Millhorn DE (1994): Early postnatal development of thyrotropin-releasing hormone (TRH) expression, TRH receptor binding, and TRH responses in neurons of rat brainstem. J Neurosci 14: 821–833.

Bernard DG, Li A, Nattie EE (1996): Evidence for central chemoreception in the midline raphe. J Appl Physiol 80: 108–115.

Boulant JA, Dean JB (1986): Temperature receptors in the central nervous system. Annu Rev Physiol 48: 639–654.

Bruce EN, Cherniack NS (1987): Central chemoreceptors. J Appl Physiol 62: 389–402.

Buckler KJ, Vaughan-Jones RD, Peers C, Lagadic-Gossmann D, Nye PC (1991): Effects of extracellular pH, P_{CO_2} and HCO_3^- on intracellular pH in isolated type-I cells of the neonatal rat carotid body. J Physiol 444: 703–721.

Buckler KJ, Vaughan-Jones RD (1994): Effects of hypercapnia on membrane potential and intracellular calcium in rat carotid body type I cells. J Physiol 478: 157–171.

Carpenter DO, Hubbard JH, Humphrey DR, Thompson HK, Marshall WH (1974): Carbon dioxide effects on nerve cell function. In Nahas G, Schaefer KE (eds): Carbon Dioxide and Metabolic Regulation. New York: Springer, pp 49–62.

Caspers J, Speckmann EJ (1971): Membrane potential and postsynaptic potentials of spinal motoneurones in seizure inhibition by CO_2. Electroenceph Clin Neurophysiol 30: 265.

Cherniack NS, von Euler C, Homma I, Kao FF (1979): Graded changes in central chemoreceptor input by local temperature changes on the ventral surface of the medulla. J Physiol (Lond) 287: 191–211.

Chonan T, Adams EM, von Euler C, Cherniack NS (1990): Effects of focal cooling in the ventrolateral medulla on chemoresponsiveness in dogs. Respir Physiol 80: 45–54.

Coates EL, Li AH, Nattie EE (1991): Acetazolamide on the ventral medulla of the cat increases phrenic output and delays the ventilatory response to CO_2. J Physiol (Lond) 441: 433–451.

Coates EL, Li A, Nattie EE (1993): Widespread sites of brain stem ventilatory chemoreceptors. J Appl Physiol 75: 5–14.

Cohen MI (1968): Discharge patterns of brain-stem respiratory neurons in relation to carbon dioxide tension. J Neurophysiol 31: 142–165.

Connelly CA, Ellenberger HH, Feldman JL (1989): Are there serotonergic projections from raphe and retrotrapezoid nuclei to the ventral respiratory group in the rat? Neurosci Lett 105: 34–40.

Cragg P, Patterson L, Purves MJ (1977): The pH of brain extracellular fluid in the cat. J Physiol (Lond) 272: 137–166.

Cunningham DJC, Robbins PA, Wolff CB (1986): Integration of respiratory responses to changes in alveolar partial pressures of CO_2 and O_2 and in arterial pH. In: Handbook of Physiology, Section: The Nervous System IV, editor FE Bloom, Baltimore: Williams & Wilkins, pp 475–528.

De Jong RH, Wagman IH, Prince DA (1967): Effect of carbon dioxide on the cortical seizure threshold to lidocaine. Exp Neurol 17: 221–232.

Dean JB, Bayliss DA, Erickson JT, Lawing WL, Millhorn DE (1990): Depolarization and stimulation of neurons in nucleus tractus solitarii by carbon dioxide does not require chemical synaptic input. Neuroscience 36: 207–216.

Dean C, Marson L, Kampine JP (1993): Distribution and co-localization of 5-hydroxytryptamine, thyrotropin-releasing hormone and substance-P in the cat medulla. Neuroscience 57: 811–822.

Dekin MS (1993): Inward rectification and its effects on the repetitive firing properties of bulbospinal neurons located in the ventral part of the nucleus tractus solitarius. J Neurophysiol 70: 590–601.

Dekin MS, Getting PA (1987): In vitro characterization of neurons in the ventral part of the nucleus tractus solitarius. II. Ionic basis for repetitive firing patterns. J Neurophysiol 58: 215–229.

Dekin MS, Richerson GB, Getting PA (1985): Thyrotropin-releasing hormone induces rhythmic bursting in neurons of the nucleus tractus solitarius. Science 229: 67–69.

Dillon GH, Waldrop TG (1992): In vitro responses of caudal hypothalamic neurons to hypoxia and hypercapnia. Neuroscience 51: 941–950.

Eldridge FL., Gill-Kumar P, Millhorn DE (1981): Input-output relationships of central neural circuits involved in respiration in cats. J Physiol (Lond) 311: 81–95.

Feldman JL (1986): Neurophysiology of breathing in mammals. In Bloom FE (ed): Handbook of Physiology; Section: The Nervous System IV. Baltimore, Williams & Wilkins, pp 463–524.

Filiano JJ, Kinney HC (1992): Arcuate nucleus hypoplasia in the sudden infant death syndrome. J Neuropathol Exp Neurol 51: 394–403.

Filiano JJ, Choi JC, Kinney HC (1990): Candidate cell populations for respiratory chemosensitive fields in the human infant medulla. J Comp Neurol 293: 448–465.

Folgering H, Kuyper F, Kille JF (1979): Primary alveolar hypoventilation (Ondine's curse syndrome) in an infant without external arcuate nucleus. Case report. Bull Eur Physiopathol Respir 15: 659–665.

Forslid A (1987): Transient neocortical, hippocampal and amygdaloid EEG silence induced by one minute inhalation of high concentration CO_2 in swine. Acta Physiol Scand 130: 1–10.

Fukuda Y, Honda Y (1975): pH-sensitive cells at ventro-lateral surface of rat medulla oblongata. Nature 256: 317–318.

Fukuda Y, Loeschcke HH (1977): Effect of H^+ on spontaneous neuronal activity in the surface layer of the rat medulla oblongata in vitro. Pflügers Arch 371: 125–134.

Fukuda Y, Honda Y, Schlaefke ME, Loeschcke H (1978): Effect of H^+ on the membrane potential of silent cells in the ventral and dorsal surface layers of the rat medulla in vitro. Pflügers Arch 376: 229–235.

Fukuda Y, See WR, Honda Y(1980): H^+-sensitivity and pattern of discharge of neurons in the chemosensitive areas of the ventral medulla oblongata of rats in vitro. Pflüegers Arch 388: 53–61.

Gill PK, Kuno M (1963): Properties of phrenic motoneurones. J Physiol (Lond) 168: 258–273.

Gronroos M, Pertovaara A (1994): A selective suppression of human pain sensitivity by carbon dioxide: Central mechanisms implicated. Eur J Appl Physiol Occ Physiol 68: 74–79.

Haddad GG, Mazza NM, Defendini R, Blanc WA, Driscoll JM, Epstein MA, Epstein RA, Mellins RB (1978): Congenital failure of automatic control of ventilation, gastrointestinal motility and heart rate. Medicine (Baltimore) 57: 517–526.

Haxiu, MA, Erokwu B, Prabhakar NR, Cherniack NS, Strohl KP (1992): Locus coeruleus neurons express c-fos immunoreactivity upon stimulation of central chemosensory system. Soc Neurosci Abstr 18: 828.

Ho K, Nichols CG, Lederer WJ, Lytton J, Vassilev PM, Kanazirska MV, Hebert SC (1993): Cloning and expression of an inwardly rectifying ATP-regulated potassium channel. Nature 362: 31–38.

Hunt CE (1992): Sudden Infant Death Syndrome. In Beckerman RC, Brouillette RT, Hunt CE (eds): Respiratory Control Disorders in Infants and Children. Baltimore: Williams & Wilkins, pp 190–211.

Jacobs BL, Azmitia EC (1992): Structure and function of the brain serotonin system. Physiol Rev 72: 165–229.

Jarolimek W, Misgeld U, Lux HD (1990): Neurons sensitive to pH in slices of the rat ventral medulla oblongata. Pflüegers Arch 416: 247–253.

Jodkowski JS, Lipski J (1986): Decreased excitability of respiratory motoneurons during hypercapnia in the acute spinal cat. Brain Res 386: 296–304.

Kawai A, Ballantyne D, Muckenhoff K, Scheid P (1996): Chemosensitive medullary neurones in the brainstem-spinal cord preparation of the neonatal rat. J Physiol (Lond) 492: 277–292.

Kinney HC, Filiano JJ, Sleeper LA, Mandell F, Valdes-Dapena M, White WF (1995): Decreased muscarinic receptor binding in the arcuate nucleus in sudden infant death syndrome. Science 269: 1446–1450.

Kogo N, Arita H (1990): In vivo study on medullary H^+ sensitive neurons. J Appl Physiol 69: 1408–1412.

Krnjevic K, Randic M, Siesjo B (1965): Cortical CO_2 tension and neuronal excitability. J Physiol (Lond) 176: 105–122.

Lalley PM (1986): Serotoninergic and non-serotoninergic responses of phrenic motoneurones to raphe stimulation in the cat. J Physiol (Lond) 380: 373–385.

Larnicol N, Wallois F, Berquin P, Gros F, Rose D (1994): c-fos-like immunoreactivity in the cat's neuraxis following moderate hypoxia or hypercapnia. J Physiol (Paris) 88: 81–88.

Lehmenkuhler A, Bingmann D, Speckmann EJ (1989): Neuronal and glial responses to hypoxia and hypercapnia. Biomed Biochim Acta 48: S155–S160.

Libscomb WT, Boyarsky LL (1972): Neurophysiological investigations of medullary chemosensitive areas of respiration. Respir Physiol 16: 362–376.

Madden JA (1993): The effect of carbon dioxide on cerebral arteries. Pharmacol Ther 59: 229–250.

Marino PL, Lamb TW (1975): Effects of CO-2 and extracellular H^+ iontophoresis on single cell activity in the cat brainstem. J Appl Physiol 38: 688–695.

McCormick DA, Pape HC (1990): Properties of a hyperpolarization-activated cation current and its role in rhythmic oscillation in thalamic relay neurones. J Physiol (Lond) 431: 291–318.

Miles R (1983): Does low pH stimulate central chemoreceptors located near the ventral medullary surface? Brain Res 271: 349–353.

Millhorn DE, Eldridge FL (1986): Role of ventrolateral medulla in regulation of respiratory and cardiovascular systems. J Appl Physiol 61: 1249–1263.

Mitchell RA, Herbert DA (1974): The effect of carbon dioxide on the membrane potential of medullary respiratory neurons. Brain Res 75: 345–349.

Mitchell RA, Loeschcke HH, Massion WH, Severinghaus JW (1963): Respiratory responses mediated through superficial chemosensitive areas on the medulla. J Appl Physiol 18: 523–533.

Moss IR, Inman JG (1989): Neurochemicals and respiratory control during development. J Appl Physiol 67: 1–13.

Nattie EE, Mills JW, Ou LC, St. John WM (1988): Kainic acid on the rostral ventrolateral medulla inhibits phrenic output and CO_2 sensitivity. J Appl Physiol 65: 1525–1534.

Neubauer JA, Gonsalves SF, Chou W, Geller HM, Edelman NH (1991): Chemosensitivity of medullary neurons in explant tissue cultures. Neuroscience 45: 701–708.

Okada Y, K Mückenhoff, G Holtermann, H Acker, P Scheid (1993): Depth profiles of pH and Po_2 in the isolated brain stem-spinal cord of the neonatal rat. Respiration Physiology 93: 315–326.

O'Regan RG, Majcherczyk S (1982): Role of peripheral chemoreceptors and central chemosensitivity in the regulation of respiration and circulation. J Exp Biol 100: 23–40.

Peers C, Green FK (1991): Inhibition of Ca^{2+}-activated K^+ currents by intracellular acidosis in isolated type I cells of the neonatal rat carotid body. J Physiol (Lond) 437: 589–602.

Penington NJ, Kelly JS (1993): Ionic dependence of a slow inward tail current in rat dorsal raphe neurones. J Physiol 464: 33–48.

Richerson GB (1992): Electrophysiologic effects of changes in Pco_2 on rat medullary neurons in vitro. Soc Neurosci Abstr 18: 348.

Richerson GB (1995): Response to CO_2 of neurons in the rostral ventral medulla in vitro. J Neurophysiol 73: 933–944.

Richerson GB (1997): Sudden infant death syndrome: The role of central chemosensitivity. Neuroscientist 3: 309–313.

Rigatto H, Fitzgerald SC, Willis MA, Yu C (1992): In search of the central respiratory neurons. II. Electrophysiologic studies of medullary fetal cells inherently sensitive to CO_2 and low pH. J Neurosci Res 33: 590–597.

Rigual R, Lopez-Lopez JR, Gonzalez C (1991): Release of dopamine and chemoreceptor discharge induced by low pH and high P_{CO_2} stimulation of the cat carotid body. J Physiol (Lond) 433: 519–531.

Rocher A, Obeso A, Gonzalez C, Herreros B (1991): Ionic mechanisms for the transduction of acidic stimuli in rabbit carotid body glomus cells. J Physiol (Lond) 433: 533–548.

Sato M, Severinghaus JW, Basbaum A (1992): I. Medullary CO_2 chemoreceptor neuron identification by c-*fos* immunocytochemistry. J Appl Physiol 73: 96–100.

Schlaefke ME, See WR, Loeschcke HH (1970): Ventilatory response to alterations of H^+ ion concentration in small areas of the ventral medullary surface. Respir Physiol 10: 198–212.

Schlaefke ME (1981): Central chemosensitivity: A respiratory drive. Rev Physiol Biochem Pharmacol 90: 171–244.

Segal M (1985): A potent transient outward current regulates excitability of dorsal raphe neurons. Brain Res 359: 347–350.

Smith JC, Morrison DE, Ellenberger HH, Otto MR, Feldman JL (1989): Brainstem projections to the major respiratory neuron populations in the medulla of the cat. J Comp Neurol 281: 69–96.

St. John WM (1977): Integration of peripheral and central chemoreceptor stimuli by pontine and medullary respiratory centers. Fed Proc 36: 2421–2427.

Suzuki M, Takahashi K, Ikeda M, Hayakawa H, Ogawa A, Kawaguchi Y, Sakai O (1994): Cloning of a pH-sensitive K^+ channel possessing two transmembrane segments. Nature 367: 642–645.

Veasey SC, Fornal, CA, Metzler CW, Jacobs BL (1995): Response of serotonergic caudal raphe neurons in relation to specific motor activities in freely moving cats. J Neurosci 15: 5346–5359.

Verloes A, Elmer C, Lacombe D, Heinrichs C, Rebuffat E, Demarquez JL, Moncla A, Adam E (1993): Ondine-Hirschsprung syndrome (Haddad syndrome). Further delineation in two cases and review of the literature. Eur J Pediatr 152: 75–77.

Wang WH, Schwab A, Giebisch G (1990): Regulation of small-conductance K^+ channel in apical membrane of rat cortical collecting tubule. Am J Physiol 259: F494–F502.

CHAPTER 27

Central Chemoreceptors, pH, and Respiratory Control

EUGENE NATTIE
Department of Physiology, Dartmouth Medical School, Lebanon, NH

THE BRAINSTEM RESPIRATORY CONTROL SYSTEM

Breathing serves the dual purpose of providing oxygen (O_2) to the body tissues for oxidative process and removing the volatile metabolite, carbon dioxide (CO_2). CO_2 regulation is also important in acid-base balance (see Chapter 20). The motor act of breathing originates via neural mechanisms in the brainstem, and breathing is regulated by a complex system that includes chemical feedback. Chemical sensors are the carotid and aortic bodies, located peripherally, and the central chemoreceptors, the subject of this chapter, located in the brainstem. Before considering central chemoreceptors, I shall outline the neural aspects of this control system. These have been the subject of a number of recent reviews (e.g., Bianchi et al., 1995; Duffin et al., 1995; Feldman and Smith, 1995; Richter, 1996).

Location and Function of Respiratory-Related Neuron Groups

Figure 1 shows the location of the major groups of brainstem neurons involved in the control of breathing (see Chapter 26). Neurons at these sites have discharge patterns related to respiration and are classified by the temporal nature of this relationship. For example, a neuron that fires during inspiration is called an Inspiratory or I neuron.

Rostrally, in the dorsal lateral pons, lies the pontine respiratory group (PRG). These neurons modulate afferent input from vagal reflexes related to lung volume. Lesions at this site in anesthetized animals change the components of ventilation, frequency and tidal volume, but have little discernible effect in chronic, unanesthetized animals, an observation that demonstrates the plasticity or duplication within this neural control system (St. John et al., 1971). Such lesions have, however, resulted in a diminished ventilatory sensitivity of CO_2 suggesting the presence of central chemoreception at this location (Johnston and Gluckman, 1989). The pons can generate a respiratory rhythm even in the absence of neural connections to the medulla (St. John and Bledsoe, 1985).

pH and Brain Function, Edited by Kai Kaila and Bruce R. Ransom
ISBN 0-471-11838-9

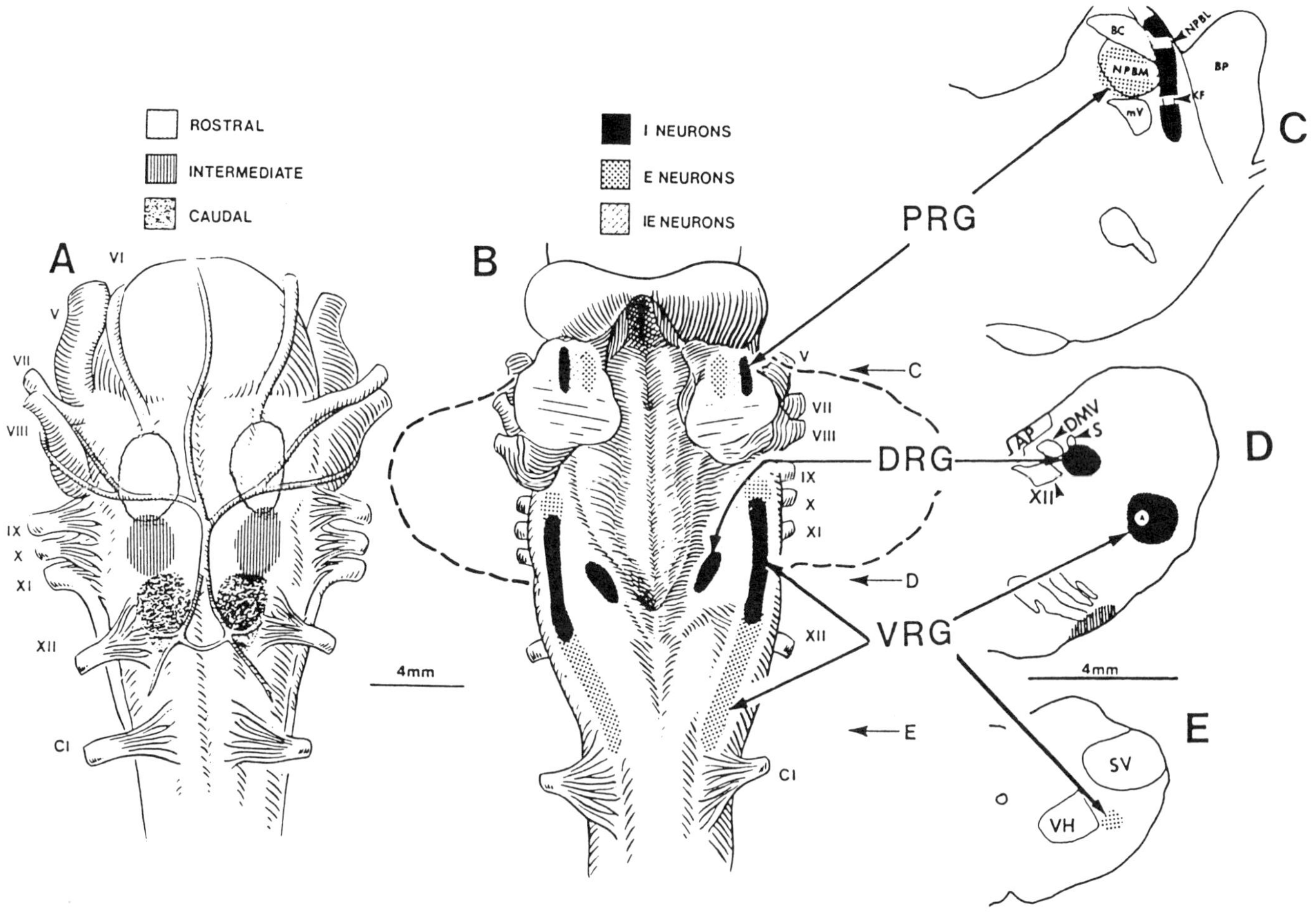
ROSTRAL
INTERMEDIATE
CAUDAL
I NEURONS
E NEURONS
IE NEURONS
A
B
C
D
E
PRG
DRG
VRG
VI
V
VII
VIII
IX
X
XI
XII
CI
4mm
BC
NPBL
NPBM
BP
KF
mV
AP
DMV
S
XII
SV
VH

In the region of the nucleus tractus solitarius lies the dorsal respiratory group (DRG). Primary afferents from peripheral mechano- and chemoreflexes synapse at this site and direct efferent connections arise to spinal motoneurons and medullary premotoneurons. Lesions here have also diminished CO_2 sensitivity, an effect that is greater during anesthesia (Berger and Cooney, 1982).

In the ventral medulla lies a long column of respiratory-related neurons, the ventral respiratory group (VRG). Most caudally in the VRG are expiratory neurons, many of which send efferents to motoneurons that control the resistance in the upper airway. More rostral sites also contain inspiratory neurons that influence inspiratory premotor and motoneurons. The PRG, DRG, and VRG are the three main clusters of respiratory neurons. However, other neurons appear to play a major, if not fully understood, role in the control of breathing.

We can, for this presentation, classify a cluster of neuron groups as the rostral ventrolateral medulla (RVLM) (see Bianchi et al., 1995; Duffin et al., 1995; Feldman and Smith, 1995; Richter, 1996; Nattie 1995a,b for references). In Fig. 1 these would extend rostrally from the VRG and dip ventrally towards the medullary surface lying ventral to the facial nucleus and lateral to the pyramidal tract (see Fig. 6 for a view of one portion of the RVLM, the retrotrapezoid nucleus, RTN). This RVLM region contains the RTN, portions of the nucleus paragigantocellularis lateralis (PGCL) and parapyramidal neurons of the midline raphe. It could also include neurons and processes of the subretrofacial and the pre-Bötzinger regions, which are just rostral to the VRG and contain many propriobulbar neurons.

FIGURE 1. Location of the traditional respiratory central chemosensitive areas (A) and the major groups of neurons involved in the control of breathing (B–E). **A.** Ventral medulla of the cat, with the location of the two surface areas traditionally associated with central chemosensitivity: the rostral area (RA), also called Mitchell's area, and the caudal area (CA), also called Loeschcke's area. Application of acidic fluids to the medullary surface at these areas produces an increase in breathing. The area labelled IA (intermediate area), also called Schläfke's area, was described as a nonchemosensitive site possibly involved in the integration of chemosensory inputs (see Schläfke, 1981). Evaluation of this site has continued (see Nattie, 1995a,b). The blood vessels shown on the ventral surface are the major supply to the medulla. Smaller vessels enter the medulla ventrally and penetrate to more dorsal regions. The Roman numerals refer to cranial nerves, C1 to the first cervical nerve root. **B.** Dorsal medulla of the cat, with the surface sites under which the major groups of respiratory neurons lie. These groups, the PRG (pontine respiratory group), DRG (dorsal respiratory group), and VRG (ventral respiratory group), are also displayed in cross sections (C–E) that show their location in that axis. Note that these neurons lie at some distance from the ventral medullary surface sites traditionally identified as the location of central chemoreceptors. Abbreviations: I, inspiratory; E, expiratory; IE, inspiratory/expiratory (these refer to the phase of breathing during which the neurons are active); BC, brachium conjunctivum; NPBL, nucleus parabrachialis lateralis; KF, Kölliker-Fuse nucleus; NPBM, nucleus parabrachialis medialis; BP, brachium pontis; AP, area postrema; DMV, dorsal motor nucleus of the vagus; S, solitary tract; XII, twelfth nerve nucleus; SV, spinal trigeminal nucleus, VH, ventral horn. (From Feldman, 1986, with permission of author and the American Physiological Society.)

Respiratory Rhythm Generation

The pre-Bötzinger region in the isolated neonatal rat brainstem preparation contains neurons with spontaneous depolarizations that are necessary for the production of the rhythmic output observed (Feldman and Smith, 1995). These and other data support the pacemaker hypothesis for the generation of respiratory rhythm, although the role of these putative pacemaker neurons in the genesis of the respiratory rhythm in the intact, adult mammal is unclear (Duffin et al., 1995; Bianchi et al., 1995; Richter, 1996). Here it seems that some type of neuronal network is needed to account for the origin of the rhythm (Cohen, 1995; Richter, 1996). In either case, the region of the RVLM is importantly involved (Duffin et al., 1995).

Respiratory Drive

Respiratory rhythm generation in the intact, adult mammal requires some type of tonic "drive" to maintain the interactions among neurons necessary to sustain the rhythm. The origin of this "drive" is unknown. It could result from ill-described reticular formation activity, input from afferents involved in respiratory control (Sullivan et al., 1978), or input related to CO_2 sensed by central (and peripheral) chemoreceptors (Cohen, 1995). Intriguing in this regard are observations showing that with step-by-step removal of various afferent inputs, ventilatory output declines (Sullivan et al., 1978; Cohen, 1995) suggesting that this tonic "drive" results from the summation of individual afferent inputs, including those from central chemoreceptors.

Arousal State

The arousal state of the animal is of crucial importance in understanding the significance of results obtained from various experimental paradigms. Anesthesia, necessary for many experimental approaches, dramatically affects the brainstem respiratory control system. Its output is diminished, the sensitivity to various stimuli, including CO_2 via central chemoreceptors, is diminished, and the distribution and number of various neuron types recorded in the PRG, DRG, and VRG can vary. Physiologically, respiration is state-dependent, and, importantly for this chapter, the role and importance of central chemoreception may vary with arousal state (see p 540, 541).

CENTRAL CHEMORECEPTION

For a recent in-depth review of this topic and other references see Nattie (1995a) and Chapter 26.

Evidence for Central Chemoreception

Infusion of acidic solutions within the brain ventricular system stimulates breathing (Fig. 2) (Fencl et al., 1966; Leuson, 1954). In these experiments, the access of the perfused acidic fluid is limited to central structures; the increase in breathing must result from a central nervous system process. Also, increased systemic P_{CO_2} or intravenous infusion of acidic solutions stimulates breathing in animals with surgical removal of peripheral chemoreceptors (e.g., see Dempsey and Forster, 1982; Nattie, 1983). Here, the response site must again be within the central nervous system.

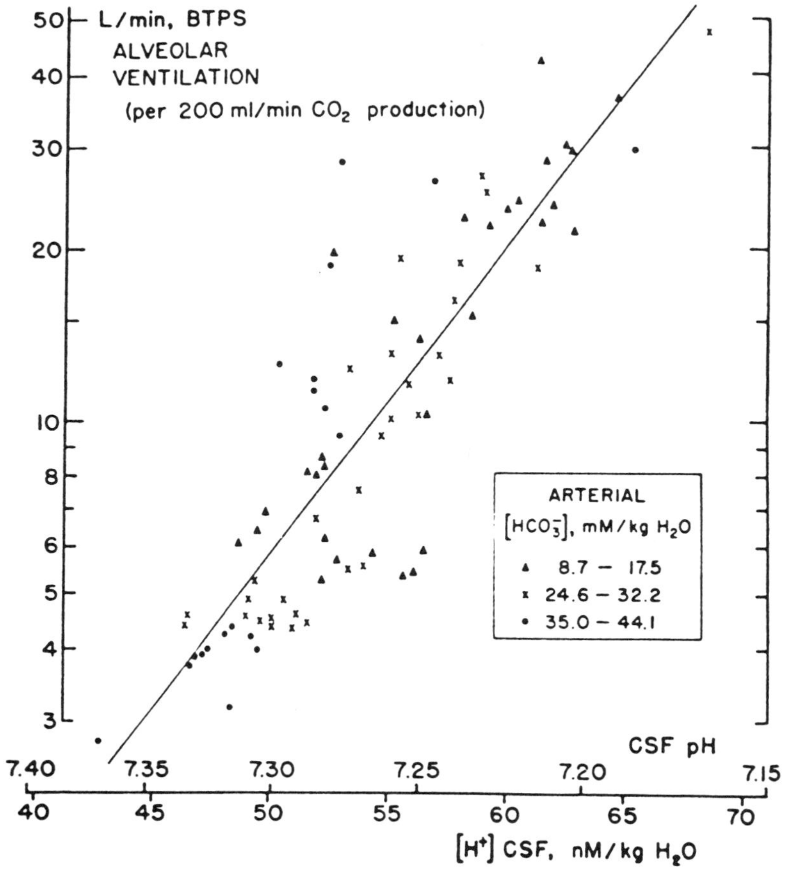

FIGURE 2. Alveolar ventilation is shown to be a linear function of $[H^+]$ in the cerebrospinal fluid (CSF) in unanesthetized goats under normal acid-base status (X) and with chronic metabolic acidosis (solid triangle) or alkalosis (solid circle) with and without CO_2 stimulation. These data show the high sensitivity of central chemoreceptors to cerebral pH changes in the unanesthetized state and have been used to support the proposal that central chemoreceptors play a dominant role in the response to CO_2 and pH. (From Fencl et al., 1966, with permission of author and the American Physiological Society.)

Function of Central Chemoreceptors

Link with CO_2 production: metabolic acid-base disorders. Chemical feedback from peripheral chemoreceptors at the carotid and aortic bodies and from central chemoreceptors involves changes in the partial pressure of CO_2 (P_{CO_2}) and pH. The peripheral receptors also sense the partial pressure of O_2 (P_{O_2}) especially levels of P_{O_2} below normal, hypoxia. Stimulation or inhibition of either set of receptors changes ventilation appropriately for homeostasis. It is a common hypothesis that CO_2-sensitive central and peripheral chemoreceptors provide ongoing and rapid feedback to the brainstem respiratory control system concerning the arterial blood levels of CO_2, and hence the adequacy of alveolar ventilation relative to metabolism. Thus they should be located to allow sampling of arterial blood (see p 541, 543). An additional hypothesis would allow central chemoreceptors to sense brain tissue pH at appropriate anatomic locations (see p 541, 543). If changes in pH result from metabolic acid-base disorders, these receptors can bring about appropriate ventilatory homeostatic mechanisms. For example, a metabolic acidosis stimulates ventilation via peripheral and central chemoreceptors, resulting in hyperventilation and lowering of arterial P_{CO_2}, a response that elevates (corrects) the arterial pH. With metabolic alkalosis, ventilation is inhibited and P_{CO_2} rises alleviating the alkalosis (see Dempsey and Forster, 1982).

Peripheral, central chemoreceptors. Central chemoreceptors appear to be quantitatively more important than do the peripheral chemoreceptors in the ventilatory response to changes in CO_2 levels. Experiments involving surgical removal of the peripheral chemoreceptors (see Dempsey and Foster, 1982) or independent stimulation of central versus peripheral chemoreceptors (Heeringa et al., 1979; Smith et al., 1984) support this conclusion. However, in some cases (e.g., intravenous infusion of metabolic acid in an unanesthetized rabbit with peripheral chemoreceptor denervation) the ventilatory response seems more dependent on the peripheral than on central chemoreceptors (Nattie, 1983). Arousal state and the ability of the central nervous system to adapt to denervation need to be considered carefully in evaluation of the relative importance of peripheral versus central chemoreceptors (Smith et al., 1984).

Hypercapnia and hypocapnia. Hypercapnia is a powerful stimulus to the respiratory control system. Both respiratory frequency and tidal volume are increased, with a predominant effect on tidal volume. Some studies suggest that there might be separate brainstem sites for these frequency and tidal volume responses (Morin-Surun et al., 1995; Issa and Remmers, 1992; Kawai et al., 1996). In hypercapnia, virtually all brainstem respiratory neurons are stimulated, a response that contrasts to the specific neuron group responses to hypoxic stimulation of the peripheral chemoreceptors (Cohen, 1968; St. John and Bianchi, 1985).

Lowering the P_{CO_2} below the normal level, hypocapnia, can result in apnea in an anesthetized animal usually at P_{CO_2} values of 15–25 mmHg, depending on the depth of anesthesia. This hypocapnic apnea is associated with a marked change in the firing pattern of brainstem respiratory neurons. Inspiratory neurons become quiescent and expiratory neurons fire tonically and with greater activity (Sears et al., 1982; Cohen, 1995). Some level of CO_2 seems needed to maintain the normal interactions among the different types of the respiratory neurons and to provide a "drive" to the system. If the hypocapnia is limited to the brainstem (i.e., the central chemoreceptors) in an anesthetized animal, some respiratory output remains until the P_{CO_2} reaches levels of 3–5 mmHg (Berkenbosh et al., 1984), suggesting that the threshold for hypocapnic apnea involves input from both the central and the peripheral chemoreceptors.

In an unanesthetized animal, hypocapnia limited to the peripheral chemoreceptors decreases ventilation without apnea in wakefulness (Daristotle et al., 1990) and in sleep (Smith et al., 1995). However, in sleeping man even mild systemic hypocapnia affecting both central and peripheral chemoreceptors results in apnea (Data et al., 1991). These data indicate that hypocapnia can result in apnea, but only in sleep or during anesthesia, and that central chemoreceptors are crucially important in this response.

Absence of central chemoreception. How important are central chemoreceptors? The congenital central hypoventilation syndrome (CCHS) is a human condition recognized within the first year of life in which central chemoreception is virtually absent. These children suffer from severe hypoventilation during sleep and require night-time ventilatory support for survival (Weese-Mayor et al., 1992). Responses to hypoxia and to exercise can be affected as well, but the most dramatic and consistent abnormalities are the loss of CO_2 sensitivity and the severe hypoventilation during sleep. Many of these children function quite well while awake, provided adequate night-time ventilation is maintained. This striking experiment of nature suggests that loss of central chemoreception is not an immediately life-threatening event, if the subject is awake, although chronic hypoventilation with its secondary pathophysiology related to hypoxemia and hypercapnia can eventually become a serious problem. During sleep, however, it appears that central chemoreception is necessary to prevent more severe life-threatening hypoventilation and apnea.

Location of Central Chemoreceptors

Traditional view: the ventrolateral medulla. The observation that perfusion of the brain ventricular system with fluids of differing pH affects breathing led to investigations of possible central chemoreceptor locations near brain surfaces. Direct application of artificial cerebrospinal fluid with altered pH resulted in the description of chemosensitive areas on the surface of the ventrolateral medulla (see Fig. 1) (Loeschcke, 1982; Mitchell et al., 1963; Schläfke, 1981). The view that central chemoreceptors are located superficially at a rostral and caudal area of the ventral medulla has been well accepted. But, critics have pointed out that the blood supply to the medulla enters the tissue from the ventral surface; substances applied to this surface may be distributed quite deeply within the tissue (Lipscomb and Boyarsky, 1972).

Cells found near the medullary surface have been suggested as putative central chemoreceptors (Trouth et al., 1979; Schläfke, 1981). Recently, Kawai et al. (1996), using the *in vitro* neonatal rat brainstem preparation, have described chemosensitive neurons in the ventral medulla with processes that extend to the ventral surface itself (Fig. 3). Others have shown such processes from neurons with cell bodies at slightly more dorsal sites in the retrofacial region (Bianchi et al., 1988; Pilowsky et al., 1990). That processes project from neurons to the ventral surface has been interpreted to indicate, in keeping with the older "traditional" view, that central chemoreception is present only, or mainly, at the ventral surface of the medulla. This location would clearly serve to sample cerebrospinal fluid (CSF) in the subarachnoid space. CSF has a slow turnover rate (see Chapter 32) and its chemical nature could represent a time-related average of events occurring within or modified by the brain. It would not serve as an adequate source of information for rapid responses to physiological perturbations.

An alternative view is that ventral neurons with surface processes sense arterial blood pH and P_{CO_2} via pial vessels or the interstitial fluid near them. Central chemoreceptors, and interstitial fluid pH (see Chapter 32), can respond quite rapidly to abrupt changes in arterial pH produced under isocapnic conditions (Borison et al., 1978; Nattie, 1983), as well as to hypercapnia.

Newer view: more widespread locations within the brainstem. A number of studies support the view that central chemoreception occurs over a more widespread distribution of sites than just the region lying close to the ventral surface. In brainstem slices, decreased pH stimulates neurons in both ventral and dorsal locations (Dean et al., 1989;

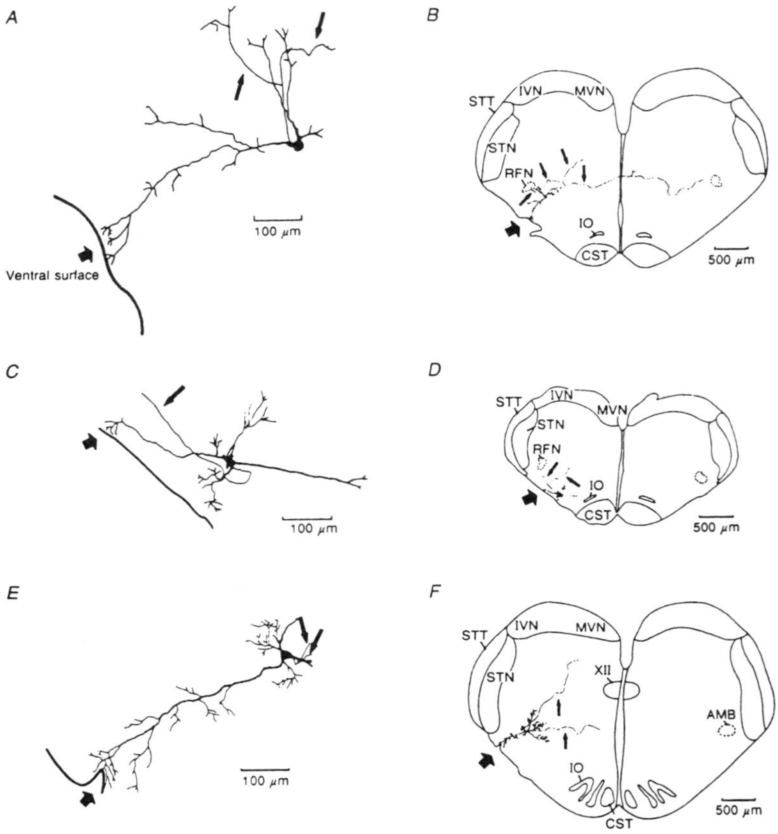

FIGURE 3. Three chemosensitive neurons injected intracellularly with Lucifer yellow are shown from the ventrolateral medulla of the isolated neonatal rat brainstem preparation. In each case, the panel on the right shows the neuron on the left under lower magnification. The thick arrows show the projections of these neurons to the ventral medullary surface; the thin arrows show the axonal projections. Note that the cell bodies are hundreds of micrometers from the ventral surface. Abbreviations: STT, spinal trigeminal tract; IVN, inferior vestibular nucleus; MVN, medial vestibular nucleus; STN, spinal trigeminal nucleus; RFN, retrofacial nucleus; IO, inferior olivary nucleus; CST, corticospinal tract; XII, twelfth nerve nucleus. (From Kawai et al., 1996, with permission of authors and the Physiological Society.)

Fukuda and Honda, 1976; Miles, 1983; Richerson, 1995) that include the midline raphe (Richerson, 1995), the locus caeruleus (Pineda and Aghajanian, 1996), hypothalamus (Dillon and Waldrop, 1992), and a region near the nucleus tractus solitarius or dorsal respiratory group (DRG) (Dean et al., 1989). Tissue explants taken from both dorsal and ventral medullary sites of the neonatal rat contain neurons responsive to acidosis (Neubauer et al., 1991; Rigatto et al., 1994). In these *in vitro* experiments, it is not known whether the responsive cells are involved in the control of breathing.

Two studies *in vivo* provide support for a widespread distribution of central chemoreceptor locations. Kogo and Arita (1990) used brief injections of CO_2-loaded saline into the vertebral artery to expose all medullary neurons to hypercapnia and stimulate a short-lived increase in respiratory output. Single neurons evaluated at many brainstem locations using extracellular recording electrodes were identified as part of the total respiratory response. Subsequent microinjection of acidic fluid limited to the vicinity of the recording pipette then also identified neurons responding to local changes in pH. Neurons responsive both to medullary hypercapnia and to focal acid injection were characterized as central chemoreceptors. They were found near the ventral surface of the medulla and deeper within the tissue near the DRG.

The second *in vivo* study used an entirely different strategy. Coates et al. (1991) observed that application to the ventral medullary surface of acetazolamide, an inhibitor of the enzyme carbonic anhydrase, increased respiratory output. This response was attributed to stimulation of central chemoreceptors by a decrease in medullary tissue pH caused by the acetazolamide (see Chapters 3, 16 and 17). They refined the approach and showed that 1-nl injections of acetazolamide (10^{-5} to 5×10^{-6} M) produced a decrease in pH at the injection center equivalent to that associated with an increase in systemic P_{CO_2} of 36–46 mmHg (Coates et al., 1993; Nattie and Li, 1996). At 100 μm from the center, the decrease in pH was like that produced by an increase in P_{CO_2} of 9 mmHg; at 300–400 μm from the center, pH was unchanged. The small-volume, low-dose acetazolamide injections produced focal regions of tissue acidosis that lasted for 90–120 minutes. At some sites, they caused an increase in integrated phrenic nerve activity in anesthetized cats and rats whose ventilation and systemic CO_2 levels were under the control of the investigator (Fig. 4). The focal regions of acidosis must be stimulating central chemoreceptors, with an effect on the control system measurable as an increase in phrenic nerve activity.

Acetazolamide injections were then used as a probe to search for medullary chemoreceptor sites. Such sites were found:

1. Along the ventral medulla within 800 μm of the surface.
2. Near the nucleus tractus solitarius.
3. Near the locus caeruleus, dorsally located but without a known role in respiratory control.
4. The medullary raphe (Bernard et al., 1996)
5. The ventral respiratory group (VRG) region (Nattie and Li, 1996).

Figure 5 shows these sites schematically, as determined in rat and cat, alongside a schematic representation of the major respiratory neuron groups.

Why such a widespread distribution? One hypothesis, alluded to above, is that central chemoreceptors may serve to sense both rapid changes in blood pH via interstitial sites near capillaries and slow averaged changes in brain pH via sites that can evaluate CSF. In any disturbance of CO_2 or pH homeostasis, such a design would allow a measured respiratory response reflecting both (i) rapid, initial and (ii) slower, modified pH changes, the latter affected by CSF pH regulation. An alternative hypothesis would attribute differing thresholds to central chemoreceptors at different locations (see Nattie and Li, 1996; this idea arises from discussions with Dr. Peter Schied). Some would react within the physiological range of pH disturbances and others only at more severe pH changes; "physiological" and "emergency" receptors. The acetazolamide injections do produce an increase in P_{CO_2} of 36–47 mmHg at the injection center, which is a strong stimulus. Experiments using a specially designed diffusion pipette with CO_2 have shown that chemoreceptors in the RTN do respond to focal changes in tissue pH within the physiological range of values (Li and Nattie, 1997). A third hypothesis is that these widespread receptor sites allow the brain to monitor interstitial pH at many locations. Stimulation at each site is capable

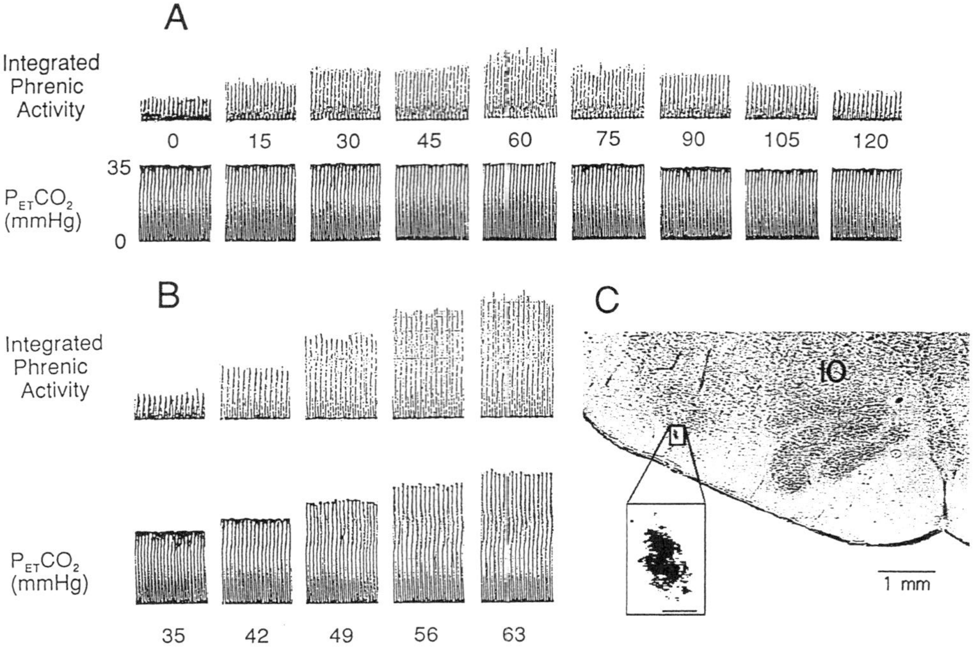

FIGURE 4. Response to focal medullary acidification. This figure shows a typical response of the respiratory control system of an anesthetized cat to the focal acidification of a very small region of the medulla. Integrated phrenic nerve activity is the measure of respiratory output. Focal acidification is produced by a 1-nl injection of acetazolamide, an inhibitor of the enzyme carbonic anhydrase. Such an injection decreases tissue pH in a region with radius of <350 μm, as measured by tissue pH electrodes. **A.** Following the injection, integrated phrenic amplitude increases over 60 min and then returns to the original baseline level by 120 min. This occurs with the end-tidal P_{CO_2} (and presumably the arterial P_{CO_2}) held constant. **B.** Integrated phrenic amplitude response to an increase in end-tidal P_{CO_2} (from 35 to 63 mmHg), which increases P_{CO_2} at all medullary and other chemoreceptor sites. **C.** Appearance and anatomical location of the fluorescent microbeads included in the injection. This injection is just lateral to the inferior olive at its rostralmost aspect. The bar within the enlargement is 100 μm. Note that the response to focal acidification of a very small medullary region is a large fraction of the response observed with acidification of the entire medulla. This acetazolamide injection technique has been used to probe for the location of central chemosensitive regions. (From Coates et al., 1993, with permission of authors the American Physiological Society.)

of increasing respiratory output by 36–56% of the change observed, with all central chemoreceptor sites stimulated by a P_{CO_2} of approximately 63 mmHg. The apparent sensitivity at each site suggests that they are of functional importance in the regulation of brain pH via the respiratory control system. Resultant system responses (e.g., increased ventilation and decreased P_{CO_2}) would tend to correct a localized interstitial acidosis but would alkalinize other sites. The widespread chemoreceptor sites may also differ in their roles according to the arousal state of the animal. Finally, it has been proposed that central chemoreceptors may be linked by low resistance gap junctions (Huang et al., 1997; see also Chapter 24).

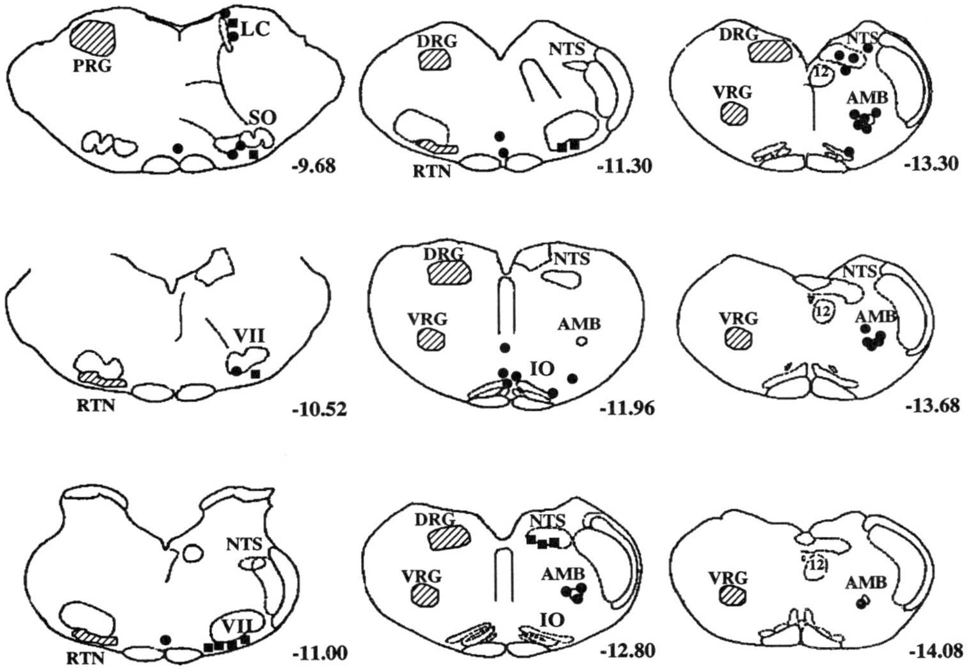

FIGURE 5. Evidence for widespread location of central chemosensitive sites. This series of cross sections of the rat medulla, labeled in millimeters caudal to the bregma, shows, on the right side of each section, the location of positive acetazolamide injection sites (i.e., sites at which the injection produced a response like that shown in Fig. 4) in cats (solid squares) and rats (solid circles). On the left side of each section are shown, schematically, the approximate locations of the major groups of respiratory neurons (see Fig. 1). Note that chemosensitive sites are located near the ventral medullary surface, as suggested by prior classic studies (Fig. 1), and that they are located more dorsally near the nucleus tractus solitarius, the locus ceruleus and PRG, the midline raphe, and the VRG. Abbreviations: PRG, pontine respiratory group; LC, locus ceruleus; SO, superior olivary nucleus; RTN, retrotrapezoid nucleus; VII, facial nucleus; DRG, dorsal respiratory group; NTS, nucleus of the solitary tract; AMB, nucleus ambiguus; 12, twelfth nerve nucleus. (Data are taken from Coates et al., 1993; Bernard et al., 1996; Nattie and Li, 1996; the cross sections are modified from those in the atlas of Paxinos and Watson, 1982).

Rostral Ventrolateral Medulla (RVLM) and Central Chemoreception

Cooling or destruction of a specific region near the RVLM surface in an anesthetized animal produces quite dramatic results: apnea ensues and the response to CO_2 virtually disappears (Fig. 6) (Loeschcke, 1982; Schläfke, 1981; Nattie 1995a,b; Nattie and Li, 1991, 1994; Forster et al., 1995; Ohtake et al., 1995). Originally labeled the intermediate area (lying intermediate between more rostral and caudal chemosensitive surface regions) (Loeschcke, 1982; Schläfke, 1981), this RVLM region was proposed to be an integrative site for information coming from other surface chemosensitive regions, hence the absence of a response to central chemoreceptor stimulation when the intermediate area is functionally disrupted. The RVLM is now known to be a site of central chemoreception, as detected by the focal acidosis produced by small injections of acetazolamide (see above).

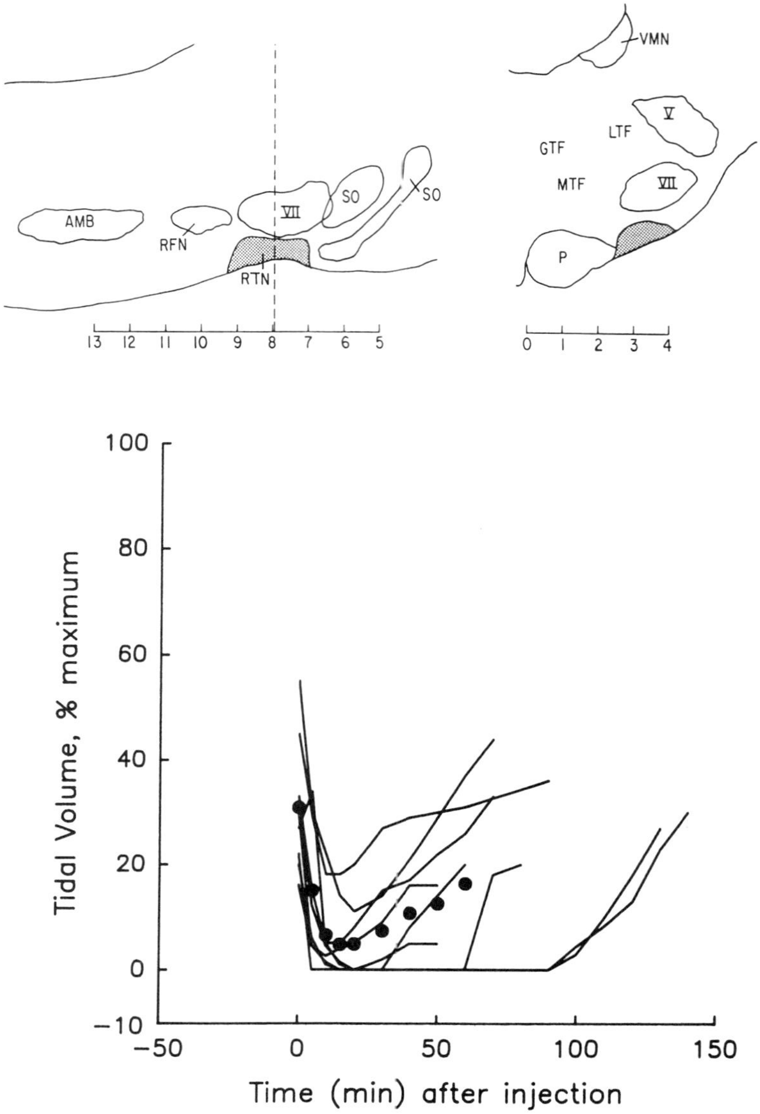

FIGURE 6. The location of the retrotrapezoid nucleus (RTN) is shown in a sagittal section (top left) and cross section taken at the dotted line (top right) of the cat brainstem. The numbered line at the bottom designates the number of millimeters caudal to the interaural line. At the bottom are shown the responses of integrated phrenic activity (labeled as tidal volume) to unilateral microinjection of 10 nl of 4.7 mM kainic acid, a neurotoxin, into the RTN of nine anesthetized cats with carotid body denervation. The solid circles represent the mean response. Note the profound depression of respiratory activity. Not shown is that the response to CO_2, a pure central chemoreceptor-mediated event in this preparation as peripheral chemoreceptors are denervated, is virtually abolished. (Data are from Nattie and Li, 1990a with permission of the American Physiological Society.)

Given the widespread nature of central chemoreceptor locations, it is puzzling that such small and focal lesions, placed well away from most of these sites, could essentially block the response to hypercapnia and produce apnea. I believe that recent results obtained in unanesthetized animals may help to explain this.

In decerebrate, unanesthetized cats (Nattie and Li, 1991), RVLM lesions in the retrotrapezoid nucleus (RTN) region diminish baseline respiratory output, but do not result in apnea, and diminish the CO_2 response, but do not abolish it. Cozine and Nagai (1967), using local procaine application in the ventral surface intermediate area, also found more profound respiratory inhibition in the anesthetized as compared to the decerebrate animal. In preinstrumented goats (Forster et al., 1995; Ohtake et al., 1995), RVLM cooling has much less dramatic effects on baseline and CO_2-stimulated breathing when the goat is awake and unanesthetized than when the goat is anesthetized (Fig. 7). In conscious, unanesthetized rats with prior unilateral RTN lesions made by injections of a neurotoxin, there is no detectable effect on baseline resting breathing, and the response to hypercapnia is reduced by 39% (Akilesh et al., 1996), a response that differs markedly from the diminished respiratory output and absent CO_2 response observed in anesthetized animals with similar lesions (Nattie and Li, 1994).

In contrast to these studies, disruption of the RVLM region produced by bilateral coagulation (Schläfke et al., 1979) or bilateral neurotoxin-induced lesions (Schäfer and Schläfke, 1995) resulted in more severe hypoventilation and reduced CO_2 sensitivity in the unanesthetized state. The greater disruptive effects in these studies may be ascribed to less discrete lesions. It does appear that the effects of RVLM disruption on breathing are much more dramatic in the anesthetized state. The remaining response to hypercapnia in the unanesthetized state with RVLM disruption must originate in the other (non-RVLM) central chemoreceptor locations. If, as hypothesized above, central chemoreception at different locations has different thresholds, the RVLM site may represent receptors operating at physiological pH levels while the other sites may operate at more severe or pathological pH levels. These thresholds and their physiological impact may depend also on arousal state. Thus, the impact on the respiratory output of stimulation of a given central chemoreceptor location could depend on (i) stimulus intensity, and (ii) the state of arousal.

RVLM cooling in the unanesthetized goat can affect responses to exercise and hypoxia as well as to CO_2 (Forster et al., 1995; Ohtake et al., 1995), suggesting that RVLM neurons play important roles beyond just central chemoreception. In sleeping unanesthetized goats, RVLM cooling has a slightly greater inhibitory effect on breathing than during wakefulness but with peripheral chemoreceptor denervation, RVLM cooling in sleep produces long-lasting apnea, suggesting that input from the carotid body and the RVLM together support respiration during sleep (Ohtake et al., 1996).

THE IDENTITY OF THE CENTRAL CHEMORECEPTOR STIMULUS: CO_2 VERSUS H^+

Responses to changes in pH under conditions of constant P_{CO_2} (isocapnia) have been attributed, reasonably, to an effect of pH. The issue here is whether effects on breathing attributed to changes in P_{CO_2} result from pH changes following CO_2 hydration and dissociation or whether there is a direct effect on CO_2. This is an old problem and it is not fully settled, although a number of current workers presume that pH is the single stimulus for chemoreceptors.

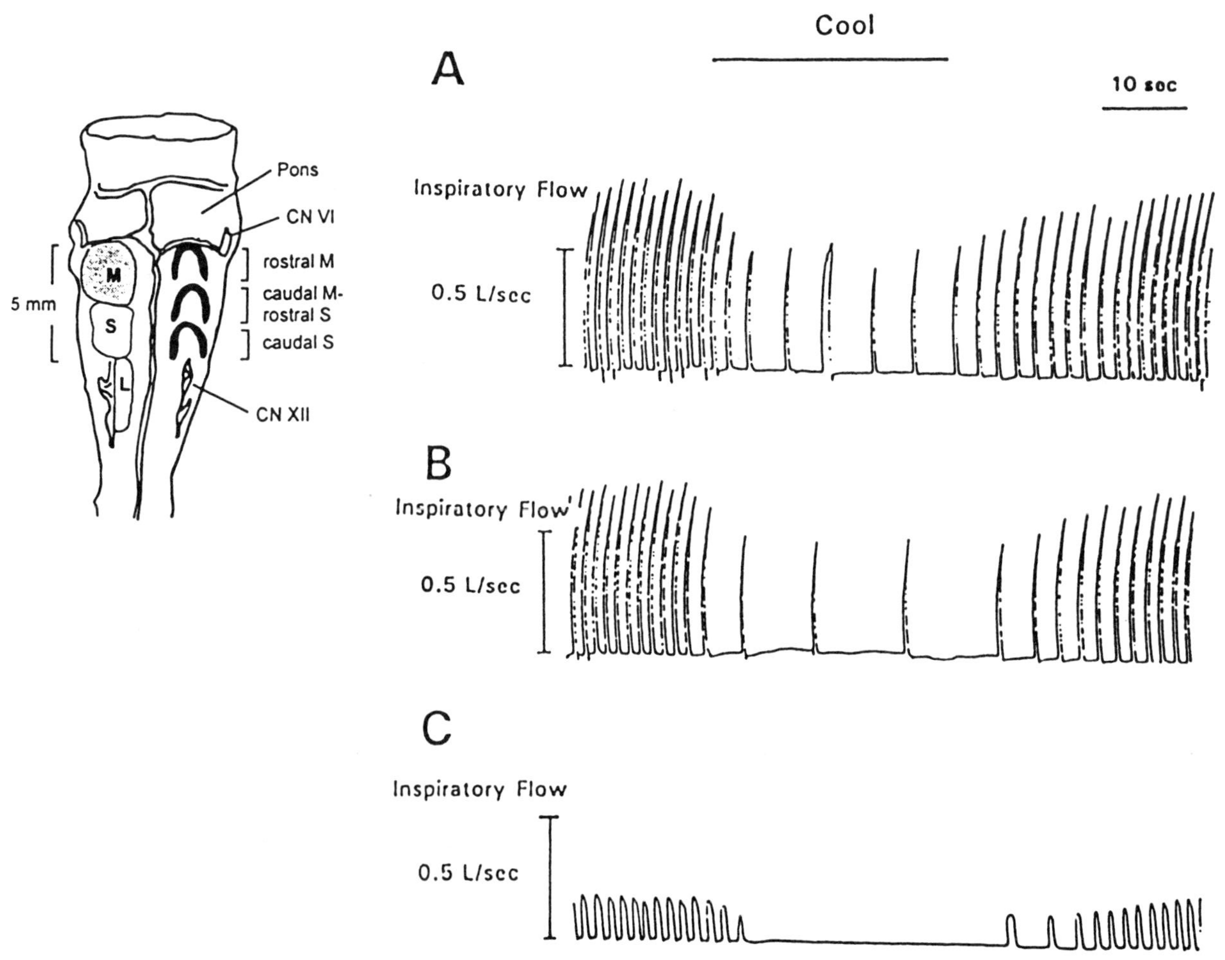
Pons
CN VI
rostral M
caudal M-
rostral S
caudal S
CN XII
M
S
L
5 mm
Cool
A
10 sec
Inspiratory Flow
0.5 L/sec
B
Inspiratory Flow
0.5 L/sec
C
Inspiratory Flow
0.5 L/sec

In Vivo Stimulus

In anesthetized, ventilated, peripherally chemodenervated animals (Eldridge et al., 1985; and see Nattie, 1995a, for other references) extracellular fluid pH on the ventral medullary surface was measured and assumed to be an index of the stimulus at the central chemoreceptor site. The ventilatory response expressed per unit change in medullary surface pH was much greater for systemic hypercapnia than for isocapnic metabolic acidosis. One concludes that either CO_2 level and pH act independently or medullary surface pH does not represent the pH at the central chemoreceptor site (i.e., the chemoreceptor site may be in the intracellular compartment or at an interstitial site with less effective pH regulation than the medullary surface). Direct perfusion via the subarachnoid space of the adult rat ventral medulla showed ventilatory responses to changes in pH produced at constant P_{CO_2} but not at constant $[HCO_3^-]$ (Tojima et al., 1991). These data suggest that P_{CO_2}-dependent pH changes are not sensed. However, the high ventral medullary blood flow could wash away the CO_2 load produced by perfusion and explain the lack of response to CO_2.

In Vitro Stimulus

Work in the isolated neonatal rat brainstem preparation (Harada et al., 1985) and in tissue explants taken from neonatal rat medulla and grown in culture (Rigatto et al., 1994) has shown responses to both pH and P_{CO_2}. Other studies of cultured neurons have shown responses to changes in P_{CO_2} but not to changes in pH produced under isocapnic conditions (Neubauer et al., 1991). There is no ready explanation for these different findings in cultured neurons.

In medullary slice preparations, decreased pH under isocapnic conditions stimulated neurons located near the ventral medullary surface (Fukuda and Honda, 1976; Miles, 1983) and at other locations (Miles, 1983). Increased P_{CO_2} has been found to stimulate neurons in the dorsally located nucleus tractus solitarius (Dean et al., 1989) and in the midline raphe (Richerson, 1995). In other experiments, a decreased pH produced via increasing P_{CO_2} with constant $[HCO_3^-]$ stimulates some neurons; a decreased pH produced by decreasing $[HCO_3^-]$ with constant CO_2 stimulates other neurons (Fukuda, 1983). This investigator concluded that neurons can respond independently to both stimuli.

Since none of these *in vitro* studies could show that the neurons examined were related to the control of breathing, the results are inconclusive with respect to central chemoreceptor function. They do serve as a warning of the possible complexities of individual neuronal responses to specific changes in P_{CO_2} or pH.

FIGURE 7. Effects of rostral ventrolateral medullary (RVLM) cooling on breathing in goats **(A)** while awake breathing air, **(B)** awake and breathing oxygen, and **(C)** under anesthesia. At left is a schematic view of the goat ventral medulla showing the position of the cooling probes. The data shown are from cooling at the rostral M–caudal M region. Abbreviations: M, Mitchell's area (see legend to Fig. 1); S, Schläfke's area; L, Loeschcke's area. Note that cooling the RVLM has a much less dramatic effect on breathing if the goat is awake (i.e., unanesthetized). The effect on RVLM cooling on the response to CO_2 is similarly less dramatic in the awake condition. (From Ohtake et al., 1995 with permission of authors and the American Physiological Society.)

MEDULLARY pH AND CHEMORECEPTION; INTRACELLULAR VERSUS EXTRACELLULAR?

For this discussion we presume that at the least some portion of the central chemoreceptor signal is pH. Where is it sensed?

Heterogeneity of Tissue pH Responses

Measurements of medullary tissue pH have been made at different locations during abrupt and transient medullary hypercapnia produced by brief intravertebral artery infusion of hypercapnic saline (Fig. 8) (Ichikawa et al., 1989). At some sites, the change in tissue pH was much greater than at other sites, presumably reflecting differences in local tissue blood flow and pH regulatory mechanisms. The investigators postulated that sites with large tissue pH changes would be central chemoreceptor locations.

Arguments for an Extracellular Site

Most studies of ventilatory output while acid-base balance is perturbed have presumed that the stimulus is in brain extracellular fluid. Historically, studies in unanesthetized goats (Fig. 2) (Fencl et al., 1966) used the strategy of first producing chronic metabolic acid-base disorders and then performing steady-state ventriculo-cisternal perfusion of artificial CSF. The use of different HCO_3^- concentrations in the cerebral perfusate allowed the investigators to examine the relationship of respiratory output and brain extracellular pH. They describe ventilation as a unique function of a calculated extracellular pH at a specific distance along a concentration gradient between the capillary and ventricular cerebrospinal fluid. This important work, a prototype of numerous similar studies, indicated a virtual location for central chemoreceptors; it did not identify their anatomical location.

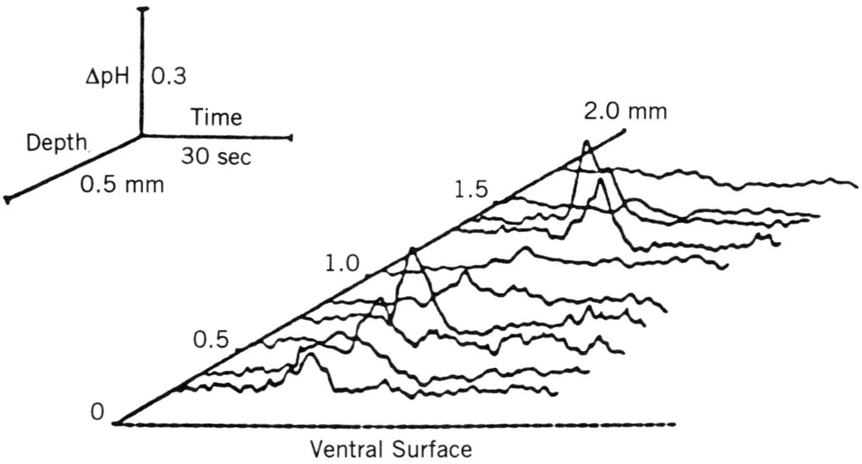

FIGURE 8. The heterogeneity of medullary pH responses to abrupt, brief medullary acidification produced by intravertebral artery injection of saline equilibrated with 100% CO_2 in the anesthetized cat. The tract shown is at the level of the caudal chemosensitive area on the VLM surface (see Fig. 1). (From Ichikawa et al., 1989 with permission of authors and the American Physiological Society.)

Arguments for an Intracellular Site

***Mammalian data* in vivo.** After reviewing data on brain pH in hypercapnia and following acetazolamide treatment, Lassen (1990) hypothesized that neuronal intracellular pH was the stimulus for central chemoreception. There is other, indirect, support for this hypothesis. Both amiloride, an inhibitor of the Na^+/H^+ antiporter, and 4,4′-diisothiocyanatostilbene-2,2′-disulfonic acid (DIDS), an inhibitor of the Cl^-/HCO_3^- antiporter, inhibit CSF pH regulation in systemic hypercapnia (see Nattie, 1995a). However, amiloride treatment has no effect on the respiratory response to hypercapnia, while DIDS treatment significantly enhances this response. In that both inhibitors produce similar effects on CSF pH regulation, one interpretation is that DIDS treatment interferes with intracellular pH regulation in central chemoreceptor neurons, resulting in a greater stimulation in response to hypercapnia.

Other data supporting an intracellular site. In studies on enzymatically dispersed neonatal rat carotid body Type I cells, intracellular pH during acid-base changes (Buckler et al., 1991) was found to have a remarkable temporal correlation with the afferent firing rate of the carotid body exposed, in separate experiments, to similar acid-base changes (obtained from Gray, 1968). While central chemoreceptors do not necessarily operate in the same fashion as these carotid body cells, this work is of great interest, in that intracellular pH was directly measured and suggested to be the stimulus. In the pulmonate snail (Erlichman and Leiter, 1993, 1994), perturbations of the acid-base environment of neurons within a small region of the nervous system can be correlated to a ventilatory response, the size of the pneumostome opening. Experiments using primary changes in P_{CO_2} versus $[HCO_3^-]$, cell alkalinization with NH_4Cl, and inhibition of Cl^-/HCO_3^- transport by DIDS supported the interpretation that in this system intracellular pH is the chemoreceptor stimulus. In rat medullary slices, measurements of pH_i using fluorescent pH-sensitive dyes have shown that neurons from chemosensitive areas regulate pH_i less well than do neurons from non-chemosensitive areas (Ritucci et al., 1997).

The Hypoxia Paradox

If extra- or intracellular pH is the stimulus for central chemoreceptors, one would expect to see ventilatory stimulation when the medulla is acidified under natural conditions such as hypoxia. In an animal with intact peripheral chemoreceptors, the ventilatory response to hypoxia is biphasic: an initial increase is followed by a waning of the response, called roll-off (see Bisgard and Neubauer, 1995). This secondary waning is especially evident in younger, less mature animals. Medullary pH has been shown to undergo an initial alkalinization (Neubauer et al., 1985; Silver, 1990), possibly due to increased blood flow (Neubauer et al., 1985), followed by a phase of acidification (Neubauer et al., 1985; Silver, 1990; see Biscard and Neubauer, 1995). This acidosis is not associated with increased ventilatory output.

In anesthetized animals with absent peripheral chemoreceptors, a number of investigators (see Kiwull-Schöne and Kiwull, 1992; Xu et al., 1992) have shown that hypoxia does not stimulate ventilation. During such hypoxia, medullary pH is known to be acidic predominantly at ventral surface areas identified with central chemosensitivity (Fig. 9) (Xu et al., 1992). Central chemoreceptors are still functional, as the response to a hypercapnic stimulus remains intact during central hypoxia and its associated respiratory depression (Kiwull-Schöne and Kiwull, 1992; Melton et al., 1988).

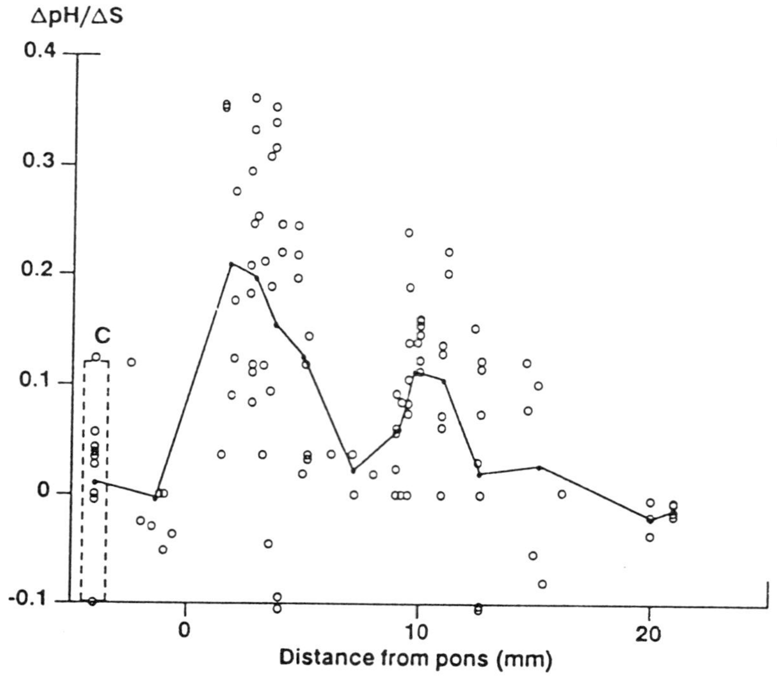

FIGURE 9. The heterogeneity of medullary surface pH response to isocapnic hypoxia in the anesthetized cat. The data are depicted as the change in pH per unit change in arterial oxygen saturation (S) shown as a function of the distance from the ponto-medullary junction (pons). Area "C" enclosed by dotted lines refers to control sites at the parietal cortex. Note that acidification is greater at two locations which coincide with the rostral and caudal chemosensitive areas (see Fig. 1). (From Xu et al., 1992, with permission of authors and the American Physiological Society.)

The absence of a ventilatory response to the acidosis induced by hypoxia in animals that still respond vigorously to increased P_{CO_2} is difficult to explain by either an extra- or intracellular central chemoreceptor sensing hypothesis. This paradox led to the development of a transmembrane H^+ gradient hypothesis for the pH sensing mechanism (Kiwull-Schöne and Kiwull, 1992; Xu et al., 1992). In this view, conditions that decreased the normal outwardly directed H^+ gradient would stimulate the chemoreceptor cells. They would include (i) hypercapnia that acidified extracellular fluid to a greater extent than intracellular fluid due to the presence of greater cell pH regulation and (ii) metabolic acidosis that originated in the extracellular environment. An acidosis that originated within the cell (e.g., the hypoxic-induced lactic acidosis) would not stimulate such a chemoreceptor mechanism.

There is an alternative explanation: In awake, unanesthetized goats with intact but separately perfused peripheral chemoreceptors, hypoxia limited to the central nervous system does not depress ventilation, it actually stimulates it (Smith et al., 1993; see Biscard and Neubauer, 1995). Thus, it is possible that central chemoreceptors do respond to the hypoxia-induced increase in lactic acid, if the neurons are not depressed by anesthesia and there is simultaneous input from the carotid body. There are also reports of neurons that are stimulated by hypoxia, both in the rostral ventrolateral medulla (Sun and Reis, 1994) and in the diencephalon (Dillon and Waldrop, 1992).

Carbonic Anhydrase

Focal application of the carbonic anhydrase inhibitor acetazolamide to the medullary surface or tissue produces a local acidosis and stimulates respiratory output (Coates et al., 1993). The acidosis is thought to result from diminished removal of metabolic end products in the form of H^+ and HCO_3^- (Severinghaus et al., 1969) or from diminished removal of the decarboxylation products as CO_2 (see Nattie and Li, 1996; see Chapters 3 and 4). Data obtained from single cells show an intracellular acidosis following carbonic anhydrase inhibition (Kaila et al., 1990), although measurements averaged for the whole brain do not (Bickler et al., 1988; Vorstrup et al., 1989).

Some experiments suggest that chemoreceptors require carbonic anhydrase to speed the reaction of molecular CO_2 to a form recognized by the receptor, presumable H^+, but the enzyme is not involved in the transduction of the stimulus. Such experiments include observations in the isolated carotid body (Iturriaga et al., 1991), investigations of the respiratory response to a step increase in medullary P_{CO_2} (Coates et al., 1993), and studies in the snail pneumostome model (Erlichman et al., 1994).

THE MECHANISM FOR SENSING pH

Cellular Mechanism

Is central chemoreception an intrinsic cell property or does it involve synaptic transmission? *In vitro* studies (Dean et al., 1989; Richerson, 1995; Pineda and Aghajanian, 1997) suggest that neuronal depolarization (or hyperpolarization) produced by hypercapnia is an intrinsic neuronal property, perhaps involving a potassium channel (Chapters 21 and 26). There is evidence that the Type I cells of the carotid body which depolarize with acidosis have a H^+-sensitive K^+ channel and/or Na^+ extrusion protein. Such proteins could conceivably be the pH sensors for these cells and, perhaps, for central chemoreceptor cells as well.

Synaptic Mechanism

It remains a plausible hypothesis that some, or all, of central chemoreception involves a pH effect on synaptic transmission. In principle, it is possible that any neurotransmitter with a pH-sensitive function could be involved in chemoreception. The pH could affect the receptor, or the release, synthesis, or uptake of the transmitter. Two neurotransmitters have been studied in some detail in this regard, acetylcholine and glutamate.

Cholinergic hypothesis. For some time, investigators have proposed a relationship between muscarinic cholinergic neurotransmission and central chemoreception (Burton et al., 1994; Loeschcke, 1982; see Nattie and Li, 1990b, for references). Microinjection of specific muscarinic receptor subtype antagonists has localized one site of these effects to the RVLM region (Nattie and Li, 1990b), with the M3 subtype antagonist having greatest efficacy in inhibiting the CO_2 response (Fig. 10) (Nattie and Li, 1990b). Binding studies have shown the presence of three muscarinic receptor subtypes in the RVLM region (Mallios et al., 1995). Data obtained in the isolated brainstem preparation of the neonatal rat (Monteau et al., 1990; Burton et al., 1994) and in the newborn opossum (Eugenin, 1994) also support an important role for cholinergic neurotransmission in the respiratory response to changes in P_{CO_2} and pH. Study of the brainstems of human victims of sudden

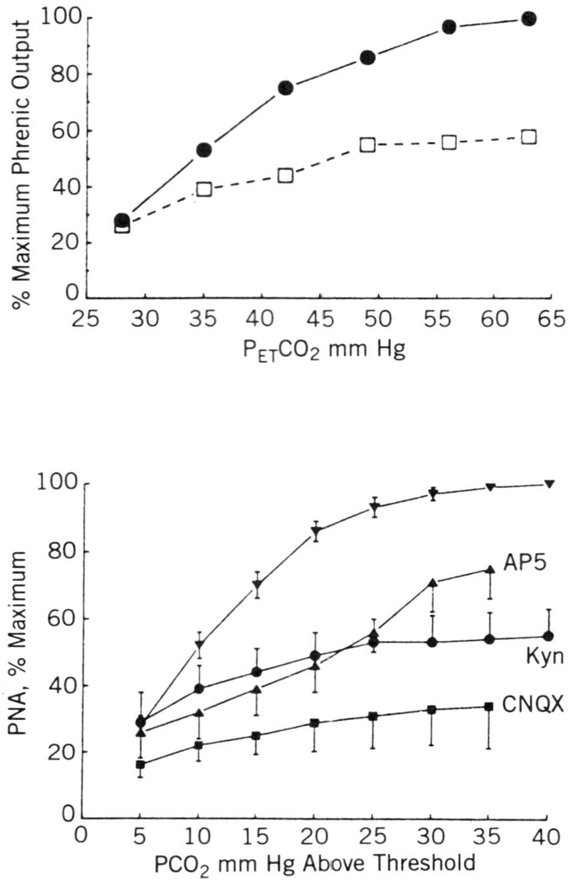

FIGURE 10. Effects on the Response to CO_2 in the anesthetized cat of the antagonism of muscarinic cholinergic receptors of the M3 subtype with DAMP (top panel open squares versus solid circles, control) and of antagonisms of glutamate receptor subtypes (bottom panel) with AP5 (an NMDA antagonist), CNQX (a non-NMDA antagonist) and kynurenate (a nondiscriminatory antagonist). The respiratory output is integrated phrenic activity (PNA) expressed as a per cent of the maximum value observed under control conditions. The agents were injected unilaterally into the RTN region (see Fig. 6). (From Nattie and Li, 1990b, and Nattie et al., 1993 with permission of American Physiological Society.

infant death syndrome indicate altered muscarinic binding in a region of the RVLM called the arcuate nucleus (Kinney et al., 1995), a possible homologue with the RVLM region studied in rat and cat.

At issue with the cholinergic hypothesis for central chemoreception is that the effects could involve RVLM neurons that are not chemosensitive but, instead, are involved in some other aspect of respiratory control. For example, it seems possible that cholinergic muscarinic input to RVLM neurons provides modulation related to sleep and/or wakefulness (Mallios et al., 1995), which, if disrupted, could alter the ventilatory system response to CO_2.

Glutamate hypothesis. Studies have also examined the role of RVLM glutamate receptors in the control of respiratory output and in central CO_2 sensitivity. Microinjection into the RVLM region of glutamate receptor antagonists decreases respiratory output and CO_2 sensitivity (Fig. 10) in a dose-dependent manner in anesthetized cats (Nattie et al., 1993) and in the unanesthetized goat (H. V. Forster, unpublished observations). Direct RVLM injection of glutamate itself increases respiratory output and the response is attributed to stimulation of ionotropic and metabotropic glutamate receptors (Nattie et al., 1993; Li and Nattie, 1995). The physiological significance of these receptors is unknown, as is their role in central chemosensitivity.

As for muscarine receptors, the effects of glutamate antagonists have not been shown to be specific for chemoreception.

Proteins and pH

While there are many mechanisms by which a protein can exhibit pH sensitivity, the role of imidazole-histidine, as predicted by the alphastat hypothesis, is of particular interest to many physiologists. This hypothesis arises from considering the effects of temperature on extracellular (Reeves, 1977) and intracellular pH (Hitzig et al., 1994; Malan et al., 1976; Marjanovic et al., 1994), both of which vary by approximately -0.015 pH unit/°C. Based on the observation that, in solution, the p*K* of imidazole-histidine had a similar temperature relationship, Reeves (1977), suggested that the fractional dissociation of imidazole-histidine (alpha-imidazole) would remain constant as temperature varied even though pH changes significantly. In contrast, as pH is altered in isothermal conditions, alpha-imidazole changes. The alphastat hypothesis implies that proteins with tertiary structure and function determined by histidine groups would be functionally unaffected by pH changes caused by temperature, while isothermal pH changes could affect them in a pH-dependent manner.

Proving this hypothesis is difficult as the precise p*K* of imidazole can be affected by its neighbor amino acids (see Nattie, 1990, for discussion), and most physiologically relevant proteins contain many histidines (Nattie, 1995a). Direct measurements of alpha imidazole using NMR techniques in the intact newt show that alpha imidazole is constant when temperature varies even though intracellular pH changes dramatically (Hitzig et al., 1994). In the clinical situation of induced hypothermia, most physicians now regulate blood acid-base status using this hypothesis. Treatment with diethylpyrocarbonate (DEPC), a substance that is relatively specific for binding to imidazole-histidine, interferes with the pH sensitivity of many proteins (see Nattie, 1995a), and application of DEPC to the ventral medullary surface diminishes baseline ventilatory output and the response to increased CO_2 (Nattie, 1986). This is very indirect evidence, and further studies are needed to show the predicted role of imidazole-histidine in physiological pH-sensing processes.

CONCLUSIONS

Stimulation of central chemoreceptors located at many sites within the brainstem increases respiratory output. The physiological role of these widespread central chemoreceptor sites is not well understood. They may differ in their threshold for responses to acidic stimuli and in their dependence on arousal state. The role of the RVLM in the control of breathing involves more than just chemoreception and seems clearly state-dependent. The central chemoreceptor stimulus appears to be pH, although an independent P_{CO_2}-specific effect remains a possibility. The pH most likely acts at an intracellular site. In the context of pH and brain function, it is intriguing that important pH sensors exist within a structure that has many levels of pH regulation: the neuron, glia, blood-brain barrier, and ventricular cerebrospinal fluid. While a broader view of central chemoreceptor location and function is presently emerging, how pH is sensed by central chemoreceptors and how this sensing process is related to brain pH regulation is unknown.

REFERENCES

Akilesh MR, Kamper M, Li A, Nattie EE (1997): Effects of unilateral lesions of the retrotrapezoid nucleus on breathing in awake rats. J Appl Physiol 32: 469–479.

Berger AJ, Cooney KA (1982): Ventilatory effects of kainic acid injection of the ventrolateral solitary nucleus. J Appl Physiol 52: 131–140.

Berkenbosch A, Van Beek JHGM, Olievier CN, De Goede J, Quanjer Ph H (1984): Central respiratory CO_2 sensitivity at extreme hypocapnia. Respir Physiol 55: 95–102.

Bernard DG, Li A, Nattie EE (1996): Evidence for central chemoreception in the medullary raphe. J Appl Physiol 80: 108–115.

Bianchi AL, Grélot L, Iscoe S, Remmers JE (1988): Electrophysiological properties of rostral medullary respiratory neurons in the cat: An intracellular study. J Physiol (Lond) 407: 293–310.

Bianchi AL, Denavit-Saubié, Champagnat J (1995): Central control of breathing in mammals: Neuronal circuitry, membrane properties, and neurotransmitters. Physiol Rev 75: 1–45.

Bickler PE, Litt L, Banville DL, Severinghaus JW (1988): Effects of acetazolamide on cerebral acid-base balance. J Appl Physiol 65: 422–427.

Bisgard GE, Neubauer JA (1995): Peripheral and central effects of hypoxia. In Dempsey JA, Pack AI (eds): Regulation of Breathing. New York: Marcel Dekker, pp 617–668.

Borison HL, Gonsalves SF, Montgomery SP, McCarthy LE (1978): Dynamics of respiratory VT response to isocapnic pHa forcing in chemodenervated cats. J Appl Physiol 45: 502–511.

Buckler KJ, Vaughn-Jones RD, Peers C, Lagadic-Grossman D, Nye PCG (1991): Effects of extracellular pH, P_{CO_2}, and HCO_3^- on intracellular pH in isolated Type-I cells of the neonatal rat carotid body. J Physiol (Lond) 444: 703–721.

Burton MD, Nouri K, Baichoo S, Samuels-Toyloy N, Kazemi H (1994): Ventilatory output and acetylcholine: Perturbations in release and muscarinic receptor activation. J Appl Physiol 77: 2275–2284.

Coates EL, Li A, Nattie EE (1991): Acetazolamide on the ventral medulla of the cat increases phrenic output and delays the ventilatory response to CO_2. J Physiol (Lond) 441: 433–451.

Coates EL, Li A, Nattie EE (1993): Widespread sites of brainstem ventilatory chemoreceptors. J Appl Physiol 75: 5–14.

Cohen MI (1995): Tonic chemoreceptor input as the background for respiratory rhythm. In Trouth CO, Millis RM, Kiwull-Schöne HF, Schläfke ME (eds): Ventral Brainstem Mechanisms and Control of Respiration and Blood Pressure. New York: Marcel Dekker, pp 797–799.

Cohen MI (1968): Discharge patterns of brain-stem respiratory neurons in relation to carbon dioxide tension. J Neurophysiol 31: 142–165.

Cozine RA, Ngai SH (1967): Medullary surface chemoreceptors and regulation of respiration in the cat. J Appl Physiol 22: 117–121.

Daristotle L, Bersenbrugge AD, Engwall MJ, Bisgard GE (1990): The effect of carotid body hypocapnia on ventilation in goats. Respir Physiol 79: 123–136.

Datta AK, Shea SA, Horner RL, Guz A (1991): The influence of induced hypocapnia and sleep on the endogenous respiratory rhythm in humans. J Physiol (Lond) 440: 17–33.

Dean JB, Lawing WL, Millhorn DE (1989): CO_2 decreases membrane conductance and depolarizes neurons in the nucleus tractus solitarius. Exp Brain Res 76: 656–661.

Dempsey JA, Forster HV (1982): Mediation of ventilatory adaptation. Physiol Rev 62: 262–346.

Dillon GH, Waldrop TG (1992): In vitro responses of caudal hypothalamic neurons to hypoxia and hypercapnia. Neuroscience 51: 941–950.

Duffin J, Ezure K, Lipski J (1995): Breathing rhythm generation: Focus on the rostral ventrolateral medulla. NIPS 10: 133–140.

Eldridge FL, Kiley JP, Millhorn DE (1985): Respiratory responses to medullary hydrogen ion changes in cats: Different effects of respiratory and metabolic acidoses. J Physiol (Lond) 358: 285–297.

Erlichman JS, Leiter JC (1993): CO_2 chemoreception cells in the pulmonate snail, *H Aspersa*. Respir Physiol 93: 347–364.

Erlichman JS, Leiter JC (1994): Central chemoreceptor stimulus in the terrestrial, pulmonate snail, *H Aspersa*. Respir Physiol 95: 209–226.

Erlichman JS, Coates EL, Leiter JC (1994): Carbonic anhydrase and CO_2 chemoreception in the pulmonate snail *Helix aspersa*. Respir Physiol 98: 27–42.

Eugenin J, Nicholls JG (1997): Chemosensory and cholinergic stimulation of fictive respiration in isolated CNS of neonatal opossum. J Physiol (Lond) 501: 425–437.

Feldman JL (1986): Neurophysiology of breathing in mammals. In Bloom FE (ed): Handbook of Physiology, Section I: The Nervous System, Vol 4, Intrinsic Regulatory Systems of the Brain. Bethesda, MD: American Physiological Society, pp 463–524.

Feldman JL, Smith JC (1995): Neural control of respiratory pattern in mammals: an overview. In Dempsey JA, Pack AI (eds): Regulation of Breathing. New York: Marcel Dekker, pp 39–69.

Fencl V, Miller TB, Pappenheimer JR (1966): Studies on the respiratory response to disturbances of acid-base balance, with deductions concerning the ionic composition of cerebral interstitial fluid. Am J Physiol 210: 459–472.

Forster HV, Ohtake PJ, Pan LG, Lowry TF (1995): Effects on breathing of ventrolateral medulla cooling in awake goats. J Appl Physiol 78: 258–265.

Fukuda Y (1983): Difference between actions of high P_{CO_2} and low $[HCO_3^-]$ on neurons in the rat medullary chemosensitive areas in vitro. Pflügers Arch 398: 324–330.

Fukuda Y, Honda Y (1976): pH sensitivity of cells located at the ventrolateral surface of the cat medulla oblongata in vitro. Pflügers Arch 364: 243–247.

Gray BA (1968): Responses of the perfused carotid body to changes in pH and P_{CO_2}. Respir Physiol 4: 580–584.

Harada Y, Kuno M, Wang YZ (1985): Differential effects of carbon dioxide and pH on central chemoreceptors in the rat in vitro. J Physiol (Lond) 368: 679–693.

Heeringa J, Berkenbosch A, DeGoede J, Olievier CN (1979): Relative contribution of central and peripheral chemoreceptors to the ventilatory response of CO_2 during hyperoxia. Respir Physiol 37: 365–379.

Hitzig BM, Perng WC, Burt T, Okunieff P, Johnson DC (1994): ^{1}H-NMR measurement of fractional dissociation of imidazole in intact animals. Am J Physiol (Regul Integ Comp Physiol) 266: R1008–R1015.

Huang R-Q, Erlichman JS, Dean JB (1997): Cell-cell coupling between CO_2-excited neurons in the dorsal medulla oblongata. Neurosci 80: 41–57.

Ichikawa K, Kuwana S, Arita H (1989): ECF pH dynamics within the ventrolateral medulla: A microelectrode study. J Appl Physiol 67: 193–198.

Issa FG, Remmers JE (1992): Identification of a subsurface area in the ventral medulla sensitive to local changes in P_{CO_2}. J Appl Physiol 72: 439–446.

Iturriaga R, Lahiri S, Mokashi A (1991): Carbonic anhydrase and chemoreception in the cat carotid body. Am J Physiol (Cell Physiol) 261: C565–C573.

Johnston BM, Gluckman PD (1989): Lateral pontine lesions affect central chemosensitivity in unanesthetized fetal lambs. J Appl Physiol 67: 1113–1118.

Kaila K, Saarikoski J, Voipio J (1990): Mechanisms of action of GABA on intracellular pH and on surface pH in crayfish muscle fibers. J Physiol (Lond) 427: 241–260.

Kawai A, Ballantyne D, Mücckenhoff, P Scheid (1996): Chemosensitive medullary neurones in the brainstem-spinal cord preparation of the neonatal rat. J Physiol (Lond) 492: 277–292.

Kinney HC, Filiano JJ, Sleeper LA, Mandell F, Valdes-Dapena M, Frost White W (1995): Decreased muscarinic receptor binding in the arcuate nucleus in sudden infant death syndrome. Science 269: 1446–1450.

Kiwull-Schöne H, Kiwull P (1992): Hypoxia and the "reaction theory" of central respiratory chemosensitivity. Adv Exp Med Biol 316: 347–357.

Kogo N, Arita H (1990): In vivo study on medullary H^+-sensitive neurons. J Appl Physiol 69: 1408–1412.

Lassen NA (1990): Is central chemoreceptor sensitivity to intracellular rather than extracellular pH? Clin Physiol 10: 311–319.

Leuson IR (1954): Chemosensitivity of the respiratory center. Influence of CO_2 in the cerebral ventricles on respiration. Am J Physiol 176: 39–44.

Li A, Nattie EE (1995): Prolonged stimulation of respiration by brainstem metabotropic glutamate receptors. J Appl Physiol 79: 1650–1656.

Li A, Nattie EE (1997): Focal central chemoreceptor sensitivity in the RTN studied with a CO_2 diffusion pipette in vivo. J Appl Physiol 83: 420–428.

Lipscomb WT, Boyarski LL (1972): Neurophysiological investigations of medullary chemosensitive areas of respiration. Respir Physiol 16: 362–376.

Loeschcke HH (1982): Central chemosensitivity and the reaction theory. J Physiol (Lond) 332: 1–24.

Malan A, Wilson TE, Reeves RB (1976): Intracellular pH in cold-blooded vertebrates as a function of body temperature. Respir Physiol 28: 29–47.

Mallios V, Lydic R, Bagdoyan HA (1995): Muscarinic receptor subtypes are differentially distributed across brain stem nuclei. Am J Physiol 268: L941–L949.

Marjanovic M, Elliot AC, Roman BB, Dawson MJ (1994): The temperature dependence of intracellular pH in frog skeletal muscle. Physiologist 37: A-50.

Melton JE, Neubauer JA, Edelman NH (1988): CO_2 sensitivity of cat phrenic neurogram during hypoxic ventilatory depression. J Appl Physiol 65: 736–743.

Miles R (1983): Does low pH stimulate central chemoreceptors located near the ventral medullary surface? Brain Res 271: 349–353.

Mitchell RA, Loeschcke HH, Massion WH, Severinghaus JW (1963): Respiratory responses mediated through superficial chemosensitive areas on the medulla. J Appl Physiol 18: 523–533.

Monteau R, Morin D, Hilaire G (1990): Acetylcholine and central chemosensitivity: In vitro study in the newborn rat. Respir Physiol 81: 241–254.

Morin-Surun MP, Boudinot E, Schäfer T, Denavit-Saubié (1995): Localization of chemosensitive structures in the isolated brainstem of adult guinea-pig. J Physiol (Lond) 485: 203–212.

Nattie EE (1983): Ventilation during acute HCl infusion in intact and chemodenervated conscious rabbits. Respir Physiol 54: 97–107.

Nattie EE (1986): Diethylpyrocarbonate (an imidazole binding substance) inhibits rostral VLM CO_2 sensitivity. J Appl Physiol 61: 843–850.

Nattie EE (1990): The alphastat hypothesis in respiratory control and acid-base balance (brief review). J Appl Physiol 69: 1201–1207.

Nattie EE (1995a): Central chemoreception. In Dempsey JA, Pack AI (eds): Regulation of Breathing. New York: Marcel Dekker, pp 473–510.

Nattie EE (1995b): Integrative aspects of ventral brainstem function. In Trouth CO, Millis RM, Kiwull-Schöne HF, Schläfke ME (eds): Ventral Brainstem Mechanisms and Control of Respiration and Blood Pressure. New York: Marcel Dekker, pp 253–266.

Nattie EE, Li A (1990a): Fluorescent location of RVLM kainate microinjections that alter the control of breathing. J Appl Physiol 68: 1157–1166.

Nattie EE, Li A (1990b): Ventral medulla sites of muscarinic receptor subtypes involved in cardiorespiratory control. J Appl Physiol 69: 33–41.

Nattie EE, Li A (1994): Retrotrapezoid nucleus lesions decrease phrenic activity and CO_2 sensitivity in rats. Respir Physiol 97: 63–77.

Nattie EE, Li A (1996): Central chemoreception in the region of the ventral respiratory group of the rat. J Appl Physiol 81: 1987–1995.

Nattie EE, Gdovin M, Li A (1993): Retrotrapezoid nucleus glutamate receptors: Control of CO_2-sensitive phrenic and sympathetic output. J Appl Physiol 74: 2958–2968.

Neubauer JA, Santiago TV, Posner MA, Edelman NH (1985): Ventral medullary pH and ventilatory responses to hyperperfusion and hypoxia. J Appl Physiol 58: 1659–1668.

Neubauer JA, Gonsalves SF, Chou W, Geller HM, Edelman NH (1991): Chemosensitivity of medullary neurons in explant tissue cultures. Neuroscience 45: 701–708.

Ohtake PJ, Forster HV, Pan LG, Lowry TF, Korducki MJ, Aaron EA, Weiss EM (1995): Ventilatory responses to cooling the ventrolateral medullary surface of awake and anesthetized goats. J Appl Physiol 78: 247–257.

Ohtake PJ, Forster HV, Pan LG, Lowry TF, Korducki MJ, Whaley A (1996): Effects of cooling the ventrolateral medulla on diaphragmatic activity during NREM sleep. Respir Physiol 104: 127–135.

Paxinos G, Watson C (1982): The Rat Brain in Stereotaxic Coordinates. New York: Academic Press.

Pilowsky PM, Jiang C, Lipski J (1990): An intracellular study of respiratory neurons in the rostral ventrolateral medulla of the cat and their relationship to catecholamine containing neurons. J Comp Neurol 301: 604–617.

Pineda J, Aghajanian GK (1997): Carbon dioxide regulates the tonic activity of locus ceruleus neurons by modulating a proton- and polyamine-sensitive inward rectifier potassium current. Neurosci 77: 723–734.

Reeves RB (1977): The interactions of body temperature and acid-base balance in ectothermic vertebrates. Annu Rev Physiol 39: 559–586.

Richerson GB (1995): Response to CO_2 of neurons in the rostral ventral medulla in vitro. J Neurophysiol 36: 207–216.

Richter DW (1996): Neural regulation of respiration: Rhythmogenesis and afferent control. In Gregor R, Windohorst U (eds): Comprehensive Human Physiology, Vol. 2. Berlin, Heidelberg: Springer-Verlag, pp 2079–2095.

Rigatto H, Fitzgerald SF, Willis MA, Yu C (1994): In search of the real respiratory neurons: Culture of medullary fetal cells sensitive to CO_2 and low pH. Biol Neonate: 149–155.

Ritucci NA, Dean JB, Putnam RW. Intracellular pH response to hypercapnia in neurons from chemosensitive areas of the medulla. Am J Physiol (Regul Integ Comp Physiol) 273: R433–R441.

Schläfke ME (1981): Central chemosensitivity: A respiratory drive. Rev Physiol Biochem Pharm 90: 171–172.

Schäfer D, Schläfke MA (1995): Cardiorespiratory regulation in a model for Ondine's curse syndrome. In Trouth CO, Millis RM, Kiwull-Schöne HF, Schläfke ME (eds): Ventral Brainstem Mechanisms and Control of Respiration and Blood Pressure. New York: Marcel Dekker, pp 253–266.

Schläfke ME, Kille JF, Loeschcke HH (1979): Elimination of central chemosensitivity by coagulation of a bilateral area on the ventral medullary surface in awake cats. Pflügers Arch 378: 231–241.

Sears TA, Berger AJ, Phillipson EA (1982): Reciprocal tonic activation of inspiratory and expiratory motoneurons by chemical drives. Nature 299: 728–730.

Severinghaus JW, Hamilton FN, Cotev S (1969): Carbonic acid production and the role of carbonic anhydrase in decarboxylation in brain. Biochem J 114: 703–705.

Silver IA, Erecinska M (1990): Intracellular and extracellular changes of [Ca^{++}] in hypoxia and ischemia in rat brain in vivo. J Gen Physiol 95: 837–866.

Smith CA, Engwall MJA, Dempsey JA, Bisgard GE (1993): Effects of specific carotid body and brain hypoxia on respiratory muscle control in the awake goat. J Physiol (Lond) 460: 623–640.

Smith CA, Jameson LC, Mitchell GS, Musch TI, Dempsey JA (1984): Central–peripheral chemoreceptor interaction in awake cerebrospinal fluid-perfused goats. J Appl Physiol 56: 1541–1549.

Smith CA, Saupe KW, Henderson KS, Dempsey JA (1995): Ventilatory effects of specific carotid body hypocapnia in dogs during wakefulness and sleep. J Appl Physiol 79: 689–699.

St. John WM, Bianchi AL (1985): Responses of bulbospinal and laryngeal respiratory neurons to hypercapnia and hypoxia. J Appl Physiol 59: 1201–1207.

St. John WM, Bledsoe TA (1985): Genesis of rhythmic respiratory activity in pons independent of medulla. J Appl Physiol 59: 684–690.

St. John WM, Glasser RI, King RA (1971): Apneustic breathing after vagotomy in cats with chronic pneumotaxic center lesions. Respir Physiol 12: 239–250.

Sullivan CE, Kozar LF, Murphy E, Phillipson EA (1978): Primary role of respiratory afferents in sustaining breathing rhythm. J Appl Physiol 45: 11–17.

Sun M, Reis DJ (1994): Hypoxia selectively excites vasomotor neurons of rostral ventrolateral medulla in rats. Am J Physiol (Regul Integr Comp Physiol) 266: R245–R256.

Tojima H, Kuriyama T, Fukuda Y (1991): Differential respiratory effects of HCO_3^- and CO_2 applied on ventral medullary surface of rats. J Appl Physiol 70: 2217–2225.

Trouth CO, Loeschcke HH, Berndt J (1979): A superficial substrate on the ventral surface of the medulla oblongata influencing respiration. Pflügers Arch 339: 135–152.

Vorstrup S, Jensen KE, Thomsen C, Henriksen O, Lassen NA, Paulson OB (1989): Neuronal pH regulation: Constant normal intracellular pH is maintained in brain during low extracellular pH induced by acetazolamide. ^{31}P NMR study. J Cereb Blood Flow Metab 9: 417–421.

Weese-Mayer DE, Silvestri JM, Menzies LJ, Morrow-Kenny AS, Hunt CE, Hauptman SA (1992): Congenital central hypoventilation syndrome: Diagnosis, management, and long-term outcome in thirty-two children. J Pediatr 120: 381–387.

Xu F, Sato M, Spellman MJ Jr, Mitchell RA, Severinghaus JW (1992): Topography of cat medullary ventral surface hypoxic acidification. J Appl Physiol 73: 2631–2637.

SECTION VII

pH and Pathophysiology of the Nervous System

CHAPTER 28

Acid-Base Metabolism in Ischemia

K. KATSURA AND BO. K. SIESJÖ
Laboratory for Experimental Brain Research, Wallenberg Neuroscience Center, University Hospital, Lund, Sweden (B.K.S.); Second Department of Internal Medicine, Nippon Medical School, Tokyo, Japan (K.K.)

INTRODUCTION

Both ischemia and hypoxia (i.e., conditions which cause bioenergetic failure) are associated with enhanced production of acid equivalents, and with a fall in intra- and extracellular pH (pH_i and pH_e, respectively). The acidosis has been quantified by various methods. For example, changes in pH_i have been estimated with pH indicators (LaManna and McCracken, 1984; Paschen et al., 1987), by application of the CO_2/HCO_3^- method (Kraig et al., 1986; Smith et al., 1986; Katsura et al., 1992a,b), or with ^{31}P-NMR techniques (e.g., Behar et al., 1989; Chopp et al., 1987). Corresponding changes in pH_e have been followed with microelectrodes (Mutch and Hansen, 1984; Kraig et al., 1986; Smith et al., 1986; Katsura et al., 1992a,b).

More than 20 years ago, results obtained in complete ischemia demonstrated that hypoglycemia reduced and hyperglycemia increased the amount of lactate accumulated during ischemia; furthermore the amounts accumulated correlated closely with the preischemic tissue concentrations of glycogen and glucose (Ljunggren et al., 1974). The intraischemic pH_i values calculated from changes in lactate concentration were 6.6, 6.4, and 6.2 in hypo-, normo-, and hyperglycemic animals, respectively. Subsequent results have amply verified these predictions, and demonstrated that preischemic hyperglycemia exaggerates the fall in pH_i and pH_e (Kraig et al., 1986; Smith et al., 1986; Katsura et al., 1992a,b; Chopp et al., 1988). However, the close correlation between preischemic plasma glucose concentration and tissue lactate content is no longer observed during incomplete ischemia (or in hypoxia), since the persistance of tissue perfusion delivers additional substrate to the ischemic tissues; for example, during incomplete ischemia tissue lactate content may rise to excessive values (30–40 mmol/kg) (Nordström et al., 1978; for corresponding data in hypoxia, see Salford and Siesjö, 1974).

Results obtained during the last 20 years have clearly shown that preischemic hyperglycemia aggravates damage due to transient global ischemia, most probably because of enhanced intra- and extracellular acidosis (Siesjö, 1988a,b; Siesjö et al., 1993). Thus, it

pH and Brain Function, Edited by Kai Kaila and Bruce R. Ransom
ISBN 0-471-11838-9

has been tacitly assumed that the aggravated cell damage is acidosis-mediated. The presumed coupling between acidosis and aggravated damage is not unique to the brain. For example, it has been proposed that flooding intolerance in plants is, at least in part, caused by acidosis (Roberts et al., 1984). Furthermore, cells in culture are readily killed by an acid load. However, it has remained a moot point whether this requires reduction of pH to values (<5) which are not observed *in vivo* (Norenberg et al., 1987; Nedergaard et al., 1991a). Recent results obtained on brain slices demonstrate, though, that if these are exposed to suspensions with moderately low pH for sustained periods, neuronal damage is indeed observed (Sipe et al., 1991).

Intraischemic changes in pH are usually large to excessive, and a reduction of pH_i to less than 6.3–6.4 may be required to trigger damage (see p 566). More moderate changes in cytoplasmic pH have been implicated in the regulation of diverse cellular processes (Busa and Nuccitelli, 1984). Alterations in pH modulate the activity of many enzymes and the function of a variety of ion channels, thereby affecting neuronal excitability (Moody, 1984; Balestrino and Somjen, 1988; Church and McLennan, 1989; Jarolimek et al., 1989; Traynelis and Cull-Candy, 1990; Chapter 20). For example, acidosis reduces the inward currents gated by N-methyl-D-aspartate (NMDA) receptors (Traynelis and Cull-Candy, 1990; Vyklicky et al., 1990; Chapter 21), or voltage-dependent Ca^{2+} channels (Iijima et al., 1986; Chapter 22). At the same time, inhibitory currents mediated by GABA receptors may be increased by a fall in pH_e (Pasternack et al., 1993). Thus H^+ ions play an important role, both in physiological processes and in pathological ones leading to intraischemic or postischemic damage.

The purpose of the present chapter is to discuss acid-base changes in ischemia, and the relationship between reduction in pH_i and/or pH_e and brain damage. We will put some emphasis on controversial or unresolved issues. These concern the sources of H^+ generated in ischemic tissues, the question of whether a critical pH value exists below which damage is exaggerated, and the mechanisms by which such damage is incurred.

SOURCES OF METABOLICALLY PRODUCED H^+

Glycolysis

As discussed by Krebs et al. (1975), H^+ ions are produced in a multitude of metabolic reactions. Our immediate concern is H^+ generation during glycolysis and during pyruvate oxidation. In general, H^+ is produced in various reactions in which ATP is used or the oxidized form of nicotinamide-adenine dinucleotide (NAD^+) is reduced (Krebs et al., 1975; Alberti and Cuthbert, 1982; see also Hochachka and Mommsen, 1983; Siesjö et al., 1993). We may take as examples the phosphorylation of glucose to glucose-6-phosphate (G-6-P) and the oxidation of glyceraldehyde 3-phosphate (GAP) to 1,3-diphosphoglycerate (1,3DPG).

$$\text{glucose} + ATP^{4-} \rightarrow \text{G-6-P}^{2-} + ADP^{3-} + H^+ \quad (1)$$

$$\text{GAP} + NAD^+ + HPO_4^{2-} \rightarrow \text{DPG} + \text{NADH} + H^+ \quad (2)$$

However, this net acid production is balanced by a corresponding consumption of H^+. For example, when phosphoenol pyruvate (PEP) is converted to pyruvate, and the latter is reduced, we get

$$PEP^{5-} + ADP^{3-} + H^+ \rightarrow \text{pyruvate}^- + ATP^{4-} + HPO_4^{2-} \quad (3)$$

$$\text{pryuvate}^- + \text{NADH} + H^+ \rightarrow \text{lactate}^- + NAD^+ \quad (4)$$

The net result of glycolysis (i.e., the anaerobic conversion of one mole of glucose to 2 moles of lactate) can be described by a series of equations.
At pH > 8.0:

$$\text{glucose} + 2HPO_4^{2-} + 2ADP^{3-} \rightarrow 2\ \text{lactate}^- + 2H_2O + 2\ ATP^{4-} \quad (5)$$

At pH < 6.0:

$$\text{glucose} + 2H_2PO^{4-} + 2\ HADP^{2-} \rightarrow 2\ \text{lactate}^- + 2H_2O + 2\ HATP^{3-} + 2\ H^+ \quad (6)$$

At physiological pH values:

$$\text{glucose} + 2[ADP] + 2[Pi] \rightarrow 2\ \text{lactate}^- + 2\ H_2O + 2[ATP] + n_1H^+ \quad (7)$$

However, the ATP formed does not accumulate but is hydrolyzed according to the equation

$$2[ATP] + 2H_2O \rightarrow 2[ADP] + 2[Pi] + n_2H^+ \quad (8)$$

Here the terms, [ADP], [ATP], and [Pi] refer to the sum of all species of ADP, ATP, and Pi, respectively. Although n_1 and n_2 are dependent constants upon pH and $[Mg^{2+}]$, the amount of H^+ released is always equal to the amount of lactate formed (Hochachka and Mommsen, 1983). Thus, the stoichiometry of proton production is independent of pH and Mg^{2+} levels, hence we can write the following equation

$$\text{glucose} \rightarrow 2\ \text{lactate}^- + 2\ H^+ \quad (9)$$

A corresponding relationship holds for the breakdown of glycogen.

$$\text{glycogen} \rightarrow 2\ \text{lactate}^- + 2\ H^+ \quad (10)$$

ATP Hydrolysis During Ischemia and Hypoxia

If ischemia and/or hypoxia lead to hydrolysis of the preexisting ATP stores, some extra H^+ will be released. The ATP available is both that stored in the tissue (about 3 mmol/kg) and that formed by the creatine kinase reaction (about 5 mmol/kg). However, since the latter consumes H^+, the total H^+ accumulated is less than 8 mmol/kg. We can write the following equations

$$ATP + HOH \rightarrow ADP + Pi + n'\ H^+ \quad (11)$$

$$PCr + ADP + n''\ H^+ \rightarrow Cr + ATP \quad (12)$$

where n' and n'' vary with pH and Mg^{2+} concentration (Lawson and Veech, 1979; Vink et al., 1988; Erecinska and Silver, 1989). Probably, hydrolysis of all PCr and ATP during ischemia yields 2–4 mmol/kg of H^+. However, ischemia also leads to the breakdown of other nucleoside triphosphates, such as uridine triphosphate (UTP), cytidine triphosphate (CTP), and guanosine triphosphate (GTP) (Kleihues et al., 1974; see also Chapman et al., 1981). Taken together, the hydrolysis of all nucleoside triphosphates probably yields 3–5 mM/kg of H^+ (see p 566).

Other Sources of H^+

Additional sources of H^+ generation during ischemia and/or hypoxia include hydrolysis of phospholipids; however, the net production is quantitatively small compared to glycolysis.

One can thus tentatively conclude that the amount of H^+ generated during ischemia and/or anoxia corresponds to that released when lactate accumulates, plus that released when the ATP stores (preischemic ATP and the ATP formed by hydrolysis of PCr) are hydrolyzed, minus the H^+ absorbed when PCr is hydrolyzed to generate ATP. These general conclusions are strongly supported by the following findings. First, although hypoglycemic coma leads to extensive ATP hydrolysis and lipolysis, pH_i is either unchanged or slightly increased, as would be expected from the decreased lactate levels (and the general consumption of metabolic acids) (Lewis et al., 1974; Feise et al., 1976; Pelligrino et al., 1981; for a recent review, see Chapter 5). Second, when complete ischemia is induced at varying preischemic plasma and tissue glucose levels, extrapolation of the lactate–pH_i curve to a normal tissue lactate content yields moderately reduced pH_i values, suggesting that 3–5 mmol/kg^{-1} of the H^+ accumulated is derived from other sources than glycolysis (Katsura et al., 1992b).

CHANGES IN pH_e AND pH_i IN ISCHEMIA AND HYPOXIA

It is clear that the conditions like ischemia, hypoxia, and status epilepticus lead to acid-base transients that are relatively dramatic, simply because the rate of acid production changes precipitously. The produced acid primarily acidifies intracellular fluids. Secondarily, nonionic or facilitated diffusion of the acid transmits the changes from one compartment to another (see Roos and Boron, 1981; Nedergaard and Goldman, 1993). There is another way in which H^+ equivalents can be translocated between intra- and extracellular fluid. Thus, translocation of H^+ via conductance channels has been discussed (Thomas, 1984), and Kaila and Voipio (1987) reported HCO_3^- translocation via Cl^- channels linked to GABA receptors.

Complete or Dense Ischemia

When ischemia is induced, pH_e is gradually reduced, reflecting acid production and accumulation of CO_2. However, at the point of ischemic depolarization (direct current potential shift), the pH record shows an alkaline shift (Kraig et al., 1983; Mutch and Hansen, 1984). It seems likely that the alkaline shift is due to influx of H^+ via one of the nonspecific cation channels gated by glutamate receptors; alternatively, the pH shift could reflect HCO_3^- flux in the opposite direction (see Siesjö, 1988a). As extensively discussed (see Chesler, 1990; Chesler and Kaila, 1992), H^+ or HCO_3^- fluxes via membrane carriers or channels commonly accompany changes in physiological activities and could well reflect the operation of a messenger system in which H^+ is an important determinant (see Chapter 17). The pH changes during ischemia could represent the pathological counterparts of such a phenomenon.

Compartmentation of H^+ during ischemia. Compartmentation of H^+ exists at several levels *in vivo*. First, the extracellular fluid (ECF) usually has a more alkaline pH than the "average" pH_i (by about 0.3 unit). Second, compartmentation within cells is obvious, since mitochondrial pH is higher than that of the cytosol, and since lysosomes as well as transmitter-containing granulae and/or vesicles have a lower pH than that of the cytosol (Roos and Boron, 1981; Busa and Nuccitelli, 1984). Third, spontaneous physiological stimulation can trigger dynamic changes of pH_i that differ between neurons and glia and also lead to local changes in pH_i (and pH_e) in the vicinity of channels or receptors (Chesler and Kaila, 1992).

Thus, in the steady state there is a clear compartmentation of H^+ between and within cells. Our present concern is the possibility that cells respond differently to ischemia and that H^+ gradients are maintained in the energy-deprived and grossly acidotic tissue of an ischemic subject. It was proposed by Kraig et al. (1986) that compartmentation of H^+ between glial cells, neurons, and ECF occurs also during ischemia. These authors based their analyses on the fact that preischemic hyperglycemia aggravates damage due to transient ischemia, and set out to correlate tissue P_{CO_2} (P_{tCO_2}), pH_e, and tissue lactate content in animals whose preischemic tissue glucose concentration varied over wide limits. Thus, over the lactate range of 9–40 mmol/kg, pH_e showed two plateau values with ΔpH of about 0.5 and 1.1, respectively, and a transition zone at lactate values of 15–20 mmol/kg. In this zone the P_{tCO_2} increased abruptly but remained constant at higher lactate values. The results hinted at the formation of lactic acid in a compartment whose HCO_3^- content was close to zero (explaining the constancy of P_{CO_2} at high lactate contents). The authors suggested that, at high tissue glucose contents, lactate is preferentially formed in glia cells that suffer damage from the lactic acidosis, leading to pannecrotic lesions.

Although the direct pH_i measurements reported by Kraig and Chesler (1990) suggest that the ischemic tissue of grossly hyperglycemic animals (plasma glucose ~40 mM) contains compartments with pH values of about 5.3, the concept of marked compartmentation of H^+ in ischemic tissue has not received additional support. Boris-Möller et al. (1988) did not detect a splitting of the Pi peak in ^{31}P-NMR studies on ischemic tissue, and Walz and Mukerji (1988, 1990) failed to find evidence that cultured astrocytes, loaded with lactic acid, could trap the acid. On the contrary, they found that lactic acid rapidly equilibrated between glia, neurons, and ECF.

Additional data on this controversy have been collected. Thus, our own group found that the relationship between lactate and either P_{CO_2} or pH_e in ischemic tissue in linear, or close to linear (see Katsura et al., 1991, 1992a,b). Such a relationship, which is illustrated in Fig. 1, does not support the notion of compartmentation of H^+, but suggests that pH_e and pH_i change in parallel when increasing amounts of lactate accumulate in the ischemic tissue.

Focal Ischemia

A crucial question in this field is: What determines the severity of intra- and extracellular acidosis? With complete or near-complete ischemia, where oxygen supply and glucose delivery are arrested, the amount of lactate accumulated depends on the perischemic glucose and glycogen content of the tissue (Ljunggren et al., 1974). On the other hand, in ischemia due to middle cerebral artery (MCA) occlusion, acidosis of comparable magnitude has only been observed in the dense ischemic focus of the lesion (Anderson and Sundt, 1983; Kobatake et al., 1984; Sako et al., 1985). Although there are relatively limited reports existing in the literature, much less severe degrees of acidosis in focal ischemia are usually reported (Hakim and Arrieta, 1986; Meyer et al., 1986, 1990; Nedergaard et al., 1991b). Nakai et al. (1988) illustrated both the flow dependence of the reduction in pH_i and the aggravating effect of hyperglycemia. However, their data showed that, unless cerebral blood flow (CBF) fell below about 20 ml 100 mg^{-1} min^{-1}, the acidosis was relatively moderate.

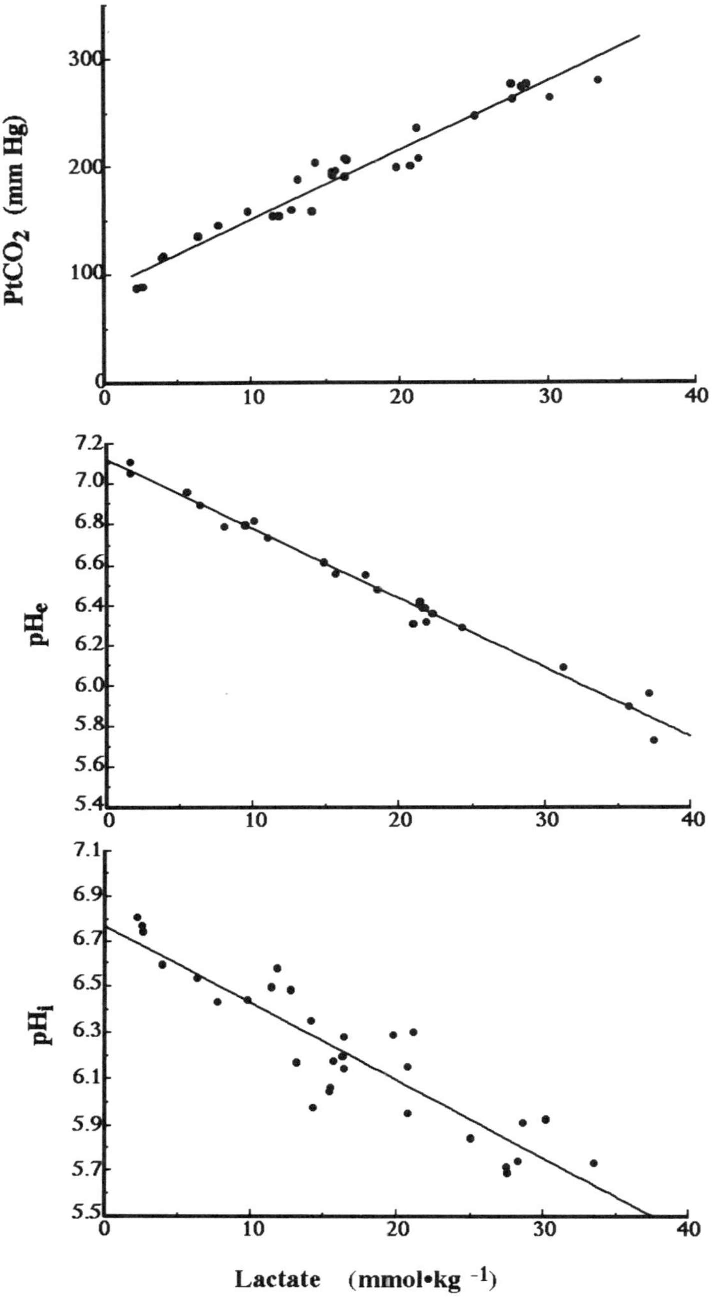

FIGURE 1. Relationships between tissue lactate content and P_tCO_2, pH_e, and pH_i in the rat brain during complete ischemia. Data from Katsura et al., 1992a,b.

There is a controversy in the literature with respect to the aggravating effects of hyperglycemia in focal ischemia (see Siesjö, 1992a,b; Siesjö et al., 1996). Thus in permanent middle cerebral artery occlusion (MCAO), induced hyperglycemia does not consistently affect infarction size. One could argue that this is because pH_i does not fall to critically low values. However, it should be recalled that another important determinant of tissue damage is the cellular ATP concentration. Thus, if ATP contents are only marginally reduced, hyperglycemia may not lead to further accumulation of lactate plus H^+, but if the ATP content is very low, most of glucose delivered will be anaerobically metabolized. In ischemia of intermediate severity hyperglycemia may well increase lactate content, but at the same time allow additional formation of ATP (Folbergrová et al., 1992). In summarizing the divergent results in the literature, the following considerations seem valid: First, if the focal ischemia is of brief duration, hyperglycemia aggravates damage, exactly as it does in transient global ischemia (Nedergaard, 1987; Yip et al., 1991). Second, if the ischemia is severe enough and prolonged enough to yield maximal infarct size, it is not likely that hyperglycemia can further aggravate the size of the lesion.

Hypoxia

Like ischemia, hypoxia reduces oxygen supply to cells; however, the important difference is that CBF is maintained, or only moderately reduced. This means that although oxygen supply is curtailed, glucose delivery is not. Thus, if the arterial Po_2 is reduced to very low values, excessive amounts of lactate can accumulate in the tissue (Salford and Siesjö, 1974; Myers, 1979; Gardiner et al., 1982). Since glucose is readily available, the amount of lactate accumulated is determined by the severity of energy failure, and not by the plasma glucose concentration (Gardiner et al., 1982). This is illustrated in Fig. 2. Caution must be exercised when predicting acidosis from the amount of lactate accumulated, since the latter may not necessary reflect the severity of acidosis. This is because H^+ can be transported independently from lactate in the open system that operates during hypoxia.

Hypoglycemia

Like hypoxia and ischemia, severe hypoglycemia ("coma") leads to brain damage and is, therefore, used as an injury model for exploring mechanisms of cell damage. Hypoglycemia offers a unique opportunity to study such mechanisms, since it leads to bioenergetic failure and cell damage, although oxygen supply and CBF are maintained, and it does so in the absence of cellular acidosis (see Chapter 5).

In summary, during hypoglycemic coma pH_i either remains unchanged, or increases somewhat, while pH_e decreases as pH_i is increased (Pelligrino et al., 1981; Behar et al., 1985; Katsura et al., 1993). During recovery, pH_e decreases further, while pH_i either remains normal or is increased above normal (Behar et al., 1985; Bengtsson et al., 1990; Katsura et al., 1993). The results suggest that hypoglycemia upregulates H^+ extrusion by Na^+/H^+ or Cl^-/HCO_3^- exchange, raising pH_i and reducing pH_e. However, the important feature is the absence of lactic acidosis. This may explain why hypoglycemia does not cause infarction but only selective neuronal necrosis (Auer and Siesjö, 1988; Siesjö, 1988b).

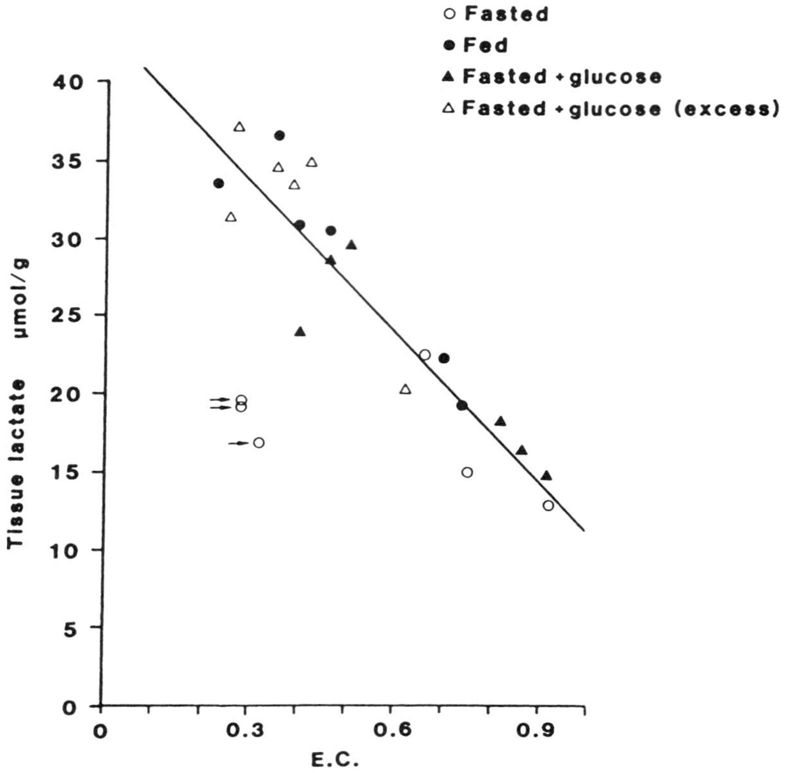

FIGURE 2. Individual values for adenylate energy charge and brain tissue lactate concentrations at the end of hypoxia. The regression line is shown for brain glucose concentrations >0.97 μmol/g ($r = 0.933$, $p < 0.001$). Arrows indicate brain glucose concentration <0.97 μmol/g. Reproduced with permission from Gardiner et al. (1982).

ACIDOSIS-RELATED BRAIN DAMAGE

Acidosis as a Major Factor for Ischemic Damage in Hyperglycemic Subjects

Although the adverse effects of preischemic hyperglycemia in global ischemia are generally accepted, the mechanisms are still not well known. It is also an open question whether or not hyperglycemia exerts its effects by aggravating tissue acidosis.

Compared to normoglycemic subjects, those animals made hyperglycemic prior to induction of transient ischemia show more rapidly maturing lesions, which also affect additional structures (e.g., cingulate cortex, thalamus, and substantial nigra) and often are pannecrotic in nature, yielding infarction (Petito et al., 1982; Pulsinelli et al., 1982; Smith et al., 1988; Inamura et al., 1987). Postischemic seizures in hyperglycemic ischemic animals are common, and the animals often die in status epilepticus (Myers and Yamaguchi, 1977; Siemkowicz and Hansen, 1978; Warner et al., 1987). Aggravation is also observed in transient focal ischemia (Nedergaard, 1987; Yip et al., 1991; L. Gisselson, M.-L. Smith, B.K. Siesjö, to be submitted), and hyperglycemia in animals subjected to permanent MCA occlusion is reportedly accompanied by destruction of all cellular elements, including capillaries, within the infarct (Nedergaard and Diemer, 1987). Thus, it is tempting to conclude that hyperglycemia (and acidosis?) accelerates processes that lead to glial and microvessel destruction and infarction (Plum, 1983; Nedergaard, 1987).

It is only recently that the effects of hyperglycemia and acidosis could be separated. Hurn et al. (1991) used excessive hypercapnia in animals subjected to transient ischemia to study how exaggerated acidosis altered the short-term recovery of metabolic and electrophysiological functions. With this experimental paradigm, hypercapnic animals showed improved short-term recovery. When the same paradigm was used in normoglycemic rats subjected to 10 min of transient ischemia, pH_e could be reduced to the same values as those recorded in normocapnic, hyperglycemic animals (Katsura et al., 1994). The further reduction in pH_e (and presumably, also in pH_i) exaggerated neuronal damage, as assessed after 1 week of recovery (see Fig. 3). Interestingly, though, only some of the animals showed pannecrotic lesions, and none had postischemic seizures.

Additional experiments shed further light on this problem. Thus, when animals subjected to hypoglycemic coma of 30 min duration were rendered hypercapnic during coma periods (Pa_{CO_2} ~200 and ~300 mmHg), pH_e was reduced from the values of about 7.1 to about 6.8 and 6.7, respectively. The hypercapnia, and the accompanying pH reduction, clearly enhanced the tissue damage (Kristián et al., 1995). These results reinforce our previous conclusion (i.e., that a reduction of pH *per se* aggravates the damage incurred by the energy-depleted tissue in ischemia and in hypoglycemic coma). The results also suggest that the effect of acidosis has two characteristics: it enhances selective neuronal vulnerability and it transforms the neuronal lesions to one of pannecrosis.

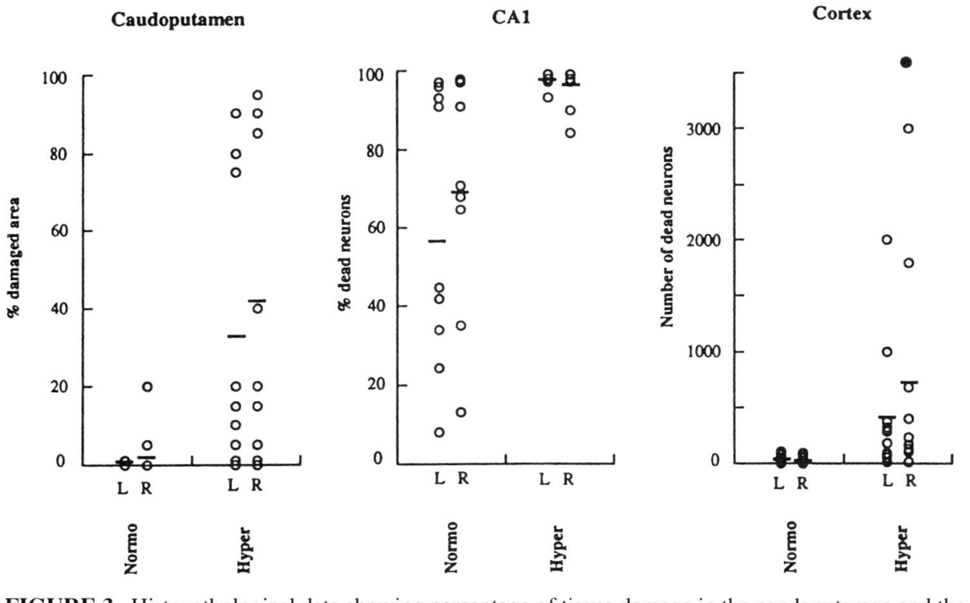

FIGURE 3. Histopathological data showing percentage of tissue damage in the caudoputamen and the percentage of dead neurons in the CA1 pyramidal cell layer, as well as the number of dead neurons in the neocortex. L and R, left and right hemisphere, respectively. Abbreviations: hyper, hypercapnic, normoglycemic ischemia ($n = 11$); normo, normocapnic, normoglycemic ischemia ($n = 9$). Horizontal bars represent the mean values. In the panel on the right, infarction of the cortex is represented by a filled circle. In all areas in the hypercapnic group, damage was significantly worse than in normocapnic animals ($p < 0.01$, Mann-Whitney test). Reproduced with permission from Katsura et al. (1994).

Mechanisms of Damage in Ischemia

Acidosis and free radical formation. Free radicals are formed in many reactions, and these radicals are activated when ischemia is followed by recirculation (Hall and Braughler, 1989; Kontos, 1989; Siesjö et al., 1989; Traystman et al., 1991). Evidence exists that ischemia with recirculation leads to hydroxylation of salicylate to 2,3- and 2,5-dihydroxybenzoic acid, and to the appearance of an electron spin resonance (ESR) signal derived from the spin trap N-*tert*-butyl-alpha-phenyl-nitrone (PBN) (Floyd, 1990; Oliver et al., 1990; Carney and Floyd, 1991; Zini et al., 1992; Christensen et al., 1994). The question arises whether the free radical formation is an epiphenomenon, or if the free radicals formed are mediators of tissue damage. Substantial evidence now exists that free radicals do in fact contribute to tissue damage following transient ischemia. Thus the free radical scavengers PBN and dimethylthiourea (DMTU) ameliorate biochemical and structural tissue lesions (Cao et al., 1988; Patt et al., 1988; Carney and Floyd, 1991; Pahlmark et al., 1993; Clough-Helfman and Phillis, 1991; Cao and Phillis, 1994; Zhao et al., 1994; Folbergrová et al., 1995).

Theoretically, a direct coupling exits between acidosis and enhanced formation of free radicals (Siesjö et al., 1985). This coupling can be exemplified by the following three reactions:

$$Fe^{3+} + HCO_3^- + H_3TF_n \rightleftarrows Fe\,TF_n\,HCO_3^- + 3\,H^+ \tag{12}$$

$$\bullet O_2^- + H_2O_2 \rightleftarrows O_2 + \bullet OH + OH^- \tag{13}$$

$$\bullet O_2^- + H^+ \rightleftarrows H\bullet O_2^- \qquad pK' = 4.7 \tag{14}$$

Equation 1 depicts the equilibrium in which iron is bound to, or released from transferrin (TF). As observed, this reaction involves H^+ and HCO_3^- (Aisen, 1979). Acidosis shifts the equilibrium to the left (i.e., displaces Fe^{3+} from the HCO_3^--TF complex), making free iron available for free radical formation. This reaction is further strengthened by the fact that in ischemia the HCO_3^- bridge in the transferrin complex is attacked by hydrogen ions to yield CO_2 and H_2O. The second reaction shown in Eq. 13, is the classical Haber-Weiss reaction. This reaction is extremely slow, but iron catalysis can increase its rate so that •OH formation becomes appreciable, and pathologically important. The third reaction, shown in Eq. 14, describes the equilibrium between $\bullet O_2^-$ and $H\bullet O_2$, the protonated form of $\bullet O_2^-$. Since the latter is a stronger pro-oxidant than $\bullet O_2^-$, and also more lipid soluble, the acidosis predictably leads to the formation of more injurous free radicals (Gebicki and Bielski, 1981).

The role of acidosis in triggering iron-catalyzed free radical production and lipid peroxidation has been shown *in vitro* (Siesjö et al., 1985). It was speculated that the effect of acidosis was to release iron from its binding to proteins like transferrin, a binding which involves HCO_3^- or CO_3^{2-} as a bridging molecule. Supporting this contention were results showing that acidosis due to a rise in CO_2 had less effect on lipid peroxidation than lactic acidosis at identical homogenate pH, conceivably because the rise in HCO_3^- concentration at high CO_2 tensions "stabilized" iron binding to proteins (Rehncrona et al., 1989). Further support for the importance of acidosis in delocalizing iron was reported by Bralet et al. (1992), who directly demonstrated pH-dependent release of low-molecular-weight iron species, and the pro-oxidant effect of iron-containing ultrafiltrates on brain membranes. It is interesting to note that if transferrin is bound to its cellular receptor, one of its iron can be released at acidic pH values (~6.2) (Sipe and Murphy, 1991). Such values can easily be encountered during ischemia in hyperglycemic animals. Results which show protection by iron depletion or chelation support this contention (Rehncrona et al., 1989).

Acidosis and cell calcium metabolism. Cellular Ca^{2+} homeostasis intimately depends on changes in intra- and extracellular pH (for a recent review, see Kristián and Siesjö, 1997; and Chapter 15–17). Calcium ion is translocated from the cell by an ATP-dependent $Ca^{2+}/2H^+$ exchanger, and similar mechanisms are responsible for sequestration of Ca_i^{2+} by intracellular organelles. Furthermore, calcium cycling across the inner mitochondrial membrane involves $Ca^{2+}/2Na^+$ exchange, the subsequent restoration of the Na^+ gradient requiring Na^+/H^+ antiport activity. Finally, Ca^{2+} and H^+ may compete for common binding sites in proteins, and H^+ can displace Ca^{2+} from the binding sites.

The relationship between H^+ and Ca^{2+} has been thoroughly studied. Results showing that acidosis is accompanied by a rise in $[Ca^{2+}]_i$ were reported by several investigators (Abercrombie and Hart, 1986; Baker and Umback, 1987; Ou-Yang et al., 1994b). These results suggest that acidosis may exaggerate the rise in $[Ca^{2+}]_i$ during ischemia, and/or delay normalization of $[Ca^{2+}]_i$ during recirculation. However, recent data on the relationship between acidosis and Ca^{2+} currents through NMDA-gated channels add complexity to this field. Thus, patch-clamp studies on neurons showed that these currents were progressively reduced when pH was decreased from values above 7.0 towards 6.0 (Morad et al., 1988; Tang et al., 1989, 1990; Traynelis and Cull-Candy, 1990; Vyklicky et al., 1990). Other results showed that acidosis reduced the influx of $^{45}Ca^{2+}$ during glutamate exposure, and that a reduction of pH in the incubation fluid ameliorated cell death due to both glutamate exposure and anoxia (Giffard et al., 1990; Tombaugh and Sapolsky, 1990). The reduction of Ca^{2+} influx is so dominating that, *in vitro,* glutamate exposure is associated with a smaller rise in $[Ca^{2+}]$ under conditions of acidosis, than in normal conditions (Ou-Yang et al., 1994a).

At first sight, these results seem contrary to those demonstrating aggravation of cell damage by acidosis in ischemia. The possibility exists that conditions *in vitro* do not mimick those prevailing *in vivo*. For example, the Ca^{2+} accumulated during glutamate exposure *in vitro* may be excessive because of the unlimited source of extracellular Ca^{2+} (Kristián et al., 1996). If this massive accumulation of Ca^{2+} is reduced by mild acidosis, reducing flux through Ca^{2+} channels, the protective effects of acidosis is easily explained. This situation seems different *in vivo,* since the extracellular Ca^{2+} content is limited during ischemia and the total Ca^{2+} content of tissues is not altered over a period of 1–2 days after ischemia (see Kristián et al., 1996). Thus, the recent arguments by Tombaugh and Sapolsky (1993) should be interpreted with some caution since the contention may be valid only under *in vitro* conditions, but not in global ischemia.

The rate of Ca^{2+} influx during ischemia becomes slower in hyperglycemic animals than in normoglycemic ones. In the former, the influx occurs in two phases, leading to a very gradual uptake of Ca^{2+} by cells (Kristián et al., 1994). Since a similarly slow influx occurs in hypercapnic animals (in which the intraischemic acidotic levels are similar to those observed in hyperglycemic animals), the effect of hyperglycemia was interpreted to be due to the fall in pH. Furthermore, since MK-801 (a noncompetitive antagonist of NMDA receptors) had similar effects, the reduction in pH must have reduced Ca^{2+} influx through NMDA-gated channels. Since hyperglycemic animals also showed a more rapid repolarization upon recirculation (Ekholm et al., 1995), the net duration of complete depolarization (i.e., the total duration of Ca^{2+} load) is shorter in hyperglycemic than in normoglycemic animals. Clearly, the aggravating effect of hyperglycemia on ischemic brain damage cannot be explained by an effect on the duration of the Ca^{2+} transient that accompanies the ischemia, nor does it seem to involve a rise in $[Ca^{2+}]_i$ of other origin.

It has been generally accepted that apoptosis, which may represent a common mechanism in cell death in various pathological situations, involves DNA breakage by endogenous endonucleuses (Orrenius et al., 1988; Barry and Eastman, 1993; Barry et al., 1993). Theoretically, lethal DNA damage precedes breakdown of plasma membrane function by many hours, which explains why cells that are destined to die look morphologically intact for some time after the insult. Although morphological studies suggest that apoptosis is not a likely mechanism in the delayed cell death after ischemia (Desphande et al., 1992), chemical analysis has shown typical laddered DNA fragmentation in vulnerable neurons in the forebrain and in focal ischemia (MacManus et al., 1993, 1994). A possible link between acidosis and apoptotic cell death was recently proposed. Results obtained on a non-neuronal system suggested the involvement of an endonuclease, DNase II, which is activated at pH values between 6.5 (Barry and Eastman 1992, 1993). These results raise the possibility that the severity of acidosis encountered during ischemia in hyperglycemic and hypercapnic animals activates endonucleases and triggers apoptotic cell death. However, there is no direct evidence at this moment showing a direct link between exaggerated acidosis and apoptosis.

Targets of Acidosis-Mediated Tissue Damage

Neurons. Selective neuronal necrosis occurs after transient global ischemia (e.g., in the CA1 region of hippocampus). If acidosis is exaggerated during ischemia by preischemic hyperglycemia or superimposed hypercapnia, new areas are recruited, such as the cingulate cortex, the CA3 sector of hippocampus, and the thalamus. This damage seems confined to the neuronal population (Katsura et al., 1994; Kristián et al., 1995; Li et al., 1995a,b). The possibility remains that an additional reduction of pH enhances reactions leading to neuronal cell death in ischemia. We then have to assume that certain neurons which are recruited in the damage in hyperglycemic or hypercapnic animals are particularly sensitive to a relatively small additional reduction of pH. Thus, a fall in pH may be a common pathogenetic factor in ischemia. However, since there is a difference between hyperglycemic and hypercapnic animals in that hyperglycemic but not hypercapnic animals develop seizures, factors other than pH should be considered.

Glial cells. The evidence that glial cells may be involved in the acidosis-related damage is circumstantial. Thus, Plum (1983) suggested that excessive acidosis "selectively" damages glial cells, and that this is what leads to infarction (see Chapter 29). However, it is not known if glial cell damage is a primary, or secondary to neuronal or microvascular damage.

In vitro studies have shown that astrocytes are strikingly resistant to both acidosis and to glutamate toxicity (Goldman et al., 1989; Nedergaard et al., 1991; Norenberg et al., 1987; Tombaugh and Sapolsky, 1990). Therefore the question has to be asked: What are the mechanisms which lead to pannecrotic lesions and to glial cell damage? Is glial cell damage secondary to dense neuronal necrosis, reflecting the adverse effects of the "toxins" released when neurons die and disintegrate? Are glial cells affected whenever microvessels are damaged?

The microvessels. Recent studies show that microvessel dysfunction may contribute to the postischemic microcirculatory disturbance that leads to pannecrotic infarction (del Zoppo, 1994; Feuerstein et al., 1994; Garcia et al., 1994). Activation of polymorphonuclear leukocytes (PMN), activation of T cells, and capillary "plugging," were suggested to be the causes of secondary damage leading to infarction, which starts some time after ischemic insults. Thus, del Zoppo et al. (1991) found 40% of capillaries obstructed by

PMNs in baboon focal ischemia with recirculation. Furthermore, antibodies against adhesion molecules, which enhance attachment between PMNs and endothelial cells (i.e., ICAM-1, CD11b, CD18, integrins, etc.), are reported to be significantly protective even though administered postischemically (Chen et al., 1994; Chopp et al., 1994; Matsuo et al., 1994). These events involve reactions observed in inflammation, which usually involves free radical formation. As discussed, this may be enhanced by acidosis (see p 572). Endothelial cells are closest to the source of acidosis because the intraluminal (intravascular) glucose is readily converted to lactate plus H^+ during ischemia. Thus, exaggerated acidosis may be one of the factors leading to functional changes of endothelial cells. In support of this view is a report showing swollen endothelial cells with nuclear chromatin changes in hyperglycemic ischemia (Paljärvi et al., 1983). However, additional information is urgently needed on this issue.

SUMMARY

It is still unexplained why preischemic hyperglycemia so dramatically aggravates damage due to dense global ischemia and why it increases infarct size in focal ischemia, particularly if followed by reperfusion. It is generally assumed that hyperglycemia worsens the damage by exaggerating intraischemic acidosis, but the aggravating effect of acidosis in the absence of hyperglycemia has only recently been established.

There is a major discrepancy between *in vivo* and *in vitro* results. The latter demonstrate that moderate acidosis ameliorates neuronal damage due to glutamate exposure or anoxia, probably because a reduction of pH inhibits Ca^{2+} influx through agonist-operated and voltage-sensitive channels, preventing excessive accumulation of Ca^{2+} during the insults. This suggests that the experiments on neuronal culture or slices *in vitro* do not reflect conditions existing under ischemic conditions *in vivo* (see p 573).

It is very likely that acidosis *in vivo* requires other factors to trigger exaggerated brain damage. Such factors encompass severe energy failure, loss of ion homeostasis with cell Ca^{2+} overload, changes in oxidation state, enhanced lipolysis, and production of free radicals.

ACKNOWLEDGMENTS

This work was supported by the Swedish Medical Research Council, the National Institute of Neurological Disorders and Stroke via the National Institutes of Health of the United States Public Health Service, and the Medical Faculty of Lund University.

REFERENCES

Abercrombie RF, Hart CE (1986): Calcium and proton buffering and diffusion in isolated cytoplasm from Myxicola axons. Am J Physiol 250: C391–C405.

Aisen P (1979): Some physicochemical aspects of iron metabolism. In: CIBA Foundation Symposium, No. 51: Iron Metabolism. Amsterdam: Elsevier, pp 1–17.

Alberti KGMM, Cuthbert C (1982): The Hydrogen Ion in Normal Metabolism: A Review. In Porter R and Lawrensen G Eds, Metabolic Acidosis, CIBA Foundation Symposium No. 87. London: Pitman Books, pp 1–19.

Anderson R, Sundt T (1983): Brian pH in focal cerebral ischemia and the protective effects of barbiturate anesthesia. J Cereb Blood Flow Metab 3: 493–497.

Auer RN, Siesjö BK (1988): Biological differences between ischemia, hypoglycemia and epilepsy. Ann Neurol 24: 699–707.

Baker PF, Umbach JA (1987): Calcium buffering in axons and axoplasm of loligo. J Physiol (Lond) 383: 369–394.

Balestrino M, Somjen GG (1988): Concentration of carbon dioxide, interstitial pH and synaptic transmission in hippocampal formation of the rat. J Physiol (Lond) 396: 247–266.

Barry MA, Eastman A (1992): Endonuclease activation during apoptosis: The role of cytosolic Ca^{2+} and pH. Biochem Biophys Res Commun 186: 782–789.

Barry MA, Eastman A (1993): Identification of deoxyribonuclease II as an endonuclease involved in apoptosis. Arch Biochem Biophys 300: 440–450.

Barry MA, Reynolds JE, Eastman A (1993): Etoposide-induced apoptosis in human HL-60 cells is associated with intracellular acidification. Cancer Res 53: 2349–2357.

Behar KL, den Hollander JA, Petroff OAC, Hetherington HP, Prichard JW, Shulman RG (1985): Effect of hypoglycemic encephalopathy upon amino acids, high-energy phosphates, and pH_i in the rat brain in vivo: Detection by sequential ^{1}H and ^{31}P NMR spectroscopy. J Neurochem 44: 1045–1055.

Behar KL, Rothman DL, Hossmann K-A (1989): NMR Spectroscopic investigation of the recovery of energy and acid-base homeostasis in the cat brain after prolonged ischemia. J Cereb Blood Flow Metab 9: 655–665.

Bengtsson F, Boris-Möller F, Hansen AJ, Siesjö BK (1990): Extracellular pH in the rat brain during hypoglycemic coma and recovery. J Cereb Blood Flow Metab 10: 262–269.

Boris-Möller F, Drakenberg T, Elmdén K, Forsén S, Siesjö BK (1988): Evidence against major compartmentalization of H^+ in ischemic rat brain tissue. Neurosci Lett 85: 113–118.

Bralet J, Schreiber L, Bouvier C (1992): Effect of acidosis and anoxia on iron delocalization from brain homogenates. Biochem Pharmacol 43: 979–983.

Busa WB, Nuccitelli R (1984): Metabolic regulation via intracellular pH. Am J Physiol 246: R409–438.

Cao W, Carney JM, Duchon A, Floyd RA, Chevion M (1988): Oxygen free radical involvement in ischemia and reperfusion injury to brain. Neurosci Lett. 88: 233–8.

Cao X, Phillis J (1994): Phenyl-t-butyl-nitrone reduces cortical infarct and edema in rats subjected to focal ischemia. Brain Res 644: 267–272.

Carney JM, Floyd RA (1991): Protection against oxidative damage to CNS by α-Phenyl-*tert*-Butyl Nitrone (PBN) and other spin-trapping agents: A novel series of nonlipid free radical scavengers. J Mol Neurosci 3: 47–57.

Chapman AG, Westerberg EV, Siesjö BK (1981): The metabolism of purine and pyrimidine nucleotides in rat cortex during insulin-induced hypoglycemia and recovery. J Neurochem 36: 179–189.

Chen H, Chopp M, Zhang RL, Bodzin G, Chen Q, Rusche JR, Todd III RF (1994): Anti-CD11b monoclonal antibody reduces ischemic cell damage after transient focal cerebral ischemia in rat. Ann Neurol 35: 458–463.

Chesler M (1990): The regulation and modulation of pH in the nervous system. Prog Neurobiol 34: 401–427.

Chesler M, Kaila K (1992): Modulation of pH by neuronal activity. Trends Neurosci 15: 396–402.

Chopp M, Frinak S, Dean BS, Walton BS, Smith MB, Welch MD (1987): Intracellular acidosis during and after cerebral ischemia: In vivo nuclear magnetic resonance study of hyperglycemia in cats. Stroke 18: 919–923.

Chopp M, Welch K, Tidwell C, Helpern J (1988): Global cerebral ischemia and intracellular pH during hyperglycemia and hypoglycemia in cats. Stroke 19: 1383–1387.

Chopp M, Zhang RL, Chen H, Li Y, Jiang N, Rusche R (1994): Postischemic administration of an Anti-Mac-1 antibody reduces ischemic cell damage after transient middle cerebral artery occlusion in rats. Stroke 25: 869–876.

Christensen T, Bruhn T, Balchen T, Diemer NH (1994): Evidence for formation of hydroxyl radicals during reperfusion after global cerebral ischemia in rats using salicylate trapping and microdialysis. Neurobiol Dis 1: 131–138.

Church J, McLennan H (1989): Electrophysiological properties of rat CA1 pyramidal neurones in vitro modified by changes in extracellular bicarbonate. J Physiol (Lond) 415: 85–108.

Clough-Helfman C, Phillis JW (1991): The free radial trapping agent N-tert-butyl-α-phenylnitrone (PBN) attenuates cerebral ischemic injury in gerbils. Free Rad Res Commun 15: 177–186.

del Zoppo GJ (1994): Microvascular changes during cerebral ischemia and reperfusion. Cerebrovasc Brain Metab Rev 6: 47–96.

del Zoppo GJ, Schmid-Schönbein GW, Mori E, Copeland BR, Chang CM (1991): Polymorphonuclear leukocytes occlude capillaries following middle cerebral artery occlusion and perfusion in baboons. Stroke 22: 1276–1283.

Desphande J, Bergstedt K, Lindén T, Kalimo H, Wieloch T (1992): Ultrastructural changes in the hippocampal CA1 region following transient cerebral ischemia: Evidence against programmed cell death. Exp Brain Res 88: 91–105.

Ekholm A, Kristián T, Siesjö BK (1995): Influence of hyperglycemia and of hypercapnia on cellular calcium transients during reversible brain ischemia. Exp Brain Res (submitted).

Ereciñska M, Silver I (1989): ATP and brain function. J Cereb Blood Flow Metab 9: 2–19.

Feise G, Kogure K, Busto R, Scheinberg P, Renmuth OM (1976): Effect of insulin hypoglycemia upon cerebral energy metabolism and EEG activity in the rat. Brain Res 126: 263–280.

Feuerstein GZ, Liu T, Barone FC (1994): Cytokines, inflammation, and brain injury: Role of tumor necrosis factor-α. Cerebrovasc Brain Metab Rev 6: 341–360.

Floyd R (1990): Role of oxygen free radicals in carcinogenesis and brain ischemia. FASEB J 4: 2587–2597.

Folbergrová J, Memezawa H, Smith M-L, Siesjö BK (1992): Focal and perifocal changes in tissue energy state during middle cerebral artery occlusion in normo- and hyperglycemic rats. J Cereb Blood Flow Metab 12: 25–33.

Folbergrová J, Zhao Q, Katsura K, Siesjö BK (1995): N-tert-Butyl-α-phenylnitrone improves recovery of brain energy state in rats following transient focal ischemia. Proc Natl Acad Sci USA 92 (in press).

Gardiner M, Smith M-L, Kågström E, Shohami E, and Siesjö BK (1982): Influence of blood glucose concentration on brain lactate accumulation during severe hypoxia and subsequent recovery of brain energy metabolism. J Cereb Blood Flow Metab 2: 429–438.

Garcia JH, Liu K-F, Yoshida Y, Chen S, Lian J (1994): Brain microvessels: Factors altering their patency after the occlusion of a middle cerebral artery (wistar rat). Am J Pathol 145: 728–740.

Gebicki JM, Bielski BHJ (1981): Comparison of the capacities of the perhydroxyl and the superoxide radicals to initiate chain oxidation of linoleic acid. J Am Chem Soc 103: 7020–7022.

Giffard RG, Monyer H, Christine CW, Choi DW (1990): Acidosis reduces NMDA receptor activation, glutamate neurotoxicity, and oxygen-glucose deprivation neuronal injury in cortical cultures. Brain Res 506: 339–342.

Goldman SA, Pulsinelli WA, Clarke WY, Kraig RP, Plum F (1989): The effects of extracellular acidosis on neurons and glia in vitro. J Cereb Blood Flow Metab 9: 471–477.

Hakim A, Arrieta M (1986): Cerebral acidosis in focal ischemia. I. A method for the simultaneous measurement of local cerebral pH with cerebral glucose utilization or cerebral blood flow in the rat. J Cereb Blood Flow Metab 6: 667–675.

Hall E, Braughler J (1989): Central nervous system trauma and stroke. II. Physiological and pharmacological evidence for involvement of oxygen radicals and lipid peroxidation. Free Radic Biol Med 6: 303–313.

Hochachka PW, Mommsen TP (1983): Protons and anaerobiosis. Science 219: 1391–1397.

Hurn PD, Koehler RC, Norris SE, Schwentker AE, Traystman RJ (1991): Bicarbonate conservation during incomplete cerebral ischemia with superimposed hypercapnia. Am J Physiol 261: H853–H859.

Iijima T, Ciani S, Hagiwara S (1986): Effects of external pH on Ca channels: Experimental studies and theoretical considerations using a two-site, two-ion model. Proc Natl Acad Sci USA 83: 654–658.

Inamura K, Olsson Y, Siesjö BK (1987):Substantia nigra damage induced by ischemia in hyperglycemic rats. A light and electron microscopic study. Acta Neuropathol (Berl) 75: 131–139.

Jarolimek W, Misgeld U, and Lux HD (1989): Activity dependent alkaline and acid transients in guinea pig hippocampal slices. Brain Res 505: 225–232.

Kaila K, Voipio J (1987): Postsynaptic fall in intracellular pH induced by GABA-activated bicarbonate conductance. Nature 330: 163–165.

Katsura K, Ekholm A, Asplund B, Siesjö BK (1991): Extracellular pH in the brain during ischemia: Relationship to the severity of lactic acidosis. J Cereb Blood Flow Metab 11: 597–599.

Katsura K, Asplund B, Ekholm A, Siesjö BK (1992a): Extra- and intracellular pH in the brain during ischemia, related to tissue lactate content in normo- and hypercapnic rats. Euro J Neurosci 4: 166–176.

Katsura K, Ekholm A, Siesjö BK (1992b): Tissue P_{CO_2} in brain ischemia related to lactate content in normo- and hypercapnic rats. J Cereb Blood Flow Metab 12: 270–80.

Katsura K, Bengrsson F, Ekholm A, Kristián T, Gidö G, Siesjö BK (1993): Recovery of mitochondrial and plasma membrane function following hypoglycemic coma: ATP synthesis, K^+ accumulation, and changes in extra- and intracellular pH. J Cereb Blood Flow Metab 13: 820–826.

Katsura K, Kristián T, Smith M-L, Siesjö BK (1994): Acidosis induced by hypercapnia exaggerates ischemic brain damage. J Cereb Blood Flow Metab 14: 243–250.

Kleihues P, Kobayashi K, Hossmann K-A (1974): Purine nucleotide metabolism in the cat brain after one hour of complete ischemia. J Neurochem 23: 417–425.

Kobatake K, Sako K, Izawa M, Yamamoto Y, Hakim A (1984): Autoradiographic determination of brain pH following middle cerebral artery occlusion in the rat. Stroke 15: 540–547.

Kontos HA (1989): Oxygen radicals in cerebral ischemia. In M.D. Ginsberg, W.D. Dietrich (eds): Cerebrovascular Diseases, Sixteenth Research (Prinston) Conference. New York, Raven Press, pp 365–371.

Kraig RP, Chesler M (1990): Astrocytic acidosis in hyperglycemic and complete ischemia. J Cereb Blood Flow Metab 10: 104–114.

Kraig RP, Ferreira-Filho CR, Nicholson C (1983): Alkaline and acid transients in the cerebellar microenvironment. J Neurophysiol 49: 831–850.

Kraig RP, Pulsinelli WA, Plum F (1986): Carbonic acid buffer changes during complete brain ischemia. Am J Physiol 250: R348–R357.

Krebs H, Woods H, Alberti K (1975): Hyperlactataemia and lactic acidosis. Essays Med Biochem 1: 81–103.

Kristián T, Katsura K, Gidö G, Siesjö BK (1994): The influence of pH on cellular calcium influx during ischemia. Brain Res 641: 295–302.

Kristián T, Gidö G, Siesjö BK (1995): The influence of acidosis on hypoglycemic brain damage. J Cereb Blood Flow Metab 15: 78–87.

Kristián T, Siesjö B (1997): Changes in ionic fluxes during cerebral ischemia in Cross A, Green A (eds): Neuroprotective agents and Cerebral Ischemia. Academic Press, New York, pp 27–45.

Kristián T, Ouyang Y, Siesjö BK (1996): Calcium-induced neuronal cell death in vivo and in vitro: are the pathophysiologic mechanisms different?. Advances in Neurology. 71: 107–113.

LaManna JC, McCracken KA (1984): The use of neutral red as an intracellular pH indicator in rat brain cortex in vivo. Anal Biochem 142: 117–125.

Lawson JW, Veech RL (1979): Effects of pH and free Mg^{2+} on the Keq of the creatine kinase reaction and other phosphate hydrolyses and phosphate transfer reactions. J Biol Chem 254: 6528–6537.

Lewis LD, Ljunggren B, Ratcheson RA, Siesjö BK (1974): Cerebral energy state in insulin-induced hypoglycemia, related to blood glucose and to EEG. J Neurochem 23: 673–679.

Ljunggren B, Ratcheson RA, Siesjö BK (1974): Cerebral metabolic state following complete compression ischemia. Brain Res 73: 291–307.

MacManus JP, Buchan AM, Hill IE, Rasquinha I, Preston E (1993): Global ischemia can cause DNA fragmentation indicative of apoptosis in rat brain. Neurosci Lett 164: 89–92.

Matsuo Y, Onodera H, Shiga Y, Shozuhara H, Ninomiya M, Kihara T, Tamatani T, Miyasaka M, Kogure K (1994): Role of cell adhesion molecules in brain injury after transient middle cerebral artery occlusion in the rat. Brain Res 656: 344–352.

Meyer FB, Anderson RE, Sundt TM Jr, Yaksh TL (1986): Intracellular brain pH, indicator tissue perfusion, electroencephalography, and histology in severe and moderate focal cortical ischemia in the rabbit. J Cereb Blood Flow Metab 6: 71–8.

Meyer FB, Anderson RE, Sundt TM Jr (1990): The novel dihydronaphthyridine Ca^{2+} channel blocker CI-951 improves CBF, brain pH_i, and EEG recovery in focal cerebral ischemia. J Cereb Blood Flow Metab 10: 97–103.

Moody W (1984): Effects of intracellular H^+ on the electrical properties of excitable cells. Annu Rev Neurosci 7: 257–278.

Morad M, Dichter M, Tang CM (1988): The NMDA activated current in hippocampal neurons is highly sensitive to $[H^+]_o$. Soc Neurosci Abstr 14: 791.

Mutch WA, Hansen AJ (1984): Extracellular pH changes during spreading depression and cerebral ischemia: Mechanisms of brain pH regulation. J Cereb Blood Flow Metab 4: 17–27.

Myers RE (1979): Lactic acid accumulation as a cause of brain edema and cerebral necrosis resulting from oxygen deprivation. In R.K. and G. Guilleminault (Eds). Advances in Perinatal Neurology. New York: Spectrum, pp 85–114.

Myers R, Yamaguchi S (1977): Nervous system effects of cardiac arrest in monkeys. Arch Neurol 34: 65–74.

Nakai H, Yamamoto Y, Diksic M, Worsley K, Takara E (1988): Triple-tracer autoradiography demonstrates effects of hyperglycemia on cerebral blood flow, pH, and glucose utilization in cerebral ischemia of rats. Stroke 19: 764–772.

Nedergaard M (1987): Transient focal ischemia in hyperglycemic rats is associated with increased cerebral infarction. Brain Res 408: 79–85.

Nedergaard M, Diemer N (1987): Focal ischemia of the rat brain, with special reference to the influence of plasma glucose concentration. Acta Neuropathol Berl 73: 131–137.

Nedergaard M, Goldman SA (1993): Carrier-mediated transport of lactic acid in cultured neurons and astrocytes. Am J Physiol 265: R282–R289.

Nedergaard M, Goldman S, Desai S, Pulsinelli W (1991a): Acid-induced death in neurons and glia. J Neurosci 11: 2489–2497.

Nedergaard M, Kraig R, Tanabe J, Pulsinelli W (1991b): Dynamics of interstitial and intracellular pH in evolving brain infarct. Am J Physiol 260: R581–R588.

Nordström C-H, Rehncrona S, Siesjö BK (1978): Restitution of cerebral energy state, as well as of glycolytic metabolites, citric acid cycle intermediates and associated amino acids after 30 min of complete ischemia in rats anaesthetized with nitrous oxide or phenobarbital. J Neurochem 30: 479–486.

Norenberg MD, Mozes LW, Gregorios JB, Norenberg L-LB (1987): Effects of lactic acid on astrocytes in primary culture. J Neuropathol Exp Neurol 46: 154–166.

Oliver CN, Starke-Reed PE, Stadtman ER, Liu GJ, Carney JM, Floyd RA (1990): Oxidative damage to brain proteins, loss of glutamine synthetase activity, and production of free radicals during ischemia/reperfusion-induced injury to gerbil brain. Proc Natl Acad Sci USA 87: 5144–5147.

Orrenius S, McConkey D, Jones D, Nicotera P (1988): Ca^{2+}-activated mechanisms in toxicity and programmed cell death. ISI Atlas Sci: Pharmacology 2: 319–324.

Ou-Yang Y, Kristián T, Mellergård P, Siesjö BK (1994a): The influence of pH on glutamate- and depolarization-induced increase of intracellular calcium concentration in cortical neurons in primary culture. Brain Res 646: 65–72.

Ou-Yang Y-B, Mellergård P, Kristián T, Kristiánová V, Siesjö BK (1994b): The influence of acid-base changes on intracellular calcium concentration of neurons in primary culture. Exp Brain Res 101: 265–271.

Pahlmark K, Folbergrová J, Smith M-L, Siesjö BK (1993): Effects of dimethylthiourea on selective neuronal vulnerability in forebrain ischemia in rats. Stroke 24: 731–737.

Paljärvi L, Rehncrona S, Söderfeldt B, Olsson Y, Kalimo H (1983): Brain lactic acidosis and ischemic cell damage: Quantitative ultrastructural changes in capillaries of rat cerebral cortex. Acta Neuropathol (Berlin) 60: 232–240.

Patt A, Harken AH, Burton LK, Rodell TC, Piermattei D, Schorr WJ, Parker NB, Berger EM, Horesh IR, Terada LS, Linas SL, Cheronis JC, Repine JE (1988): Xanthine oxidase-derived hydrogen peroxide contributes to ischemia reperfusion-induced edema in gerbil brains. J Clin Invest 81: 1556–1562.

Paschen W, Djuricic B, Mies G, Schmidt KR, Linn F (1987): Lactate and pH in the brain: Association and dissociation in different pathophysiological states. J Neurochem 48: 154–159.

Pasternack M, Voipio J, Kaila K (1993): Intracellular carbonic anhydrase activity and its role in GABA-induced acidosis in isolated rat hippocampal pyramidal neurones. Acta Physiol Scand 148: 229–231.

Pelligrino D, Almqvist L-O, Siesjö BK (1981): Effects of insulin-induced hypoglycemia on intracellular pH and impedance in the cerebral cortex of the rat. Brain Res 221: 129–147.

Petito CK, Pulsinelli WA, Jacobson G, Plum F (1982): Edema and vascular permeability in ischemia: Comparison between ischemic neuronal damage and infarction. J Neuropathol Exp Neurol 41: 423–436.

Plum F (1983): What causes infarction in ischemic brain? The Robert Wartenburg lecture. Neurology 33: 222–233.

Pulsinelli WA, Waldman S, Rawlinson D, Plum F (1982): Moderate hyperglycemia augments ischemic brain damage: A neuropathological study in the rat. Neurology 32: 1239–1246.

Rehncrona S, Hauge HN, Siesjö BK (1989): Enhancement of iron-catalyzed free radical formation by acidosis in brain homogenates: Difference in effect by lactic acid and CO_2. J Cereb Blood Flow Metab 9: 65–70.

Roberts JKM, Callis J, Jardetzky O, Walbot V, Freeling M (1984): Cytoplasmic acidosis as a determinant of flooding intolerance. Proc Natl Acad Sci USA 81: 6029–6033.

Roos A, Boron WF (1981): Intracellular pH. Physiol Rev 61: 296–434.

Sako K, Kobatake K, Yamamoto L, Diksic M (1985): Correlation of local cerebral blood flow, glucose utilization, and tissue pH following a middle cerebral artery occlusion in the rat. Stroke 16: 828–834.

Salford L, Siesjö B (1974): The influence of arterial hypoxia and unilateral carotid artery occlusion upon regional blood flow and metabolism in the rat brain. Acta Physiol Scand 92: 130–141.

Siemkowicz E, Hansen A (1978): Clinical restitution following cerebral ischemia in hypo-, normo-, and hyperglycemic rats. Acta Neurol Scand 58: 1–8.

Siesjö BK (1988a): Acidosis and ischemic brain damage. Neurochem Pathol 9: 31–88.

Siesjö BK (1988b): Hypoglycemia, brain metabolism, and brain damage. Diabetes Metab Rev 4:113–144.

Siesjö BK (1992a): Pathophysiology and treatment of focal cerebral ischemia. I. Pathophysiology. J Neurosurg 77: 169–184.

Siesjö BK (1992b): Pathophysiology and treatment of focal cerebral ischemia. II. Mechanisms of damage and treatment. J Neurosurg 77: 337–354.

Siesjö BK, Bendek G, Koide T, Westerberg E, Wieloch T (1985): Influence of acidosis on lipid peroxidation in brain tissues in vitro. J Cereb Blood Flow Metab 5: 253–258.

Siesjö BK, Agardh C-D, Bengtsson F (1989): Free radicals and brain damage. Cerebrovasc Brain Metab Rev 1: 165–211.

Siesjö BK, Katsura K, Mellergård P, Ekholm A, Lundgren J, Smith M-L (1993): Acidosis-Related Brain Damage, In Neurobiology of Ischemic Brain Damage, Kogure K, Hossmann K-A, Siesjö BK, eds, Progress in Brain Research. vol 96: pp 23–48.

Siesjö BK, Katsura K, Kristián T (1996): Acidosis-related damage. Advances in Neurology. 71: 209–236.

Sipe DM, Murphy RF (1991): Binding to cellular receptors results in increased iron release from transferrin at mildly acidic pH. J Biol Chem 266: 8002–8007.

Smith M-L, von Hanwehr R, Siesjö BK (1986): Changes in extra- and intracellular pH in the brain during and following ischemia in hyperglycemic and in moderately hypoglycemic rats. J Cereb Blood Flow Metab 6: 574–583.

Smith M-L, Kalimo H, Warner DS, Siesjö BK (1988): Morphological lesions in the brain preceding the development of postischemic seizures. Acta Neuropathol (Berl) 76: 253–264.

Tang CM, Dichter M, Morad M (1989): Mechanism of NMDA channel modulation by H^+ at near physiologic pH. Soc Neurosci Abstr 15: 325.

Tang CM, Dichter M, Morad M (1990): Modulation of the N-methyl-D-aspartate channel by extracellular H^+. Proc Natl Acad Sci 87: 6445–6449.

Thomas RC (1984): Experimental displacement of intracellular pH and the mechanisms of its subsequent recovery. J Physiol (Lond) 354: 3p-22p.

Tombaugh GC, Sapolsky RM (1990): Mild acidosis protects hippocampal neurons from injury induced by oxygen and glucose deprivation. Brain Res 506: 343–345.

Traynelis S, Cull-Candy S (1990): Proton inhibition of N-methyl-D-aspartate receptors in cerebellar neurons. Nature 345: 347–350.

Traystman R, Kirsch J, Koehler R (1991): Oxygen radical mechanisms of brain injury following ischemia and reperfusion. J Applied Physiol 71: 1185–1195.

Vink R, Faden AI, Mcintosh TK (1988): Changes in cellular bioenergetic state following graded traumatic brain injury in rats: Determination by phosphorus 31 magnetic resonance spectroscopy. J Neurotrauma 5: 315–330.

Vyklicky L, Vlachova V, Krusek J (1990): The effect of external pH changes on responses to excitatory amino acids in mouse hippocampal neurones. J Physiol (Lond) 430: 497–517.

Walz W, Mukerji S (1988): Lactate production and release in cultured astrocytes. Neurosci Lett 86: 296–300.

Walz W, Mukerji S (1990): Simulation of aspects of ischemia in cell culture: Changes in lactate compartmentation. Glia 3: 522–528.

Warner DS, Smith M-L, Siesjö BK (1987): Ischemia in normo- and hyperglycemic rats: Effects on brain water and electrolytes. Stroke 18: 464–471.

Yip P, He Y, Hsu C, Garg N, Marangos P, Hogan E (1991): Effect of plasma glucose on infarct size in focal cerebral ischemia-reperfusion. Neurology 41: 899–905.

Zhao Q, Pahlmark K, Smith M-L, Siesjö BK (1994): Delayed treatment with the spin trap a-phenyl-N-tert-butyl nitrone (PBN) reduces infarct size following transient middle cerebral artery occlusion in rats. Acta Physiol Scand 152: 349–350.

Zini I, Tomasi A, Grimaldi R, Vannini V, Agnati LF (1992): Detection of free radicals during brain ischemia and reperfusion by spin trapping and microdialysis. Neurosci Lett 138: 279–282.

CHAPTER 29

Astroglial pH During and After Global Ischemia

CHRISTOPHER D. LASCOLA AND RICHARD P. KRAIG
Committee on Neurobiology, The University of Chicago, Chicago, IL (C.D.L.); Departments of Neurology and Pharmacological and Physiological Sciences, Committee on Neurobiology, The University of Chicago, Chicago, IL (R.P.K.)

INTRODUCTION

Direct measurements of astroglial acid-base shifts during and following physiological depolarization strongly indicate astroglial pH is actively regulated and not simply a reflection of interstitial pH (Chapter 13). During cortical stimulation and spreading depression (SD), for example, depolarization-dependent astroglial alkalosis appears to arise in part from the active uptake of interstitial HCO_3^- and/or CO_3^{2-}. Measurements of astroglial pH also suggest astrocytes maintain the capacity to undergo separate acid-base shifts during injurious perturbations such as global ischemia. Depending on the form and extent of global ischemia, astroglial pH changes may either be an extension of those observed during and following physiological depolarization, or an exaggeration of those observed throughout brain tissue during the most severe global ischemic injury associated with infarction.

This chapter will first detail the history of astroglial acidosis during hyperglycemic global ischemia and discuss the hypothesis of compartmentalized astroglial acidosis and the mechanisms that may lead to this phenomenon. We will next examine astroglial alkalosis during normoglycemic global ischemia and consider the implications of astroglial alkalosis for the pathogenesis of selective neuronal necrosis. Finally, we will look at astroglial acid-base shifts during reperfusion after normoglycemic ischemia, and then speculate about how these rebound acid-base shifts might further enhance injury to vulnerable neurons.

ASTROGLIAL ACIDOSIS AND INFARCTION

History

Friede and Van Houten (1961) first suggested a relationship between brain injury and glial acidosis. In anoxic cerebellar brain slices, they enhanced tissue pathology by incubating their

pH and Brain Function, Edited by Kai Kaila and Bruce R. Ransom
ISBN 0-471-11838-9

slices in elevated concentrations of glucose. Regions of high neuroglial density, such as the aggregations of glial cell bodies between the perikarya of Purkinje cells, demonstrated the most pronounced cellular pathology. In the presence of agents that blocked glycolysis, however, glial pathology and tissue acidosis were dramatically reduced. These results led Friede and Van Houten to speculate that glycolytic acidosis during anoxia can cause tissue injury in brain, and that the primary target of acid-induced injury as well as the source of hydrogen ions may in both cases be neuroglial cells.

Later experiments by others extended the concept of acid-induced injury, but did not emphasize the specific role of glial cells in generating and succumbing to anaerobic acidosis. Ljunggren et al. (1974) and others (Myers, 1979; Rehncrona et al., 1981) demonstrated that tissue pH drops and lactate content rises during global ischemia, and that lactate concentrations are significantly higher in animals made hyperglycemic before ischemia. Siemkowiscz and Hansen (1981) showed, using pH-selective microelectrodes, that the interstitium in ischemic brain is markedly more acidotic (pH 6.1) in hyperglycemic animals than in saline-treated controls (pH 6.6). Myers and Yamaguchi (1977) provided the first direct *in vivo* evidence of the harmful effects of ischemic lactate by showing that animals infused with glucose before ischemia suffered far more brain damage than normoglycemic controls. Pulsinelli et al. (1982) demonstrated that preischemic hyperglycemia shifted pathological outcome from selective neuronal death to infarction.

By the early 1980s, it was clear that two very different pathological outcomes could follow global ischemia: one in which only selectively vulnerable neurons die and the other in which all brain cells die (infarction) (Plum, 1983). Although these two outcomes appeared to be dependent on preischemic carbohydrate levels, it became evident that the degree of brain injury is discontinuously correlated with brain lactate level. Below a threshold of 16–20 mmol/kg (Myers, 1979), variations in lactate concentration did not appear to influence neuronal injury (Pulsinelli et al., 1983). Above this threshold, however, the brain is suddenly beset by astroglial death, widespread neuronal death, and edema (i.e., infarction) (Petito and Babiak, 1982). Particularly striking were the marked histological changes seen in astrocytes when viewed by both light and electron microscopy (Petito and Babiak, 1982).

The above evidence not only suggested a threshold lactate level for infarction, but also a threshold lactate level for astroglial injury. This led Plum, drawing on the work of Friede and Van Houten, to propose that the presence or absence of lactic acidosis damage to astrocytes is *the* critical factor determining whether brain ischemia limits its damage to selectively distributed neurons or produces regional necrosis of all cellular elements (Plum, 1983). With moderate to low lactate, therefore, select neurons are injured by a process independent of pH or osmolal change, and astrocytes are spared. High lactate concentrations, however, are associated with excessive acidosis to which astrocytes seem most sensitive. Plum also speculated that astrocytes were the glycolytic "engines" in brain most responsible for generating the lactic acidosis levels associated with infarction.

The rationale implicating astrocytes in "acidosis-induced" infarction was as follows (Plum, 1983): Histological and biochemical data demonstrate that astrocytes contain more glycogen than other brain cells (Lowry et al., 1964; Koizumi, 1974; Phelps, 1975). During ischemia, this store could be utilized to drive glycolysis as astrocytes attempt to reestablish transmembrane ion gradients. If astroglial glycogen stores are elevated, which might occur with preischemic hyperglycemia, then more glycogen-derived glucosyl units would be available for conversion to lactate via anaerobic glycolysis, and more ATP would be available for hydrolysis. Indeed, during hyperglycemic global ischemia in rats, interstitial

$[K^+]$ can be maintained within a physiological range for a period twice that seen in normoglycemic animals (Hansen, 1978; Ekholm et al., 1993). Elevated glucose also reduces the rise in interstitial $[K^+]$ during anoxia in the isolated rat optic nerve (Ransom et al., 1992). Thus, energy-dependent homeostatic mechanisms and proton-generating ATP hydrolysis (Hochachka and Mommsen, 1983; Chapter 5)) may also occur for this extended period. Past a threshold level of lactate, however, acidosis irreversibly injures anoxic astrocytes, damaging critical astroglial ionic, nutritional, and neurotransmitter homeostatic support mechanisms. This ultimately leads to generalized necrosis, blood-brain barrier breakdown, and progressive edema. It should be noted, however, that enhanced glycolysis may protect some regions of CNS white matter such as the optic nerve (Ransom et al., 1992), but not other white matter such as that in the cerebellum (Friede and Van Houten, 1961). The consequences of runaway glycolysis during anoxia may therefore vary regionally.

Measurement of Astroglial Acidosis

Studies in whole animals and tissue culture experiments have since added considerable support for the astroglial acidosis hypothesis of infarction. The studies *in vivo* began with a group of experiments simultaneously measuring brain lactate content, tissue P_{CO_2}, interstitial $[HCO_3^-]$ and pH during hyperglycemic and normoglycemic complete ischemia. In hyperglycemic animals, interstitial pH, recorded with pH-sensitive microelectrodes (pH-ISMs), remained fixed at various tissue lactate levels above a critical threshold of ~17 mmol/kg (Kraig et al., 1985a,b; Kraig et al., 1986). This implied that a separate compartment, perhaps a population of cells, could retain the hydrogen ions cogenerated with continued production of lactate. Perhaps more importantly, tissue P_{CO_2} also remained fixed above this critical threshold, suggesting this increasingly acidic compartment had lost its titratable store of HCO_3^-. Estimates of ischemic total brain and interstitial $[HCO_3^-]$ confirmed that the level of this anion remained unchanged at lactate levels above those associated with infarction (Kraig et al., 1986). Further calculations demonstrated that the HCO_3^--free compartment was approximately 36% of the total brain volume (Kraig et al., 1986), a value close to that estimated for glial cell space in the neocortex (Hertz and Shousboe, 1975).

Although some studies questioned whether astroglial acid-compartmentation occurs with severe hyperglycemic global ischemia (Katsura et al., 1991, 1992), direct pH-ISM measurements of astrocytic intracellular pH (pH_i) in rat neocortex (Kraig and Chesler, 1990) have corroborated the indirect evidence detailed above. In these experiments, two cellular foci at least an order of magnitude more acidic than the interstitial space (pH ~ 6.2) were observed during both hyperglycemic and complete ischemia. The first and most extreme (pH 4.3 ± 0.19) was seen only at the onset of anoxic depolarization, while the second, and more typical, (pH 5.3 ± 0.07) was measured 5–46 min after cardiac arrest (Fig. 1). High input resistance and small, residual membrane potentials strongly suggested that these foci were from intracellular space. When the pH-ISMs were removed from these high-impedance foci, they consistently recorded typical values of interstitial pH (~6.2). Horseradish peroxidase injections via the reference barrels of the pH-ISMs and histological analysis added further evidence that these extreme levels of acidosis occurred in astroglial cells. Subsequent pH-ISM experiments in the neocortex of rats undergoing short-term, transient (20–30 min) hyperglycemic global ischemia have since provided additional support for severe astroglial acidosis (Kraig, unpublished observations).

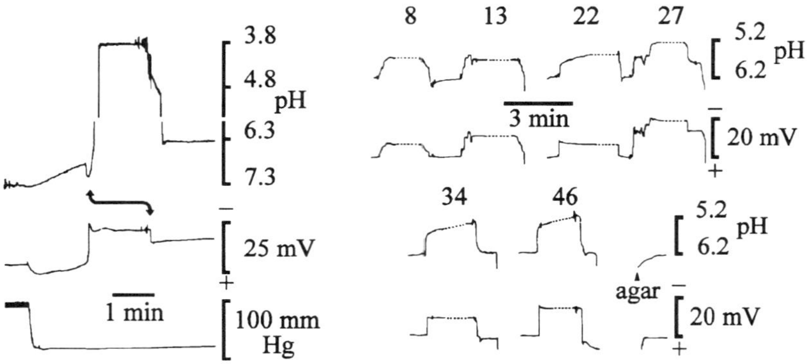

FIGURE 1. Acid shifts associated with hyperglycemic and complete ischemia. pH-ISMs are positioned approximately 400–800 μm below the parietal pial surface. Following cardiac arrest, systolic blood pressure rapidly falls to a minimum, as indicated in the lower left-hand record. The trace at the top left shows first a slow interstitial acid shift, and then, at the onset of anoxic depolarization (left middle trace), indicates a pH shift (upward arrow) that is more acidic than the interstitial space by orders of magnitude. This suggests entry of the electrode into an extremely acidic environment. In this example, pH_i reaches a peak value of 3.82. It is believed that the electrode enters glial cells as they swell with depolarization. This notion is supported by the positive shift in DC potential (downward arrow) and return of pH to 6.3 when the electrode is raised by 200 μm. The pH values observed are expected values of potential and pH in the interstitial space of hyperglycemic and completely ischemic brain. The record of pH is interrupted to allow visualization of the lower and uppermost portions of the record. On the right, the top and third traces show pH changes, and the second and fourth traces show simultaneous DC changes. The numbers above each pH change indicate the number of minutes after ischemia. Breaks in pH and DC traces (e.g., at 34 min and 46 min) correspond to withdrawal of the electrode from the brain. After a final measurement, the electrode was removed from the brain and advanced into a phosphate-buffered (6.20 pH) agar gel (3.5%), containing 150 mM NaCl. The dotted lines in each record indicate a 3-minute period during which horseradish peroxidase was electrophoretically injected for later morphological identification. (With permission from Lascola and Kraig, 1995).

Other acid-base studies using techniques that average pH_i are quantitatively consistent with the finding of compartmentalized acidosis. For example, using the fluorescence pH indicator dye umbelliferone, Meyer et al. (1986) demonstrated that average pH_i in rabbit superficial cortex falls as low as 5.3 pH during complete ischemia. Since glial fibrillary acidic protein (GFAP) immunohistochemical staining is concentrated in the superficial layers of neocortex (Petito et al., 1990), astrocytes may predominate there, and these low values may reflect primarily astroglial pH_i. In addition, Smith et al. (1986) calculated (using the Van Slyke technique) an average pH_i of 5.9 during hyperglycemic and complete ischemia. And Chopp et al. (1987), using ^{31}P NMR spectroscopy, also measured an average pH_i of 5.9 under similar ischemic conditions. Given that astroglial, neuronal, and interstitial volumes are roughly 33%, 52%, and 15% of total neocortical volume, respectively (Hertz and Shousboe, 1975), an average of pH_i of 5.9 is still consistent with an average astroglial pH_i of 5.3, and neuronal and interstitial pH of 6.2.

Acidosis Kills Astrocytes

A series of recent experiments in culture have shown that astrocytes die following exposure to acid, and perhaps more importantly, that they are more sensitive to the toxic effects of acidosis than neurons (Goldman et al., 1989; Giffard et al., 1990; Nedergaard et al., 1991). These results support the idea that astrocytes are uniquely susceptible to acidosis *in vivo*, and that irreversible astroglial injury from acidosis may be a critical event precipitating the death of all brain cells. The precise molecular mechanisms of acid-induced injury, however, remain to be defined. Pulsinelli et al. (1985; see also Chapter 28) proposed an iron-mediated free radical mechanism for cellular injury (see also Siesjo et al., 1985) that depends on acid-induced release of iron from organic stores. Sequestration of iron may require bicarbonate. In an excessively acidic environment (e.g., extremely low [HCO_3^-]), free iron may be liberated to catalyze the generation of free radicals with subsequent irreversible injury to membranes by lipid peroxidation (Siesjo et al., 1985). However, other injurious mechanisms such as mitochondrial dysfunction (Hillared et al., 1984) most likely also interact in an extremely acidic environment to destroy astrocytes (Kraig et al., 1995; Siesjo et al., 1995).

Characteristics of Compartmentalized Acidosis

If astrocytes generate and maintain a compartmentalized pH of 5.3 or less, these cells must exhaust their HCO_3^- store as protons are produced in excess (Kraig and Chesler, 1990; Chapter 4). It follows then that proton accumulation proceeds beyond a point where astrocytes can absorb HCO_3^- from the interstitium, since interstitial HCO_3^- remains at ~12 mM throughout ischemia (Kraig et al., 1986). Although specific mechanisms have not yet been clarified in astrocytes, a wealth of data from astrocytes and other cellular preparations suggest likely metabolic and biophysical processes underlying this phenomenon. First, astrocytes must produce excess protons. Substantial *in vivo* and *in vitro* evidence indicates cortical astrocytes contain sizable amounts of glycogen (Cataldo and Broadwell, 1986; Rosenberg and Dichter, 1985; Swanson and Choi, 1993), the major energy reserve in brain (Lowry et al., 1964), whereas neurons from adult (Koizumi, 1974; Phelps, 1975; Cataldo and Broadwell, 1986) and fetal (Bruekner and Bisolde, 1981) brains contain little glycogen. In addition, although neurons possess some enzymes for glycogen metabolism, their levels are relatively low compared to those of glia (Knull and Khandelwahl, 1982; Reinhart et al., 1990). Finally, brain glucose levels are also considerably elevated in hyperglycemic animals (Welsh et al., 1980; Mason et al., 1992), providing additional substrate for anaerobic glycolysis in the first few minutes of ischemia. Under hyperglycemic conditions, therefore, astrocytes could generate a considerable intracellular acid load relative to neurons and the interstitium, as anaerobic glycolysis and ATP hydrolysis yield protons when pH falls below 6.8. An important question, however, is how astrocytes sustain anaerobic glycolysis in the face of acidosis, which has been shown to reduce the rate of glycolysis (Hochachka and Mommsen, 1983). Although decreasing pH inhibits the rate-limiting glycolytic enzyme phosphofructokinase, this inhibition can be prevented in part by the accumulation of products of ATP hydrolysis (e.g., ADP, AMP, and Pi), which lower the pH past the optimum for this enzyme (Hochachka and Mommsen, 1983; Erecinska and Silver, 1989). As experiments in frog and rat muscle undergoing anoxia (i.e., glycolytic energy supply but no oxygen) have already shown, some cells are capable of sustaining high and unchanging rates of anaerobic glycolysis as proton production continues and pH_i decreases (Fitts and Holloszy, 1976; Sahlin et al., 1981).

Second, astrocyte membranes must be impermeable to HCO_3^-, despite a substantial inward driving force for this anion during hyperglycemic global ischemia (intracellular $[HCO_3^-] < 2$ mM vs interstitial $[HCO_3^-] \sim 12$ mM; Kraig et al., 1986). Several studies have documented potential pathways for HCO_3^- permeability in astrocytes. Each, however, appears to be inhibited by pathological levels of acidosis. For example, HCO_3^- transport occurs via Cl^-/HCO_3^- exchange, Na^+-dependent Cl^-/HCO_3^- exchange, and electrogenic Na^+/HCO_3^- cotransport (Boyarski et al., 1993; Shrode and Putnum, 1994; O'Connor et al., 1994; Grichtenko and Chesler, 1994a,b). Yet, at an extracellular pH of 6.8 or lower, these HCO_3^--dependent pH regulatory mechanisms may be inhibited by protons (Mellergard et al., 1994a,b). Another potential mechanism for HCO_3^- transport is via glutamate transport (Bouvier et al., 1992), but this is also inhibited by ischemic levels of acidosis in cultured astrocytes (Swanson et al., 1995). Together, these results complement earlier data from barnacle muscle experiments showing that acid extrusion (e.g., HCO_3^- uptake) falls almost three orders of magnitude when bath pH drops to 6.8 (Boron et al., 1979).

Transport of HCO_3^- however, could still potentially occur by other means such as through astroglial Cl^- channels (Gray and Ritchie, 1986; Barres et al., 1990; Fraser et al., 1995; Lascola and Kraig, 1996). Although this form of HCO_3^- transport (Bormann et al., 1987; Kaila and Voipio, 1987; Tabcharani et al., 1989; Hille, 1992; Poulsen et al., 1994; Chapter 16) may be important under more physiological conditions, astroglial Cl^- channels are also likely to be inhibited by extreme acidosis. For example, the Cl^- conductance of frog skeletal muscle membranes is a steep sigmoidal function of external pH, and appears completely inhibited at a pH of ~6.0 (Hutter and Warner, 1967). In addition, the probability that single-channel Cl^- currents in tracheal epithelial cells are open is significantly reduced in acid solutions (Duszyk et al., 1995). Finally, weakly voltage-dependent Cl^- channels in lobster nerves are irreversibly suppressed at a cytoplasmic pH of 6.0 (Lukacs and Moczydlowski, 1990). Indeed, inhibition of electrogenic HCO_3^- transport, whether via channels or transporters, should be reflected more generally as an increase in cell input resistance, which is precisely what is measured *in vivo* during hyperglycemic and complete ischemia (Kraig and Chesler, 1990), and in cultured astrocytes following either exposure to acid (Walz and Wuttke, 1989) or metabolic injury (Harold and Walz, 1992).

Glial input impedance *in vivo,* however, may be more a function of syncytial and high glial K^+ conductances. Therefore, increased membrane resistance is also likely to reflect combined acidotic inhibition of K^+ channels (Coulter et al., 1995), other electrogenic mechanisms (Walz and Hinks, 1987), and gap junctions (Connors et al., 1984; Anders et al., 1988; DeVries and Schwartz, 1992), in addition to decreased anion conductance. In fact, acid inhibition of ion channels and transporters seems to be a general phenomenon, blocking both anion and cation (including proton) flux (Lang et al., 1988; Chapter 21–24). Extracellular acid even appears to inactivate the Na^+/H^+ exchanger (Aronson, 1985; Vaughan-Jones and Wu, 1990). With intact, high-impedance membranes, astrocytes would nonetheless remain permeable to nonionized organic acids (HA) such as lactate. At first glance, this may appear to be still another potential mechanism for dissipating proton gradients between astrocytes and the interstitium. However, if astrocytes are impermeable to anions and cations as suggested above, and the concentration of pH-related anions (e.g., HCO_3^- or lactate) is greater outside astrocytes than inside, then once intracellular and interstitial nonionized acids equilibrate, the proton gradient across the astroglial membrane must be equal to the *inverse* of the trans-membrane pH-related anion ratio (Kraig and Chesler, 1990). This analysis is demonstrated in Fig. 2.

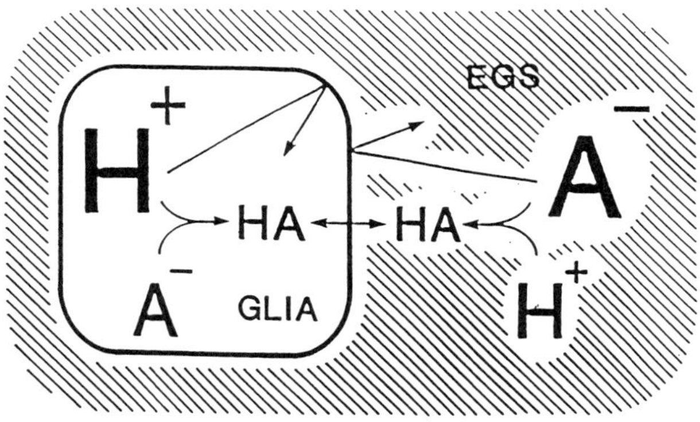

$$\text{(1) } pH_i = pK_a + \log [A^-]_i/[HA]_i$$

$$\text{(2) } pH_o = pK_a + \log [A_o^-]_o/[HA]_o$$

$$\text{(3) } [HA]_i = [HA]_o$$

$$\text{(4) } [H^+]_i/[H^+]_o = [A^-]_o/[A^-]_i$$

FIGURE 2. Extreme astroglial acidosis is sustained as protonated forms of pH-related anions (e.g., HCO_3^- or lactate) equilibrate across cell membranes. During hyperglycemic and complete ischemia, astroglial membranes are still permeable to nonionized organic acids (HA) such as lactic acid and carbonic acid. The concentrations of these acids should then be equal within astrocytes and the extraglial space (EGS). If the EGS is a restricted compartment, and the concentration of pH-related anions is greater than that found in astrocytes, then the proton gradient across astroglial membranes must be equal to the inverse of the transmembrane anion ratio. In Eqs. 1 and 2, A_i^- and A_0^- represent intracellular and interstitial conjugate bases, respectively. Assuming the ionization constant for each acid is the same in both the interstitial and astroglial compartments, and that Eq. 3 holds, then substituting Eqs. 1 and 2 into 3, yields Eq. 4, where H_1^+ and H_0^+ are the intracellular and interstitial proton concentrations. (With permission from Kraig and Chesler, 1990).

ASTROGLIAL ALKALOSIS AND SELECTIVE NEURONAL VULNERABILITY

Astroglial pH_i During Normoglycemic Ischemic Depolarization

Astrocytes appear extremely resistant to injury following normoglycemic global ischemia (Petito and Babiak, 1982) while select populations of neurons are highly vulnerable (Pulsinelli, 1985). Plum (1983) speculated that under these circumstances astrocytes may instead defend neuronal cells pH-ISM measurements of astroglial and interstitial pH during reversible normoglycemic ischemia support this hypothesis (Kraig and Petito, 1989), especially when considered in light of recent data suggesting mild acidosis protects neurons from excitotoxicity in culture (Giffard et al., 1990; Tombaugh and Sapolski, 1990). During normoglycemic global ischemia, astrocytes not only maintain pH_i despite a mild interstitial acidosis (pH ~ 6.8), but may even alkalinize slightly relative to their own resting baseline pH_i (Fig. 3). Before ischemia, interstitial pH is 7.34 ± 0.01 and astroglial pH 7.03 ± 0.01; during ischemic depolarization, the interstitium acidifies to pH 6.72 + 0.03 as glial pH shifts more alkaline to 7.11 + 0.04 (Fig. 3).

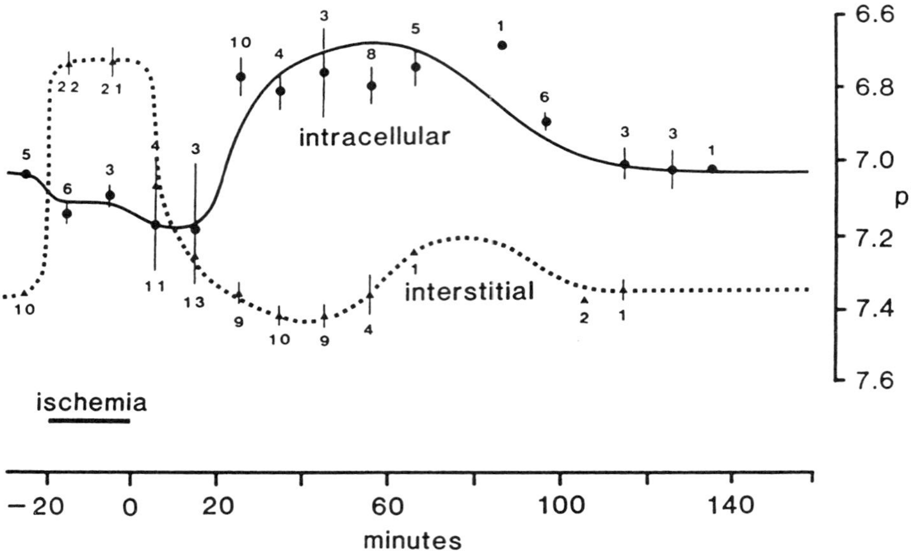

FIGURE 3. Composite of interstitial and presumed intraglial pH recordings associated with reversible, nearly complete global ischemia under normoglycemic conditions. Recordings were made with pH-ISMs and cells were identified as glial (prior to and several minutes after ischemia) by their high membrane potential, low membrane impedance, and absence of synaptic or injury discharges. Methods for identifying glial cells during ischemic depolarization are described in Kraig and Chesler, 1990. Dots indicate intracellular recordings, and triangles interstitial measurements. Vertical bar through points represents standard error of mean of recordings averaged over 10-minute intervals. Numbers above and below symbols indicate number of recordings for each 10-min interval. Solid and dotted lines were drawn by sight. Note that astrocytes are more alkaline than normal during ischemia, when the interstitial space was most acidic. Following reperfusion, astrocytes become most acidic just after the peak rebound alkalosis in the interstitium. Interstitial and astroglial pH returned to normal approximately 2 hours after reperfusion. (With permission from Kraig and Petito, 1989.)

Astroglial alkalosis during normoglycemic ischemic depolarization may at first seem surprising given the massive acidosis seen during hyperglycemic global ischemia. One might expect that during global ischemia astrocytes acidify along a continuum, with the greatest acidosis reached during the most severe and complete hyperglycemic ischemia. That astrocytes are relatively alkaline, however, suggests that these cells can maintain some active transport processes moving proton equivalents across astrocyte membranes. If this is true, then alkalosis during normoglycemic global ischemia happens in an "open" rather than "closed" compartment: pH shifts in the glial compartment are not due to intracellular isolation as in hyperglycemic ischemia, but are rather the reciprocation in part of those shifts occurring in the interstitium.

Depolarization-Induced Astroglial Alkalosis

The relative astroglial alkalosis during normoglycemic ischemia is strongly reminiscent of astroglial pH shifts observed during physiological depolarization both *in vivo* (Chesler and Kraig, 1987, 1989) and *in vitro* (Boyarsky et al., 1993; Gritchenko and Chesler, 1994a; Pappas and Ransom, 1994; Chapters 14 and 17). The experiments *in vivo* using pH-ISMs first demonstrated that astrocytes can alkalinize with depolarization (Chesler and Kraig, 1987, 1989). Astroglial alkalosis was measured during cortical stimulation and SD (Fig. 4), which produces a documented rise in brain lactate (Krivanek and Fifkova, 1965), CO_2 (Kraig and Chesler, 1988), and interstitial acid (Kraig et al., 1983). These experiments showed that (i) a depolarizing shift in membrane potential is essential for alkalosis, and (ii) the alkalosis arises (at least partly) from the active transport of acid-base equivalents across the glial membrane. The dependence on membrane potential is indicated by the direct correlation between the magnitude of glial alkalotic shifts and the degree of depolarization following stimulation at various intensities (Fig. 4). Cortical stimulation of increasing strength produced a graded rise in glial pH_i (~0.1–0.3) (Fig. 4). During maximal depolarization with SD, astroglial pH rises by as much as 0.80 pH unit (Figs. 4 and 5). When Ba^{2+} is added to block astroglial depolarization from cortical stimulation, no alkalosis is seen.

With astroglial pH at 7.0 and interstitial pH at 7.3, and average membrane potentials between −70 mV and −80 mV, a considerable outward driving force exists for pH-related anions (and an equivalent, but inward, driving force for pH-related cations). Therefore, active transport of pH-related ions, rather than the passive influx of base or efflux of acid through ion channels, is strongly implicated in depolarization-dependent alkalosis. Intracellular measurements of strong electrolytes (i.e., Na^+, K^+, and Cl^-; Stewart, 1981) in glial cells of olfactory slices suggests that depolarization produces an anion deficit that can be filled by HCO_3^-. Calculations of intracellular and interstitial $[HCO_3^-]$ indicate astroglial $[HCO_3^-]$ may rise an order of magnitude during SD (Fig. 5). At any given depolarization, a rise in $[HCO_3^-]_i$ could account for the astroglial intracellular anion deficit. One likely mechanism underlying alkalosis during depolarization with brain activity, spreading depression, and normoglycemic global ischemia, therefore, is the active transport of HCO_3^- into astrocytes.

Electrogenic Na^+/HCO_3^- Cotransport and Acid Secretion by Astrocytes

Data from nonmammalian glial preparations (Deitmer and Schlue, 1989; Newman, 1991, 1996), mammalian brain slices (Grichtchenko and Chesler, 1994a), and cultured astrocytes (Shrode and Putnum, 1994; O'Connor et al., 1994; Brune et al., 1994; Brookes and Turner, 1994) suggest the presence of an electrogenic carrier that transports HCO_3^- and/or CO_3^{2-} by coupling the movement of these anions to the electrochemical gradient of Na^+. Electrogenicity arises from the net transfer of at least one negative charge (as either HCO_3^- or CO_3^{2-}) moving in the same direction as Na^+. The transporter is electrogenic, and depolarization itself can also drive HCO_3^- anion flux. Details of the transporter are reviewed elsewhere in this volume (see Chapters 13–14).

Experiments in neuron-depleted and normal hippocampal slices (Grichtchenko and Chesler, 1994a,b) and in isolated retinal glial cells (Newman, 1996) provide convincing evidence that Na^+/HCO_3^- cotransport, while in part responsible for depolarization-induced alkalosis, may be the most important mechanism underlying depolarization-dependent acid-secretion from glial cells. In the slice work, Grichtchenko and Chesler (1994b) measured interstitial acid shifts that were abolished in Na^+-free media, and significantly diminished in HEPES (N-[2-hydroxyethyl]piperazine-N′-[2-ethanesulfonic acid]) buffered, HCO_3^--free solutions. Notably, these acid shifts were not affected by amiloride, zero-Cl^- media, high or low glucose, lactate transport inhibitors, or zero interstitial Ca^{2+}.

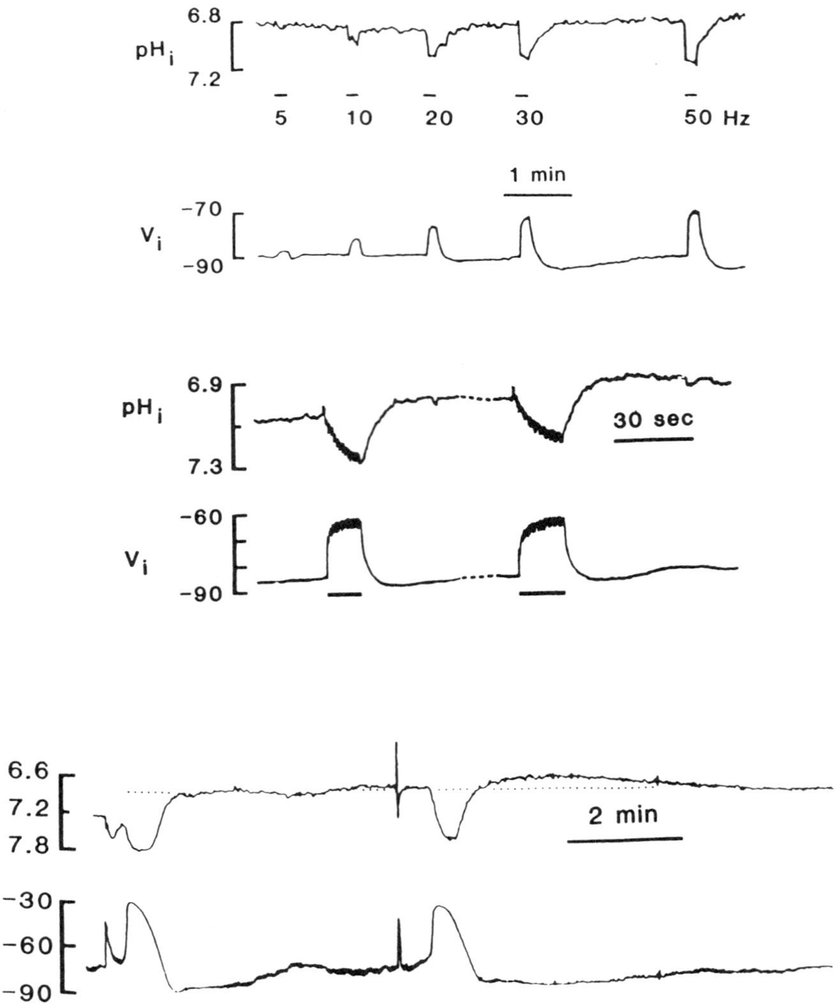

FIGURE 4. Astroglial pH transients with cortical stimulation and spreading depression (SD) measured *in vivo* with pH-ISMs. The top two traces demonstrate the correlation between stimulus frequency and astroglial alkaline shifts. In these experiments, alkaline shifts became greater as stimulus frequency was increased. Depolarizations and alkalinizations were positively correlated ($r = 0.68$, $n = 50$). In the middle two traces, 20-Hz stimulation produced a 20-mV depolarization as astroglial pH shifted to 7.25 alkaline. After stimulation, membrane potential hyperpolarized before returning to baseline. A second 20-Hz stimulation, delivered 4 minutes after the first, produced another stereotypic alkaline shift and acid rebound. Following this second stimulus, astroglial pH reached an acidic plateau of 6.85. These measurements demonstrate that recurrent stimuli sufficiently close in time can progressively acidify astrocytes. The bottom two traces represent astroglial pH changes with successive spreading depressions. A 100-Hz initiating stimulus caused first a brief depolarization and concomitant alkalosis, and then a wave of spreading depression, giving rise to a pH approaching 7.8. With recovery, a prolonged acid shift is sustained as membrane potential returns to baseline. After 3 minutes, a second 100-Hz stimulus produced another spreading depression, alkaline shift, and acid rebound. The net effect of these recurrent spreading depressions was a progressive acidification of astroglial pH to 6.6. This acidification could arise from the loss of HCO_3^- through astroglial Cl^- channels (see text). (With permission from Chesler and Kraig, 1989.)

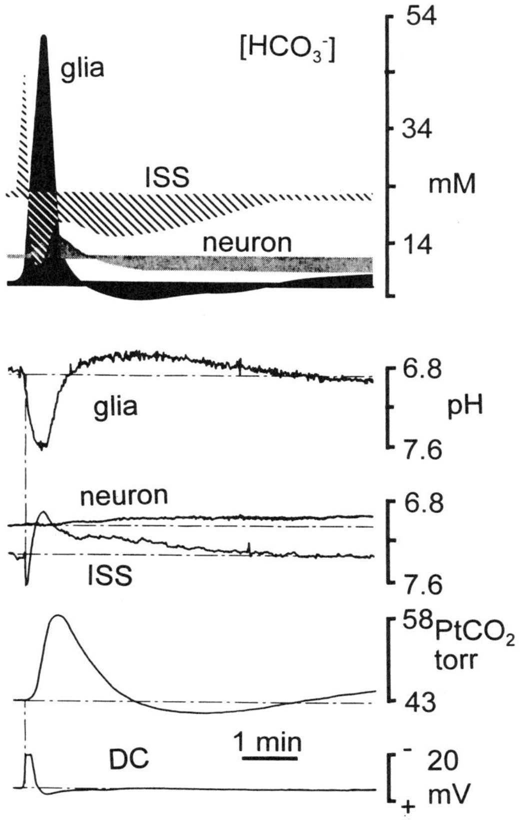

FIGURE 5. Acid-base and calculated $[HCO_3^-]$ changes during spreading depression (SD). The bottom four traces are from different experiments but represent typical recordings during spreading depression. The top figure is a composite of $[HCO_3^-]$ in all three brain compartments, calculated via the Henderson-Hasselbach equation from the data shown below it. The DC (bottom trace) and interstitial (ISS) pH changes were recorded simultaneously. The glial and ISS pH changes show that the peak interstitial acid shift (pH to 6.93) was correlated in time with the peak alkaline glial transient (pH to 7.61). Neuronal pH (7.08) was comparatively unchanged during SD but became slightly acidic (pH 6.92) with cell repolarization. Calculation $[HCO_3^-]$ was based on the assumption that carbon dioxide levels and the first ionization equilibrium constant of carbonic acid (pK_a 6.13) were the same in all brain compartments. Glial $[HCO_3^-]$ (solid area) was 7 mM before SD and reached a peak of 51 mM during SD and maximum depolarization. In contrast, interstitial $[HCO_3^-]$ (hatched area) predominantly fell to <11 mM before progressively returning to baseline. After increasing slightly from 12 mM to 16 mM during SD, neuronal $[HCO_3^-]$ (stippled area) eventually dropped to <8 mM. (With permission from Lascola and Kraig, 1996.)

These results were mirrored recently in an elegant series of experiments in freshly dissociated Müller cells of the salamander retina (Newman, 1996). Again, depolarization-induced acid secretion from Müller cells appears to be almost exclusively due to electrogenic cotransport, whereas potential-dependent alkalosis is due only in part to this cotransport plus some additional and as-yet undefined mechanisms.

Defense of Normoglycemic Brain Cells During Ischemic Depolarization: Astroglial Na^+/HCO_3^- Cotransport

During global ischemia, Na^+/HCO_3^- cotransport may protect astrocytes from the damaging effects of acidosis that may otherwise occur without access to titratable base. Compared to neurons, cultured astrocytes appear to maintain their Na^+ electrochemical gradient by utilizing predominantly glycolytic rather than oxidatively-derived ATP to drive Na^+/K^+ ATPase activity (Kauppinen et al., 1988; Harold and Walz, 1992). If this holds true *in vivo,* then during ischemic depolarization, hydrolysis of glycolytic ATP should occur not only at the onset of ischemia but also throughout the ischemic period to the extent that substrate is available for glycolysis. Indeed, during incomplete transient global ischemia of 15 minutes duration (Blomqvist et al., 1985) or 60 (Yoshida et al., 1985), neocortical and hippocampal glucose and glycogen concentrations drop an order of magnitude, but do not fall to zero. Astrocytes may attempt to buffer the resultant intracellular acid load in part by transporting HCO_3^- coupled to an inwardly directed Na^+ gradient.

One remaining issue that requires examination, however, is the extent to which astrocytes can maintain any transmembrane Na^+ gradient during ischemic depolarization. During *physiologic* depolarization, glia exhibit a drop in intracellular Na^+ activity (Ballanyi et al., 1987). In addition, cerebral astrocytes acutely isolated either have a relatively low (Barres et al., 1990) or unmeasurable (Tse et al., 1992) Na^+ channel density. These facts, combined with the results of Goldberg et al. (1984) that indicate ATPase activity may be elevated in ischemia, suggest that astrocytes with low membrane Na^+ permeability should be capable of generating at least some Na^+ gradient sufficient to drive the continued uptake of HCO_3^- during ischemic depolarization. Direct measurements of intraglial Na^+ activity and membrane electrical properties during transient normoglycemic global ischemia would help resolve this issue. Nonetheless, what is now clear is that some brain compartment is capable of maintaining a relatively more alkalotic pH during normoglycemic global ischemia (Sunagawa et al., 1994), and direct electrophysiological measurements suggest strongly that this compartment is astroglial (Kraig and Petito, 1989).

Transport of HCO_3^- into astrocytes during normoglycemic ischemia may also be an important mechanism by which astrocytes contribute acid to the interstitium (Grichtchenko and Chesler, 1994b). As detailed above, Na^+/HCO_3^- cotransport appears to be the major mechanism by which astrocytes expel protons in a depolarization-dependent fashion. Experiments in tissue culture (Giffard et al., 1990; Tombaugh and Sapolski, 1990) and brain slices (Tombaugh, 1994) have demonstrated a neuroprotective effect of mild extracellular acidosis against hypoxic and excitotoxic injury. Acid protection is also seen with hypercarbic acidosis *in vivo* following focal cerebral ischemia (Simon et al., 1993), although it is notable that the reduction in infarct size is maximal at brain pH 6.8 and then lost at pH 6.5 (see also Katsura et al., 1994). Acidic pH dramatically reduces the probability that N-methyl-D-aspartate (NMDA) channels will be open (Tang et al., 1990; Chapter 22) and blocks voltage-dependent Ca^{2+} channels (Prod'hom et al., 1989; Barnes and Bui, 1991; Chapter 21), both of which are major neuronal pathways for Ca^{2+} influxes associated with excitotoxicity. Indeed, neuroprotection from interstitial acidosis appears to arise from the more generalized phenomenon of acid-inhibited ion permeability detailed above for hyperglycemic global ischemia. Acid-inhibited ion permeability may therefore facilitate both acidotic limitation of excitotoxicity associated with selective neuronal injury, and when taken to an extreme, massive compartmentalized astroglial acidosis associated with infarction.

ASTROGLIAL ACIDOSIS AND SELECTIVE NEURONAL VULNERABILITY

Rebound Acidosis with Reperfusion

Although mild interstitial acidosis and intraglial alkalosis may protect neurons and astrocytes from injury in normoglycemic ischemia, this protection *in vivo* can be relevant only during ischemic depolarization. With postischemic reperfusion, astrocytes, the interstitial space, and neurons undergo measurable acid-base rebound shifts that may instead enhance selective neuronal injury. After recovery of cell membrane potentials, interstitial pH rises to a peak of 7.41 ± 0.02 (range 7.35–7.60) 30–40 min following normoglycemic ischemia (Fig. 3) (Kraig and Petito, 1989). This compares to a preischemic interstitial pH of 7.34 ± 0.01 and ischemic interstitial pH of 6.72 ± 0.03. In contrast, astroglia become markedly acidotic (pH 6.75 ± 0.05, range 6.62–6.92) 60–70 minutes after the start of reperfusion (Fig. 3). Normal intraglial pH is ~7.04 (Chesler and Kraig, 1989; Swain et al., 1991). Finally, measurements of average brain intracellular pH indicate a transient alkalosis within the first hour after reperfusion (Mabe et al., 1983; Yoshida et al., 1985). Since direct pH-ISM measurements in astrocytes measure a distinct acidosis after ischemia, these average pH_i results must reflect a volume-weighted, predominant intracellular alkalosis occurring in other brain cells.

As with ischemic depolarization, the intraglial acidosis observed after reperfusion appears to be an exaggeration both in magnitude and time of astroglial shifts following cortical stimulation and SD. In pH-ISM experiments discussed previously (Chesler and Kraig, 1987, 1989), surface stimulation produces an intraglial alkaline shift that in 38% of cells is followed by an acid rebound (Fig. 4). In addition, repetitive periods of stimulation close in time progressively acidify astrocytes (Fig. 4). The size of the acid rebound, therefore, appears to depend on the duration of stimulation as well as the frequency of stimulation periods. With SD, the pattern of astroglial acid-base transients is further enhanced. For 26 SD events, the peak alkaline shift with depolarization is 0.32 + 0.04 pH unit (range 0.11–0.78) and the final acid shift above baseline is 0.19 + 0.02 pH unit (range 0.03–0.39). Moreover, astrocytes become progressively acidotic with successive SDs (Fig. 4).

In contrast, an alkalotic rebound in the interstitium analogous to that observed after reperfusion is not discernable after physiologic stimulation or SD (Chesler and Kraig, 1989). It is also not observed in similar experiments where interstitial pH was measured with glass-based pH-ISMs following both normoglycemic and hyperglycemic ischemia (Siemkowicz and Hansen, 1981) or with liquid exchanger ISMs following low-flow ischemia (Silver and Erecinska, 1992), although the disparate results may have to do with differences in experimental techniques. Underlying mechanisms for interstitial alkalotic rebound are therefore not immediately obvious. However, NMR measurements of average brain pH_i reveal two distinct cycles of alkalosis after ischemic insult. The first occurs and resolves within 60 minutes after reperfusion (Mabe et al., 1983; Yoshida et al., 1985), and the second occurs approximately 48 hours later (Chopp et al., 1989). The time course for the first cycle is similar to the time course for interstitial alkalosis measured with pH-ISMs. Since slow interstitial and neuronal acid-base shifts appear to parallel one another during physiologic stimulation (Chesler, 1990) and during ischemia (Silver and Erecinska, 1992), perhaps they also follow one another during reperfusion after normoglycemic ischemia.

Interstitial acid-base shifts are also often inversely related to astroglial pH shifts. Thus the interstitial alkalosis may in turn arise in part from loss of base or uptake of acid from astrocytes. One potential mechanism for loss of HCO_3^- base is through astroglial Cl^- channels permeable to this anion (Chapter 16). In culture experiments, stellate Type I neocortical astrocytes express Cl^- conductances open at resting potentials that are enhanced by elevated intracellular calcium (Lascola and Kraig, 1996). This enhancement seems to be dependent on Ca^{2+}-induced depolymerization of cytoskeletal actin. Since elevated extracellular K^+ and neurotransmitters (Finkbeiner, 1993), as well as SD *in vivo* (Kraig and Lascola, 1994), all elevate intraglial Ca^{2+}, it is likely that intraglial Ca^{2+} also rises during ischemic depolarization, inducing breakdown of the actin cytoskeleton in astrocytes and perhaps increasing astroglial anion conductances. When cells are returned to resting potential with reperfusion, HCO_3^- anions would then leave astrocytes, moving down their electrochemical gradient through opened anion channels, resulting in cytoplasmic acidification and interstitial alkalinization. As the concentration of ATP, which aids repolymerization of actin (Pollard and Cooper, 1986), is slowly restored with recirculation (Mabe et al., 1983), astroglial anion conductances would eventually decrease, reducing HCO_3^- efflux and allowing intraglial pH to normalize.

Acid-Base Shifts with Reperfusion: Enhanced Injury

The consequences of interstitial alkalosis with reperfusion are not known, but several possibilities can be generated by extending the same reasoning applied to acid-base shifts occurring during ischemic depolarization. If interstitial acidosis protects neurons from excitotoxicity by reducing the probability of open neuronal NMDA and Ca^{2+} channels, then conversely, excitotoxicity may be increased by alkalotic enhancement of these probabilities. For example, Tang et al. (1990) have shown that the pK_a of NMDA channel modulation is 6.9, rendering the channels highly sensitive to both alkalotic and acidotic pH shifts between pH 6.0 and 8.0. Acid decreases whereas alkalosis increases NMDA currents. Alkalosis may therefore make neurons susceptible to greater influxes of Ca^{2+} during cellular activity following reperfusion (Ou-Yang et al., 1994).

The rise in interstitial pH (7.41), albeit prolonged, is nevertheless slight, and greater harm for vulnerable neurons may arise indirectly from the considerable rebound acidosis (6.75) occurring in astrocytes after reperfusion. With restoration of cell membrane potentials, ionic gradients, and aerobic metabolism, ATP hydrolysis is presumably reduced. Under these circumstances, aerobic glycolysis is likely to be highly susceptible to inhibition by acidosis (Roos and Boron, 1981). Gemba et al. (1994) have demonstrated that specific inhibition of glycolysis in cultured astrocytes completely reduces glutamate uptake, in part by partial collapse of the transmembrane Na^+ gradient, and in part by acidosis-driven reversal of the Na^+-dependent glutamate transporter. Although the consequences of reduced glial capacity for glutamate uptake combined with NMDA receptor potentiation alone could devastate selectively vulnerable neurons, intraglial acidosis is also likely to affect not only other glial functions dependent on aerobic glycolysis but also cell functions directly dependent on intraglial pH.

CONCLUSION

The measurements of astroglial acid-base shifts during and following both physiological and ischemic depolarization provide strong evidence that astroglial pH is compartmentalized and not a simply a direct reflection of interstitial pH. The concept of acid-base "compartmentalization" is practical since it emphasizes the fact that astrocytic pH changes during physiological and pathological conditions are partitioned from those occurring in neuronal and interstitial space. Under circumstances in which astrocytes remain permeable to pH-related ions, astrocytic space may be thought of as an "open compartment." In this case, changes in astroglial pH yield inverse changes in interstitial pH as proton equivalents are actively transported across astroglial membranes. Depolarization-induced astroglial alkalosis, as seen with cortical stimulation, spreading depression, and normoglycemic ischemia, seems to occur in such an open compartment. Furthermore, it may arise in part from active uptake of interstitial HCO_3^- and/or CO_3^{2-} into astrocytes. In contrast, astroglial space may be viewed as a "closed compartment" under conditions in which astrocytes lose most, if not all, permeability to pH-related ions. Severe astroglial acidosis during hyperglycemic and complete ischemia, for example, must occur in a "closed compartment," since astrocytes accumulate protons independent of pH changes in neurons or the interstitium.

The progression from an open to a closed astroglial pH compartment during global ischemia may mark the pathological transition from selective neuronal vulnerability to infarction. This progression depends on preischemic brain carbohydrate content. Under normoglycemic ischemic depolarization, astrocytes alkalinize their interior and release acid to the interstitium, buffering ischemic acid loads while contributing to a mild interstitial acidosis that can limit neuronal excitotoxicity. After reperfusion and repolarization, astrocytes experience cytplasmic acidosis as the interstitium swings to slightly alkaline. The rebound pH shifts may enhance selective neuronal injury, but do not lead to infarction. With hyperglycemic ischemia, however, excessive interstitial acidosis can be associated with a closed compartment and severe astroglial acidosis. Thus, the general phenomenon of brain cell ion channel inhibition by acidosis can have a dual effect. It retards neuronal injury by reducing an influx of Ca^{2+} under normoglycemic conditions and global ischemia. However, with hyperglycemia and an analogous period of ischemia, acid accumulation within astrocytes transforms the membranes of these cells into an impenetrable barrier for at least an hour. Deciphering how acid differentially creates these effects in ischemia may shed light on how astrocytes modulate brain infarction from ischemia.

ACKNOWLEDGMENTS

This work was supported by a grant from the National Institute of Neurological Disorders and Stroke (NS-19108), A Zenith Award from the Alzheimer's Association and the Brain Research Foundation of The University of Chicago. C.D.L. was supported by an MD/PhD Training Grant in Growth and Development (HD-07009) from the NICHD and an NRSA (F31-MH11126) from the N1MH.

REFERENCES

Anders JJ (1988): Lactic acid inhibition of gap junctional intercellular communication in vitro astrocytes as measured by fluorescence recovery after laser photobleaching. Glia 1: 371–379.

Aronson PS (1985): Kinetic properties of the plasma membrane Na^+-H^+ exchanger. Annu Rev Physiol 47: 545–560.

Ballanyi K, Grafe P, ten Bruggencate G (1987): Ion activities and potassium uptake mechanisms of glial cells in guinea-pig olfactory cortex slices. J Physiol (Lond) 273: 295–316.

Barnes S, Bui Q (1991): Modulation of calcium-activated chloride current via pH-induced changes of calcium channel properties in cone photoreceptors. J Neurosci 11: 4015–4023.

Barres BA, Chun LLY, Corey DP (1990): Ion channels in vertebrate glia. Annu Rev Neurosci 13: 441–474.

Blomqvist P, Lindvall O, Stenevi U, Wieloch T (1985): Cyclic AMP concentrations in rat neocortex and hippocampus during and following incomplete ischemia: Effects of central noradrenergic neurons, prostaglandins, and adenosine. J Neurochem 44: 1345–1353.

Bormann J, Hamill OP, Sakmann B (1987): Mechanism of anion permeation through channels gated by glycine and γ-aminobutyric acid in mouse cultured spinal neurons. J Physiol (Lond) 385: 243–286.

Boron WF, McCormick WC, Roos A (1979): pH regulation in barnacle muscle fibers: Dependence on intracellular and extracellular pH. Am J Physiol 237: C185–C193.

Bouvier M, Szatkowski M, Amato A, Attwell D (1992): The glial glutamate uptake carrier countertransports pH-changing anions. Nature 360: 471–474.

Boyarsky G, Ransom B, Schlue WR, Davis MBE, Boron WF (1993): Intracellular pH regulation in single cultured astrocytes from rat forebrain. Glia 8: 241–248.

Brookes N, Turner RJ (1994): K^+-induced alkalinization in mouse cerebral astrocytes mediated by reversal of electrogenic Na^+-HCO_3^- cotransport in cortical neurons in primary cultures. Am J Physiol 267: C1633–1640.

Bruckner R, Biesold D (1981): Histochemistry of glycogen deposition in perinatal rat brain: Importance of radial glial cells. J Neurocytol 10: 749–757.

Brune T, Fetzer S, Backus KH, Deitmer JW (1994): Evidence for electrogenic sodium-bicarbonate cotransport in cultured rat cerebellar astrocytes. Pflügers Arch 429: 64–71.

Cataldo AM, Broadwell RD (1986): Cytochemical identification of cerebral glycogen and glucose-6-phosphate activity under normal and experimental conditions. I. Neurons and glia. J Electron Microsc Tech 3: 413–437.

Chesler M (1990): The regulation and modulation of pH in the nervous system. Prog in Neurobiol 34: 401–427.

Chesler M, Kraig RP (1987): Intracellular pH of astrocytes increases rapidly with cortical stimulation. Am J Physiol 253: 666–670.

Chesler M, Kraig RP (1989): Intracellular pH transients of mammalian astrocytes. J Neurosci 9: 2011–2019.

Chopp M, Walton DR, Smith MB, Welch KMA (1987): Intracellular acidosis during and after cerebral ischemia: In vivo nuclear magnetic resonance study of hyperglycemia in cats. Stroke 18: 919–923.

Chopp M, Vande Linde AMQ, Chen H, Knight R, Helpern JA, Welch KMA (1989): Chronic cerebral intracellular alkalosis following forebrain ischemic insult in rats. Stroke 21: 463–466.

Connors BW, Bernardo LS, Prince DA (1984): Carbon dioxide sensitivity of dye coupling among glia and neurons of the neocortex. J Neurosci 4: 1324–1330.

Coulter KL, Perier F, Radeke CM, Vandenberg CA (1995): Identification and molecular localization of pH-sensing domain for the inward rectifier potassium channel HIR. Neuron 15: 1157–1168.

Deitmer JW, Schlue WR (1989): An inwardly directed electrogenic sodium-bicarbonate co-transport in leech glial cell. J Physiol (Lond) 411: 179–194.

Devries SH, Schwartz EA (1992): Hemi-gap-junction channels in solitary horizontal cells of the catfish retina. J Physiol (Lond) 445: 201–230.

Duszyk M, Liu D, French AS, Man SFP (1995): Evidence that pH-titratable grouped control the activity of a large epithelial chloride channel. Biochem Biophys Res Commun 215: 355–360.

Ekholm A, Katsura K, Siesjo BK (1993): Coupling of energy failure and dissipative K^+ flux during ischemia: Role of preischemic plasma glucose concentration. J Cereb Blood Flow Metab 13: 193–200.

Erecinska M, Silver IA (1989): ATP and brain function. J Cereb Blood Flow Metab 9: 2–19.

Finkbeiner SM (1993): Glial calcium. Glia 9: 83–104.

Fitts RH, Holoszy JO (1976): Lactate and contractile force in frog muscle during development of fatigue and recovery. Am J Physiol 231: 430–433.

Fraser DD, Duffy S, Angelides KJ, Perez-Velazquez JL, Kettenmann H, MacVicar BA (1995): GABA/benzodiazepine receptors in acutely isolated hippocampal astrocytes. J Neurosci 15: 2720–2732.

Friede RL, Van Houten WH (1961): Relations between post-mortem alterations and glycolytic metabolism in the brain. Exp Neurol 4: 197–204.

Gemba T, Oshima T, Ninomiya M (1994): Glutamate efflux via the reversal of the sodium-dependent glutamate transporter caused by glycolytic inhibition in rat cultured astrocytes. Neuroscience 63: 789–795.

Giffard RG, Monyer H, Choi DW (1990): Selective vulnerability of cultured cortical glia to injury by extracellular acidosis. Brain Res 530: 138–141.

Goldberg WJ, Watson BD, Busto R, Kurchner H, Santiso M, Ginsberg MD (1984): Concurrent measurement of (Na^+, K^+)-ATPase activity and lipid peroxides in rat brain following reversible global ischemia. Neurochem Res 9: 1737–1747.

Goldman SA, Pulsinelli WA, Clarke WY, Kraig RP, Plum F (1989): The effects of extracellular acidosis on neurons and glia in vitro. J Cereb Blood Flow Metab 9: 471–477.

Gray PTA, Ritchie JM (1986): A voltage-gated chloride conductance in rat cultured astrocytes. Proc R Soc Lond (Biol) 228: 267–288.

Grichtchenko II, Chesler M (1994a): Depolarization-induced alkalinization of astrocytes in gliotic hippocampal slices. Neuroscience 62: 1071–1078.

Grichtchenko II, Chesler M (1994b): Depolarization-induced acid secretion in gliotic hippocampal slices. Neuroscience 62: 1057–1070.

Hansen AJ (1978): The extracellular potassium concentration in brain cortex following ischemia in hypo- and hyperglycemic rats. Acta Physiol Scand 102: 324–329.

Harold DE, Walz W (1992): Metabolic inhibition and electrical properties of type-1-like cortical astrocytes. Neuroscience 47: 203–211.

Hertz L, Shousboe A (1975): Ions and energy metabolism of the brain at the cellular level. In Pfeiffer CC, Smythies JR (eds): International Review of Neurobiology, Vol 10. New York: Academic Press, pp 141–211.

Hille B (1992): Ionic Channels of Excitable Membranes, 2nd ed. Sunderland MA: Sinauer.

Hillered L, Ernster L, Siesjo BK (1984): Influence of in vitro lactic acidosis and hypercapnia on respiratory activity of isolated rat brain mitochondria. J Cereb Blood Flow Metab 4: 430–437.

Hochachka PW, Mommsen TP (1983): Protons and anaerobiosis. Science 219: 1391–1397.

Hutter OF, Warner AE (1967): The pH sensitivity of the chloride conductance of frog skeletal muscle. J Physiol (Lond) 189: 403–425.

Kaila K, Voipio J (1987): Postsynaptic fall in intracellular pH induced by GABA-activated bicarbonate conductance. Nature 330: 163–165.

Katsura K, Ekholm A, Asplund B, Siesjo BK (1991): Extracellular pH in the brain during ischemia: Relationship to the severity of lactic acidosis. J Cereb Blood Flow Metab 11: 597–599.

Katsura K, Ekholm A, Siesjo BK (1992): Tissue P_{CO_2} in brain ischemia related to lactate content in normo- and hypercapnic rats. J Cereb Blood Flow Metab 14: 243–250.

Katsura K, Kristian T, Smith ML, Siesjo BK (1994): Acidosis induced by hypercapnia exaggerates ischemic brain damage. J Cereb Blood Flow Metab 14: 243–250.

Kauppinen RA, Enkvist K, Holopainen I, Akerman KEO (1988): Glucose deprivation depolarizes plasma membrane of cultured astrocytes and collapses transmembrane potassium and glutamate gradients. Neuroscience 1: 283–289.

Knull HR, Khandelwahl RL (1982): Glycogen metabolizing enzymes in the brain. Neurochem Res 7: 1307–1317.

Koizumi J (1974): Glycogen in the central nervous system. Prog Histochem Cytochem 6: 1–35.

Kraig RP, Ferreira-Filho CR, Nicholson C (1983): Alkaline and acid transients in cerebellar microenvironment. J Neurophys 49: 831–850.

Kraig RP, Pulsinelli WA, Plum F (1985a): Hydrogen ion buffering during complete brain ischemia. Brain Res 342: 281–290.

Kraig RP, Pulsinelli WA, Plum F (1985b): Heterogenous distribution of hydrogen and bicarbonate ions during complete brain ischemia. Prog in Brain Res 63: 155–166.

Kraig RP, Pulsinelli WA, Plum F (1986): Carbonic acid buffer changes during complete brain ischemia. Am J Physiol 250: R348–R357.

Kraig RP, Chesler M (1988): Dynamics of volatile buffers in brain cells during spreading depression. In Somjen GG (ed): Mechanisms of Cerebral Hypoxia and Stroke. New York: Plenum, pp 279–289.

Kraig RP, Chesler M (1990): Astrocytic acidosis in hyperglycemic and complete ischemia. J Cereb Blood Flow Metab 10: 104–114.

Kraig RP, Lascola CD (1994): Astrocytic H^+ and Ca^{2+} changes: Implications for ischemic brain injury. In Hartmann A, Yatsu F, Kuchisky W (eds): Cerebral Ischemia and Basic Mechanisms. Berlin: Springer-Verlag, pp 64–76.

Kraig RP, Lascola CD, Caggiano A (1995): Glial response to brain ischemia. In Kettenmann H, Ransom B (eds): Neuroglial Cells. New York: Oxford University Press, pp 965–976.

Kraig RP, Petito CK (1989): Interrelation of proton and volume regulation in astrocytes. In Ginsberg MD, Dietrich WD (eds): Cerebrovascular Disease. New York: Raven Press, pp 239–246.

Krnjevic K, Fifkova E (1965): The value of ultramicroanalysis of lactic acid in tracing the presentation of Leao's cortical spreading depression. J Neurol Sci 2: 385–392.

Lang F, Oberleithner H, Kolb HA, Paulmichl M, Vokl H, Wang W (1988): Interaction of intracellular pH and cell membrane potential. In Haussinger D (ed): pH Mechanisms and control. New York: Academic Press, pp 27–42.

Lascola CD, Kraig RP (1995): Brain cell acid-base changes measured with ion-selective microelectrodes. In Kraicer J, Dixon SJ (eds): Measurements and Manipulation of Intracellular Ions. Academic Press, New York, pp 214–232.

Lascola CD, Kraig RP (1996): Whole-cell chloride currents in rat astrocytes accompany changes in cell morphology. J Neurosci 16: 2532–2545.

Ljunnggren B, Norberg K, Siesjo BK (1974): Influence of tissue acidosis upon restitution of brain energy metabolism following total ischemia. Brain Res 77: 173–183.

Lowry OH, Passonneau JV, Hasselberger FH, Shulz DW (1964): Effect of ischemia on known substrates and cofactors of the glycolytic pathway in brain. J Biol Chem 239: 18–30.

Lukacs GL, Moczydlowski E (1990): A chloride channel from lobster walking leg nerves. Characterization of single-channel properties in planar bilayers. J Gen Physiol 96: 707–733.

Mabe H, Blomqvist P, Siesjo BK (1983): Intracellular pH in the brain following transient ischemia. J Cereb Blood Flow Metab 3: 109–114.

Mason GF, Behar KL, Rothman DL, Shulman RG (1992): NMR determination of intracerebral glucose concentration and transport kinetics in rat brain. J Cereb Blood Flow Metab 12: 448–455.

Mellergard PE, Ouyang YB, Siesjo BK (1994a): The regulation of intracellular pH is strongly dependent on extracellular pH in cultured rat astrocytes and neurons. Acta Neurochir Suppl (Wien) 60: 34–37.

Mellergaard P, Ou-yang Y, Siesjo BK (1994b): Relationship between intra- and extracellular pH in primary cultures of rat astrocytes. Am J Physiol 267: 581–589.

Meyer FB, Anderson RE, Sundt TM, Yaksh T (1986): Intracellular brain pH, indicator tissue perfusion, electroencephalography, and histology in severe and moderate focal cortical ischemia in the rabbit. J Cereb Blood Flow Metab 6: 71–78.

Myers ME, Yamaguchi S (1977): Nervous system effects of cardiac arrest in monkeys. Arch Neurol 35: 64–75.

Myers RE (1979): Lactic acid accumulation as a cause of brain edema and cerebral necrosis resulting from oxygen deprivation. In Korobkin R, Guilleminault C (eds): Advances in Perinatal Neurology. New York: Spectrum, pp 88–114.

Nedergaard M, Goldman SA, Desai S, Pulsinelli WA (1991): Acid-induced death in neurons and glia. J Neurosci 11: 2489–2497.

Newman EA (1991): Sodium-bicarbonate cotransport in retinal Muller (glial) cells of the salamander. J Neurosci 11: 3972–3983.

Newman EA (1996): Acid efflux from retinal glial cells generated by sodium bicarbonate cotransport. J Neurosci 11: 3972–3983.

O'Connor ER, Sontheimer H, Ransom BR (1994): Rat hippocampal astrocytes exhibit electrogenic sodium-bicarbonate co-transport. J Neurophysiol 72: 2580–2589.

Ou-Yang Y, Kristian T, Mellergard P, Siesjo BK (1994): The influence of pH on glutamate- and depolarization-induced increases of intracellular calcium concentration. Brain Res 646: 65–72.

Pappas CA, Ransom BR (1994): Depolarization-induced alkalosis (DIA) in rat hippocampal astrocytes. J Neurophysiol 72: 2816–2825.

Petito CK, Babiak T (1982): Early proliferative changes in astrocytes in post-ischemic noninfarcted rat brain. Ann Neurol 11: 510–518.

Phelps CH (1975): An ultrastructural study of methionine sulphoximine-induced glycogen accumulation in astrocytes of the mouse cerebral cortex. J Neurocytol 4: 479–490.

Plum F (1983): What causes infarction in ischemic brain? The Robert Wartenberg lecture. Neurology 33: 222–233.

Pollard TD, Cooper JA (1986): Actin and actin-binding proteins: A critical evaluation of mechanisms and functions. Annu Rev Biochem 55: 987–1035.

Poulsen JH, Fischer H, Illek B, Machen TE (1994): Bicarbonate conductance and pH regulatory capability of cystic fibrosis transmembrane conductance regulator. Proc Natl Acad Sci USA 91: 5340–5344.

Prod'hom B, Pietrobon D, Hess P (1989): Interactions of protons with single open L-type calcium channels. Location of protonation site and dependence of proton-induced current fluctuations on concentration and species of permeant ion. J Gen Physiol 94: 23–42.

Pulsinelli WA, Waldman S, Rawlinson D, Plum F (1982): Hyperglycemia converts ischemic neuronal damage into brain infarction. Neurology 32: 1239–1246.

Pulsinelli WA, Duffy TE (1983): Regional energy balance in rat brain after transient forebrain ischemia. J Neurochem 40: 1500–1503.

Pulsinelli WA (1985): Selective neuronal vulnerability: Morphological and molecular characteristics. Prog Brain Res 63: 29–37.

Pulsinelli WA, Kraig RP, Plum F (1985): Hyperglycemia, cerebral acidosis, and ischemic brain damage. In Plum F, Pulsinelli W (eds): Cerebrovascular Diseases. New York: Raven Press, pp 201–205.

Rehncrona S, Rosen I, Siesjo B (1981): Brain lactic acidosis and ischemic cell damage. I. Biochemistry and neurophysiology. J Cereb Blood Flow Metab 1:297–311.

Reinhart PH, Pfeiffer B, Spengler S, Hamprecht B (1990): Purification of glycogen phosphorylase from bovine brain and immunocytochemical examination of rat glial primary cultures using monoclonal antibodies raised against this enzyme. J Neurochem 54: 1474–1483.

Ransom BR, Walz W, Davis PK, Carlini WG (1992): Anoxia-induced changes in extracellular K^+ and pH in mammalian central white matter. J Cereb Blood Flow Metab 12: 593–602.

Roos A, Boron WF (1981): Intracellular pH. Physiol Rev 61: 296–433.

Rosenberg PA, Dichter MA (1985): Glycogen accumulation rat cerebral cortex in dissociated cell culture. J Neurosci Methods 15: 101–112.

Sahlin K, Edstrom L, Sjoholm H, Hultman E (1981): Effects of lactic acid accumulation and ATP decrease on muscle tension and relaxation. Am J Physiol 240: C121–C126.

Shrode LD, Putnam RW (1994): Intracellular pH regulation in primary rat astrocytes and C6 glioma cells. Glia 12: 196–210.

Siesjo BK, Bendek G, Koide TY, Westberg E, Weiloch T (1985): Influences of acidosis on lipid peroxidation in brain tissue *in vitro*. J Cereb Blood Flow Metab 5: 253–260.

Siesjo BK, Katsura KI, Zhao Q, Folbergrova J, Pahlmark K, Siesjo P, Smith ML (1995): Mechanisms of secondary brain damage in global and focal ischemia: A speculative synthesis. J Neurotrauma 12: 943–956.

Siemkowicz E, Hansen A (1981): Brian extracellular ion composition and EEG activity following 10 min ischemia in normo- and hypoglycemic rats. Stroke 12: 236–240.

Silver IA, Erecinska M (1992): Ion homeostasis in rat brain in vivo: Intra- and extracellular [Ca^{2+}] and [H^+] in the hippocampus during recovery from short-term, transient ischemia. J Cereb Blood Flow Metab 12: 759–772.

Simon RP, Niro M, Gwinn R (1993): Brain acidosis induced by hypercarbic ventilation attenuates focal ischemic injury. J Pharm Exp Ther 267: 1428–1431.

Smith ML, von Hanwehr R, Siesjo BK (1986): Changes in extra- and intracellular pH in the brain during and following ischemia in hyperglycemic and moderately hypoglycemic rats. J Cereb Blood Flow Metab 6: 574–583.

Stewart, PA (1981): How to Understand Acid-Base. Elsevier Press, New York.

Sunagawa S, Buist RJ, Hruska FE, Sutherland GR, Pelling J (1994): Hydrogen ion compartmentation during and following cerebral ischemia evaluated by ^{31}P NMR spectroscopy. Brain Res 641: 328–332.

Swain MS, Blei AT, Butterworth RF, Kraig RP (1991): Intracellular pH rises and astrocytes swell after portacaval anastomasosis in rats. Am J Physiol 261: 1491–1496.

Swanson RA, Choi DW (1993): Glial glycogen stores affect neuronal survival during glucose deprivation in vitro. J Cereb Blood Flow Metab 13: 162–169.

Swanson RA, Farrell K, Simon RP (1995): Acidosis causes failure of astrocytes glutamate uptake during hypoxia. J Cereb Blood Flow Metab 15: 417–424.

Tabcharani JA, Jensen JR, Riordan JR, Hanrahan JW (1989): Bicarbonate permeability of the outwardly rectifying anion channel. J Membr Biol 112: 109–122.

Tang CM, Dichter M, Morad M (1990): Modulation of the N-methyl-D-aspartate channel by extracellular H^+. Proc Natl Acad Sci USA 87: 6445–6449.

Tombaugh GC, Sapolski RM (1990): Mild acidosis protects hippocampal neurons from injury induced by oxygen and glucose deprivation. Brain Res 506: 343–345.

Tombaugh GC (1994): Mild acidosis delays hypoxic spreading depression and improves neuronal recovery in hippocampal slices. J Neurosci 14: 5635–5643.

Tse FW, Fraser DD, Duffy S, MacVicar BA (1992): Voltage-activated K^+ currents in acutely isolated hippocampal astrocytes. J Neurosci 12: 1781–1788.

Vaughan-Jones RD, Wu ML (1990): Extracellular H^+ inactivation of Na^+/H^+ exchange in the sheep cardiac Purkinje fibre. J Physiol (Lond) 428: 441–466.

Walz W, Hinks EC (1987): Extracellular hydrogen ions influence channel-mediated and carrier-mediated K^+ fluxes in cultured mouse astrocytes. Neuroscience 20: 341–346.

Walz W, Wuttke WA (1989): Resistance of astrocyte electrical membrane properties to acidosis changes in the presence of lactate. Brain Res 504: 82–86.

Welsh FA, Ginsberg MD, Rieder W, Budd WW (1980): Deleterious effect of glucose pretreatment on recovery from diffuse cerebral ischemia in the cat. II. Regional metabolite levels. Stroke 11: 353–363.

Yoshida S, Busto R, Martinez E, Scheinberg P, Ginsberg MD (1985): Regional brain energy metabolism after complete versus incomplete ischemia in the rat in the absence of severe lactic acidosis. J Cereb Blood Flow Metab 5: 490–501.

CHAPTER 30

Use of NMR Spectroscopy in Monitoring Cerebral pH and Metabolism During Systemic and Focal Acid-Base Disturbances

RISTO A. KAUPPINEN AND STEPHEN R. WILLIAMS
NMR Research Group, A. I. Virtanen Institute for Molecular Sciences, University of Kuopio, FIN-70211 Kuopio, Finland (R.A.K.); RCS Unit of Biophysics, Institute of Child Health, London, United Kingdom (S.R.W)

INTRODUCTION

Systemic acid-base disturbances have strong effects on brain function (Chapter 20). The precise role of brain energy metabolism and brain intracellular pH (pH_i) in mediation of these effects is unclear, since neither can be observed easily by traditional techniques. Recently developed NMR methods allow direct observation of both and can provide new insights into the role of pH regulation and into the functional consequences of its failure in pathological conditions. In this chapter we discuss NMR measurements of brain pH_i both *in vivo* and *in vitro* during systemic acid-base disturbances and relate this to information about cerebral physiology and metabolism under these conditions. We also discuss "focal" acid-base disturbances and describe NMR studies of pH_i in human brain disorders such as ischemia, head trauma, cancer, and epilepsy.

Compared with other methods (Chapters 6–8), NMR spectroscopy possesses several advantages for measuring pH_i. It is noninvasive, and allows direct monitoring of pH_i without the need to introduce exogenous probes, while simultaneously providing information about bioenergetic status (see Chapter 9). On the other hand, NMR provides less data with temporal and spatial resolution than the use of microelectrodes or optical probes to monitor $[H^+]$, and both of these techniques are superior in studying pH_i in single cells. Despite these penalties, ^{31}P NMR is now the method of choice to study regulation of pH_i *in vivo* in the brain of both animals and humans. The most commonly used indicator of pH_i is inorganic phosphate (Pi, see Fig. 1A). Unlike muscle, brain tissue has no endogenous indicator of pH_i that can be detected by ^1H NMR (Williams et al., 1985), since adult brain contains anserine and carnosine at concentrations too low to be utilized as a pH indicator by ^1H NMR.

pH and Brain Function, Edited by Kai Kaila and Bruce R. Ransom
ISBN 0-471-11838-9

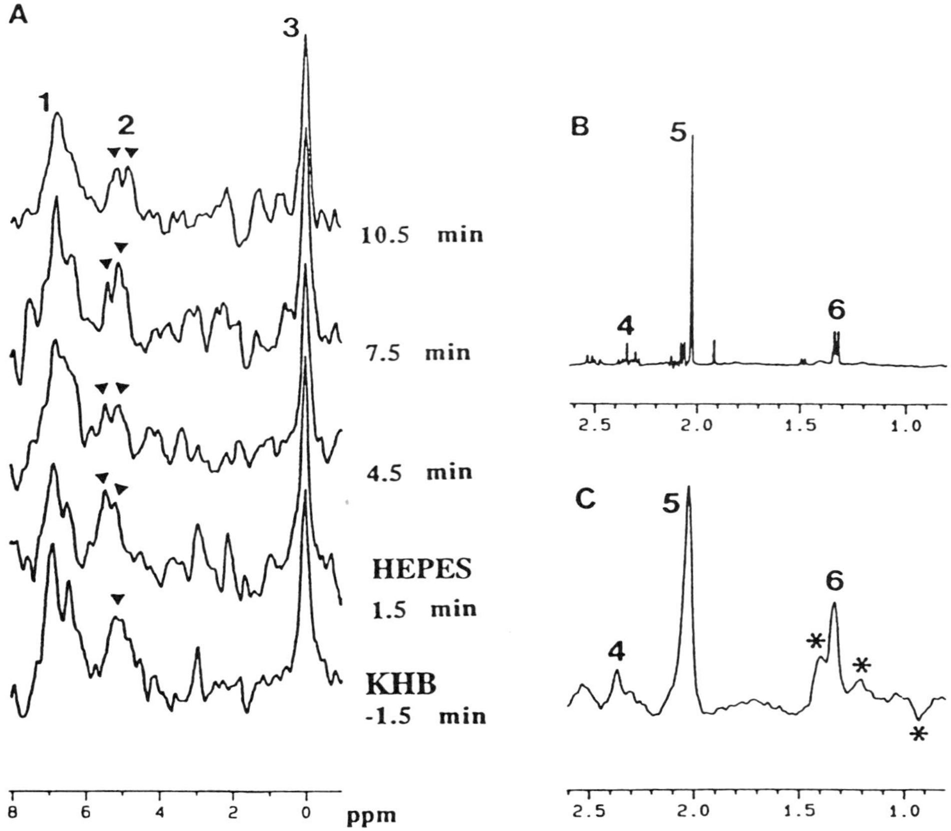

FIGURE 1. ^{31}P (**A**) and ^{1}H NMR (**C**) spectra of a superfused brain preparation and a ^{1}H NMR spectrum of their acid extract (**B**).
A cortical brain slice preparation was initially superfused in KHB, and HCO_3^--free buffer (HEPES) was introduced as indicated. Peaks are assigned as follows: 1, phosphomonoesters; 2, inorganic phosphate; 3, phosphocreatine; 4, glutamate; 5, N-acetylaspartate; 6, lactate. Triangles in conjunction with peak 2 in (A) indicate split Pi, and asterisks in (C) indicate signals from macromolecules. Details in Pirttilä and Kauppinen (1994). Reproduced with permission from The International Society for Neurochemistry.

The resonance frequency of Pi is sensitive to $[H^+]$, which rapidly equilibrates between dihydrogen and monohydrogen phosphate in the cell. Pi is particularly useful as an indicator of pH_i as its pK_a is 6.7. In order to determine the chemical shift of the pH_i indicator, one has to acquire signal from a volume of several milliliters over a period of several minutes. Therefore pH_i measured by ^{31}P NMR is both time and volume-averaged. It is generally accepted in the NMR literature that the Pi signal in the brain ^{31}P NMR spectrum originates only from cytoplasmic phosphate and that phosphate either from other subcellular compartments such as mitochondria or from extracellular space does not contribute to the peak. In some species, however, such as in sheep, the cerebral Pi peak consists of two components and the one at higher frequency (at more alkaline pH) has been assigned to Pi in extracellular space (Portman et al., 1991; see also Chapter 9). If this is the case, it would allow simultaneous monitoring of pH_i and extracellular pH (pH_o).

The main difficulty in interpreting of pH data from NMR lies in heterogeneity. As mentioned above the measurement of pH is volume-averaged, and is biased to those regions with high Pi, which contribute most to the NMR signal. Thus in a population of cells that respond differently to an insult, correlations between parameters such as concentrations of ATP and lactate with pH may be misleading. As a simple example, consider a situation in which half the cells maintain normal ATP levels, with low Pi and lactate, while the other half have undergone complete energy failure and contain no ATP, with high levels of Pi and lactate. In the volume-averaged spectrum, all the ATP signal will come from the normal cells, whereas most of the Pi and lactate signals will be from the ischemic population. It would be biologically incorrect to associate the measured pH with the levels of either ATP or lactate estimated from the spectra, since the former would be in a different compartment, and the latter would be underestimated by dilution from cells at normal pH_i with low lactate. These issues are discussed further below.

SYSTEMIC ACID-BASE DISTURBANCES

Brain pH_i and Metabolism During Changes in Blood pH

Microelectrode studies have shown that during prolonged isocapnic acidosis, when the blood pH is ~7, pH in the extracellular compartment (ECF) is reduced from normal values (Javaheri et al., 1983). However, the pH changes in ECF are only about 20% of those encountered in the blood. The small response in ECF following isocapnic acidosis is thought to reflect selective permeability of the blood-brain barrier (BBB; see Chapters 27, 32) and indeed, it has been recently shown, using NMR imaging methods, that permeability to gadolinium remains unaffected even at a blood pH of 7 (Adler et al., 1993). Thus the failure to detect a drop in brain pH_i by ^{31}P NMR during blood acidification is not unexpected (Adler et al., 1990, 1993). However, changes in the volume-averaged and time-averaged pH_i of less than 0.05 pH unit may go undetected if the signal-to-noise ratio or spectroscopic resolution are inadequate. In any event, such changes, if present during isocapnic acidosis, are expected to be of no major pathophysiological significance.

Levels of brain high-energy phosphates remain unaffected even after prolonged metabolic acidosis. It appears, therefore, that the mammalian brain is protected from acidosis provided that P_{CO_2} is maintained. Based on the NMR data, it has been suggested that *in vivo* there are three pH compartments (Portman et al., 1991) in the brain and that these correspond to intracellular pH, extracellular pH, and blood pH. Furthermore, all these compartments have different mechanisms to maintain stable $[H^+]$. The extracellular compartment can be further divided into two subcompartments: bulk cerebrospinal fluid (CSF) and interstitial fluid. Studies of animal models have also shown that dysfunctions of the nervous system associated with metabolic acidosis, and possibly also with metabolic alkalosis, are not caused directly by energy failure but rather by more specific mechanisms.

Effects of Altered P_{CO_2} on Cerebral pH and Metabolism

Hypocapnia. It appears that at either extreme hypocapnia ($P_{CO_2} < 10$ mmHg) or extreme hypercapnia ($P_{CO_2} > 200$ mmHg) the responses in the CNS are equivalent, namely, there is a severe depression of electrical activity. However, modest changes of P_{CO_2} are associated with increased excitation (hypocapnia) or decreased neuronal activity (hypercapnia). ^{31}P NMR studies have indicated that there is a large variation in responses to severe hypocapnia ($P_{CO_2} < 20$ mmHg or less) from species to species. Cerebral pH_i in anaesthetised sheep (Cady et al., 1987), rat (Adler et al., 1988; Nioka et al., 1987) or in dog (Sieber et al., 1992) remained unaffected under these conditions but became slightly alkaline by 0.15 pH unit in rabbit brain following reduction of P_{CO_2} to 15 mmHg (Petroff et al., 1985). In lamb, cerebral blood flow (CBF) is reduced by about 40% to ~30 ml 100 g^{-1} min^{-1} (Gardiner, 1980) and in dog by a similar percentage to 22 ml 100 g^{-1} min^{-1} during hypocapnia (Sieber et al., 1992).

Hypocapnia *in vivo* does not influence cerebral high-energy phosphates, phosphocreatine (PCr) and ATP, provided that the blood glucose concentration is maintained above 2 mM (Sieber et al., 1992). This is consistent with observations that oxygen and glucose consumption in the brain do not change during hypocapnia (Gardiner, 1980; Sieber et al., 1992). It has been reported that brain lactate increases ~100% at $P_{CO_2} < 15$ mmHg as detected by ^{1}H NMR spectroscopy *in vivo* (Petroff et al., 1985) (See Fig. 1B for an example of a proton spectrum). Lactate accumulation under these conditions (see Chapter 5) is not caused by energy failure, which could result from reduced CBF, but can plausibly be argued to be a result of the activation of glycolysis. As pH_i changes at reduced P_{CO_2}, both free cytoplasmic ADP ($[ADP]_f$) and AMP ($[AMP]_f$) increase (Petroff et al., 1985; Nioka et al., 1987). Taken together, these changes in $[ADP]_f$, $[AMP]_f$, and pH_i, will stimulate glycolysis via effects on phosphofructokinase.

Hypocapnia renders brain energy metabolism susceptible to reduced blood glucose levels. It has been previously shown that at a blood glucose concentration of 1 mM, superimposed on hypocapnia, ($P_{CO_2} < 15$ mmHg) leads to severe energy failure (Sieber et al., 1992). This combination of hypoglycemia and hypocapnia did not influence CBF, so that the energy failure was not caused by hypoxia. Previous NMR studies of superfused cortical brain slices have indicated that the energy state of the preparation is not compromised at low CO_2 tension in the absence of HCO_3^- (Pirttilä and Kauppinen, 1994), even at a pH_o of 7.8 (Brooks and Bachelard, 1992), but that brain lactate concentration increases at alkaline pH_i (Pirttilä and Kauppinen, 1993). It may be therefore that at alkaline pH_i *in vivo*, such as that observed during hypocapnia, the supply of glycolytic substrate for mitochondrial oxidation becomes limiting at glucose concentrations higher than normal due to an increased production of lactate.

Hypercapnia. Acute exposure of animals to hypercapnia (P_{CO_2} of >50 mmHg) inevitably increases CBF in a dose-dependent manner so that, at a P_{CO_2} of 120 mmHg, CBF has doubled relative to normocapnia (Tomlinson et al., 1993). ^{31}P NMR studies have uniformly demonstrated that hypercapnia causes brain pH_i to decrease and therefore the difference between blood and brain pH_i is reduced (Cady et al., 1987; Nioka et al., 1987; Adler et al., 1988). Cerebral acidification by elevated P_{CO_2} is not influenced by the anesthesia used in most animal studies (Barrere et al., 1990). Following a return to normocapnia, brain pH_i recovers faster than blood pH, and an overshoot of cerebral pH_i is a consistent finding (Nioka et al., 1987; Adler et al., 1988).

Changes in pH_i during and following hypercapnia have been used in calculations of intrinsic buffering power (B_i) in brain *in vivo*. Estimates of B_i of 36 and 45 mM buffer base per pH unit in the brain of rabbit (Petroff et al., 1985) and rat (Adler et al., 1988), respectively have been obtained. It is evident from the ^{31}P NMR data that brain "buffering capacity" follows P_{CO_2} and brain $[HCO_3^-]$. As P_{CO_2} is elevated to as high as 130 mmHg, the apparent buffering ability of the brain increases to 140 mM buffer base per pH unit in rat brain (Adler et al., 1988). In anesthetized rats evolution of "full buffering capacity" requires some 30 min to develop during hypercapnia and subsequent recovery (Adler et al., 1988). This slow phase of $[H^+]$ "buffering" may represent operations of transmembrane acid extrusion and/or alkali loading mechanisms in brain. Both acute and chronic hyponatremia (plasma $[Na^+] < 110$ mM) have been shown to reduce cerebral buffering as calculated from responses of pH_i to hypercapnia measured by ^{31}P NMR (Adler et al., 1993).

In the majority of species studied by ^{31}P NMR, the low concentration of extracellular Pi, together with the small relative volume of the extracellular compartment means that the contribution from intracellular Pi masks any signal from extracellular Pi (see Chapter 9). Thus only pH_i is obtained. In sheep, however, two Pi peaks are detected, as the Pi concentration in cerebrospinal fluid and the interstitial space is high enough to rise to a strong NMR signal, and appears to make it possible to monitor simultaneously both pH_i and pH_o (Portman et al., 1991; Swain et al., 1991). The two Pi peaks have been assigned as intracellular (with smaller chemical shift) and extracellular (with larger chemical shift, see Fig. 1A for a split Pi peak). Therefore a unique opportunity exists to study individual pH compartments noninvasively in sheep. It was shown that the three separate brain pH compartments corresponding to blood, extracellular space, and the intracellular compartment, behaved differently during exposure to altered P_{CO_2} (Portman et al., 1991). Arterial pH and pH_o were related more directly to each other than was either to pH_i, indicating that the intracellular compartment has an efficient means of regulating $[H^+]$. The pH_o determined by ^{31}P NMR was higher than previously reported from invasive microelectrode studies. This result may suggest that tissue damage following opening of the skull can influence pH in the extracellular space.

In contrast to hypocapnia, hypercapnia in anesthetized animals causes a decrease in cerebral PCr and an increase in Pi that is dependent on the degree of hypercapnia, but even at a P_{CO_2} of 140 mmHg, ATP remains unaffected (Petroff et al., 1985; Nioka et al., 1987). The decrease in PCr can be interpreted to reflect a shift in the creatine kinase equilibrium caused by an increase in $[H^+]$. However, other factors, including anesthesia, may also be involved, since a previous ^{31}P NMR study on awake rats showed unchanged energy state by ^{31}P NMR despite acidosis induced by hypercapnia (P_{CO_2} of 90 mmHg; Barrere et al., 1990). This finding is consistent with ^{31}P NMR results from superfused neonatal cortical brain slices showing that the PCr level remained constant until external P_{CO_2} was in excess of physiological levels (i.e., >200 mmHg; Espanol et al., 1992). These data can be reconciled with the requirement that creatine kinase remains at equilibrium if there are compensating changes in free ADP in the latter experiments conducted without anesthesia.

Studies using ^1H NMR have shown that the cerebral lactate level does not change during hypercapnia (P_{CO_2} of about 140 mmHg) *in vivo* (Petroff et al., 1985; Behar et al., 1987). However, accumulation of lactate has been shown during severe hypercapnic energy failure of cerebral cortex *ex vivo* (Espanol et al., 1992). An unexpected decrease in the ^1H NMR detected glutamate peak during prolonged hypercapnia *in vivo* occurs (Behar et al., 1987). In the steady state, there is a carbon flux from glucose through the metabolic glutamate pool, and in recent years NMR methods have been used to probe this noninvasively (Gruetter et al., 1994; Kauppinen et al., 1994). Taking this into account, hypercapnia may divert the supply of glycolytic substrates from mitochondrial oxidation towards lactate production and thereby cause a need for oxidation of endogenous substrates (Agardh et al., 1981). If this were the case, cerebral dysfunction caused by hypercapnia may be partially due to "hypoglycaemia-like" conditions, which would result in an inability to maintain membrane polarization despite an apparently normal energy state (Kauppinen and Nicholls, 1986).

NMR STUDIES OF HUMAN BRAIN DURING ALTERED P_{CO_2}

The behavior of pH_i and the energy metabolism in human brain during altered P_{CO_2} are of major clinical interest. In studies of awake humans using ^{31}P NMR, it has been shown that mild hypercapnia ($P_{CO_2} > 50$ mmHg) reduces cerebral pH_i, whereas mild hypocapnia (P_{CO_2} ranging from 19–25 mmHg) appears to be ineffective (Jensen et al., 1988). In this study, cerebral high-energy phosphates were not quantified. Using anatomically localised ^{31}P NMR spectroscopy, Cadoux-Hudson and associates (Cadoux-Hudson et al., 1990a) showed greater acidification at a given elevated P_{CO_2} level in human white matter than in grey matter. The reasons for this apparent difference in pH_i during hypercapnia may reflect distinct responses of CBF in grey and white matter. Alternatively pH_i regulation, buffering power, or both may vary between white matter and grey matter. An alternative way of studying the role of the bicarbonate-dependent system on pH_i is to inhibit carbonic anhydrase (CA). In a recent study, cerebral pH_i in humans did not change following inhibition of CA with acetazolamide, despite the continuous production of CO_2 and the reduction of pH_o (Vorstrup et al., 1989). Therefore, compensatory internal H^+ neutralizing mechanisms must operate in the presence of acetazolamide.

NMR STUDIES OF CEREBRAL pH_i AND METABOLISM USING BRAIN SLICES

In order to study the contribution of the transmembrane mechanisms that are responsible for pH_i regulation in the intact brain, *ex vivo* superfused brain slice preparations have been used as models. In this preparation, the extracellular environment can be controlled and manipulated to mimick acid-base and P_{CO_2} changes encountered *in vivo,* and drugs that specifically affect transmembrane mechanisms of net H^+ movements are easily applied. In the *ex vivo* preparations, complications associated with anesthetics, as well as effects of the BBB, are eliminated. In the ^{31}P NMR studies of *ex vivo* brain preparations, a contribution from Pi in extracerebral tissues is also avoided.

The pH_i is around 7.2–7.3 in cortical brain slices from guinea pig or rat superfused in Krebs-Hensleit bicarbonate buffer (KHB) with an external P_{CO_2} of about 40 mmHg (pH_o of 7.5) (Brooks et al., 1989; Brooks and Bachelard, 1992; Pirttilä and Kauppinen, 1992, 1993, 1994). This is some 0.1–0.2 pH units higher than in the cerebral cortex *in vivo* (see Chapter 8). Applications of either amiloride or 4,4′-diisothiocyanatostilbene-2,2′-disulfonic acid (DIDS) (Pirttilä and Kauppinen, 1992, 1994) to the superfusions with KHB did not influence pH_i, indicating that neither the amiloride-sensitive Na^+/H^+ exchangers nor the Cl^-/HCO_3^- exchanger is significant in maintaining pH_i under normal conditions (see Chapters 12 and 14). Reduction of external Na^+ to 20 mM or below resulted in acidification of brain slices (Pirttilä and Kauppinen, 1992).

Elevation of external CO_2 to about 100 mmHg in the presence of 26 mM HCO_3^-, which reduces pH_o to 7.0, results in a drop of pH_i to 7.0, without a change in the concentration of high-energy phosphates (Pirttilä and Kauppinen, 1993). Further increases in external P_{CO_2} up to 540 mmHg (P_{O_2} 236 mmHg) acidifies pH_i down to 6.3 with concomitant energy failure, though it would be unwise to ascribe the effects of such an extreme intervention solely to effects on extracellular pH (Espanol et al., 1992). Both pH_i and the energy state of slice preparations recover completely following restoration of external P_{CO_2} to 40 mmHg (Espanol et al., 1992). A previous study has indicated that the recovery from acid load by reduced pH_o of 6.4 may be slower in the presence of amiloride, DIDS, or acetazolamide (Brooks and Bachelard, 1992) so that intact brain may use a number of different mechanisms to restore pH_i.

The experiments carried out in the nominal absence of CO_2 and HCO_3^- have suggested heterogeneity of pH_i in cortical brain slices (Pirttilä and Kauppinen, 1994). Following removal of CO_2, the Pi resonance splits, and the two peaks indicate that there is a pH difference of about 0.4 pH unit (Fig. 1A). The intensities of the two signals are fairly similar, a feature that tends to suggest that each peak corresponds to a given cytoplasmic origin, possibly in different cell types. Interestingly enough, both DIDS and the absence of external Cl^- prevent the splitting of Pi in conjunction with HCO_3^-/CO_2 removal, but do not prevent the increase of pH_i (Pirttilä and Kauppinen, 1994). Under these conditions the restoration of pH_i to about 7.2 is retarded. As there is no external phosphate in the medium that would contribute to the split of Pi peak, these data strongly indicate a differential response of neural cell population to the absence of the HCO_3^-/CO_2 system.

CEREBRAL pH_i AND METABOLISM AS MONITORED BY NMR SPECTROSCOPY DURING ENERGY FAILURE

A common theme throughout this chapter is the use of NMR to measure pH in conjunction with measurements of other metabolites, and this becomes particularly relevant in the study of ischemia (Chapters 28, 29, 31). For example, numerous studies have confirmed the primary role of anaerobic glycolysis in generating acid during ischemia, by revealing a close correlation between lactate production, energy failure, and acidosis (Crockard et al., 1987; Bolas et al., 1988; Naritomi et al., 1988; Kauppinen and Williams, 1990). Although a close association between pH and lactate concentration is generally seen during ischemia, the precise relationship is obviously a function of brain buffering capacity, and contributions from the HCO_3^-/CO_2 system will influence this. There are also effects dependent on anesthesia, since a strong correlation between lactate concentration and pH has been reported during anoxia in conscious rats; the anoxia was weaker under Nembutal anesthesia, and abolished using chloral hydrate (Lotito et al., 1989). Another example of lactate concentration–pH uncoupling, demonstrated by NMR occurs in mild hypoxia, where lactate production is stimulated before there are any changes in pH or high energy phosphate (i.e., at higher P_{O_2}; Allen et al., 1992). Such an effect has not been reported in studies of ischemia and may represent a relatively unstable threshold region, in which failure to fully meet ATP requirements by oxidative metabolism can be compensated for by stimulation of anaerobic glycolysis. The mismatch between ATP production rates from aerobic and anaerobic metabolism is so large that even a further slight decline in rates of oxidative ATP production would push the system into complete energy failure. However, while ATP synthesis is maintained, the extra H^+ resulting from lactate production can be extruded by energy-dependent mechanisms such as the Na^+/H^+ exchanger.

As described in the introduction to this chapter and elsewhere (Brooks et al., 1989; Portman et al., 1991), interpretation of pH changes is complicated by the issues of heterogeneity and compartmentation. Under some circumstances, heterogeneity can be addressed directly, as two Pi resonances (and thus two pH compartments) can sometimes be identified. Nioka and coworkers (1990) and Sunagawa and associates (1994) have both reported split Pi resonances in studies of animal models of ischemia and recovery. The authors of these two papers offer quite different interpretations: Nioka and coworkers favor metabolic heterogeneity resulting from different levels of hypoxic stress, whereas Sunagawa and associates postulate that the two compartments correspond to neuronal and glial cells. Although the latter hypothesis is attractive, and would fit in with other (non-NMR) data indicating differential responses of neurons and glia to ischemia (Kraig et al., 1986; see Chapter 29), glia to ischemia (Kraig et al., 1986), the authors of this chapter tend to accept the former explanation, since, in our own studies of hypoxia in superfused brain slices (where there is no flow heterogeneity), we have never detected a split Pi peak during recovery from energy failure. Readers may care to disregard our opinion since they may argue that we have simply not looked hard enough or long enough, or are looking for heterogeneity in the wrong anatomical areas of the brain.

Recovery of pH_i Following Energy Failure

The lactic acidosis that results from oxygen deprivation is generally considered to be harmful to cells and one of a number of factors that ultimately lead to neuronal death. It is of interest, therefore, to find out how the cell copes with a severe acid load once flow is restored. Using ^{23}Na NMR in the dog, it was shown that restoration of blood flow following a brief (12 min) ischemic period led to rapid recovery of the Na^+ gradient within 10–15 minutes, followed by full recovery of ATP level, but a delayed normalization of pH at about 40 min (Eleff et al., 1991). Similar observations have been reported following bilateral common carotid artery occlusion in the gerbil: recovery of ATP within 10 min, followed by pH recovery at 20 minutes, with lactate level returning more slowly to control in about 40 min. This dissociation of lactate washout and recovery of pH has also been observed in superfused brain slices (Pirttilä and Kauppinen, 1992), in which a transient alkalosis is observed, while the lactate level remains elevated. Taken together, these data present a picture in which reflow leads to rapid activation of mitochondrial ATP synthesis, which in the first instance is used to reestablish the Na^+ electrochemical gradient. Until intracellular Na^+ is normalized, the drain on ATP that overstimulation of the Na^+/K^+-ATPase produces, clamps ATP levels below normal. Once the ATP level returns to control, presumably energy-dependent H^+-export mechanisms (including the Na^+/H^+ exchanger) lead to recovery of pH independently of lactic acid washout. A consequence of severe ischemia is subsequent development of vasogenic edema. Recent NMR data suggest that edematous regions are high in lactate but normal in pH_i (Mun-Bryce et al., 1993). Slow development of edema is unlikely to be the cause of the delay in the washout of lactate following acute ischemia, since at 1 hour lactate levels are normal. However, it may explain the observations of Houkin et al. (1990), who have reported elevated lactate levels 24 hours after an infarct. In summary, NMR measurements confirm that acidosis is always associated with elevated lactate, but the reverse is not necessarily true.

FOCAL ALTERATIONS IN ACID-BASE BALANCE

Metabolism and pH_i During Brain Activation

Under normal conditions there is tight coupling between blood flow and metabolism in the brain, with delivery of oxygen and glucose both being closely matched to the tissue's oxidative requirements. Under these conditions, about 85% of glucose consumption is directed towards mitochondrial oxidation. However, under conditions of metabolic stress, ranging from mild stimuli such as physiological task activation to severe stimuli such as epileptic seizures, there is evidence of the uncoupling of blood flow from oxygen consumption, with a disproportionate increase in anaerobic glucose consumption (Fox and Raichle, 1986; Fox et al., 1988). Since anaerobic glycolysis is known to be both very sensitive to regulation by pH_i, and to be a significant source of acid generation, NMR methods have been used to investigate cerebral pH_i in a variety of models of brain activation.

Early studies of seizures in experimental animals indicated a degree of energy failure (reduced PCr, elevated Pi), associated with mild acidosis and lactate elevation (for review see Prichard, 1994). Subsequent studies of bicuculline-induced seizures in the rabbit indicated that at low seizure levels lactate production could be stimulated without energy failure or detectable acidosis (Petroff et al., 1984), while at slightly higher doses of the drug there was a transient acidosis with sustained elevation of lactate levels. Similar results for pH_i and lactate have been reported for kainate-induced seizures (Meric et al., 1994), and in domoic acid-induced seizures no changes in phosphorus energy metabolites or pH_i were observed during 30–50 minutes of seizure activity (Sutherland et al., 1993). It appears from these data that there is a range of effects of seizure activity, with stimulation of anaerobic glycolysis being the first response. (Although lactate was not measured in the domoic acid study, it seems reasonable to suppose that it would have been elevated.) Increased levels of activity lead to acidosis and ultimately energy failure. Thus, it seems that brain pH_i would not necessarily change in response to the demands of physiological, as opposed to pathological brain activation. This contention is supported by studies of metabolism during amphetamine stimulation of rat brain *in vivo,* in which large increases in flow were unaccompanied by changes in the levels of phosphorus metabolites, pH_i, or lactate concentration (Detre et al., 1990).

NMR has also been brought to bear on human epilepsy, including early studies of seizures in neonates (Younkin et al., 1986). Consistent with the animal data, energy failure was observed during seizures with changes in pH_i. Studies of chronic epilepsy have produced conflicting results, with suggestions of alkalosis at the seizure focus from one group (Laxer et al., 1992), whereas another group has reported normal pH_i but reduced PCr in the focus (Kuzniecky et al., 1992). The findings in chronic epilepsy probably indicate neuronal loss or damage resulting from seizure activity over many years, and in this context can be related to ^{1}H NMR data showing loss of N-acetylaspartate (NAA), generally considered to be a neuronal marker, in the temporal lobe ipsilateral to the seizure focus (Connelly et al. 1994, Vainio et al., 1994). We are unaware of any adult data recorded ictally using NMR spectroscopy, so the question as to whether pathological brain activation in man is associated with acidosis or lactate generation remains unanswered. There are, however, studies on visual stimulation in human cortex. Stimulation of lactate production has been reported in two laboratories (Prichard et al., 1991; Sappey-Marinier et al., 1992b), with a third group indicating an inconsistent pattern, in which lactate increased in some subjects, but not in others (Merboldt et al., 1992). Sappey-Marinier et al.

(1992b) also recorded ^{31}P NMR spectra, and described significant reductions in PCr together with a slight, but statistically insignificant, alkalosis. This seems to fit with the animal data, in suggesting stimulation of anaerobic glycolysis at a level which does not produce a large acid load, but is sufficient to fuel an increased demand for ATP. It seems unlikely that physiological activation in human brain is associated with significant acidosis.

Brain pH_i in Focal Human Brain Disease

The acute pH changes observed in animal models of disease, such as stroke and head injury would only be expected to be detected in man if observations were made soon after the cerebral insult. It is very often difficult to undertake an NMR spectroscopy exam on an acutely ill patient within a few hours of arrival at the hospital. Nevertheless, there is information on subacute and chronic disturbances in acid-base balance in a number of diseases, including epilepsy, stroke, hemorrhage, hematomas and cancer. The first ^{31}P NMR study of human stroke (Bottomley et al., 1986), undertaken on a group of patients studied between 10 weeks and several years after a stroke, did not show significant differences in phosphorus metabolite ratios or pH compared to normal subjects. However, there were significant overall reductions in NMR signal, suggesting that necrotic tissue contributes little if any signal to the ^{31}P spectrum. Improvements in spectroscopic localization methods and in instrumentation have revealed that there are chronic alterations in both phosphorus metabolites and pH_i in stroke. Weiner and colleagues (Hugg et al., 1992; Sappey-Mariner et al., 1992a) have reported that subacute and chronic stroke lesions are often alkaline (pH_i 7.2 compared to 7.05 in control contralateral hemispheres), and that elevated lactate concentration is always associated with the alkalosis. Welch and coworkers (Levine et al., 1992; Helpern et al., 1993) have been able to study stroke patients starting some hours after the onset of symptoms. They reported that an early acidosis associated with energy failure in the stroke lesion (1–2 days) was followed by a transition through to normal pH (2–3 days), and subsequently by alkalosis between 4 and 10 days. The subacute and/or chronic alkalosis, which has also been observed following transient forebrain ischemia in the rat (van de Linde et al., 1993), cannot be indicative of chronic ischemia, despite the presence of lactate, but may indicate the presence of phagocytotic cells and/or gliosis in the infarcted region. In support of this, ^{1}H NMR data in chronic stroke reveal a disproportionate loss of NAA (Bruhn et al., 1989) indicating selective loss of neurons in the infarct.

In stroke, a correlation between pH_i or energy metabolism and neurological status has not been demonstrated (Levine et al., 1992). This probably reflects the fact that location of the lesion, rather than its size or relative severity is the primary determinant of neurological abnormalities. However, in both subarachnoid haemorrhage (Brooke et al., 1994) and subdural hematomas (Yoshida et al., 1994), clinical status has been shown to correlate with brain pH_i. In the subarachnoid haemorrhage study, regions of low pH_i ($pH_i < 6.8$) were associated in 4 of 5 patients with impaired consciousness. Over several days, both pH_i and consciousness returned to normal. Other patients, with normal pH_i throughout the brain, generally displayed fewer or no neurological abnormalities. Low pH_i probably indicates an area of ongoing ischemia, which has not developed to infarction and therefore may either spontaneously resolve, or be amenable to treatment. Patients with severe subdural hematomas all had low pH_i preoperatively. This was normalized after surgery to evacuate the hematoma brought relief of neurological symptoms. A group of patients with chronic hematomas and only mild symptoms proved to have normal pH_i throughout the brain.

In all the examples above in which there was chronic pH_i disturbance, that disturbance was alkaline. It would appear that tissue cannot sustain an acid pH_i chronically; either pH_i normalizes through recovery, or increases as a result of necrosis. However, studies of head injury by ^{31}P NMR contradict this assertion. A small group of patients were studied 6–18 months after severe head injury (Cadoux-Hudson et al., 1990b), and in each case, pH_i decreased to below control values (minimum 6.85) on going deeper into the brain (predominantly white matter). The worst affected hemisphere (contralateral to the worst spasticity) had a lower pH_i in all cases. In addition to the pH_i changes, there was a general reduction in metabolite levels. The cause of this persistent acidosis is not known—although reduced oxygenation can cause acidosis, this is not a stable situation, as discussed above. The authors postulate that accumulation of lysosomes, which is known to occur following head injury, may explain the acidification, since the lysosome is more acidic than cytosol.

Another disease in which pH regulation might be expected to be perturbed, is cancer. Cancer cells are generally poorly perfused and are expected to be hypoxic in comparison to normal tissue. Furthermore, since the days of classical biochemistry, it has been suggested that tumors rely to a large extent on anaerobic glycolysis to meet their energy requirements. It was therefore something of a surprise to discover, using ^{31}P NMR spectroscopy that many human tumours were, in fact, alkaline. A systematic study of human brain tumours has shown that both astrocytomas and meningiomas are significantly more alkaline (pH 7.08 and 7.19 respectively) than normal brain (Cadoux-Hudson et al., 1989). A plausible explanation for the alkalosis is that ion transport processes, and particularly Na^+/H^+ exchange are differently regulated in transformed cells.

On the whole, it appears that focal human brain disease is most commonly associated with a chronic elevation of pH_i in affected regions. Acutely, acidification may occur, which is usually an indication of ongoing ischemia, and is often accompanied by impaired consciousness. Chronic acidification has only been observed following severe head injury, when it occurs in white matter and is associated with major neurological disabilities.

CONCLUSIONS

Multinuclear NMR spectroscopy provides a versatile picture of cerebral metabolism and pH_i *in vivo*. One can obtain unique information concerning brain pathology and physiology under different conditions and this information may be useful in understanding the relationships between pH_i and brain function. We hope that we have succeeded in demonstrating that cerebral metabolism responds to systemic acid-base disturbances in an unpredictable manner and that the NMR data assist in determining possible causative factors that underlie associated neurological symptoms. These include the "hypoglycemia-like" metabolic condition that occurs in hypercapnia. We wish to emphasize that focal alterations in brain CO_2 status, for instance following neuronal activation, may have "feedback" effects on metabolism. ^1H NMR spectroscopic evidence already exists to corroborate this notion. It appears that under some clinical brain conditions, an abnormal pH_i may be involved in potentiation of focal or even global brain dysfunction.

REFERENCES

Adler S, Simplaceanu V, Ho C (1988): Determination of rat brain buffering in vivo by ^{31}P-NMR. J Appl Physiol 64: 1829–1836.

Adler S, Simplaceanu V, Ho C (1990): Brain pH in acute isocapnic metabolic acidosis and hypoxia: A ^{31}P-nuclear magnetic resonance study. Am J Physiol 258: F34–F40.

Adler S, Williams D, Verbalis JG (1993): Effect of acute and chronic hyponatremia on blood-brain barrier function in the rat. NMR Biomed 6: 119–124.

Agardh C-D, Chapman AG, Nilsson B, Siesjö BK (1981): Endogenous substrates utilized by rat brain in severe insulin-induced hypoglycemia. J Neurochem 36: 490–500.

Allen K, Busza AL, Crockard HA, Gadian DG (1992): Brain metabolism and blood flow in acute cerebral hypoxia studied by NMR spectroscopy and hydrogen clearance. NMR Biomed 5: 48–52.

Barrere B, Meric P, Borredon J, Berenger G, Beloeil J-C, Seylaz J (1990): Cerebral intracellular pH regulation during hypercapnia in unanesthetized rats: A ^{31}P nuclear magnetic resonance spectroscopy study. Brain Res 516: 215–221.

Behar KL, Rothman DL, Fitzpatrick SM, Hetherington HP, Shulman RG (1987): Effects of altered intracellular pH on metabolism. Ann NY Acad. Sci 508: 81–88.

Bolas NM, Rajagopalan B, Mitsumori F, Radda GK (1988): Metabolic changes during experimental cerebral ischemia in hyperglycemic rats, observed by ^{31}P and ^{1}H magnetic resonance spectroscopy. Stroke 19: 763–766.

Bottomley PA, Drayer BP, Smith LS (1986): Chronic adult cerebral infarction studied by phosphorus NMR spectroscopy. Radiology 160: 763–766.

Brooke NSR, Ouwerkerk R, Adams CBT, Radda GK, Ledingham JGG, Rajagopalan B (1994): Phosphorus-31 magnetic resonance spectra reveal prolonged intracellular acidosis in the brain following subarachnoid hemorrhage. Proc Natl Acad Sci USA 91: 1903–1907.

Brooks KJ, Bachelard, HS (1992): The regulation of intracellular pH studied by ^{31}P- and ^{1}H-NMR spectroscopy in superfused guinea-pig cerebral cortex slices. Neurochem Int 21: 379–383.

Brooks KJ, Kauppinen RA, Williams SR, Bachelard HS, Bates TE, Gadian DG (1989): Ammonia causes a drop in intracellular pH in metabolising cortical brain slices. A ^{1}H and ^{31}P nuclear magnetic resonance study. Neuroscience 33: 185–192.

Bruhn H, Frahm J, Gyngell ML, Merboldt KD, Hänicke W, Sauter R (1989): Cerebral metabolism in man after acute stroke: New observations using localized proton NMR spectroscopy. Magn Reson Med 9: 126–131.

Cadoux-Hudson TAD, Blackledge MJ, Rajagopalan B, Taylor DJ, Radda GK (1989): Human primary brain tumour metabolism in vivo—a phosphorus magnetic resonance spectroscopy study. Br J Cancer 60: 430–436.

Cadoux-Hudson TAD, Rajagopalan B, Ledingham JGG, Radda GK (1990a): Response of the human brain to a hypercapnic acid load *in vivo*. Clin Sci 79: 1–3.

Cadoux-Hudson TAD, Wade D, Taylor DJ, Rajagopalan B, Ledingham JGG, Briggs M, Radda GK (1990b): Persistent metabolic sequelae of severe head injury in humans in vivo. Acta Neurochir 104: 1–7.

Cady EB, Chu A, Costello AMDL, Delpy DT, Gardiner RM, Hope PL, Reynolds EOR (1987): Brain intracellular pH and metabolism during hypercapnia and hypocapnia in the new-born lamb. J Physiol 382: 1–14.

Connelly A, Jackson GD, Duncan JS, King MD, Gadian DG (1994): Magnetic resonance spectroscopy in temporal lobe epilepsy. Neurology 44: 1411–1417.

Crockard HA, Gadian DG, Frackowiak RSJ, Proctor E, Allen K, Williams SR, Russell RWR

(1987): Acute cerebral ischaemia: Concurrent changes in cerebral blood flow, energy metabolites, pH, and lactate measured with hydrogen clearance and ^{31}P and ^{1}H nuclear magnetic resonance spectroscopy. II. Changes during ischaemia. J Cereb Blood Flow Metab 7: 394–402.

Detre JA, Williams DS, Koretsky AP (1990): Nuclear magnetic resonance determination of flow, lactate, and phosphate metabolites during amphetamine stimulation of the rat brain. NMR Biomed 3: 272–278.

Eleff SM, Maruki Y, Monsein LH, Traystman RJ, Bryan RN, Koehler RC (1991): Sodium, ATP, and intracellular pH transients during reversible complete ischemia of dog cerebrum. Stroke 22: 233–241.

Espanol MT, Litt L, Yang G-y, Chang L-H, Chan PH, James TL, Weinstein PR (1992): Tolerance of low intracellular pH during hypercapnia by rat cortical brain slices: A $^{31}P/^{1}H$ NMR study. J Neurochem 59: 1820–1828.

Fox PT, Raichle ME (1986): Focal physiological uncoupling of cerebral blood flow and oxidative metabolism during somatosensory stimulation in human subjects. Proc Natl Acad Sci USA 83: 1140–1144.

Fox PT, Raichle ME, Mintun MA, Dence C (1988): Nonoxidative glucose consumption during focal physiologic neural activity. Science 241: 462–464.

Gardiner RM (1980): Cerebral blood flow and oxidative metabolism during hypoxia and asphyxia in the new-born calf and lamb. J Physiol (Lond) 305: 357–376.

Gruetter R, Novotny EJ, Boulware SD, Mason GF, Rothman DL, Shulman GI, Prichard JW, Shulman RG (1994): Localized ^{13}C NMR spectroscopy in the human brain of amino acid labeling from D-[1-^{13}C]glucose. J Neurochem 63: 1377–1385.

Helpern JA, Vande Linde AMQ, Welch KMA, Levine SR, Schultz LR, Ordidge RJ, Halvorson HR, Hugg JW (1993): Acute elevation and recovery of intracellular [Mg^{2+}] following human focal cerebral ischemia. Neurology 43: 1577–1581.

Houkin K, Kwee IL, Nakada T (1990): Persistent high lactate level as a sensitive MR spectroscopy indicator of completed infarction. J Neurosurg 72: 763–766.

Hugg JW, Matson GB, Twieg DB, Maudsley AA, Sappey-Marinier D, Weiner MW (1992): Phosphorus-31 MR spectroscopic imaging (MRSI) of normal and pathological human brains. Magn Reson Imaging 10: 227–243.

Javaheri S, DeHemrtinne A, Van Heel B, Leusen I (1983): Changes in brain ECF pH during metabolic acidosis and alkalosis: A microelectrode study. J Appl Physiol, 55: 1849–1853.

Jensen KE, Thomsen C, Henriksen O (1988): In vivo measurement of intracellular pH in human brain during different tensions of carbon dioxide in arterial blood. A ^{31}P-NMR study. Acta Physiol Scand 134: 295–298.

Kauppinen RA, Nicholls DG (1986): Synaptosomal bioenergetics: Role of glycolysis and pyruvate oxidation and responses to hypoglycaemia. Eur J Biochem 158: 159–165.

Kauppinen RA, Williams SR (1990): Cerebral energy metabolism and intracellular pH during severe hypoxia and recovery. A study using ^{1}H, ^{31}P and $^{1}H\{^{13}C\}$ nuclear magnetic resonance spectroscopy in the guinea pig cerebral cortex in vitro. J Neurosci Res 26: 359–369.

Kauppinen RA, Pirttila TRM, Auriola SOK, Williams SR (1994): Compartmentation of cerebral glutamate in situ as detected by H-1/C-13 NMR. Biochem J 298: 121–127.

Kraig RP, Pulsinelli WA, Plum F (1986): Carbonic acid buffer changes during complete brain ischemia. Am J Physiol 250: R348–R357.

Kuzniecky R, Elgavish GA, Hetherington HP, Evanochko WT, Pohost GM (1992): In vivo ^{31}P nuclear magnetic resonance spectroscopy of human temporal lobe epilepsy. Neurology 42: 1586–1590.

Laxer KD, Hubesch B, Sappey-Marinier D, Weiner MW (1992): Increased pH and inorganic phosphate in temporal seizure foci demonstrated by [^{31}P]MRS. Epilepsia 33: 618–623.

Levine SR, Helpern JA, Welch KMA, van de Linde AMQ, Sawaya KL, Brown EE, Ramadan NM (1992): Human focal cerebral ischemia—evaluation of brain pH and energy metabolism with ^{31}P NMR spectroscopy. Radiology 185: 537–544.

Lotito S, Blondet P, François A, von Kienlin M, Rémy C, Albrand JP, Décorps M, Benabid AL (1989): Correlation between intracellular pH and lactate levels in the rat brain during potassium cyanide induced metabolism blockade: A combined ^{31}P-^{1}H in vivo nuclear magnetic spectroscopy study. Neurosci Lett 97: 91–96.

Merboldt KD, Bruhn H, Hänicke W, Michaelis T, Frahm J (1992): Decrease of glucose in the human visual cortex during photic stimulation. Magn Res Med 25: 187–194.

Meric P, Barrere B, Peres M, Gillet B, Berenger G, Beloeil JC, Seylaz J (1994): Effects of kainate-induced seizures on cerebral metabolism—a combined H-1 and P-31 NMR study in rat. Brain Res 638: 53–60.

Mun-Bryce S, Kroh FO, White J, Rosenberg GA (1993): Brain lactate and pH dissociation in edema: ^{1}H and ^{31}P NMR spectroscopy in collagenase-induced hemorrhage in rats. Am J Physiol 265: R697–R702.

Naritomi H, Sasaki M, Kanashiro M, Kitani M, Sawada T (1988): Flow thresholds for cerebral energy disturbance and Na^+ pump failure as studied by ^{31}P and ^{23}Na NMR spectroscopy. J Cereb Blood Flow Metab 8: 16–23.

Nioka S, Chance B, Hilberman M, Subramanian HV, Leigh JSJ, Veech RL, Forster RE (1987): Relationship between intracellular pH and energy metabolism in dog brain as measured by ^{31}P-NMR. J Appl Physiol 62: 2094–2102.

Nioka S, Smith DS, Chance B, Lockard S (1990): Metabolic heterogeneity in brain tissue during incomplete ischemia and reperfusion. NMR Biomed 3: 239–247.

Petroff OAC, Prichard JW, Behar KL, Alger JR, Shulman RG (1984): In vivo phosphorus nuclear magnetic resonance spectroscopy in status epilepticus. Ann Neurol 16: 169–177.

Petroff OAC, Prichard JW, Behar KL, Rothman DL, Alger JR, Shulman RG (1985): Cerebral metabolism in hyper- and hypocarbia: ^{31}P and ^{1}H nuclear magnetic resonance studies. Neurology 35: 1681–1688.

Pirttilä T-R, Kauppinen RA (1992): Recovery of intracellular pH in cortical brain slices following anoxia studied by nuclear magnetic resonance spectroscopy: Role of lactate removal, extracellular sodium and sodium/hydrogen exchange. Neuroscience 47: 155–164.

Pirttilä T-RM, Kauppinen RA (1993): Extracellular pH and buffering power determine intracellular pH in cortical brain slices during and following hypoxia. Neuroreport 5: 213–216.

Pirttilä TRM, Kauppinen RA (1994): Regulation of intracellular pH in guinea pig cerebral cortex ex vivo studied by P-31 and H-1 nuclear magnetic resonance spectroscopy—role of extracellular bicarbonate and chloride. J Neurochem 62: 656–664.

Portman MA, Lassen NA, Cooper TG, Sills AM, Potchen EJ (1991): Intra- and extracellular pH of the brain in vivo studied by ^{31}P-NMR during hyper- and hypocapnia. J Appl Physiol 71: 2168–2172.

Prichard JW (1994): Nuclear magnetic resonance spectroscopy of seizure states. Epilepsia 35: S14–S20.

Prichard J, Rothman D, Novotny E, Petroff O, Kuwabara T, Avison M, Howseman A, Hanstock C, Shulman R (1991): Lactate rise detected by ^{1}H NMR in human visual cortex during physiologic stimulation. Proc Natl Acad Sci USA 88: 5829–5831.

Sappey-Mariner D, Hubesch B, Matson GB, Weiner MW (1992a): Decreased phosphorus metabolite concentrations and alkalosis in choronic cerebral infarction. Radiology 182: 29–34.

Sappey-Marinier D, Calabress G, Fein G, Hugg JW, Biggins C, Weiner MW (1992b): Effect of photic stimulation on human visual cortex lactate and phosphates using ^{1}H and ^{31}P magnetic resonance spectroscopy. J Cereb Blood Flow Metabol 12: 584–592.

Sieber FE, Derrer SA, Eleff SM, Koehler RC, Traystman RJ (1992): Hypocapnic-hypoglycemic interactions on cerebral high-energy phosphates and pH in dogs. Am J Physiol 263: H1864–H1871.

Sunagawa S, Buist RJ, Hruska FE, Sutherland GR, Peeling J (1994): Hydrogen ion compartmentation during and following cerebral ischemia evaluated by P-31 NMR spectroscopy. Brain Res 641: 328–332.

Sutherland GR, Ross BD, Lesiuk H, Peeling J, Pillay N, Pinsky C (1993): Phosphate energy metabolism during domoic acid-induced seizures. Epilepsia 34: 996–1002.

Swain JA, McDonald TJ, Robbins RCJ, Balaban RS (1991): Relationship of cerebral and myocardial intracellular pH to blood pH during hypothermia. Am J Physiol 260: H1640–H1644.

Tomlinson FH, Anderson RE, Meyer FB (1993): Effect of superhypercapnia on cortical pH_i and cortical blood flow. Am J Physiol 265: R974–R981.

Vainio P, Usenius JP, Vapalahti M, Partanen K, Kalviainen R, Rinne J, Kauppinen RA (1994): Reduced N-acetylaspartate concentration in temporal lobe epilepsy by quantitative ^{1}H MRS in vivo. Neuroreport 5: 1733–1736.

van de Linde AMQ, Chopp M, Lee SA, Schultz LR, Welch KMA (1993): Postischemic brain tissue alkalosis supressed by U74006F. J Neurol Sci 114: 36–39.

Vorstrup S, Jensen KE, Thomsen C, Henriksen O, Lassen NA, Paulson OB (1989): Neuronal pH regulation: Constant normal intracellular pH is maintained in brain during low extracellular pH induced by acetazolamide-^{31}P NMR study. J Cereb Blood Flow Metabol 9: 417–421.

Williams SR, Gadian DG, Proctor E, Sprague DB, Talbot DF, Young IR, Brown FF (1985): Proton NMR studies of muscle metabolites *in vivo*. J. Mag. Reson 63: 406–412.

Yoshida K, Furuse M, Izawa A, Hirano T, Kuchiwaki H, Inao S, Sugita K (1994): Dynamics of cerebral metabolism in patients with chronic subdural hematoma evaluated with phosphorus-31 MR spectroscopy before and after surgery. Am J Neuroradiol 15: 1681–1686.

Younkin DP, Delivoria-Papadopoulos MJM, Donlon E, Clancy R, Chance B (1986): Cerebral metabolic effects of neonatal seizures measured with in vivo P-31 NMR spectroscopy. Ann Neurol 20: 513–519.

CHAPTER 31

Abnormal Cytoplasmic pH in Diseases of Nerve and Muscle

PETER GRAFE
Department of Physiology, University of Munich, Munich, Germany

INTRODUCTION

Cytoplasmic pH has not been considered as a key factor in the pathophysiology of neuromuscular disorders. However, in the last few years, an abnormal cytoplasmic pH in nerve and muscle has been observed by means of pH-sensitive microelectrodes, pH-sensitive fluorescent dyes, and magnetic resonance spectroscopy. Furthermore, it is known that changes in extra- and/or intracellular pH can alter membrane conductances, energy metabolism, and mechanical properties of nerve and muscle. This raises the question of whether an abnormal cytoplasmic pH might contribute to some of the functional deficits in diseases of nerve and muscle.

DISEASES OF NERVE

Activity-Dependent Changes in Nerve pH

Activity-dependent changes in extracellular pH have been observed during repetitive electrical stimulation of nonmyelinated axons in isolated rat vagus nerve (Endres et al., 1986) and of thin myelinated axons in isolated rat optic nerve (Davis et al., 1987). Acidic pH shifts reached 0.2 pH unit in the optic nerve, whereas a relatively small, transient acidification of 0.03 pH unit was found in vagus nerve. Repetitive electrical stimulation of isolated rat spinal roots (Bostock and Grafe, 1985) and rat sciatic nerves (Hoppe et al., 1991), however, did not result in changes of extracellular pH. Intracellular pH was measured in frog sciatic nerve by means of a fluorescent pH-sensitive dye (Valkina et al., 1993; Khodorov et al., 1994). Under control conditions, repetitive stimulation caused only a very small decrease in pH_i (see also Chapter 16). After modification of Na^+ channels by veratridine, however, stimulus-evoked internal acidification was greatly increased (Valkina et al., 1993). Optical recordings of intracellular pH in a peripheral nerve have

pH and Brain Function, Edited by Kai Kaila and Bruce R. Ransom
ISBN 0-471-11838-9

been performed after transmembrane loading of axons and/or glia with pH-sensitive fluorescent dyes from the bathing solution (e.g., Wächtler et al., 1996b). Confocal images can be used to determine whether the light emission is coming from the axons and/or the adjacent glial cells (Kriegler and Chiu, 1993; Lev-Ram and Ellisman, 1995).

Nerve Ischemia, Diabetic Neuropathy

Abnormalities of cytoplasmic pH in diseases of human nerve, to my knowledge, have not been described. However, clear changes in tissue pH were found in isolated nerves exposed to hypoxia and/or ischemia, a condition possibly involved in some of the neuropathies. One of the first recordings of changes in cytoplasmic pH during metabolic inhibition of axons has been performed on squid axons (Boron and De Weer, 1976). A fall in pH_i of 0.3 pH unit was observed after 50 minutes of exposure to cyanide. Effects of hypoxia on tissue pH of mammalian nerves have been measured in peripheral nerves and the optic nerves of rats. In isolated peroneal nerves and spinal roots (Strupp et al., 1991; Schneider et al., 1993a), an extracellular acidification of up to 0.3 pH unit was found after 30 min of exposure to hypoxia. A similar acidification was observed in rat optic nerves after a much shorter period of hyoxia (Ransom et al., 1992). The extent of hypoxia-induced acidification was dependent on the buffering power of the bathing solution (see Chapters 2–4). Hypoxia-induced acidification was increased in N-2-Hydroxyethylpiperazine-N′-2-ethane sulfonic acid (HEPES)-buffered solutions compared to HCO_3^-/CO_2-buffered solutions (Strupp et al., 1991).

Diabetic neuropathy is a condition in which nerve hypoxia is combined with hyperglycemia. Therefore, the importance of glucose availability for hypoxia-induced acidification was tested in peroneal nerve (Strupp et al., 1991), spinal roots (Schneider et al., 1993a), and optic nerves (Ransom et al., 1992) of rats. It was found in all these preparations that a high glucose concentration enhanced hypoxia-induced tissue acidification. Inhibition of glycolysis, on the other hand, prevented hypoxia-induced nerve acidosis (Strupp et al., 1991; Schneider et al., 1993a). Recently, such findings were confirmed when a fluorescent dye was used to follow changes in intracellular pH of isolated rat spinal roots (Wächtler et al., 1996b). These observations indicate that tissue acidification might contribute to the pathophysiology of diabetic neuropathy. Typical symptoms in this disease are alterations in nerve excitability and the liability to ischemic lesions (see following section).

Cytoplasmic pH and Pathophysiology of Nerve

Abnormal excitability. A link between axonal irritability and *extracellular* pH is known for the axonal membrane of the nerve trunk and of the receptive area. Tissue alkalosis in the region of the nerve trunks can lead to tetany and paraesthesias (Layzer, 1985). This phenomenon can be explained by the surface-potential hypothesis (Hille, 1968, 1984; Chapter 21): Some of the plasma protons and Ca^{2+} ions are bound to fixed negative charges on the surface of membranes and affect the Na^+ conductance by altering the local potential gradient around the ion channel. A reduction of protons would enhance Na^+ channel excitability by activation of the voltage sensor in this channel either directly, or indirectly, by binding of free Ca^{2+} ions to plasma proteins. Acidosis, on the other hand, has been described to induce strong excitation of nociceptors (see Chapter

23) and might be involved in the generation of pain in human skin or muscle (Steen et al., 1995; Issberner et al., 1996). In this hypothesis, excitation of nociceptive nerve terminals by low extracellular pH is explained by a proton-induced opening of unspecific cation channels (Bevan and Geppetti, 1994; Waldmann et al., 1997).

Changes in *cytoplasmic* pH also may lead to abnormal excitability of peripheral nerves. This is likely through an effect at the cytoplasmic face of axonal ion channels (Roos and Boron 1981; Moody, 1984). Giant axons of the squid allow internal perfusion and have been used to study the effects of cytoplasmic acidification on axonal K^+ channels. In these experiments an inhibition of K^+ conductance was observed (Wanke et al., 1979; Clay, 1990). This effect was confirmed in rat spinal root axons by measurements of whole nerve electrotonus (Grafe et al., 1994) and of multichannel currents from excised inside-out patches (Schneider et al., 1993b). It was found that lowering of intracellular pH (pH_i) to 6.8 was sufficient to block 50% of one of the voltage dependent K^+ conductances present in mammalian axons (Schneider et al., 1993b). Such an intracellular acidification can be reached in peripheral nerve by hyperglycemic hypoxia (see Fig. 1). Inhibition of axonal K^+ channels by cytoplasmic acidification could explain hypoexcitability of axons. This effect would be the result of an inhibition of K^+ conductances involved in the generation of the resting potential. Whether a pH-dependent inhibition of repolarizing K^+ currents could result in hyperexcitability of axons is unclear. Experiments on dorsal root ganglions indicate that the proton-induced depolarization does not result in excitation of the neurons (Tegeder and Reeh, 1993).

Ischemic lesions in diabetic neuropathy. Peripheral nerves of diabetic patients or animals show an increased susceptibility to ischemic damage (Nukada, 1986, 1992; Low, 1987; Lindström et al., 1997). A similar phenomenon has been observed in isolated peripheral nerves of frog (Lorente de No, 1947) and rat (Strupp et al., 1991; Schneider et al., 1992). The combination of hyperglycemia and hypoxia was found to result in an impairment of posthypoxic electrophysiological recovery (Fig. 1). Further experimental findings on these isolated nerves indicate that enhanced cytoplasmic acidification might contribute to the increased susceptibility to hypoxic damage. First, recovery from hyperglycemic hypoxia was greatly improved in bicarbonate- as compared to HEPES-buffered bathing solutions (Schneider et al., 1992). Second, recovery was also much improved when reoxygenation was combined with passive cytoplasmic alkalinization (Euchner-Wamser et al., 1994) or after inhibition of glycolysis (Schneider et al., 1993a).

The mechanism(s) underlying the deleterious effect of enhanced cytoplasmic acidification for recovery of peripheral nerves from hyperglycemic hypoxia are not yet clear. This is similar to the situation in the central nervous system, in which hyperglycemia impairs recovery from transient cerebral ischemia (Siesjö et al., 1993; see Chapter 28). One factor involved might be inhibition of energy metabolism. It was found that the recovery rate of ATP and phosphocreatine levels was reduced when reoxygenation of peripheral nerves was combined with acidification in high P_{CO_2} (Stewart and Moonsammy, 1966). Another factor might be failure of ionic homeostasis. Cytoplasmic acidification has been reported to inhibit active Na^+ extrusion from squid axons (Breitwieser et al., 1987). It is interesting to note that, in contrast to peripheral nerves, increased extracellular glucose concentrations actually improved functional recovery from anoxia in the optic nerve (Ransom et al., 1992). The reason for this discrepancy is not clear. However, it is known that central and peripheral axons differ strongly in their ability to recover from hypoxia (Lehning et al., 1996).

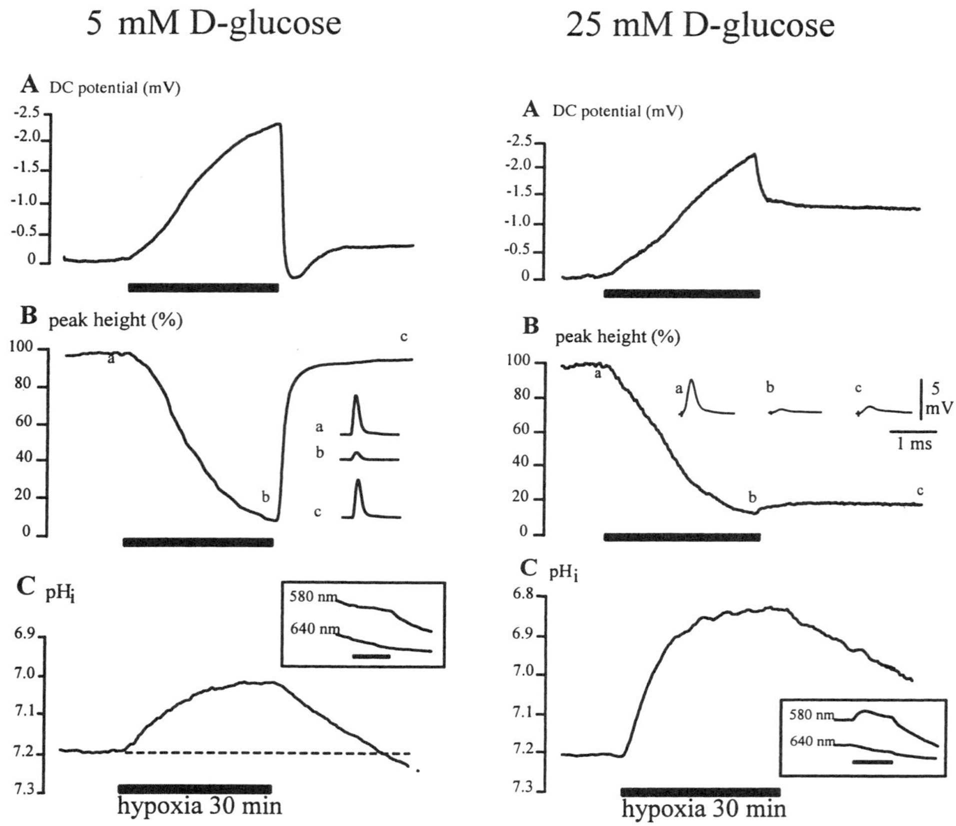

FIGURE 1. Cytoplasmic acidification and lack of neurophysiological recovery from hyperglycemic hypoxia. Illustrated are changes in extracellular DC potential, action potential, and cytoplasmic pH of isolated rat dorsal spinal roots exposed to hypoxia in normal (5 mM) and high (25 mM) glucose concentrations. The electrophysiological parameters were recorded as previously described (Schneider et al., 1993b). Cytoplasmic pH was determined photometrically by means of the pH-sensitive fluorescent dye seminaphthorhodafluor SNARF-1 (Buckler and Vaughan-Jones, 1990). The isolated spinal roots were excited at 515 nm, emission was measured at 580 and 640 nm, and changes in cytoplasmic pH were plotted as changes in the ratio of emission at 580 nm to emission at 640 nm. Note a correlation between lack of functional recovery from hyperglycemic hypoxia and greater cytoplasmic acidification (Liedl E, Euchner-Wamser I, Schneider U, Rucker F and Grafe P; unpublished; see also Wächtler et al., 1996b).

DISEASES OF MUSCLE

Activity-Dependent Changes in Myoplasmic pH

pH-sensitive microelectrodes were the first direct method used to study cytoplasmic pH in mammalian skeletal muscle (Aickin and Thomas, 1977); however, changes in pH_i in a contracting muscle could not be recorded. The latter parameter was investigated by means of ^{31}P magnetic resonance spectroscopy (MRS), a noninvasive method for measurement of cytoplasmic pH. This technique has been used to study intracellular pH in exercising normal and diseased human skeletal muscle. The results have been reviewed by Kent-Braun et al., (1994). A classic study on human muscle was reported by Taylor et al. (1986). Intensive exercise was found to lower intracellular pH to less than 6.2. Similarly, Miller et al. (1987) observed that pH_i fell from 7.1 to 6.4 during muscle fatigue. Recovery from this acidification was dependent on the intensity of exercise but was complete within several minutes.

Vascular Diseases, Myopathies, Neuromuscular Diseases

The characteristics of activity-dependent changes in cytoplasmic pH are altered in conditions of reduced oxygen supply to the muscle and in myopathies (review by Kent-Braun et al., 1994) (Table 1). Muscle exercise in patients suffering from heart failure, peripheral vascular disease, and respiratory failure led to lower pH_i as compared to controls. In contrast, lack of myophosphorylase (McArdle's syndrome) and deficiency in phosphofructokinase are accompanied by a failure of acidification during exercise. Impairment of aerobic metabolism in mitochondrial myopathies, on the other hand, resulted in more acidification and slowing of pH recovery. The latter findings were also made in Duchenne's and Becker's muscular dystrophy (Kemp et al., 1993; Kent-Braun et al., 1994). Although resting pH was elevated in these conditions, a greater than normal reduction of pH during exercise and slower metabolic recovery were found. Abnormalities of myoplasmic pH were also found in some other neuromuscular diseases (Taylor et al., 1993; Kent-Braun et al., 1994). In the interpretation of these observations, Kent-Braun et al., (1994) came to the conclusion that only the changes found in the metabolic myopathies can be regarded as a specific result of the underlying pathophysiology. Unspecific factors, such as physical deconditioning, might explain the abnormal cytoplasmic pH in most of the other diseases.

Cytoplasmic pH and Pathophysiology of Muscle

Fatigue and recovery from exercise. An unusually fast decline of force and slowing of force recovery during and after muscle exercise are common symptoms in some of

TABLE 1 Observations on Myoplasmic pH in Diseases of Muscle

Disease	pH at Rest	Acidosis During Exercise	Recovery of pH
Impaired circulation			
Heart failure	Normal	Greater	Impaired
Peripheral vascular disease	High	Greater	Impaired
Chronic respiratory failure	Normal	Greater	—
Metabolic Myopathies			
McArdle's disease	Normal	Reduced	Normal
Phosphofructokinase deficiency	Normal	Reduced	Normal
Mitochondrial myopathies	Normal	Greater	Impaired
Thyroid disease	Low	Greater	Impaired
Muscular dystrophy	High	Greater	Impaired
Neuromuscular disease	High	Greater	Impaired

Modified from Kent-Braun et al., 1994, with permission.

the myopathies (Layzer, 1986). Several experimental studies, described below, led to the suggestion that myoplasmic acidification might contribute to these phenomena not only in diseased but also in normal skeletal muscle. The results, however, are not yet conclusive, and the exact quantitative link between myoplasmic pH and other metabolic alterations to the development of muscle fatigue and slowing of recovery remains a matter of discussion (Miller et al., 1987; Bendahan et al., 1990; Lännergren and Westerblad, 1991; Kent-Braun et al., 1994).

A key observation was a negative inotropic effect of low pH_i (see Roos and Boron, 1981). In skinned fibers, two effects of low pH_i were observed (Fabiato and Fabiato, 1978). First, more Ca^{2+} was required for the development of 50% of maximum tension. This indicates a reduced sensitivity of actin and/or myosin interaction to Ca^{2+}. On the other hand, in an acid environment, more Ca^{2+} was released by the sarcoplasmic reticulum. In general accordance with this view are several recent studies. Using a Ca^{2+}-sensitive dye, Westerblad and Allen (1993) found that tetanic tension was decreased at acidic pH_i in spite of an increase in tetanic $[Ca^{2+}]_i$. This finding was explained by the reduced Ca^{2+} sensitivity. Acidosis was also found to slow the rate of sarcoplasmic reticulum Ca^{2+} pumping. In addition, acid- and alkaline-going shifts in pH_i induced changes in the kinetics of posttetanic relaxation. Comparative studies about the effects of cytoplasmic pH on excitation-contraction coupling were performed on toad (Lamb et al., 1992) and rat muscle (Lamb and Stephenson, 1994). In both preparations depolarization at pH 6.2 still elicited massive Ca^{2+} release, sufficient to produce near-maximal force. Thus, the ability of the voltage sensors to activate Ca^{2+} release appeared to be little affected by lowering the pH to 6.2.

Another effect of myoplasmic acidification that might contribute to fatigue is inhibition of muscle glycolysis. According to Roos and Boron (1981), at least two loci on the glycolytic pathway are inhibited by low pH_i. One is the conversion of (inactive) phosphorylase *b* into (active) phosphorylase *a*. The second locus is phosphofructokinase, the enzyme that phosphorylates fructose-6-phosphate.

In summary, cytoplasmic acidification might be one of the components involved in the development of muscle fatigue and in the abnormal slowing of recovery following exercise. Other components involved are pH-independent alterations in cellular metabolism, excitability of cell membrane, and excitation-contraction coupling.

REFERENCES

Aickin CC, Thomas RC (1977): Micro-electrode measurement of the intracellular pH and buffering power of mouse soleus muscle fibres. J Physiol (Lond) 267: 791–810.

Bendahan D, Confort-Gouny S, Kozak-Ribbens G, Cozzone PJ (1990): Heterogeneity of metabolic response to muscular exercise in humans. FEBS Lett 272: 155–158.

Bevan S, Geppetti P (1994): Protons: Small stimulants of capsaicin-sensitive sensory nerves. Trends Neurosci 17: 509–512

Boron WF, De Weer P (1976): Intracellular pH transients in squid axons caused by CO_2, NH_3 and metabolic inhibitors. J Gen Physiol 67: 91–112.

Bostock H, Grafe P (1985): Activity-dependent excitability changes in normal and demyelinated rat spinal root axons. J Physiol (Lond) 365: 239–257.

Breitwieser GE, Altamirano AA, Russell JM (1987): Effects of pH changes on sodium pump fluxes in squid giant axon. Am J Physiol 253: C547–C554.

Buckler KJ, Vaughan-Jones RD (1990): Application of a new pH-sensitive fluoroprobe (carboxy-SNARF-1) for intracellular pH measurement in small, isolated cells. Pflügers Arch 417: 234–239.

Clay JR (1990): IK inactivation in squid axons is shifted along the voltage axis by changes in the intracellular pH. Biophys J 58: 797–801.

Davis PK, Carlini WG, Ransom BR, Black JA, Waxman SG (1987): Carbonic anhydrase activity develops postnatally in the rat optic nerve. Dev Brain Res 31: 291–298.

Endres W, Grafe P, Bostock H, ten Bruggencate G (1986): Changes in extracellular pH during electrical stimulation of isolated rat vagus nerve. Neurosci Lett 64: 201–205.

Euchner-Wamser I, Grafe P, Sennefelder E, Schneider U (1994): Alkalinization during re-oxygenation prevents functional damage by hyperglycaemic hypoxia. Neuroreport 5: 1574–1576.

Fabiato A, Fabiato F (1978): Effects of pH on the myofilaments and the sarcoplasmic reticulum of skinned cells from cardiac and skeletal muscle. J Physiol (Lond) 276: 233–255.

Grafe P, Bostock H, Schneider U (1994): The effects of hyperglycaemic hypoxia on rectification in rat dorsal root axons. J Physiol (Lond) 480: 297–307.

Hille B (1968): Charges and potentials at the nerve surface. Divalent ions and pH. J Gen Physiol 51: 221–236.

Hille B (1984): Ionic Channels of Excitable Membranes. Sunderland, MA: Sinauer.

Hoppe D, Chvatal A, Kettenmann H, Orkand RK, Ransom BR (1991): Characteristics of activity-dependent potassium accumulation in mammalian peripheral nerve in vitro. Brain Res 552: 106–112.

Issberner U, Reeh PW, Steen KH (1996): Pain due to tissue acidosis: A mechanism for inflammatory and ischemic myalgia? Neurosci Lett 208: 191–194

Kemp GJ, Taylor DJ, Dunn JF, Frostick SP, Radda GK (1993): Cellular energetics of dystrophic muscle. J Neurol Sci 116: 201–206.

Kent-Braun JA, Miller RG, Weiner MW (1994): Magnetic resonance spectroscopy studies of human muscle. Radiol Clin North Am 32: 313–335.

Khodorov B, Valkina O, Turovetsky V (1994): Mechanisms of stimulus-evoked intracellular acidification in frog nerve fibres. FEBS Lett 341: 125–127.

Kriegler S, Chiu SY (1993): Calcium signaling of glial cells along mammalian axons. J Neurosci 13: 4229–4245.

Lamb GD, Stephenson DG (1994): Effects of intracellular pH and [Mg^{2+}] on excitation-contraction coupling in skeletal muscle fibres of the rat. J Physiol (Lond) 478: 331–339.

Lamb GD, Recupero E, Stephenson DG (1992): Effect of myoplasmic pH on excitation-contraction coupling in skeletal muscle fibres of the toad. J Physiol (Lond) 448: 211–224.

Layzer RB (1985): Neuromuscular Manifestations of Systemic Disease. Philadelphia: FA Davis.

Layzer B (1986): Muscle pain, cramps, and fatigue. In Engel AG, Banker BQ (eds): Myology. New York: McGraw-Hill, pp 1907–1922.

Lännergren J, Westerblad H (1991): Force decline due to fatigue and intracellular acidification in isolated fibres from mouse skeletal muscle. J Physiol (Lond) 434: 307–322.

Lehning EJ, Stys PK, LoPachin RM Jr (1996): Reoxygenation of anoxic peripheral nerve myelinated axons promotes re-establishment of normal elemental composition. Brain Res 715: 189–196.

Lev-Ram V, Ellisman MH (1995): Axonal activation-induced calcium transients in myelinating Schwann cells: Sources and mechanisms. J Neurosci 15: 2628–2637.

Lindström P, Lindblom U, Brismar T (1997): Delayed recovery of nerve conduction and vibratory sensibility after ischaemic block in patients with diabetes mellitus. J Neurol Neurosurg Psychiat 63: 346–350.

Lorente de No R (1947): A Study of Nerve Physiology. New York: The Rockefeller Institute for Medical Research.

Low PA (1987): Recent advances in the pathogenesis of diabetic neuropathy. Muscle Nerve 10: 121–128.

Miller RG, Giannini D, Milner-Brown HS, Layzer RB, Koretsky AP, Hooper D, Weiner MW (1987): Effects of fatiguing exercise on high energy phosphates, force, and EMG: Evidence for three phases of recovery. Muscle Nerve 10: 810–821.

Moody WJ (1984): Effects of intracellular H^+ on the electrical properties of excitable cells. Annu Rev Neurosci 7: 257–278.

Nukada H (1986): Increased susceptibility to ischemic damage in streptozocin-diabetic nerve. Diabetes 35: 1058–1061.

Nukada H (1992): Mild ischemia causes severe pathological changes in experimental diabetic nerve. Muscle Nerve 15: 1116–1122.

Ransom BR, Walz W, Davis PK, Carlini WG (1992): Anoxia-induced changes in extracellular K^+ and pH in mammalian central white matter. J Cereb Blood Flow Metab 12: 593–602.

Roos A, and Boron WF (1981): Intracellular pH. Physiol Rev 61: 296–434.

Schneider U, Jund R, Nees S, Grafe P (1992): Differences in sensitivity to hyperglycemic hypoxia of isolated rat sensory and motor nerve fibers. Ann Neurol 31: 605–610.

Schneider U, Niedermeier W, Grafe P (1993a): The paradox between resistance to hypoxia and liability to hypoxic damage in hyperglycemic peripheral nerves: Evidence for glycolysis involvement. Diabetes 42: 981–987.

Schneider U, Quasthoff S, Mitrovic N, Grafe P (1993b): Hyperglycaemic hypoxia alters afterpotential and fast K^+ conductance of rat axons by cytoplasmic acidification. J Physiol (Lond) 465: 679–697.

Siesjö BK, Katsura K, Mellergård P, Ekholm A, Lundgren J, Smith ML (1993): Acidosis-related brain damage. Prog Brain Res 96: 23–48.

Steen KH, Issberner U, Reeh PW (1995): Pain due to experimental acidosis in human skin: Evidence for non-adapting nociceptor excitation. Neurosci Lett 199: 29–32.

Stewart MA, Moonsammy GI (1966): Substrate changes in peripheral nerve recovering from anoxia. J Neurochem 13: 1433–1439.

Strupp M, Jund R, Schneider U, Grafe P (1991): Glucose availability and sensitivity to anoxia of isolated rat peroneal nerve. Am J Physiol 261: E389–E394.

Taylor DJ, Styles P, Matthews PM, Arnold DA, Gadian DG, Bore P, Radda GK (1986): Energetics of human muscle: Exercise-induced ATP depletion. MRM 3: 44–54.

Taylor DJ, Kemp GJ, Woods CG, Edwards JH, Radda GK (1993): Skeletal muscle bioenergetics in myotonic dystrophy. J Neurol Sci 116: 193–200.

Tegeder C, Reeh PW (1993): Protons induce sustained depolarization but no excitation in all cell types of the intact rat spinal ganglion. Pflügers Arch 422: R51

Valkina ON, Vergun OV, Turovetsky VB, Khodorov BI (1993): Effects of repetitive stimulation, veratridine and ouabain on cytoplasmic pH in frog nerve fibres: Role of internal Na^+. FEBS Lett 334: 83–85.

Wächtler J, Mayer C, Quasthoff S, Adelsberger H, Grafe P (1996a): P_2 purinoceptor–mediated intracellular Ca^{2+} transients in human sural nerve. Neuroreport 7: 1289–1292.

Wächtler J, Mayer C, Rucker F, Grafe P (1996b): Glucose availability alters ischaemia-induced changes in intracellular pH and calcium of isolated rat spinal roots. Brain Res 725: 30–36.

Waldmann R, Champigny G, Bassilana F, Heurteaux C, Lazdunski M (1997): A proton-gated cation channel involved in acid sensing. Nature 386: 173–177.

Wanke E, Carbone E, Testa PL (1979): K^+ conductance modified by a titratable group accessible to protons from the intracellular side of the squid axon membrane. Biophys J 26: 319–324.

Westerblad H, Allen DG (1993): The influence of intracellular pH on contraction, relaxation and $[Ca^{2+}]_i$ in intact single fibres from mouse muscle. J Physiol (Lond) 466: 611–628.

CHAPTER 32

Control and Disturbances of Cerebrospinal Fluid pH

EUGENE NATTIE
Department of Physiology, Dartmouth Medical School, Lebanon NH

INTRODUCTION

The System

The central nervous system can be considered as having two extracellular compartments, the ventricular compartment containing cerebrospinal fluid (CSF) and the parenchymal extracellular fluid (ECF) in the tissue matrix itself. These two compartments communicate without diffusional restriction but have different volumes; 140 ml or 10% of brain weight for CSF and 280 ml or 20% brain weight for tissue ECF. CSF provides hydraulic protection for the brain and both CSF and ECF provide a relatively constant chemical environment for neuronal function (see Fishman, 1990; Johanson, 1995)

Specialized cells in the choroid plexus secrete the CSF. It is a clear, colorless liquid with low protein content (typically 25 mg/100 ml) in comparison to plasma and body interstitial fluid (typically 6.5 mg/100 ml in plasma). It circulates through the large ventricular cavities and exits from the brain into the venous system via the arachnoid villi (Fig. 1). The brain, because of the very tight junctions in the capillary endothelium (Chapter 14), has no Starling's type bulk flow fluid exchange between the capillary and the ECF space. However, a fluid similar to CSF is formed by secretory processes at the capillary endothelial cells. This secreted ECF, estimated to be one third of the total CSF flow (Pollay and Curl, 1967), moves by bulk flow through perivascular channels and along nerve fiber tracts to the ventricular spaces or to cervical lymphatics (see Fig. 5 and discussion below).

Figure 2 shows schematically the main elements of this system that are involved in cerebral pH regulation. Normal cerebral function requires that pH inside and outside the neuron be maintained within a normal range of values. Both neurons and glia have well-developed mechanisms for the regulation of intracellular pH (pH_i) (see Chapters 10–15 in this volume). In this chapter, we shall outline processes involved in the regulation of pH in the fluids lying outside the brain cells, brain ECF and CSF. The thesis is that this regulation occurs in parallel to that of pH_i and it provides a stable environment for pH_i regulation (see Chapter 17).

pH and Brain Function, Edited by Kai Kaila and Bruce R. Ransom
ISBN 0-471-11838-9

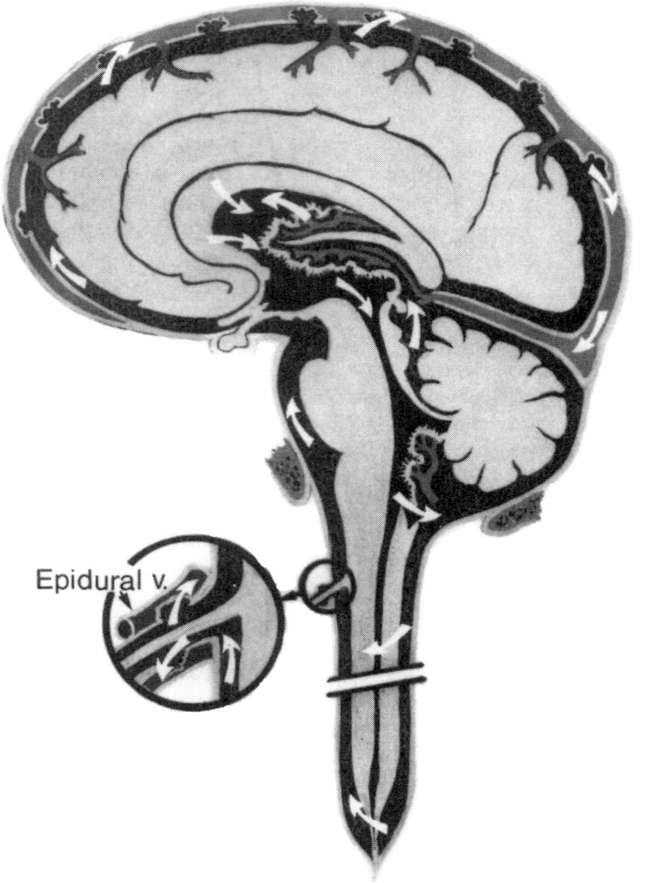

FIGURE 1. Sites of CSF formation at both choroid plexus and non–choroid plexus sources, the circulation of this fluid through the large ventricular system to the subarachnoid space, and the reabsorption of the fluid, via the arachnoid villi, into the venous sinuses at both intracranial and spinal root sites. (from Fishman, 1990, with permission of author and WB Saunders, Co.)

To understand how the pH of brain ECF is regulated is not an easy task. There are many measurements of CSF pH in various acid-base disorders, and the regulation of CSF pH has been the subject of recent, thorough reviews (Kazemi and Johnson, 1986; Nattie, 1983; Siesjö, 1972; Davson and Segal, 1996). The purpose of this chapter is to provide a general understanding of the processes of CSF formation, the movement of CSF through the brain, and how CSF pH is regulated. The relationship between ECF and CSF, in the context of pH regulation, will also be considered.

Importance of Blood-Brain Barrier Processes

An important issue is the extent to which ion secretory or exchange processes at the capillary endothelium can participate in, or even dominate, ECF pH and ion regulation. One thesis of this chapter is that such blood-brain barrier mechanisms are of crucial importance; they pro-

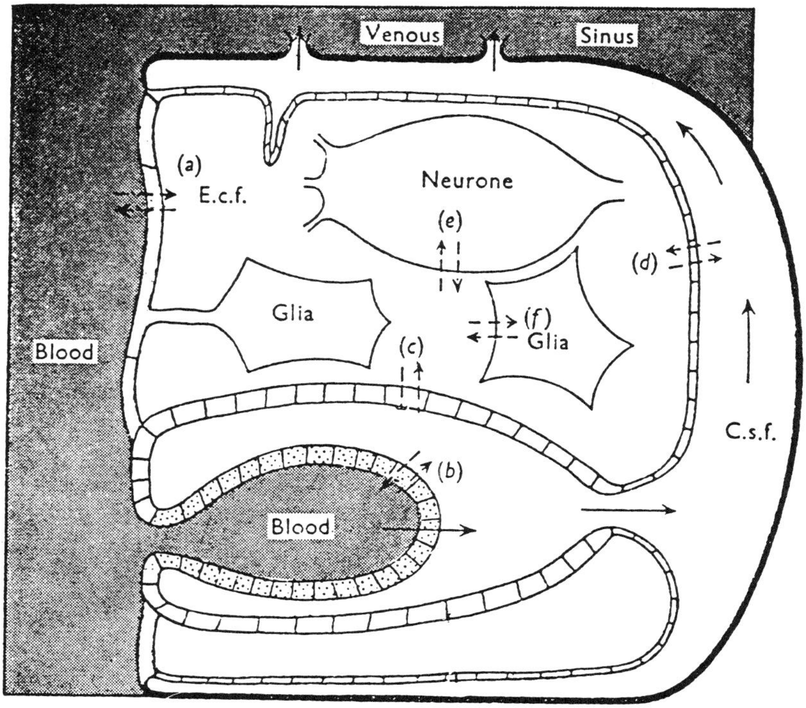

FIGURE 2. The production, circulation, and reabsorption of cerebrospinal fluid (CSF) and the compartments of the brain. At (*a*) the blood-brain barrier separates capillary blood from brain extracellular fluid (ECF); (*b*) represents the choroid plexus secretory site; (*c*) and (*d*) show the sites of free exchange at the ependyma between large-cavity CSF and brain tissue ECF; (*e*) and (*f*) show sites of ion exchange between brain cells and ECF. Ion exchange between cells and brain ECF involved in cell acid-base balance must, in the steady state, be balanced by ion exchange at blood-brain barrier (*a*) or choroid plexus (*b*) or by removal of fluid at venous sinuses or via the lymph (not shown here). (From Davson and Welch, 1971, with permission of Academic Press.)

vide a constant ECF environment for neuronal and glial membrane mechanisms of pH_i regulation. The ability of brain cells to exchange ions (Na^+/H^+, Cl^-/HCO_3^-) with the ECF will depend on how well ECF ion exchange occurs with fluids outside the brain. For example, in intact rats exposed to 11%, 27.5%, or 42.5% CO_2 for 90 minutes, Katsura et al., (1994) evaluated brain pH_i, ECF pH, and concentrations of Na^+, K^+, Cl^-, and HCO_3^-. With each increase in CO_2, ECF pH decreased initially, followed quickly by recovery toward the normal value. The magnitude of the initial decreases in ECF and pH_i were dependent on the level of CO_2. However, pH_i recovery was observed only in the group exposed to 11% CO_2. Exposure to higher CO_2 levels, with consequent greater ECF acidity, interfered with pH_i recovery. In these experiments, pH_i regulation depended on the degree of ECF pH change. The greater the degree of ECF acidosis the less effective was pH_i regulation.

The Chemistry of CSF

The chemical characteristics of brain ECF and CSF have relevance to considerations of pH regulation. Stewart (1981) formalized, by solving multiple simultaneous equations, the relationships of pH, P_{CO_2}, bicarbonate, carbonic acid, ions, and proteins (see Chapter 4). In summary, he determined that in physiological solutions there are, in a chemical sense, three independent variables: (i) the P_{CO_2}, (ii) the strong ion difference (SID) (the difference in the sum of all essentially fully dissociated cations and anions), and (iii) A_{TOT}, the concentration of non-HCO_3^- weak acid buffers (in blood primarily made up of charged protein) that can buffer protons generated by hydrolysis of CO_2. This approach has not been universally adopted, although it is chemically correct (see Cameron, 1989, for discussion of the pros and cons of the Stewart approach). Without debating its general utility, it has been of value in trying to understand CSF pH regulation (see Nattie, 1983). Applying the principles of Stewart's analysis to CSF, the pH in this fluid can be determined by the P_{CO_2} and the SID; A_{TOT} is not involved, as the protein concentration is so low. Further, any process involved in the regulation of CSF HCO_3^- must involve changes in strong ions; a fluid with no non-HCO_3^- weak acid buffers (A_{TOT}) has little detectable change in HCO_3^- when P_{CO_2} is changed (in a pure $NaHCO_3^-$ solution at 40 mM an increase in P_{CO_2} from 40 to 80 mmHg will increase the [HCO_3^-] by 40 nM, a virtually undetectable amount.) Thus, ion exchange with cells within the brain or with blood is necessary for the regulation of CSF and ECF HCO_3^-.

Before the emphasis on the role of strong ions in the determination of the CSF [HCO_3^-], very few studies of CSF pH and HCO_3^- regulation included measurement of these ions. It seems likely that the processes involved in this regulation use some or all of the transport proteins shown in Figs. 3 and 4, the same proteins used by cells for pH_i regulation. The challenge is to determine which proteins are involved and at what sites. In the regulation of brain ECF pH and [HCO_3^-], does exchange between blood and ECF dominate? Does exchange between ICF and ECF dominate? Are both involved?

OVERVIEW OF CEREBROSPINAL FLUID PRODUCTION AND FLOW

Role of Choroid Plexus as a Source for CSF

A current model for the process of CSF formation in choroid plexus epithelium appears in Fig. 3 (Johanson and Murphy, 1990). The key element is the Na^+/K^+ ATPase at the CSF facing (apical) membrane, which decreases cell Na^+ concentration and maintains the driving Na^+ concentration gradient from interstitial fluid to cell interior. At the apical surface, Na^+ leaves the cell via the ATPase and possibly via a $NaHCO_3$ pathway although this would require movement against a concentration gradient. Both K^+ and Cl^- exit via channels. At the interstitial fluid (basolateral) facing membrane, the Na^+/H^+ exchange protein enhances Na^+ entry and removes protons, and the Cl^-/HCO_3^- exchange protein enhances Cl^- entry in exchange for HCO_3^-. Carbonic anhydrase catalyzes the rapid production within the cell of the HCO_3^- and H^+ used for these ion exchanges. There appears to be sodium chloride cotransport at both the apical and basolateral membranes as well (not shown on Fig. 2) (see Johanson et al., 1989).

At the right of Fig. 3 are shown typical averaged ion concentrations and acid-base values. Note that the secreted CSF has a higher [Na^+] and [Cl^-] and lower [K^+] than does plasma (labeled interstitial fluid in the figure). Note also that CSF pH is acidic relative to plasma.

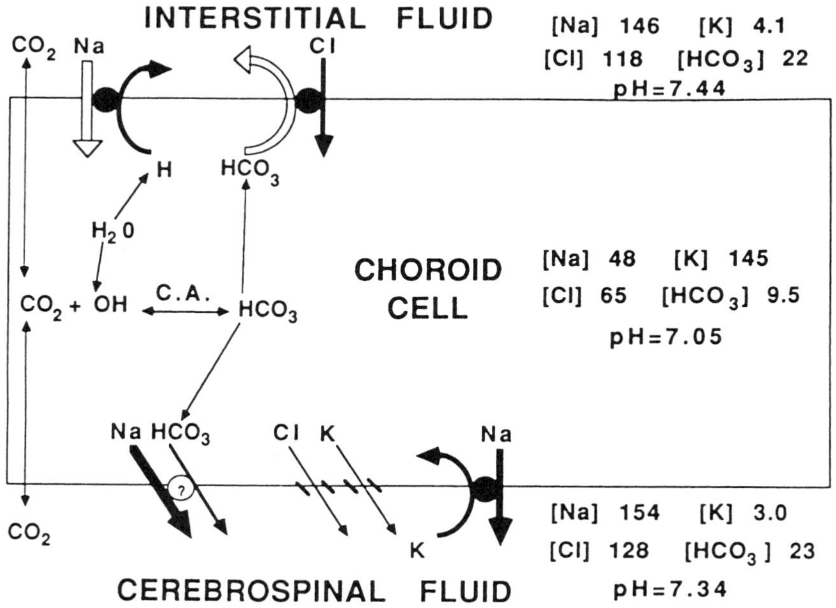

FIGURE 3. A model for the proteins involved in the secretion of CSF at the choroid plexus. At the blood-facing membrane there is evidence for N^+/H^+ and Cl^-/HCO_3^- exchange proteins. Within the cell, the presence of carbonic anhydrase speeds the production of bicarbonate that accompanies sodium secretion by a process not fully understood. At the CSF membrane, a Na^+/K^+-ATPase accounts for Na^+ movement into the ventricle. K^+ and Cl^- channels exist at this membrane. There is also evidence for a cation-coupled movement of Cl^- from blood to CSF. Any of these proteins could be involved in the regulation of pH effected by changes in choroid plexus secretion. (From Johanson and Murphy, 1990, with permission of author and the American Physiological Society.)

Brain Capillaries as a Source of CSF

In 1967, Pollay and Curl perfused the cerebral aqueduct and anterior portion of the fourth ventricle of the rabbit after removing the choroid plexus. Dilution of marker molecules in their perfusate indicated the presence of a fluid added from the brain. This fluid must have originated at the capillary endothelial cells that form the blood-brain barrier. These endothelial cells have low ion permeabilities (e.g., for Na^+ it is 0.004 that observed in muscle); low hydraulic conductivity, 0.012 that in muscle; and high electrical resistance, 67 times that seen in muscle (see Bradbury, 1979, 1985; Schielke and Betz, 1992). Thus ion movement to account for this observed fluid flow must be via a secretory process. In their study (Pollay and Curl, 1967), the flow attributed to the nonchoroidal source was one third of the total CSF flow measured with choroid plexus intact. More recent work showing the presence of various ion transport proteins at the blood-brain barrier supports the secretion of brain ECF by these cells (Bradbury, 1985).

A schematic of possible ion transport proteins at the blood-brain barrier appears in Fig. 4 (Schielke and Betz, 1992). Evidence from *in vivo* and *in vitro* studies (reviewed by Schielke and Betz, 1992) supports the existence in brain capillaries of an amiloride-sensitive nonselective cation channel (*a* in Fig. 4) (Betz, 1983a; Vigne et al., 1989). The Na^+/Cl^- cotransport

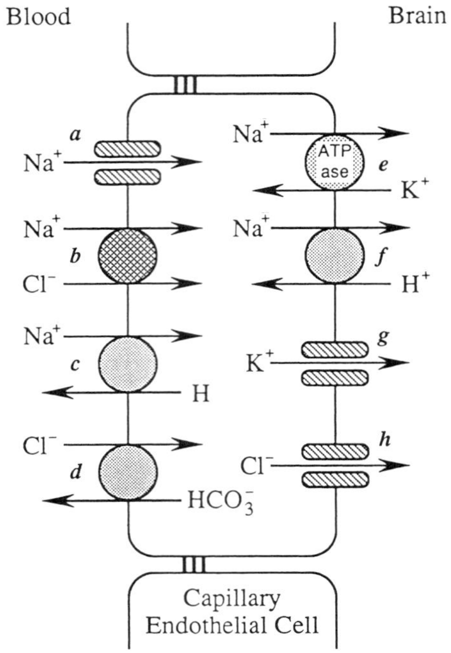

FIGURE 4. A model for the proteins possibly involved in the movement of ions and water between blood and brain ECF at the capillary (blood-brain barrier). At the brain ECF side, a Na^+/H^+ exchange protein could be sensitive to ECF pH. The Na^+/H^+ and Cl^-/HCO_3^- exchanger proteins at the blood-facing membrane would have to respond to changes in endothelial cell pH (from Schielke and Betz, 1992, with permission of author and Springer-Verlag).

protein (*b* in Fig. 4) is less well substantiated. The evidence for an Na^+/H^+ exchanger (*c* in Fig. 4) is largely from *in vivo* studies (Murphy and Johanson, 1989), as is the evidence for the Cl^-/HCO_3^- exchanger (Smith and Rapaport, 1984). At the CSF facing membrane, there is clear evidence *in vivo* and *in vitro* for a N^+/K^+-ATPase (Betz and Goldstein, 1981; Bradbury and Stulcová, 1970). This protein has kinetics, different from the ATPase of neuronal membranes, that would explain observed K^+ regulation at the blood-brain barrier (Schielke et al., 1990; see below). The Na^+/H^+ transport exchange protein (*f* in Fig. 4) has been demonstrated in isolated cerebral capillaries (Betz, 1983b) and the K^+ (*g* in Fig. 4) and Cl^- (*h* in Fig. 4) channels have been suggested based on observations made in other analogous systems. This model not only accounts for an actively secreted brain ECF originating at the brain capillary endothelium, but explains how these capillaries might be involved in brain ECF pH regulation.

Outflow of CSF

The CSF exits from the subarachnoid space, not the ventricles, into the venous sinuses through the arachnoid villi. The fluid leaves and enters the venous system by a pressure-dependent mechanism.

ECF Movement and Lymphatics

More recent work (Cserr and Patlak, 1992) has expanded on the hypothesis that brain ECF is secreted by capillary endothelial cells. This work shows that fluid can move by a pressure-dependent bulk flow through ECF spaces into ventricular CSF via the perivascular space (see Fig. 5). Cserr and her colleagues microinjected tracer molecules of different sizes into the brain parenchyma and analyzed the time course and pathways involved in their clearance. The time course for clearance of a 900-Da and a 4000-Da polyethylene glycol and of the 69,000-Da serum albumin were similar, results that can be explained by bulk flow movement but not by diffusion (Cserr et al., 1981). The maximum rate of this clearance is 0.19 μl/min for a 1.8-g rat brain. Part of this fluid movement can be traced into the large ventricular cavity CSF; part of it is cleared by other pathways (e.g., cervical lymphatics; Szentistványi et al., 1984). ECF turnover via bulk flow is slow under normal experimental conditions; the half time is 6–16 hr (Szentistványi et al., 1984). But in other conditions (e.g., increased plasma osmolality), ECF bulk flow can be faster with half-times of 30 minutes (Pullen et al., 1987). In contrast, large-cavity CSF turnover is approximately 8 hours. The ECF turnover in acid-base abnormalities (e.g.,hypercapnia) is unknown.

The Relationship of CSF to Brain Extracellular Fluid

This flow pathway for brain ECF from capillary to large-cavity CSF or lymphatics raises an important question. What is the role of ion concentrations in ventricular CSF in the

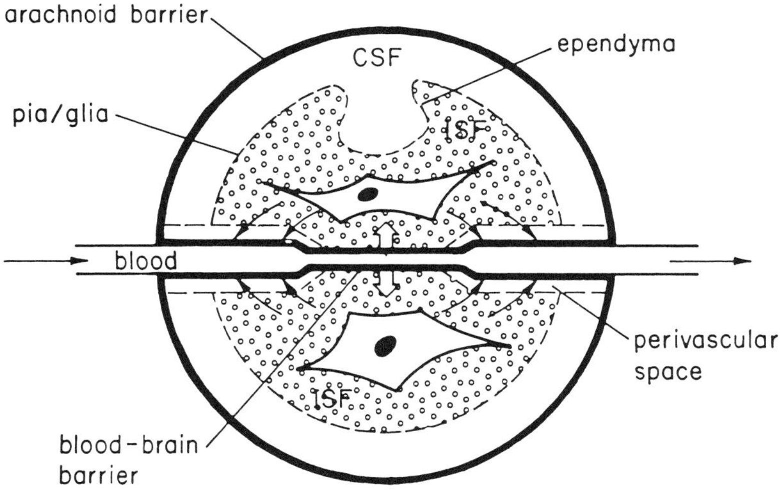

FIGURE 5. A model of fluid movement within the brain parenchyma via bulk flow. The perivascular spaces provide a pathway for the movement of brain ECF from tissue to large-cavity CSF. This hypothesis says that brain ECF, secreted by capillary endothelial cells, moves by bulk flow through the interstitial and perivascular spaces into the CSF. This would provide a method for continual renewal of the ECF and its ionic and acid-base constituents. (From Cserr and Patlak, 1992, with permission of author and Springer-Verlag.)

regulation of ECF? To what extent does the pH of freshly secreted CSF at the choroid plexus affect ECF pH? This is a subset of the larger issue concerning the function of ventricular CSF. Is CSF merely an effluent drainage system or does it modify brain ECF? The studies of ECF bulk flow cited above suggest that brain ECF normally moves from the endothelial secretory site into large-cavity CSF. It is only with increased CSF pressure or decreased vascular pressure that this flow reverses direction. In contrast, there are studies on the perfusion of tracer molecules via the cerebral ventricular system that show these molecules do enter the parenchyma (e.g., Borison et al., 1980). The mechanism could be diffusion or bulk flow if the normal flow-determining pressure gradients are disturbed. It seems reasonable to conclude that ECF, near the ventricles in most cases and deeper within the tissue in some cases, is exposed to the ionic constituents of the CSF as secreted by the choroid plexus.

K^+ as an Example of Ion Homeostasis Within the Brain

In CSF, the concentrations of the ions K^+, Ca^{2+}, and Mg^{2+}, vary little when the blood levels of these ions are altered (Bradbury, 1979). These ions can affect neuronal function at membranes and synapses. The maintenance of their concentrations within a narrow range of values optimizes central nervous system function even when their balance in body or blood is abnormal. There is evidence that part of this regulation takes place at the choroid plexus secretory cells. This will maintain appropriate ion concentrations in the CSF that circulates through the brain ventricular system. Such regulation must also take place at the blood-brain barrier to insure appropriate ion concentrations within the parenchyma.

We shall briefly consider, as an example, the evidence for the regulation of potassium by both choroid plexus and blood-brain barrier (Keep et al., 1993). The general problem in the evaluation of the relative roles of blood-brain barrier and choroid plexus in the regulation of ions or pH is technical. *In vivo* approaches (e.g., the study of radiolabeled ion entry into brain) cannot always dissociate between effects at the two sites. Recent *in vitro* approaches have disadvantages as well. For example, isolated microvessels have depleted ATP levels and a collapsed lumen; cultured endothelial cells can coalesce to form a barrier but this barrier has a resistance lower than that observed *in vivo* (Keep et al., 1993).

Studies using ion-selective electrodes placed into brain tissue in rats of different ages have shown that ECF [K^+] regulation is present, developmentally, before large-cavity CSF [K^+] regulation (Jones and Keep, 1987). These results show a clear role for the blood-brain barrier in the regulation of ECF [K^+] and suggest that the similar regulation of [K^+] at the choroid plexus may have but a secondary, modulatory effect.

Na^+/K^+-ATPases have been described at the brain-facing surface of both choroid plexus and blood-brain barrier cells and at both sites the K_m for K^+ transport is 3 mM, close to the normal brain ECF and CSF concentration of K^+. Cotransport of potassium with chloride is present at the choroid plexus and, perhaps, at the blood-brain barrier as well (not shown on Figs. 3, 4). Also, many K^+ channels have been identified in membranes at both sides of capillary endothelial cells (see Keep et al., 1993). These data show that mechanisms exist at both the choroid plexus and at the blood-brain barrier for the regulation of K^+ in brain ECF and CSF. Such a model should also apply to the regulation of brain ECF and CSF pH.

OBSERVED VALUES FOR CSF pH IN PERIPHERAL ACID-BASE DISTURBANCES

Primary Changes in Arterial P_{CO_2} (Respiratory Disorders)

Many authors have pooled CSF data obtained from humans with steady-state acid-base disorders (see Chapter 30). One example is shown in Fig. 6 (Fencl, 1971). At the right are shown the ranges of mean values from many studies of normal humans, patients with CO_2 retention secondary to respiratory disease, and patients with alveolar hyperventilation caused by hypoxic stimulation of the peripheral chemoreceptors. Note first that normal CSF pH is slightly more acidic than that of blood, due to a slightly higher P_{CO_2} in CSF. With CO_2 retention, blood pH is acidic as is CSF pH, but the change in CSF is less than in blood. These data are for patients with chronic disorders, so adaptive mechanisms in blood, kidney, and brain have had time to regulate pH. In an acute case of CO_2 retention, CSF pH rapidly becomes acidic, as the CO_2 can diffuse freely across blood-brain and blood-CSF barriers. With time, processes remove H^+ ions (in exchange with Na^+) or add HCO_3^- (in exchange with Cl^-) and the pH is corrected. Figure 7 shows data obtained from rats exposed for varying times to an increase in P_{CO_2} (Messeter and Siesjö, 1971). The top panel shows that the CSF P_{CO_2} is approximately 85 mm Hg for the entire duration of the experiment. Note that within the initial 45 minutes there is a significant increase in CSF

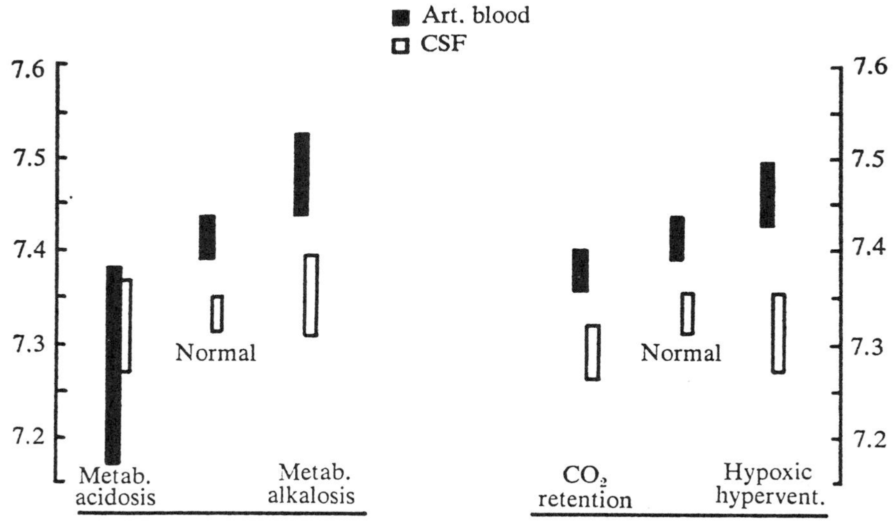

FIGURE 6. Summary of pooled human blood and CSF acid-base data taken from patients with various acid-base abnormalities. *Left:* Metabolic acidosis and alkalosis. Note that in the normal condition, CSF pH is slightly acidic relative to blood. In acidosis or alkalosis, CSF pH changes much less than does the blood value. *Right:* In respiratory acidosis produced by CO_2 retention, CSF pH does become slightly acidic. In respiratory alkalosis produced by hypoxic hyperventilation, CSF pH remains more constant than does blood pH. In general, CSF pH in these conditions is well regulated and changes less than does the blood value. (From Fencl, 1971, with permission of author and Academic Press.)

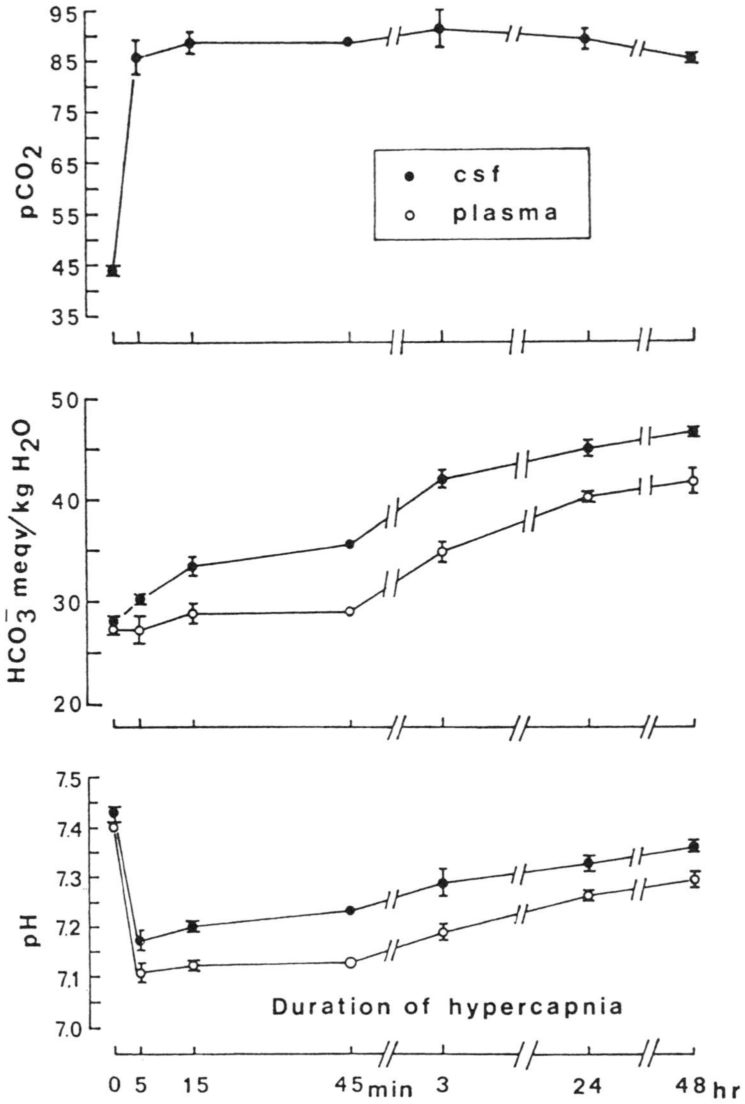

FIGURE 7. The time course of the changes in blood and CSF P_{CO_2}, HCO_3^-, and pH in rats exposed to an increase in P_{CO_2} for 48 hours. *Top:* Constant elevation of P_{CO_2}. *Bottom:* The pH falls initially in both compartments but slowly increases toward the original normal control value with time. In blood, this reflects, in large part, renal mechanisms for extrusion of H^+ with an accompanying increase in blood [HCO_3^-]. (Middle panel)—In CSF, H^+ is also removed and [HCO_3^-] increases due to ion exchange processes at choroid plexus and the blood-brain barrier (middle panel). Note that CSF [HCO_3^-] increases within 5 minutes and 15 minutes of the initial increase in P_{CO_2} before changes in plasma [HCO_3^-]. After 15 minutes, [HCO_3^-] increases in both plasma and CSF. (From Messeter and Siesjö, 1971, with permission of author and Academic Press.)

[HCO_3^-]. This change is evident at 5 minutes and 15 minutes, although plasma bicarbonate changes minimally at those times. The result is a much smaller decrease in CSF pH during this initial phase of the increased P_{CO_2}. Over 3 hours to 48 hours of exposure, plasma and CSF [HCO_3^-] both increase slowly but steadily, with corresponding changes in pH occurring in both compartments. As shown in Fig. 6, the resultant pH in a group of patients with chronic respiratory acidosis is, on average, quite close to normal.

In hypoxic hyperventilation, the blood pH becomes alkaline as the arterial P_{CO_2} is decreased. Initially, CSF and brain P_{CO_2} also decrease. With time, processes of ion exchange correct this, bringing the pH back toward the normal value, as for the respiratory acidosis shown in Fig. 7.

Primary Changes in Arterial [HCO_3^-] (Metabolic Disorders)

At the left on Fig. 6 are shown pooled data from patients with metabolic acid-base disorders (Fencl, 1971). The decrease in CSF pH in metabolic acidosis is less than in the arterial blood, as is the increase in pH in metabolic alkalosis. One can view a metabolic acid-base disorder as consisting of an initial change in the HCO_3^- and H^+ concentrations, produced by the addition of metabolic acid and titration of the existing HCO_3^- or produced by the addition of base. This is followed by a secondary change in the P_{CO_2} caused by changes in ventilation.

As an example, Fig. 8 shows data obtained in conscious goats with chronic metabolic acid-base disorders induced over a period of days (Fencl et al., 1966). The CSF pH, CSF and arterial PCO_2, and the CSF [HCO_3^-] are shown as a function of the plasma [HCO_3^-], the index of the acid-base state. In metabolic acidosis (low plasma HCO_3^- values), the CSF [HCO_3^-] is also decreased, but the change is much less than that observed in the plasma. CSF [HCO_3^-] appears to be regulated. The arterial and CSF P_{CO_2} are both decreased, a change reflecting increased alveolar ventilation caused by the acidic stimulation of peripheral and central chemoreceptors (Chapters 26 and 27). The CSF pH is slightly acidic, much less so than plasma pH (data not shown), reflecting the decreases in [HCO_3^-] and P_{CO_2}. Were there no increase in ventilation (and decrease in P_{CO_2}), the CSF pH would be more acidic. Were the change in CSF bicarbonate more equivalent to that in plasma, the CSF pH would also be more acidic. Thus, both processes are important in CSF pH regulation.

In metabolic alkalosis (higher plasma HCO_3^- concentrations), CSF [HCO_3^-] is increased, as is CSF (and arterial) P_{CO_2}, the latter due to the hypoventilation induced by the alkalosis. The resultant increase in CSF pH is much less than that observed in plasma and reflects, as in metabolic acidosis, the dual effects of the secondary P_{CO_2} changes and the processes that keep the changes in plasma HCO_3^- from being fully reflected within the CSF.

CSF [HCO_3^-] Regulation

The relationship of CSF to arterial [HCO_3^-] is plotted in Fig. 9 (Fencl, 1971) for the two types of acid-base disorder, respiratory and metabolic. Note that, in both cases, the change in CSF is much less than in plasma and that the relationship differs in the two cases. In a respiratory acid-base disorder, the change in CSF [HCO_3^-] is a larger fraction of the change in plasma [HCO_3^-]. It is important to note that in both acid-base disorders when the [HCO_3^-] concentration is increased above normal, the P_{CO_2} is also increased; the P_{CO_2} change is the primary event in the respiratory acidosis, the secondary event in the metabolic alkalosis. Similarly, when the plasma [HCO_3^-] is decreased below normal, the P_{CO_2}

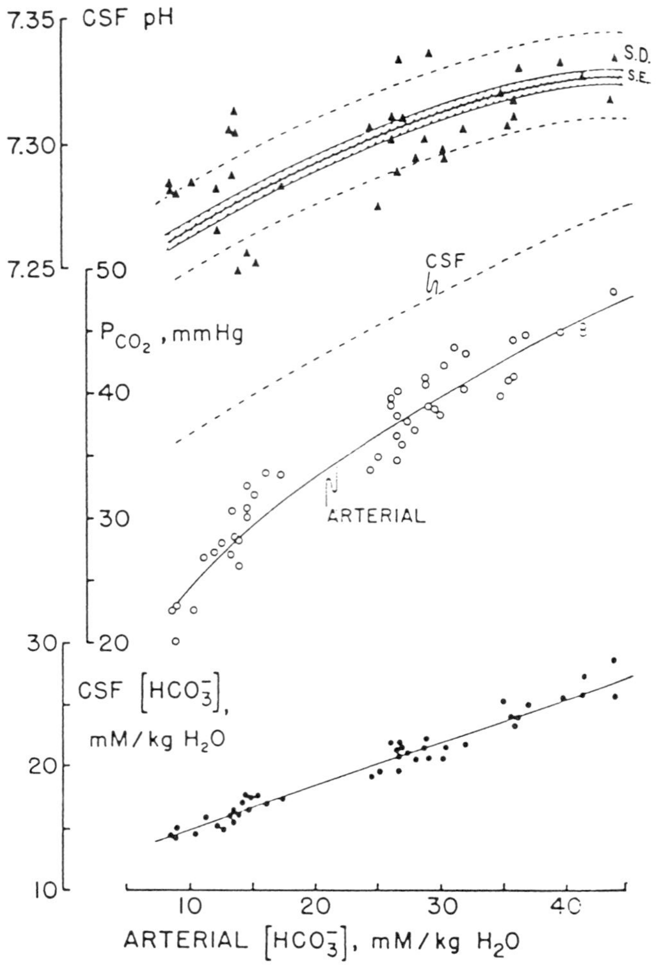

FIGURE 8. CSF pH (top panel), P_{CO_2} (middle panel), and $[HCO_3^-]$ (bottom panel) shown as a function of arterial $[HCO_3^-]$. Data obtained in conscious goats with metabolic acid-base disturbances induced chronically. In these carefully controlled experimental conditions, as in the human data above, CSF pH varies much less than does blood pH. Note that CSF P_{CO_2} decreases in metabolic acidosis, as does arterial P_{CO_2}, because of the increase in ventilation induced by acidic stimulation of respiratory chemoreceptors. In metabolic alkalosis, arterial and CSF P_{CO_2} both increase, reflecting the hypoventilation caused by inhibition of respiratory chemoreceptors by the alkaline pH. These P_{CO_2} changes are in a direction such as to reduce the initial pH insult. For example, with metabolic acidosis, the initial decrease in pH is modified by the subsequent lowering of P_{CO_2} from the hyperventilation. In addition, there are changes in CSF $[HCO_3^-]$ that are specific and reproducible. Note here that CSF $[HCO_3^-]$ is $= 11.3 + 0.352$ arterial $[HCO_3^-]$. The mechanism of this apparent CSF $[HCO_3^-]$ regulation is discussed in the text. (From Fencl et al., 1966, with permission of author and the American Physiological Society.)

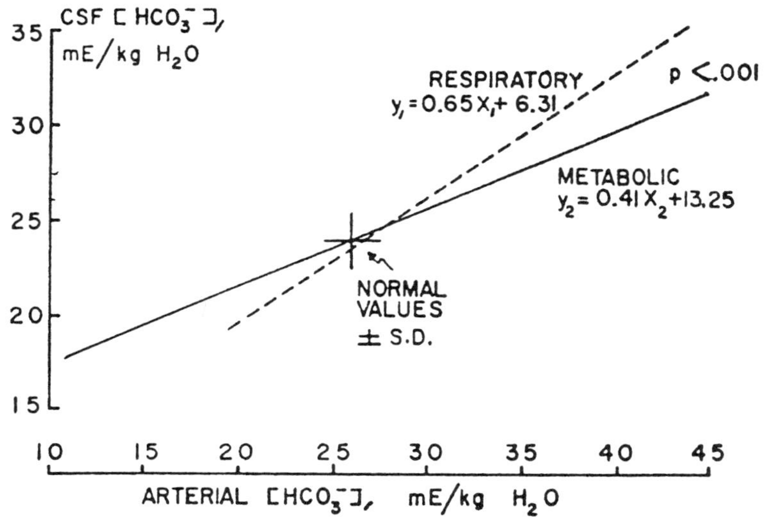

FIGURE 9. Plot of CSF versus arterial [HCO$_3^-$] for metabolic and respiratory acid-base disturbances. Data pooled from human patients with these disorders. Note the different relationship in respiratory and metabolic acid-base disorders. In the case with high arterial [HCO$_3^-$], a respiratory acidosis is compared to a metabolic alkalosis. In each, the P_{CO_2} and the blood [HCO$_3^-$] are both increased, but in slightly different proportions. The same holds true, in reverse, for the case with low arterial [HCO$_3^-$]. (From Fencl, 1971, with permission of author and Academic Press.)

is also decreased; it is the primary event in the respiratory alkalosis, the secondary event in the metabolic acidosis. This and other observations (e.g., Fig. 7, and discussed above) led to the proposal that the CSF [HCO$_3^-$] in acid-base disorders could be determined by (i) a P_{CO_2}-dependent process, and (ii) a process dependent on or related to the plasma [HCO$_3^-$].

Results of one experiment designed to test this hypothesis as shown in Fig. 10 (Nattie and Romer, 1978). Here the plasma [HCO$_3^-$] and the P_{CO_2} were independently controlled under isosmotic conditions by peritoneal dialysis and artificial ventilation; CSF [HCO$_3^-$] was the response measure. Solid circles show the results from control animals with no change in plasma [HCO$_3^-$] or P_{CO_2} and no effect on CSF [HCO$_3^-$]. When the P_{CO_2} was increased and plasma [HCO$_3^-$] kept at the normal level by dialysis (open circles), the CSF [HCO$_3^-$] increased significantly within 2 hr and did not change after that. The isolated increase in plasma [HCO$_3^-$] produced by dialysis with the P_{CO_2} kept constant (solid squares) resulted in a slow and steady increase of CSF [HCO$_3^-$] over the 6-hour experimental period. The combination of the two stimuli, an increase in the P_{CO_2} and the plasma [HCO$_3^-$] (open squares), produced both the initial rapid and the slower prolonged increases in CSF [HCO$_3^-$]. Other experiments performed at the same time produced similar results and led Kazemi and co-workers to formulate a "dual contribution hypothesis" for CSF HCO$_3^-$ regulation in acid-base disorders (Hasan and Kazemi, 1976; see also Kazemi and Johnson, 1986).

These descriptive data show that CSF pH is regulated in peripheral acid-base disorders. As a result, CSF pH remains close to normal despite large changes in blood pH. They suggest that at least two types of processes might be involved: one dependent on the P_{CO_2}, the other dependent on the level of plasma HCO$_3^-$ or on a related variable. They do not show the mechanisms involved.

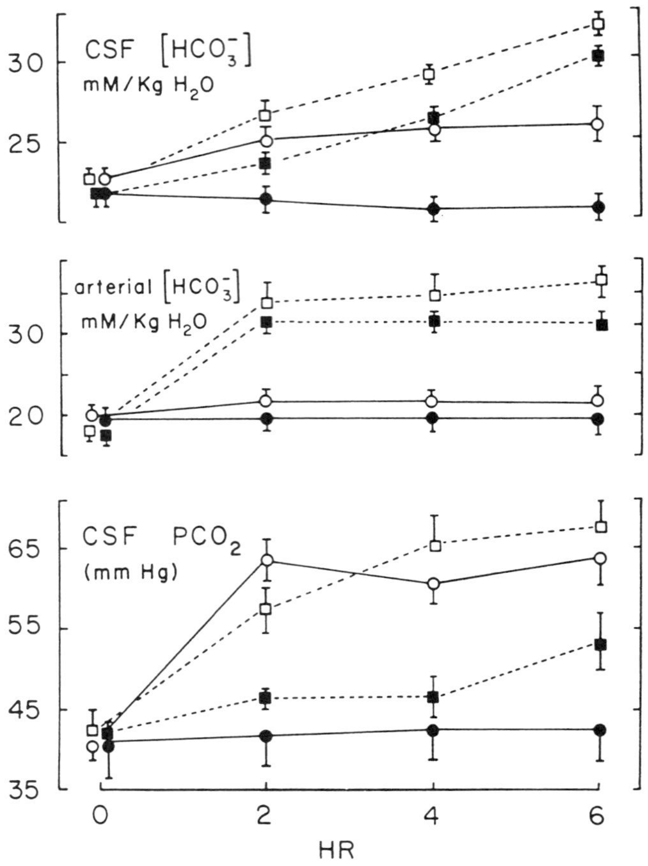

FIGURE 10. Separate effects of P_{CO_2} (bottom) and plasma $[HCO_3^-]$ (middle) on CSF $[HCO_3^-]$ (top). In anesthetized cats, the time course and magnitude of the increase in CSF $[HCO_3^-]$ depend on the relative changes in arterial $[HCO_3^-]$ and the P_{CO_2}. In the control (solid circles), with arterial $[HCO_3^-]$ and CSF P_{CO_2} kept constant over 6 hours, no change in CSF $[HCO_3^-]$ is observed. With an increase in CSF P_{CO_2} alone (open circles), and arterial $[HCO_3^-]$ kept constant, CSF $[HCO_3^-]$ is increased within 2 hours and remains at that level for 6 hours. If arterial $[HCO_3^-]$ is increased while CSF P_{CO_2} is kept constant (solid squares), CSF $[HCO_3^-]$ increases steadily throughout the 6-hour period of the experiment. With an increase in both arterial $[HCO_3^-]$ and CSF P_{CO_2} (open squares), CSF $[HCO_3^-]$ shows an early increase at 2 hours and a steady further increase over 6 hours. Increases in P_{CO_2} stimulate production of a CSF with a greater $[HCO_3^-]$; increases in plasma $[HCO_3^-]$ result in a slower increase in CSF $[HCO_3^-]$. (From Nattie and Romer, 1978, with permission of author and Elsevier Science).

MECHANISTIC DETERMINANTS OF CSF pH

Introduction

To understand mechanisms that might explain the observed regulation of CSF pH shown in the previous section, it is useful to consider here the Henderson-Hasselbalch relationship:

$$pH = pK' + \log\left(\frac{[HCO_3^-]}{S \times P_{CO_2}}\right)$$

where S is the solubility of CO_2, and pK' a combined dissociation constant. Isolated changes in P_{CO_2} or bicarbonate can affect the pH. To understand brain ECF and CSF, pH regulation requires separate analysis of physiological processes that can affect these two variables. The situation becomes complex in that the P_{CO_2} can in turn be a direct stimulus for changes in the amount of HCO_3^- in cerebral fluids via effects on ion secretory processes. The [HCO_3^-], or a related variable, also appears to be associated with regulatory responses.

The P_{CO_2} in brain tissue and CSF is determined by (i) alveolar ventilation, (ii) metabolism, and (iii) cerebral blood flow. Changes in cerebral metabolic rate will affect tissue P_{CO_2}, all else being equal. Because cerebral P_{CO_2} is related to arterial P_{CO_2}, the level of alveolar ventilation is a key determinant. An increase in alveolar ventilation decreases P_{CO_2} in arteries, brain tissue, and CSF; a decrease in alveolar ventilation increases P_{CO_2} in all three compartments. Chemoreceptors involved in the control of breathing (see Chapters 26 and 27) sense changes in pH and change the level of ventilation. This system can help to regulate cerebral ECF pH. For example, a metabolic acidosis stimulates chemoreceptors, increases alveolar ventilation, and decreases arterial and tissue P_{CO_2}. This decrease in P_{CO_2}, via mass action, tends to correct the pH. Cerebral blood flow can also affect tissue and CSF P_{CO_2}. If blood flow decreases, tissue and CSF P_{CO_2} increase; if blood flow increases, they decrease. This can operate to regulate cerebral pH, as cerebral blood flow is quite sensitive to ECF pH. For example, acidosis decreases cerebral vascular resistance, thus increasing blood flow. The increased flow enhances CO_2 elimination so that tissue and CSF P_{CO_2} decreases. This decrease in P_{CO_2} tends to correct the initial acidosis.

We now discuss how P_{CO_2} can stimulate ionic secretory processes that change [H^+] and/or [HCO_3^-] and how HCO_3^- enters the brain in exchange for Cl^-.

Role of P_{CO_2}

Choroid plexus. Studies of isolated choroid plexus fluid and of CSF collected in the cisterna magna show that increased P_{CO_2} changes the ionic composition of the secreted fluid; the [HCO_3^-] increases and the [Cl^-] decreases (Kazemi and Johnson, 1986; Nattie, 1983). Some studies also show an increase in CSF [Na^+] in the immediate response (Nattie, 1980; Katsura et al., 1994). These changes in strong ions that accompany the increased [HCO_3^-] suggest ionic mechanisms are involved in the pH regulation. As [HCO_3^-] increases, the pH also increases. With a decrease in P_{CO_2} the reverse occurs; [HCO_3^-] decreases and [Cl^-] increases (Javaheri et al., 1994). The mechanisms for these changes in the ionic character of the secreted CSF are not understood. They appear to involve Na^+/H^+ and Cl^-/HCO_3^- exchange proteins. Examples that support this contention are:

1. Infusion of amiloride or DIDS can inhibit the response of CSF [HCO_3^-] to increased P_{CO_2} (e.g., see Nattie and Adams, 1988; Nattie and Giddings, 1988).
2. Part of the change in CSF [HCO_3^-] in hypercapnia requires the presence of carbonic anhydrase (Wichser and Kazemi, 1975).
3. Changes in pH can affect choroid plexus cell ion concentrations and Na^+ entry from blood to CSF via an amiloride-sensitive process (Murphy and Johanson, 1990).

Blood-brain barrier. While ionic events at the blood-brain barrier that help to regulate brain ECF pH will also affect the composition of ventricular CSF, we are most interested in their influence on the ECF. As discussed above, CSF ion concentrations change dramatically and reliably when the P_{CO_2} is altered above or below the normal level. Unfortunately, there are very few studies that measure ECF ions directly with changes in P_{CO_2}.

Attempts have been made to detect changes in ion entry into the brain by measurement of ions in whole brain tissue and CSF. The rationale is that ion entry will be reflected in whole tissue measurements. However, the brain ECF space is only between 15–20% of brain volume, so substantial changes in ECF ion concentrations can be masked. In the absence of changes in brain cell ion content, it can be quite difficult to detect the small changes in tissue ion concentrations caused entirely by changes in ECF ion concentration. Nonetheless, changes in tissue total CO_2 content are clearly demonstrable in response to changes in P_{CO_2} (Wichser and Kazemi, 1975).

Two studies have measured ECF ions directly in response to changes in P_{CO_2}. Ahmad and Loeschcke (1982a) (Fig. 11) measured cortical surface pH, P_{CO_2}, and $[Cl^-]$ during a sudden, sustained increase in P_{CO_2} in anesthetized cats. Cortical P_{CO_2} increased and pH decreased. The ECF $[HCO_3^-]$ increased and the $[Cl^-]$ decreased, both with surprising speed. The time constant for the chloride response was 30.5 seconds, that for the bicarbonate response, 30.7 seconds. The interpretation was that the responses reflected

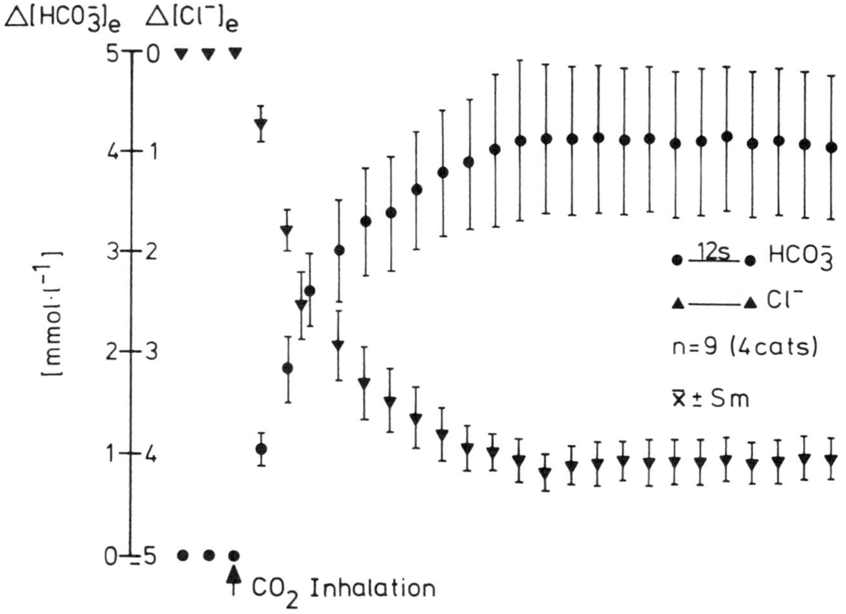

FIGURE 11. Rapid changes in cortical ECF $[HCO_3^-]$ and $[Cl^-]$ in response to abrupt sustained increases in P_{CO_2}. In anesthetized cats, electrodes measured cortical surface pH, P_{CO_2}, and $[Cl^-]$ during a sudden, sustained increase in P_{CO_2}. The ECF $[HCO_3^-]$ was calculated from the pH and P_{CO_2} data. The symbols are separated by 12-second intervals. Note the immediate increase in ECF $[HCO_3^-]$ and the immediate equimolar decrease in ECF $[Cl^-]$. The interpretation was that these responses reflected glia-ECF exchange. Such an exchange would leave the glia depleted of HCO_3^- and severely acidotic as a result of the increase in P_{CO_2}. An alternate explanation could be the exchange of HCO_3^- and Cl^- between blood and ECF via capillary endothelial cells. (From Ahmad and Loeschcke, 1982a, with permission of author and Springer-Verlag.)

glia–ECF exchange of Cl^- and HCO_3^-; it was a rapid cell response that worked to regulate ECF pH. Such an exchange would leave the glia depleted of HCO_3^- and more severely acidotic. An alternate explanation is that the exchange of HCO_3^- and Cl^- was between blood and ECF via capillary endothelial cells. The transport processes shown in Fig. 4 could cause these changes. In an accompanying paper, Ahmad and Loeschcke (1982b) attributed similarly timed changes in [Cl^-] and [HCO_3^-] to ion movement across the blood-brain barrier. And measurements of cisternal CSF ions, at some distance from the capillary endothelial cell site, show ion responses to increased CO_2 within minutes (Nattie, 1980).

Katsura et al. (1994) used ion-selective electrodes to measure rat cortex ECF ions in response to increased P_{CO_2}. They also measured brain tissue ions and calculated [HCO_3^-] in intracellular (ICF) and ECF. They reported significant changes in ECF [HCO_3^-], [Na^+], and [Cl^-] within minutes of the onset of hypercapnia. Total tissue bicarbonate content was also increased significantly within minutes, a result also found by others (Nattie, 1980; Wichser and Kazemi, 1975). These investigators interpret the data to reflect an exchange of Cl^- for HCO_3^- and Na^+ for H^+ between ICF and ECF. That is, the cell is losing HCO_3^-, or gaining H^+, relative to the ECF. Or, ECF pH is regulated at the expense of pH_i. The major observation to support this interpretation is the absence of any measurable change in whole-tissue Na^+ and Cl^- content. But, because of the small ECF space and the distribution of Na^+ and Cl^-, both being of greater concentration in ECF than in ICF, it may be quite difficult to detect significant changes in ECF sodium or chloride by these whole-tissue measurements.

It seems likely that rapid ion movements at the blood-brain barrier play a very important role in ECF pH regulation in hypercapnia. Such regulation of ECF pH by ion transport processes at the capillary endothelium would maintain an extracellular environment that would optimize the processes at the neuronal and glial cell membranes that focus on pH_i regulation. The alternate view, that brain cells lose HCO_3^- and gain H^+ during hypercapnia to regulate ECF pH would require a greater acidification of the cells involved. Simultaneous measurement of ICF and ECF pH in the brain of the anesthetized sheep using ^{31}P NMR show that after 10 minutes of hypercapnia, pH_i decreases less than does ECF pH (Portman et al., 1991). In a separate study, following 50 minutes of increased P_{CO_2}, pH_i measured by ^{31}P NMR changes much less than blood pH, with exposure to both 5% and 10% CO_2 (Nishimura et al., 1989). Similar data have been reported in rats using tissue and CSF measurements (Kjallquist et al., 1969). These observations of a minimally changed average brain pH_i in acute hypercapnia do not support the idea of ICF acidification to protect or regulate ECF pH. There remains the possibility that some brain compartments (e.g., glial cells) could perform such a function, while other compartments do not.

At issue, then, is where do the rapid changes in brain ECF [Na^+], [HCO_3^-], and chloride come from during the response to hypercapnia? One view is that these ion changes occur between ICF and ECF; the other is that they occur between ECF and blood. Both sites could be involved.

Role of Bicarbonate Entry from Blood

Ahmad and Loeschcke (1982b) measured on the cortical surface of anesthetized cats the pH, P_{CO_2}, and [Cl^-] (and calculated the ECF [HCO_3^-]) in response to a bolus infusion of $NaHCO_3$ (3 M). This was given intravenously while the arterial P_{CO_2} was kept constant. The results of the experiment, shown in Fig. 12, show that there is a rapid increase in ECF

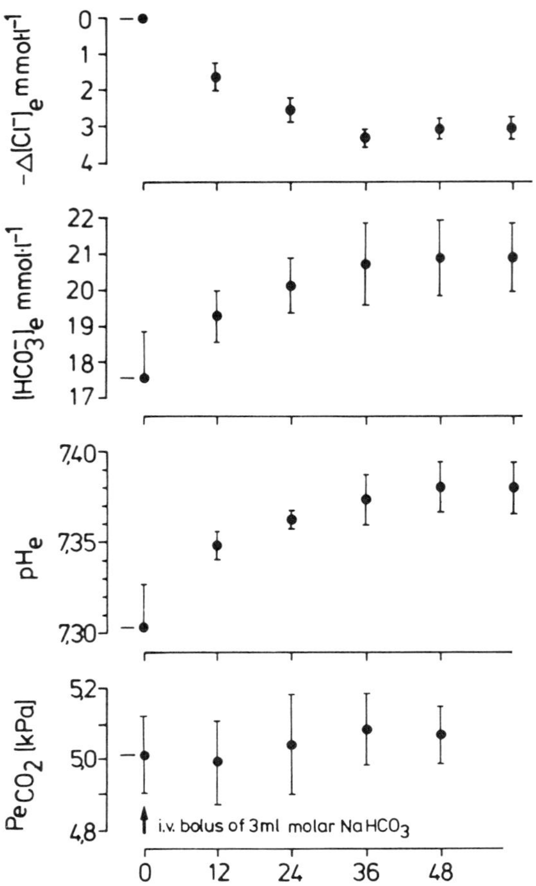

FIGURE 12. Rapid changes in cortical ECF (top) [Cl^-] and [HCO_3^-] (second from top) in response to intravenous infusion of $NaHCO_3$, with P_{CO_2} kept constant (bottom). The time scale is in seconds. In anesthetized cats, electrodes measured cortical surface pH (second from bottom), P_{CO_2}, and [Cl^-]. The ECF [HCO_3^-] was calculated from the pH and P_{CO_2} data. Note the time (in seconds) and the rapid increase in ECF [HCO_3^-] with the accompanying decrease in ECF [Cl^-]. This is a surprisingly rapid exchange of HCO_3^- and Cl^- between plasma and ECF, given the known characteristics of the blood-brain barrier. (From Ahmad and Loeschcke, 1982b, with permission of author and Springer-Verlag.)

[HCO_3^-] accompanied by a decrease in ECF [Cl^-]. Many other studies have shown rapid changes in ECF pH following changes in arterial pH caused by intravenous infusion of acid or base when arterial P_{CO_2} was maintained at a constant value. Flat pH electrodes placed on the ventral surface of the medulla (see Javaheri et al., 1981; Teppema et al., 1983) show ECF pH changes within minutes of the change produced in blood. In the steady state these are approximately one third of the pH change in blood. Tissue pH electrodes placed into the ventral medulla and thalamus (Davies and Nolan, 1982) and cortex (Javaheri et al., 1983) also show such responses, although of a smaller magnitude. The mechanism of the rapid pH response in these studies is not known although the HCO_3^-/Cl^- exchange protein shown on Fig. 4 is a likely participant.

In contrast, measurement of average brain cell pH using the ^{31}P NMR technique in isocapnic metabolic acidosis (Adler et al., 1990) failed to show any acid shift during the first hour of acid infusion. There are three possible explanations for this absence of NMR-detectable pH changes. First, the NMR study used a less severe arterial acidosis than did the studies in which surface or tissue pH was measured by electrode. It is possible that pH_i changes were present but undetectable due to their small magnitude. Second, pH_i may not change over the period of 1 hour of an isocapnic metabolic acidosis or alkalosis, even when ECF pH does change. Third, the pH response within brain to this stress may be heterogeneous. There is evidence for regional heterogeneity of medullary pH responses when CO_2-saturated saline is infused as a bolus via the vertebral arteries (Arita et al., 1989). Some regions showed no pH change; others showed large changes. It is not known whether such a regional response also pertains to an isocapnic metabolic acidosis.

Relationship of CSF to ECF pH

In respect to the relationship of brain ECF to ventricular CSF pH during isocapnic intravenous infusion of acid or base with a demonstrable steady-state change in brain ECF pH, no change is CSF pH is observed for the 60-minute duration of the experiments (e.g., Davies and Nolan, 1982; Javaheri et al., 1983). In response to a change in blood [HCO_3^-] (or [H^+]) produced in isocapnic conditions, ECF [HCO_3^-] (or [H^+]) changes quite a bit faster than does the fluid measured in the cisterna magna.

In sum, rapid changes in the ionic composition of secreted CSF (and ECF) can occur in response to both respiratory and peripheral metabolic acid-base disorders. These ionic events, not fully understood at present, affect pH regulation in the fluids surrounding the neurons.

CSF pH IN ACID-BASE DISTURBANCES THAT ARISE WITHIN THE BRAIN

Endogenous Lactic Acidosis

The conditions discussed above, hypercapnia and a metabolic acidosis or alkalosis that originates in blood, provide a very different type of acid-base stress to the brain than does acidosis that arises within the brain itself. A lactic acidosis induced by hypoxia or ischemia would be an example of such a situation. Adler et al. (1990), using ^{31}P NMR, measured brain and blood pH and lactate concentration in rats before, during, and after 30-minute exposure to 8% inspired O_2 (arterial Po_2 = 25 to 36 mmHg). During the 30 minutes of hypoxia, mean blood pH decreased from 7.45 to 7.00 and mean brain pH decreased from 7.12 to 6.77. The change in arterial [H^+] was 37 nM, while that in brain was 94 nM. Brain H^+ ion regulation was not apparent during the hypoxia, in contrast to the virtual absence of any NMR-detectable brain pH change in peripheral metabolic acid-base disorders (see above). In the recovery period, however, brain pH returned to normal within 12 minutes, then showed an alkaline overshoot, while arterial pH or blood lactate levels did not recover over 40 minutes. This rapid recovery of brain pH and the subsequent overshoot speak to active cell pH_i regulation following a brief hypoxic insult such as this. It is likely that ECF regulatory mechanisms, most likely blood-brain barrier processes given the speed of the response are also involved in this rapid recovery.

Under more severe stress, a sustained altered pH that overwhelms pH regulatory mechanisms, can be important in the subsequent pathology (see Siesjö et al., 1993).

SUMMARY

CSF is formed by ionic exchange processes at the choroid plexus and circulates through the brain ventricular system. Ion exchange occurs at the blood-brain barrier, and a poorly understood secretory process creates an ECF that moves by bulk flow through brain parenchyma via the perivascular spaces. CSF pH changes very little in chronic respiratory and peripheral metabolic acid-base disturbances. The mechanisms of this CSF pH regulation involve system responses (ventilation and cerebral blood flow) to control P_{CO_2} and ion exchange processes to regulate $[HCO_3^-]$. A thesis of this chapter is that brain ECF pH regulation occurs quickly via ion exchanges that take place across brain capillaries. ECF pH changes quickly in response to a peripheral isocapnic metabolic acid-base disorder, but the ECF pH change is less than in blood. Part of this apparent regulation could be due to the low permeability of the blood-brain barrier to HCO_3^- (or H^+); part could be due to ion transport regulation. Brain ECF pH and ionic regulation, via ion exchange with blood, provides a stable environment for neuronal pH_i regulation via ion exchange with ECF.

REFERENCES

Adler S, Simplaceanu V, Ho C (1990): Brain pH in acute isocapnic metabolic acidosis and hypoxia: A ^{31}P- nuclear magnetic resonance study. Am J Physiol 258: F34–F40.

Ahmad HR, Loeschcke HH (1982a): Fast bicarbonate-chloride exchange between brain cells and brain extracellular fluid in respiratory acidosis. Pflügers Arch 395: 293–299.

Ahmad HR, Loeschcke HH (1982b): Fast bicarbonate-chloride exchange between plasma and brain extracellular fluid at maintained P_{CO_2}. Pflügers Arch 395: 300–305.

Arita H, Ichikawa K, Huwana S, Kogo N (1989): Possible locations of pH-dependent central chemoreceptors: Intramedullary regions with acidic shift of extracellular fluid pH during hypercapnia. Brain Res 485: 285–293.

Betz AL (1983a): Sodium transport from blood to brain: Inhibition by furosemide and amiloride. J Neurochem 41: 1158–1164.

Betz AL (1983b): Sodium transport in capillaries isolated from rat brain. J Neurochem 41: 1150–1157.

Betz AL, Goldstein GW (1981): The basis for active transport at the blood-brain barrier. In Eisenberg HM, Suddith RL (eds): Advances in Experimental Medicine Biology. New York: Plenum, pp 5–15.

Borison HL, Borison R, McCarthy LE (1980): Brain stem penetration by horseradish peroxidase from cerebrospinal fluid spaces in the cat. Exp Neurol 69: 271–289.

Bradbury MWB (1979): The Concept of a Blood Brain Barrier. Chichester: John Wiley & Sons.

Bradbury MWB (1985): The blood-brain barrier. Transport across the cerebral endothelium. Circ Res 57: 213–222.

Bradbury MWB, Stulcová B (1970): Efflux mechanism contributing to the stability of the potassium concentration in cerebrospinal fluid. J Physiol (Lond) 208: 415–430.

Cameron J (1989): Acid-base homeostasis: Past and present perspectives. Physiol Zoology 62: 845–865.

Cserr HF, Cooper DN, Suri PK, Patlak CS (1981): Efflux of radiolabeled polyethylene glycols and albumin from rat brain. Am J Physiol 240: F319–F328.

Cserr HF, Patlak CS (1992): Secretion and bulk flow of interstitial fluid. In Bradbury MWB (ed): Physiology and Pharmacology of the Blood-Brain Barrier. New York: Springer-Verlag, pp 245–261.

Davies DG, Nolan WF (1982): Cerebral interstitial fluid acid-base status follows arterial acid-base perturbations. J Appl Physiol 53: 1551–1555.

Davson H, Welch K (1971): The relations of blood, brain and cerebrospinal fluid. In Siesjö BK, Sorenson SC (eds): Ion Homeostasis of the Brain. New York: Academic Press, p 10.

Davson H, Segal MB (1996): Acid-base relations in the central nervous system. In Davson H, Segal MB: Physiology of the CSF and Blood-brain Barriers. New York: CRC Press, pp 459–488.

Fencl V (1971): Distribution of H^+ and HCO_3^- in cerebral fluids. In Siesjö BK, Sorenson SC (eds): Ion Homeostasis of the Brain. New York: Academic Press, pp 175–185.

Fencl V, Miller TB, Pappenheimer JR (1966): Studies on the respiratory response to disturbances of acid-base balance, with deductions concerning the ionic composition of cerebral fluids. Am J Physiol 210: 459–472.

Fishman R (1990): Cerebrospinal Fluid in Diseases of the Nervous System. Philadelphia: WB Saunders.

Hasan FM, Kazemi H (1976): The dual contribution theory of regulation of CSF HCO_3^- in respiratory acidosis. J Appl Physiol 40: 559–567.

Javaheri S, Clendenning A, Papadakis N, Brody JS (1981): Changes in brain surface pH during acute isocapnic metabolic acidosis and alkalosis. J Appl Physiol 51: 276–281.

Javaheri S, De Hemptine A, Vanheel B, Leusen I (1983): Changes in brain ECF pH during metabolic acidosis and alkalosis: A microelectrode study. J Appl Physiol 55: 1849–1853.

Javaheri S, Corbett W, Wagner K, Adams JM (1994): Quantitative cerebrospinal fluid acid-base balance in acute respiratory alkalosis. Am J Respir Crit Care Med 150: 78–87.

Johanson CE (1995): Ventricular and cerebrospinal fluid. In Conn PM (ed): Neuroscience in Medicine. Philadelphia, JB Lippincott, pp 171–196.

Johanson CE, Murphy VA (1990): Acetazolamide and insulin alter choroid plexus epithelial cell [Na^+], pH, and volume. Am J Physiol 258 (Renal Fluid Electrolyte Physiol 27): F1538–F1546.

Johanson, CE, Murphy VA, Bairmian D, Epstein MH (1989): Exchange vs cotransport of Na^+ and Cl^- across the barriers interfacing cerebrospinal fluid and brain with blood. In Hoff J, Betz AL (eds): Intracranial Pressure VII. New York: Springer-Verlag, pp 303–311.

Jones HC, Keep RF (1987): The control of potassium concentration in the cerebrospinal fluid and brain interstitial fluid of developing rats. J Physiol (Lond) 383: 441–453.

Katsura K, Kristián T, Nair R, Siesjö BK (1994): Regulation of intra- and extracellular pH in the rat brain in acute hypercapnia: A re-appraisal. Brain Res 651: 47–56.

Kazemi H, Johnsen DC (1986): Regulation of cerebrospinal fluid acid-base balance. Physiol Rev 66: 953–1037.

Keep RF, Xiang J, Betz AL (1993): Potassium transport at the blood-brain and blood-CSF barriers. In Drewes LR, Betz AL (eds): Frontiers in Cerebral Vascular Biology: Transport and Its Regulation. New York: Plenum Press, pp 43–54.

Kjallquist A, Nardini M, Siesjö BK (1969): The regulation of extra- and intracellular acid-base parameters in the rat brain during hyper- and hypocapnia. Acta Physiol Scand 76: 485–494.

Messeter K, Siesjö BK (1971): Electrochemical gradients for H^+ and HCO_3^- between blood and CSF during sustained acid-base changes. In Siesjö BK, Sorenson SC (eds): Ion Homeostasis of the Brain. New York: Academic Press, pp 190–200.

Murphy VA, Johanson CE (1989): Acidosis, acetazolamide, and amiloride: Effects on $^{22}Na^+$ transfer across the blood-brain and blood-CSF barriers. J Neurochem 52: 1058–1063.

Murphy VA, Johanson CE (1990): Na^+-H^+ exchange in choroid plexus and CSF in acute metabolic acidosis or alkalosis. Am J Physiol 258: F1528–F1537.

Nattie EE (1980): Brain and cerebrospinal fluid ionic composition and ventilation in acute hypercapnia. Respir Physiol 40: 309–322.

Nattie EE (1983): Ionic mechanisms of cerebrospinal fluid acid-base regulation (brief review). J Appl Physiol 54: 3–12.

Nattie EE, Adams JM (1988): DIDS decreases CSF HCO_3^- and increases breathing in response to CO_2 in awake rabbits. J Appl Physiol 64: 397–403.

Nattie EE, Giddings B (1988): Effects of amiloride and diethylpyrocarbonate on CSF HCO_3^- and ventilation in hypercapnia. J Appl Physiol 65: 242–248.

Nattie EE, Romer L (1978): CSF HCO_3 regulation in isosmotic conditions: The role of brain P_{CO_2} and plasma HCO_3^-. Respir Physiol 33: 177–198.

Nishimura M, Johnson DC, Hitzig BM, Okunieff P, Kazemi H (1989): Effects of hypercapnia on brain pH_i and phosphate metabolite regulation by ^{31}P-NMR. J Appl Physiol 66: 2181–2188.

Pollay M, Curl F (1967): Secretion of cerebrospinal fluid by the ventricular ependyma of the rabbit. Am J Physiol 213: 1031–1038.

Portman MA, Lassen NA, Cooper TG, Sills AM, Potchen EJ (1991): Intra- and extracellular pH of the brain in vivo studies by ^{31}P-NMR during hyper- and hypocapnia. J Appl Physiol 71: 2168–2172.

Pullen RGL, DePasquale M, Cserr HF (1987): Bulk flow of cerebrospinal fluid into brain in response to acute hyperosmolality. Am J Physiol 253: F358–F545.

Schielke GP, Betz AL (1992): Electrolyte transport. In Bradbury MWB (ed): Physiology and Pharmacology of the Blood-Brain Barrier. New York: Springer-Verlag, pp 221–243.

Schielke GP, Moises HC, Betz AL (1990): Potassium activation of the Na,K-pump in isolated brain microvessels and synaptosomes. Brain Res 524: 291–296.

Siesjö BK (1972): The regulation of cerebrospinal fluid pH. Kidney Intern 1: 360–374.

Siesjö BK, Katsura K, Mellergård P, Ekholm A, Lundgren KJ, Smith M (1993): Acidosis-related brain damage. In Kogure H, Hossman K-A, Siesjö BK (eds): New York: Elsevier: Progress in Brain Research 96: 23–48.

Smith QR, Rapoport SI (1984): Carrier-mediated transport of chloride across the blood-brain barrier. J Neurochem 42: 754–763.

Stewart PA (1981): How To Understand Acid-Base: A Quantitative Acid-Base Primer For Biology And Medicine. New York: Elsevier.

Szentistványi I, Patlak CS, Ellis RA, Cserr HF (1984): Drainage of interstitial fluid from different regions of rat brain. Am J Physiol 246: F835–F844.

Teppema LJ, Barts PWJA, Folgering HtH, Evers JAM (1983): Effects of respiratory and (isocapnic) metabolic arterial acid-base disturbances on medullary extracellular fluid pH and ventilation in cats. Respir Physiol 53: 379–395.

Vigne P, Champigny G, Marsault R, Barbry P, Frelin C, Lazdunski M (1989): A new type of amiloride-sensitive cationic channel in endothelial cells of brain microvessels. J Biol Chem 264: 7663–7668.

Wichser J, Kazemi H (1975): CSF bicarbonate regulation in respiratory acidosis and alkalosis. J Appl Physiol 38: 504–511.

CHAPTER 33

Brain Edema and pH

HAROLD K. KIMELBERG AND PAUL J. FEUSTEL
Department of Surgery and Department of Pharmacology and Neuroscience (H.K.K.), Department of Surgery and Department of Physiology and Cell Biology, Albany Medical College, Albany, NY (P.J.F.)

INTRODUCTION

Two forms of brain swelling or edema are currently recognized based on the basis of their location and mechanisms (Klatzo, 1967). One is termed *vasogenic edema,* which essentially involves breakdown of the blood-brain barrier (BBB) and the extravasation of blood plasma into the brain. This is associated with increased brain volume, raised intracranial pressure (ICP), increased extracellular space and tissue displacement. The second form of edema is termed *cytotoxic* or, perhaps preferably, *cellular edema* (Kimelberg, 1995). Its main characteristics are cellular swelling in the absence of any measurable breakdown of the BBB. The cell type usually swollen is the astrocyte, especially the "endfeet" of their processes found around blood vessels. It should be emphasized, however, that both forms of edema often coexist. Astrocytic swelling is associated with pathological metabolic states such as cerebral ischemia, when there is also acidification. Extracellular acidification in the presence of ionizable species such as lactate or HCO_3^-, which have a permeant undissociated form, can lead to astrocytic swelling *in vitro* (Kempski et al., 1988; Kimelberg and Ransom, 1986). Several HCO_3^-, H^+, and anion transporters are present on astrocytes that could mediate such swelling and pH changes (Chapter 14). Measurements of intracellular pH (pH_i) in astrocytes *in situ* have been done in pathological conditions such as ischemia, where astrocytic swelling is common, and a marked acidification has been found (Kraig and Chesler, 1990). In brain slices or astrocyte cultures where interstitial K^+ concentration, $[K^+]_o$, is raised and astrocytic swelling would be expected to occur, a marked depolarization-induced alkalinization is observed (see Chapters 10 and 11). However, there have been no measurements where pH_i has been directly correlated with cell swelling. In general there is a clear need for further studies before the cause-and-effect relationships between both forms of brain edema and associated pH changes can be resolved.

Recently a third form of edema, *ionic edema,* has been proposed (Young and Constantini, 1994). In this concept it is proposed that there is net entry of ions and water into the

pH and Brain Function, Edited by Kai Kaila and Bruce R. Ransom
ISBN 0-471-11838-9

CNS from the plasma across an intact BBB. Although the routes of ion entry were not defined, they most likely involve ion channels or specific transporters. Since these transporters likely include HCO_3^--and H^+-driven uptake of Na^+ and Cl^-, pH changes may also be expected (Fig.1.). We need to emphasize, however, that there has been little experimental work to refute or support the occurrence of this type of edema or many of these pH-related transport possibilities.

VASOGENIC EDEMA

Characteristics

In vasogenic edema water enters the brain secondary to a breakdown of the BBB. The BBB is primarily due to the blood vessels and capillaries of the central nervous system (CNS) being lined with endothelial cells that have tight junctions between them and to low pinocytotic activity (Crone, 1986; Chapter 32).

Movement of water across a membrane occurs by diffusion. Diffusion of water can be measured by the unidirectional flux of radio-labeled water. Labeled water injected into blood crosses the BBB and comes into equilibrium with cerebral water within minutes (Eichling et al., 1974). The ease with which water crosses the BBB means that edema fluid will be nearly isotonic with respect to plasma except under very transient conditions. In contrast, diffusion of ions through the normal BBB is very slow. Under normal conditions, when there is no edema, equilibration of K^+, Na^+, or H^+, takes up to 5 hr (Crone and Levitt, 1984).

Potential Effects of Vasogenic Edema on Brain pH

Changes in ionic composition of the interstitial space due to breakdown of the BBB will potentially influence pH. The pathophysiological changes following edema-inducing

FIGURE 1. Hypothetical scheme for the orientation of transport systems in endothelial cells (ENDO) and astrocytes (AST) leading to NaCl movement from blood into brain and involving linked movements of HCO_3^-. Movements of Na^+ and Cl^- from the blood (Cap., capillary) across the luminal membrane of the endothelial cell are shown as being principally due to Na^+/H^+ and Cl^-/HCO_3^- exchange with net movement in the directions shown because of formation of H^+ and HCO_3^- due to carbonic anhydrase-catalyzed intracellular hydration of CO_2. CL^- diffuses passively across the abluminal surface of the endothelial cell through Cl^- channels, while Na^+ is pumped out of the cell into the CNS in exchange for K^+ by the $(Na^+ + K^+)$ATPase. Now in the extracellular space of the CNS Na^+ and Cl^- can either diffuse extracellularly or enter the perivascular astrocyte foot processes by the same exchangers as postulated for the luminal surface of the endothelial cell. Na^+ and Cl^- that enter the astrocyte can then diffuse through the intracellular space of the astrocytic syncytium and likely exit by the same mechanisms as for the abluminal membrane of the endothelial cell. There is evidence for some aspects of these overall schema, as noted in the text on p (001068). Other abbreviations; T.J., tight junction; B.M., basement membrane; ECS, extracellular space; N., neuron.

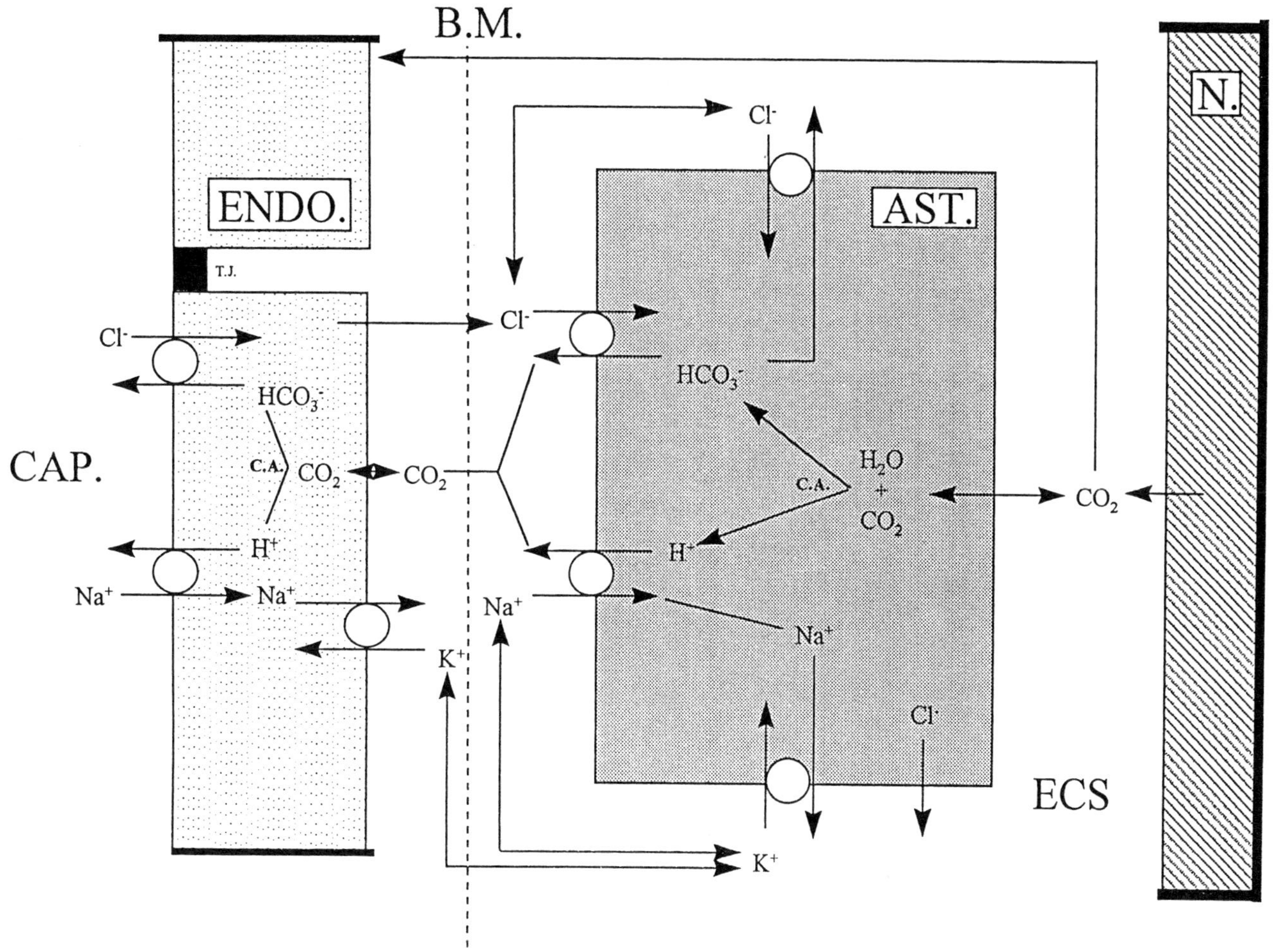
B.M.
N.
ENDO.
T.J.
AST.
CAP.
ECS
Cl-
HCO3-
C.A.
CO2
H+
Na+
K+
H2O
+
CO2
Cl-

events such as stroke, uncontrolled seizure activity, or traumatic brain injury can alter ionic composition via BBB breakdown and extravasation of plasma, or by alterations in ion transporters and receptor-gated ion channels within the BBB. The tissue P_{CO_2} may also change depending upon the blood flow relative to the metabolic rate in the tissue, as well as to the degree of anaerobiosis (Chapters 5, 28–32). Although vasogenic edema formation and pH changes in pathological situations have been independently studied, the two are rarely determined in the same experiment. In the absence of experimental data, the only thing that can currently be said is that vasogenic edema can alter pH dependent upon how ions in CSF, interstitial fluid and intracellular space by the increased interactions with the vascular compartment.

Respiratory Changes, Intracranial Pressure, Edema, and pH

Intracranial pressure (ICP) is often used as an indicator of the extent of brain swelling. The ratio ($\Delta V/\Delta$ICP) of the change in intracranial volume (ΔV) to the change in intracranial pressure (ICP) is termed the compliance of the brain. When compliance is low, small volumes of edema will lead to large ICP changes. Normally, compliance is relatively high and volume added to the CNS within the closed confines of the skull and meninges displaces CSF and venous blood. This allows brain tissue volume to increase without large changes in ICP. As swelling progresses further, compliance decreases, displaceable volume is depleted and increases in brain volume cause larger changes in ICP (Popp et al., 1996). Such increases in ICP lead to life-threatening cessation of blood supply and displacement of and damage to brain tissue. This sequence of events is a frequent cause of death and disability following cerebral ischemia, traumatic brain injury, hypoxia, or uncontrolled seizure activity (Long, 1982).

As an indicator of edema, however, ICP is limited because (i) volume and ICP are not linearly related, since compliance decreases as ICP increases, and (ii) ICP is influenced by vascular and CSF volume changes as well as by the extent and severity of edema. Indeed, the treatments for high ICP act, at least initially and perhaps solely, by decreasing vascular volume; these include mannitol infusion (Muizelaar et al., 1986; Bouma and Muizelaar, 1992) and hyperventilation (Fortune et al., 1995a). Hyperventilation will affect brain pH by reducing arterial CO_2 tension (Pa_{CO_2}) and thus brain P_{CO_2}, but also will reduce blood flow, which in turn will increase the brain-to-arterial P_{CO_2} gradient. Normally the P_{CO_2} in CSF is about 7 mmHg greater than the arterial P_{CO_2}, and this difference increases as blood flow decreases (Caronna et al.,1974). However, this increase is a small effect and the net result is that CSF P_{CO_2} changes essentially follow the Pa_{CO_2} changes caused by hyperventilation.

During acute respiratory changes, ^{31}P NMR studies (see Chapters 9 and 30) have shown that extracellular pH changes are about 50% of the arterial pH changes. Intracellular pH changes are smaller, about 20% of arterial pH changes (Portman et al., 1991). How such respiratory acid-base changes influence edema formation is unclear. In sheep that developed cerebral edema when subjected to hypoxia, Yang et al., (1994) found that ICP elevations and brain water content increases were not different in eucapnic compared to hypocapnic animals, despite higher P_{CO_2} and higher cerebral blood flow (CBF) in the eucapnic group. In contrast, Bakay and Lee (1968) found, by electron microscopy, enlargement of the extracellular space and marked astrocytic swelling (i.e., cellular edema) when hypercapnia was combined with hypoxia in the rat.

pH and Edema in Ischemia and Traumatic Brain Injury

In cerebral ischemia, edema fluid increases in brain tissue within several hours of arterial occlusion. The amount of fluid increases progressively for 2 to 3 days and then resolves (Gotoh et al., 1985; Hatashita and Hoff, 1990). Although the water and Na^+ content increase within 12 hr, increased BBB permeability, as measured by appearance of marker proteins, is not apparent until after this edema has begun to form. In order to assess where the increased water resides, a form of magnetic resonance imaging (MRI) termed diffusion-weighted imaging or DWI was used to distinguish between changes in intracellular and extracellular water. Based on DWI, common carotid artery ligation in the rat showed an early phase of water shifts occurring within 2 hr, which seemed exclusively intracellular, presumably representing cellular edema. Extracellular, vasogenic edema occurred later, and only with prolonged, superimposed hypoxia (Rumpel et al., 1995). The early increases in water content appear before the BBB has been shown to open to macromolecules (Kógure et al., 1981), but at the same time as an increased brain Na^+ content (Young et al., 1987; Gotoh et al., 1985). Gotoh et al. (1985) and Menzies et al. (1993) found a poor correlation between water content and albumin content of brain tissue and attributed edema formation to increasing ionic content of the tissue. Furthermore, an increased permeability of the BBB to labeled albumin (Gotoh et al., 1985) or sucrose (Hatashita and Hoff, 1990) generally occurs 2 to 8 days after occlusion. Thus, the current opinion regarding early edema formation in ischemia is that it arises from a translocation of ions from the blood into brain tissue, rather than an increase in protein permeability (i.e., an observed breakdown of the BBB; Menzies et al., 1993), and has been proposed to represent a third form of edema, termed ionic edema (Young and Constantini, 1994).

In focal ischemia, pH changes are rapid, occurring within seconds of occlusion (Menzies et al., 1993; Siesjo, 1992). For days after stroke, lactate concentration has been found to be elevated in CSF (Zupping et al., 1971) and in tissue using magnetic resonance spectroscopy (MRS) (Graham et al., 1992). Pacshen et al. (1987) found that the brain lactate level was still elevated after 24 hr of recirculation following a 30-min episode of global cerebral ischemia when pH had returned to control levels.

In contrast to ischemia, following traumatic brain injury (TBI) injury, there is an immediate transient increased permeability of the BBB to proteins, which decreases over the course of an hour (Fukuda et al., 1995; Tanno et al., 1992). In a rat model of head trauma plus hypoxia and hypotension, it was found that a decreased apparent diffusion constant (ADC) of water measured by DWI occurred at 1 hr after injury, indicating that cellular edema was a relatively early event (Ito et al., 1996).

In a rat model of traumatic brain injury intracellular pH changes measured with MRS, showed little or no changes during 5 hr following injury (Yamakami et al., 1995), except when injury was accompanied by hypoxia (Ishige et al., 1987) or hypoventilation (Andersen et al., 1988) when acidosis results. Cellular edema, defined as astrocytic swelling, has been observed by electron microscopy, after experimental head and spinal cord injury in cats and rats (Barron et al., 1988; Dietrich et al., 1994), and in contused areas after head injury in human patients (Bullock et al., 1991). In an animal model it was more pronounced when hypoxia was added as an additional insult (Barron et al., 1988).

As in ischemia, elevations of CSF lactate level are also a common finding after TBI in humans (DeSalles et al., 1987) and animals (Rosner and Becker, 1984). Following traumatic brain injury in humans, the use of the buffer tris-(hydroxymethyl)aminomethane

(THAM) is thought to reduce edema formation. Evidence for this is the reduction in the requirement to use other modalities to treat intracranial hypertension in humans (Marmarou, 1992). Following brain injury, THAM has been found to decrease brain levels of lactate in animals, to increase the ratio of creatinine phosphate to inorganic phosphate, and to reduce brain water content (Yoshido and Marmarou, 1991). How much of the beneficial effect, however, is due to the alkalinization produced by THAM is unclear. A similar alkalinization secondary to hyperventilation may be ineffective, either because it is accompanied by vasoconstriction (Fortune et al., 1995b) and increased brain lactate production secondary to relative ischemia (Yoshida and Marmarou, 1991), or because the respiratory brain alkalosis is not sustained for more than 12 hr (Muizelaar et al, 1988).

Conclusions

Any breakdown of the BBB that leads to an increased permeability to proteins would be expected to also increase permeability to small ions. However, increased ion transport need not necessarily indicate a breakdown of the BBB, since ion transport across the BBB can also involve specific transporters located in cells forming the interface between the blood and the interstitial fluid of the brain (Fig.1) (Crone, 1986; Bradbury, 1985; Betz, 1983). Thus the concept that a physical opening of the BBB, sufficient to allow proteins to enter the CNS is alone responsible for vasogenic edema may be too narrow. Gotoh et al. (1985) proposed that early in edema the BBB has selective increases in Na^+ permeability without a measurable effect on protein permeability. Based on similar results, Young and Constantini (1994) have proposed that a third type of edema, ionic edema, should be defined correspond to the observed ionic changes, even though, at present, specific transport mechanisms have not been demonstrated. The fundamental observation is that the content of Na^+ in ischemic tissue increases more than tissue K^+ content decreases. Since brain fluid must remain isosmotic with plasma, water accompanies the increased [Na^+]. Young and Constantini (1994) proposed that movements of ions into the brain are driven because of glial uptake of K^+ and Cl^-, together with net cellular uptake of Na^+, so that Na^+, Cl^-, and water are drawn into the extracellular space of the brain. Possible ion transport mechanisms for these movements are shown in Fig. 1, but it must be emphasized that the identification of these requires considerable experimental study. If HCO_3^- and H^+ are also transported according to some of the schemes shown in Fig.1, specific pH changes should be part of the increased ionic transport in ionic edema. Otherwise, the consequences of vasogenic edema would be limited to equilibration with the plasma pH, or be secondary to alterations in Pa_{CO_2} caused by ventilatory changes. There have been few studies on pH changes in vasogenic edema, and although pathological situations such as traumatic brain injury or ischemia lead to both vasogenic edema and acidosis, it is unclear to what degree this is a cause and effect relationship, and what are likely to be the causes and what the effects.

CELLULAR EDEMA

Astrocytic Edema in Ischemia and Trauma

Unlike vasogenic edema, cellular edema or cell swelling can, in principle, occur without an increase in brain volume if water and solute enter the cell only from the extracellular space, which thus shrinks. The commonest cell type observed to swell is the astrocyte. In experimental ischemia, astrocytic swelling is seen as an early event, prominently involving swelling of processes around both neurons and capillaries. The astrocytic cell body is also swollen (Kimelberg and Ransom, 1986; Garcia, 1984; Jenkins et al., 1979). Similar astrocytic swelling is also seen after trauma (Barron et al., 1988; Dietrich et al., 1994; Bullock et al., 1991). Such swelling is distinguished by a pale, watery cytoplasm in electron micrographs. This contrasts with the increased cell size (hypertrophy) seen in reactive astrogliosis, which occurs considerably later than astrocytic swelling, and where the cell cytoplasm is packed with glial fibrillary acidic protein (GFAP) positive intermediate filaments (Kimelberg, Norenberg, 1994). Van Harreveld (1966) showed that in cerebellar tissue examined by electron microscopy after rapid fixation by freeze-substitution, swelling of Bergmann glia fibers was several fold greater than swelling of Purkinje cell bodies and apical dendrites, after 6–8 min of asphyxiation in the rat and rabbit.

In another study, the extent and frequency of altered neurons and astrocytic swelling in the cat cerebral cortex was seen to be a function of the duration of complete ischemia and reperfusion (Jenkins et al., 1982). There was an increased frequency of astrocytic swelling when the ischemia was increased from 5 min to 15 min, or when 5 min of ischemia was followed by reperfusion. It was further noted that oligodendrocytes never showed volume changes, while the response of neurons was generally shrinkage. However, significant numbers of dark and shrunken neurons only occurred with the most severe insults.

Astrocytic Swelling and Transport of HCO_3^-

The transport of HCO_3^- into or out of cells has implications not only for cell swelling but for pH regulation as well. The first suggestion of a role for astrocytes in movements of HCO_3^- in the CNS was made by Tschirgi (1958). He sought to explain the marked inhibition of CSF production by the carbonic anhydrase (CA) inhibitor acetazolamide (see Chapters 3 and 32). CA markedly increases the hydration of CO_2 to HCO_3^- and H^+. Since he proposed that HCO_3^- and H^+ exchanged for Cl^- and Na^+ in the plasma, he considered that the most likely site would be an intracellular location around blood vessels and postulated that this was the perivascular astrocyte endfoot. Such a scheme is depicted in Fig. 1., but CA can also be located in the endothelial cell.

Bourke et al., (1975) showed that exposure of cat brain cortical slices to isosmotic solutions with high $[K^+]$ resulted in increased slice wet weight. A portion of this increase was dependent on media $[HCO_3^-]$ and was associated with increased Na^+ and Cl^- content of the slice, which was in addition to the increased uptake of K^+ and Cl^- due to Donnan swelling. Such conditions were seen by electron microscopy to be associated with marked astrocytic swelling in the slice (Bourke et al., 1975). It was suggested that such astrocytic swelling was due to the same mechanism as proposed by Tschirgi (1958): namely, intracellular hydration of the increased CO_2 within the astrocyte to produce H^+ and HCO_3^-, and cyclical exchange of these continuously generated intracellular components for extracellular Na^+ and Cl^- respectively, resulting in accumulation of NaCl plus water intracellularly (Fig. 2C). This scheme was supported by finding Na^+/H^+ and Cl^-/HCO_3^- exchangers and CA in primary rat astrocyte cultures (Kimelberg et al., 1979; Church et al., 1980; Chapter 3). It was also suggested that a preferential localization of the Cl^-/HCO_3^- and Na^+/H^+ exchangers on the membranes of the perivascular astrocytic foot process, and their regulation in response to varying acid-base conditions within the CNS, could help to independently regulate brain pH (see Fig. 1).

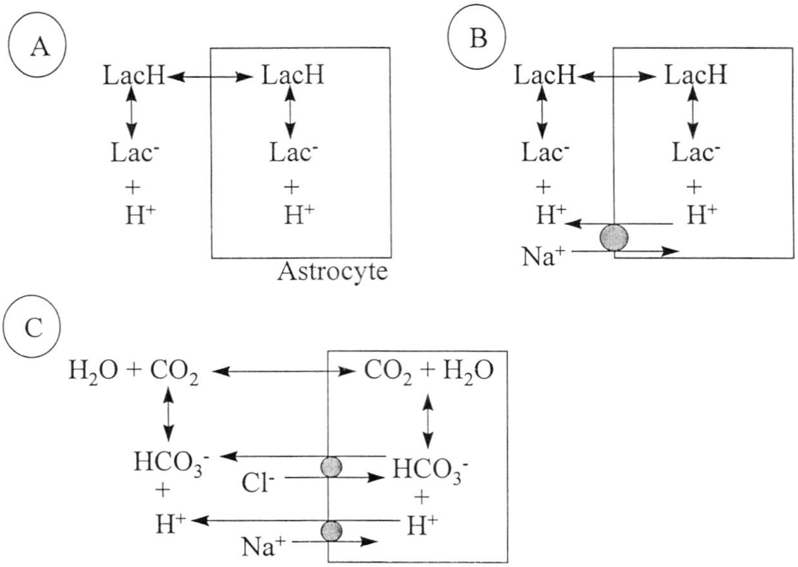

FIGURE 2. Mechanisms of pH-linked cellular edema. **A.** Permeability of undissociated lactic acid (or other membrane permeant weak acids) only leads to increased osmolarity inside the cell if $pH_i > pH_o$. This is because $\frac{[lac^-]_i}{[lac^-]_o} = \frac{[H^+]_o}{[H^+]_i}$ or taking logs $\log\left(\frac{[lac^-]_i}{[lac^-]_o}\right) = \log\,[H^+]_o - \log\,[H^+]_i$ or $pH_i - pH_o$ pH_i is usually about 0.3 pH units lower than pH_o and therefore the intracellular lactate will be about one half the concentration outside. The other basic principal is that $[lacH]_o = [lacH]_i$ (Kraig and Chesler, 1990). In hyperglycemic ischemia, pH_i is thought to be around 4.0 when pH_o is ~6.0 ((Kraig and Chesler, 1990) and Chapter 18). This will further reduce $[lac^-]_i$. With increased neural activity alone, pH_i of astrocytes is found to increase (see text and chapter 18), which will raise $[lac^-]_i$ by the ratio given.
B. With the addition of a Na^+/H^+ exchanger the H^+ can be continuously removed to the extracellular space. This will proceed until Eq. 1 (see text) is satisfied by the intracellular Na^+ and the H^+ changes. Since $[Na^+]_o$ usually decreases in conditions such as ischemia, sodium lactate can accumulate intracellularly up to about one third the outside Na^+ concentration, or 20–40 mM. Extrusion of Na^+ will likely be limited by inhibition of the ATP-driven Na^+/K^+ pump due to the ATP deficit. Since, in ischemia, pH_i of astrocytes is lower than pH_o, theoretically the exchanger can lead to increased $[Na^+]_i$. However, as noted in the text, the Na^+/H^+ exchanger is inhibited if $pH_o \leq 6$.
C. With the further addition of an anion exchanger, such as the Cl^-/HCO_3^- exchanger, the anion can also be removed, in this case, HCO_3^-. This will have the effect of accelerating the intracellular hydration, as well as increasing the extent of hydration. The equilibrium value for these anions is given by Eq. 2 in the text. Location of these transporters to different cell regions such as the astrocytic endfeet could have the very important function of exporting H^+ and conserving HCO_3^- in the CNS to reduce the fall of pH (Kimelberg and Bourke, 1982).

Pathologically high $[K^+]_o$ (> 20 mM and up to 100 mM), seen during ischemia and following stroke and head injury (Walz et al., 1993; Chapter 18), is likely to lead to astrocytic swelling by an uptake of KCl due to Donnan forces (Barres et al., 1990; Hodgkin and Horowicz, 1959; Walz et al., 1993). K^+ uptake would be via K^+ channels, of which there are several types in astrocytes (Barres et al., 1990; Jalonen and Holopainen, 1989), and Cl^- through Cl^- channels. HCO_3^- can usually also enter via Cl^- and nonspecific anion channels. The anion channels in astrocytes are usually not open at the resting membrane potential (Kimelberg, 1990), but a large conductance anion channel and smaller conductance Cl^- channels in astrocytes are voltage-sensitive (Jalonen, 1993; Barres et al., 1990; Walz et al., 1993). Massive release of K^+ from neurons via increased opening of K^+ channels may trigger the uptake of K^+, Cl^-, and HCO_3^- by astrocytes, which enter through their appropriate channels, activated and driven by the $[K^+]_o$-induced depolarization of the astrocytes (Kimelberg, 1990; Walz et al., 1993).

Much *in vitro* experimental work is done in nonbicarbonate buffers, since experimental manipulations are easier to handle in chambers that can be kept open because there is no need to equilibrate with CO_2. However, a number of transport systems in both neurons and astrocytes involve HCO_3^-. An important HCO_3^- cotransporter has been identified in almost all glial cells so far examined. This is the $Na^+ + HCO_3^-$ electrogenic cotransporter which operates at a stoichiometry of 1 Na^+ to 2 HCO_3^- and usually transports HCO_3^- into the cell (Boyarsky et al., 1993; O'Connor et al., 1994; Shrode and Putnam, 1994; Newman, 1991; Brune et al., 1994). The HCO_3^- taken up could remain intracellular or could exchange for extracellular Cl^- via the HCO_3^-/Cl^- exchanger. This system is driven by reduction of the membrane potential, an inwardly directed Na^+ gradient and intracellular acidity. Thus it is likely to be activated by depolarization caused by increased $[K^+]_o$ and may be responsible for the intracellular alkalinization observed in identified astrocytes *in situ* during cortical stimulation (Chesler and Kraig, 1987). Pappas and Ransom (1994) showed in cultured rat hippocampal astrocytes that this alkalinization could occur with extracellular K^+ concentrations as low as 6 mM, and that it was increased when the cells were grown in serum-free medium for approximately 24hr before the experiment.

Excitatory Amino Acids (EAAs), Cell Edema and pH

Extracellular levels of the excitatory amino acids (EAAs) glutamate, and aspartate, are known to increase in ischemia and other pathological states. These elevated levels are thought to cause neuronal death due to EAA receptor activation that leads to increased $[Ca^{2+}]_i$ by increasing Ca^{2+} influx or release from intracellular stores (Olney et al., 1991; Choi and Rothman, 1990). Glutamate, or the glutamate analogue kainic acid (KA), causes swelling of astrocytic retinal Müller cells and of astrocytes when directly injected into the CNS (Van Harreveld and Fifkova, 1971; Herndon et al., 1980; Van Harreveld, 1982; Matyja, 1986). These compounds may stimulate neurons and increase metabolic rate and production of CO_2 and H^+, causing astrocytic swelling by activation of the $Na^+ + HCO_3^-$ transporter. However, astrocytes are also known to take up glutamate by a Na^+-dependent mechanism (Hertz, 1979). This transporter is also dependent on K^+ efflux. Probably the stoichiometry involves movement into the cell of 1 glutamate + 2 Na^+ while 1 K^+ + 1 OH^- (or HCO_3^-) move out (Attwell et al., 1993; Bouvier et al., 1992). Net uptake of Na^+ and glutamate will thus lead to both swelling and intracellular acidification. It has now been established that there are *at least* three distinct glutamate/aspartate transporters in the mammalian CNS, of which two are found on astrocytes (Rothstein et al., 1994; Lehre et al., 1995):

1. The GLT-1 transporter found widely in the CNS on astrocytes.
2. The GLAST transporter found mainly in the cerebellum on Bergmann glia.
3. The EAAC1 transporter found in neurons. Other forms of the transporters are currently being found (Arriza et al., 1997; Nagas et al., 1997; Yamada et al., 1996).

Excitatory amino acid receptors are also present on astrocytes and could be involved in astrocytic swelling. As noted above, kainic acid or glutamate causes astrocytic swelling *in situ,* but since [^{3}H]-kainic acid is not taken up by primary astrocyte cultures (Kimelberg et al., 1989), this indicates that glutamate can cause swelling by other mechanisms such as receptor activation. Sontheimer et al., (1988) first showed that the depolarization of primary cortical astrocyte cultures by both L-glutamate and KA is best described by the activation of an amino-3-hydroxy-5-methyl-4-isoxazolepropionic acid (AMPA) ionotropic Kainate (KA) excitatory amino acid receptor (see also Robert and Magistretti, 1997). On this basis, swelling could be due to direct receptor activation leading to Na^+ entry. Cl^- or HCO_3^- could presumably enter through anion channels, which may also open under these conditions (Kimelberg, 1990; Chapter 16–17). The intracellular pH change due to activation of EAA transporters and/or ionotropic receptors in primary astrocyte cultures from rat hippocampus have been recently described in detail by Rose and Ransom 1996 and in Chapter 14. It would be of great interest to know how cell volume changes correlate with such pH changes and determine which of them occur *in vivo.*

Volume and pH Changes in Astrocyte Cell Cultures

In its simplest form, a likely basis of the CO_2/HCO_3^--driven swelling discussed above is that a membrane-permeant, undissociated acid, in this case LacH, diffuses into the cell where it dissociates into H^+ and the base (see Fig. 2A). However, this will only lead to accumulation of Lac^- and H^+ and swelling if $pH_i >> pH_o$. Several workers have tested the ability of such models *in vitro* to support swelling in glial cells utilizing not only CO_2/H_2CO_3, but also other membrane-permeant acids. As shown in Fig. 2B, when a Na^+/H^+ exchanger is present on the cell membrane, the increased intracellular $[H^+]$ can then exchange for extracellular Na^+, leading to the intracellular accumulation of the sodium salt of the base (Grinstein and Furuya, 1984). The continuous removal of the H^+ will pull the reaction in the direction of product leading to increased dissociation of the acid to H^+ and the conjugate base and greater accumulation and swelling. Thus the existence of the Na^+/H^+ exchanger is a critical factor in generating significant accumulation of osmoles inside the cell. Since the cell membrane is likely to be relatively impermeable to Na^+ and the base, this intracellular accumulation of the sodium salt can in theory continue until there is no driving force on the Na^+/H^+ exchanger, which occurs when the following relation holds:

$$\frac{[Na^+]_i}{[Na^+]_o} = \frac{[H^+]_i}{[H^+]_o} \tag{1}$$

Since the Na^+ concentration ratio starts at around 20/140 (inside to out) about 120 mM Na^+ with the conjugate base can be accumulated inside the cell when, pH is approximately the same outside and inside the cell. Changes in the pH or Na^+ gradient will alter the ratio according to Eq. 1. Thus if $[H^+]_i$ increases more than $[H^+]_o$, more sodium salt can accumulate intracellularly with greater swelling. However, this will be limited by the inhibition of the Na^+/H^+ exchanger at decreasing pH_o (Grinstein et al., 1984).

This model has been tested in primary astrocyte cultures and C_6 glioma cells for permeant alkyl carboxyl acids such as propionic acid (Jakubovicz et al., 1987). Swelling was seen that was inhibited by amiloride, indicating the involvement of the Na^+/H^+ exchanger. The high concentrations of the propionic acid used, and the unlikely generation of such concentrations of fatty acids make the operation of this system *in situ* unlikely. Swelling and moderate intracellular acidosis of C6 glioma cells or primary astrocyte cultures has been shown with arachidonic acid and other long-chain unsaturated fatty acids, but not with saturated fatty acids (Staub et al., 1994). The swelling was Na^+-dependent, but it is not clear whether the Na^+/H^+ exchanger was involved, since amiloride and other Na^+/H^+ exchange inhibitors were not used. The lack of effect of the saturated fatty acids indicates that as the authors suggested, other mechanisms occurred, such as membrane damage due to lipid peroxides formed by oxidation of the unsaturated fatty acid side chains.

More relevant to *in vivo* astrocytic swelling is the demonstration of permeant acid-driven swelling *in vitro* with CO_2/HCO_3^-, where H_2CO_3 is the permeant, undissociated acid. Amiloride-, $[HCO_3^-]_o$- and $[Na^+]_o$, -sensitive swelling has been seen in C6 glioma cells (Kempski et al., 1988). As shown in Fig. 2C, with a HCO_3^-/Cl^- transporter present, NaCl can now accumulate (Kimelberg et al., 1979). The relevant relation for the anions equilibrating on the exchanger is (Kimelberg and Bourke, 1982; Chapter 10):

$$\frac{[Cl^-]_i}{[Cl^-]_o} = \frac{[HCO_3^-]_i}{[HCO_3^-]_o} \tag{2}$$

Large amounts of lactate are produced in ischemia, anoxia, TBI or severe hypoxia (<20 mmHg O_2), due to inhibition of oxidative metabolism and greatly increased glycolysis, which produces the lactate anion. Total lactate can reach levels of 15–20 mM in ischemias of varying severity (Paschen et al., 1987), and 10 mM in traumatic brain injury (Rosner and Becker, 1984). The increased $[H^+]$ usually associated with increased lactate concentration is not thought to be due to lactic acid, since it is the lactate anion which is the end-product of glycolysis. It is rather considered to be principally due to the greater hydrolysis of creatinine phosphate and ATP than can be resynthesized by glycolysis (Paschen et al., 1987). The pK_a of lactic acid is 4.5 so that the proportion of the total lactate that will exist as undissociated lactic acid at the pH values of around 6.0 encountered in severe ischemia will be $\sim 3\%$. Experimentally, uptake of lactate as lactic acid is followed by Na^+/H^+ exchange. Thus, the maximum amount of Na^+ that could be transported into the cell after lactic acid uptake is equivalent to the intracellular concentration of lactate accumulated. There is, as yet, no direct evidence for a lactate transporter *in situ,* it has been shown that there is a lactate + H^+ cotransporter or, which would be indistinguishable, a lactate/OH^- exchanger in astrocytes *in vitro* (Walz et al., 1993). In either case these should not affect the final equilibrium, and therefore the degree of swelling, but only the routes by which lactate + H^+ will be transported.

In vitro, increased lactic acid concentrations have been shown to cause swelling and intracellular acidification of primary astrocyte cultures, C6 glioma cells, and a human astrocyte-derived cell line (Jakubovicz and Klip, 1989; Staub et al., 1990; Lomneth et al.,

1990). In the C6 glioma cells, the swelling, but not the intracellular acidification, was inhibited by amiloride (Staub et al., 1990). This nicely demonstrates that the swelling depends on the operation of the Na^+/H^+ exchanger, while intracellular acidity depends only on lactate plus H^+ entering the intracellular compartment. Without the Na^+/H^+ exchanger, the extracellular concentration of lactic acid and the intracellular buffering capacity will determine how much lactate and H^+ will accumulate intracellularly (Fig. 2A). As pointed out by Siesjo (1985), swelling of cells can be thought of as an attempt by the cells pH-controlling reactions to correct a pathologically large acid load.

Measurements of Astrocytic pH *In Situ*

In spite of all the measurements already described, there have been no concurrent measurements of both astrocytic swelling and pH *in situ*. However, a few astrocytic intracellular pH measurements in astrocytes have been made under conditions such as increased neuronal activity, spreading depression, and ischemia, when astrocytic swelling is known to occur (Kimelberg and Ransom, 1986). This work was done using the demanding technique of inserting pH electrodes into astrocytes *in situ* (see Chapter 14). Such studies have shown an alkalinization of astrocytes in response to increased activity, spreading depression or raised $[K^+]_o$ in a gliotic hippocampal slice (Grichtchenko and Chesler, 1994). This can be most easily explained on the basis of raised $[K^+]_o$ depolarizing the astrocyte and activating inward transport of Na^+ and HCO_3^- on the electrogenic cotransporter, producing an intracellular alkalinization, as observed *in vitro*. However, transmitters will also be released by high $[K^+]_o$, which could activate anion or H^+ channels and lead to electrochemically driven movements of these pH equivalents. For example, the $GABA_A$ channel is known to be very permeable to HCO_3^- as well as Cl^- (see Chapters 16 and 17).

In contrast to the intracellular alkalinization seen upon cortical stimulation a very considerable intracellular acidification of astrocytes was measured in rat cortex when the animals were subjected to the severe insult of complete ischemia with hyperglycemia (Kraig and Chesler, 1990). Intracellular pH levels as low as 3.8 were recorded, associated with greatly depolarized membrane potentials of -3 to -12 mV and a 10-fold increase in the membrane time constant, indicating a marked increase in membrane resistance. The extracellular pH was 6.2. Some of these cells were identified histologically by horseradish peroxidase injection using the recording electrode, and resembled astrocytes. The electrical characteristics associated with these cells were then used to identify the other cells studied as also being astrocytic. Interestingly for this discussion, dramatic acid pH changes were suddenly seen with some electrodes that were not being moved. This is consistent with the concept that a considerable swelling of the astrocytes had caused an astrocyte to impale itself on the microelectrode and the pH_o being measured then changed to the very acidic pH_i of an ischemic, swollen astrocyte. If this is the correct explanation, then this constitutes the only published measured value for the pH_i of a swollen astrocyte *in situ*. This is, however, purely speculative.

The explanation given for the marked acidity under these conditions of ischemia plus hyperglycemia was that, at pH_o values of 6.5 to 6.0, the astrocyte membrane becomes impermeant to H^+, so that H^+ generated by the lactic acid diffusing into the astrocyte is trapped therein. The increase in the membrane resistance seen is certainly consistent with decreased conductance to H^+, but decreased conductance to K^+ or uncoupling of the astrocytic syncytium (see Chapter 24) would also be sufficient to explain the resistance changes. Decreased efflux of K^+ through channels at low pH (see Chapter 21) will also increase cell swelling. An alternative explanation for the cell being unable to get rid of the accumulated H^+ is that the Na^+/H^+ exchanger, activated to maximal levels by a fall in pH_i to at least 6.0, is inhibited by a comparable drop in pH_o. Evidence for this is that in lymphocytes at a pH_i of 6.3 (the lowest pH_i values obtainable with the nigericin technique used) Na^+-induced H^+ extrusion was completely inhibited at pH_o values of 6.0 or less (Grinstein et al., 1984). Since in the severe hyperglycemic conditions used in the *in vivo* brain studies of Kraig and Chesler (1990) pH_o was around 6.0, one likely explanation of the severe astrocytic acidosis is that the increased $[H^+]_i$ cannot be removed by the Na^+/H^+ exchanger. However, why this is principally a problem for astrocytes and not neurons, and why the astrocytic pH_i cannot equilibrate with ECF, offset by whatever fixed buffering capacity is inside the cell, via permeant undissociated lactic acid, remain unanswered questions (Siesjö, 1985).

If the Na^+/H^+ exchanger is the mechanism for acid induced swelling of astrocytes, then the Na^+/H^+ exchanger-driven component of astrocytic swelling must occur while the pH_o is still above 6.2–6.4. Of course, the other mechanisms of swelling, such as Donnan driven KCl uptake, may still operate at a pH $\leq$ 6.0 if the K^+ and Cl^- conductances remain sufficiently high. Astrocytic swelling, from the ultrastructural studies done so far, seems to represent the major component of cellular edema, and, because of increased knowledge of astrocytic functions, it is now thought that astrocytic swelling may be an important event predisposing the brain to further damage (Kimelberg and Ransom, 1986; Kimelberg, 1995).

METHODOLOGICAL CONSIDERATIONS REGARDING RELATING EDEMA TO pH *IN VIVO*

As can be discerned from this chapter, a major problem in attempting to draw conclusions regarding the relationships between the different forms of edema and pH changes is the absence of experimental data, especially measurements of both swelling and pH, done in the same experiments. This is in part due to the difficulties of separately identifying the two forms of edema, and also making concurrent edema and pH measurements. Ultrastructural analysis of properly fixed tissue allows one to determine whether extracellular or intracellular swelling has occurred, but does not allow one to determine a continuous time course to correlate with pH measurements made in parallel experiments. Impedance measurements were developed by Van Harreveld (1966) to continuously measure decreases in the extracellular space due to ischemia or spreading depression. More recently, systems have been developed for measuring extracellular impedance noninvasively (Holder, 1992; Van Lookeren Campagne et al., 1994). To distinguish between extracellular and cellular edema noninvasively, a form of magnetic resonance imaging (MRI), called diffusion weighted MRI, is beginning to be used (Ebisu et al., 1993; Rumpel et al., 1995; Verheul et al., 1994; Ito et al., 1996; Niendorf et al., 1996). This uses additional strong magnetic field gradients to distinguish the proton signal of the more diffusion-restricted intracellular water as compared to extracellular water. However, it will only give an average value if vasogenic and cellular edema are occurring, and will also not indicate which cell types are involved in cellular edema. This *might* be clarified to some extent by a one-time ultrastructural analysis of the tissue immediately at the conclusion of any experimental study.

Other averaging but invasive methods for estimating changes in the ECS used experimentally are determination of concentration changes of a radiolabelled membrane-impermeant dye perfused through the extracellular space (ECS) by microdialysis (Katayama et al., 1992), or detection of changes in a similarly membrane impermeant dye injected into the ECS, whose extracellular activity can be recorded with an ion-specific microelectrode (ISM) *in situ* (Ransom et al., 1985). This method has also been used for measuring cell volume *in vitro* after cellular uptake of TMA^+ (tetramethyl ammonium), whose activity is then measured with an intracellular K^+ ISM that is very sensitive to TMA^+ (Ballanyi and Grafe, 1988; Ballanyi et al., 1990).

Application of these methods in future work, both in experimental animals and clinically, are critically needed to resolve the still unanswered questions of the precise relationships between brain pH and brain extracellular and cellular edema.

ACKNOWLEDGMENTS

Support from NINDS National Institute of Neurological Diseases and Stroke, NS 30303 and NS35205 are gratefully acknowledged.

REFERENCES

Andersen BJ, Unterberg AW, Clarke GD, Marmarou A (1988): Effect of posttraumatic hypoventilation on cerebral energy metabolism. J Neurosurg 68: 601–607.

Attwell D, Barbour B, Szatkowski M (1993): Nonvesicular release of neurotransmitter. Neuron 11: 403–407.

Arriza JL, Eliasof S, Kavanaugh MP, Amara SG (1997): Excitatory amino acid transporter 5, a retinal glutamate transporter coupled to a chloride conductance. Proc Natl Acad Sci U S A 94: 4155–4160.

Bakay L, Lee JC (1968): The effect of acute hypoxia and hypercapnia on the ultrastructure of the central nervous system. Brain 91: 697–706.

Ballanyi K, Grafe P (1988): Cell volume regulation in the nervous system. Renal Physiol Biochem 3-5: 142–157.

Ballanyi K, Grafe P, Serve G, Schlue W-R (1990): Electrophysiological measurements of volume changes in leech neuropile glial cells. Glia 3: 151–158.

Barres BA, Chun LLY, Corey DP (1990): Ion channels in vertebrate glia. Annu Rev Neurosci 13: 441–474.

Barron KD, Dentinger MP, Kimelberg HK, Nelson LR, Bourke RS, Keegan S, Mankes R, Cragoe EJ Jr (1988): Ultrastructural features of a brain injury model in cat. I. Vascular and neuroglial changes and the prevention of astroglial swelling by a fluorenyl (aryloxy) alkanoic acid derivative (L-644, 711). Acta Neuropathol (Berl) 75: 295–307.

Betz AL (1983): Sodium transport from blood to brain: Inhibition by furosemide and amiloride. J Neurochem 41: 1158–1164.

Bourke RS, Kimelberg HK, West CR, Bremer AM (1975): Effect of HCO_3^- on the swelling and ion uptake of monkey cerebral cortex under conditions of raised extracellular potassium. J Neurochem 25: 323–328.

Bouma GJ, Muizelaar JP (1992): Cerebral blood flow, cerebral blood volume, and cerebrovascular reactivity after severe head injury. J Neurotrauma 9 (Suppl 1): S333–S348.

Bouvier M, Szatkowski M, Amato A, Attwell D (1992): Glial cell glutamate uptake carrier countertransports pH-changing anions. Nature 360: 471–474.

Boyarsky G, Ransom B, Schlue W, Davis MBE, Boron WF (1993): Intracellular pH regulation in single cultured astrocytes from rat forebrain. Glia 8: 241–248.

Bradbury MWB (1985): Transport across the cerebral endothelium. Circ Res 57: 213–222.

Brune T, Fetzer S, Backus KH, Deitmer JW (1994): Evidence for electrogenic sodium-bicarbonate cotransport in cultured rat cerebellar astrocytes. Pflügers Arch 429: 64–71.

Bullock R., Maxwell WL, Graham DI, Teasdale GM, Adams JH (1991): Glial swelling following human cerebral contusion: An ultrastructural study. J Neurol Neurosurg Psychiatry 54: 427–434.

Caronna JJ, Plum F, Siesjö BK (1974): P_{CO_2} gradients between blood and CSF in rat during alterations of acid-base balance. Am J Physiol 227: 1173–1177.

Chesler M, Kraig RP (1987): Intracellular pH of astrocytes increases rapidly with cortical stimulation. Am J Physiol 253: R666–R670.

Choi DW, Rothman SM (1990): The role of glutamate neurotoxicity in hypoxic ischemic neuronal death. Annu Rev Neurosci 13: 171–182.

Church GA, Kimelberg HK, Sapirstein VS (1980): Stimulation of carbonic anhydrase activity and phosphorylation in primary astroglial cultures by norepinephrine. J Neurochemistry 34: 873–879.

Crone C (1986): The blood-brain barrier: A modified tight epithelium. In Sirkling AJ, Rumsby MG, Bradbury MWB (eds): Blood-Brain Barrier in Health and Disease. Chichester: Ellis Howard, pp 17–39.

Crone C, Levitt DG (1984): Capillary permeability to small solutes. In Renkin EM, Michel CC (eds): Handbook of Physiology, Section 2: The Cardiovascular System, Bethesda, MD: American Physiological Society, pp 411–466.

DeSalles AA, Muizelaar JP, Young HF (1987): Hyperglycemia, cerebrospinal fluid lactic acidosis, and cerebral blood flow in severely head-injured patients. Neurosurgery 21: 45–50.

Dietrich WD, Alonso O, Halley M (1994): Early microvascular and neuronal consequences of traumatic brain injury: A light and electron microscopic study in rats. J Neurotrauma 11: 289–301.

Ebisu T, Naruse S, Horikawa Y, Ueda S, Tanaka C, Uto M, Umeda M, Higuchi TI (1993): Discrimination between different types of white matter edema with diffusion-weighted MR imaging. J Magn Res Imag 3: 863–868.

Eichling JO, Raichle ME, Grubb RL, Ter-Pogossian MM (1974): Evidence of the limitations of water as a freely diffusible tracer in brain of the rhesus monkey. Circ Res 35: 358–364.

Fortune JB, Feustel PJ, DeLuna C, Graca L, Hasselbarth J, Kupinski AM (1995a): Cerebral blood flow and blood volume in response to O_2 and CO_2 changes in normal humans. J Trauma 39: 463–472.

Fortune JB, Feustel PJ, Graca L, Hasselbarth J, Kuehler DH (1995b): Effect of hyperventilation, mannitol, and ventriculostomy drainage on cerebral blood flow after head injury. J Trauma 39: 1091–1099.

Fukuda K, Tanno H, Okimura Y, Nakamura M, Yamaura A (1995): The blood-brain barrier disruption to circulating proteins in the early period after fluid percussion brain injury in rats. J Neurotrauma 12: 315–324.

Garcia JH (1984): Experimental ischemic stroke: A review. Stroke 15: 5–14.

Gotoh O, Asano T, Koide T, Takakura K (1985): Ischemic brain edema following occlusion of the middle cerebral artery in the rat. I. The time courses of brain water, sodium and potassium contents and blood brain permeability to ^{125}I-albumin. Stroke 21: 582–588.

Graham GD, Blamire AM, Howseman AM, Rothman DL, Fayad PB, Brass LM, Petroff OA, Shulman RG, Prichard JW (1992): Proton magnetic resonance spectroscopy of cerebral lactate and other metabolites in stroke patients. Stroke 23: 333–340.

Grichtchenko II, Chesler M (1994): Depolarization-induced alkalinization of astrocytes in gliotic hippocampal slices. Neurosci 62: 1071–1078.

Grinstein S, Furuya W (1984): Amiloride-Sensitive Na^+/H^+ exchange in human neutrophils: Mechanism of activation by chemotactic factors. Biochem Biophys Res Commun 122: 755–762.

Grinstein S, Cohen S, Rothstein A (1984): Cytoplasmic pH regulation in thymic lymphocytes by an amiloride-sensitive Na^+/H^+ antiport. J Gen Physiol 83: 341–369.

Hatashita S, Hoff JT (1990): Brain edema and cerebrovascular permeability during cerebral ischemia in rats. Stroke 21: 582–588.

Herndon RM, Coyle JT, Addicks E (1980): Ultrastructural analysis of kainic acid lesion to cerebellar cortex. Neuroscience 5: 1015–1026.

Hertz L (1979): Functional interactions between neurons and astrocytes. I. Turnover and metabolism of putative amino acid transmitters. Progress in Neurobiology, 13: (Anonymous London: Pergamon Press, pp 277–323.

Hodgkin AL, Horowicz P (1959): Influence of potassium and chloride ions on the membrane potential of single muscle fibres. J Physiol (Lond) 148: 127–160.

Holder DS (1992): Electrical impedance tomography (EIT) of brain function. Brain Topography 5: 87–93.

Ishige N, Pitts LH, Pogliani L, Hashimoto T, Nishimura MC, Bartkowski HM, James TL (1987): Effect of hypoxia on traumatic brain injury in rats. II. Changes in high energy phosphate metabolism. Neurosurgery 20: 854–858.

Ito J, Marmarou A, Barzó P, Fatouros P, Corwin F (1996): Characterization of edema by diffusion-weighted imaging in experimental traumatic brain injury. J Neurosurg 84: 97–103.

Jakubovicz DE, Klip A (1989): Lactic acid-induced swelling in C6 glial cells via Na^+/H^+ exchange. Brain Res 485: 215–224.

Jakubovicz DE, Grinstein S, Klip A (1987): Cell swelling following recovery from acidification in C6 glioma cells: An in vitro model of postischemic brain edema. Brain Res 435: 138–146.

Jalonen T (1993): Single-channel characteristics of the large-conductance anion channel in rat cortical astrocytes in primary culture. Glia 9: 227–237.

Jalonen T, Holopainen I (1989): Properties of single potassium channels in cultured primary astrocytes. Brain Res 484: 177–183.

Jenkins LW, Povlishock JT, Becker DP, Miller JD, Sullivan HG (1979): Complete cerebral ischemia: An ultrastructural study. Acta Neuropathol (Berl) 48: 113–125.

Jenkins LW, Becker DP, Coburn TH (1982): Quantitative analysis of glial swelling and ischemic neuronal injury following complete cerebral ischemia. In Go TG, Baethmann A (eds): Recent Progress in the Study and Therapy of Brain Edema. New York: Plenum Press, pp 523–537.

Katayama Y, Tamura T, Becker DP, Tsubokawa T (1992): Early cellular swelling during cerebral ischemia in vivo is mediated by excitatory amino acids released from nerve terminals. Brain Res 577: 121–126.

Kempski O, Staub F, Jansen M, Schodel F, Baethmann A (1988): Glial swelling during extracellular acidosis in vitro. Stroke 19: 385–392.

Kimelberg HK (1990): Chloride transport across glial membranes. In Alvarez-Leefmans FJ, Russell J (eds): Chloride Channels and Carriers in Nerve, Muscle, and Glial Cells. New York: Plenum Press, pp 159–192.

Kimelberg HK (1995): Current concepts of brain edema. J Neurosurg 83: 1051–1059.

Kimelberg HK, Bourke RS (1982): Anion transport in the nervous system. In Lajtha A (ed): Handbook of Neurochemistry. New York: Plenum Press, pp 31–67.

Kimelberg HK, Norenberg MD (1994): Astrocytic responses to central nervous system trauma. In Salzman SK, Faden AI (eds): The Neurobiology of Central Nervous System Trauma. New York, Oxford: Oxford University Press, pp 193–208.

Kimelberg HK, Ransom BR (1986): Physiological and pathological aspects of astrocytic swelling. In Fedoroff S, Vernadakis A (eds): Astrocytes,Vol. 3. Orlando, FL: Academic Press, pp 129–166.

Kimelberg HK, Biddlecome SM, Bourke RS (1979): SITS-inhibitable Cl^- transport and Na^+-dependent H^+ production in primary astroglial cultures. Brain Res 173: 111–124.

Kimelberg HK, Pang S, Treble DH (1989): Excitatory amino acid-stimulated uptake of $^{22}Na^+$ in primary astrocyte cultures. J Neurosci 9: 1141–1149.

Klatzo I (1967): Presidential Address: Neuropathological aspects of brain edema. J Neuropath Exp Neurol 26: 1–13.

Kogure K, Busto R, Scheinberg P (1981): The role of hydrostatic pressure in ischemic brain edema. Ann Neurol 9: 273–282.

Kraig RP, Chesler M (1990): Astrocytic acidosis in hyperglycemic and complete ischemia. J Cereb Blood Flow Metab 10: 104–114.

Lehre KP, Levy LM, Ottersen OP, Storm-Mathisen J, Danbolt NC (1995): Differential expression of two glial glutamate transporters in the rat brain: Quantitative and immunocytochemical observations. J Neurosci 15: 1835–1853.

Lomneth R, Medrano S, Gruenstein EI (1990): The role of transmembrane pH gradients in the lactic acid induced swelling of astrocytes. Brain Res 523: 69–77.

Long DM (1982): Traumatic brain edema. Clin Sci 29: 174–202.

Marmarou A (1992) Intracellular acidosis in human and experimental brain injury. Neurotrauma 9 (Suppl 2): S551–S562.

Matyja E (1986): Morphologic evidence of a primary response of glia to kainic acid administration into the rat neostriatum: Studied in vivo and in vitro. Exp Neurol 92: 609–623.

Menzies SA, Betz AL, Hoff JT (1993): Contributions of ions and albumin to the formation and resolution of ischemic brain edema. J Neurosurg 78: 257–266.

Muizelaar JP, Wei EP, Kontos HA, Becker DP (1986): Cerebral blood flow is regulated by changes in blood pressure and in blood viscosity alike. Stroke 17: 44–48.

Muizelaar JP, van der Poel HG, Li Z, Kontos HA, Levasseur JE (1988): Pial arteriolar vessel diameter and CO_2 reactivity during prolonged hyperventilation in the rabbit. J Neurosurg 69: 923–927.

Nagao S, Kwak S, Kanazawa I (1997): EAAT4, a glutamate transporter with properties of a chloride channel, is predominantly localized in Purkinje cell dendrites, and forms parasagittal compartments in rat cerebellum. Neurosci 78: 929–933.

Newman EA (1991): Sodium-bicarbonate cotransport in retinal Müller (glial) cells of the salamander. J Neurosci 11: 3972–3983.

Niendorf T, Dijkhuizen RM, Norris DG, Campagne MV, Nicolay K (1996): Biexponential diffusion attenuation in various states of brain tissue: Implications for diffusion-weighted imaging. Magn Reson Med 36: 847–857.

O'Connor ER, Sontheimer H, Ransom BR (1994): Rat hippocampal astrocytes exhibit electrogenic sodium-bicarbonate co-transport. J Neurophysiol 72: 2580–2589.

Olney JW, Labruyere J, Wang G, Wozniak DF, Price MT, Sessler FM (1991): NMDA antagonist neurotoxicity: Mechanism and prevention. Science 254: 1515–1518.

Pappas CA, Ransom BR (1994): Depolarization-induced alkalinization (DIA) in rat hippocampal astrocytes. J Neurophysiol 72: 2816–2826.

Paschen W, Djuricic B, Mies G, Schmidt-Kastner R, Linn F (1987): Lactate and pH in the brain: Association and dissociation in different pathophysiological states. J Neurochem 48: 154–159.

Popp AJ, Feustel PJ, Kimelberg HK (1996): Pathophysiology of traumatic brain injury. In Wilkins RH, Rengachary SS (eds): Neurosurgery, Vol 2, New York: McGraw-Hill, pp 2623–2637.

Portman MA, Lassen NA, Cooper TG, Sills AM, Potchen EJ (1991): Intra- and extracellular pH of the brain in vivo studied by ^{31}P-NMR during hyper- and hypocapnia. J Appl Physiol 71: 2168–2172.

Ransom BR, Yamate CL, Connors BW (1985): Activity-dependent shrinkage of extracellular space in rat optic nerve: A developmental study. J Neurosci 5: 532–535.

Robert A, Magistretti PJ (1997): AMPA/kainate Receptor Activation Blocks K^+ Currents Via Internal Na^+ Increase In Mouse Cultured Stellate Astrocytes. Glia 20: 38–50.

Rose CR, Ransom BR (1996): Mechanisms of H^+ and Na^+ changes induced by glutamate, kainate, and D-aspartate in rat hippocampal astrocytes. J Neurosci 16: 5393–5404.

Rosner MJ, Becker DP (1984): Experimental brain injury: Successful therapy with the weak base, tromethamine. With an overview of CNS acidosis. J Neurosurg 60: 961–971.

Rothstein JD, Martin L, Levey AI, Dykes-Hoberg M, Jin L, Wu D, Nash N, and Kuncl RW (1994): Localization of neuronal and glial glutamate transporters. Neuron 13: 713–725.

Rumpel H, Buchli R, Gehrmann J, Aguzzi A, Illi O, Martin E (1995): Magnetic resonance imaging of brain edema in the neonatal rat: A comparison of short and long term hypoxia-ischemia. Pediatr Res 38: 113–118.

Shrode LD, Putnam RW (1994): Intracellular pH regulation in primary rat astrocytes and C6 glioma cells. Glia 12: 196–210.

Siesjö BK (1985): Acid-base homeostasis in the brain: Physiology, chemistry, and neurochemical pathology. In Kogure K, Hossmann KA, Siesjo BK, Welsh FA (eds): Progress in Brain Research, Elsevier Science Publishers B.V. (Biomedical Division) Vol 63. pp 121–154.

Siesjö BK (1992): Pathophysiology and treatment of focal cerebral ischemia. I. Pathophysiology. J Neurosurg 77: 169–184.

Sontheimer H, Kettenmann H, Backus KH, Schachner M (1988): Glutamate opens Na^+/K^+ channels in cultured astrocytes. Glia 1: 328–336.

Staub F, Baethmann A, Peters J, Weigt H, Kempski O (1990): Effects of lactacidosis on glial cell volume and viability. J Cereb Blood Flow Metab 10: 866–876.

Staub F, Winkler A, Peters J, Kempski O, Kachel V, Baethmann A (1994): Swelling, acidosis, and irreversible damage of glial cells from exposure to arachidonic acid in vitro. J Cereb Blood Flow Metab 14: 1030–1039.

Steward PA (1981): How to Understand Acid-Base. New York: Elsevier.

Tanno H, Nockels RP, Pitts LH, Noble LJ (1992): Breakdown of the blood-brain barrier after fluid percussive brain injury in the rat. I. Distribution and time course of protein extravasation. J Neurotrauma 9: 21–32.

Tschirgi RD (1958): The blood-brain barrier. In Windle WF (ed): Biology of Neuroglia, Springfield, IL. Charles C. Thomas, pp 130–138.

Van Harreveld A (1966): Brain Tissue Electrolytes. London: Butterworths.

Van Harreveld A (1982): Swelling of the Muller fibers in the chicken retina. J Neurobiol 13: 519–536.

Van Harreveld A, Fifkova E (1971): Light-and electron-microscopic changes in central nervous tissue after electrophoretic injection of glutamate. Exp Molec Path 15: 61–81.

Van Lookeren Campagne M, Verheul JB, Nicolay K, Balázs R (1994): Early evolution and recovery from excitotoxic injury in the neonatal rat brain: A study combining magnetic resonance imaging, electrical impedance, and histology. J Cereb Blood Flow Metab 14: 1011–1023.

Verheul HB, Balázs R, Berkelbach van der Sprenkel JW, Tulleken CAF, Nicolay K, Tamminga KS,

Van Lookeren M, Campagne M (1994): Comparison of diffusion-weighted MRI with changes in cell volume in a rat model of brain injury. NMR Biomed 7: 96–100.

Walz W, Klimaszewski A, Paterson IA (1993): Glial swelling in ischemia: A hypothesis. Dev Neurosci 15: 216–225.

Yamada K, Watanabe M, Shibata T, Tanaka K, Wada K, Inoue Y (1996): EAAT4 is a post-synaptic glutamate transporter at Purkinje cell synapses. NeuroReport 7: 2013–2017.

Yamakami I, Vink R, Faden AI, Gennarelli TA, Lenkinski R, McIntosh TK (1995): Effects of acute ethanol intoxication on experimental brain injury in the rat: Neurobehavioral and phosphorus-31 nuclear magnetic resonance spectroscopy studies. J Neurosurg 82: 813–821.

Yang SP, Bergo GW, Krasney E, Krasney JA (1994): Cerebral pressure-flow and metabolic responses to sustained hypoxia: effect of CO_2. J Appl Physiol 76: 303–313.

Yoshida K, Marmarou A (1991): Effects of tromethamine and hyperventilation on brain injury in the cat. J Neurosurg 74: 87–96.

Young W, Constantini S (1994): Ionic and water shifts in injured central nervous tissue. In Salzman SK, Faden AI (eds): The Neurobiology of Central Nervous System Trauma. New York: University Press, pp 123–130.

Young W, Rappaport ZH, Chalif DJ, Flamm ES (1987): Regional brain sodium, potassium, and water changes in the rat middle cerebral artery occlusion model of ischemia. Stroke 18: 751–759.

Zupping R, Kaasik AE, Raudam E (1971): Cerebrospinal fluid metabolic acidosis and brain oxygen supply. Arch Neurol 25: 33–38.

INDEX